World Geomorphological Landscapes

Series Editor

Piotr Migoń, Institute of Geography and Regional Development, University of Wrocław, Wrocław, Poland

Geomorphology – 'the Science of Scenery' – is a part of Earth Sciences that focuses on the scientific study of landforms, their assemblages, and surface and subsurface processes that moulded them in the past and that change them today. Shapes of landforms and regularities of their spatial distribution, their origin, evolution, and ages are the subject of geomorphology. Geomorphology is also a science of considerable practical importance since many geomorphic processes occur so suddenly and unexpectedly, and with such a force, that they pose significant hazards to human populations. Landforms and landscapes vary enormously across the Earth, from high mountains to endless plains. At a smaller scale, Nature often surprises us creating shapes which look improbable. Many geomorphological landscapes are so immensely beautiful that they received the highest possible recognition – they hold the status of World Heritage properties. Apart from often being immensely scenic, landscapes tell stories which not uncommonly can be traced back in time for millions of years and include unique events. This international book series will be a scientific library of monographs that present and explain physical landscapes across the globe, focusing on both representative and uniquely spectacular examples. Each book contains details on geomorphology of a particular country (i.e. The Geomorphological Landscapes of France, The Geomorphological Landscapes of Italy, The Geomorphological Landscapes of India) or a geographically coherent region. The content is divided into two parts. Part one contains the necessary background about geology and tectonic framework, past and present climate, geographical regions, and long-term geomorphological history. The core of each book is however succinct presentation of key geomorphological localities (landscapes) and it is envisaged that the number of such studies will generally vary from 20 to 30. There is additional scope for discussing issues of geomorphological heritage and suggesting itineraries to visit the most important sites. The series provides a unique reference source not only for geomorphologists, but all Earth scientists, geographers, and conservationists. It complements the existing reference books in geomorphology which focus on specific themes rather than regions or localities and fills a growing gap between poorly accessible regional studies, often in national languages, and papers in international journals which put major emphasis on understanding processes rather than particular landscapes. The World Geomorphological Landscapes series is a peer-reviewed series which contains single and multi-authored books as well as edited volumes.

World Geomorphological Landscapes – now indexed in Scopus® !

Amos Frumkin · Nurit Shtober-Zisu
Editors

Landscapes and Landforms of Israel

Springer

Editors
Amos Frumkin
The Fredy & Nadine Herrmann Institute of Earth Sciences, The Hebrew University of Jerusalem
Jerusalem, Israel

Nurit Shtober-Zisu
School of Environmental Sciences
University of Haifa
Haifa, Israel

ISSN 2213-2090 ISSN 2213-2104 (electronic)
World Geomorphological Landscapes
ISBN 978-3-031-44763-1 ISBN 978-3-031-44764-8 (eBook)
https://doi.org/10.1007/978-3-031-44764-8

This Springer imprint is published by the registered company Springer Nature Switzerland AG
The registered company address is: Gewerbestrasse 11, 6330 Cham, Switzerland

Paper in this product is recyclable.

Series Editor's Preface

Landforms and landscapes vary enormously across the Earth, from high mountains to endless plains. At a smaller scale, Nature often surprises us by creating shapes which look improbable. Many physical landscapes are so immensely beautiful that they have received the highest possible recognition—they hold the status of World Heritage properties. Apart from often being immensely scenic, landscapes tell stories which not uncommonly can be traced back in time for tens of million years and include unique events. In addition, many landscapes owe their appearance and harmony not solely to natural forces. For centuries, or even millennia, they have been shaped by humans who modified hillslopes, river courses, and coastlines and erected structures which often blend with the natural landforms to form inseparable entities.

These landscapes are studied by geomorphology—"the Science of Scenery"—a part of Earth Sciences that focuses on landforms, their assemblages, the surface, and subsurface processes that moulded them in the past and that change them today. The shapes of landforms and the regularities of their spatial distribution, their origin, evolution, and ages are the subject of research. Geomorphology is also a science of considerable practical importance since many geomorphic processes occur so suddenly and unexpectedly, and with such a force, that they pose significant hazards to human populations and not uncommonly result in considerable damage or even casualties.

To show the importance of geomorphology in understanding the landscape, and to present the beauty and diversity of the geomorphological sceneries across the world, we have launched a new book series *World Geomorphological Landscapes*. It aims to be a scientific library of monographs that present and explain physical landscapes, focusing on both representative and uniquely spectacular examples. Each book will contain details on geomorphology of a particular country or a geographically coherent region. This volume covers Israel—a country which is rather small in terms of area, but extremely diverse in terms of geomorphology and placed in a critical point at the junction of continents. Its territory includes one of the most important tectonic structures globally—the Dead Sea transform covers a range of climates from humid to hyper-arid, changing across very short distances, and varies between flat coastal plains and steep, high-energy escarpments. The geomorphology of Israel also bears evidence of protracted human impact, one of the longest lasting on Earth. All these cases are comprehensively illustrated in this volume.

The *World Geomorphological Landscapes* series is produced under the scientific patronage of the International Association of Geomorphologists—a society that brings together geomorphologists from all around the world. The IAG was established in 1989 and is an independent scientific association affiliated with the International Geographical Union and the International Union of Geological Sciences. Among its main aims are to promote geomorphology and to foster dissemination of geomorphological knowledge. I believe that this lavishly illustrated series, which sticks to the scientific rigour, is the most appropriate means to fulfil these aims and to serve the geoscientific community. To this end, my great thanks go to the editors of the volume, Profs. Nurit Shtober-Zisu and Amos Frumkin, who agreed to coordinate the book and ensured that the final product meets all quality requirements. I am also grateful to all individual contributors who agreed to add the task of writing chapters to their busy agendas and delivered such interesting stories to read.

On a more personal note, a visit to Israel thirty years ago was one of my first geomorphological field trips abroad and my continuing fascination with desert landscapes began in Israel, where I was expertly guided across the Negev and along the Dead Sea by Prof. Dan Bowman, whose generosity and kindness will always be remembered. Repeating part of that trip this year, in 2023, prior and during the conference of the IAG Working Group of Denuchange so efficiently organized by Nurit Shtober-Zisu, was a great experience, and I strongly recommend to all geomorphologists to visit Israel and enjoy its breathtaking landscapes. Let the current volume be both an introduction and the guidebook to such a visit.

Wrocław, Poland

Piotr Migoń
Series Editor

Preface

Israel is a land of contrasts, where the ancient and the modern coexist. From the desert to the mountains, the landscapes of Israel are diverse and spectacular, celebrating natural beauty and cultural richness. Climate variability and geologic events are the foundation of present and past geomorphological evolution, responsible for most morphogenetic processes and landforms currently observed in Israel—fluvial, coastal, aeolian, karstic, volcanic, or gravitational. The natural geomorphological processes in Israel have created highly scenic and diverse landscapes; by their side, anthropic landscapes of enormous cultural value are embedded as a heritage of humanity. These landscapes are the background for much of the history and myths of the main religions of the Western world. A wide range of these landscapes is described in this volume, which contains over twenty examples, representative of all natural and human morphogenetic environments (Fig. 1).

This book is the result of a joint venture established among Israeli geomorphologists, which has also included the participation of valuable experts from other disciplines. To our knowledge, this is the first attempt to put together such a collection of papers on the landscapes and landforms of Israel. More than 25 authors from five universities and three research centres have contributed to the book. Furthermore, each chapter has undergone three rounds of thorough peer review before being accepted for publication. This book, therefore, offers a collection of contributions that can be valuable to a wide readership, ranging from professionals who study the landscapes of Israel, students, teachers, tour guides, tourists, or anyone in the international public who wants to expand their knowledge and learn about this country from first-class researchers.

We are very grateful to Piotr Migoń, Series Editor, for having invited us to join the important editorial project of the "World Geomorphological Landscapes" and for his precious suggestions, constant availability, and continuous support. We would like to acknowledge Dr. Robert K. Doe, Editorial Director and the assistance of the Springer book Project Coordinators who took care of this book with remarkable dedication and patience, in particular Ms. Banu Dhayalan who finalized the volume production.

Last but not least, our special thanks go to the individual authors for the enthusiasm with which they responded to our invitation and for the outstanding efforts made for the success of this valuable editorial initiative.

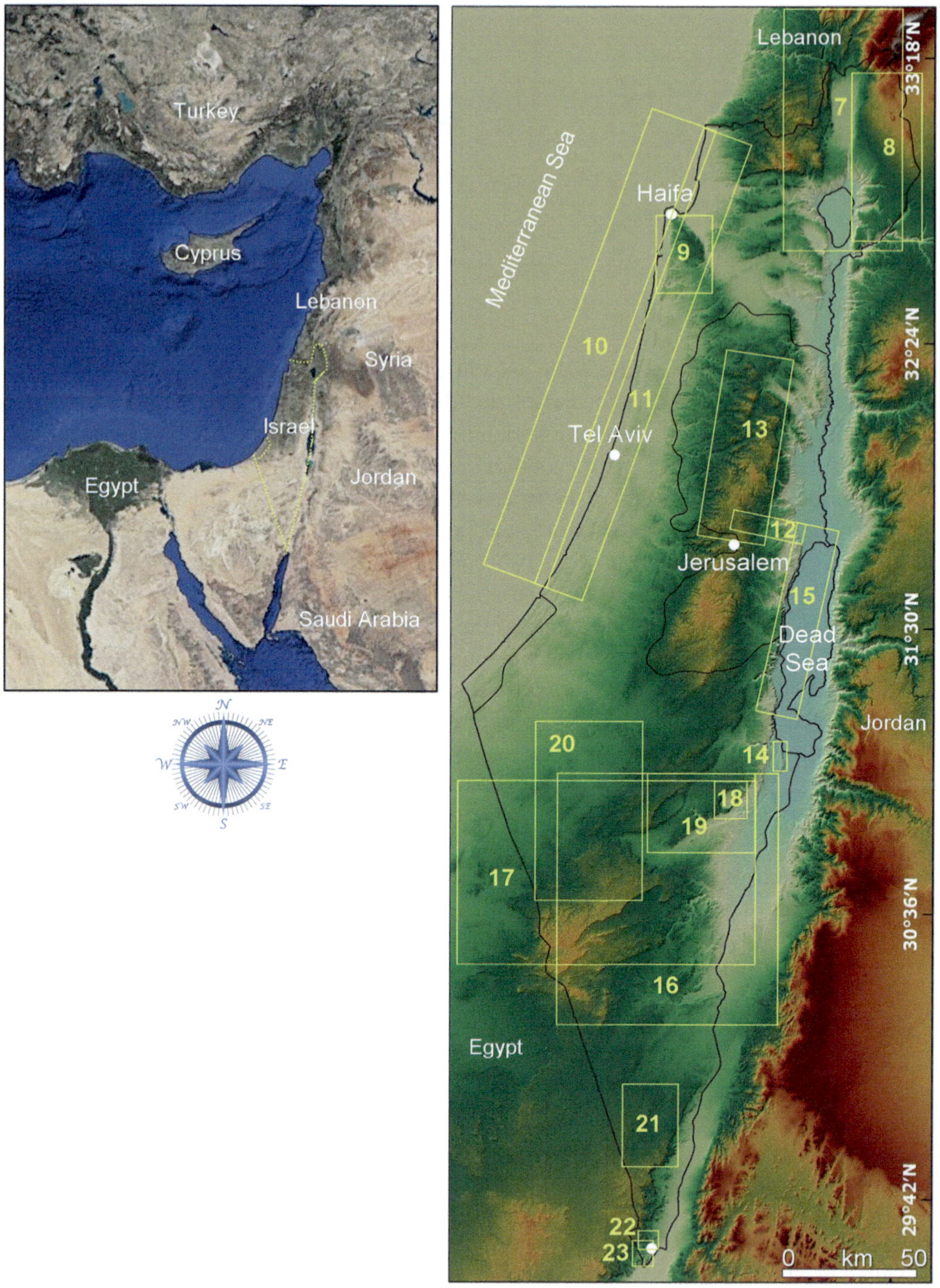

Fig. 1 Approximate areas covered by chapters, shown on the relief map of Israel

Jerusalem, Israel — Amos Frumkin
Haifa, Israel — Nurit Shtober-Zisu

Contents

Editors and Contributors

About the Editors

Amos Frumkin is a professor Emeritus at the Hebrew University of Jerusalem. His interests cover karst geomorphology and speleology, cave sediments as indicators of palaeoclimate, palaeohydrology, the development of karst aquifers, geoarchaeology, and ancient water supply systems. His research is mostly associated with underground (speleological) features studied using Earth Sciences methods, such as geomorphology, radiometric dating, and stable isotopes. He has written and edited several books and about 200 scientific papers. He founded and directs the Israel Cave Research Center.

Nurit Shtober-Zisu is a geomorphologist at the School of Environmental Sciences at the University of Haifa. She has a background in morphotectonics and landscape evolution. Her research interest includes weathering processes and rock decay mechanisms, in various temporal and spatial scales—from bacterial erosion in stalagmites to mechanical decay of rocks during forest fires, and chemomechanical decay in carbonate terrains. In recent years, she leads several projects in geoarchaeology and palaeoenvironmental reconstruction.

Contributors

Oren Ackermann The Department of Land of Israel Studies and Archaeology, Ariel University, Ariel, Israel

Yoav Avni Geological Survey of Israel, Jerusalem, Israel

Anne Bernhardt Institute of Geological Sciences, Freie Universität Berlin, Berlin, Germany

Hendrik J. Bruins Ben-Gurion University of the Negev, Jacob Blaustein Institutes for Desert Research, Swiss Institute for Dryland Environmental and Energy Research (SIDEER), Midreshet Ben-Gurion, Israel

Ran Calvo Geological Survey of Israel, Jerusalem, Israel

Elad Dente School of Environmental Sciences, University of Haifa, Haifa, Israel

Amir Eyal Be'er Sheva, Israel

Amos Frumkin Institute of Earth Sciences, The Hebrew University of Jerusalem, Jerusalem, Israel

Hanan Ginat Dead Sea and Arava Science Center, Eilat, Israel

Noam Greenbaum School of Environmental Sciences, University of Haifa, Haifa, Israel

Tamir Grodek The Fredy and Nadine Herrmann Institute of Earth Sciences, The Hebrew University of Jerusalem, Jerusalem, Israel

Liora Kolska Horwitz National Natural History Collections, Faculty of Life Sciences, The Hebrew University, Jerusalem, Israel

Moshe Inbar School of Environmental Sciences, University of Haifa, Haifa, Israel

Haim Kutiel School of Environmental Sciences, University of Haifa, Haifa, Israel

Hanoch Lavee Department of Geography and Environment, Laboratory of Soil and Geomorphology, Bar Ilan University, Ramat Gan, Israel

Judith Lekach The Fredy & Nadine Herrmann Institute of Earth Sciences, The Hebrew University of Jerusalem, Jerusalem, Israel

Noam Levin Department of Geography, The Hebrew University of Jerusalem, Jerusalem, Israel; Remote Sensing Research Centre, School of the Environment, The University of Queensland, Brisbane, Australia

Doron Mor Kibbutz Merhavia, Israel

Joel Roskin Department of Geography and Environment, Bar-Ilan University, Ramat-Gan, Israel

Uri Schattner School of Environmental Sciences, University of Haifa, Haifa, Israel

Uri Schwartz School of Environmental Sciences, University of Haifa, Haifa, Israel

Nurit Shtober-Zisu School of Environmental Sciences, University of Haifa, Haifa, Israel

Adi Torfstein The Fredy and Nadine Herrmann Institute of Earth Sciences, Hebrew University of Jerusalem, Israel, and Interuniversity Institute for Marine Sciences, Eilat, Israel

Ram Weinberger Geological Survey of Israel, Jerusalem, Israel; Department of Earth and Environmental Sciences, Ben Gurion University of the Negev, Beersheba, Israel

Yishai Weinstein Department of Geography and Environment, Bar-Ilan University, Ramat Gan, Israel

Nimrod Wieler Research Department, Israel Antiquities Authority, Jerusalem, Israel

Israel Zak (Deceased) Institute of Earth Sciences, The Hebrew University of Jerusalem, Jerusalem, Israel

Ezra Zilberman Geological Survey of Israel, Jerusalem, Israel

Motti Zohar School of Environmental Sciences, University of Haifa, Haifa, Israel

Part I
Israel Background

Landscapes and Landforms of Israel—An Overview

1

Amos Frumkin and Nurit Shtober-Zisu

Abstract

Three geographical factors determine the character of Israel: its location in the Mediterranean zone, at the crossroads of three continents and two oceans, and on the border between the desert and the sown. Like other Mediterranean countries, Israel has a predominantly hilly topography, a bedrock foundation in which carbonate rocks predominate, and a sunny climate characterized by a sharp seasonal division between a rainy winter and an arid summer. The combination of these factors together create the country's distinctive landscapes, soils, and vegetation. This chapter aims to draw a brief picture of the four lengthwise strips, running north to south, and their subdivision into major physiographic units that together form the landscapes of this country.

Keywords

Physiographic units · Relief · Coastal plain · Central mountain belt · The Dead Sea transform

1.1 Introduction

Israel is a small country, measuring only 470 km from north to south and 116 km from east to west at its widest point, extending over an area of ~20,000 km^2. Israel is based at the border region between three major tectonic plates: the Eurasian plate to the north, the Arabian plate to the east and the African plate to the south-west, forming therefore a bridge between continents and acting as a connecting corridor for animals and human passage. Zooming in, Israel is located in the north-eastern part of the Sinai micro-plate. Israel is also a climatic meeting point. Located between the Mediterranean and the desert belt, it boasts three climate zones characterized by a sharp seasonal division between a rainy winter and a dry summer: the Mediterranean at the north, the semi-arid (Irano-Turanian) intermediate belt, and the extremely arid to the south-east, which belongs to the Saharo-Arabian climate region (Fig. 1.1).

The rock record of Israel includes igneous and metamorphic crystalline basement rocks from the Precambrian, overlain by a lengthy sequence of sedimentary rocks extending up to the Quaternary, in which marine carbonate rocks prevail, partly covered by volcanic rocks, alluvium, sand dunes, and playa deposits. The combination of various parent rocks and three different climate zones culminates in highly variable resultant soil and vegetation types, resulting in incredible diversity of geomorphological landscapes.

Overall, the relief is characterized by diverse landscapes and ecosystems, ranging from fertile coastal plains to rugged mountains, shallow depressions, and desert regions. Four main physiographic belts are determined by the north–south trend of the Dead Sea transform and the strike of the Mediterranean coastline. From west to east these are: (a) the coastal plain, a narrow area with an altitude rising gradually from sea level at the coast to about 100 m asl, extending from the Mediterranean Sea to the foothills; (b) the central mountain belt, including the Galilee, Samaria, Judea, and the Negev mountains, with elevations generally from 100 to 1200 m asl; (c) the Rift Valley or the Dead Sea transform—an elongated tectonic depression trending north–south; and (d) the eastern side of the Dead Sea transform—a high plateau dissected by ridges and river canyons, reaching its highest elevation at Mt. Hermon. Only the Golan Heights and southern Mt. Hermon are within Israel's borders (Fig. 1.2).

A. Frumkin (✉)
The Hebrew University of Jerusalem, Institute of Earth Sciences, 9190401 Jerusalem, Israel
e-mail: amos.frumkin@mail.huji.ac.il

N. Shtober-Zisu
School of Environmental Sciences, University of Haifa, Abba Khoushy Ave, 3498838 Haifa, Israel

A. Frumkin and N. Shtober-Zisu (eds.), *Landscapes and Landforms of Israel*,
World Geomorphological Landscapes, https://doi.org/10.1007/978-3-031-44764-8_1

Fig. 1.1 Oblique Google Earth view of the Mediterranean climate zone of Israel, with the arid zone in the background

1.2 Physiographic Units

1.2.1 Coastal Plain

The coastal belt (Fig. 1.2, no. 1) is tens of km wide in the south, and narrow to completely absent in northern Israel. For example, at Rosh Hanikra, at the northern extremity of the country, there is no coastal plain at all, and the mountainous ridge falls directly to the sea through steep cliffs (Fig. 1.3). The continental shelf also narrows northward and is dissected by submarine canyons and slump scars (Almagor and Garfunkel 1979; Schattner and Bernhardt, this volume). The coastal plain has mainly acted as a depositional area during the Plio-Pleistocene, accumulating sand supplied by the Nile, alluvium from the backbone hills, and dust (Levin, this volume). Therefore, the coast is primarily devoid of bays or estuaries. The Haifa Bay, interrupting an otherwise almost unbroken shoreline, and the adjoining Zebulun Valley differ from the rest of the Coastal Plain, being a subsiding graben responsible for subsequent inland sedimentation. Aeolianite (locally termed 'kurkar') developed during the Pleistocene, parallel to the Mediterranean coastline and in relation to past sea levels (Sivan et al. 2011), and build low ridges within the generally flat landscape. The valleys separating the ridges attracted water, forming wetlands which were mostly drained artificially during modern times (Cohen-Seffer et al. 2005). Sand dunes are common along the sea shore in the south, becoming rare in the north.

1.2.2 The Central Mountain Range

The central mountain range is a diversified north–south trending hilly terrain, uplifted as an elongated arch along the Dead Sea rift during its development. Its highest point is Mt. Meron in the Upper Galilee, reaching 1208 m in elevation. Several transverse depressions segment the mountain range; the prominent ones are the Be'er Sheva and Jezreel valleys. The central mountain range is discussed below according to its sub-regions from south to north.

The Eilat Mountains (Fig. 1.2, no. 2) are a mountain massif at the southernmost tip of the central mountain range, on the shores of the Red Sea. It reaches its highest elevation of 838 m asl west of the city of Eilat. The Eilat Mountains are the only region of Israel where the Arabian-Nubian Massif is exposed, surrounding the Red Sea. The more pronounced the uplift, the greater was the exposure to denudation and erosion. The younger strata have been removed by erosion, exposing older and deeper rocks. Thus, Neoproterozoic magmatic and metamorphic rock units crop out only near Eilat and these are covered by the Paleozoic Nubian sandstones (Fig. 1.4); all are faulted and deformed by the Dead Sea transform system. The mountains reach the Red Sea coast; thus, the sea-level eustatic changes during the Quaternary induced base-level changes and associated geomorphic variability (Bar et al. 2018). This is also the most arid region of Israel, where the average annual rainfall is 20 mm (Israel Meteorological Service). Yet the short rainfall events may be intensive, and therefore flash-floods play

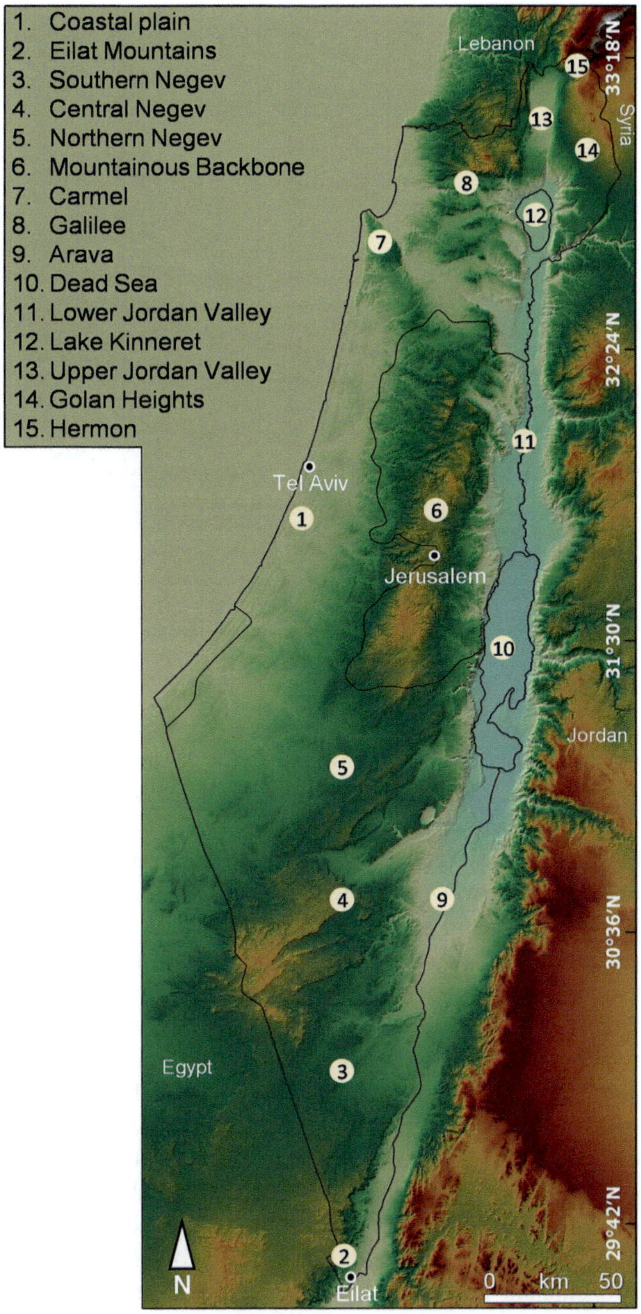

Fig. 1.2 Topographic map of Israel and neighbouring countries, presenting the main physiographic units. The south basin of the Dead Sea is shown inundated (as in the late 1970s, before major human interference)

a major geomorphic role in shaping the slopes and valleys of this area (Grodek, this volume; Lekah, this volume).

To the north of the Eilat mountains, the southern Negev (Fig. 1.2, no. 3) comprises carbonate hills and plateaus, dissected by major tectonic lines, the most prominent of which are associated with the Dead Sea Rift Valley. Most catchments drain to the NNE, ultimately reaching the Dead Sea base level. Some of these drainage systems demonstrate wide areas of ancient desert pavement, mantled by a veneer of mixed sand and gravel. These may be some of the longest-lived landforms in the world (Matmon et al. 2009). An Early Pleistocene fluvio-lacustrine sequence found in this area indicates that climatic conditions during that period were humid enough to support the formation of shallow lakes (Ginat et al. 2003, 2018).

The Central Negev (Fig. 1.2, no. 4) is the highest part of the Negev and reaches up to ~1000 m asl. It comprises mainly carbonate ridges, the spatial pattern of which is controlled by tectonic lines, NW-trending folds of the Syrian Arc (Shahar, 1994) and east-trending faults of the Negev–Sinai Shear Zone (Weinberger et al. 2020). These structures were later affected by the tectonic subsidence of the Dead Sea Rift Valley. The highest plateaus belong to the Oligocene erosion surface (Avni et al. 2012; Avni this volume), cut by various stages of entrenchment (Fig. 1.5). Some of the higher anticlines were entrenched by deep erosion cirques (locally termed 'makhteshim'), where various Triassic to Cretaceous rocks crop out (Zilberman 2000; Zilberman this volume) (Fig. 1.6). These sites have a major geodiversity and geoheritage value (Finzi et al. 2019).

The northern Negev (Fig. 1.2, no. 5) shows diverse topography. The south-eastern part comprises ridges and valleys associated with the asymmetric folds of the Syrian Arc. The north-western part of the northern Negev is covered by a veneer of late Quaternary aeolian sediments, mainly loess and sand, where gully erosion processes are quite common (Fig. 1.7) (Crouvi et al. 2008; Roskin this volume). The relatively low Beer Sheva Valley, separating the northern Negev and Judea, is also mostly covered by loess. The north-eastern Negev desert continues into the Judean Desert formed by rain-shadow effect, with similar erosional and weathering features.

The Central mountainous backbone (Fig. 1.2, no. 6) is a wide NNE-trending anticlinorium. Its southern part, the historic Judea, is divided into three longitudinal belts: (1) the Judean Shephela at the west is a syncline, where the Eocene chalk comprises most of the outcrops, forming undulating hills, reaching an altitude of up to ~500 m (Fig. 1.8a); (2) the Judea Mountains backbone ridge is roughly associated with an uplifted anticline, where mechanically resistant carbonates dominate (Shtober-Zisu and Zissu 2018). These form a continuous ridge whose mild hilly topography is mainly shaped by dissolution (Fig. 1.8b) (Ryb et al. 2014a; Frumkin this volume). The ridge is bordered by monoclines, dipping to both its sides; (3) the eastern belt, occupied by the Judean desert, demonstrates the steepest climatic gradient in Israel, on a west–east transect (Lavee et al. 1998; Lavee this volume). It comprises a hilly plateau in which mostly Senonian chalk crops out. Some ephemeral streams have cut into the underlying harder carbonates, forming steep canyons (Fig. 1.8c). These become deeply entrenched close to the Dead Sea base level, with high

Fig. 1.3 The northern termination of Israel's coastal plain: foreground–Pleistocene Kurkar eroded by the sea; Background–Rosh Hanikra: Cenomanian chalk cliff falls into the sea

vertical escarpments and associated knickpoints (Haviv et al. 2010). The rain-shadow desert conditions promote entrenchment by flash-floods, which are among the most destructive natural disasters in Israel (Morin et al. 2009). They are also eroding deep potholes in the streambeds which may retain water for the entire year. In contrast with physical erosion, dissolution play a less important role (Ryb et al. 2015; Shtober-Zisu et al. 2017). The city of Jerusalem is located in a tectonic saddle on the northern backbone of Judea mountains, north of which lies the Binyamin plateau, formed mainly by karst dissolution over Late Cretaceous carbonates (Frumkin 1993).

The northern part of the mountainous backbone, the historic Samaria, is more tectonically complicated. Comprising several sub-parallel NNE-trending folds of the Syrian Arc, it is also dissected by NW-trending faults. Thus, the continuity of the water divide is interrupted, crossing hills (up to 940 m altitude) and valleys. The combination of an anticline and faults exposes Jurassic and early Cretaceous rocks in NE Samaria, but most outcrops consist of late Cretaceous to Eocene carbonates. Some of the deep valleys have cut into complex aquifers (Khayat et al. 2018), forming large springs which promoted human habitation during ancient times, such as the cities of Tirza and Shechem (Roman *Flavia Neapolis*) (Frumkin 2017).

Mt. Carmel (Fig. 1.2, no. 7) is a NW extension of the fold system of Samaria (Shtober-Zisu, this volume). This was uplifted as a horst which extends towards the Mediterranean (Steinberg et al. 2010). The proximity to the Mediterranean and the topographic high, above 500 m, promotes more precipitation and lush vegetation. The projection of Mt. Carmel ridge into the Mediterranean Sea is associated with the largest bay along Israel's coast, where the harbour city of Haifa was built.

The Galilee region (Fig. 1.2, no. 8) is located in northern Israel, between the Mediterranean Sea and the Dead Sea Fault system, and is divided into two major domains, the Lower and Upper Galilee. The Jezreel Valley crosses the mountainous backbone, separating Samaria in the south from the Lower Galilee in the north. The Lower Galilee, rising to 600 m in elevation, comprises horsts (Fig. 1.9) and grabens which determine the relief. In the west, they trend eastward, whereas in the east, they trend to southeast, while the blocks themselves are tilted to south-west (Freund 1970; Matmon et al. 2003). These blocks contain diverse Cenozoic outcrops, including Eocene limestones, Neogene aquatic sediments, and a cover of Pliocene basalt (Rozenbaum et al. 2022). Further to the north, the Upper Galilee reaches over 1200 m in elevation. The relief of the western part is dominated by late Cretaceous carbonates cut by east-trending faults. The eastern part is more diversified, with outcrops of Eocene carbonates present in synclines and rocks of Cretaceous age in uplifted areas. The structure and recent topography did not evolve contemporaneously throughout the Galilee, and its various parts experienced different geological histories. The pre-Eocene N–S oriented anticlines and synclines (belonging to the Syrian Arc fold belt) have been truncated during the Oligocene and Miocene and subsequently faulted. Thus, the Syrian Arc fold belt does not influence the present morphology of the Galilee. The recent relief is rather determined by Plio-Pleistocene arching and normal faulting (Matmon et al.

Fig. 1.4 The magmatic and sandstone landscapes of the Eilat Mountains. **A** The igneous Mt. Shlomo in the background and the upper tributaries of Nahal Shlomo, incised into the Nubian sandstone; **B** steep slopes in magmatic rocks of Nahal Netafim; **C** Reddish sandstone of the Shani canyon; **D** Tafoni in the sandstones of the valley of Nahal Netafim (‘*nahal*’ = stream)

Fig. 1.5 Erosional surfaces in the central Negev (Nahal Ashosh), controlled by lithologic properties and stages of entrenchment

Fig. 1.6 General view of Makhtesh Hatira. The cliffs in the background form the southern rim of the Makhtesh

Fig. 1.7 Loess badlands in the northern Negev; severe bank erosion along the Nahal Besor exposes a water cistern dated to the fifth–sixth centuries CE (Shtober-Zisu et al. 2024)

2003). While patches of early Pleistocene volcanic rocks cover some areas in the eastern Upper Galilee, most areas of the Upper Galilee demonstrate well-developed fluvio-karstic terrain on carbonate rocks (Frumkin et al. 2021a), with associated karst aquifers (Magal et al. 2013).

1.2.3 The Dead Sea Transform-Rift Valley

The Dead Sea rift, also acting as transform, is a down-faulted plate boundary along the eastern side of the backbone hills (Garfunkel 2009). The tectonic framework includes a major strike-slip fault active since the early Miocene (Nuriel et al. 2017), with a present slip rate of ~5 mm/yr, and commonly normal faults along which the internal parts are down-faulted (Sneh 1996). In some areas, the western border of the rift is a monocline. The rift valley comprises several deep pull-apart basins with thick sedimentary fill, and intermediate higher relief saddles. The mild topography and climate as well as the relatively common water sources have promoted its use as a corridor for the dispersion of animals and humans between Africa and

Fig. 1.8 The mountainous backbone of Judea; **A** Central Judean Shephela looking south; **B** deeply incised valley of Nahal Soreq, Judean Mountains, looking east; **C** The Judean desert and the deeply incised canyon of Nahal Kidron looking west

Eurasia. The northern part is freshwater-rich, including the Hula Valley and the Kinneret Lake (Sea of Galilee), but the southern part is arid including the hypersaline Dead Sea, which is the lowest terrestrial point on Earth, presently (August 2023) 438 m below sea level. The rift valley will be discussed according to its sub-regions from south to north.

The Arava Valley (Fig. 1.2, no. 9) extends from the Gulf of Eilat of the Red Sea in the south to the Dead Sea in the north. It includes endorheic basins with sabkhas in the south (Fig. 1.10), a saddle rising to ~250 m elevation, similar in altitude to the low area of the southern Negev to the west, and a longer catchment drained into the Dead Sea through Nahal HaArava (Dente et al. 2017). The sedimentary fill of the Arava crops out along its entire length and reaches its maximal thickness, over 8 km, close to the Dead Sea. It dates from the Miocene until present times (Zilberman and Calvo 2013). Most of the northern Arava is covered by the sediments of last glacial Lake Lisan (Fig. 1.11A). Since the lake retreat, these soft sediments have been eroded into badlands and piping-rich landscape (Hamawi et al. 2022; Dente this volume).

The ~300 m deep Dead Sea (Fig. 1.2, no. 10) is a hypersaline lake occupying a pull-apart basin within the lowest area of the rift valley (Garfunkel and Ben-Avraham 1996). The lake shoreline has been the lowest terrestrial place in the world at least throughout the late Holocene (Bookman et al. 2004). As an endorheic lake in a hyper-arid region, the Late Pleistocene and Holocene lake levels serve as a gauge for climatic fluctuations (Fig. 1.11A) (Frumkin 1997; Enzel et al. 2003). Lake water is a residual brine, originating in a late-Miocene marine lagoon, whose water has evolved chemically ever since. The falling water level has revealed crescent-like landforms associated with mass movements (Fig. 1.11B). These landslide scarps were formed underwater during the late Quaternary, and became exposed during recent lake-level fall (Lensky et al, 2014). The Dead Sea basin is bordered by escarpments a few hundred metres high, genetically related to active normal faults. The bottom of the valley is covered mainly by recent lacustrine sediments, while older lake deposits are hardly exposed (Stein 2001; Lisker et al. 2010). These include subsurface shallow salt layers, which promote intensive dissolution and sinkhole formation during recent decades, following an anthropogenic rapid fall of the Dead Sea level and associated aquifers (Fig. 1.11C). The shallow south basin of the Dead Sea dried up when the lake level fell below −400 m and is used today for the potash industry. Deep layers of salt, attributed to late-Miocene marine intrusion, gave rise to several diapirs along the Dead Sea basin. However, the only salt diapir actively penetrating the surface is Mt. Sedom, with its unique landscape and dissolution features (Fig. 1.11D) (Frumkin et al. 2021b; Frumkin and Zak this volume).

The Lower Jordan Valley (Fig. 1.2, no. 11) is climatically diverse, stretching from the Mediterranean climate of Lake Kinneret to the hyper-arid climate of the Dead Sea. The valley bottom is covered mostly by the sediments of Lake Lisan, the glacial period predecessor of the Dead Sea (Torfstein this volume). Following the lake-level fall, the Jordan River flowed from the Sea of Galilee to the Dead Sea along the previous lake-level bed, forming an elongated depression with the Jordan floodplain at the bottom. The Jordan River is meandering along this sub-valley (Fig. 1.12). Since the anthropogenic extraction of upstream waters during the twentieth century, the river stopped inundating the floodplain, and its route has stabilized. The Bet-She'an Valley, where the Harod Valley bifurcates from the

Fig. 1.9 Mt. Tavor horst rising above the Jezreel Valley

Fig. 1.10 The Sabha of Yotvata after heavy rain; 23.04.2023. Photo credit Shahar Harel

Dead Sea rift towards the Jezreel Valley, is a deep and wide basin filled with sediments. It boasts many large springs which deposited tufa across parts of the valley bottom. Less abundant springs occur in other parts of the Lower Jordan Valley, affected by the major change in hydrogeologic configuration since the last glacial period (Levy et al. 2020).

Lake Kinneret (Fig. 1.2, no. 12), also known as the 'Sea of Galilee', is a freshwater lake, ~30 m deep at a level of ~−210 m (Tal 2019), occupying a tectonic basin within the Jordan Valley (Hazan et al. 2004; Ben-Gai 2009). Towards the south of the basin, thick Plio-Pleistocene deposits were accumulated (Davis et al. 2011; Segev

Fig. 1.11 The Dead Sea region. **A** Deformation of lake Lisan sediments; **B** Crescent-like bays along the Dead Sea shore, associated with underwater mass movement; **C** A sinkhole formed above a salt layer, Nahal Hever; **D** Vertical salt layers of the Mt. Sedom diapir

Fig. 1.12 The Jordan River meanders across its floodplain which is carved into the Lisan beds of the Jordan Valley

2017), bearing important faunal evidence for the connection between Africa and Eurasia (Belmaker 2010). Lake Kinneret is partly surrounded by water-rich valleys which attracted humans since the early Pleistocene. The lake served as an important water source for Israel until being mostly replaced by desalination projects during the twenty-first century. The main water recharge of the lake is from the Upper Jordan River, accompanied by several smaller streams and springs (Inbar this volume). Saline deposits and brines underlying the lake cause high salinity of some spring-water, part of which is artificially diverted to the Lower Jordan River (Hurwitz et al. 1999). Since the early 1930th, the Degania dam restricts the outflow of Kinneret water into the Lower Jordan River.

North of Lake Kinneret, the Jordan River flows through diverse freshwater-rich landscapes of the Upper Jordan Valley (Fig. 1.2, no. 13). Its altitude ranges from ~ +300 m in the north, to ~ −210 at Lake Kinneret in the south. In the northernmost part, three main perennial tributaries of the Jordan flow through low plateaus in basalt, tufa and alluvium (Rimmer and Salingar 2006; Shtober-Zisu and Inbar 2014). Their confluence marks the northern border of the Hula Valley–a deep pull-apart basin filled with ~4 km deep sediments (Fig. 1.13) (Zilberman et al. 2000). The valley was occupied by the Hula Lake and marsh at an elevation of ~70 m, until it was artificially drained during the 1950s. Today, only small patches of water still exist, preserving some wetland ecosystems and important bird migration stopover sites. South of the Hula Valley, the higher Korazim block is the main area in the Dead Sea rift covered with basalt, a part of the Harrat Ash-Shaam volcanic field (Heimann and Ron 1993; Weinstein et al. 2020). The Jordan River has entrenched its course along the main transform fault at the eastern side the of Korazim plateau.

1.2.4 The Eastern Shoulder of the Dead Sea Transform

The area to the east of the Dead Sea Transform in Israel includes two physiographic unit: the Golan Heights and Mt. Hermon. The Golan Heights plateau (Fig. 1.2, no. 14) is the westernmost part of the Bashan region, part of the Harrat Ash-Shaam volcanic field (Mor 1993; Mor et al. this volume). The Golan slopes from ~1300 m asl in the north-east to ~300 m asl in the south. It is covered by lava flows of Pliocene age in the south, whereas in the north, effusive activity was active until the Late Pleistocene. The basalt plateau is dotted by tephra cones (Fig. 1.14A), comprising mainly scoria, as well as volcanic depressions, such as maars (Shaanan et al. 2011) (Fig. 1.14B) and pit craters (Frumkin and Naor 2019). The edges of the plateau are dissected by deep canyons draining into the down-faulted Jordan Valley (Fig. 1.14C) (Shtober-Zisu et al. 2018). Paleosols separating lava flows give rise to perched aquifers which supply abundant water to springs and streams (Dafny et al. 2006).

Fig. 1.13 Hula Valley, view to south-east. The Golan Heights (in the background) border the graben to the east

Fig. 1.14 **A** The Golan Heights basaltic plateau, dotted with tephra cones (in the background) and incised by bedrock channels; **B** Birket Ram maar lake, 1 km in diameter, in the northern Golan Heights. Mt Hermon is on the left (Photo courtesy of Shai Abramson); **C** Nahal Daliyot canyon entrenches the basaltic plateau and flows towards Lake Kinneret

North of the Golan Heights, the southernmost part of Mt. Hermon (Fig. 1.2, no. 15) forms the highest mountain range in Israel, reaching 2224 m asl (Fig. 1.15). It comprises a faulted anticlinal ridge whose major uplift is attributed to a combination of a Syrian Arc fold and a right turn in the sinistral strike-slip of the Dead Sea transform. Deep stream incision occurs along major faults. Jurassic limestones, cropping out at the bulk of the ridge, as well as high precipitation including heavy snow, give rise to abundant karst features (Frumkin et al. 1998). Infiltration of much of this water through the karstified fracture system recharges the aquifer, giving rise to the large karstic springs (Fig. 1.16) feeding the Jordan River.

1.3 Anthropogenic Impact

In many regions of the country, the diversity of the landscape is deeply connected with human presence since prehistoric early times. Humans appeared in this region very early, in the early Paleolithic times, some 1.5 million years before present. Their traces are among the earliest found outside Africa. The use of fire ~780,000 years ago on a site near the Jordan River marks a turning point in the development of mankind.

Human interference in the Mediterranean environment exacerbates the negative natural biophysical processes, and the results are more frequent and more severe geomorphic events, such as floods, landslides and soil erosion (Nir 1983).

All rivers in Israel were affected by human activities and 95% of the rivers in the Mediterranean climate region are artificially regulated. Many streams were polluted for various lengths of time, while mitigation efforts take place with some success (e.g. Greenbaum 2007). The environmental impact of humans on drainage systems caused major anthropic landform changes, which are manifested by various geoarchaeologic and geomorphic indicators (Ackermann 2007; Ackermann and Kolska Horwitz this volume). Fluvial systems are changing from natural networks into altered systems. Human interference started in the early hydraulic civilizations, which perfected the major distributional systems of irrigation and water supply. Human activities increase the seasonality of flowing streams, with higher runoff rates during winter floods and less percolation into groundwater, decreasing spring discharge and lowering exploitable aquifers. The increased activity causes more flood damage and erosion, making water management a complex and expensive task. Sewage recycling for irrigation increases soil salinization, as salt flushing by the natural river system was impeded.

Forest fires have increased, and on average 5% of the forests in Israel are burnt each year. Urbanization changes much of the surface to an impervious one, increasing the runoff/rainfall ratio from few per cent under natural conditions to more than 50%. The coastal area of Israel is a narrow and fragile strip, with enormous human pressure for intense exploitation.

Large-scale effects are observed in the Dead Sea region. The south basin of the Dead Sea dried up during the early 1980s, and the lake level is falling by >1 m per year. Thousands of sinkholes have developed along the Dead Sea during recent decades, demonstrating the instability of coastal areas overlying aquifers which adjust to the falling lake levels over salt layers (Frumkin and Raz 2001; Yechieli et al. 2006; Frumkin et al. 2011).

Fig. 1.15 The northern Golan Heights dotted by cinder cones and the anticlinal structure of Mt. Hermon in the background

Fig. 1.16 A karst spring (En Bared) at the bottom of Mt. Hermon (background), flowing into the Jordan River

Sand dunes in the coastal belt and NW Negev have been highly impacted by anthropogenic effects such as grazing, agriculture, and building (Roskin et al. 2013).

Thus, humans have altered the landscapes since the beginnings of civilization, first as hunter-gatherers living in the open. Later on, with the beginning of animal domestication, humans became herdsmen and finally tillers of the soils. In the twentieth century and especially since the establishment of the State of Israel, humans have altered the landscape and landforms through modernization, extensive agriculture, dam building, roads, tunnels, settlements, wildfires and other practices that all result in the transformation of nature and the environment.

References

Ackermann O (2007) Reading the field: geoarchaeological codes in the Israeli landscape. Isr J Earth Sci 56:87–106

Almagor G, Garfunkel Z (1979) Submarine slumping in continental margin of Israel and Northern Sinai. AAPG Bull 63:324–340

Avni Y, Segev A, Ginat H (2012) Oligocene regional denudation of the northern Afar dome: Pre-and syn-breakup stages of the Afro-Arabian plate. GSA Bull 124:1871–1897

Bar N, Agnon A, Yehudai M, Lazar B, Shaked Y, Stein M (2018) Last interglacial sea levels and regional tectonics from fossil coral reefs in the northeast Gulf of Aqaba. Quat Sci Rev 191:41–56

Belmaker M (2010) Early Pleistocene faunal connections between Africa and Eurasia: an ecological perspective. In: Fleagle J, Grine, F, Beiden A, Leakey R, Shea JJ (Eds). Out of Africa I: The First Hominin Migration Out of Africa. Springer, pp 183–205.

Ben-Gai Y (2009) Subsurface geology of the southern Lake Kinneret (Sea of Galilee), Dead Sea transform-evidence from seismic reflection data. Isr J Earth Sci 58:163–175

Bookman (Ken-Tor) R, Enzel Y, Agnon A, Stein M (2004) Late Holocene lake levels of the Dead Sea. Geol Soc Am Bull 116(5–6):555–571

Cohen-Seffer R, Greenbaum N, Sivan D, Barmeir E, Croitoru S, Inbar M (2005) Late Pleistocene-Holocene marsh episodes along the Carmel coast, Israel. Quat Int 140–141:103–120

Crouvi O, Amit R, Enzel Y, Porat N, Sandler A (2008) Sand dunes as a major proximal dust source for late Pleistocene loess in the Negev Desert, Israel. Quat Res 70:275–282

Dafny E, Burg A, Gvirtzman H (2006) Deduction of groundwater flow regime in a basaltic aquifer using geochemical and isotopic data: the Golan Heights, Israel case study. J Hydrol 33:506–524

Davis M, Matmon A, Fink D, Ron H, Niedermann S (2011) Dating Pliocene lacustrine sediments in the central Jordan Valley, Israel—Implications for cosmogenic burial dating. Earth Planet Sc Lett 305:317–327

Dente E, Lensky NG, Morin E, Grodek T, Sheffer NA, Enzel Y (2017) Geomorphic response of a low-gradient channel to modern, progressive base-level lowering: Nahal HaArava, the Dead Sea. J Geophys Res Earth Surf 122:2468–2487

Enzel Y, Bookman R, Sharon D, Gvirtzman H, Dayan U, Ziv B, Stein M (2003) Late Holocene climates of the Near East deduced from Dead Sea level variations and modern regional winter rainfall. Quat Res 60:263–273
Finzi Y, Avni S, Maroz A, Avriel-Avni N, Ashckenazi-Polivoda S, Ryvkin I (2019) Extraordinary geodiversity and geoheritage value of erosional craters of the Negev Craterland. Geoheritage 11:875–896
Freund R (1970) The geometry of faulting in the Galilee. Isr J Earth Sci 19:117
Frumkin A (1993) Karst origin of the upper erosion surface in the Northern Judean Mountains, Israel. Isr J Earth Sci 41:169–176
Frumkin A (1997) The holocene history of the dead sea levels. In: Ben-Avraham Z, Gat Y, Niemi TM (eds) The dead sea—the lake and its setting: oxford monographs on geology and geophysics 36, Oxford Univ Press, pp 237–248
Frumkin A (2017) The underground water systems of Ma'abarta—Flavia Neapolis, Israel. Geoarchaeology 33:127–140
Frumkin A, Raz E (2001) Collapse and subsidence associated with salt karstification along the Dead Sea. Carbonates Evaporites 16(2):117–130
Frumkin A, Shimron AE, Miron Y (1998) Karst morphology across a steep climatic gradient, southern Mount Hermon, Israel. Z Geom Supp 109:23–40
Frumkin A, Ezersky M, Al-Zoubi A, Abueladas AR (2011) The Dead Sea sinkhole hazard: geophysical assessment of salt dissolution and collapse. Geomorphology 134:102–117
Frumkin A, Naor R (2019) Formation and modification of pit craters—example from the Golan volcanic plateau, southern Levant. Z Geom 62:163–181
Frumkin A, Barzilai O, Hershkovitz I, Ullman M, Marder O (2021a) Karst terrain in the western upper Galilee, Israel: Speleogenesis hydrogeology and human preference of Manot Cave. J Hum Evol 160:102618
Frumkin A, Pe'eri S, Zak I (2021b) Development of banded terrain in an active salt diapir: potential analog to Mars. Geomorphology 389:107824
Garfunkel Z (2009) The long-and short-term lateral slip and seismicity along the Dead Sea Transform: An interim evaluation. Isr J Earth Sci 58:217–235
Garfunkel Z, Ben-Avraham Z (1996) The structure of the Dead Sea basin. Tectonophysics 266:155–176
Ginat H, Zilberman E, Saragusti I (2003) Early Pleistocene lake deposits and Lower Paleolithic finds in Nahal (wadi) Zihor, Southern Negev desert, Israel. Quat Res 59:445–458
Ginat H, Opitz S, Ababneh L, Faershtein G, Lazar M, Porat N, Mischke S (2018) Pliocene-Pleistocene waterbodies and associated deposits in southern Israel and southern Jordan. J Arid Environ 148:14–33
Greenbaum N (2007) Assessment of dam failure flood and a natural, high-magnitude flood in a hyperarid region using paleoflood hydrology, Nahal Ashalim catchment, Dead Sea Israel. Water Resources 43:W02401
Hamawi M, Goren L, Mushkin A, Levi T (2022) Rectangular drainage pattern evolution controlled by pipe cave collapse along clastic dikes, the Dead Sea Basin, Israel. Earth Surf Proc Land 47:936–954
Haviv I, Enzel Y, Whipple KX, Zilberman E, Matmon A, Stone J, Fifield KL (2010) Evolution of vertical knickpoints (waterfalls) with resistant caprock: Insights from numerical modeling. J Geophys Res Earth Surf 115(F3)
Hazan N, Stein M, Marco S (2004) Lake Kinneret levels and active faulting in the Tiberias area. Israel J Earth Sci 53:199–205
Heimann A, Ron H (1993) Geometric changes of plate boundaries along part of the northern Dead Sea transform: geochronologic and paleomagnetic evidence. Tectonics 12:477–491
Hurwitz S, Goldman M, Ezersky M, Gvirtzman H (1999) Geophysical (time domain electromagnetic model) delineation of a shallow brine beneath a freshwater lake, the Sea of Galilee, Israel. Water Resour Res 35:3631–3638
Khayat S, Marei A, Geyer S, Rödiger T (2018) Investigating the complex hydrogeological settings in the northeastern slope of the West Bank to the Jordan Graben (Malih and Jeftlik). Euro-Med J Environ Integr 3:1–15
Lavee H, Imeson AC, Sarah P (1998) The impact of climate change on Geomorphology and desertification along a Mediterranean-Arid transect. Land Degrad Dev 9:407–422
Lensky NG, Calvo R, Sade AR, Gavrieli I, Katz O, Hall JK, Enzel Y, Mushkin A (2014) The scarred slopes of the dead sea–evidence for intensive subsea landsliding. In: Israel geological society proceedings, p 81
Levy Y, Burg A, Yechieli Y, Gvirtzman H (2020) Displacement of springs and changes in groundwater flow regime due to the extreme drop in adjacent lake levels: the Dead Sea rift. J Hydrol 587:124928
Lisker S, Porat R, Frumkin A (2010) Late Neogene rift valley fill sediments preserved in caves of the Dead Sea Fault Escarpment (Israel): palaeogeographic and morphotectonic implications. Sedimentology 57:429–445
Magal E, Arbel Y, Caspi S, Glazman H, Greenbaum N, Yechieli Y (2013) Determination of pollution and recovery time of karst springs, an example from a carbonate aquifer in Israel. J Contam Hydrol 145:26–36
Matmon A, Wdowinski S, Hall JK (2003) Morphological and structural relations in the Galilee extensional domain, northern Israel. Tectonophysics 371(1–4):223–241
Matmon A, Simhai O, Amit R, Haviv I, Porat N, McDonald E, Benedetti L, Finkel R (2009) Desert pavement–coated surfaces in extreme deserts present the longest-lived landforms on Earth. Geol Soc Am Bull 121(5–6):688–697
Mor D (1993) A time-table for the Levant Volcanic Province, according to K-Ar dating in the Golan Heights. Israel. J Afr Earth Sci 16(3):223–234
Morin E, Jacoby Y, Navon S, Bet-Halachmi E (2009) Towards flash-flood prediction in the dry Dead Sea region utilizing radar rainfall information. Adv Water Resour 32(7):1066–1076
Nir D (1983) Man, a geomorphological agent: an introduction to anthropic geomorphology. Springer Science & Business Media.
Nuriel P, Weinberger R, Kylander-Clark ARC, Hacker BR, Craddock JP (2017) The onset of the Dead Sea transform based on calcite age-strain analyses. Geology 45:587–590
Rimmer A, Salingar Y (2006) Modelling precipitation-streamflow processes in karst basin: the case of the Jordan River sources, Israel. J Hydrol 331:524–542
Roskin J, Katra I, Blumberg DG (2013) Late Holocene dune mobilizations in the northwestern Negev dunefield, Israel: a response to combined anthropogenic activity and short-term intensified windiness. Quat Int 303:10–23
Rozenbaum AG, Stein M, Zilberman E, Gelband DS, Starinsky A, Sandler A (2022) Sr isotopes in the Tortonian-Messinian Lake Bira and Gesher marshes, Northern Valleys of Israel: implications for hydroclimate changes in East Mediterranean-Levant margins. GSA Bull 134:762–775
Ryb U, Matmon A, Erel Y, Haviv I, Katz A, Starinsky A, Angert A (2014a) Controls on denudation rates in tectonically stable Mediterranean carbonate terrain. GSA Bull 126:553–568
Ryb U, Matmon A, Erel Y, Haviv I, Benedetti L, Hidy AJ (2014b) Styles and rates of long-term denudation in carbonate terrains under a Mediterranean to hyper-arid climatic gradient. Earth Planet Sci Lett 406:142–152
Ryb U, Matmon A, Haviv I, Benedetti L (2015) Exhumation and uplift coupled with precipitation along the western Dead Sea Rift margin. Geology 43:483–486

Segev A (2017) Zemah-1, a unique deep oil well on the Dead Sea fault zone, northern Israel: a new stratigraphic amendment. Geol Surv Isr Rep GSI/21/2017

Shahar J (1994) The Syrian arc system—an overview. Paleaogeogr Palaeocl 112:125–142

Shaanan U, Porat N, Navon O, Weinberger R, Calvert A, Weinstein Y (2011) OSL dating of a Pleistocene maar: Birket Ram, the Golan heights. J Volcanol Geotherm Res 201:397–403

Shtober-Zisu N, Inbar M (2014) Chapter 6: geomorphology. In: Zohary T, Sukenik A, Berman T, Nishri A, (eds) Lake Kinneret: ecology and management, vol 6. Springer, pp 69–77

Shtober-Zisu N, Amasha H, Frumkin A (2017) Inland notches: lithological characteristics and climatic implications of subaerial cavernous landforms in Israel. Earth Surf Proc Land 42:1820–1832

Shtober-Zisu N, Inbar M, Mor D, Jicha BR, Singer BS (2018) Drainage development and incision rates in an upper pleistocene basalt-limestone boundary channel: the sa'ar stream, golan heights, Israel. Geomorphology 303:417–433

Shtober-Zisu N, Zissu B (2018) Lithology and the distribution of Early Roman-era tombs in Jerusalem's necropolis. Prog Phys Geogr Earth Environ 42:628–649

Shtober-Zisu N, Brook A, Zissu B (2024) Soil denudation in the northwestern Negev (Israel) following the Late Byzantine–Early Islamic period. Geomorphology 446: 108983

Sivan D, Greenbaum N, Cohen-Seffer R, Sisma-Ventura G, Almogi-Labin A (2011) The origin and disappearance of the late Pleistocene–early Holocene short-lived coastal wetlands along the Carmel coast, Israel. Quat Res 76:83–92

Sneh A (1996) The Dead Sea Rift: lateral displacement and downfaulting phases. Tectonophysics 263:277–292

Stein M (2001) The sedimentary and geochemical record of Neogene-Quaternary water bodies in the Dead Sea Basin-inferences for the regional paleoclimatic history. J Paleolimn 26:271–282

Steinberg J, Gvirtzman Z, Folkman Y (2010) New age constraints on the evolution of the Mt Carmel structure and its implications on a Late Miocene extensional phase of the Levant continental margin. J Geol Soc 167:203–216

Tal A (2019) The implications of climate change driven depletion of Lake Kinneret water levels: the compelling case for climate change-triggered precipitation impact on Lake Kinneret's low water levels. Sci Total Environ 664:1045–1051

Weinberger R, Nuriel P, Kylander-Clark AR, Craddock JP (2020) Temporal and spatial relations between large-scale fault systems: evidence from the Sinai-Negev shear zone and the Dead Sea Fault. Earth Sci Rev 211:103377

Weinstein Y, Nuriel P, Inbar M, Jicha BR, Weinberger R (2020) Impact of the Dead Sea transform kinematics on adjacent volcanic activity. Tectonics 39: 2019TC005645

Yechieli Y, Abelson M, Bein A, Crouvi O, Shtivelman V (2006) Sinkhole 'swarms' along the Dead Sea coast: reflection of disturbance of lake and adjacent groundwater systems. Geol Soc Am Bull 118:1075–1087

Zilberman E (2000) Formation of makhteshim unique erosion cirques in the Negev, southern Israel. Isr J Earth Sci 49:127–141

Zilberman E, Amit R, Heimann A, Porat N (2000) Changes in Holocene paleoseismic activity in the Hula pull-apart basin, Dead Sea Rift, northern Israel. Tectonophysics 321:237–252

Zilberman E, Calvo R (2013) Remnants of Miocene fluvial sediments in the Negev Desert, Israel, and the Jordanian Plateau: evidence for an extensive subsiding basin in the northwestern margins of the Arabian plate. J Afr Earth Sci 82:33–53

2 Geology and Relief Development of Israel

Amos Frumkin

Abstract

Until the Eocene, Israel region was mostly affected by the passive margin conditions prevailing at the northwest edge of the African Plate, associated with epeirogenic movements and eustatic fluctuations close to the sea-land transition. The Syrian Arc folding, active since the Santonian, has some implications on the present-day relief. During the late Eocene–Oligocene, an extensive regression and regional uplift shifted the coast to the NW exposing the area to severe denudation. Moderate tectonism occurred since the Oligocene and the early Miocene and increased through the break-up of the Arabian–African plates along the Dead Sea transform since the early Miocene (Nuriel et al., 2017). Since the Miocene until today, the relief of Israel was enhanced mainly by deep depressions along the Dead Sea fault system, accompanied by arching of the hilly backbone of Israel, where a new water divide was established. Erosion, denudation and deposition have played a mixed role of shaping the landscape over geological timescales. The tectonic-based relief has been reduced and modified to gentle convex hills under Mediterranean environment. Fluvial erosion responding to backbone uplift coeval with lowering base levels has dissected the landscape and created canyons and cliffs, in particular under dry environments and in cohesive rocks, mainly along the Dead Sea transform margin.

A. Frumkin (✉)
Institute of Earth Sciences, The Hebrew University of Jerusalem, 9190401 Jerusalem, Israel
e-mail: amos.frumkin@mail.huji.ac.il

2.1 Introduction

The geology of Israel is briefly discussed below, starting with the Neoproterozoic Era when the regional crust was formed as part of the Pan-African orogeny. The presented timeline is roughly defined, because numerous events, in particular tectonic deformations, recurred intermittently over long periods, while only the main ones can be mentioned in this short review.

The stratigraphic column of Israel can be subdivided by major unconformities into lithostratigraphic and chronostratigraphic units from the Neoproterozoic basement onwards. The basement is covered in most of the country by a wedge of deposits commonly thickening towards the NW, the main direction of the Tethys-Neotethys-Mediterranean basins. Several major unconformities mark periods of significant uplift, erosion and truncation (Fig. 2.1).

Neoproterozoic and Paleozoic rock units crop out only near Eilat in southernmost Israel, while they are more common in Jordan and Sinai (Egypt). Mesozoic to Eocene rocks comprise most outcrops in the hilly lands of Israel. The lowlands and depressions are filled with late Cenozoic sediments. Several volcanic episodes occurred across the geological history (Segev and Rybakov 2010), as observed in the columnar section (Fig. 2.2). The most recent event, of Plio-Pleistocene age, covered most of the Golan Heights and part of the eastern Galilee with basalt and some pyroclasts.

All these units were subjected to tectonic deformation, which generally increased the relief throughout the late Cenozoic. These tectonic processes are associated with the location of Israel on the Israel-Sinai microplate, bordered by the Dead Sea transform margin in the east, which separates it from the Arabian plate. The Arabian plate moves away from the African plate along the Red Sea spreading margin and collides with the Eurasian plate along the Zagros collision margin orogenic belt (Fig. 2.3).

A. Frumkin and N. Shtober-Zisu (eds.), *Landscapes and Landforms of Israel*,
World Geomorphological Landscapes, https://doi.org/10.1007/978-3-031-44764-8_2

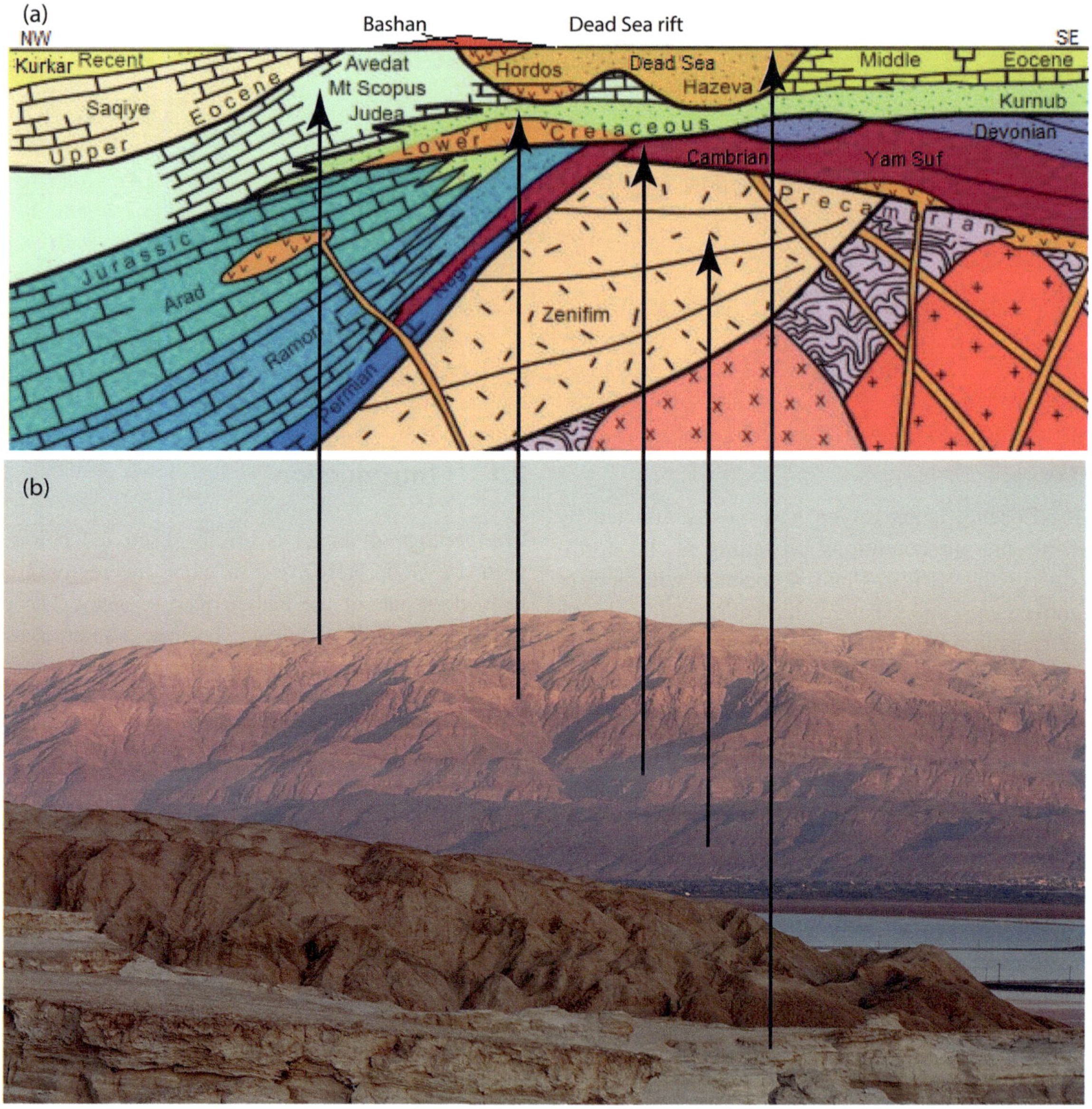

Fig. 2.1 **a** Schematic stratigraphic section of Israel from the Mediterranean (NW) to Wadi Ram, east of Eilat (SE), modified after Freund (1977); Flexer et al. (2005). Heavy lines represent regional unconformities. Important ages and associated groups/formations, respectively: Un-named Precambrian Neoproterozoic basement rocks (metamorphic, granites and volcanics, lower right corner of a); Neoproterozoic Zenifim Formation (arkose, conglomerate, igneous); Cambrian Yam-Suf Group (sandstone, shales, limestone); Devonian (sandstone, exposed in Jordan); Permian Negev Group (sandstone, shales, limestone, exposed in Jordan); Triassic Ramon Group (limestone, shales, anhydrite, sandstone); Jurassic Arad Group (limestone, sandstone, shales); lower Cretaceous Kurnub Group (sandstone, shales, limestone, igneous); Albian-Turonian Judea Group (limestone, dolomite, shales); Senonian-Paleocene Mount Scopus Group (chalk, marl, shales, chert, phosphates); early to middle Eocene Avedat Group (limestone, chalk, chert); late Eocene to Pliocene Saqiya Group (shales, sandstone, limestone, anhydrite, basalt); Quaternary Kurkar Group ('kurkar' aeolianite, sands, loess); Miocene Hordos Formation (sandstone, conglomerate, marl, chalk, volcanics); early Miocene Hazeva Group (sandstone, conglomerate, marl, chalk); Plio-Pleistocene Dead Sea Group (lacustrine, conglomerates, sandstone, volcanics). Plio-Pleistocene Bashan Group (volcanics). **b** The major rock units exposed at the south Dead Sea region, with arrows showing their stratigraphic setting (looking east). The Precambrian to Mesozoic units are exposed at the eastern fault of the Dead Sea transform (background), while the late Cenozoic Dead Sea Group is exposed in the western side of the Dead Sea (foreground) (*photo* A. Tsabar)

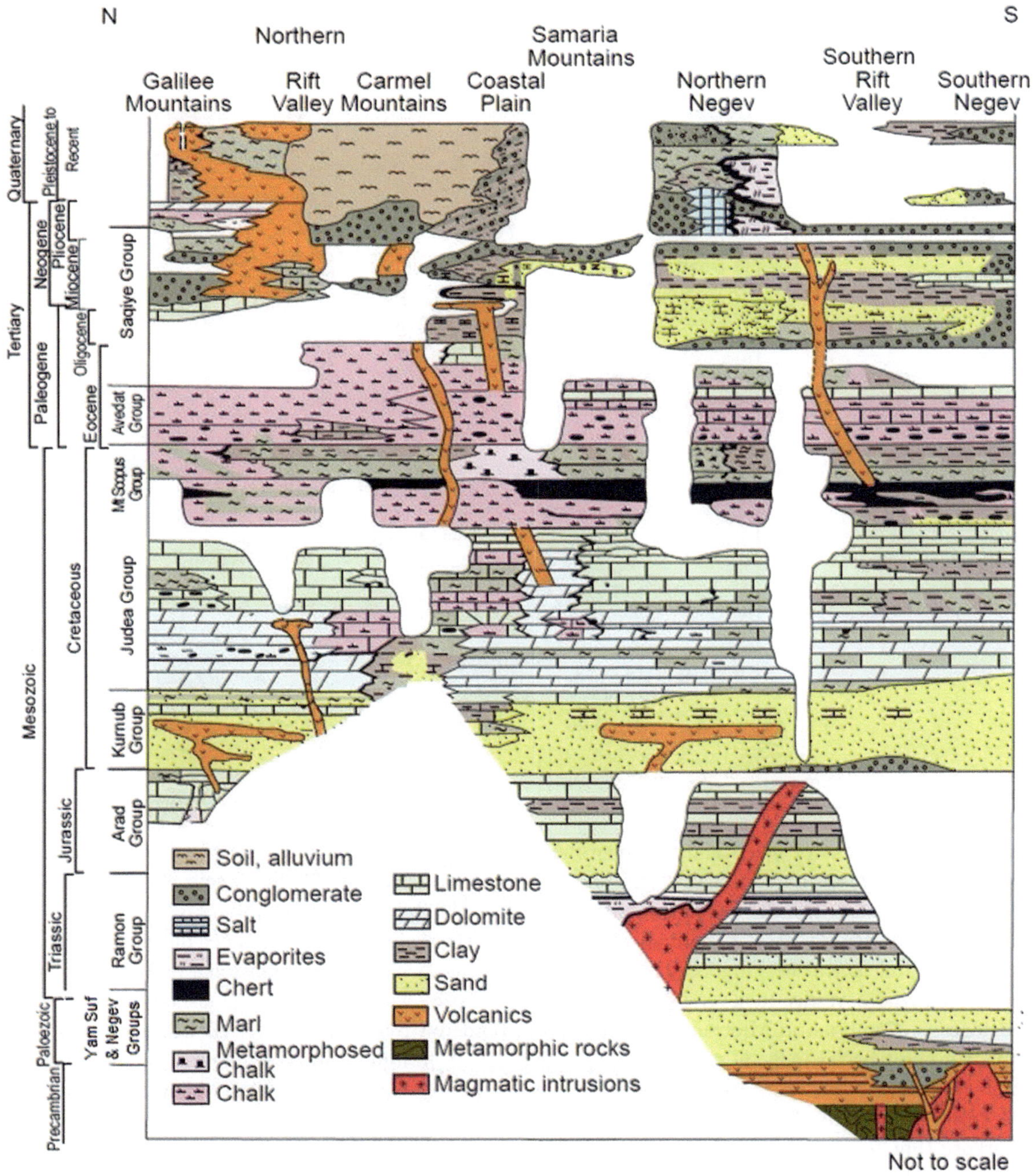

Fig. 2.2 Schematic columnar section of Israel (modified after Bartov and Arkin 1980; Gvirtzman 2004). For major regional unconformities see Fig. 2.1

2.2 The Neoproterozoic–Paleozoic Era

Israel is situated in the northern part of the Arabian–Nubian Shield. The Neoproterozoic shield is exposed along the shoulders of the Red Sea spreading axis (Fig. 2.3) which have been rising since the Oligocene (Avni et al. 2012). The shield is exposed in the Eilat region of southern Israel, while further north it is buried under north-west thickening Phanerozoic deposits (Figs. 2.1 and 2.4). The Arabian-Nubian shield evolved by complex tectonomagmatic, metamorphic, and volcanic processes during the Neoproterozoic/early Cambrian as a part of the pan-African orogeny (Fig. 2.5a) (Garfunkel 1999). The uplifted orogen was eroded and truncated by a 532 Ma peneplain (Fig. 2.5b), whose age is constrained by the youngest underlying dykes. This early Cambrian peneplain was associated with extensive erosion of ~2000 m of the section and by intensive chemical weathering (Beyth et al. 2014). The

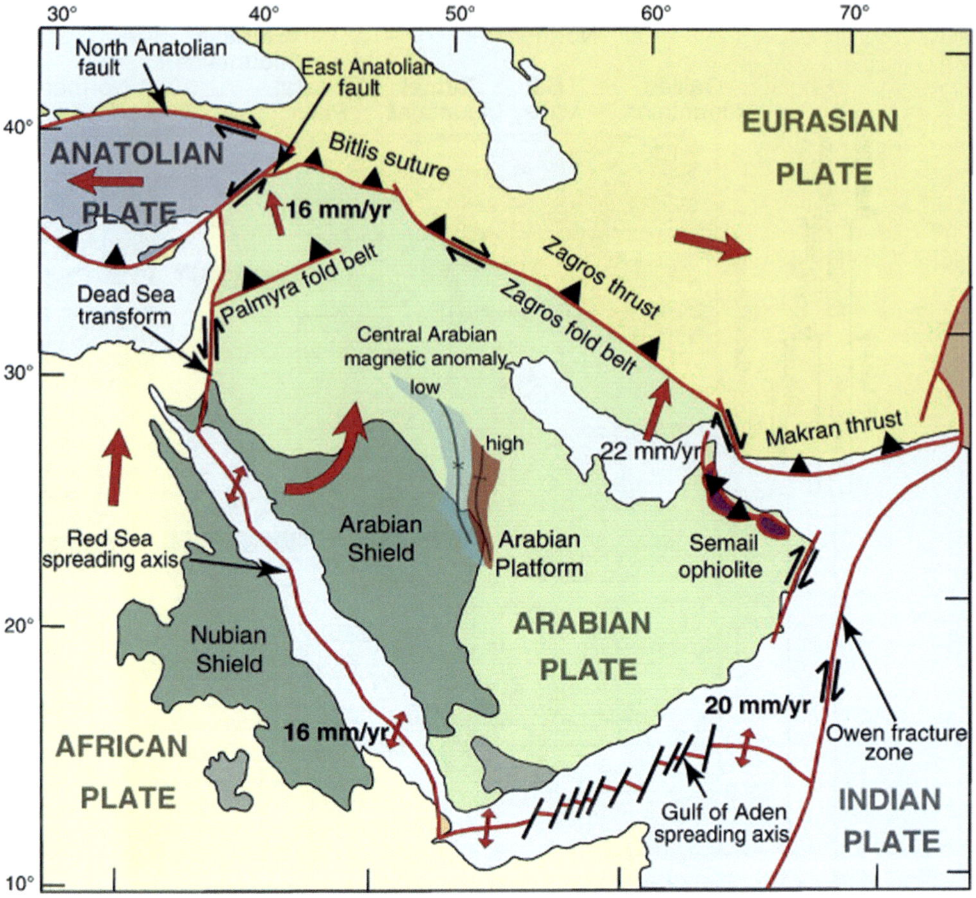

Fig. 2.3 The tectonic configuration of Israel and the Middle East. Israel forms part of the Israel-Sinai subplate, bordered by the Dead Sea transform margin from the Arabian plate. The Arabian plate moves away from the African plate along the Red Sea spreading margin, and collides with the Eurasian plate along the Zagros collision margin orogeny (Stern and Johnson 2010)

Precambrian crystalline rocks of the Arabian-Nubian massif were exposed and eroded in additional stages, the latest of which was associated with the shoulder uplift along the Red Sea rift system (Picard 1943; Zilberman and Calvo 2013) that has occurred since the Oligocene. The eroded and transported material gave rise to several generations of quartz sandstones of Cambrian to Miocene age, preserved mainly within subsiding basins in the north-western Arabian plate. Some of these deposits were exposed, eroded and redeposited, forming recycled clastic sequences. Cambrian sandstone, arkose (Fig. 2.5c) and dolostone are exposed in Eilat region, followed by a major unconformity (Fig. 2.1).

The post-Cambrian part of the Paleozoic sequence, comprising mostly sandstone, is partly preserved at the NW Arabian platform in the Arabian Peninsula, but is not exposed in Israel, mostly due to late Carboniferous–early Cretaceous truncation events.

2.3 Mesozoic to Eocene

Mesozoic and Cenozoic sedimentation and volcanism formed most of the sedimentary sequence exposed in Israel (Garfunkel and Derin 1984). During the Mesozoic–early Cenozoic times, Israel was located at the northern edge of the African plate, where the coast of the Neotethys Ocean, followed by the Mediterranean Sea, intermittently submerged the plate margins.

The disintegration of Pangea during the Triassic–Jurassic was associated with extensional deformation in the region. Triassic evaporites were deposited in extensional basins bordered by normal faults (Druckman 1974; Bialik et al. 2012), forming a thick sequence of gypsum and anhydrite, currently exposed only in the Makhtesh Ramon (Fig. 2.6).

A passive plate margin setting dominated the region during the rest of the Mesozoic. A wedge of marine sediments was formed across the crustal transition zone underlying Israel (Schattner and Ben-Avraham 2007), thickening towards the subsiding area of the present Eastern Mediterranean (Fig. 2.1). The marine carbonates occasionally interfinger with continental clastic deposits originating from erosion of the Arabian-Nubian shield in the south-east (Flexer et al. 1989). Today, the highlands' best-exposed units are thick marine sediments of the Judea, Mt. Scopus and Avedat Groups, comprising massive limestone and dolostone with marls and chalk. The common depositional environment was a wide shallow carbonate platform with barrier reefs (Sass and Bein 1978). Local reef patches produced scattered massive limestone bodies, partly dolomitized and silicified (Fig. 2.7).

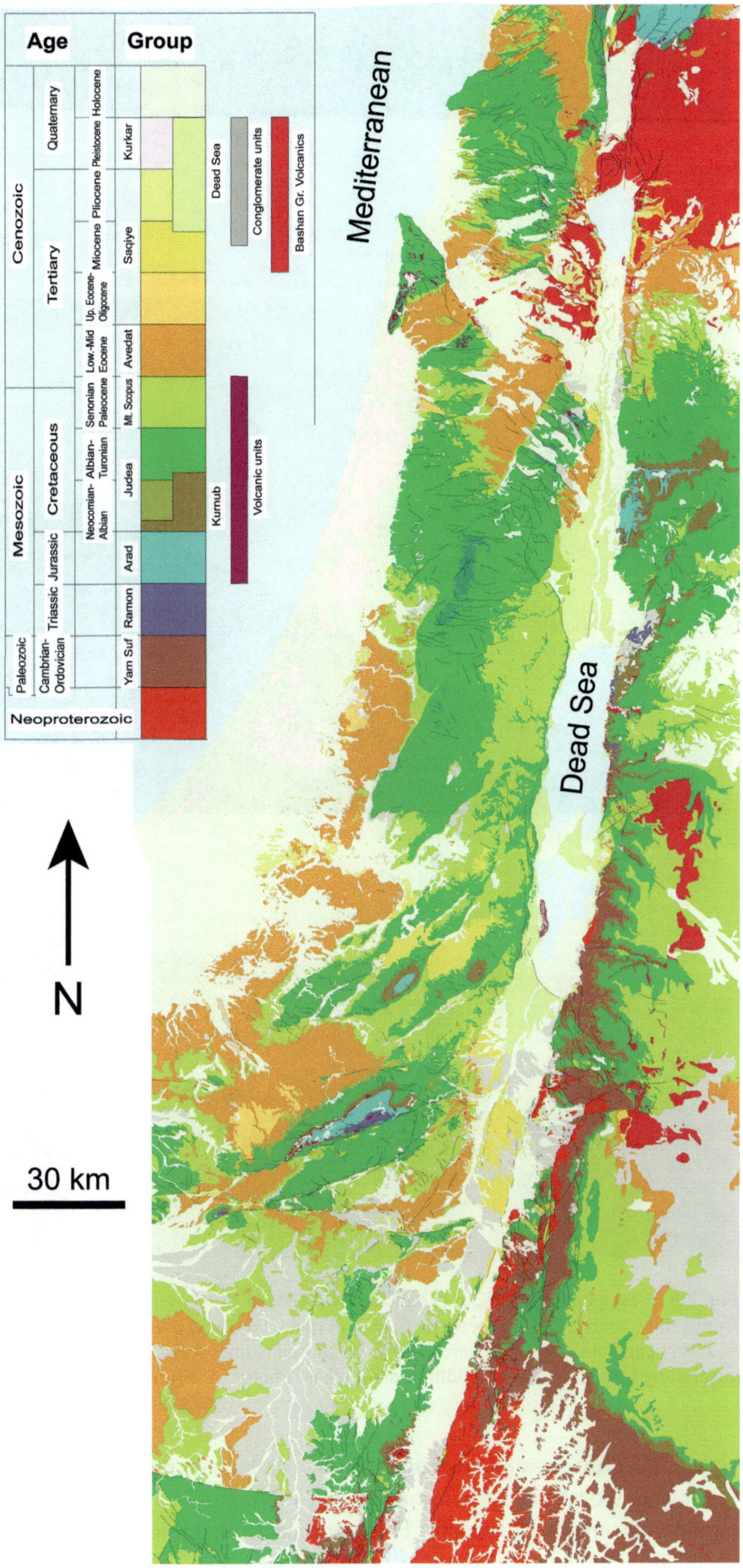

Fig. 2.4 General geological map of Israel (modified after Srebro 2011, and the Geological Survey of Israel)

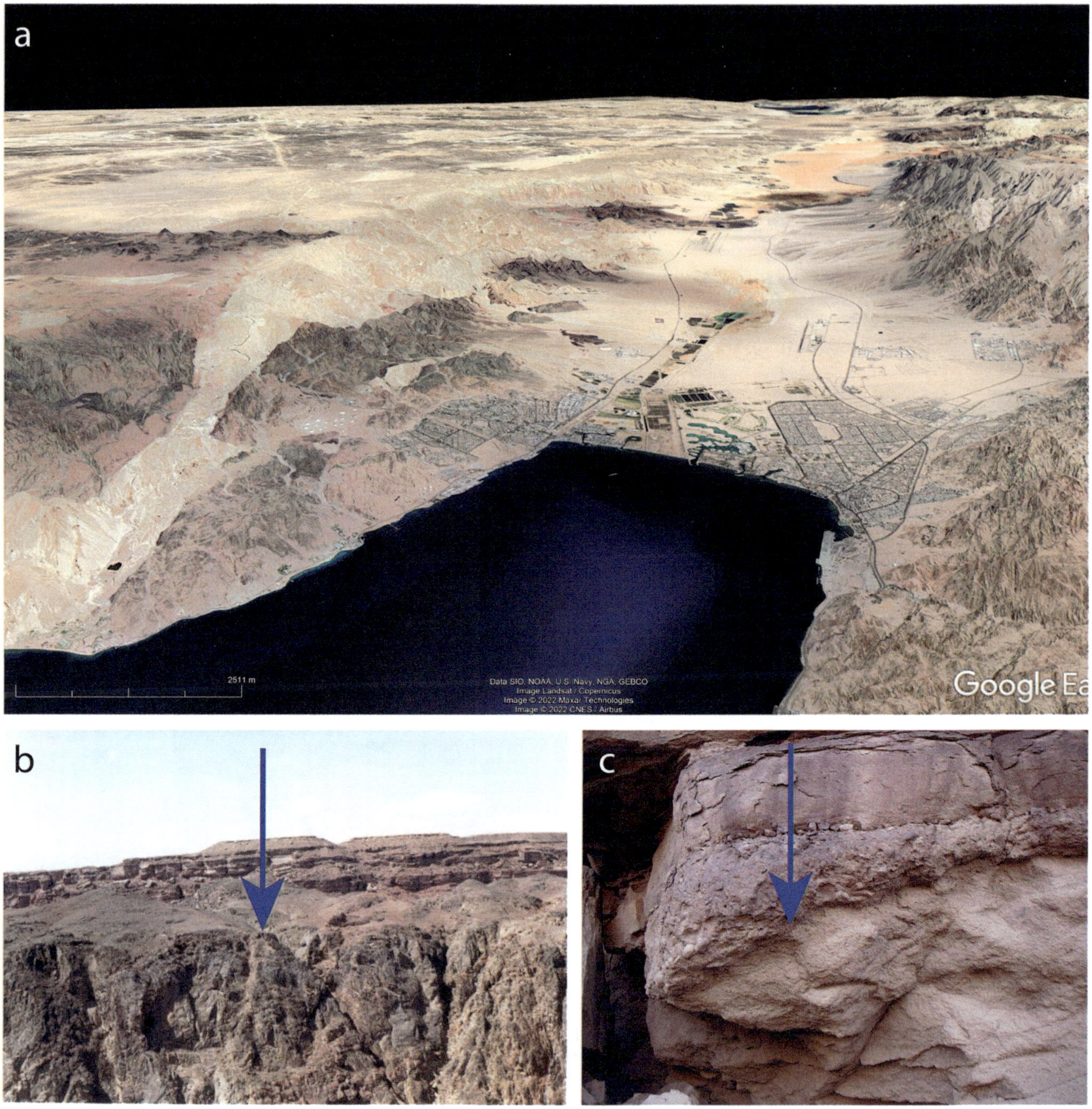

Fig. 2.5 Basement magmatic rocks of the Pan-African orogeny and overlying Cambrian clastics at Eilat region. **a** Oblique northward view from Eilat (Aqaba) Gulf. In general, basement rocks are dark, overlying sandstones are reddish, and carbonates are whitish-yellowish; **b** The 532 Ma peneplain (arrow), covered by Cambrian sandstones west of Eilat, width of photo ~1 km; **c** Close up of the 532 Ma peneplain (arrow), covered by Cambrian arkose, width of photo ~2 m. Photos: A. Frumkin

Marine transgression over the shallow carbonate platform was occasionally interrupted by regressions and periods of subaerial exposure. Following each regression, eogenetic karstification (prior to burial of the rock, *sensu* White and White 2013) developed to varying degrees (Fig. 2.6d), shallow braided rivers formed and pedogenesis occurred. These are recorded as paleokarst features within the bedrock sequence, which in some occasions have been reactivated by later telogenetic karstification (post-burial and compaction).

During the early Cretaceous, a magmatic hot spot was active in the Levant, causing uplift, tectonics, volcanism (Fig. 2.8) and subsequent wide-scale erosion and unconformities (Gvirtzman and Garfunkel 1998).

Fig. 2.6 Geologic 'window' exposing geological features of the Negev, formed by erosion of Makhtesh Ramon. **a** Google Earth view of Makhtesh Ramon, scale (lower left) is 7.15 km; **b** Triassic gypsum and anhydrite. **c** Jurassic sediments, with the northern wall of Makhtesh Ramon in the background; **d** Paleokarst doline on Triassic limestone, later filled by shales, and exposed by recent quarrying. Photos: A. Frumkin

The Syrian Arc structure, commonly observed as asymmetric folds extending across Israel from SW to NE, formed during the late Cretaceous above and along deep faults (Shahar 1994). The compressional stress field of the late Cretaceous/early Cenozoic, manifested as reverse faults and associated Syrian Arc folds, was accompanied by intraplate shortening and reactivation of ancient normal faults (Freund et al. 1975; Segev and Rybakov 2010). These folds were tectonically rejuvenated, uplifted and subsequently erosionally truncated following exposure since the late Eocene (Fig. 2.9). The Syrian Arc is until today the dominant fold structure of Israel (Figs. 2.10 and 2.11).

East-west trending faults, commonly referred to as the Negev–Sinai Shear Zone (Bartov 1974; Weinberger et al. 2020), have been active intermittently since the late Cretaceous to post-Miocene times. This system of mostly right-lateral strike-slip faults extends from central Sinai eastward through the Negev, and apparently also into the Galilee, Jordan and Lebanon. The displacement developed between the Triassic and the late Pleistocene (Bartov 1974; Matmon et al. 2003), forming push-up structures and depressions.

The Senonian-Paleocene Mount Scopus Group comprises mainly chalk and marl. An intercalated chert unit is

Fig. 2.7 Cliff of dolomitized Albian reef at Nahal Me'arot, Mount Carmel. **a** View southward; **b** Close up of rudist fossils within the reef (Frumkin and Weinstein-Evron, 2021). Photos: A. Frumkin

much more resistant to erosion, forming distinct landforms following exposure (Fig. 2.12). The last regional marine transgression during the middle Eocene deposited mainly chalk and limestone with chert nodules (Fig. 2.13). Eocene outcrops remain uneroded mainly at the Syrian Arc synclines, while on anticlines they have been truncated (Figs. 2.4 and 2.10). Indications of the pre-Oligocene relief, which was subsequently eroded, is mainly observed within the geological sequence, while since the Oligocene the present landscape started to evolve.

2.4 Late Eocene to Early Miocene and Emergence of Moderate Relief

A late Eocene regression exposed the anticlinal crests of Israel, which had been previously submerged under the Neotethys Ocean (Fig. 2.14), while the coastline shifted to the west and north (Buchbinder et al. 1993, 2005; Gvirtzman et al. 2011).

The regression and associated uplift exposed much of southern and central Israel, as well as a vast Afro-Arabian dome to the south. The following Oligo-Miocene erosion formed a regional relatively flat, low-relief, truncated terrain (Figs. 2.10 and 2.15), extending until today throughout the eastern Mediterranean and Levant (Quennell 1958; Garfunkel 1988; Frumkin 1992; Segev et al. 2011; Avni et al. 2012). Anticline crests may have been exposed and truncated already at Middle Eocene times (Mimran 1984; Lewy et al. 1995), but the associated evidence may be interpreted as submarine erosion.

The subaerial Oligo-Miocene truncation resulted from long-term dissolution and erosion. The topographic surface was deformed by subsequent tectonic phases of uplift and faulting since the middle Miocene (e.g. Zilberman 1991; Avni et al. 2000; Bar et al. 2016). Total structural uplift

Fig. 2.8 Early Cretaceous volcanics (centre and low left) filling a graben within Jurassic limestone (upper right). Wadi Malih, north-east Samaria. Photo: A. Frumkin

during various phases amounted up to 3000 m in some areas of Israel. Concurrently, tectonic depressions have continued to accumulate sediments until today, mainly in association with the evolution of the Dead Sea transform and the development of the Mediterranean coastal plain.

During the early Miocene the regional fluvial systems still drained the uplifted northern margins of the Red Sea rift across the NW Arabian platform towards the Mediterranean, as indicated by far-field early Miocene allochthonous sediments interfingering with lacustrine deposits in the Negev, termed the Hazeva Group (Fig. 2.1a) (Calvo and Bartov, 2001). Some of these sediments are preserved over the high plateaus of central and northern Negev and in the Dead Sea transform (Zilberman and Calvo 2013). While the early Miocene large-scale fluvial systems crossed the Negev, the central Israel highlands seem to have been elevated without traces of fluvial systems crossing them from Arabia, since the late Oligocene.

2.5 Dynamic Base Levels, Break-Up and Volcanism Since Early Miocene

2.5.1 Early Miocene Erosion

During early Miocene times, the east Mediterranean area was set as a deep marine basin. The deep water conditions combined with a steep slope and ample sediment supply from large streams arriving from Arabia set the stage for marine erosion processes as a result of slumping, sliding and turbidity currents, eventually resulting in the incision of submarine canyons (Druckman et al. 1995). The largest canyon crossing the ancient shelf-edge area was the Afiq Channel (near Ashqelon) (Gvirtzman and Buchbinder 1978), which is presently filled by Oligocene to recent sediments.

2.5.2 Oligocene to Miocene Tectonism

The tectonic deformation rates intensified since the Oligocene, exceeding the rates of highland denudation and lowland sedimentation. The result is the present relief, amounting to ~3 km between Mt. Hermon and the Dead Sea (Fig. 2.11).

The tectonic break-up of the African plate along the Red Sea rift began in the Oligocene and was intensified in the early Miocene. During the Miocene (18–14 Ma), the N-S trending Dead Sea transform started acting with a left-lateral motion, separating the Israel-Sinai subplate from the Arabian plate (Quennell 1958; Freund et al. 1970; Garfunkel, 1997; Nuriel et al. 2017). The Dead Sea transform was accompanied by a sub-parallel uplift belt, 60–80 km wide, observed in the Negev (Avni 2017), Galilee, and the Judean Hills (Fig. 2.15c). Until the middle Miocene, the pure strike-slip motion of the Dead Sea transform did not block streams originating in the Arabian plate from flowing to the Mediterranean Sea, although local pull-apart depressions developed along the transform. Rates of sedimentation in these depressions still prevailed over subsidence.

Fig. 2.9 Flexures of the Syrian Arc within Cretaceous carbonates, exposed by consequent canyons. Cliff-forming Turonian limestone is covered by softer Senonian chalk; **a** Eastern flexure of Ramallah Anticline, entrenched by Wadi Wahita canyon; **b** The flexure between Naftali anticline and Wadi Duba syncline entrenched by Nahal Dishon. Photos: A. Frumkin

Since the late Miocene, a directional change in plate motion by a few degrees introduced an extension component to the Dead Sea transform, and triggered significant subsidence within the developing rift and uplift along its margins (Garfunkel 1981). Thus, an elongated rift formed along the Dead Sea transform, between the Red Sea in the south and Lebanon in the north. The Dead Sea transform-rift is part of a >1000 km long fault system dividing the Arabian plate from the Israel–Sinai microplate. Several markers demonstrate a 105 km left-lateral strike-slip motion along this plate boundary (Quennell 1958; Freund et al. 1970; Weinberger et al. 2009).

2.5.3 Miocene to Pliocene Development of the Dead Sea Basin and Formation of the Current Water Divide

During the late Miocene, a temporary connection between the Dead Sea basin and the Mediterranean Sea was established, via the Yizre'el and the Jordan valley structural basins. The Bira and Sedom lagoons existed along the Dead Sea transform (Rozenbaum et al. 2016). This connection was severed not later than the early Pliocene due to the arching west of the Jordan valley and/or by the Messinian salinity crisis during which the Mediterranean Sea level fell

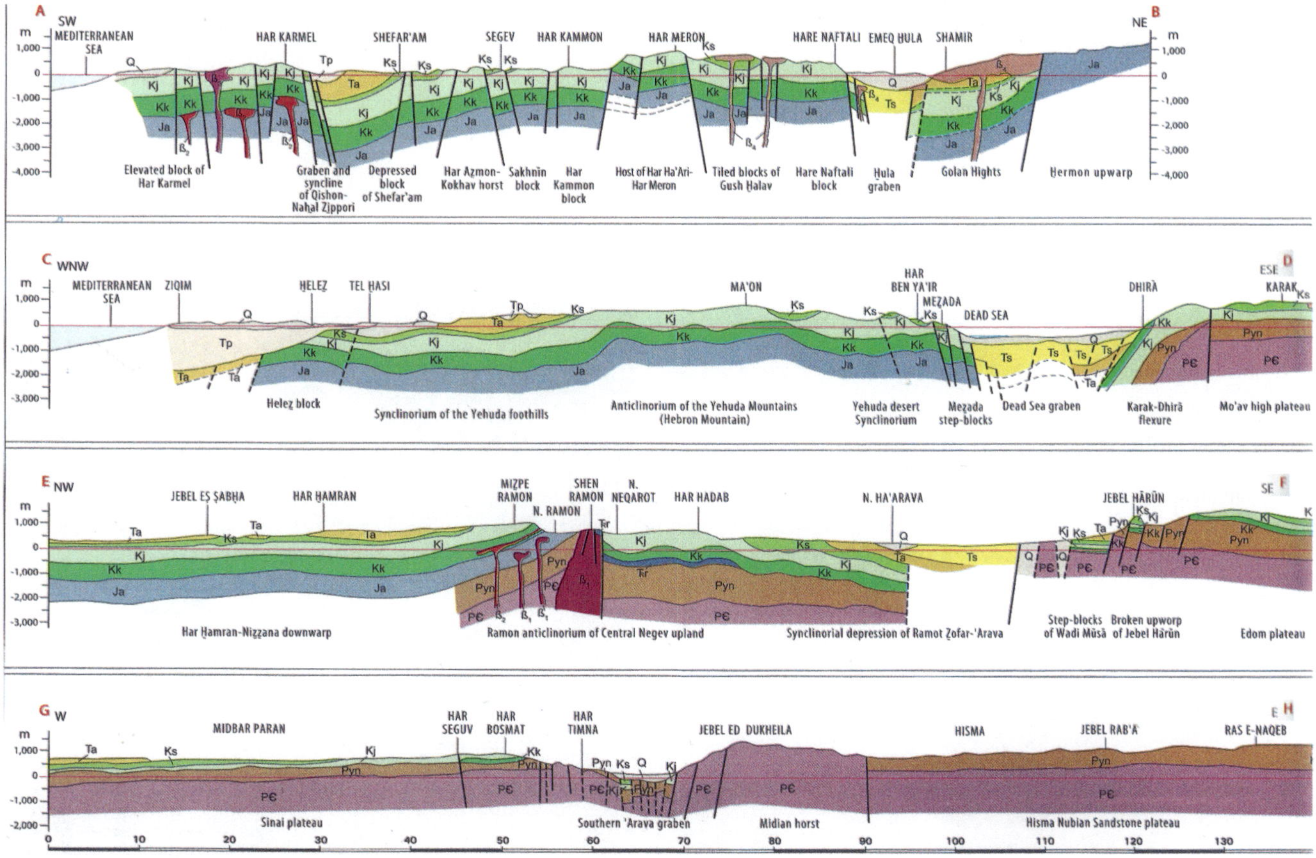

Fig. 2.10 Geological sections across Israel (modified after Srebro 2011). Note the Syrian Arc folds which are partly reflected in present landscape, depending on local rejuvenation. The dominant faults belong to the Dead Sea Transform extending from Hula graben in the north to Southern Arava graben in the south. Further notes are underscored below. Section A-B shows the highly faulted character of northern Israel. The two main horsts are Mt. Carmel and Mt. Meron, but many other faults exist, still active in the Plio-Pleistocene. The most uplifted and denuded block is the Hermon upwarp, combining an anticline and a faulted block. Section C-D shows the Judea anticlinorium bounded by the synclinoriums at the west and east, and the deep young deposits along the Mediterranean coastal belt and Dead Sea transform. Section E-F shows the central Negev Ramon anticlinorium with its early Mesozoic outcrops. Section G-H shows the faulted basement associated with the Dead Sea transform, and the relatively unfaulted sedimentary plateaus away from the rift near Eilat at the southern Negev

dramatically (Frumkin et al. 2020). The early Miocene–Pliocene water bodies and fluvial systems along the Dead Sea transform occupied a relatively shallow depression, compared with the Quaternary (Frumkin 2001; Lisker et al. 2010a; Matmon et al. 2014).

The displacement along the Dead Sea transform was accompanied by a transpression component (Butler et al. 1998) forming a large-scale (i.e. 50–100 km wide) uplifted structural arch (Fig. 2.15c), extending sub-parallel to the Dead Sea transform along its western margin (Picard, 1943; Wdowinski and Zilberman, 1997). The arching was affected also by the weight of the Nile sediment cone in the Levant basin, causing tilting of western Israel: the basin floor subsided while the land area uplifted through isostatic equilibrium (Segev et al. 2006; Gvirtzman and Steinberg 2012). The tilt axis is located roughly along the Israeli coastline (Ginzburg and Gvirtzman 1979).

These deformations had tremendous effects on Israel's physiography (Fig. 2.11). As tectonic subsidence started to prevail over the filling of the Dead Sea transform, the depth of its sub-basins has increased since the late Miocene, forming a rift valley which began acting as an endorheic basin (Bar and Zilberman, 2016; Bar et al. 2016) (Fig. 2.16). This elongated basin became the most prominent physiographic feature in Israel (Fig. 2.17). The subsidence accompanied by uplift of the transform shoulders disabled most Arabian plate catchments from flowing into the Mediterranean Sea, and new water divides were established to the east and west of the Dead Sea transform. Large rivers in northern Arabia began flowing into local endorheic basins within the peninsula, while the smaller ones on the Jordanian plateau were captured by the Dead Sea transform, as were the newly established eastern catchments of Israel.

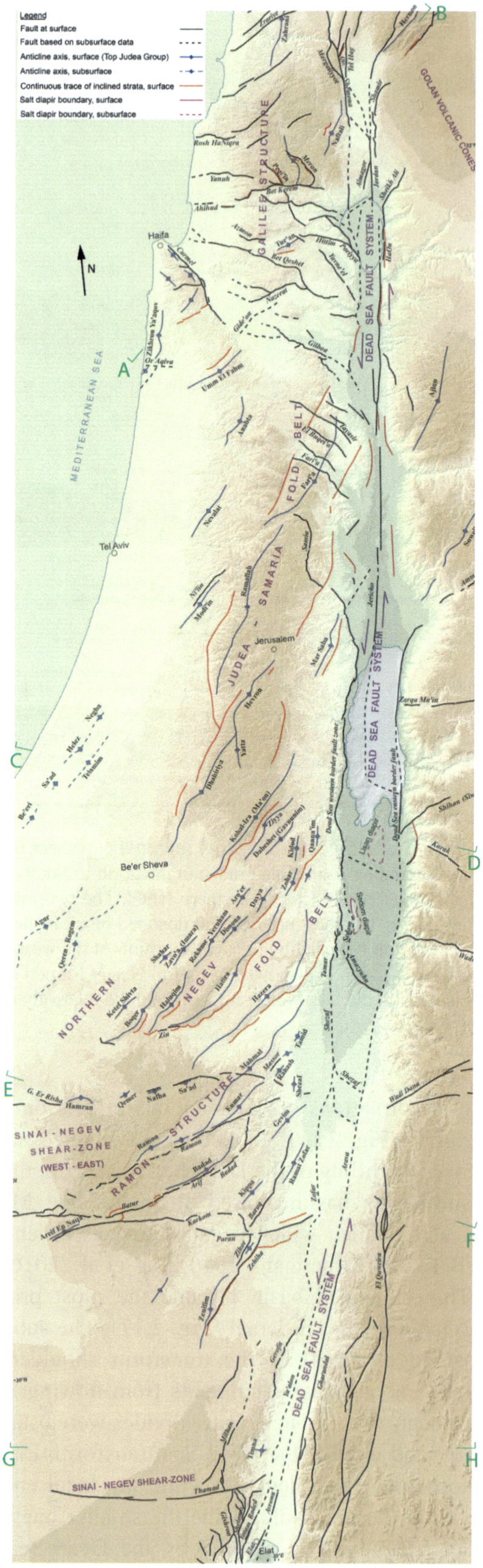

Fig. 2.11 Present relief of Israel with the major structural elements. Structure is modified after Sneh and Weinberger (2014)

Fig. 2.12 Resistant Senonian chert within soft chalk forming distinct landforms following exposure and differential erosion. **a** Table hill in the Negev, south of Makhtesh Ramon; **b** Dry waterfall, northern Judean Desert. Photos: A. Frumkin

The current water divide between the Mediterranean Sea and the Dead Sea was established roughly along the central arched ridge (Fig. 2.16). The associated tectonic deformations induced considerable relief to the eastern flank of the mountainous backbone, achieving an ~1400 m altitude difference between the Dead Sea surface and the Judean Hills. The associated water table dropped dramatically in

Fig. 2.13 Avedat Group, Eocene age, exposed in synclines; **a** Chalk at the Sartaba syncline, east Samaria; **b** Limestone cliff at Achbara, Zefat syncline. Photos: A. Frumkin

response to the uplift and the falling Dead Sea base level, as indicated by caves and their deposits (Vaks et al. 2013; Frumkin et al. 2017).

Extensive NW-striking faults within Samaria and Galilee and the NE-striking faults within the Golan Heights (Fig. 2.18) resulted from the interaction between the Dead Sea transform and the Carmel–Yizre'el–Irbid rift zone (Segev et al. 2014). The NW-oriented tectonic depression of the Carmel–Yizre'el Valley developed into a large basin which accumulated sediments and volcanic materials.

Dated vadose speleothems in caves were used to show the long-term extension of the desert belt from the Negev into the Dead Sea transform due to rain shadow effect (Frumkin et al. 2000; Vaks et al. 2003, 2006; Lisker et al. 2010b; Keinan et al. 2019; Bar-Matthews et al. 2017). In contrast, phreatic speleothems are used for dating paleo-water tables associated with uplift (Vaks et al. 2013). Chaldekas et al. (2022) used this method to suggests stable groundwater table and limited vertical tectonic motions between 14 and 6 Ma, followed by an abrupt drop in water table starting ~6 Ma, associated with uplift and folding of Israel's backbone hills, coupled with a morphotectonic base level subsidence in the Dead Sea area.

2.5.4 Miocene to Pleistocene Volcanism

A series of NW-trending faults parallel to the Red Sea were associates with its rifting. During the early Miocene, magmatic dykes penetrated into some of these faulted zones, indicating regional extension following rifting of the Red Sea (Eyal et al. 1981).

Fig. 2.14 The late Eocene-early Oligocene land-sea configuration, when a major regression exposed the anticlines crests of Israel to terrestrial denudation, active until today (modified after Rögl 1999)

Fig. 2.15 The major truncation surface formed after the emergence above sea level and transformed later by further erosion; **a** The truncation surface at NW Galilee, dissected by recent stream valleys; **b** The truncation surface at western Judean Hills, cutting across the Hevron flexure (Fig. 2.11), dissected by a recent stream valley **c** The truncation surface at the south Hevron hills (background), showing the later arching of the hilly backbone of Israel, where a new regional water divide (between the Mediterranean and Dead Sea) was established. Photos: A. Frumkin

While several episodes of volcanism occurred in Israel during the last 250 Ma (Garfunkel 1989), only the late Cenozoic volcanic phase left distinct imprint on the present-day landscape. These younger volcanics of Miocene-Pleistocene age are widespread across the Middle East, and in some areas, the onset of volcanism occurred even earlier, during the Oligocene. The largest volcanic field is Harrat ash Shaam, extending roughly along a NW-trending rift zone, parallel to the Red Sea. It includes large areas of basalt lava flows, interspersed with cinder cones, mainly east of the Dead Sea transform, within Saudi Arabia, Jordan, Syria and Israel. Within Israel, the main volcanic area of this age is the Golan Heights (Fig. 2.19a,b), where volcanism terminated ~100 ka (Weinstein et al. 2013; Shtober-Zisu et al. 2018). The Harrat ash Shaam field extended also into eastern Galilee in NE Israel, where eruptions were active since the early-Miocene and terminated during the early Pleistocene. While the Golan Heights retained its mostly plateau-like topography, dominated by lava flows, the Galilee has been deformed by tectonic faulting following its older volcanism (Fig. 2.19c).

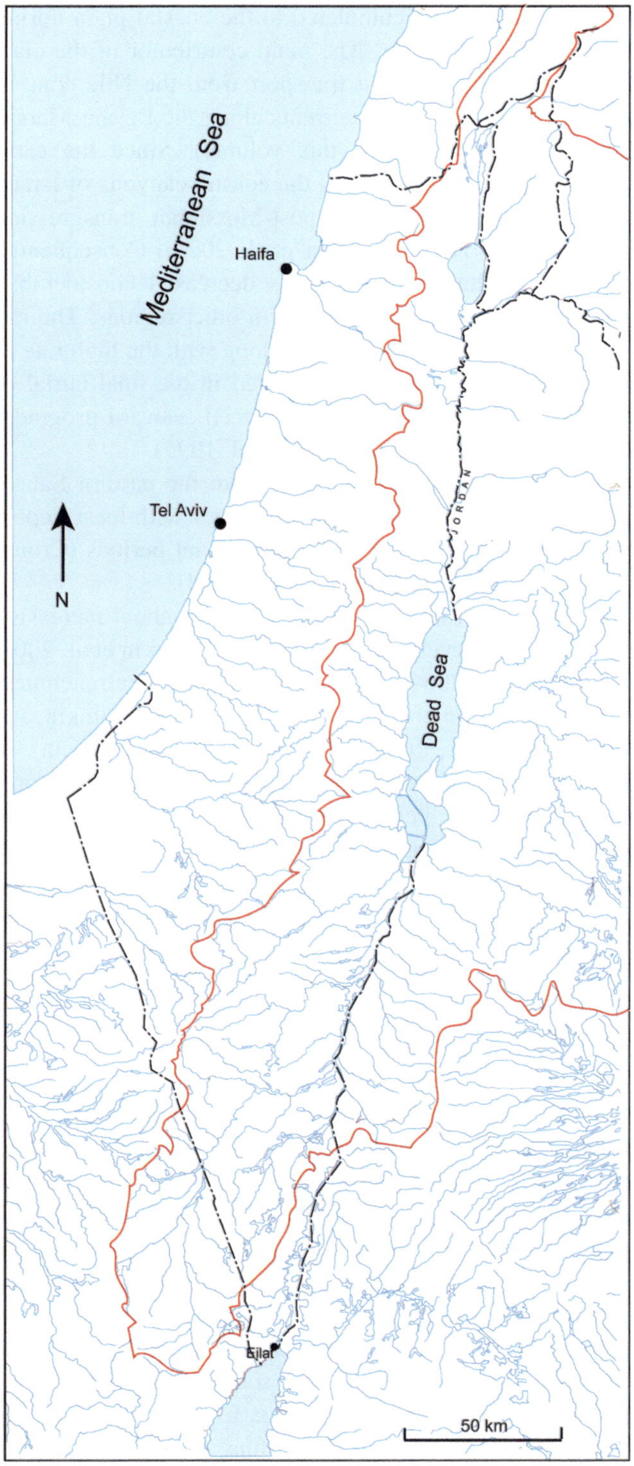

Fig. 2.16 The present hydrographic system of Israel, showing streambeds (blue) and main water divides (red). The major Dead Sea Transform has formed the lowest global depression acting as a base level to an endorheic basin of ~42,000 km^2. The current central water divide reflects the present relief, roughly following the arched backbone hills along the centre of the Israel

Fig. 2.17 The steep topography of the western shoulder of the Dead Sea fault system controlled by normal faults; **a** The normal fault at the NW edge of the Dead Sea; **b** The fault escarpment above Jericho, NW of the Dead Sea; **c** The Berenike escarpment north of Poriyya (Fig. 2.11), Sea of Galilee. Photos: A. Frumkin

2.5.5 The Plio-Pleistocene Development of the Relief

The intermittent tectonic intensification since the late Miocene was associated with rebound following the development of deep basins along the Dead Sea Transform in the east and the Nile cone submarine accumulation in the west.

Fig. 2.18 Faults affecting present relief west of the Dead Sea Rift; **a** NE Samaria; **b** Judean Desert. Photos: A. Frumkin

Erosion associated with these tectonic deformations sculpted the main present-day physiographic features of Israel. Coeval with the development of depressions along the Dead Sea transform, thick evaporites and clastic deposits, termed the Dead Sea Group, accumulated along the these basins. Of these, the late Pleistocene Lisan Formation is most widely exposed along the Jordan Valley (Fig. 2.20), the Dead Sea shores, and within the lower reaches of canyons at the Judean Desert and in the caves (Stein 2001; Lisker et al. 2009).

Accumulation in the uplifted parts of the country has been minor. Some fluvial deposits were preserved as alluvial terraces along the main drainage channels, and in tectonic or sedimentary traps, such as caves. Ryb et al. (2013) dated the base of a landslide and alluvial surfaces located ~100 m and <10 m above the channel of Nahal Soreq, to ~1.1 and ~0.5 Ma, respectively. Consequently, Ryb et al. (2013) suggested that since ~0.5 Ma, uplift and associated fluvial entrenchment in the western Judean Hills has been insignificant.

Clastic deposits accumulated in the coastal plain during the Pliocene-Quaternary. The main contributor of the clastics is the Nile. Sediment transport from the Nile cone to Israel operates by marine currents along the Levant Margin (Schattner and Bernhardt this volume). Since the early Pliocene, these deposits filled the coastal canyons of Israel, which developed until the post-Messinian transgression (Laskow et al. 2011; Frumkin et al. 2020). Consequently, the relief of the coastal belt has decreased considerably, coeval with the increase of relief in other regions. The rising sea level during the Pliocene along with the high rate of Nilotic clastic sedimentation resulted in the final burial of the coastal canyons, and in pronounced seaward progradation of the shelf-edge (Druckman et al. 1995).

Aeolian dust penetrated Israel from the eastern Sahara and Sinai, draping mainly the NW Negev with loess deposits, in particular during Pleistocene glacial periods (Crouvi et al. 2009, 2008; Frumkin and Stein 2004).

The transition to the Holocene brought about increasing aridity and low levels of the Dead Sea (Frumkin et al. 2001; Bookman et al. 2004), promoting increasing entrenchment of streams within the Dead Sea transform (Frumkin and Ford 1995; Dente, this volume, and references therein). In the Mediterranean coastal plain, the rising sea level promoted inland dune penetration and erosion of the aeolianite (kurkar) ridges along the coastline (Fig. 2.21), which had been formed by calcification of dunes. Wetlands developed between aeolianite ridges, accumulating clay and biogenic sediments.

2.6 Conclusions

Following the Pan-African orogeny and until the Eocene period, Israel was mostly under passive margin conditions at the NW edge of the African Plate, bordering the Tethys-Neotethys Sea. Thus, Israel's setting was associated with epeirogenic movements and eustatic fluctuations close to the sea-land transition. The Syrian Arc, resulting from the regional compression and Tethys closure, is manifested as extensive folding across the Levant that developed since the Santonian. These structures have some implications on the present-day relief, since some of its folds were rejuvenated by later processes, such as the arching parallel to the Dead Sea transform (Fig. 2.15c).

During the late Eocene–Oligocene, a major regression and uplift shifted the coast to the NW margins of the plate. Consequent denudation produced an extensive planation surface across the Middle East.

Moderate tectonism occurred between the Oligocene and the early Miocene. While the Judean Hills were

Fig. 2.19 Pleistocene volcanics of northern Israel; **a** The Golan Heights volcanic plateau and cinder cones, looking east across the Hula graben; **b** Late Pleistocene Birket Ram Maar, northern Golan; **c** Mt. Dalton extinct volcano and volcanic depression of early Pleistocene age, NE Galilee. Photos: A. Frumkin

uplifted, depressions formed in southern Israel and were filled with allochthonous alluvial and lacustrine deposits during the early Miocene. The break-up of the Arabian–African plate along the Dead Sea transform was associated with rifting of the Red Sea and the development of Yizre'el Valley and the Lower Galilee tectonic basins. Since the middle Miocene, the relief of Israel was enhanced mainly by deep depressions formed along the Dead Sea fault system, accompanied by arching of the hilly backbone of Israel, where a new water divide was established (Fig. 2.16).

The Galilee's and Samaria's normal faults increased the relief, and associated canyons were incised, mostly since 5 Ma. The establishment of the Dead Sea transform endorheic base level resulted in disconnection of fluvial networks which had drained northern Arabia into the Mediterranean Sea.

By the Pleistocene, the present-day drainage network of Israel was mostly established. Similarly, the present-day topography was established in the north-eastern Negev. The Judean Hills have been uplifted until the middle Pleistocene, ~0.5 Ma, and since then, fluvial, karst and aeolian processes are the main factor shaping the landscapes of Israel.

Fig. 2.20 The morphology of Lisan Formation, deposited by late Pleistocen Lake Lisan, and dissected by Holocene erosion; **a** North of the Dead Sea; **b** NW of the Dead Sea, with early Holocene redeposited red terra rossa soil, transported from east Samaria slopes; **c** Piping-dominated terrain above Amazyahu fault, south of the Dead Sea. Photos: A. Frumkin

Fig. 2.21 Aeolianite (locally termed kurkar) ridge along the Mediterranean coastline. The ridge is being eroded by littoral processes; **a** The western kurkar ridge between Tel Aviv and Haifa. Sea wave erosion has consumed much of the ridge, forming bays and eroded hills; **b** Microsculpturing at the coastline between Tel Aviv and Haifa. Complex processes involving wave quarrying, bioerosion, salt weathering, mixing corrosion, etc., form dynamic morphologic belts along the intertidal area. Photos: A. Frumkin

References

Avni Y, Bartov Y, Garfunkel Z, Ginat H (2000) Evolution of the Paran drainage basin and its relations to the Plio-Pleistocene history of the Arava Rift western margin, Israel. Israel J Earth Sci 49:215–238

Avni Y, Segev A, Ginat H (2012) Oligocene regional denudation of the northern Afar dome: pre-and syn-breakup stages of the Afro-Arabian plate. Geol Soc Am Bull 124:1871–1897. https://doi.org/10.1130/B30634.1

Avni Y (2017) Tectonic and physiographic settings of the Levant. In: Enzel Y, Bar-Yosef O (eds) Quaternary of the Levant (environment. Cambridge University Press, Climate Change and Humans), pp 3–16

Bar O, Zilberman E (2016) Subsidence and conversion of the Dead Sea basin to an inland erosion base level in the early middle Miocene as inferred from geomorphological analysis of its ancient western fluvial outlet. Geomorphology 261:147–161. https://doi.org/10.1016/j.geomorph.2016.02.028

Bar O, Zilberman E, Feinstein S, Calvo R, Gvirtzman, (2016) The uplift history of the Arabian Plateau as inferred from geomorphologic analysis of its northwestern edge. Tectonophysics 671:9–23. https://doi.org/10.1016/j.tecto.2016.01.004

Bar-Matthews M, Ayalon A, Vaks A, Frumkin A (2017) Climate and environment reconstructions based on speleothems from the Levant. In: Bar-Yosef O (ed) Enzel Y. Quaternary of the Levant, Cambridge University Press, pp 151–164

Bartov Y (1974) A structural and paleogeographical study of the central Sinai faults and domes. [Ph.D. Thesis]: Jerusalem, The Hebrew University, pp 143 (in Hebrew, English abstract).

Bartov Y, Arkin Y (1980) Geological photomap 1:500000, 2nd edn. Geological Survey of Israel, Jerusalem

Beyth M, Eyal Y, Garfunkel Z (2014) The geology of the northern tip of the Arabian–Nubian Shield. J African Earth Sci 99:332–341. https://doi.org/10.1016/j.jafrearsci.2014.03.028

Bialik OM, Korngreen D, Benjamini C (2012) Lithofacies and cyclicity of Mohilla evaporite basins on the rifted margin of the Levant in the Late Triassic, Makhtesh Ramon, southern Israel. Sedimentology 59(7):2097–2124. https://doi.org/10.1111/j.1365-3091.2012.01336.x

Bookman R, Enzel Y, Agnon A, Stein M (2004) Late Holocene lake levels of the Dead Sea. Geol Soc Am Bull 116:555–571. https://doi.org/10.1130/B25286.1

Buchbinder B, Martinotti GM, Siman Tov R, Zilberman E (1993). Temporal and spatial relationships in Miocene reef carbonates in Israel. Paleaogeogr Palaeocl 97–116. https://doi.org/10.1016/0031-0182(93)90154-B

Buchbinder B, Calvo R, Siman-Tov R (2005) The Oligocene in Israel: a marine realm with intermittent denudation accompanied by massflow deposition. Isr J Earth Sci 54:63–85

Butler RWH, Spencer S, Griffiths H (1998) The structural response to evolving plate kinematics during transpression: evolution of the Lebanese restraining bend of the Dead Sea Transform. Geologic Soc London Spec Publicat 135:81–106

Calvo R, Bartov Y (2001) Hazeva group, southern Israel: new observations, and their implications for itsstratigraphy, paleogeography, and tectono-sedimentary regime. Israel J Earth Sci 50:71–99

Chaldekas O, Vaks A, Haviv I, Gerdes A, Albert R (2022) U-Pb speleothem geochronology reveals a major 6 Ma uplift phase along the western margin of Dead Sea Transform. GSA Bull 134(5–6):1571–1584. https://doi.org/10.1130/B36051.1

Crouvi O, Amit R, Enzel Y, Porat N, Sandler A (2008) Sand dunes as a major proximal dust source for late Pleistocene loess in the Negev Desert, Israel. Quat Res 70:275–282. https://doi.org/10.1016/j.yqres.2008.04.011

Crouvi O, Amit R, Porat N. Gillespie AR, McDonald EV, Enzel Y (2009) Significance of primary hill top loess in reconstructing dust chronology, accretion rates, and sources: An example from the Negev Desert. Israel. J Geophys Res Earth 114:F02017. https://doi.org/10.1029/2008JF001083

Druckman Y (1974) The stratigraphy of the Triassic sequence in southern Israel. Bull. Geol Surv Israel 64:94

Druckman Y, Buchbinder B, Martinotti GM, Siman Tov RS, Aharon P (1995) The buried Afiq Canyon (eastern Mediterranean, Israel): a case study of a Tertiary submarine canyon exposed in Late Messinian times. Marine Geol 123(3–4):167–185. https://doi.org/10.1016/0025-3227(94)00127-7

Eyal M, Eyal Y, Steinitz BY, G, (1981) The tectonic development of the western margin of the Gulf of Elat (Aqaba) rift. Tectonophysics 80(1–4):39–66. https://doi.org/10.1016/0040-1951(81)90141-4

Flexer A, Gilat A, Hirsch F, Honigstein A, Rosenfeld A, Reuffer T (1989) Late Cretaceous evolution of the Judean Mountains as indicated by ostracodes. In: Hall JK, Krasheninnikov VA, Hirsch F, Benjamini C, Flexer A (eds) Geological Framework of the Levant -, vol 2. The Levantine Basin and Israel: Jerusalem. Israel, Historical Productions-Hall, pp 349–358

Flexer A, Hirsch F, Hall JK (2005) Israel—Introduction. In: Hall JK, Krasheninnikov VA, Hirsch F, Benjamini C, Flexer A (eds) Geological Framework of the Levant—, vol 2. The Levantine Basin and Israel: Jerusalem. Israel, Historical Productions-Hall, pp 215–268

Freund R (1977) Updating the stratigraphic framework of Israel. Isr J Earth Sci 26:30–33, Jerusalem.

Freund R, Garfunkel Z, Zak I, Goldberg M, Weissbrod T, Derin B, Bender F, Wellings FE, Girdler RW (1970) The shear along the Dead Sea Rift. Philos T Roy Soc Lond 267(1181):107–130. https://www.jstor.org/stable/73612

Freund R, Goldberg M, Weissbrod T, Druckman Y, Derin B (1975) The Triassic-Jurassic structure of Israel and its relation to the origin of the eastern Mediterranean. Geol Survey Isr Bull 65:26

Frumkin A (1992) Karst origin of the upper erosion surface in the Northern Judean Mountains. Israel: Isr J Earth Sci 41:169–176

Frumkin A (2001) The cave of the letters sediments—indication of an early phase of the Dead Sea depression? J Geol 109(1):79–90. https://doi.org/10.1086/317962

Frumkin A, Ford DC (1995) Rapid entrenchment of stream profiles in the salt caves of Mount Sedom. Israel: Earth Surf Proc Land 20:139–152. https://doi.org/10.1002/esp.3290200205

Frumkin A, Ford DC, Schwarcz HP (2000) Paleoclimate and vegetation of the last glacial cycles in Jerusalem from a speleothem record. Global Biogeochem Cy 14:863–870. https://doi.org/10.1029/1999GB001245

Frumkin A, Kadan G, Enzel Y, Eyal Y (2001) Radiocarbon chronology of the Holocene Dead Sea: attempting a regional correlation. Radiocarbon 43(2C):1179–1190. https://doi.org/10.1017/S0033822200038479

Frumkin A, Stein M (2004) The Sahara—East Mediterranean dust and climate connection revealed by strontium and uranium isotopes in a Jerusalem speleothem. Earth Planet Sci Lett 217:451–464.https://doi.org/10.1017/S0033822200038479

Frumkin A, Langford B, Lisker S, Amrani A (2017) Hypogenic karst at the Arabian platform margins: implications for far-field groundwater systems. Bull Geologic Soc Am 129(11–12):1636–1659. https://doi.org/10.1130/B31694.1

Frumkin A, Dimentman C, Naaman I (2020) Biogeography of living fossils as a key for geological reconstruction of the East Mediterranean: Ayyalon-Nesher Ramla system, Israel. Quat Int 624:168–180. https://doi.org/10.1016/j.quaint.2020.11.036

Frumkin A, Weinstein-Evron M (2021) Nahal Me'arot caves: archive of human evolution against the background of prolonged karstic processes. Z Geomorphol Supp 62:283–300. https://doi.org/10.1127/zfg_suppl/2021/0699

Garfunkel Z (1981) Internal structure of the Dead Sea leaky transform (rift) in relation to plate kinematics. Tectonophysics 80:81–108. https://doi.org/10.1016/0040-1951(81)90143-8

Garfunkel Z (1988) Relations between continental rifting and uplifting: evidence from the Suez rift and northern Red Sea. Tectonophysics 150:33–49

Garfunkel Z (1989) Tectonic setting of Phanerozoic magmatism in Israel. Israel J Earth Sci 38:51–74. https://doi.org/10.1016/0040-1951(88)90294-6

Garfunkel Z (1997) The history and formation of the Dead Sea Basin. In: Niemi TM, Ben-Avraham Z, Gat Y (eds) The Dead Sea—the lake and its setting: Oxford, Oxford University Press, pp 36–55

Garfunkel Z (1999) History and paleogeography during the Pan-African orogen to stable platform transition: reappraisal of the evidence from the Elat area and the northern Arabian-Nubian Shield. Israel J Earth Sci 48(3):135–157

Garfunkel Z, Derin B (1984) Permian-early Mesozoic tectonism and continental margin formation in Israel and its implications for the history of the Eastern Mediterranean. Geologic Soc London Spec Publ 17(1):187–201

Ginzburg A, Gvirtzman G (1979) Changes in the crust and in the sedimentary cover across the transition from the Arabian Platform to the Mediterranean Basin: evidence from seismic refraction and sedimentary studies in Israel and in Sinai. Sed Geol 23:19–36. https://doi.org/10.1016/0037-0738(79)90004-6

Gvirtzman G, Buchbinder B (1978) The late tertiary of the coastal plain and continental shelf of israel and itsbearing on the history of the eastern mediterranean. Initial reports of the deep sea drilling project/1975/Istanbul, Turkey to Istanbul, Turkey; USA; Washington: U.S. government printing office; da 42(2):1195–1222

Gvirtzman Z (2004) Chronostratigraphic table and subsidence curves of southern Israel. Israel J Earth Sci 53(1).

Gvirtzman Z, Garfunkel Z (1998) The transformation of southern Israel from a swell to a basin: stratigraphic and geodynamic implications for intracontinental tectonics. Earth Planet Sc Lett 163(1–4):275–290. https://doi.org/10.1016/S0012-821X(98)00193-9

Gvirtzman Z, Steinberg J, Bar O, Buchbinder B, Zilberman E, Siman-Tov R, Calvo R, Grossowicz L, Almogi-Labin A, Rosensaft M (2011) Retreating Late Tertiary shorelines in Israel: Implications for the exposure of north Arabia and Levant during Neotethys closure. Lithosphere 3:95–109. https://doi.org/10.1130/L124.1

Gvirtzman Z, Steinberg J (2012) Inland jump of the Arabian northwest plate boundary from the Levant continental margin to the Dead Sea Transform. Tectonics 31: TC4003. https://doi.org/10.1029/2011TC002994

Keinan J, Bar-Matthews M, Ayalon A, Zilberman T, Agnon A, Frumkin A (2019) Paleoclimatology of the Levant from Zalmon Cave speleothems, the northern Jordan Valley, Israel. Quat Sci Rev 220:142–153. https://doi.org/10.1016/j.quascirev.2019.07.018

Laskow M, Gendler M, Goldberg I, Gvirtzman H, Frumkin A (2011) Deep confined karst detection, analysis and paleo-hydrology reconstruction at a basin-wide scale using new geophysical interpretation of borehole logs. J Hydrol 406:158–169. https://doi.org/10.1016/j.jhydrol.2011.06.011

Lewy Z, Almogi-Labin A, Siman-Tov R (1995) The paleoecologic and paleogeographic significance of upper Maastrichtian and lower Eocene rocks recently discovered in Jerusalem, Israel: Israel J Earth Sci 44:25–32

Lisker S, Vaks A, Bar-Matthews M, Porat R, Frumkin A (2009) Stromatolites in caves of the Dead Sea Fault Escarpment: implications to latest Pleistocene lake levels and tectonic subsidence. Quat Sci Rev 28:80–92. https://doi.org/10.1016/j.quascirev.2008.10.015

Lisker S, Porat R, Frumkin A (2010a) Late Neogene rift valley fill sediments preserved in caves of the Dead Sea Fault Escarpment (Israel): palaeogeographic and morphotectonic implications. Sedimentology 57:429–445. https://doi.org/10.1111/j.1365-3091.2009.01089.x

Lisker S, Vaks A, Bar-Matthews M, Porat R, Frumkin A (2010b) Late Pleistocene palaeoclimatic and palaeoenvironmental reconstruction of the Dead Sea area (Israel) based on speleothems and cave stromatolites. Quat Sci Rev 29:1201–1211. https://doi.org/10.1016/j.quascirev.2010.01.018

Matmon A, Wdowinski S, Hall J (2003) Morphological and structural relations in the Galilee extensional domain, northern Israel. Tectonophysics 371:223–241. https://doi.org/10.1016/S0040-1951(03)00237-3

Matmon A, Fink D, Davis M, Niedermann S, Rood D, Frumkin A (2014) Unraveling rift margin evolution and escarpment development ages along the Dead Sea fault using cosmogenic burial ages. Quat Res 82:281–295. https://doi.org/10.1016/j.yqres.2014.04.008

Mimran YA (1984) Unconformities on the eastern flank of the Fari'a anticline, and their implications on the structural evolution of Samaria (central Israel). Israel J Earth Sci 33:1–11

Nuriel P, Weinberger R, Kylander-Clark ARC, Hacker BR, Craddock JP (2017) The onset of the Dead Sea transform based on calcite age-strain analyses. Geology 45(7):587–590. https://doi.org/10.1130/G38903.1

Picard L (1943) Structure and evolution of Palestine: Jerusalem, Hebrew University, pp 187

Quennell AM (1958) The structural and geomorphic evolution of the Dead Sea Rift. Q J Geol Soc London 114:1–24. https://doi.org/10.1144/gsjgs.114.1.0001

Rögl F (1999) Mediterranean and Paratethys: facts and hypotheses of an Oligocene to Miocene paleogeography (short overview). Geol carpathica 50(4):339–349

Rozenbaum AG, Sandler A, Zilberman E, Stein M, Jicha BR, Singer BS (2016) 40Ar/39Ar chronostratigraphy of late Miocene–early Pliocene continental aquatic basins in SE Galilee. Israel. GSA Bull 128(9–10):1383–1402. https://doi.org/10.1130/B31239.1

Sass E, Bein A (1978) Platform carbonates and reefs in the Judean Mountains, Carmel and Galilee proceedings of the tenth international congress on sedimentology. Int Assoc Sed 2:239–274

Schattner U, Ben-Avraham Z (2007) Transform margin of the northern Levant, eastern Mediterranean: From formation to reactivation. Tectonics 26:1–17. https://doi.org/10.1029/2007TC002112

Segev A, Rybakov M, Lyakhovsky V, Hofstetter A, Tibor G, Goldshmidt V, Ben Avraham Z (2006) The structure, isostasy and gravity field of the Levant continental margin and the southeast Mediterranean area. Tectonophysics 425:137–157. https://doi.org/10.1016/j.tecto.2006.07.010

Segev A, Rybakov M (2010) Effects of Cretaceous plume and convergence, and Early Tertiary tectonomagmatic quiescence on the central and southern Levant continental margin. J Geol Soc Lond 167:731–749. https://doi.org/10.1144/0016-76492009-118

Segev A, Schattner U, Lyakhovsky V (2011) Middle-late eocene structure of the southern Levant continental margin—tectonic motion versus global sea-level change. Tectonophysics 499:165–177. https://doi.org/10.1016/j.tecto.2011.01.006

Segev A, Lyakhovsky V, Weinberger R (2014) Continental transform–rift interaction adjacent to a continental margin: the Levant case study. Earth Sci Rev 139:83–103. https://doi.org/10.1016/j.earscirev.2014.08.015

Shahar J (1994) The Syrian arc system - an overview. Paleaogeogr Palaeocl 112:125–142. https://doi.org/10.1016/0031-0182(94)90137-6

Shtober-Zisu N, Inbar M, Mor D, Singer JBR, BS, (2018) Drainage development and incision rates in an upper Pleistocene basalt-limestone boundary channel: The Sa'ar Stream, Golan Heights, Israel. Geomorphology 303:417–433. https://doi.org/10.1016/j.geomorph.2017.12.002

Sneh A, Weinberger R (2014) Major geological structures of Israel and environs. Map, Geol Surv Israel.
Srebro H (2011) The New Atlas of Israel: The National Atlas. Survey of Israel and The Hebrew University of Jerusalem.
Stein M (2001) The sedimentary and geochemical record of Neogene-Quaternary water bodies in the Dead Sea Basin — inferences for the regional paleoclimatic history. J Paleolimnol 26:271–282. https://doi.org/10.1023/A:1017529228186
Stern RJ, Johnson PR (2010) Continental lithosphere of the Arabian Plate: a geologic, petrologic, and geophysical synthesis. Earth-Sci Rev 101:29–67. https://doi.org/10.1016/j.earscirev.2010.01.002
Ryb U, Matmon A, Porat N, Katz O (2013) From mass-wasting to slope stabilization—putting constrains on a tectonically induced transition in slope erosion mode: a case study in the Judea Hills, Israel. Earth Surf Proc Land 38:551–560. https://doi.org/10.1002/esp.3291
Vaks A, Bar-Matthews M, Ayalon A, Schilman B, Gilmour M, Hawkesworth CJ, Frumkin A, Kaufman A, Matthews A (2003) Paleoclimate reconstruction based on the timing of speleothem growth, oxygen and carbon isotope composition from a cave located in the rain shadow in Israel. Quat Res 59:182–193. https://doi.org/10.1016/S0033-5894(03)00013-9
Vaks A, Bar-Matthews M, Ayalon A. Matthews A, Frumkin A, Dayan U, Halicz L, Almogi-Labin A, Schilman B (2006) Paleoclimate and location of the border between Mediterranean climate region and the Saharo–Arabian Desert as revealed by speleothems from the northern Negev Desert, Israel. Earth Planet Sc Lett 249:384–399. https://doi.org/10.1016/j.epsl.2006.07.009
Vaks A, Woodhead J, Bar-Matthews M, Ayalon A, Cliff R, Zilberman T, Matthews A, Frumkin A (2013) Pliocene-Pleistocene climate of the northern margin of Saharan-Arabian Desert recorded in speleothems from the Negev Desert, Israel. Earth Planet Sc Lett 368:88–100. https://doi.org/10.1016/j.epsl.2013.02.027
Wdowinski S, Zilberman E (1997) Systematic analysis of the large scale topography and structure across the Dead Sea Rift. Tectonics 16:409–424. https://doi.org/10.1029/97TC00814
Weinberger R, Gross MR, Sneh A (2009). Evolving deformation along transform plate boundary: Example from the Dead Sea fault in northern Israel. Tectonics 28:TC5005. https://doi.org/10.1029/2008TC002316
Weinberger R, Nuriel P, Kylander-Clark AR, Craddock JP (2020) Temporal and spatial relations between large-scale fault systems: Evidence from the Sinai-Negev shear zone and the Dead Sea Fault. Earth Sci Rev 211:103377. https://doi.org/10.1016/j.earscirev.2020.103377
Weinstein Y, Weinberger R, Calvert A (2013) High-resolution 40Ar/39Ar study of Mount Avital, northern Golan: Reconstructing the interaction between volcanism and a drainage system and their impact on eruptive styles. B Volcanol 75:1–12. https://doi.org/10.1007/s00445-013-0712-7
White WB, White EL (2013) Karst landforms: Scope and processes in the early twenty-first century. In: Frumkin A, Shroder JF (eds) Treatise on geomorphology, vol 6. Academic Press, San Diego, California, pp 14–21
Zilberman E (1991) Landscape evolution in the central, northern and northwestern Negev during the Neogene and the Quaternary. Geol Surv Israel Rep GSI/45/90.
Zilberman E, Calvo R (2013) Remnants of Miocene fluvial sediments in the Negev Desert, Israel, and the Jordanian Plateau: evidence for an extensive subsiding basin in the northwestern margins of the Arabian plate. J African Earth Sci 82:33–53. https://doi.org/10.1016/j.jafrearsci.2013.02.006

3 Climate of Israel

Haim Kutiel

Abstract

The main characteristics of the complexity of the climate in Israel are presented. Despite its very moderate size, Israel presents a variability of climates which includes three main climatic regions; hot summer Mediterranean, hot semi-arid and hot arid climates. The wind regime comprises Easterlies mainly in winter, whereas the summer is characterized by a sea breeze. Relative humidity tends to be higher in winter and lower in summer in all parts of Israel except along the coastal plain due to the sea breeze which in turn affects the seasonality of forest fires. Air temperature regime displays a larger annual range than diurnal range which is characteristic of high latitudes. Finally, the rainfall regime across the country presents a very steep rainfall gradient from the mountain ridge in the west to the Jordan Rift Valley in the east. All parts of the country experience both temporal (year to year) and spatial uncertainty which increases southeastwards.

Keywords

Wind regime · Relative humidity · Air temperature · Rainfall regime · Israel

3.1 Introduction

Israel lies between latitudes 29.56 and 33.21 N and longitudes 34.57 and 35.57 E. Its climate is determined primarily by its location at the junction of three continents: Europe in the North-West; Asia in the East and Africa in the South-West (Fig. 3.1). Together with its proximity to the Mediterranean Sea and the large variability in terrain elevation ranging from 2236 m a.s.l. at Mt. Hermon to −434 m a.s.l. at the Dead Sea which is also the world's lowest point on land, causes that despite its very modest size (22,145 km^2), Israel demonstrates a large variability of climates.

Three major climatic regions according to the Köppen climate classification can be identified and there are four additional tiny regions (Fig. 3.2c).

These regions are:

- Csa–Hot summer Mediterranean climate.
- BSh–Hot semi-arid climate.
- BWh–Hot desert climate.

Most of northern Israel has a hot summer Mediterranean climate, whereas most of southern Israel has a hot desert climate, with a narrow intermediate band of hot semi-arid climate. Summer is dry in all regions and the main differences among these three are a decrease in mean annual rainfall. Roughly the mean annual rainfall in the Mediterranean region is>400 mm, 200–400 mm in the semi-arid region, and<200 mm in the arid one (compare Fig. 3.2b, c). Most rain events occur in winter with a large variability from year to year (e.g., Aviad et al. 2013).

In addition to these three regions, four tiny regions, BWk (Cold desert climate); Csb (Warm summer Mediterranean climate); Dsb (Dry warm continental climate); Dsc (Dry cool continental climate) can also be found as a result of the local topography reaching elevations of over 1000 m a.s.l. in the case of BWk and over 2000 m a.s.l. in the remaining cases.

Roughly, Israel can be divided into three main longitudinal bands from west to east, namely, the coastal plain, the mountain ridge and the Jordan Rift Valley (JRV hereafter, Fig. 3.2a). Each of these three bands differs a great deal from the other two in its climatic characteristics.

H. Kutiel (✉)
School of Environmental Sciences,
University of Haifa, Haifa, Israel
e-mail: kutiel@geo.haifa.ac.il

A. Frumkin and N. Shtober-Zisu (eds.), *Landscapes and Landforms of Israel*,
World Geomorphological Landscapes, https://doi.org/10.1007/978-3-031-44764-8_3

Fig. 3.1 Location map of Israel (red rectangle) at the joining point of three continents

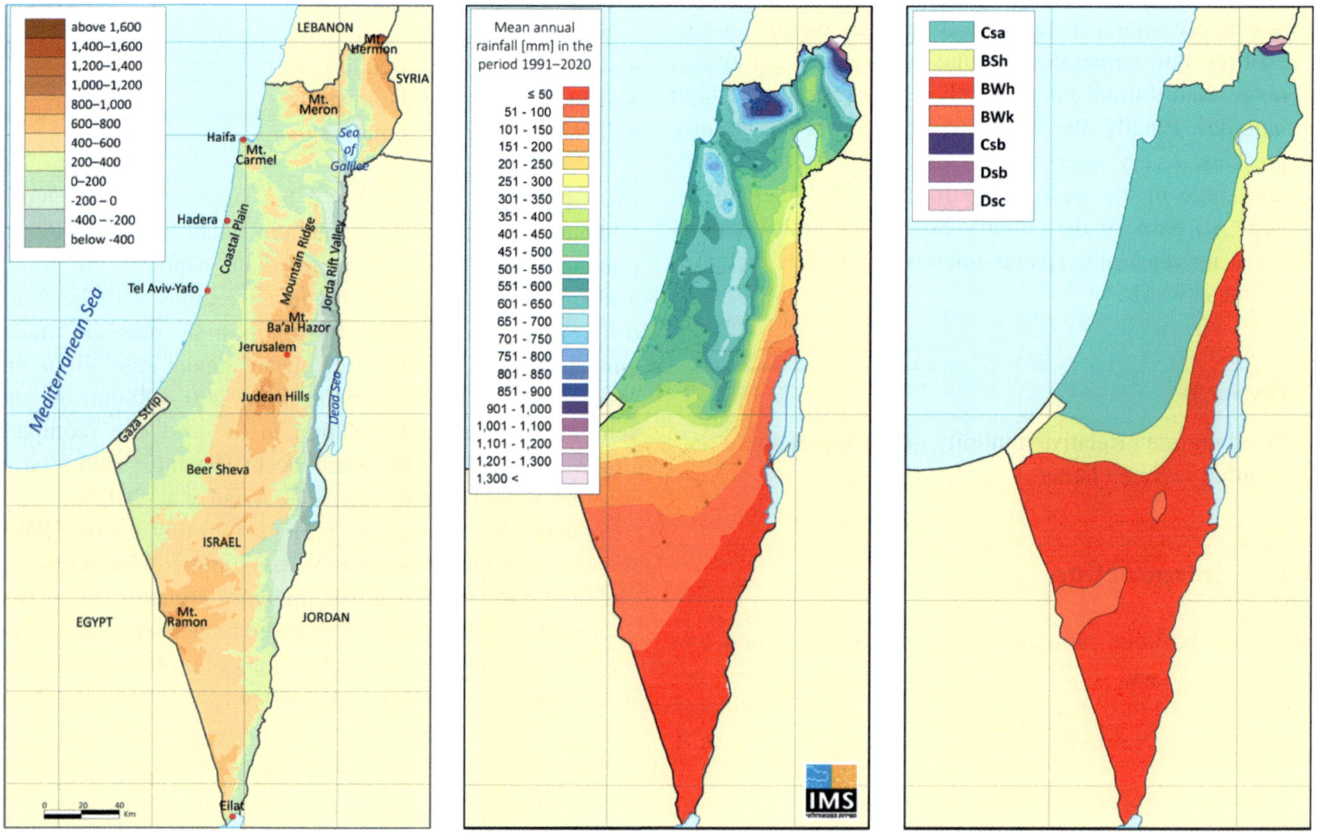

a b c

Fig. 3.2 Climate of Israel. **a** Location map of major physiographic regions in Israel. **b** Mean annual rainfall totals 1991–2020 adapted from the Israel Meteorological Service. **c** Köppen climate classification of Israel

The coastal plain is the westernmost region of the country bordering the Mediterranean with elevations spanning from sea level up to 300 m a.s.l. The mountain ridge, located east of the coastal plain, is characterized by elevations ranging from 700 to 900 m a.s.l. and a few peaks over 1000 m a.s.l. The peaks are Mt. Meron 1208 m a.s.l. in the north, Ba'al Hazor 1016 m a.s.l. in the center and Mt. Ramon 1033 m a.s.l. in the south. The JRV, located on the east side of the mountain ridge, mostly lies below sea level. Elevations ranging from 210 m b.s.l. at the Sea of Galilee, dropping to 434 m b.s.l. at the Dead Sea and then raising steadily to the sea level at Eilat, the southernmost city in Israel. The proximity of the JRV to the mountain ridge presents a dramatic topographic slope which affects the local climate of that region.

3.2 Wind Regime

The wind regime anywhere on the globe depends on the pressure fields in a certain region at various levels and influences other climatic variables in that region, such as temperature, relative humidity, rainfall occurrence and its characteristics, and many others. Therefore, it is crucial to study the wind regime if one is willing to understand the climatic regimes of a certain region.

The wind, unlike other meteorological parameters, is a vector, which means that in addition to its magnitude it is crucial to know its azimuth. Sometimes, it is even more important to know wherefrom the wind is blowing than its intensity (see for example Kutiel 2019). Furthermore, the magnitude and direction of wind close to the surface varies a great deal both in space and time, which makes it more difficult to be represented on a map in a way similar to the other weather variables, but rather is presented by means of wind roses.

Due to its location within the westerlies band of the northern hemisphere, Israel is affected mainly by winds from the west sector (SW to NW). Easterlies are frequent mainly in winter, when they may be sometimes very intense, and to a lesser extent, in the transitional seasons. In summer and during many days in the transitional seasons, the wind regime is relatively stable with westerly winds along most hours of the day. This is due to a combination of the Etesian blowing from Greece toward the eastern Mediterranean and the seasonal thermal low over the Persian Gulf which intensify the sea breeze Fig. 3.3). This moist sea breeze starts to blow along the coastal plain at around 08:00–09:00 and ends in the late afternoon-early evening at an average speed of 2.8–3.0 m/s (10–11 km/h) with increasing speed around noon time.

During most of the daytime from March to September and in certain cases even until October, the temperature near the ground over land is higher than the sea surface temperatures in the eastern Mediterranean, meaning that for most of the time the sea breeze brings cool moist air. In its trajectory over Israel, this breeze ascends the mountain ridge and then descends to the JRV. The air masses cool (on ascending) and warm (on descending) adiabatically. As most of the JRV is below sea level, the heating is greater than the cooling and therefore when reaching this region, the air masses are hotter than at the coastal plain. Due to the coastal line shape, the distance from the sea to the JRV varies from north to south, being shorter in the north and increasingly longer as one moves southwards (Fig. 3.2a). This implies that in the northern part of the JRV, the descending hot air masses reach the region around noon time which coincides with the timing of the radiative maximum causing an additional increase of the daily maximum. This wind reaches the central JRV after the timing of the radiative maximum which results in a longer duration of the daily maximum temperature. In the Dead Sea region, the sea breeze arrives in the afternoon, after the air temperature started to decrease causing a secondary maximum at that time of the day (Bitan 1974, 1977, 1981). Afterward, the wind shifts southwards toward the town of Eilat and the Red Sea. This hot and very dry wind reaches Eilat before midnight. In many cases, the air temperature in Eilat may even rise at that time of the night. This is noticeable, if we compare the mean air temperature at various hours along the day. Whereas in Hadera Port (which represents a Mediterranean-type climate station), the average temperatures at 08:00 and 23:00 LT are almost identical, in Eilat, the temperature at 23:00 LT is significantly higher than at 08:00 LT (Table 3.1).

3.3 Humidity Regime

The spatial variability of relative humidity (RH) depends on the proximity of a location to a major water body, i.e., RH decreases as the distance from the water body increases. However, the wind regime may alter this scheme. For example, in Eilat, the southernmost town in Israel, one would expect high RH rates as Eilat is located on the coastline of the Red Sea. But, as most of the time Eilat is affected by northerly dry winds, the RH in Eilat is one of the lowest in the country, despite its proximity to a major water body (Table 3.2).

Regarding the temporal variability, normally there is an opposite course of the RH and the temperature, with RH decreasing as the temperature increases and vice versa. Therefore, one may expect the RH to be at its maximum in winter and at its minimum in summer for the annual cycle and to present a nocturnal maximum and an early afternoon minimum for the daily cycle.

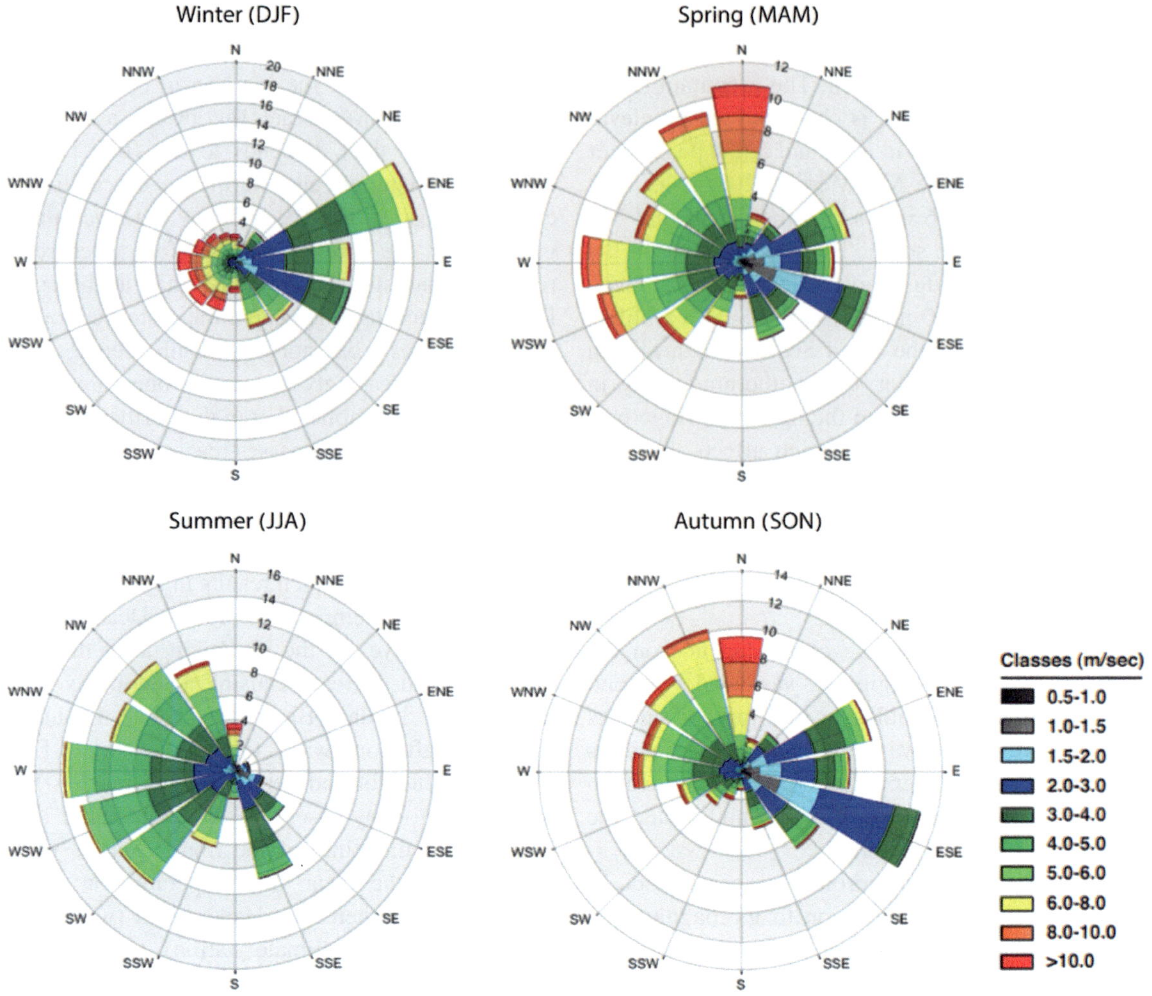

Fig. 3.3 Wind roses for Hadera Port (2001–2012). Data after the IMS (https://ims.gov.il)

Here again this scheme is altered by the wind regime as described earlier. For the annual cycle, only along the mountain ridge and the Jordan Rift Valley the RH annual courses follow the expected pattern, i.e., maximum in winter and minimum in summer, whereas along the coastal plain, the course presents an opposite cycle with its maximum in summer and minimum in winter. This regime is the main reason for the discrepancy between the Judean Hills and Mt. Carmel in the timing of major forest fires (>50 ha). Whereas in the former, these fires are concentrated in summer, and in the latter, they occur mainly in autumn (Visnovaty 2021). This difference in timing has a crucial implication for soil erosion in both regions, as a major forest fire leaves the soil bare without tree cover and a considerable amount of ash at different time from the first significant rainfall of the next rainy season. Figure 3.4 presents the annual course of the mean daily RH along the three main longitudinal bands.

3.4 Temperature Regime

The temperature regime anywhere on the globe presents two cycles, the diurnal cycle and the annual cycle. These two cycles are not even everywhere. Whereas near the equator the diurnal cycle is dominant, and the annual cycle does not exist or is insignificant, as one moves poleward on both hemispheres, the annual cycle becomes more evident, and in parallel, there is a clear reduction of the diurnal cycle until at the poles, where the diurnal cycle does not exist at all and only the annual cycle is identifiable. Another factor that reduces both the diurnal and the annual ranges is the proximity to the ocean or any other major water body. Maritime locations benefit from milder temperatures as compared to inland stations.

Israel lies in the midway between the equator and the North Pole, much closer to the equator. One could expect the diurnal cycle would be larger than the annual, however

Table 3.1 Hourly and monthly average temperatures in Eilat (2001–2020) and Hadera Port (2003–2020)

Eilat

	Jan	Feb	Mar	Apr	May	Jun	Jul	Aug	Sep	Oct	Nov	Dec	Average
Average	15.9	18.1	21.4	25.2	29.6	32.7	34.4	34.3	31.9	28.0	22.3	17.5	
2:00	12.9	14.7	17.8	21.1	25.2	28.1	30.2	30.5	28.2	24.5	19.1	14.3	22.2
5:00	11.8	13.5	16.3	19.5	23.4	25.9	28.0	28.3	26.5	23.1	17.8	13.4	20.6
8:00	13.1	15.3	19.0	23.3	27.8	30.5	32.2	31.8	29.4	25.9	20.4	15.1	23.6
11:00	18.6	20.7	23.9	28.0	32.6	35.3	36.7	36.4	33.9	30.6	25.6	20.6	28.6
14:00	21.0	23.3	26.6	30.6	35.2	38.5	39.8	39.7	37.2	33.3	27.8	22.9	31.3
17:00	19.1	21.7	25.3	29.5	34.1	38.0	39.2	39.1	36.0	31.1	25.1	20.3	29.9
20:00	16.4	18.8	22.4	26.4	30.9	34.4	36.0	36.0	33.3	28.7	22.6	17.6	27.0
23:00	14.4	16.5	19.9	23.5	27.9	31.0	32.9	32.9	30.4	26.3	20.5	15.6	24.3

Legend: ≤15.0 °C; 15.1-20.0 °C; 20.1-25.0 °C; 25.1-30.0 °C; 30.1-35.0 °C; >35.0 °C

Hadera Port

	Jan	Feb	Mar	Apr	May	Jun	Jul	Aug	Sep	Oct	Nov	Dec	Average
Average	14.3	14.9	16.7	18.6	21.5	24.4	26.8	27.7	26.7	23.8	20.0	16.1	
2:00	12.9	13.4	15.1	17.2	20.2	23.5	26.1	27.0	25.5	22.2	18.0	14.4	19.6
5:00	12.3	12.8	14.4	16.5	19.6	22.6	25.1	26.0	24.5	21.5	17.4	13.9	18.9
8:00	12.6	13.5	16.0	18.8	21.8	24.5	26.8	27.7	26.7	23.7	18.7	14.3	20.4
11:00	15.7	16.4	18.1	19.8	22.4	24.9	27.4	28.3	27.5	25.6	22.2	17.9	22.2
14:00	16.6	17.1	18.3	19.7	22.3	25.2	27.6	28.5	27.7	25.7	22.7	18.9	22.5
17:00	16.1	16.7	18.0	19.6	22.3	25.3	27.7	28.6	27.7	25.4	21.9	18.0	22.3
20:00	14.5	15.4	17.2	19.0	21.8	24.8	27.2	28.1	27.3	23.3	20.1	16.1	21.2
23:00	13.6	14.2	16.1	18.2	21.3	24.4	26.8	27.8	26.6	23.3	18.8	15.0	20.5

Legend: ≤15.0 °C; 15.1-20.0 °C; 20.1-25.0 °C; 25.1-30.0 °C

Table 3.2 Mean daily and annual RH in selected stations

	Average Daily Relative Humidity [%]												Annual Mean
	Jan	Feb	Mar	Apr	May	Jun	Jul	Aug	Sep	Oct	Nov	Dec	
University of Haifa	68	67	63	61	63	74	80	82	74	67	59	65	69
Ein Carmel	66	68	67	66	70	76	77	76	69	68	60	64	69
Ein-ha-Shofet	75	75	72	67	67	74	77	76	69	67	63	71	71
Hadera Port	65	67	67	69	75	77	76	74	69	68	61	65	69
Coastal Plain	**69**	**69**	**67**	**66**	**69**	**75**	**78**	**77**	**70**	**68**	**61**	**66**	**70**
Har Cnaan	74	72	62	55	48	49	50	58	60	60	65	70	60
Jerusalem	65	63	56	48	42	47	49	57	59	57	55	61	55
Arad	65	65	58	48	48	54	56	61	59	59	55	61	57
Mitzpe Ramon	62	59	53	44	43	45	44	52	61	63	57	56	53
Mountain Ridge	**67**	**65**	**57**	**49**	**45**	**49**	**50**	**57**	**60**	**60**	**58**	**62**	**56**
Havat Eden	75	73	66	56	51	52	52	54	53	54	57	68	59
Gilgal	70	68	57	48	43	44	44	48	50	53	55	65	54
Sdom	48	46	41	37	33	34	34	38	40	42	42	49	40
Eilat	47	42	37	32	27	26	27	31	37	41	41	46	36
Rift Valley	**60**	**57**	**50**	**43**	**39**	**39**	**39**	**43**	**45**	**48**	**49**	**57**	**47**

Legend: ≤30%; >30 - ≤40%; >40 - ≤50%; >50 - ≤60%; >60 - ≤70%; >70 - ≤80%; >80%

Data after the IMS (https://ims.gov.il)

this is not the case, and the annual cycle is more significant than the diurnal cycle (Tables 3.1 and 3.3). Furthermore, despite the very small size of the country, the annual range increases as one moves inland. This range is about 13–14 °C along the coastal plain, around 15.5 °C along the mountain ridge and about 18 °C along the JRV (Tables 3.1 and 3.3). Figure 3.5 presents the annual course of the daily temperature range along the three main longitudinal bands.

Fig. 3.4 The annual course of the mean daily relative humidity along the three main longitudinal bands

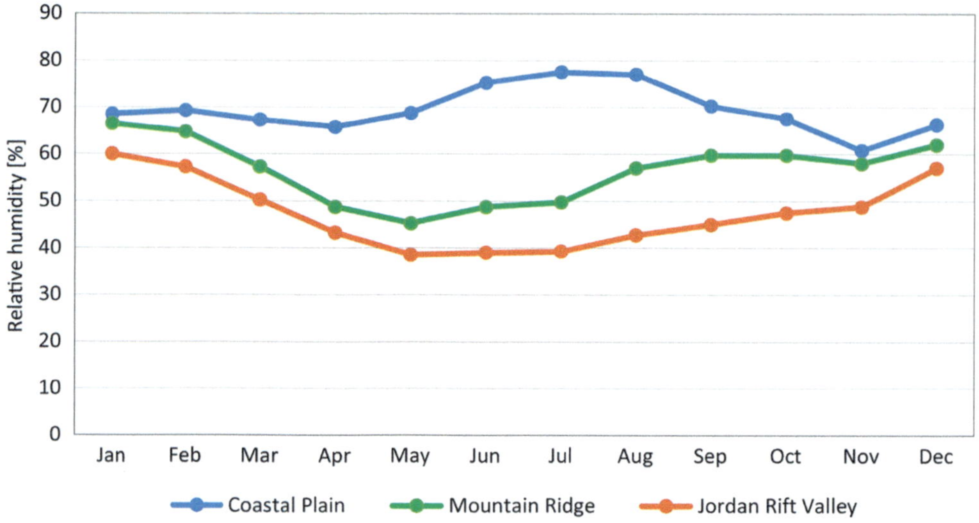

During the transitional seasons, several episodes of extremely hot and dry weather known as *Sharav* may occur. In spring, *Sharav* conditions are related mainly with a dynamical depression moving rapidly eastward from North Africa. In Israel, the passage of such a depression is rapid and may last between half a day and a day and half. The passage is associated with a major temperature rise, a decrease in relative humidity and a dramatic visibility reduction due to severe dust storms caused by the movement of the depressions over the Sahara Desert (Winstanley 1972; Alpert and Ziv 1989; Kutiel and Furman 2003; Barkan et al. 2004). The southern parts of the country are the most affected by the *Sharav* depressions. Such events tend also to enhance airborne allergenic pollen concentrations (Helfman-Hertzog et al. 2018).

Hot and dry conditions may be also provoked by an active Red Sea Trough (RST hereafter) mainly in autumn but sometimes also in spring. In such a case, the RST penetrates deep north along the Jordan Rift Valley and like the *Sharav* depressions, cause a temperature rise and a relative humidity drop. These episodes are less extreme as compared with the *Sharav* depression; however, they may last up to one week and their accumulated impact is considerable. The eastern parts of the country, namely along the JRV, are the most affected during such conditions.

Table 3.3 Mean daily and annual temperature ranges in selected stations

	Mean Daily Range [°C]												Annual Range
	Jan	Feb	Mar	Apr	May	Jun	Jul	Aug	Sep	Oct	Nov	Dec	
Eilon	7.4	8.2	8.9	9.9	10.5	10.0	9.5	9.5	9.5	9.2	8.7	7.5	14.0
Haifa Port	7.8	8.4	8.8	9.1	8.3	7.6	7.2	6.5	7.0	8.4	9.3	8.0	14.5
Ein Carmel	7.5	8.5	9.6	10.5	10.0	8.5	7.5	7.6	8.6	9.4	9.5	7.9	12.8
Hadera Port	6.9	6.8	6.6	6.5	5.2	4.1	3.7	3.7	4.7	6.1	7.6	7.0	12.6
TA Sde Dov	7.4	7.4	7.6	7.7	6.9	6.1	5.7	5.6	6.3	7.1	8.2	7.5	13.9
Negba	9.3	9.9	10.8	12.3	12.6	11.4	10.4	9.9	10.2	10.2	10.8	9.8	14.0
Coastal Plain	**7.7**	**8.2**	**8.7**	**9.3**	**8.9**	**8.0**	**7.3**	**7.1**	**7.7**	**8.4**	**9.0**	**8.0**	**13.6**
Har Cnaan	5.1	5.9	7.4	8.9	10.5	11.1	10.8	10.6	10.3	8.9	6.7	5.1	17.6
Neve Yaar	9.7	10.9	12.0	12.8	13.3	12.7	12.1	11.8	12.1	12.6	11.8	9.9	15.4
Jerusalem	5.9	6.7	7.8	9.4	9.9	10.2	9.8	9.9	9.7	8.5	7.1	6.2	15.3
Netiv ha-Lamed Hei	12.0	12.4	14.2	16.4	18.1	18.0	17.3	16.5	16.4	15.7	14.6	12.6	14.4
Arad	7.4	8.2	10.2	12.3	14.3	15.0	14.5	14.2	13.2	11.4	8.8	7.6	15.7
Beer Sheva	10.6	11.0	12.2	13.7	14.5	14.1	13.4	13.2	13.3	13.0	12.8	11.2	15.7
Mitzpe Ramon	6.7	7.7	9.4	11.1	11.5	11.7	11.3	11.4	10.8	9.6	7.7	6.8	15.5
Mountain Ridge	**8.2**	**9.0**	**10.5**	**12.1**	**13.2**	**13.3**	**12.7**	**12.5**	**12.3**	**11.4**	**9.9**	**8.5**	**15.7**
Kefar Blum	11.0	11.7	13.1	14.2	16.0	15.7	15.2	14.7	15.5	15.5	14.3	11.6	16.5
Ayelet ha-Shahar	9.0	9.8	11.6	13.2	14.9	15.0	14.5	13.9	13.7	13.0	11.4	9.3	16.9
Gilgal	11.2	12.2	13.8	15.0	16.5	16.4	15.9	15.0	14.5	13.9	13.3	11.5	18.0
Sdom	7.5	7.4	8.0	8.7	9.5	9.9	10.2	9.1	8.1	7.1	7.1	7.3	18.5
Hatzeva	10.5	10.8	11.9	12.8	13.8	14.0	14.0	12.7	11.5	10.4	10.7	10.5	18.8
Eilat	10.9	11.2	11.8	12.6	13.2	13.7	13.1	12.6	12.1	11.3	11.4	11.1	18.0
Rift Valley	**10.0**	**10.5**	**11.7**	**12.8**	**14.0**	**14.1**	**13.8**	**13.0**	**12.6**	**11.9**	**11.4**	**10.2**	**17.8**

≤6.0 °C
6.1-9.0 °C
9.1-12.0 °C
12.1-15.0 °C
15.1-18.0 °C
>18.0 °C

Data after the IMS (https://ims.gov.il)

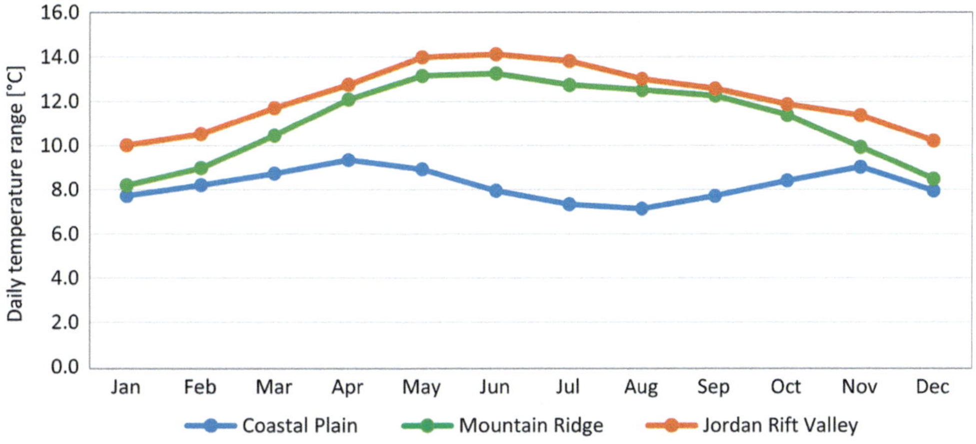

Fig. 3.5 The annual course of the daily temperature ranges along the three main longitudinal bands

In summer, subsidence of air masses as part of the Hadley cell may result in very hot and dry conditions along the mountain ridge and the JRV, whereas along the coastal plain, temperatures are average but the relative humidity is very high resulting in a very high discomfort index. This is caused mainly as the moisture brought by the sea breeze is trapped in a narrow zone due to the subsidence that disable the air masses to move inland. This explains the unusual annual course of the relative humidity described in the previous paragraph.

3.5 Rainfall Regime and Uncertainty

The rainfall regime in Israel is of a Mediterranean type, i.e., most rains are concentrated in the winter season with some sporadic events during the transitional seasons and the summer is dry. A typical seasonal summer drought lasts for 4–5 months which may extend up to 8–9 months in extreme dry years.

Anywhere in the world, mean annual rainfall increases with elevation (up to a certain level), decreases with increasing distance from the sea and is reduced on the lee side of a mountain or a hill. These three rules apply also in Israel and can be detected when looking at the mean annual rainfall map. However, in Israel, a fourth rule determines the spatial distribution; rainfall decreases from north to south. This is the most evident feature regarding the spatial rainfall distribution. This is because most depression trajectories in the Eastern Mediterranean, the so-called *Cyprus Low* (which is the main rainfall mechanism in Israel), are located north of the country and their influence diminishes as one moves southwards (Fig. 3.2b). During the transitional seasons, rainfall may originate from an active RST. Not all active RSTs cause rainfall in our region. Only the RST associated with a deep narrow upper level (500 hPa) trough located west of Israel may result in intense rains, mainly in the southern parts of the country and along the JRV (Tsvieli and Zangvil 2007).

Climatic uncertainty refers to the inability to forecast accurately when, where and how intense a certain climatic event will occur. Among all climatic parameters, rainfall is one of the most variable and therefore less predictable compared to, e.g., air temperature or atmospheric pressure.

In a series of articles, Reiser and Kutiel (2008, 2009, 2010a, b, 2011, 2012) analyzed many components of the rainfall regime in the Mediterranean basin including Israel. Among the analyzed components were the rainy season duration, its start and end, the number of rain-spells (consecutive rainy days), their average yield, various components of dryness and many others. One of the most evident characteristics of the rainfall regime in Israel as compared to other parts of the Mediterranean with similar annual totals, is that there are relatively fewer rain-spells but with higher yields. For example, Jerusalem, Marseille (France) and Brindisi (Italy) have a similar mean annual rainfall total (slightly above 500 mm), but while this amount in Brindisi and Marseille is obtained by an annual average of 37 and 30 rain-spells, respectively, yielding each 14.8–17.7 mm, respectively, in Jerusalem the annual total is obtained in 18 rain-spells, yielding 28.4 mm each (Reiser and Kutiel 2011). Furthermore, the average rainy season length in Jerusalem lasts for 129 days whereas, in Marseille (276 days) and Brindisi (254 days) is two or more times longer. This causes that the effective rainfall in Jerusalem is much lower, as a large percentage of it becomes runoff and does not percolate. The same characteristics are true anywhere within the Mediterranean zone of the country.

A person who drives from the Judean Hills (where average annual rainfall is 690 mm) to the Dead Sea (with <100 mm) will notice a dramatic change in landscape over about 25 km. The associated rainfall gradient is one of the steepest in the world, with an average decrease of more than 22 mm/km. In a drive of less than half an hour, one will ride from a humid Mediterranean to an extremely arid climate, crossing in between a narrow semi-arid region. This rainfall gradient is a result of a decrease, both in the mean annual

Table 3.4 A sample of countries with elevation differences between their highest and lowest points relative to their total area

Country	Area (km^2)	Highest (m asl)	Lowest (m asl)	Difference (m)	D/A (cm/km^2)
Israel	22,145	2236	−434	2670	12.1
Portugal	92,226	2351	0	2351	2.5
Greece	131,957	2917	0	2917	2.2
Italy	301,339	4748	0	4748	1.6
France	543,940	4807	−2	4809	0.9
Spain	505,992	3718	0	3718	0.7

number of rain-spells (consecutive days with a measurable rainfall amount) from 16 to 6 and in mean rain-spell yield from 39 to <15 mm. However, the mean rain-spell duration along this gradient remains constant, about 2 days (Aviad et al. 2013).

Aviad et al. (2004) analyzed the changes in mean annual rainy season length (RSL) along the same climatic transect from the Judean Hills to the Dead Sea. The RSL is the time elapsed from the first day when the measured rain equals or exceeds a daily rainfall threshold (DRT) until the last day with a DRT. The RSL depends on the DRT; RSL shortens as DRT increases. The RSL shortens with increasing aridity. For example, for a DRT = 10 mm, the average RSL in the Mediterranean part of the transect varies between 120 and 135 days and drops to less than 70 days in the arid part of this transect. All these rainfall characteristics are crucial for hydrologic and geomorphic processes.

3.6 Conclusions and Summary

Israel, despite its very modest size, demonstrates a large variability of climates. This variability can be noticed in almost all climatic parameters (temperature, relative humidity, wind regime, rainfall, etc.). The variability is observed not only in the magnitude of the parameters (hotter or cooler, moister or dryer) but also in the timing, for example when is the moister season (summer vs. winter), when is the onset of the rainfall season and so on.

One of the main reasons for this variability is due to the tremendous topographic differences relative to the country size. Table 3.4 presents a sample of few Mediterranean countries showing their highest and lowest points and the difference between the two, relative to the country size. While in France, Italy and Spain, the differences are larger than in Israel, due to their very large area the relative topographic relief per square km is by far smaller than in Israel as it can be noticed in the last column of Table 3.4.

Acknowledgements The author thanks Ms. Noga Yoselevich for preparing the drawings.

References

Alpert P, Ziv B (1989) The Sharav Cyclone: observations and some theoretical considerations. J Geophys Res Atmos 941(D15):18495–18514

Aviad Y, Kutiel H, Lavee H (2004) Characteristics of the rainy season along a Mediterranean-Arid climate transect: Start, end & length – a geomorphic approach. J Arid Environ 59:189–204

Aviad Y, Kutiel H, Lavee H (2013) Empirical models of rain-spells characteristics—a case study of a Mediterranean-arid climatic transect. J Arid Environ 97:84–91

Barkan J, Kutiel H, Alpert P (2004) Climatology of dust sources in North Africa and the Arabian Peninsula, based on TOMS data. Indoor Built Environ 13:407–419

Bitan A (1974) The wind regime in the North-West section of the dead-sea. Arch Met Geoph Biokl Ser B 22:313–335

Bitan A (1977) The influence of the special shape of the dead-sea and its environment on the local wind system. Arch Met Geoph Biokl Ser B 24:283–301

Bitan A (1981) Lake Kinneret (Sea of Galilee) and its exceptional wind system. Bound- Lay Meteorol 21:477–487

Helfman-Hertzog I, Kutiel H, Levetin E, Galán C, Hefer T (2018) The impact of Sharav weather conditions on airborne pollen in Jerusalem and Tel Aviv (Israel). Aerobiologia 34:497–511

IMS. https://ims.gov.il

Kutiel H (2019) Climate uncertainty in the Mediterranean basin and its possible relevance to important economic sectors. Atmosphere 10(1):10

Kutiel H, Furman H (2003) Dust storms in the middle east: sources of origin and their temporal characteristics. Indoor Built Environ 12:419–426

Reiser H, Kutiel H (2008) Rainfall uncertainty in the Mediterranean: definition of the rainy season–a methodological approach. Theor Appl Climatol 94:35–49

Reiser H, Kutiel H (2009) Rainfall uncertainty in the Mediterranean: definition of the daily rainfall threshold (DRT) and the rainy season length (RSL). Theor Appl Climatol 97:151–162

Reiser H, Kutiel H (2010a) Rainfall uncertainty in the Mediterranean: intra-seasonal rainfall distribution. Theor Appl Climatol 100:105–121

Reiser H, Kutiel H (2010b) Rainfall uncertainty in the Mediterranean: dryness distribution. Theor Appl Climatol 100:123–135

Reiser H, Kutiel H (2011) Rainfall uncertainty in the Mediterranean: time series, uncertainty and extreme events. Theor Appl Climatol 104:357–375

Reiser H, Kutiel H (2012) The dependence of the annual total on the number of rain-spells and their yield in the Mediterranean. Geografiska Ann 94A:285–299

Tsvieli Y, Zangvil A (2007) Synoptic climatological analysis of red sea trough and non-red sea trough rain situations over Israel. Adv Geosci 12:137–143

Visnovaty B (2021) Weather conditions and major forest fires. MA thesis, Department of Geography and Environmental Studies, University of Haifa

Winstanley D (1972) Sharav. Weather 27:146–160

The Quaternary Climate of Israel

4

Adi Torfstein

Abstract

Israel is located in the southern sector of the Levant, spanning the climatic transition between the hyperarid Sahara Desert in the south and the moderate Mediterranean climate in the north. Past changes in the locus of this climatic transition zone have been recorded at numerous sites in Israel and its surroundings, reflecting major Quaternary climate changes that have impacted regional fauna, flora, and hominid migration patterns, and influenced the cultural history of this region. Importantly, two uniquely resolved, continuous and long records—the Dead Sea sediments and concomitant cave deposits—serve as globally prominent archives of subtropical continental climate, which together with numerous additional regional climate reconstructions provide a highly detailed record of Quaternary climate change in the Levant. In this chapter, the main archives and observations that provide the basis for regional climate reconstruction are described, and used to provide an overview of climate patterns in the southern Levant throughout the Quaternary.

Keywords

Levant · Quaternary climate · Glacial–interglacial cycles

4.1 Introduction

Israel is located in the southern sector of the Levant, which extends between Egypt in the south and Anatolia and modern Iraq in the north, serving as the triple junction of Asia, Europe, and Africa (Fig. 4.1). Past climate change patterns in the Levant are recognized as highly important for two primary reasons. First, the Levant is located on the fringe of the Sahara Desert, and thus even small global climate perturbations have an immense impact on the regional climate conditions. This is particularly relevant for Israel, located in the southern Levant, where a sharp climatic gradient differentiates between the hyperarid Sahara Desert and moderate Mediterranean zones. In light of anthropogenic climate changes that include a rise in CO_2 levels and global warming over the last century, which are predicted to persevere and even accelerate in future decades (Masson-Delmotte et al. 2021), it is important to understand the potential impact of these processes on regional climate and to provide policy makers with tools to prepare for the changing conditions. The combination of regional geopolitical unrest and water scarcity (Kelley et al. 2015) warrants this mission to be of the highest priority. Second, the Levant has served as a major pathway for the dispersion of fauna and flora between Africa and Eurasia during the Quaternary. Importantly, it has also been argued that the main waves of mid- and late Quaternary hominid dispersion were modulated by wet intervals that provided the possibility for humans to cross the hyperarid Sahara and the Negev Desert on their way "out of Africa" (Vaks et al. 2007; Waldmann et al. 2010; Frumkin et al. 2011; Beyer et al. 2021), though this has also been discussed in the context of being driven by either dry or wet climate conditions (Tierney et al. 2017). Understanding the triggers of hominid dispersion is therefore critical for understanding important aspects of human evolution.

Israel, located in the southern Levant, is thus extremely sensitive to global and regional climate change, but at the

A. Torfstein (✉)
The Fredy and Nadine Herrmann Institute of Earth Sciences, Hebrew University of Jerusalem, Israel, and Interuniversity Institute for Marine Sciences, Eilat, Israel
e-mail: adi.torf@mail.huji.ac.il

A. Frumkin and N. Shtober-Zisu (eds.), *Landscapes and Landforms of Israel*,
World Geomorphological Landscapes, https://doi.org/10.1007/978-3-031-44764-8_4

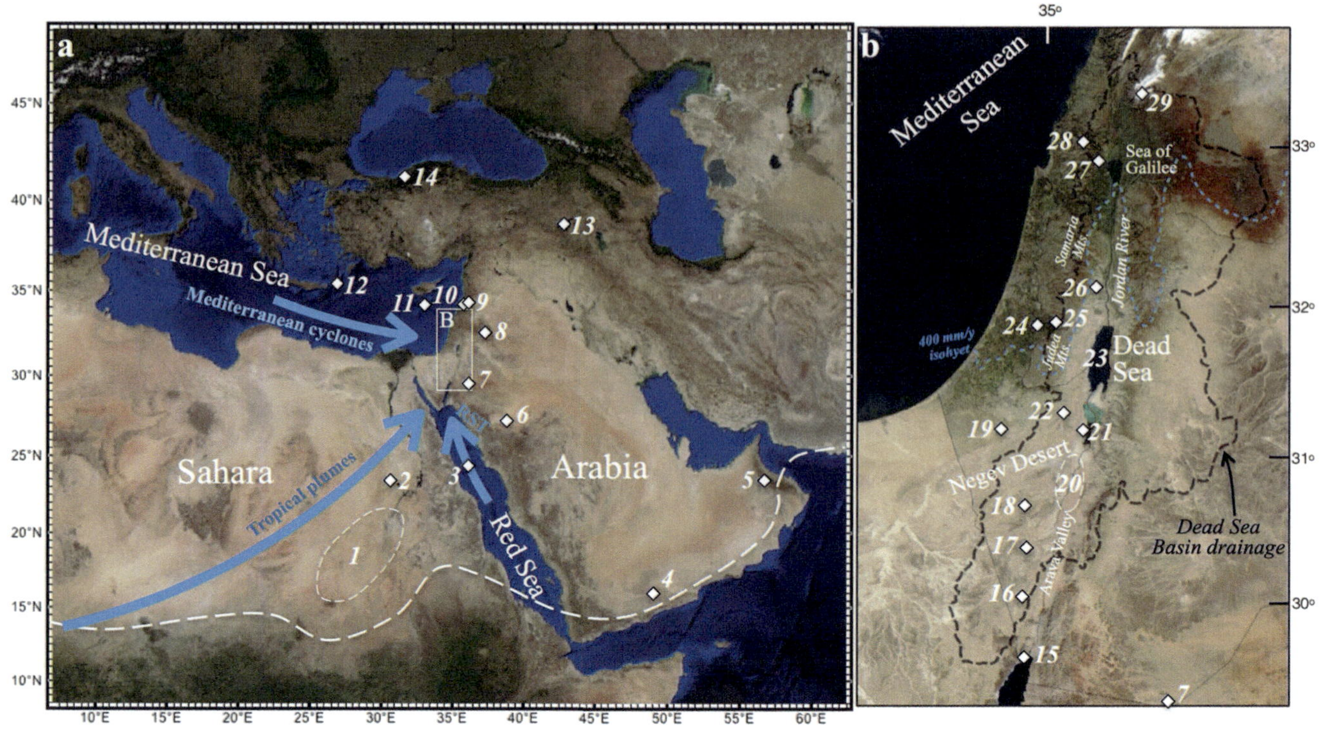

Fig. 4.1 **a Regional location map and climate archives**. The Levant is located between the Saharo–Arabia desert belt and the Mediterranean climate zones. The modern northern boundary of the equatorial summer rains is marked by a thick-dashed white line. The main systems contributing precipitation to the southern Levant are marked by blue arrows (Mediterranean cyclones, Red Sea Troughs (RST) and Tropical plumes); **b Southern Levant climate archives**. A dashed blue line marks the approximate locus of the 400 mm/y isohyet, beyond which a steep climate gradient leads to an extreme desert environment. Currently, most of the precipitation in northern Israel comes from the Mediterranean during the winter. The Negev Desert and Arava Valley receive minor amounts of localized rain, primarily associated with active RST systems and tropical plumes (Ziv et al. 2006). During the last interglacial the areas that currently receive rains from eastern Mediterranean sources were significantly drier. Locations of prominent geological archives in the region: (1) Eastern Sahara freshwater deposits (Szabo et al. 1995; Smith et al. 2007), (2) Western Desert–Egypt travertines (Crombie et al. 1997), (3) Core KL23 in the northern Red Sea (Palchan et al. 2013; Hartman et al. 2020), (4) Yemen Cave (Fleitmann et al. 2011), (5) Oman Cave (Fleitmann et al. 2003), (6) Nafud Desert (Rosenberg et al. 2013), (7) South Jordan (Petit-Maire et al. 2010), (8) Azraq Oasis–Jordan (Cordova et al. 2013), (9) Lake Yammouneh (Gasse et al. 2011), (10) Jeita Cave (Cheng et al. 2015), (11) ODP967 (Grant et al. 2017), (12) Core LC21 (Grant et al. 2012), (13) Lake Van (Litt et al. 2014; Stockhecke et al. 2016), (14) Sofular Cave (Fleitmann et al. 2009), (15) Gulf of Aqaba corals (Lazar and Stein 2011; Yehudai et al. 2017), (16) Ketora Cracks (Vaks et al. 2007), (17) Ma'ale-ha-Meyshar (Vaks et al. 2007), (18) Even-Sid (Vaks et al. 2007), (19) Ashalim Cave (Vaks et al. 2007), (20) Arava Valley travertines (Livnat and Kronfeld 1985; Waldmann et al. 2010), (21) Amora Formation (Zak 1967; Torfstein et al. 2009), (22) Tzavoa Cave (Vaks et al. 2006), (23) Dead Sea (see text for references), (24) Soreq Cave (Bar-Matthews et al. 2003), (25) Har Nof Cave (Frumkin et al. 2022), (26) Ma'ale Efrayim Cave (Vaks et al. 2003), (27) Zalmon Cave (Keinan et al. 2019), (28) Peqi'in Cave (Bar-Matthews et al. 2003), (29) Mt. Hermon (Ayalon et al. 2013)

same time also hosts some of the most detailed Quaternary climatic records in subtropical continental zones, allowing for a robust reconstruction of Quaternary climate change in the region.

4.2 Modern Climate Framework

Current climate patterns across Israel display sharp spatial gradients, with the northern and central regions characterized by Mediterranean climate (warm dry summers, cool wet winters), while the southern parts (i.e., the Negev Desert) experience hot, hyperarid desert climates (Goldreich 2003). Mean late twentieth century temperatures range between 24 °C (summer) and 7 °C (winter) in northern Israel, and 33 °C (summer) and 17 °C (winter) in southern Israel (Goldreich 2003). Present-day precipitation, stream flow, aquifer recharge, and spring discharge in the southern Levant are generated primarily (>90%) by winter Mediterranean cyclones, whose latitude, depth, and frequency are also the main controls over regional hydrology patterns on geological timescales (Enzel et al. 2003; Ziv et al. 2006; Kushnir and Stein 2010; Armon et al. 2019; Morin et al. 2019). By contrast, summers are dry in the Levant, in part because of air subsidence driven by the monsoon convection over northern India and Bay of Bengal (Goldreich 2003).

The Mediterranean winter systems yield a maximum of>1000 mm/y and>75 rainy days in the northern Galilee mountains, but their impact diminishes toward the south, reaching minimum values in the southernmost parts of the Negev Desert of<40 mm/y and less than 10 rainy days each year (Goldreich 2003). Another important feature is a pronounced orographic effect of the central Samaria and Judea mountains that shade the Dead Sea Basin (DSB), resulting in a sharp west to east precipitation decrease from>500 to less than 100 mm/y, respectively, over a distance of several tens of km (Armon et al. 2019). Other regional precipitation systems produce lower amounts of rain or are restricted in area. The Red Sea trough is the second most important source of rainfall in the region, active during the fall–winter (and occasionally spring), primarily affecting areas in the south and east of the Levant with localized storms producing limited amount of runoff. Subtropical jet activity forms well-defined cloud bands that originate in tropical Africa and extend across the Sahara Desert. These "tropical plumes" are active during winter–spring and bring larger precipitation to the arid Negev Desert, mostly south of the latitude of the Dead Sea (Ziv 2001; Kahana et al. 2002; Rubin et al. 2007; Tubi et al. 2017; Armon et al. 2019). Summer monsoon rains from either the African or the Indian monsoons do not reach the region.

4.3 Climate Archives

The Levant in general, and Israel in particular, host some of the most remarkable climate archives available in mid-latitude continental settings. These include the lacustrine records along the DSB (e.g., Zak 1967; Begin et al. 1974; Bartov et al. 2003; Hazan et al. 2005; Stein 2014; Torfstein et al. 2015a, b, 2009) (Fig. 4.2), as well as from the Soreq Cave (Bar-Matthews et al. 2003), and a series of cave deposits spread between the Galilee and the Negev Desert (Frumkin and Stein 2004; Vaks et al. 2018, 2006, 2003; Keinan et al. 2019; Orland et al. 2019) (Fig. 4.3). In addition to these detailed, long and mostly continuous records, other archives in the region include vast areas of the northern Negev Desert covered with loess deposits (Crouvi et al. 2008; Ben-Israel et al. 2015), soil and desert pavement records (Amit et al. 2006; Matmon et al. 2009; Porat et al. 2010), fluvial deposits (Greenbaum et al. 2006; Faershtein et al. 2016) (Fig. 4.4), and ample archeological sites and historical records that contribute to the understanding of the interplay between hominid behavior, cultural developments and climate conditions (e.g., Hirschfeld 2006; Frumkin et al. 2011; Ellenblum 2012). All the above should be considered in the context of additional archives in the region such as cave deposits across the Arabian Peninsula (Fleitmann et al. 2003) and northeastern Sahara Desert (El-Shenawy et al. 2018), lacustrine and fluvial deposits (Petit-Maire et al. 2010; Rosenberg et al. 2013; Gasse et al. 2015), as well as marine deposits across the eastern Mediterranean (Almogi-Labin et al. 2009; Box et al. 2011; Grant et al. 2017) and the Red Sea (Legge et al. 2006; Arz et al. 2007, 2003; Lazar and Stein 2011; Palchan et al. 2013; Bar et al. 2018; Hartman et al. 2020). Although this is clearly not an exhaustive list of the available regional paleoclimate archives, combined, the studies provide us with a detailed reconstruction of Quaternary climate in the southern Levant, as will be discussed hereafter.

Although pre-Quaternary climate is not discussed here, it is nonetheless worth noting that two large-scale processes must have had a major impact on regional climate throughout the Pliocene–Quaternary. The first is the Northern Hemisphere (NH) glaciation whose onset corresponds to the Pliocene—Quaternary transition, and which, based on the sensitive regional climate response to global trends during the Quaternary, could be postulated to have caused a gradual change in the Levant hydroclimate, as indeed is reflected by the Plio-Pleistocene aridification of the Sahara (deMenocal 2004; Trauth et al. 2009). The second process is the long-term regional landscape evolution associated with the development of the Dead Sea rift valley (Matmon and Zilberman 2017), which resulted in the development of the central mountain ridge (e.g., Judea and Samaria Mountains). This imposed a strong local orographic effect and determined the location and nature of the climate records that are preserved across Israel. Unfortunately, the lack of appropriate archives inhibits a more detailed evaluation of the impacts of both processes on pre-Quaternary Levant climate.

4.4 Quaternary Paleoclimate

4.4.1 Early-Pleistocene (2.6–0.78 Ma)

Climate conditions in the Levant during the early and middle Quaternary are poorly constrained relative to later stages. Nevertheless, several records span this period and provide information on regional climate. A long-term climatic reference is provided by soils found across the currently hyperarid Negev Desert (Fig. 4.4). Amit et al. (2011) concluded that relative to present, slightly wetter conditions prevailed across the Negev Desert between 2 and 1 Ma, allowing for sparse vegetation cover that in turn supported the formation of moderately developed calcic soils. Widespread formation of waterbodies across the Negev Desert and southern Jordan during the Pliocene and early Pleistocene further reflects an overall relatively wet period,

Fig. 4.2 **Dead Sea lacustrine records**. **a** Aerial view of the Lisan Peninsula, **b** The Perazim Valley PZ1 type section of the Lisan Formation. The alternating aragonite detritus (aad) laminae are magnified in upper right. U-Th ages are superimposed over their corresponding sampling spots (Haase-Schramm et al. 2004) and reveal three hiatuses in the section, **c** The sedimentary record of the Dead Sea during the last four decades (Sirota et al. 2021). The top of this sequence precipitated underwater ca. 2016 and has since been exposed due to continued lake-level drop, **d** Retreating shorelines of the Dead Sea that formed since the 1970s due to the modern lake-level drop. Each strand line represents the lake level during its deposition (Bookman et al. 2004)

though age constraints remain poor (Ginat et al. 2018). Following a tectonic uplift of ca. 200 m of the Negev highland, along with a regression of the eastern Mediterranean coast to the northwest (Matmon and Zilberman 2017), less moisture reached the Negev Desert from north-west, triggering the development of more hyperarid conditions in the Negev Desert after ~1 Ma.

During the first half of the Quaternary, evidence from the DSB is mostly limited to a series of deep boreholes, where palynological records were argued to correlate with global climate fluctuations, though they lacked temporal resolution and absolute chronology (Horowitz 1987). Nevertheless, the lithostratigraphy of the Sedom-1 deep borehole suggests a late Pliocene to early Pleistocene dominance of a sandy lithology (i.e., *Melekh Sedom Sand* Member) marking fluvial conditions. The morphology of the DSB, however, is not well reconstructed for that time, again limiting the ability to interpret the deposits in the context of lake configuration and water depth.

Additional details on the hydroclimatic regime during the early and middle Quaternary can be derived from cave deposits in the northern Negev Desert, which were dated using the U–Pb method (Vaks et al. 2018). These indicate that this region was exposed to intermittent wet pulses ("Negev Humid Periods," NHP) between ~1.27 and 0.5 Ma (Fig. 4.5). Speleothem $\delta^{18}O$ values (6.9–11.2‰) were corrected for rainfall $\delta^{18}O$ shifts associated with ice volume and temperature effects, yielding an observed relative constancy of speleothem $\delta^{18}O$ values between the Pliocene and Late Pleistocene. This was interpreted by Vaks et al. (2018) to indicate a common source of rainfall originating from the eastern Mediterranean. These authors postulated

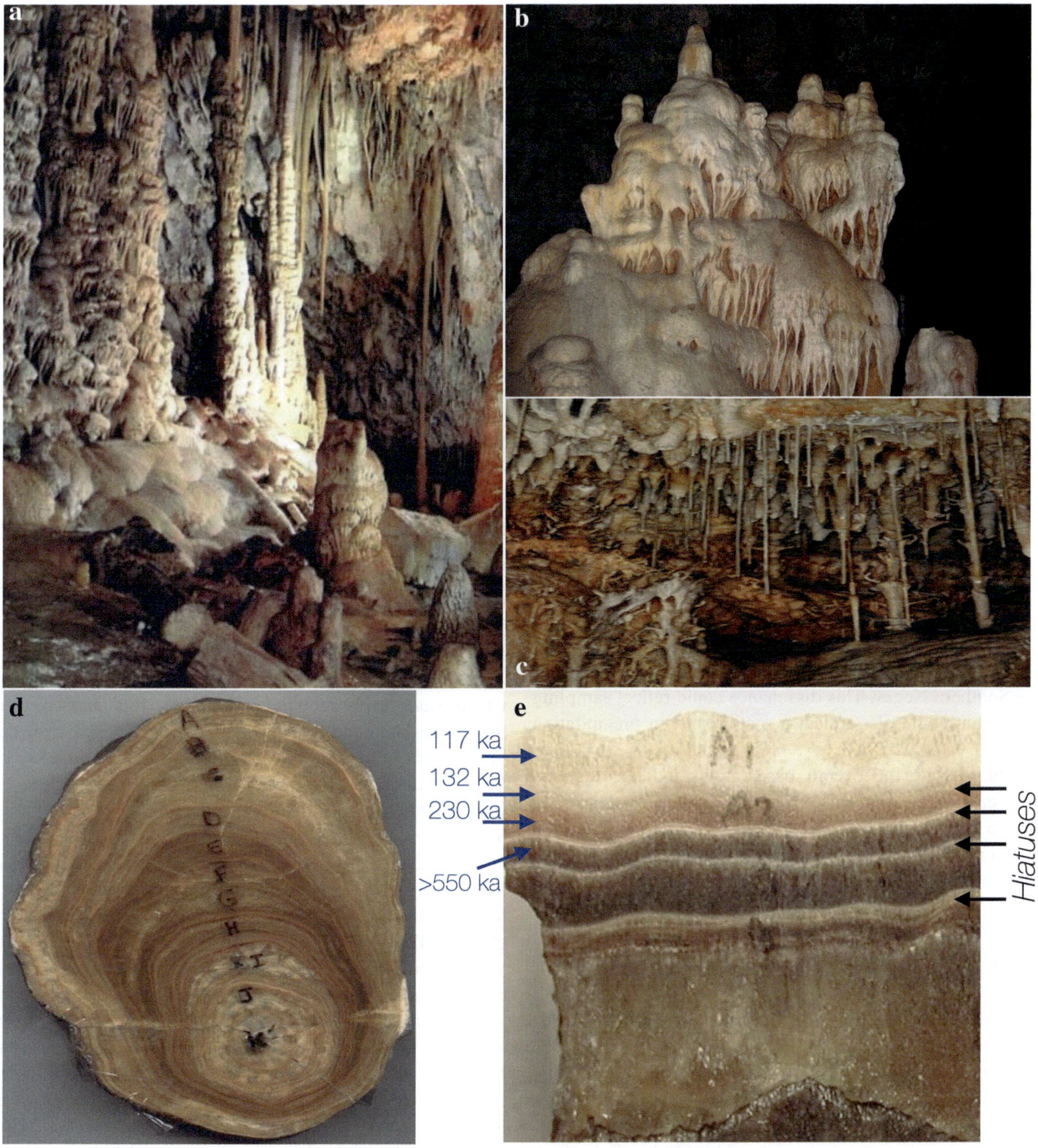

Fig. 4.3 **Cave deposits in Israel.** **a, b** Soreq cave, **c** Needle speleothems, Soreq cave, **d** Speleothem growth bands, Soreq Cave, **e** Ashalim Cave stalactite in the northern Negev Desert with U-Th ages (blue) marking deposition during interglacials (modified after Vaks et al. 2018). **a, c, d** courtesy of M. Bar-Matthews and A. Ayalon, **d** courtesy of E. Kagan, M. Bar-Matthews and A. Ayalon

that the humid Pliocene and early Pleistocene conditions could be generated by a more southerly position of the Mediterranean coast (allowing the access of Mediterranean precipitation to the Negev), as well as by warmer Atlantic Sea Surface Temperatures that weakened the Azores High Pressure Cell. It should be noted that *a-priori*, the formation of cave deposits in the northern Negev lags the early Pleistocene wet phase discussed by Amit et al. (2011) in

Fig. 4.4 Desert pavement, soils, fluvial terraces and loess from the Negev Desert. **a** Overview and **b** close view of the Desert Pavement of the Paran Plains (modified after Amit et al. 2011), **c** Soil Profile (Paran Plains) with OSL ages (modified after Amit et al. 2011), **d** Shehoret fluvial terraces, Southern Negev Desert. OSL ages are from Gellman et al. (2018), **e** General view of Mount Harif area, showing the primary loess mantle and the fluvial/colluvial terraces at the bottom (modified after Crouvi et al. 2009).

the Southern Negev. This dissonance could reflect sampling biases but could also reflect a spatial shift in precipitation patterns.

Thus, while the overall observations indicate that the southern Negev was wet during the early Pleistocene relative to later stages, knowledge of the climatic configuration is still very limited due to the poor, or lack of, temporal resolution of related evidence, warranting further investigation.

4.4.2 Mid-Pleistocene (780–130 ka)

One of the most prominent archives of mid-Pleistocene climate in Israel is a sedimentary sequence exposed at the Arubotaim Cave site, along the eastern margins of Mount Sedom (Fig. 4.6). There, the rising Sedom salt diapir uplifted and tilted overlaying sediments, exposing deposits of the Amora Formation dated to ~740–130 ka (Zak 1967; Torfstein et al. 2009). This sequence, comprised of alternating intervals of marls, gypsum, aragonite, and detritus, displays cyclic shifts between dry-low-stand and wet-high-stand conditions. During ~740–550 ka, the lake experienced frequent but limited amplitude level fluctuations, after which water levels declined and the lake was characterized by sabkha-like conditions. This period terminated with the instigation of the exceptionally warm Marine Isotope Stage (MIS) 11 at ca. 420 ka, during which a prominent salt layer was deposited in Lake Amora, reflecting a major regional hyperarid pulse. MIS7 was characterized by a low-stand and sporadic intensive flooding events, while MIS6 was a high-stand period that was followed by a significant lake-level decline corresponding with the onset of the last interglacial (MIS5).

The shift from frequent lake stand fluctuations experienced prior to 550 ka to a more stable lake thereafter, corresponds with a transition in the global climate patterns from a cyclicity of 41–100 ka (i.e., the mid-Pleistocene transition; Mudelsee and Schulz 1997). During MIS7 (and to a lesser extent during MIS17, 15 and 9), $\delta^{18}O$ excursions toward lighter values are interpreted to reflect short but intense freshwater flooding of the lake, possibly caused by rain storms originating from southern, low latitudes, during an otherwise arid period (Torfstein et al. 2009). This conclusion is further supported by studies of younger intervals and of the modern Dead Sea (Gat 1984; Kolodny et al. 2005), as will be discussed below.

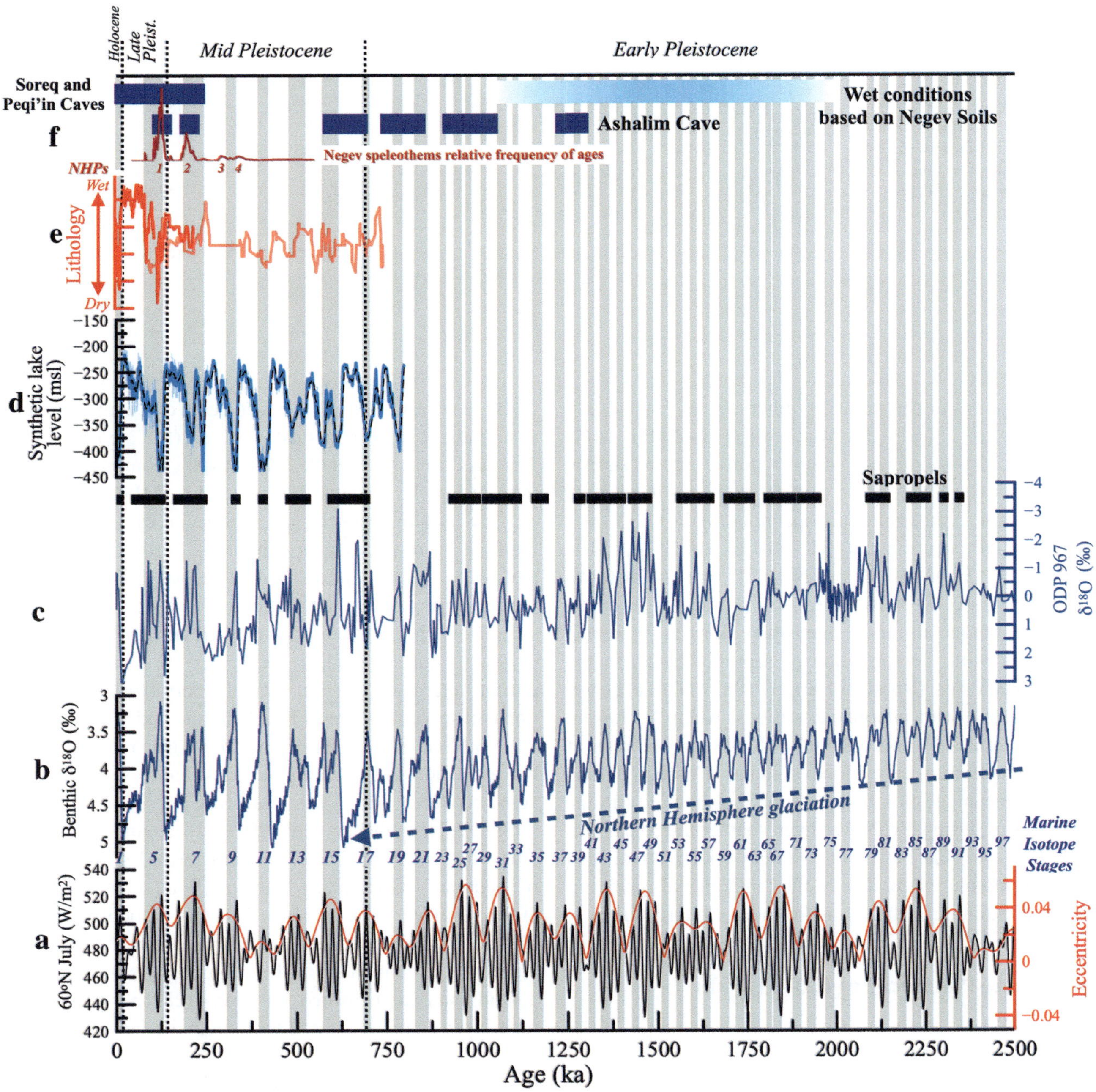

Fig. 4.5 **Climate patterns since 2.5 Ma.** **a** Eccentricity and northern hemisphere summer insolation (Laskar et al. 2004), **b** Stacked benthic $\delta^{18}O$ record (Lisiecki and Raymo 2005), **c** Eastern Mediterranean $\delta^{18}O$ and timing of sapropels (ODP967, Kroon et al. 1998), **d** Dead Sea synthetic lake level (Torfstein 2019), **e** Lithological facies of the Dead Sea deposits. A running curve reflects varying limnologic–climatic conditions, ranging between *aad* intervals, through laminated and massive marls, gypsum and fine sand, to sand, pebbles and halite, reflecting wet, intermediate, arid, and hyperarid climate conditions, respectively; the DSDDP (Torfstein et al. 2015a, red curve) and the Amora Formation Arubotaim Cave section (Torfstein et al. 2009, faint red curve), **f** Timing of Negev Humid Periods (NHPs) (Vaks et al. 2010), timing of deposition at Ashalim, Soreq and Peqi'in Caves (Bar-Matthews et al. 2003; Vaks et al. 2007, 2003), and formation of Negev soils (Amit et al. 2011). Gray rectangles mark interglacial periods

Fig. 4.6 The Amora Formation at the Arubotaim Cave (AC) site. The lacustrine section has been uplifted and tilted to a vertical position by the rising Sedom Diapir that forms Mount Sedom (Zak, 1967). The AC-1 section consists of sediments deposited between MIS11-5 (Torfstein et al. 2009)

Cave deposits in central and northern Israel extend only as far back as ~250 ka (Bar-Matthews et al. 2003), while speleothem deposition has been reported from several caves across the Negev Desert, as far south as the Ketura Cracks (Fig. 4.1), intermittently spanning the last 350 kyrs, during NHPs occurring at 350–310 ka (NHP-4), 310–290 ka (NHP-3), 220–190 ka (NHP-2), and 142–109 ka (NHP-1) (Vaks et al. 2010) (Figs. 4.3, 4.5 and 4.7). NHPs are contemporaneous with intensified African monsoon activity and with the formation of sapropel layers in the Mediterranean Sea (Rossignol-Strick 1983). These generally correspond with interglacial periods, when the Levant was characterized by arid conditions due to a weakening of Atlantic–Mediterranean cyclone systems. Combined, this reinforces the suggestion that the source of humidity in the Negev Desert during NHPs was from southern, tropical latitudes. However, the best studied interglacial interval, and the most extreme expression of the "interglacial-type" climatic configuration in the Levant, is the last interglacial.

4.4.3 Late Pleistocene (130–11.7 ka)

4.4.3.1 Last Interglacial

The last interglacial (MIS5; Fig. 4.7) is globally considered to be an analogue for a warm future Earth due to its temperatures that were, on average, approximately 1 °C higher than pre-industrial values (Otto-Bliesner et al. 2021). In the Levant, this period is particularly interesting, as abundant evidence suggests a unique switch in climate systems and a superposition of tropical humid incursions over an extreme hyperarid climate imposed by a weakening of NH systems. For the purpose of discussing this unique climatic configuration, we first consider the record of cave deposits across Israel. Caves in northern (Peqi'in and Zalmon) and central (Soreq) Israel show continuous deposition over the last ~200 kyrs (Bar-Matthews et al. 2003; Vaks et al. 2003; Keinan et al. 2019), assumingly reflecting above-threshold precipitation levels during this period.

By contrast, nearby caves located down the precipitation gradient in the rain shadow desert (Ma'ale Efrayim and Tzavoa Caves; Vaks et al. 2006) display continuous deposition during the last two glacial cycles (MIS6,4–2), bracketed by long hiatuses during interglacial cycles (MIS7,5,1) marking corresponding arid periods. Further to the south, the ages of cave deposits spread across the Negev Desert cluster during interglacial cycles (Vaks et al. 2007). This spatial switch in climatic patterns—relatively wet-north and dry-south during glacials versus dry-north and wet-south during interglacials (Fig. 4.7)—is corroborated by additional observations across Israel and the wider region.

Dead Sea lake levels during MIS5 generally lack absolute elevation anchors though thick halite sequences in a core recovered during the Dead Sea Deep Drilling Project (DSDDP; Neugebauer et al. 2014; Torfstein et al. 2015a), along with marginal sections characterized by coarse clastic beds and sedimentary hiatuses (Waldmann et al. 2010, 2009), reflect extreme low stands, particularly during MIS5e. During this interval, the thick sequence of massive halite that accumulated in the deep basin is interrupted by a mud interval deposited between ~128 and 122 ka. This mud layer contains intermittent aragonite laminae that display strong negative $\delta^{18}O$ excursions. Combined, these observations have been interpreted to reflect extreme, albeit limited in quantitative impact, flooding events in the Dead Sea watershed, mostly in its southern parts (Torfstein et al. 2015a; Kiro et al. 2020) (Fig. 4.8). This shift in the water balance across the Dead Sea watershed hindered a full drydown of the Dead Sea during MIS5e (Torfstein et al. 2015a). After the effective cessation of southern moisture incursions at the end of MIS5e, the Dead Sea water levels dropped dramatically, as reflected by a deposition sequence of mud–halite–beach pebbles, suggesting that the bottom of the lake was at least partially exposed during MIS5d, resulting in a short sedimentary hiatus represented by the pebble layer (~116–110 ka) (Torfstein et al. 2015a). While this suggestion has been debated (Katz and Starinsky 2015; Torfstein et al. 2015b), a corresponding depositional hiatus in the Har Nof Cave in Jerusalem (Frumkin et al. 2022), as

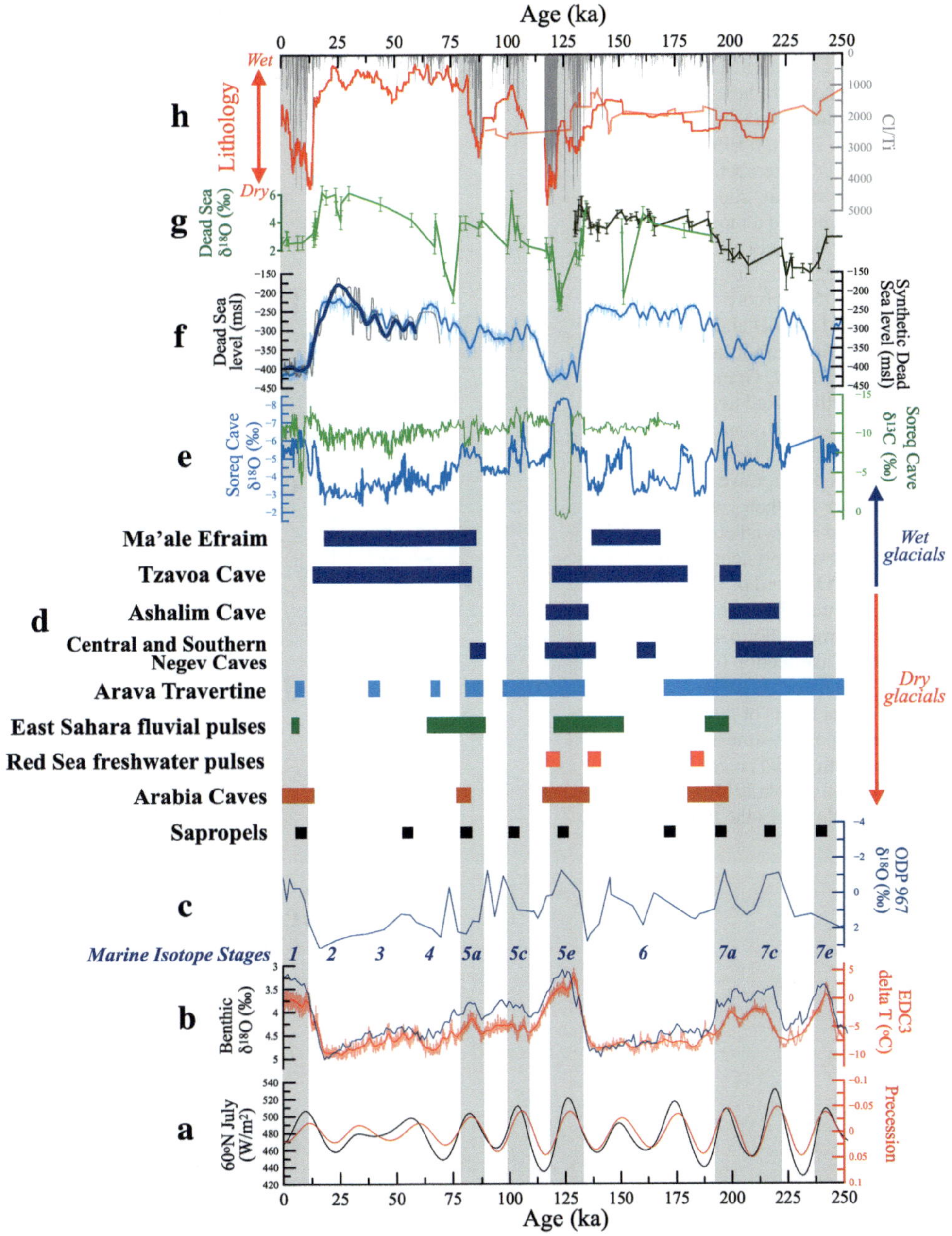

Fig. 4.7 Climate patterns over the last 250 kyrs. a Precession and northern hemisphere summer insolation (Laskar et al. 2004), **b** Stacked benthic $\delta^{18}O$ record (Lisiecki and Raymo 2005) and EPICA Dome C3 (EDC3) temperature changes (Jouzel et al. 2007), **c** Eastern Mediterranean $\delta^{18}O$ and timing of sapropels (ODP967, Kroon et al. 1998), **d** Timing of wet pulses in the region (see text for details and Fig. 4.1 for locations), **e** Soreq Cave $\delta^{18}O$ and $\delta^{13}C$ (Bar-Matthews et al. 2003), **f** Dead Sea absolute lake level (dark blue; Torfstein et al. 2013) and extended synthetic curve (navy blue; Torfstein 2019), **g** Dead Sea $\delta^{18}O$ record of DSDDP core (light green; Torfstein et al. 2015a) and Amora Formation Arubotaim Cave section (dark green; Torfstein et al. 2009), **h** Lithological facies of the Dead Sea deposits (same as in Fig. 4.5e) with a Cl/Ti record (grey; Torfstein et al. 2015a) representing pulses of halite deposition during hyperarid intervals. Gray rectangles mark interglacial periods

well as a slightly lagging depositional hiatus (<111 ka) in a stalagmite record from Kanaan Cave Lebanon (Nehme et al. 2015), strongly support the occurrence of extreme regional hyperaridity during MIS5d, presumably reflecting the shutdown of the southern rain systems after MIS5e, while eastern Mediterranean systems continued to remain exceptionally weak. Nevertheless, it is readily acknowledged that the evidence for the magnitude, timing, and temporal extent of this important arid spell is still very much lacking and requires further investigation.

Multiple additional lines of evidence indicate that the incursion of precipitation into the region during MIS5, and in particular MIS5e, originated from the south. These include evidence for massive flooding of the northern Red Sea (Hartman et al. 2020) and enhanced freshwater flow into the northern Gulf of Aqaba (Lazar and Stein 2011), development of fluvial deposits and water bodies in Jordan, Arabia, and the northeastern Sahara Desert (Szabo et al. 1995; Petit-Maire et al. 2010; Cordova et al. 2013; Rosenberg et al. 2013), travertine deposition along the Arava Valley (Livnat and Kronfeld 1985; Waldmann et al. 2010), speleothem formation across Arabia, northeastern Sahara Desert and the Negev Desert (Fleitmann et al. 2003; Vaks et al. 2007, 2006; El-Shenawy et al. 2018; Henselowsky et al. 2021), and deposition of sapropel layers in the Mediterranean Sea (Thunell et al. 1977).

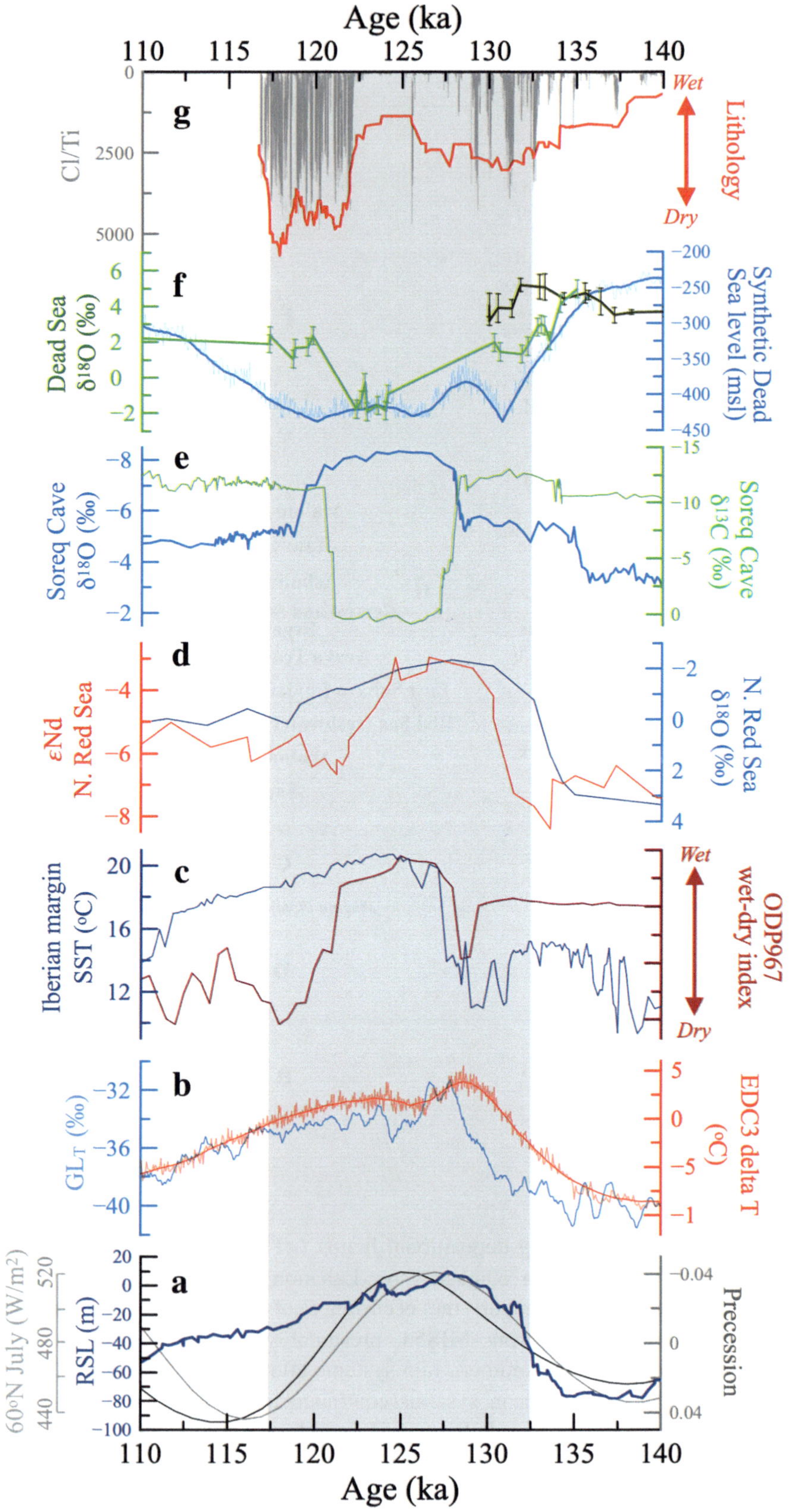

Fig. 4.8 Climate patterns during MIS5e. **a** Red Sea Relative Sea Level (RSL) (Grant et al. 2012), northern hemisphere summer insolation and precession (Laskar et al. 2004), **b** EPICA Dome C3 (EDC3) temperature changes (Jouzel et al. 2007) and Greenland synthetic temperature record (GL_T; Barker et al. 2011), **c** Iberian Margin Sea Surface Temperature (SST) (core md012443, Martrat et al. 2007) and eastern Mediterranean wet-dry index (ODP967; Grant et al. 2017), **d** Northern Red Sea $\delta^{18}O$ and δNd curves from core KL23 (Palchan et al. 2013; Hartman et al. 2020), **e** Soreq Cave $\delta^{18}O$ and $\delta^{13}C$ (Bar-Matthews et al. 2003), **f** Dead Sea synthetic lake-level curve (Torfstein 2019), Dead Sea $\delta^{18}O$ record of DSDDP core (light green; Torfstein et al. 2015a) and Amora Formation Arubotaim Cave section (dark green; Torfstein et al. 2009), **g** Lithological facies of the Dead Sea deposits (same as in Fig. 4.5e) with a Cl/Ti record (grey; Torfstein et al. 2015a). Gray rectangle marks the timing of MIS5e

The latter organic-rich layers reflect the impact of the Nile River overflow during peak equatorial insolation cycles that drive enhanced monsoon rain activity in east Africa and the Ethiopian Highlands (Rossignol-Strick 1983). Moreover, a composite index of humidity and aridity for the wider North African region has been developed using core ODP967 in the eastern Mediterranean (Grant et al. 2017), and interpreted to reflect increased relative regional wetness during MIS5e (Figs. 4.1 and 4.8C). The flooding of the eastern Mediterranean by terrestrial freshwater resulted in the eastern Mediterranean surface waters becoming more dilute with a shift in their $\delta^{18}O$ signature toward lighter values, which is reflected by the cave records in Israel, whose isotopic composition is modulated by shifts in the water source (Kolodny et al. 2005). Accordingly, the peak of MIS5e is expressed by significant $\delta^{18}O$ and $\delta^{13}C$ excursions in the Soreq and Peqi'in Caves (Bar-Matthews et al. 2003). While the $\delta^{18}O$ excursion is associated with a shift in eastern Mediterranean source waters, the shift in $\delta^{13}C$ compositions toward higher values reflects a wash-out of the covering soil and organic matter that serve as a low $\delta^{13}C$ end member, and the dominance of residual inorganic Cenozoic limestone as a source of bicarbonate.

It is emphasized here that the intensification of tropical monsoon rains in Africa during MIS5e is not thought to have a direct impact on the Levant (i.e., the monsoon rains did not actually reach the Levant), in agreement with the interpreted nature of precipitation during MIS5e, which was abrupt and "violent," most likely reflecting more frequent incursion of rainstorms associated with tropical plumes or Red Sea troughs (Waldmann et al. 2010; Enzel et al. 2015; Torfstein et al. 2015a; Hartman et al. 2020). Thus, while the Dead Sea regressed to extreme minimum water levels imposed by the unprecedented weakening of Mediterranean cyclones during the warm MIS5e, a second, concomitant process of increased incursions of southerly sourced rainstorms nourished this region, particularly its southern part. As mentioned, the primary sources of precipitation in the southern Levant are winter Mediterranean cyclones. By contrast, Red Sea troughs are active mainly during fall–winter (and occasionally spring), while tropical plumes are active during winter–spring (Armon et al. 2019). Thus, weakening of the main rain system in winter together with a relative intensification of fall and spring rain resulted in a change in seasonal rain patterns, such that the rain season was longer relative to present. This scenario can explain shifts in the regional vegetation cover (e.g., Chen and Litt 2018), but does not require the occurrence of monsoon-related summer rains. Indeed, global climate model reconstructions of Levant climate during MIS5e fail to show significant summer precipitation, and generally imply summer rain rates that are limited to less than 1 mm/day (Herold and Lohmann 2009; Otto-Bliesner et al. 2020; Kutzbach et al. 2020). Given the present-day extreme hyperarid conditions in this region, with negligible precipitation between fall and spring and no summer precipitation, it is difficult to envision that the very minor modeled summer precipitation increase relative to present could explain the above-mentioned observations during MIS5e. Rather, an increase in the frequency and intensity of abrupt storm-like events that occur between fall and spring and originate in lower latitudes is a much likelier mechanism to explain the compiled observations.

Another interesting aspect regarding MIS5e conditions stems from pollen records from the Dead Sea, which indicate the presence of grasslands and expanded woodlands across the Dead Sea watershed, as well as a large increase in the charcoal-to-pollen ratio (Chen and Litt 2018). The relative increase in charcoal marks an increased frequency of fires in the region, presumably driven by the buildup of vegetation "fuel" and possibly combined with a high intensity of lightning that ignited the dry vegetation at the end of the dry season, when the fires are particularly widespread due to the mostly dead biomass (Keeley and Bond 2001; Frumkin and Comay 2019; Frumkin et al. 2022).

4.4.3.2 Last Glacial

The last glacial period (MIS4-2; Fig. 4.9) was characterized by a gradual drop in temperatures, growth of the polar ice sheets, and drop in sea level that peaked during the Last Glacial Maximum (LGM) (~25 ka), followed by a warming phase that resulted in the resumption of interglacial conditions and the onset of the Holocene. During glacial times, the expansion of the NH ice sheet, including the European snow cover, resulted in increased albedo and a strong temperature gradient that deflected the jet stream southward. This caused an overall southward shift in the route of westerly storm tracks over the Mediterranean region, resulting in an intensification and higher frequency of eastern Mediterranean cyclones that increased the direct delivery of moisture to the southern Levant, which was accordingly characterized by extremely wet conditions during the last glacial cycle (Fig. 4.9).

Concomitantly, the drop in sea level during MIS2 to ~120 m below the current sea level (mbsl) drove the Israeli Mediterranean coastline to advance ~30–70 km northwest of its present-day position (Figs. 4.9 and 4.11). This caused areas currently receiving ~200 mm/y or less to be located much farther from the coast, thereby further weakening their rain supply. Subsequently, while northern and central Israel were significantly wetter during the glacial, the desert boundary toward the south remained in place or could have even migrated slightly northwards, resulting in a steeper, relative to present, rainfall gradient in the northern Negev (Enzel et al. 2008).

Another effect of the northward migration of the coastline due to the glacial sea-level drop was that the newly

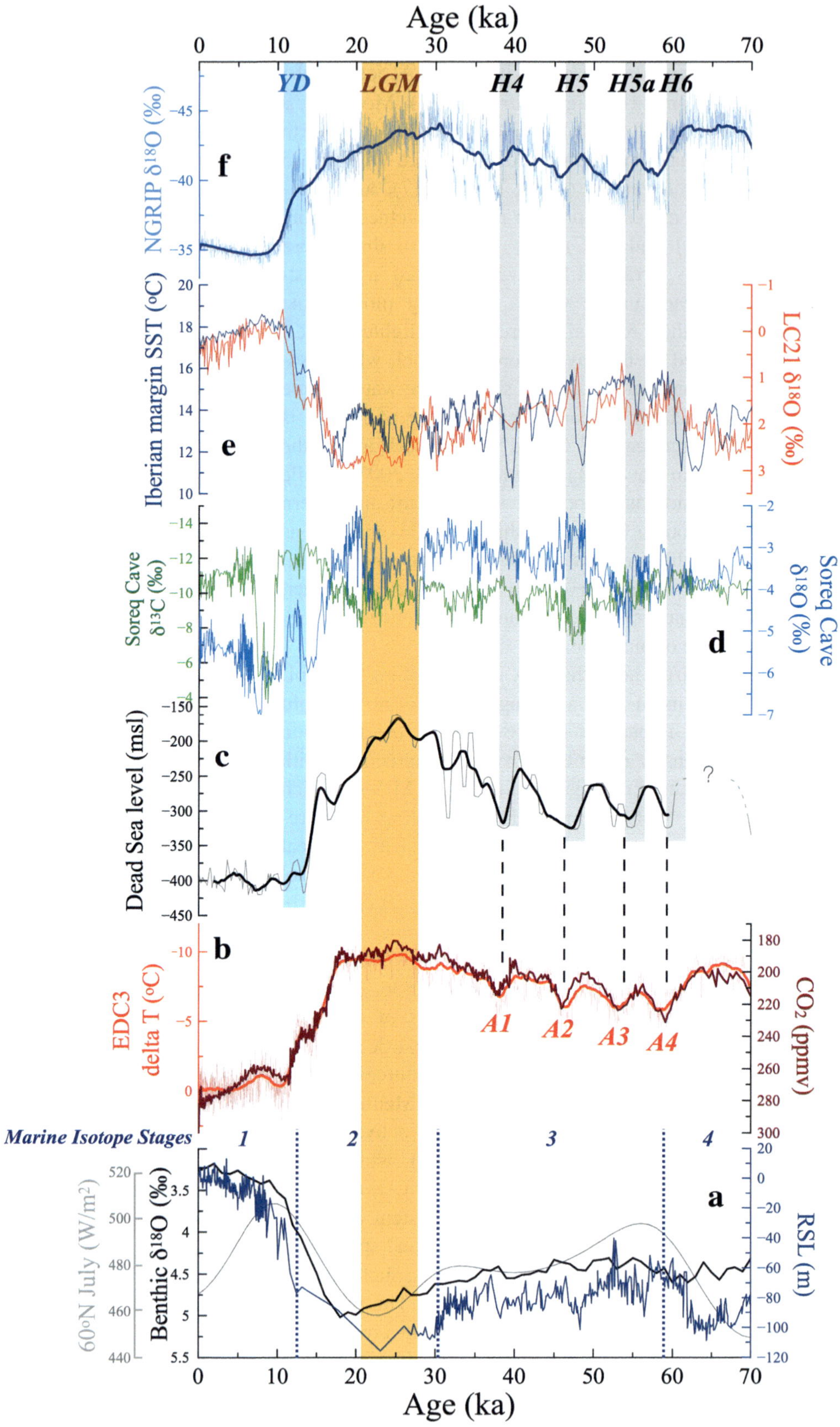

Fig. 4.9 Climate patterns over the last 70 kyrs. a Stacked benthic $\delta^{18}O$ record (Lisiecki and Raymo 2005), Red Sea Relative Sea Level (RSL) (Grant et al. 2012) and northern hemisphere summer insolation (Laskar et al. 2004), **b** EPICA Dome C3 (EDC3) temperature changes (Jouzel et al. 2007) and atmospheric CO_2 concentrations (Kawamura et al. 2007), **c** Observed (gray) and 2-kyr smoothed (black) Lake Lisan level (Torfstein et al. 2013), **d** Soreq Cave $\delta^{18}O$ and $\delta^{13}C$ (Bar-Matthews et al. 2003), **e** Eastern Mediterranean $\delta^{18}O$ (core LC21, Grant et al. 2012) and Iberian Margin Sea Surface Temperature (SST) (core md012443, Martrat et al. 2007), **f** NGRIP $\delta^{18}O$ (‰) plotted on the GICC05 timescale (Wolff et al. 2010)

exposed large areas became a potential regional dust source. During high sea-level intervals (interglacials), the Mediterranean shelf served as a sink for Sahara dust and Nilotic sediments. Once exposed, the winds associated with the eastern Mediterranean cyclones that penetrated southward from the Mediterranean but delivered little precipitation to Sinai and the Negev Desert, entrained and transported large amounts of sedimentary material from the newly exposed Nile Delta and Mediterranean shelf areas. As a result, large amounts of dust penetrated into the northern Negev and produced a wide blanket of loess deposits that fingers out to the south and east. To the north, dust penetration was constrained by the strong precipitation gradient that imposed dust wet deposition, though finer dust grains were able to continue farther north to form dust-born soils in central and northern Israel (Crouvi et al. 2008; Enzel et al. 2008) (Figs. 4.4 and 4.11b).

Speleothems formed continuously during the last glacial in central and northern Israel, but not further to the north (Lebanon), or to the south (Negev Desert). The last glacial, and in particular the LGM, were on average much colder, by approximately 6–7 °C relative to the present (Affek et al. 2008), and high mountain ranges such as Lebanon, anti-Lebanon, and Hermon probably experienced a freezing periglacial environment, which persisted not only during winter months but even year round (e.g., Ayalon et al. 2013), inhibiting the infiltration of meltwater from local snow caps and resulting in depositional gaps in local speleothem records. In less extreme environments such as the Galilee and central Israel, continuous deposition of speleothems in the Peqi'in and Soreq Caves reflects an ongoing flow of freshwater throughout the glacial period (Figs. 4.3 and 4.7).

Today, speleothems do not form in areas where annual precipitation is below ~500 mm (Enzel et al. 2008), indicating this to be a minimum threshold across wide areas of the southern Levant during the last glacial. Considering that speleothems formed across the last glacial in the Ma'ale Efraim and Tzavoa Caves (Figs. 4.7 and 4.11), where current precipitation is ~300 and ~150 mm/y, respectively, the increase in precipitation and its inland incursion were significant relative to the present. Indeed, speleothems did not form in either of these caves during interglacial periods, reflecting again the impact of the shutdown of Mediterranean climate systems on northern and central Israel, and indicating that the southern humid incursions recorded during interglacial cycles probably had a limited effect in central and northern Israel.

The absolute Dead Sea level curve over the past 70 kyrs (Figs. 4.2 and 4.9) indicates that water levels of the last glacial precursor of the Dead Sea, Lake Lisan, were significantly higher during the last glacial, relative to interglacial periods (Bartov et al. 2003; Torfstein et al. 2013). Hydrological changes in the DSB watershed are associated with orbital timescales, where interglacials are characterized by lower levels (e.g., the Holocene Dead Sea at 430–370 mbsl) and glacial times by higher levels. In Lake Lisan, the highest stand of ~160 mbsl was reached ca. 26–24 ka (Bartov et al. 2003; Torfstein et al. 2013), coinciding with minimum sea levels (Peltier and Fairbanks 2006) during the coldest stage of the last glacial cycle (MIS 2) (Fig. 4.9). At this time, the lake extended between the Sea of Galilee in the north and Hazeva (northern Arava Valley) in the south. Compared with the relatively stable, high-stand conditions characterizing most of MIS 2, the lake was significantly less stable during the warmer MIS 3, and experienced repeated millennial-scale vertical perturbations of up to 100 m. The details of the lake levels during MIS 4 are not as well resolved, although the sedimentary facies indicate relatively high lake levels, perhaps similar to those that existed during MIS 2, reflecting wet conditions in the DSB watershed. Overall, this pattern is supported by pollen records in the Dead Sea area that display a mixture of Irano-Turanian steppe, Saharo-Arabian desert vegetation, and Mediterranean woodland elements. In particular, an increased proportion of Irano-Turanian steppe vegetation and Mediterranean woodland elements during the last glacial period suggests consistently high effective moisture during late MIS4, MIS3, and MIS2 (Miebach et al. 2019).

Globally, the last glacial period is characterized by repeated millennial-timescale climatic perturbations (Blunier and Brook 2001) that had strong impacts on the Levant climate. For example, during Heinrich stadials, large armadas of icebergs collapsed into the North Atlantic (e.g., Hemming 2004) causing an abrupt southward expansion of the polar cold front that resulted in abrupt drops in sea surface temperatures (SSTs) in the mid-latitude Atlantic, western Mediterranean Sea (Cacho et al. 1999) and even the Red Sea (Arz et al. 2007). The lower SSTs, in combination with additional cooling by the air masses originating in the North Atlantic and bound for the Levant, weakened the air–water thermal gradient over the Mediterranean. Consequently, cyclogenesis over the eastern Mediterranean and precipitation over the southern Levant were considerably reduced, leading to severe arid spells across the southern Levant, exemplified by sharp drops in water levels and deposition of massive gypsum layers in Lake Lisan (Bartov et al. 2003; Torfstein et al. 2013) (Fig. 4.9). As discussed, reduced temperatures that caused freezing in the Hermon and Lebanon Mountains, some of the primary water sources in the DSB, might have further contributed to this drop. With the dissipation of the Heinrich iceberg armada and subsequent interstadial warming, the frequency and possibly intensity of Mediterranean cyclones recovered, and relatively wet glacial conditions resumed in the region, resulting in a rise in Lake Lisan levels.

Soreq and Peqi'in Caves display continuous deposition throughout the last glacial, with characteristic heavy $\delta^{18}O$

compositions of −2 to −3‰. These values, about ~3–4‰ heavier than typical interglacial values, reflect the same variability in eastern Mediterranean surface waters, the source of humidity to the region. Normalizing the Soreq Cave $\delta^{18}O$ record to the eastern Mediterranean record yields a constant difference ca. 5–6‰ over the last 70 kyrs (Kolodny et al. 2005). A similar comparison between the Dead Sea and Soreq Cave record yields a constant difference ca. 8–9‰. The constant difference over time between these three records, resulting from the large glacial–interglacial variability relative to second-order sub-orbital variations, reflects the dominance of the source water (eastern Mediterranean surface water) composition on that of meteoric water in Israel (Kolodny et al. 2005).

4.4.3.3 The Last Deglaciation

The shift into the current interglacial (Holocene), and the retreat of the NH ice cover during the last deglaciation, coincided with a major ~200 m Dead Sea lake-level decline to its common Holocene elevation of ~400 m bmsl (Fig. 4.9), a shift in $\delta^{18}O$ compositions of Soreq Cave deposits toward lighter values (Bar-Matthews et al. 2003), and a shift in regional vegetation cover (Langgut et al. 2011; Miebach et al. 2019). The turbulent climate of the last deglaciation spans the last Heinrich event (H1) arid spell at 17.5–15.5 ka, the arid Bolling/Allerod (~14 ka), and a brief resumption of wet conditions during the first part of the Younger Dryas (YD; 12.9–11.7 ka) (Stein et al. 2010). The latter represents a relapse of glacial conditions, i.e., a short cooling within the overall warming trend. The structure of the YD is well expressed in the Soreq Cave, with a shift toward heavier $\delta^{18}O$ values, while in the Dead Sea it resulted in a brief lake-level rise, in accord with the resumption of glacial conditions further north. This again reflects the overwhelming forcing of the NH ice-sheet cover in determining the latitude and path of westerly storm tracks, whose more northerly position during the Holocene resulted in weaker and less frequent cyclones over the eastern Mediterranean and in turn, less precipitation in the DSB drainage basin (Enzel et al. 2003, 2008). Accordingly, the regional response to millennial-scale perturbations during the Holocene and glacial intervals was different, as reflected for example by the lake-level rise during the NH cold pulse of the YD, in contrast with its response to glacial Heinrich stadials (Stein et al. 2010; Torfstein et al. 2013).

4.4.4 The Holocene (11.7 ka—Present)

The Holocene climate in the Levant is characterized by overall arid conditions relative to the last glacial cycle, whereby Dead Sea levels are considerably lower, at ca. 400 mbsl, and the $\delta^{18}O$ and $\delta^{13}C$ compositions of cave deposits in northern and central Israel are −6 to −5‰ and −12 to −10‰, respectively (Figs. 4.9 and 4.10). An exception to the latter is identified during the early Holocene, between 9.4 and 7.0 ka, a period also referred to as the "African Humid Period," for which the evidence suggests a northward excursion of tropical precipitation across the Sahel (Gasse 2000; deMenocal 2004; Revel et al. 2015; Shanahan et al. 2015; Palchan and Torfstein 2019), though the magnitude and geographic span of this excursion are debated, with compiled evidence suggesting that the so-called "Green Sahara" was more likely a "pale-green and discontinuously wet Sahara" (Quade et al. 2018; Drake et al. 2022).

In Israel, the Soreq Cave during the early Holocene was characterized by low speleothem $\delta^{18}O$ values coupled with the very high $\delta^{13}C$ values that supposedly reflect a washout of the covering soil and organic matter that serve as a low $\delta^{13}C$ end member, rendering the residual inorganic Cenozoic limestone as a dominant source of bicarbonate (Bar-Matthews et al. 2003). Regionally, the latter $\delta^{13}C$ excursion coincides with the deposition of sapropel S1 in the eastern Mediterranean (Bar-Matthews et al. 2000), travertine deposition in the Arava Valley (Waldmann et al. 2010), shifts in the mineralogy, oxygen isotopic compositions and a drop in atmospheric dust fluxes in the northern Red Sea (Arz et al. 2003; Palchan and Torfstein 2019), and with additional observations, some of which have been mentioned earlier in this discussion, that support increased humidity across the Sahara Desert and Arabian Peninsula. It is noted, however, that the available evidence supporting such an incursion—only one reported age of travertine in the Arava Valley (Waldmann et al. 2010), absence of Negev Desert speleothems (Bar-Matthews 2014), and only brief occurrence of lacustrine conditions in the northeastern Sahara Desert (Szabo et al. 1995)—suggests that the early Holocene incursion of wetness was weaker relative to the last interglacial. Indeed, the Dead Sea water levels do not display notable fluctuations during the early Holocene, implying that even if a wet incursion reached the southern Levant, its quantitative impact on the hydrological mass balance was quite limited.

By contrast, millennial-timescale climatic perturbations have been identified throughout the Holocene in the Dead Sea water levels and Soreq Cave oxygen isotopic compositions. In the Dead Sea, Kushnir and Stein (2010) discussed level drops on the order of at least 10 m at ~8.4, 8.1, 5.7, 4.1, 3.2, and 1.4 ka that reflect marked Holocene arid events, which correspond with pronounced cooling episodes recorded in eastern Mediterranean winter SSTs (reconstructed from planktonic foraminifera in marine cores) and with cold events in the northern latitudes (e.g., Bond ice-rafting events). More recently, Weber et al. (2021) suggested that the ages of gypsum structures along

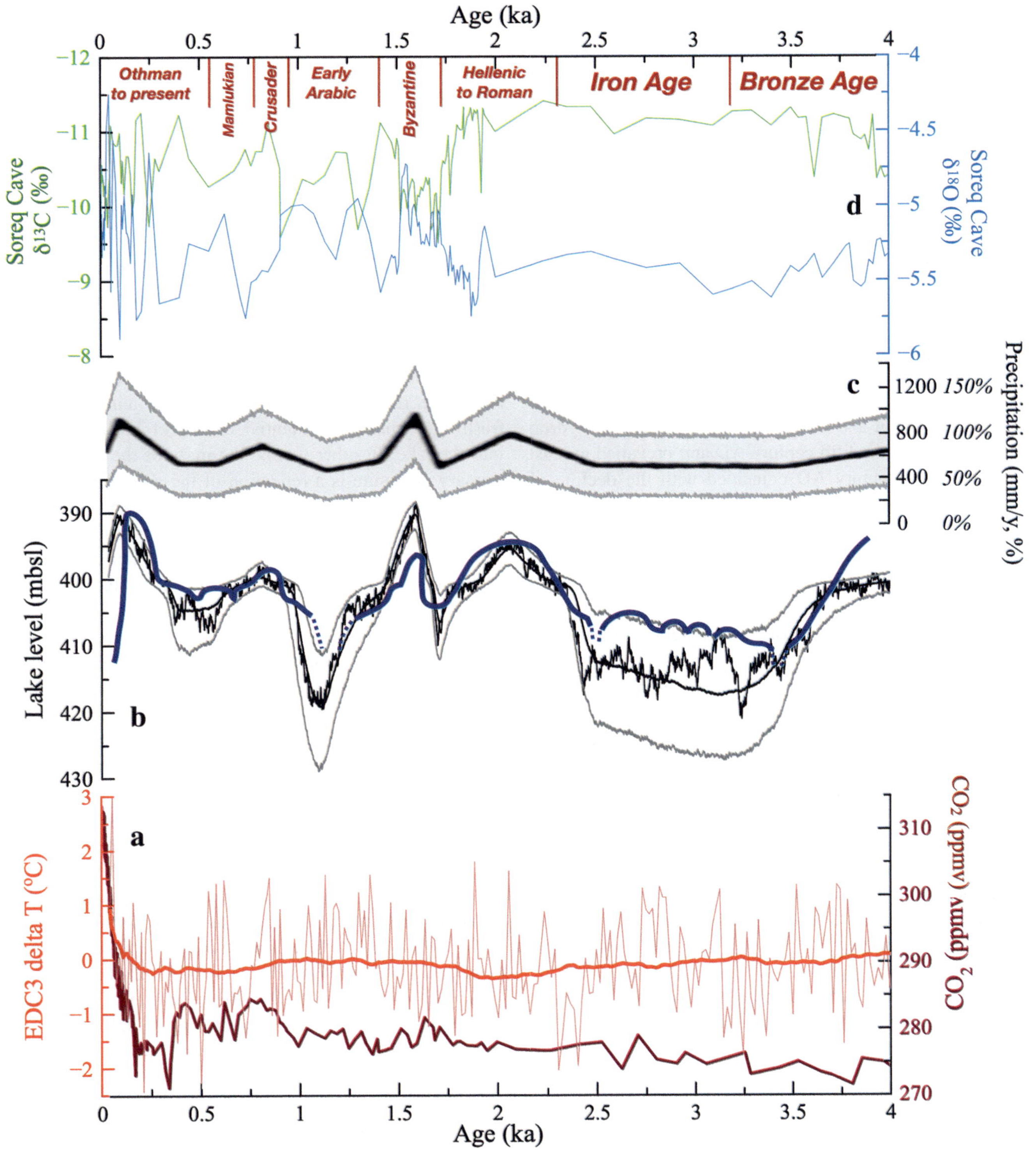

Fig. 4.10 **Climate patterns since 4 ka**. **a** EPICA Dome C3 (EDC3) temperature changes (Jouzel et al. 2007) and atmospheric CO_2 concentrations (Kawamura et al. 2007), **b** Reconstructed curve of the Dead Sea lake level from field observations (blue, Bookman (Ken-Tor) et al. 2004), and median of modeled levels (black) and uncertainty envelope (95%, gray) (Morin et al. 2019), **c** Reconstructed late Holocene annual precipitation at Kfar Giladi (northern Galilee) (Morin et al. 2019). Ensemble median and 95% range are shown as black curve and gray envelope, respectively, **d** Soreq Cave $\delta^{18}O$ and $\delta^{13}C$ (Bar-Matthews et al. 2003)

the shores of the Dead Sea coincide with times of North Atlantic cooling events and grand solar minima, pointing to a direct impact of the latter on regional hydrology. Oxygen and carbon isotope records in the Soreq Cave revealed several short-lived decadal-to centennial-scale climatic events during the mid-Holocene, at 6.65–6.60, 6.25–6.18, 5.70–5.6, 5.25–5.17, and 4.2–4.05 ka. The last two events coincided with the cultural collapses of the Uruk society in Mesopotamia and the Akkadian Empire (Bar-Matthews and Ayalon 2011).

Late Holocene Dead Sea level fluctuations are correlated to regional cultural shifts (Bookman et al. 2006) (Fig. 4.10). The high-stands of the second and first centuries BC, the fourth century AD, and the eleventh and twelfth centuries AD coincide with the Roman, Byzantine, and Crusader periods, respectively, reflecting an association between cultural expansion and wet climatic pulses. The lake drop that started after the fifth century AD and prevailed at least until the ninth century AD coincided with the decline of the Byzantine regime and invasion of the Arab tribes into the area. At the end of the nineteenth century, the lake level reached a temporal high-stand of ca. 390 mbsl, which persisted until roughly 1929 (Bookman et al. 2006). Thereafter, and particularly since the 1960s, the Dead Sea level has been retreating drastically mostly due to human interference with its water balance.

Morin et al. (2019) developed a model based on the correlation between Dead Sea levels and regional present-day precipitation that yielded a quantitative assessment of late Holocene precipitation in the Levant, including mean and variance of annual precipitation and their trends (Fig. 4.10c). The model, which is based on the strong correlation between annual precipitation in Kfar Giladi (northern Israel), precipitation across the entire Levant, and Dead Sea level changes, shows that mean annual precipitation during the late Holocene ranged between −37 and + 25% of the present-day conditions, and that the twentieth century was substantially wetter than most of the late Holocene, i.e., the recent drying of the region can be partially explained as a return to the "normal" drier conditions, with anthropogenic processes strengthening this trend.

4.4.5 Was the Last Glacial Drier or Wetter Than Today?

Despite decades of research, a fundamental aspect of hydro-climate in the southern Levant, pertaining in particular to the last glacial cycle, and by inference, to the entire Quaternary climate history of the region, remains debated. During the last glacial, Lake Lisan's water levels were up to 240 m higher than typical levels in the Holocene Dead Sea, implying that the last glacial was significantly wetter than the Holocene (see Sect. 4.3.2). Following the above discussion, it is important to discuss an alternative explanation that the high glacial lake levels were driven by decreased evaporation rates (due to colder temperatures) rather than by an increase in precipitation. In this context, a fundamental observation is that the deposition of primary aragonite from the Ca-chloride hypersaline Lake Lisan brine requires an addition of bicarbonate from external sources, i.e., runoff. Calcium concentrations in the modern Dead Sea are extremely high (~18,000 ppm), which drastically limits any possibility of bicarbonate buildup in the lake. Rather, the brine composition imposes a rapid removal from the water column of any influxing bicarbonate through aragonite precipitation (Stein et al. 1997). Thus, the deposition of the thick sequences of aragonite laminae that form the Lisan Formation requires a large and persistent annual influx of freshwater, the only potential carrier of bicarbonate ions into the lake. In other words, the mass of the precipitating primary aragonite is a reflection of the amount of bicarbonate entering the lake (Stein et al. 1997; Barkan et al. 2001) and hence of freshwater fluxes and regional rainfall (Enzel et al. 2008). In the modern setting of the southern Levant, such increases in runoff are only driven by increased precipitation (Enzel et al. 2003) and calculations show that average aragonite deposition rates in Lake Lisan were about six-fold higher than those in the modern Dead Sea (Barkan et al. 2001), requiring an approximate doubling of mean annual precipitation in northern and central Israel during the last glacial (Enzel et al. 2008). Thus, although decreased evaporation could in principle support a lake-level rise, it could not support the precipitation of the thick glacial aragonite sequences found in the Lisan Formation, confirming that regional precipitation was significantly higher during the last glacial relative to present. It is further noted that prominent cold-spells that occurred during Heinrich stadials within the last glacial cycle are associated with significant Lake Lisan level drops, indicating that reduced temperatures alone cannot drive a rise in lake levels (Enzel et al. 2008).

Several lines of evidence indicate that evaporation rates during the glacial period were likely as high as, or possibly higher than, the present-day ones, which emphasizes the requirement for significant freshwater input as the main driver of lake-level high-stands. For example, wind stress in the Levant was persistent throughout the late Pleistocene, as indicated by dune movement and aeolian loess accretion in the nearby northern Negev desert (e.g., Crouvi et al. 2008); increased wind gustiness would enhance the advection of moisture over the lake and support increased evaporation (Rohling 2013). Similarly, the larger Lake Lisan volume is associated with a more dilute salt composition and a higher water activity, which in turn allows for increased evaporation rates (Krumgalz 2000). Indeed, evaporation rates in Lake Lisan were estimated to be much higher (~3–5 m/y)

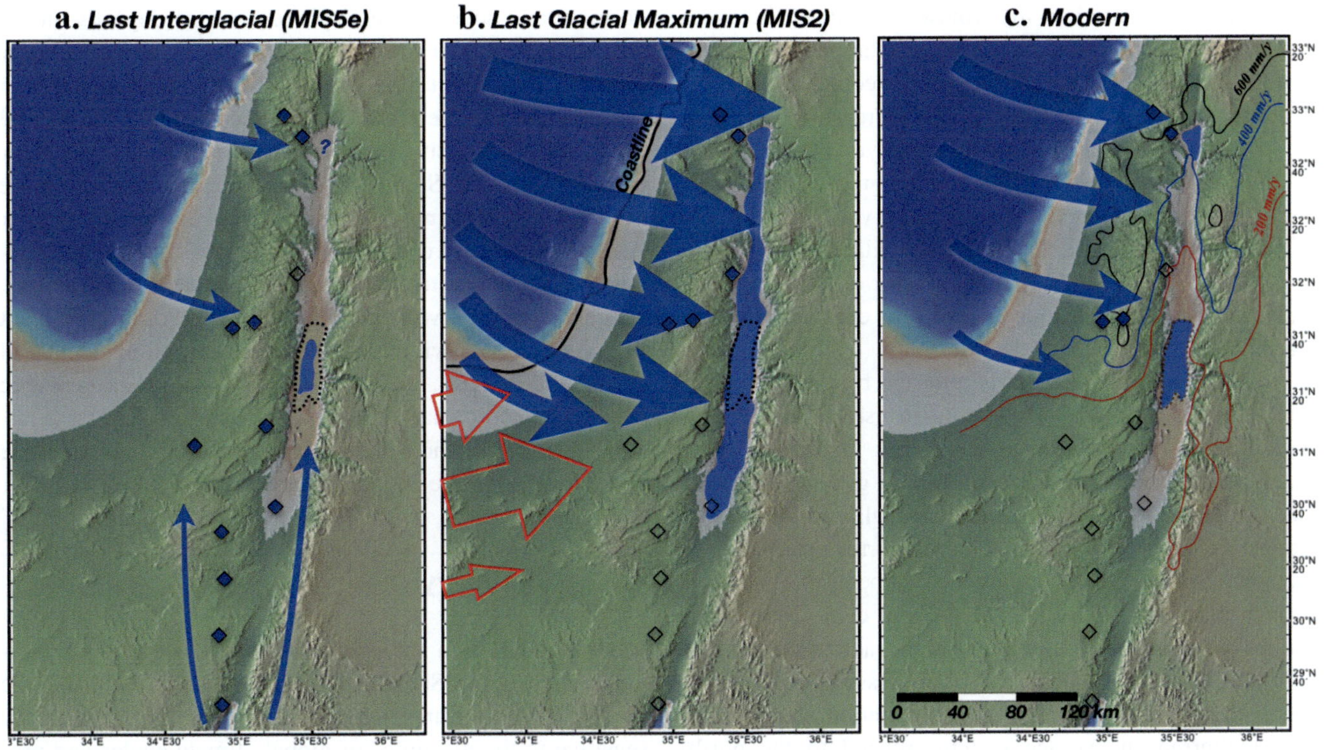

Figure. 4.11. Schematic representations of late Quaternary climate patterns in the southern Levant. Blue arrows represent the magnitude and pathway of the main rain systems and the blue and empty symbols mark sites that indicate relatively wet and dry conditions, respectively (for details on the sites see Fig. 4.1 and the main text). For reference, a dashed black curve marks the size of the modern Dead Sea in all three panels. **a** During MIS5e, weak eastern Mediterranean cyclones imposed overall hyperarid conditions in the region, resulting in particularly low Dead Sea stands. There are no records of the history of the Sea of Galilee during this time by contrast, coeval incursions from the south, most likely in the form of strong but episodic storms associated with tropical plumes and the Red Sea trough, are recorded by cave and spring deposits across the Negev Desert, together with additional archives discussed in the text. Shortly after MIS5e, during MIS5d (not shown), the activity of the southern rain systems stopped and the regional hydrological balance was based only on the weak eastern Mediterranean systems, resulting in an extreme hyperarid spell. **b** The Last Glacial Maximum (MIS2) in northern and central Israel was significantly wetter than interglacial periods due to an intensification of eastern Mediterranean cyclone systems, as reflected for example by the widespread expansion of Lake Lisan between the Sea of Galilee in the north and the Arava Valley in the south. At this stage, southern rain systems were not effective and the north–south precipitation gradient became even sharper as a result of coastline advancement of several tens of km to the northwest (because of the global sea-level drop) that limited inland penetration of EM rain systems. In turn, this supported increased dust fluxes from Sinai and the exposed Mediterranean shelf into the northern Negev Desert (marked by red arrows). **c** Modern climate is dominated by eastern Mediterranean rain systems, with minor contributions of southern systems. Currently, precipitation of cave deposits is limited to areas where precipitation exceeds ~500 mm/y

(Stein et al. 1997) than the present-day rates (~1.5 m/y) (even though it is recognized that the former rates probably overestimate the true values; Ben Dor et al. 2021), despite the overall colder glacial conditions. It is therefore clear that Lisan lake-level fluctuations are overwhelmingly controlled by the precipitation component of the regional hydrological cycle, rendering the lake-level curve to be a robust recorder of the regional hydrological history (Fig. 4.9). By contrast, the dominance of the isotopic source effect in regional meteoric waters renders cave and lake $\delta^{18}O$ compositions to be tracers of the common marine reservoir (i.e., eastern Mediterranean surface waters) rather than local rainfall amounts (Frumkin et al. 1999; Kolodny et al. 2005), and hence as an ambiguous proxy for the reconstruction of past hydrology in this region (Goldsmith et al. 2017).

An additional "smoking gun" for a wet glacial in Israel arises from the reconstruction of weathering rates in central Israel. Ryb et al. (2014) found that the long-term ^{36}Cl-based chemical weathering rates and long-term denudation rates in the Judea Mountains and the Arugot watershed (located in the Judea Desert, proximal to the Dead Sea) correlate linearly with mean annual precipitation and agree with the linear trend of contemporary chemical weathering rates. By contrast, chemical weathering rates from the western face of the Judea Range (proximal to the Soreq Cave) diverge from the linear trend toward higher values. This is explained by a higher, relative to the present, effective precipitation across the western face of the Judea Range over the time scales averaged by ^{36}Cl method (10^4–10^5 years), suggesting that on a glacial–interglacial time scale precipitation increased on the windward flank of the Judea Range, while on the leeward flank precipitation did not vary significantly and arid conditions persisted. The observed weathering rates require an effective increase in the average precipitation over the last tens of kyrs of ~40%, with absolute mean annual precipitation ranging between 600 and > 1200 mm (Ryb et al. 2014). Such a value can be considered to be a reasonable glacial–interglacial average, as an increase in regional rainfall by a factor of two has previously been suggested for the LGM (Vaks et al. 2006; Enzel et al. 2008).

A similar rainfall gradient is reflected by the spatiotemporal distributions of speleothems in Israel (see details in Sect. 4.3.2), as well as in the Negev region (southern Israel) where the distribution of calcic, gypsic, and salic soils indicates that while significant glacial–interglacial changes in precipitation existed in the northwestern Negev, the southeastern Negev maintained hyperarid conditions through glacial–interglacial cycles (Amit et al. 2006). This implies a stronger precipitation spatial gradient during glacial periods, which is attributed, as discussed above, to the intensification, but limited inland penetration, of the eastern Mediterranean cyclone systems related to the northward advance of the coastline location due to sea-level drop (Enzel et al. 2008).

In summary, the evidence reviewed here and in previous studies (e.g., Enzel et al. 2008) supports a wetter glacial period in northern and central Israel relative to the Holocene and the present, with the Negev Desert experiencing coeval hyperarid conditions that are briefly perturbed by limited humid incursions from low latitudes during interglacial peaks. The wet-dry glacial–interglacial patterns discussed here are variably challenged, it should be noted, by some proxy studies (e.g., Gasse et al. 2011; Langgut et al. 2011; Miebach et al. 2019; Tierney et al. 2022) as well as by global climate model results (e.g., Kageyama et al. 2021; Ludwig and Hochman 2022), warranting further investigation of the mechanistic drivers of climate change in this region.

4.5 Global Connections

A comparison to a range of regional and global climate records fleshes out a tight correlation between Dead Sea water levels and global temperatures and atmospheric CO_2 concentrations (Torfstein 2019). This indicates that global temperature changes are the underlying process that modulates Levant hydro-climate even though the mechanisms supporting this connection are both complex and ambiguous, involving inter-latitudinal climatic teleconnections on orbital and millennial timescales, whose understanding requires the determination of leads and lags between the northern and southern hemisphere but also depends on the sensitivities of each of the records and proxies used. For example, the impact of Heinrich events is very prominent in the Dead Sea in the form of dramatic lake-level drops and deposition of massive primary gypsum layers, while their expression in the Soreq Cave $\delta^{18}O$ record is much more mild and they cannot be easily distinguished from other stadial–interstadial cycles (Fig. 4.9). Torfstein (2019) suggested that Antarctic temperatures and atmospheric CO_2 concentrations provide the best proxy records of Levant paleohydrology (with a prominent exception identified during the MIS5e), and used this to establish a synthetic lake-level record that extends as far back as 800 ka (Figs. 4.5, 4.7 and 4.8) and serves as a regional climate reference curve for the middle and late Quaternary.

4.6 Summary

Israel is located in the southern Levant, along the transition between the Sahara and Arabia Desert belt and the moderate Mediterranean climate zone. As such, it is highly sensitive to the dynamics of global climate cycles and provides exquisite records of climate history throughout the Quaternary. The overarching modulator of southern Levant hydro-climate changes is global temperature and atmospheric CO_2. In more detail, regional climate is controlled by the interplay between NH westerly systems and incursions of low latitude precipitation systems. The collective observations show that the Mediterranean winter cyclone activity patterns imposed arid interglacials and wetter glacials in the region. During warm interglacials, tropical African moisture, delivered by Red Sea troughs and tropical plumes, moderated the otherwise expected hyperaridity. In the absence of these tropical effects, the southern Levant would have been extremely more hyperarid during warm interglacial peaks.

During the early Quaternary the Negev Desert experienced relatively wet conditions, though coeval climate records from central and northern Israel are limited. Negev Humid Periods (NHPs) are recorded in the northern Negev between ~1.27 and 0.5 Ma. Later, NHPs are recorded across the Negev Desert as well as in the DSB during warm interglacial peaks younger than ~300 ka. Accumulated evidence suggests that climatic configuration that supported a dramatic weakening of Mediterranean westerly storm systems concurrent with incursions of humidity from more southern latitudes was most pronounced during the last interglacial but also existed during earlier interglacials, as well as during the early Holocene. By contrast, although water availability was affected by lower glacial temperatures (less evaporation), several lines of evidence require that glacial precipitation was increased relative to the present, and hence glacial periods in general, and the last glacial in particular, were significantly wetter than interglacial periods. Moreover, a northward shift of the Mediterranean coastline by several tens of km due to the global sea-level drop resulted in an intensification of the precipitation gradient between north and south Israel during glacial periods, relative to interglacials, and in enhanced dust penetration into the region, particularly into the northern Negev Desert. Repeated millennial-timescale perturbations during the last glacial imposed abrupt climatic shifts, such as hyperarid spells associated with Heinrich stadials.

The Holocene was overall warmer and more arid than the last glacial, but was superimposed by millennial-timescale fluctuations associated with changes in solar activity. Finally, proxy- and model-based climate reconstructions indicate that the twentieth century was substantially wetter than most of the late Holocene, i.e., the recent drying of the region can be partially explained as a return to the "normal" drier conditions, which overlays and strengthens the effects of anthropogenic processes.

References

Affek HP, Bar-Matthews M, Ayalon A, Matthews A, Eiler JM (2008) Glacial/interglacial temperature variations in Soreq cave speleothems as recorded by 'clumped isotope' thermometry. Geochim Cosmochim Acta 72:5351–5360. https://doi.org/10.1016/j.gca.2008.06.031

Almogi-Labin A, Bar-Matthews M, Shriki D, Kolosovsky E, Paterne M, Schilman B, Ayalon A, Aizenshtat Z, Matthews A (2009) Climatic variability during the last ~90ka of the southern and northern Levantine Basin as evident from marine records and speleothems. Quat Sci Rev 28:2882–2896. https://doi.org/10.1016/j.quascirev.2009.07.017

Amit R, Enzel Y, Sharon D (2006) Permanent quaternary hyperaridity in the Negev, Israel, resulting from regional tectonics blocking mediterranean frontal systems. Geology 34:509. https://doi.org/10.1130/G22354.1

Amit R, Simhai O, Ayalon A, Enzel Y, Matmon A, Crouvi O, Porat N, McDonald E (2011) Transition from arid to hyper-arid environment in the southern Levant deserts as recorded by early Pleistocene cummulic Aridisols. Quat Sci Rev 30:312–323. https://doi.org/10.1016/j.quascirev.2010.11.007

Armon M, Morin E, Enzel Y (2019) Overview of modern atmospheric patterns controlling rainfall and floods into the Dead Sea: implications for the lake's sedimentology and paleohydrology. Quat Sci Rev 216:58–73. https://doi.org/10.1016/j.quascirev.2019.06.005

Arz HW, Lamy F, Ganopolski A, Nowaczyk N, Pätzold J (2007) Dominant Northern Hemisphere climate control over millennial-scale glacial sea-level variability. Quat Sci Rev 26:312–321. https://doi.org/10.1016/j.quascirev.2006.07.016

Arz HW, Lamy F, Pätzold J, Muller PJ, Prins M (2003) Mediterranean moisture source for an early-Holocene humid period in the northern Red Sea. Science 300:118–121. https://doi.org/10.1126/science.1080325

Ayalon A, Bar-Matthews M, Frumkin A, Matthews A (2013) Last Glacial warm events on Mount Hermon: the southern extension of the Alpine karst range of the east Mediterranean. Quat Sci Rev 59:43–56. https://doi.org/10.1016/j.quascirev.2012.10.047

Bar-Matthews M (2014) History of Water in the Middle East and North Africa. In: Treatise on geochemistry, 2nd edn. Elsevier, pp 109–128

Bar-Matthews M, Ayalon A (2011) Mid-Holocene climate variations revealed by high-resolution speleothem records from Soreq Cave, Israel and their correlation with cultural changes. Holocene 21:163–171. https://doi.org/10.1177/0959683610384165

Bar-Matthews M, Ayalon A, Kaufman A (2000) Timing and hydrological conditions of Sapropel events in the Eastern Mediterranean, as evident from speleothems, Soreq cave, Israel. Chem Geol 169:145–156. https://doi.org/10.1016/S0009-2541(99)00232-6

Bar-Matthews M, Ayalon A, Gilmour M, Matthews A, Hawksworth CJ (2003) Sea–land oxygen isotopic relationships from planktonic foraminifera and speleothems in the Eastern Mediterranean region and their implication for paleorainfall during. Geochim Cosmochim Acta 67:3181–3199. https://doi.org/10.1016/S0016-7037(02)01031-1

Bar N, Agnon A, Yehudai M, Lazar B, Shaked Y, Stein M (2018) Last interglacial sea levels and regional tectonics from fossil coral reefs in the northeast Gulf of Aqaba. Quat Sci Rev 191:41–56. https://doi.org/10.1016/J.QUASCIREV.2018.04.031

Barkan E, Luz B, Lazar B (2001) Dynamics of the carbon dioxide system in the Dead Sea. Geochim Cosmochim Acta 65:355–368

Barker S, Knorr G, Edwards RL, Parrenin F, Putnam AE, Skinner LC, Wolff E, Ziegler M (2011) 800,000 years of abrupt climate variability. Science 334(80):347–351. https://doi.org/10.1126/science.1203580

Bartov Y, Goldstein SL, Stein M, Enzel Y (2003) Catastrophic arid episodes in the Eastern Mediterranean linked with the North Atlantic Heinrich events. Geology 31:439–442

Begin ZB, Ehrlich A, Nathan Y (1974) Lake Lisan, the Pleistocene precursor of the Dead Sea. Geol Surv Isr Bull 63:1–30

Ben-Israel M, Enzel Y, Amit R, Erel Y (2015) Provenance of the various grain-size fractions in the Negev loess and potential changes in major dust sources to the Eastern Mediterranean. Quat Res 83:105–115. https://doi.org/10.1016/j.yqres.2014.08.001

Ben Dor Y, Flax T, Levitan I, Enzel Y, Brauer A, Erel Y (2021) The paleohydrological implications of aragonite precipitation under contrasting climates in the endorheic Dead Sea and its precursors revealed by experimental investigations. Chem Geol 576:120261. https://doi.org/10.1016/j.chemgeo.2021.120261

Beyer RM, Krapp M, Eriksson A, Manica A (2021) Climatic windows for human migration out of Africa in the past 300,000 years. Nat Commun 12:4889. https://doi.org/10.1038/s41467-021-24779-1

Blunier T, Brook EJ (2001) Timing of millennial-scale climate change in Antarctica and Greenland during the last glacial period. Science 291:109–112. https://doi.org/10.1126/science.291.5501.109

Bookman (Ken-Tor) R, Enzel Y, Agnon A, Stein M (2004) Late Holocene lake levels of the Dead Sea. Geol Soc Am Bull 116:555–571https://doi.org/10.1130/B25286.1

Bookman R, Bartov Y, Enzel Y, Stein M (2006) Quaternary lake levels in the Dead Sea basin: Two centuries of research. In: New Frontiers in Dead Sea Paleoenvironmental Research. Geological Society of America, pp 155–170

Box MR, Krom MD, Cliff R, a., Bar-Matthews M, Almogi-Labin a., Ayalon a., Paterne M, (2011) Response of the Nile and its catchment to millennial-scale climatic change since the LGM from Sr isotopes and major elements of East Mediterranean sediments. Quat Sci Rev 30:431–442. https://doi.org/10.1016/j.quascirev.2010.12.005

Cacho I, Grimalt JO, Pelejero C, Canals M, Sierro FJ, Flores JA, Shackleton N (1999) Dansgaard-Oeschger and Heinrich event imprints in Alboran Sea paleotemperatures. Paleoceanography 14:698–705. https://doi.org/10.1029/1999PA900044

Chen C, Litt T (2018) Dead Sea pollen provides new insights into the paleoenvironment of the southern Levant during

MIS 6–5. Quat Sci Rev 188:15–27. https://doi.org/10.1016/j.quascirev.2018.03.029

Cheng H, Sinha A, Verheyden S, Nader FH, Li XL, Zhang PZ, Yin JJ, Yi L, Peng YB, Rao ZG, Ning YF, Edwards RL (2015) The climate variability in northern Levant over the past 20,000 years. Geophys Res Lett 42(20):8641–8650. https://doi.org/10.1002/2015GL065397

Cordova CE, Nowell A, Bisson M, Ames CJH, Pokines J, Chang M, Al-Nahar M (2013) Interglacial and glacial desert refugia and the Middle Paleolithic of the Azraq Oasis, Jordan. Quat Int 300:94–110. https://doi.org/10.1016/j.quaint.2012.09.019

Crombie MK, Arvidson RE, Sturchio NC, El Alfy Z, Zeid KA (1997) Age and isotopic constraints on Pleistocene pluvial episodes in the Western Desert Egypt. Palaeogeogr Palaeoclimatol Palaeoecol 130:337–355. https://doi.org/10.1016/S0031-0182(96)00134-4

Crouvi O, Amit R, Porat N, Gillespie AR, McDonald EV, Enzel Y (2009) Significance of primary hilltop loess in reconstructing dust chronology, accretion rates, and sources: an example from the Negev Desert, Israel. J Geophys Res Earth Surf 114:F02017. https://doi.org/10.1029/2008JF001083

Crouvi O, Amit R, Enzel Y, Porat N, Sandler A (2008) Sand dunes as a major proximal dust source for late Pleistocene loess in the Negev Desert, Israel. Quat Res 70:275–282. https://doi.org/10.1016/j.yqres.2008.04.011

DeMenocal PB (2004) African climate change and faunal evolution during the Pliocene-Pleistocene. Earth Planet Sci Lett 220:3–24. https://doi.org/10.1016/S0012-821X(04)00003-2

Drake NA, Candy I, Breeze P, Armitage SJ, Gasmi N, Schwenninger JL, Peat D, Manning K (2022) Sedimentary and geomorphic evidence of Saharan megalakes: a synthesis. Quat Sci Rev 276:107318. https://doi.org/10.1016/j.quascirev.2021.107318

El-Shenawy MI, Kim ST, Schwarcz HP, Asmerom Y, Polyak VJ (2018) Speleothem evidence for the greening of the Sahara and its implications for the early human dispersal out of sub-Saharan Africa. Quat Sci Rev 188:67–76. https://doi.org/10.1016/j.quascirev.2018.03.016

Ellenblum R (2012) The collapse of the eastern Mediterranean: climate change and the decline of the East, 950–1072. Cambridge University Press

Enzel Y, Bookman (Ken Tor) R, Sharon D, Gvirtzman H, Dayan U, Ziv B, Stein M (2003) Late Holocene climates of the Near East deduced from Dead Sea level variations and modern regional winter rainfall. Quat Res 60:263–273. https://doi.org/10.1016/j.yqres.2003.07.011

Enzel Y, Amit R, Dayan U, Crouvi O, Kahana R, Ziv B, Sharon D (2008) The climatic and physiographic controls of the eastern Mediterranean over the late Pleistocene climates in the southern Levant and its neighboring deserts. Glob Planet Change 60:165–192. https://doi.org/10.1016/j.gloplacha.2007.02.003

Enzel Y, Kushnir Y, Quade J (2015) The middle Holocene climatic records from Arabia: Reassessing lacustrine environments, shift of ITCZ in Arabian Sea, and impacts of the southwest Indian and African monsoons. Glob Planet Change 129:69–91. https://doi.org/10.1016/j.gloplacha.2015.03.004

Faershtein G, Porat N, Avni Y, Matmon A (2016) Aggradation–incision transition in arid environments at the end of the Pleistocene: an example from the Negev Highlands, southern Israel. Geomorphology 253:289–304. https://doi.org/10.1016/j.gca.2003.07.016

Fleitmann D, Burns SJ, Neff U, Mangini A, Matter A (2003) Changing moisture sources over the last 330,000 years in Northern Oman from fluid-inclusion evidence in speleothems. Quat Res 60:223–232. https://doi.org/10.1016/S0033-5894(03)00086-3

Fleitmann D, Cheng H, Badertscher S, Edwards RL, Mudelsee M, Göktürk OM, Fankhauser a., Pickering R, Raible CC, Matter a., Kramers J, Tüysüz O (2009) Timing and climatic impact of Greenland interstadials recorded in stalagmites from northern Turkey. Geophys Res Lett 36(19):L19707. https://doi.org/10.1029/2009GL040050

Fleitmann D, Burns SJ, Pekala M, Mangini A, Al-Subbary A, Al-Aowah M, Kramers J, Matter A (2011) Holocene and Pleistocene pluvial periods in Yemen, southern Arabia. Quat Sci Rev 30:783–787. https://doi.org/10.1016/j.quascirev.2011.01.004

Frumkin A, Stein M (2004) The Sahara-East Mediterranean dust and climate connection revealed by strontium and uranium isotopes in a Jerusalem speleothem. Earth Planet Sci Lett 217:451–464. https://doi.org/10.1016/S0012-821X(03)00589-2

Frumkin A, Comay O (2019) The last glacial cycle of the southern Levant: Paleoenvironment and chronology of modern humans. J Hum Evol 2019:102609. https://doi.org/10.1016/j.jhevol.2019.04.007

Frumkin A, Ford D, Schwarcz H (1999) Continental Oxygen Isotopic Record of the Last 170,000 Years in Jerusalem. Quat Res 327:317–327

Frumkin A, Bar-Yosef O, Schwarcz HP (2011) Possible paleohydrologic and paleoclimatic effects on hominin migration and occupation of the Levantine Middle Paleolithic. J Hum Evol 60:437–451. https://doi.org/10.1016/j.jhevol.2010.03.010

Frumkin A, Stein M, Goldstein SL (2022) High resolution environmental conditions of the last interglacial (MIS5e) in the Levant from Sr, C and O isotopes from a Jerusalem stalagmite. Palaeogeogr Palaeoclimatol Palaeoecol 586:110761. https://doi.org/10.1016/j.palaeo.2021.110761

Gasse F (2000) Hydrological changes in the African tropics since the Last Glacial Maximum. Quat Sci Rev 19:189–211. https://doi.org/10.1016/S0277-3791(99)00061-X

Gasse F, Vidal L, Develle AL, Van Campo E (2011) Hydrological variability in the Northern Levant: A 250 ka multiproxy record from the Yammoûneh (Lebanon) sedimentary sequence. Clim past 7:1261–1284. https://doi.org/10.5194/cp-7-1261-2011

Gasse F, Vidal L, Van Campo E, Demory F, Develle AL, Tachikawa K, Elias A, Bard E, Garcia M, Sonzogni C, Thouveny N (2015) Hydroclimatic changes in northern Levant over the past 400,000 years. Quat Sci Rev 111:1–8. https://doi.org/10.1016/j.quascirev.2014.12.019

Gat JR (1984) The stable isotope composition of Dead Sea waters. Earth Planet Sci Lett 71:361–376. https://doi.org/10.1016/0012-821X(84)90103-1

Gellman Y, Matmon A, Mushkin A, Porat N (2018) Drainage system reorganization and late Quaternary tectonic deformation along the southern Dead Sea Transform. Quatern Res 90:380–393

Ginat H, Opitz S, Ababneh L, Faershtein G, Lazar M, Porat N, Mischke S (2018) Pliocene-Pleistocene waterbodies and associated deposits in southern Israel and southern Jordan. J Arid Environ 148:14–33. https://doi.org/10.1016/j.jaridenv.2017.09.007

Goldreich Y (2003) The climate of Israel. Springer, US, Boston, MA

Goldsmith Y, Polissar PJ, Ayalon A, Bar-Matthews M, deMenocal PB, Broecker WS (2017) The modern and Last Glacial Maximum hydrological cycles of the Eastern Mediterranean and the Levant from a water isotope perspective. Earth Planet Sci Lett 457:302–312. https://doi.org/10.1016/j.epsl.2016.10.017

Grant KM, Rohling EJ, Bar-Matthews M, Ayalon A, Medina-Elizalde M, Ramsey CB, Satow C, Roberts AP (2012) Rapid coupling between ice volume and polar temperature over the past 150,000 years. Nature 491:744–747. https://doi.org/10.1038/nature11593

Grant KM, Rohling EJ, Westerhold T, Zabel M, Heslop D, Konijnendijk T, Lourens L (2017) A 3 million year index for North African humidity/aridity and the implication of potential pan-African humid periods. Quat Sci Rev 171:100–118. https://doi.org/10.1016/j.quascirev.2017.07.005

Greenbaum N, Porat N, Rhodes E, Enzel Y (2006) Large floods during late oxygen isotope stage 3, southern Negev desert Israel. Quaternary Sci Rev 25:704–719. https://doi.org/10.1016/j.quascirev.2005.07.008

Haase-Schramm A, Goldstein SL, Stein M (2004) U-Th dating of Lake Lisan (late Pleistocene Dead Sea) aragonite and implications for glacial East Mediterranean climate change. Geochim Cosmochim Acta 68:985–1005. https://doi.org/10.1016/j.gca.2003.07.016

Hartman A, Torfstein A, Almogi-Labin A (2020) Climate swings in the northern Red Sea over the last 150,000 years from εNd and Mg/Ca of marine sediments. Quat Sci Rev 231:106205. https://doi.org/10.1016/j.quascirev.2020.106205

Hazan N, Stein M, Agnon a., Marco S, Nadel D, Negendank JFW, Schwab MJ, Neev D, (2005) The late Quaternary limnological history of Lake Kinneret (Sea of Galilee), Israel. Quat Res 63:60–77. https://doi.org/10.1016/j.yqres.2004.09.004

Hemming SR (2004) Heinrich events: massive late Pleistocene detritus layers of the North Atlantic and their global imprint. Rev Geophys 42:2003RG000128. https://doi.org/10.1029/2003RG000128

Henselowsky F, Eichstädter R, Schröder-Ritzrau A, Herwartz D, Almoazamy A, Frank N, Kindermann K, Bubenzer O (2021) Speleothem growth phases in the central Eastern Desert of Egypt reveal enhanced humidity throughout MIS 5. Quat Int. https://doi.org/10.1016/j.quaint.2021.05.006

Herold M, Lohmann G (2009) Eemian tropical and subtropical African moisture transport: an isotope modelling study. Clim Dyn 33:1075–1088. https://doi.org/10.1007/s00382-008-0515-2

Hirschfeld Y (2006) The archaeology of the Dead Sea valley in the Late Hellenistic and Early Roman periods. Spec Pap Soc. https://doi.org/10.1130/2006.2401(14).

Horowitz A (1987) Palynological evidence for the age and rate of sedimentation along the Dead Sea rift, and structural implications. Tectonophysics 141:107–115. https://doi.org/10.1016/0040-1951(87)90178-8

Jouzel J, Masson-Delmotte V, Cattani O, Dreyfus G, Falourd S, Hoffmann G, Minster B, Nouet J, Barnola JM, Chappellaz J, Fischer H, Gallet JC, Johnsen S, Leuenberger M, Loulergue L, Luethi D, Oerter H, Parrenin F, Raisbeck G, Raynaud D, Schilt A, Schwander J, Selmo E, Souchez R, Spahni R, Stauffer B, Steffensen JP, Stenni B, Stocker TF, Tison JL, Werner M, Wolff EW (2007) Orbital and millennial Antarctic climate variability over the past 800,000 years. Science 317(80):793–796. https://doi.org/10.1126/science.1141038

Kageyama M, Harrison SP, Kapsch M-L, Lofverstrom M, Lora JM, Mikolajewicz U, Sherriff-Tadano S, Vadsaria T, Abe-Ouchi A, Bouttes N, Chandan D, Gregoire LJ, Ivanovic RF, Izumi K, LeGrande AN, Lhardy F, Lohmann G, Morozova PA, Ohgaito R, Paul A, Peltier WR, Poulsen CJ, Quiquet A, Roche DM, Shi X, Tierney JE, Valdes PJ, Volodin E, Zhu J (2021) The PMIP4 Last Glacial Maximum experiments: preliminary results and comparison with the PMIP3 simulations. Clim past 17:1065–1089. https://doi.org/10.5194/cp-17-1065-2021

Kahana R, Ziv B, Enzel Y, Dayan U (2002) Synoptic climatology of major floods in the Negev Desert, Israel. Int J Climatol 22:867–882. https://doi.org/10.1002/joc.766

Katz A, Starinsky A (2015) No drawdown and no hyperaridity in the ancient Dead Sea (Comments to Torfstein's et al. (2015) paper, EPSL 412, 235–244). Earth Planet Sci Lett 427:303–305. https://doi.org/10.1016/j.epsl.2015.07.006

Kawamura K, Parrenin F, Lisiecki L, Uemura R, Vimeux F, Severinghaus JP, Hutterli M a, Nakazawa T, Aoki S, Jouzel J, Raymo ME, Matsumoto K, Nakata H, Motoyama H, Fujita S, Goto-Azuma K, Fujii Y, Watanabe O (2007) Northern Hemisphere forcing of climatic cycles in Antarctica over the past 360,000 years. Nature 448:912–6. https://doi.org/10.1038/nature06015

Keeley JE, Bond WJ (2001) On incorporating fire into our thinking about natural ecosystems: a response to Saha and Howe. Am Nat 158:664–670. https://doi.org/10.1086/323594

Keinan J, Bar-Matthews M, Ayalon A, Zilberman T, Agnon A, Frumkin A (2019) Paleoclimatology of the Levant from Zalmon Cave speleothems, the northern Jordan Valley, Israel. Quat Sci Rev 220:142–153. https://doi.org/10.1016/j.quascirev.2019.07.018

Kelley CP, Mohtadi S, Cane MA, Seager R, Kushnir Y (2015) Climate change in the Fertile Crescent and implications of the recent Syrian drought. Proc Natl Acad Sci 112:3241–3246. https://doi.org/10.1073/pnas.1421533112

Kiro Y, Goldstein SL, Kushnir Y, Olson JM, Bolge L, Lazar B, Stein M (2020) Droughts, flooding events, and shifts in water sources and seasonality characterize last interglacial Levant climate. Quat Sci Rev 248:106546. https://doi.org/10.1016/j.quascirev.2020.106546

Kolodny Y, Stein M, Machlus M (2005) Sea-rain-lake relation in the Last Glacial East Mediterranean revealed by δ18O-δ13C in Lake Lisan aragonites. Geochim Cosmochim Acta 69:4045–4060. https://doi.org/10.1016/j.gca.2004.11.022

Kroon D, Alexander I, Little M, Lourens LJ, Matthewson A, Robertson AHF, Sakamoto T (1998) Oxygen isotope and sapropel stratigraphy in the Eastern Mediterranean during the last 3.2 million years. Proc Ocean Drill Program, Sci Results 160:181–189

Krumgalz B (2000) Thermodynamic constraints on Dead Sea evaporation: can the Dead Sea dry up? Chem Geol 165:1–11. https://doi.org/10.1016/S0009-2541(99)00156-4

Kushnir Y, Stein M (2010) North Atlantic influence on 19th–20th century rainfall in the Dead Sea watershed, teleconnections with the Sahel, and implication for Holocene climate fluctuations. Quat Sci Rev 29:3843–3860. https://doi.org/10.1016/j.quascirev.2010.09.004

Kutzbach JE, Guan J, He F, Cohen AS, Orland IJ, Chen G (2020) African climate response to orbital and glacial forcing in 140,000-y simulation with implications for early modern human environments. Proc Natl Acad Sci 201917673 https://doi.org/10.1073/pnas.1917673117

Langgut D, Almogi-Labin a., Bar-Matthews M, Weinstein-Evron M (2011) Vegetation and climate changes in the South Eastern Mediterranean during the Last Glacial-Interglacial cycle (86 ka): new marine pollen record. Quat Sci Rev 30:3960–3972.https://doi.org/10.1016/j.quascirev.2011.10.016

Laskar J, Robutel P, Joutel F, Gastineau M, Correia ACM, Levrard B (2004) A long term numerical solution for the insolation quantities of the Earth. Astron Astrophys 428:261–285. https://doi.org/10.1051/0004-6361:20041335

Lazar B, Stein M (2011) Freshwater on the route of hominids out of Africa revealed by U-Th in Red Sea corals. Geology 39:1067–1070. https://doi.org/10.1130/G32257.1

Legge HL, Mutterlose J, Arz HW (2006) Climatic changes in the northern Red Sea during the last 22,000 years as recorded by calcareous nannofossils. Paleoceanography 21. https://doi.org/10.1029/2005PA001142

Lisiecki LE, Raymo ME (2005) A Pliocene-Pleistocene stack of 57 globally distributed benthic δ18O records. Paleoceanography 20https://doi.org/10.1029/2004PA001071

Litt T, Pickarski N, Heumann G, Stockhecke M, Tzedakis PC (2014) A 600,000 year long continental pollen record from Lake Van, eastern Anatolia (Turkey). Quat Sci Rev 104:30–41. https://doi.org/10.1016/j.quascirev.2014.03.017

Livnat A, Kronfeld J (1985) Paleoclimatic implications of U-series dates for lake sediments and travertines in the Arava Rift Valley, Israel. Quat Res 24:164–172. https://doi.org/10.1016/0033-5894(85)90003-1

Ludwig P, Hochman A (2022) Last glacial maximum hydro-climate and cyclone characteristics in the Levant: a regional modelling perspective. Environ Res Lett 17:014053. https://doi.org/10.1088/1748-9326/ac46ea

Martrat B, Grimalt JO, Shackleton NJ, de Abreu L, Hutterli MA, Stocker TF (2007) Four Climate Cycles of Recurring Deep and Surface Water Destabilizations on the Iberian Margin. Science (80-) 317:502–507. https://doi.org/10.1126/science.1139994

Masson-Delmotte V, Zhai P, Priani A, Connors S, Péan C, Berger S, Caud N, Chen Y, Goldfarb L, Gomis M, Huang M, Leitzell K, Lonnoy E, Matthews J, Maycock T, Waterfield T, Yelekçi O, Yu R, Zhou B (eds) (2021) IPCC, 2021: Climate Change 2021: The Physical Science Basis. Contribution of Working Group I to the Sixth Assessment Report of the Intergovernmental Panel on Climate Change. Cambridge University Press

Matmon A, Zilberman E (2017) Landscape evolution along the Dead sea fault and its margins. In: Quaternary of the levant: environments, climate change, and humans. Cambridge University Press, pp 17–30.

Matmon A, Simhai O, Amit R, Haviv I, Porat N, McDonald E, Benedetti L, Finkel R (2009) Desert pavement–coated surfaces in extreme deserts present the longest-lived landforms on Earth. Geol Soc Am Bull 121:688–697

Miebach A, Stolzenberger S, Wacker L, Hense A, Litt T (2019) A new Dead Sea pollen record reveals the last glacial paleoenvironment of the southern Levant. Quat Sci Rev 214:98–116. https://doi.org/10.1016/j.quascirev.2019.04.033

Morin E, Ryb T, Gavrieli I, Enzel Y (2019) Mean, variance, and trends of Levant precipitation over the past 4500 years from reconstructed Dead Sea levels and stochastic modeling. Quat Res 91:751–767. https://doi.org/10.1017/qua.2018.98

Mudelsee M, Schulz M (1997) The Mid-Pleistocene climate transition: onset of 100 ka cycle lags ice volume build-up by 280 ka. Earth Planet Sci Lett 1:117–123

Nehme C, Verheyden S, Noble SR, Farrant AR, Sahy D, Hellstrom J, Delannoy JJ, Claeys P (2015) Reconstruction of MIS 5 climate in the central Levant using a stalagmite from Kanaan Cave, Lebanon. Clim past 11:1785–1799. https://doi.org/10.5194/cp-11-1785-2015

Neugebauer I, Brauer A, Schwab MJ, Waldmann N, Enzel Y, Kitagawa H, Torfstein A, Frank U, Dulski P, Agnon A, Ariztegui D, Ben-Avraham Z, Goldstein SL, Stein M (2014) Lithology of the long sediment record recovered by the ICDP Dead Sea Deep Drilling Project (DSDDP). Quat Sci Rev 102:149–165. https://doi.org/10.1016/j.quascirev.2014.08.013

Orland IJ, He F, Bar-Matthews M, Chen G, Ayalon A, Kutzbach JE (2019) Resolving seasonal rainfall changes in the Middle East during the last interglacial period. Proc Natl Acad Sci 201903139. https://doi.org/10.1073/pnas.1903139116

Otto-Bliesner BL, Brady EC, Zhao A, Brierley CM, Axford Y, Capron E, Govin A, Hoffman JS, Isaacs E, Kageyama M, Scussolini P, Tzedakis PC, Williams CJR, Wolff E, Abe-Ouchi A, Braconnot P, Ramos Buarque S, Cao J, de Vernal A, Guarino MV, Guo C, LeGrande AN, Lohmann G, Meissner KJ, Menviel L, Morozova PA, Nisancioglu KH, O'ishi R, Salas y Mélia D, Shi X, Sicard M, Sime L, Stepanek C, Tomas R, Volodin E, Yeung NKH, Zhang Q, Zhang Z, Zheng W (2021) Large-scale features of Last Interglacial climate: results from evaluating the lig127k simulations for the Coupled Model Intercomparison Project (CMIP6)–Paleoclimate Modeling Intercomparison Project (PMIP4). Clim Past 17:63–94https://doi.org/10.5194/cp-17-63-2021

Otto-Bliesner BL, Brady EC, Tomas RA, Albani S, Bartlein PJ, Mahowald NM, Shafer SL, Kluzek E, Lawrence PJ, Leguy G, Rothstein M, Sommers AN (2020) A Comparison of the CMIP6 midHolocene and lig127k Simulations in CESM2. Paleoceanogr Paleoclimatology 35.https://doi.org/10.1029/2020PA003957

Palchan D, Stein M, Almogi-Labin A, Erel Y, Goldstein SL (2013) Dust transport and synoptic conditions over the Sahara-Arabia deserts during the MIS6/5 and 2/1 transitions from grain-size, chemical and isotopic properties of Red Sea cores. Earth Planet Sci Lett 382:125–139. https://doi.org/10.1016/j.epsl.2013.09.013

Palchan D, Torfstein A (2019) A drop in Sahara dust fluxes records the northern limits of the African Humid Period. Nat Commun 10:3803. https://doi.org/10.1038/s41467-019-11701-z

Peltier WR, Fairbanks RG (2006) Global glacial ice volume and Last Glacial Maximum duration from an extended Barbados sea level record. Quat Sci Rev 25(23–24):3322–3337

Petit-Maire N, Carbonel P, Reyss JL, Sanlaville P, Abed a., Bourrouilh R, Fontugne M, Yasin S, (2010) A vast Eemian palaeolake in Southern Jordan (29°N). Glob Planet Change 72:368–373. https://doi.org/10.1016/j.gloplacha.2010.01.012

Porat N, Amit R, Enzel Y, Zilberman E, Avni Y, Ginat H, Gluck D (2010) Abandonment ages of alluvial landforms in the hyperarid Negev determined by luminescence dating. J Arid Environ 74:861–869. https://doi.org/10.1016/j.jaridenv.2009.10.018

Quade J, Dente E, Armon M, Ben Dor Y, Morin E, Adam O, Enzel Y (2018) Megalakes in the Sahara? a review. Quat Res 1–23. https://doi.org/10.1017/qua.2018.46

Revel M, Ducassou E, Skonieczny C, Colin C, Bastian L, Bosch D, Migeon S, Mascle J (2015) 20,000 years of Nile River dynamics and environmental changes in the Nile catchment area as inferred from Nile upper continental slope sediments. Quat Sci Rev 130:200–221. https://doi.org/10.1016/j.quascirev.2015.10.030

Rohling E (2013) Quantitative assessment of glacial fluctuations in the level of Lake Lisan, Dead Sea rift. Quat Sci Rev 70:63–72

Rosenberg TM, Preusser F, Risberg J, Plikk A, Kadi K, a., Matter A, Fleitmann D, (2013) Middle and Late Pleistocene humid periods recorded in palaeolake deposits of the Nafud desert, Saudi Arabia. Quat Sci Rev 70:109–123. https://doi.org/10.1016/j.quascirev.2013.03.017

Rossignol-Strick M (1983) African monsoons, an immediate climate response to orbital insolation. Nature 304:46–49

Rubin S, Ziv B, Paldor N (2007) Tropical Plumes over Eastern North Africa as a Source of Rain in the Middle East. Mon Weather Rev 135:4135–4148. https://doi.org/10.1175/2007MWR1919.1

Ryb U, Matmon A, Erel Y, Haviv I, Katz A, Starinsky A, Angert A, and ASTER Team (2014) Controls on denudation rates in tectonically stable Mediterranean carbonate terrain. Geol Soc Am Bull 126:553–568https://doi.org/10.1130/B30886.1

Shanahan TM, Mckay NP, Hughen KA, Overpeck JT, Otto-Bliesner B, Heil CW, King J, Scholz CA, Peck J (2015) The time-transgressive termination of the African humid period. Nat Geosci 8:140–144. https://doi.org/10.1038/ngeo2329

Sirota I, Enzel Y, Mor Z, Ben Moshe L, Eyal H, Lowenstein TK, Lensky NG (2021) Sedimentology and stratigraphy of a modern halite sequence formed under Dead Sea level fall. Sedimentology 68:1069–1090. https://doi.org/10.1111/sed.12814

Smith JR, Hawkins AL, Asmerom Y, Polyak V, Giegengack R (2007) New age constraints on the Middle Stone Age occupations of Kharga Oasis, Western Desert. Egypt. J Hum Evol 52:690–701. https://doi.org/10.1016/j.jhevol.2007.01.004

Stein M (2014) The evolution of neogene-quaternary water- bodies in the Dead Sea Rift Valley. In: Garfunkel Z, Ben-Avraham Z, Kagan E (eds) Dead sea transform fault system: reviews. Springer, Netherlands, Dordrecht, pp 279–316

Stein M, Starinsky A, Katz A, Goldstein SL, Machlus M, Schramm A (1997) Strontium isotopic, chemical, and sedimentological evidence for the evolution of Lake Lisan and the Dead Sea. Geochim Cosmochim Acta 61:3875–3992. https://doi.org/10.1016/S0016-7037(97)00191-9

Stein M, Torfstein A, Gavrieli I, Yechieli Y (2010) Abrupt aridities and salt deposition in the post-glacial Dead Sea and their North Atlantic connection. Quat Sci Rev 29(3–4):567–575. https://doi.org/10.1016/j.quascirev.2009.10.015

Stockhecke M, Timmermann A, Kipfer R, Haug GH, Kwiecien O, Friedrich T, Menviel L, Litt T, Pickarski N, Anselmetti FS (2016) Millennial to orbital-scale variations of drought intensity in the

Eastern Mediterranean. Quat Sci Rev 133:77–95. https://doi.org/10.1016/j.quascirev.2015.12.016
Szabo BJ, Haynes CV, Maxwell TA (1995) Ages of Quaternary pluvial episodes determined by uranium-series and radiocarbon dating of lacustrine deposits of Eastern Sahara. Palaeogeogr Palaeoclimatol Palaeoecol 113:227–242. https://doi.org/10.1016/0031-0182(95)00052-N
Thunell RC, Williams DF, Kennett JP (1977) Late Quaternary paleoclimatology, stratigraphy and sapropel history in eastern Mediterranean deep-sea sediments. Mar Micropaleontol 2:371–388. https://doi.org/10.1016/0377-8398(77)90018-4
Tierney JE, DeMenocal PB, Zander PD (2017) A climatic context for the out-of-Africa migration. Geology 45:1023–1026. https://doi.org/10.1130/G39457.1
Tierney JE, Torfstein A, Bhattacharya T (2022) Late Quaternary hydroclimate of the Levant: The leaf wax record from the Dead Sea. Quat Sci Rev 289:107613. https://doi.org/10.1016/j.quascirev.2022.107613
Torfstein A (2019) Climate cycles in the southern Levant and their global climatic connections. Quat Sci Rev 221:105881. https://doi.org/10.1016/j.quascirev.2019.105881
Torfstein A, Haase-Schramm A, Waldmann N, Kolodny Y, Stein M (2009) U-series and oxygen isotope chronology of the mid-Pleistocene Lake Amora (Dead Sea basin). Geochim Cosmochim Acta 73:2603–2630. https://doi.org/10.1016/j.gca.2009.02.010
Torfstein A, Goldstein SL, Stein M, Enzel Y (2013) Impacts of abrupt climate changes in the Levant from Last Glacial Dead Sea levels. Quat Sci Rev 69:1–7
Torfstein A, Goldstein SL, Kushnir Y, Enzel Y, Haug G, Stein M (2015a) Dead Sea drawdown and monsoonal impacts in the Levant during the last interglacial. Earth Planet Sci Lett 412:235–244. https://doi.org/10.1016/j.epsl.2014.12.013
Torfstein A, Goldstein SL, Kushnir Y, Enzel Y, Haug G, Stein M (2015b) Response to comment on: "Dead Sea drawdown and monsoonal impacts in the Levant during the last interglacial" [EPSL, 412, 235–244, 2015]. Earth Planet Sci Lett 427:306–308. https://doi.org/10.1016/j.epsl.2015.07.014
Trauth MH, Larrasoaña JC, Mudelsee M (2009) Trends, rhythms and events in Plio-Pleistocene African climate. Quat Sci Rev 28:399–411. https://doi.org/10.1016/j.quascirev.2008.11.003
Tubi A, Dayan U, Lensky IM (2017) Moisture transport by tropical plumes over the Middle East: a 30-year climatology. Q J R Meteorol Soc 143:3165–3176. https://doi.org/10.1002/qj.3170
Vaks A, Bar-Matthews M, Ayalon A, Schilman B, Gilmour M, Hawkesworth CJ, Frumkin A, Kaufman A, Matthews A (2003) Paleoclimate reconstruction based on the timing of speleothem growth and oxygen and carbon isotope composition in a cave located in the rain shadow in Israel. Quat Res 59:182–193. https://doi.org/10.1016/S0033-5894(03)00013-9
Vaks A, Bar-Matthews M, Ayalon A, Matthews A, Frumkin A, Dayan U, Halicz L, Almogi-Labin A, Schilman B (2006) Paleoclimate and location of the border between Mediterranean climate region and the Saharo-Arabian Desert as revealed by speleothems from the northern Negev Desert, Israel. Earth Planet Sci Lett 249:384–399. https://doi.org/10.1016/j.epsl.2006.07.009
Vaks A, Bar-Matthews M, Ayalon A, Matthews A, Halicz L, Frumkin A (2007) Desert speleothems reveal climatic window for African exodus of early modern humans. Geology 35:831–834. https://doi.org/10.1130/G23794A.1
Vaks A, Bar-Matthews M, Matthews A, Ayalon A, Frumkin A (2010) Middle-Late Quaternary paleoclimate of northern margins of the Saharan-Arabian Desert: reconstruction from speleothems of Negev Desert, Israel. Quat Sci Rev 29:2647–2662. https://doi.org/10.1016/j.quascirev.2010.06.014
Vaks A, Bar-Matthews M, Ayalon A, Matthews A, Frumkin A (2018) Pliocene-Pleistocene palaeoclimate reconstruction from Ashalim Cave speleothems, Negev Desert, Israel. Geol Soc London, Spec Publ 466:201–216. https://doi.org/10.1144/SP466.10
Waldmann N, Stein M, Ariztegui D, Starinsky A (2009) Stratigraphy, depositional environments and level reconstruction of the last interglacial Lake Samra in the Dead Sea basin. Quat Res 72:1–15. https://doi.org/10.1016/j.yqres.2009.03.005
Waldmann N, Torfstein A, Stein M (2010) Northward intrusions of low- and mid-latitude storms across the Saharo-Arabian belt during past interglacials. Geology 38:567–570. https://doi.org/10.1130/G30654.1
Weber N, Lazar B, Gavrieli I, Yechieli Y, Stein M (2021) Gypsum deltas at the holocene dead sea linked to grand solar minima. Geophys Res Lett 48:1–10. https://doi.org/10.1029/2020GL091034
Wolff EW, Chappellaz J, Blunier T, Rasmussen SO, Svensson A (2010) Millennial-scale variability during the last glacial: the ice core record. Quat Sci Rev 29:2828–2838. https://doi.org/10.1016/j.quascirev.2009.10.013
Yehudai M, Lazar B, Bar N, Kiro Y, Agnon A, Shaked Y, Stein M (2017) U-Th dating of calcite corals from the Gulf of Aqaba. Geochim Cosmochim Acta 198:285–298. https://doi.org/10.1016/j.gca.2016.11.005
Zak I (1967) The Geology of Mount Sedom, PhD dissertation. The Hebrew Univ of Jerusalem
Ziv B (2001) A subtropical rainstorm associated with a tropical plume over Africa and the Middle-East. Theor Appl Climatol 102:91–102
Ziv B, Dayan U, Kushnir Y, Roth C, Enzel Y (2006) Regional and global atmospheric patterns governing rainfall in the southern Levant. Int J Climatol 26:55–73. https://doi.org/10.1002/joc.1238

The Paleo-Anthropocene and the Genesis of the Current Landscape of Israel

5

Oren Ackermann and Liora Kolska Horwitz

Abstract

This chapter explores the nature of the Paleo-Anthropocene of Israel, starting with the earliest human occupation in the region that served as a catalyst for subsequent changes in the physical and biotic landscapes. Focusing on human impacts to the natural and physical environment, we detail four key stages that served as pivotal events in the evolution of anthropogenic activities: (1) the "Natural Ecosystem and Landscape" that characterized the landscape inhabited by early hunter-gatherers starting ~ 1.5 Ma years BP; (2) the advent and progress of "Human Niche Construction" beginning ~12,500 BCE; (3) the establishment of an "Anthropogenic Ecosystem" starting ~ 3300 BCE; and (4) the transformation of the country to a "Total Anthropogenic Ecosystem and Landscape" ~ 1200 CE. Each stage marked a progression, that was characterized by an acceleration in the pace and extent of human intervention. Thus, the present-day Israeli landscape and ecosystems are not 'natural', but should be considered as the direct outcome of a series of anthropogenic interventions.

Keywords

Anthropocene · Landscape history · Human geography · Human–environment interaction · Anthropogenic geomorphology

The current research started when O.A. was a staff member at Bar-Ilan University.

O. Ackermann (✉)
The Department of Land of Israel Studies and Archaeology, Ariel University, Ariel, Israel
e-mail: orenac@ariel.ac.il

L. K. Horwitz
National Natural History Collections, Faculty of Life Sciences, The Hebrew University, Jerusalem, Israel
e-mail: lix1000@gmail.com

5.1 Introduction

The etiology of the term Anthropocene lies in the Greek words for human (*Anthropos*) and new (*cene*). The irreparable and global-scale influence that the collective activities of human beings have exerted on our planet's geology, ecology, and natural cycles, lies at the core of the proposal by Crutzen and Stoermer (2000; see also Crutzen 2002a, b) to define the Anthropocene as a new geological epoch. They suggested that the onset of this era coincided with the Industrial Revolution i.e., the latter part of the eighteenth century (Crutzen and Stoermer 2000). Since then, other scholars (e.g., Steffen et al. 2015; Hajdas et al. 2015; Gibbard et al. 2022) have proposed a more recent commencement in the mid-twentieth century, marked by the "Great Acceleration" that began with the detonation of the first nuclear bomb in 1945. In contrast, Smith and Zeder (2013) place the boundary much earlier i.e., at the start of the Holocene and coincident with the onset of the so-called Neolithic Revolution some 12,000 years ago.

Although it is important to establish a starting point for the Anthropocene, it is more critical to disseminate the message behind the term, namely "that humans are an integral part of the Earth system rather than merely an external forcing factor" (Foley et al. 2013: 83). This is the concept underlying the term the "Paleo-Anthropocene" which was coined by Foley et al. (2013) to encompass the full 2.5 million years of human impact on our environment, since the first appearance of the genus *Homo* in Africa.

The Paleo-Anthropocene archive is based on data assembled from the geosciences and environmental archaeology, that document important events in Earth's history which can be explicitly linked to human agency and engendered long-term and irreversible changes (Weiberg et al. 2019; Ellis et al. 2021). However, in the distant past, it is often difficult to disentangle climate-driven changes from those resulting from human agency, as exemplified by the ongoing debate surrounding the cause of Late Pleistocene mega-faunal

A. Frumkin and N. Shtober-Zisu (eds.), *Landscapes and Landforms of Israel*, World Geomorphological Landscapes, https://doi.org/10.1007/978-3-031-44764-8_5

extinctions (e.g., Barnofsky and Lindsey 2010; Prates and Perez 2021). Consequently, temporal studies undertaken at a local scale can be very useful in separating natural trends and those driven by human activities.

In this chapter, we describe some of the key events and features of the Paleo-Anthropocene of Israel from the earliest hominin occupation ~ 1.5 million years ago, through to the Islamic period, 628–1517 CE. This region is eminently suitable for such a temporal study, given its history as the main corridor for biotic exchanges and movements of people, plants, and animals, between Africa and Eurasia as attested in its paleobiological and cultural records (Zohary 1973; Horowitz 1979; Tchernov 1988a, b; Mazar 1990; Enzel and Bar-Yosef 2017; Barash et al. 2022). Sitting strategically at the interface of diverse biota, the north of Israel experiences a sub-humid Mediterranean climate while the south and east are semi-arid to hyper-arid (Kafle and Bruins 2009). The phytogeography of the region is diverse, comprising the Mediterranean biome in the north, trending to the Irano-Turanian (Asian steppes) and Saharo-Arabian (Sahara, Sinai, and Arabian deserts) vegetation zones in the south and east, with Sudanian enclaves (subtropical savannas of Africa) (Danin 1988). The present-day faunal community contains Palaearctic, Oriental, and Ethiopian elements (Tchernov 1988a, b). In terms of history and archaeology (prehistoric through historical periods), the area has been comprehensively studied and provides a well-dated record of various sites (for main sites and regions that are mentioned in the text, see Fig. 5.1) and of different population groups that inhabited or passed through the Levantine corridor (e.g., Mazar 1990; Levy 1995; Killebrew and Steiner 2014; Enzel and Bar-Yosef 2017), as well their interactions with the paleoenvironment (e.g., Ashkenazi et al. 2012; Gadot et al. 2016; Ackermann et al. 2017; Bar-Oz and Weissbrod 2017).

As elsewhere in the world, the pace and enormity of the scale of human impacts on the Israeli landscape have exponentially increased since the twentieth century. This is illustrated in Fig. 5.2a, which shows an endgame photo taken in 1909, showing the founding of the city of Tel Aviv on April 11. A small crowd of formally dressed men and women are captured standing in a cluster against the backdrop of a tall sand dune. Other dunes can be seen in the distance. The sand in the foreground of the photo is marked by myriad footprints, a harbinger of things to come, made by the people as they hurried to join the crowd, while the dune behind the group remains unmarked.

The photograph captures a crowd—the future inhabitants of the city—who had gathered to allocate land by lottery (using seashells). By 1914, some 1500 people had settled in this area, which currently (2023) serves as the core of a metropolis with an estimated population of 5,261,883 persons and a built-up area density of 7606 people for every square kilometer (https://worldpopulationreview.com/world-cities/tel-aviv-population). Today, there are no sand dunes visible in the Tel Aviv landscape (Fig. 5.2b), which is dominated by apartment buildings and office high-rises (Fig. 5.2c), while open spaces are constructed urban parks (Fig. 5.2d), and the modified seafront includes a built promenade, some 13 monitored municipal beaches, and breakwaters. The image in Fig. 5.2a offers a glimpse of a natural landscape and its associated ecosystem, which was totally obliterated within the space of just over 100 years.

With recourse to archaeological and paleoenvironmental proxy data, we have tracked the evolution of the Paleo-Anthropocene in Israel and highlighted the key milestones in anthropogenic activities which have shaped the physical and biological landscape of this region over time. Although the nature and extent of human presence in the region has fluctuated temporally and spatially (either by being transient, or by processes of settlement emergence, blossoming, and collapse), the human imprint has been a constant factor in modifying the natural processes and landscapes over the past ~ 1.5 million years (Ackermann 2007). Over time, the ecosystem has continually adapted to changing human management (Perevolotsky 2005; Perevolotsky et al. 2013), an ongoing process that has led to the evolution of an Israeli landscape that is now a "Total Anthropogenic Landscape and Ecosystem" (Fig. 5.3).

Based on the available data, we outline here a four-phase trajectory (Fig. 5.3) for the Paleo-Anthropocene of Israel, each stage a milestone marked by steadily increasing intensification in the interplay between humans and their environment:

(1) "Natural Ecosystem and Landscape" represents the primal environment occupied by the first inhabitants of the region, who were hunter-gatherers, starting at ~ 1.5 million years BP.
(2) A period of "Human Niche Construction", comprising a series of decisive events including the "Broad Spectrum Revolution" associated with the earliest sedentary villages of the Natufian culture (beginning ~ 12,500 BCE), followed by the so-called Neolithic Revolution (~ 10,500–6400 BCE) through to the adoption of metallurgy in the Late Chalcolithic period (~ 4500–3800 BCE).
(3) The third stage is marked by the establishment of a large-scale "Anthropogenic Ecosystem" beginning with the earliest urban centers in the Early Bronze Age II (~ 3300 BCE) and peaking in the Roman and Byzantine periods (~ 63 BCE–628 CE), an epoch characterized by extensive agricultural expansion.
(4) The foundation of a "Total Anthropogenic Landscape and Ecosystem" characterized by the introduction of terracing and industrialization beginning in the late Islamic period ~ 800 years ago.

Fig. 5.1 Main sites and regions in Israel that are mentioned in the text

5.2 The Natural Ecosystem and Landscape

The genesis of human-landscape interactions in Israel started in the Early Pleistocene at ~ 1.5 Ma, the date of the Lower Paleolithic site of 'Ubeidiya, the earliest hominin occupation in the country (Tchernov 1986, 1988a; Bar-Yosef 1994; Bar-Yosef and Belmaker 2017). Located in the Jordan Rift Valley, the site comprises a sequence of faunal and artifact-bearing layers, the latter attributed to the Early Acheulean industrial complex, that accumulated on the shoreline of a freshwater lake. Human fossils, probably belonging to *Homo* cf. *erectus/ergaster*, have been recovered at the site (Bar-Yosef and Belmaker 2017; Barash et al. 2022). From this point in time, human populations in the region flourished and their activities increased in extent and intensity. This trend is manifest even in early prehistory, by the increased number of Late Acheulean and Acheulo-Yabrudian sites (~ 400–200 ka) relative to those of the preceding Early and Middle Acheulean (~ 1.5 Ma to ~ 500 ka respectively) (see chapters in Enzel and Bar-Yosef 2017).

In the Early Acheulean, as attested to at 'Ubeidiya, faunal taphonomy suggests hominin acquisition of meat from scavenged carcasses (Gaudzinski 2004). The first convincing evidence for hunting, an activity that may have had an impact on wild animal communities, is only evident in the later, Lower Paleolithic as attested to at the site of

Fig. 5.2 Changes in the landscape of Tel Aviv. **a** The first families of Tel Aviv gathered in the sand dunes on April 11, 1909 for the lottery of plot allocation where the city would be built (KKL archives). **b** The same general location of Fig. 5.1a, Tel Aviv, 2021 (from Google Earth). **c** Highway and office buildings, Tel Aviv, 2023 (photo by Kerem Ackermann). **d** Edith Wolfson Park, an urban park in modern city of Tel Aviv (photo by Oren Ackermann)

Gesher Benot Ya'aqov, dated to ~ 780 ka (Rabinovich et al. 2012). This marks the onset of a traditional hunter-gatherer way of life that lasted in the Southern Levant until the first appearance of domestic animals and plants with the Pre-Pottery Neolithic B period, ~ 10,500 years BCE. From Ubeidiya onwards, the fauna of Israel comprised numerous mammalian species that became extinct before the Holocene, including mega-herbivores (e.g., elephants extinct by ~ 250/200 ka, rhinoceros by ~ 20,000 BCE), several species of small mammals and rodents (e.g., pika extinct by ~ 200 ka), and carnivores (e.g., African hunting dog extinct by ~ 250–190 ka; spotted hyaena by ~ 20,000 BCE) (Tchernov 1975, 1988b; Bar-Oz and Weissbrod 2017). However, there is no indication that humans played a role in any of these Pleistocene extinctions[1] since this process affected a range of different faunal families and genera, as well as taxa inhabiting different ecological niches and occurred at different times during the Pleistocene, such that it appears to have been climate-driven.

For the Lower Paleolithic, evidence of direct human impact on the landscape and ecosystem is difficult to detect as, most probably, it was ephemeral and of limited scale and duration or has been erased or covered over time by subsequent biotic and abiotic activities in the landscape. This is partly due to the fact that most early hominin activities occurred in open-air sites since caves were only regularly occupied from the Middle Paleolithic (Mousterian) onwards (> 250,000 BCE). The earliest constructed dwellings in Israel, from the site of Ohalo II, are much later and date to the Early Epipaleolithic (Kebaran) period (~ 21,000 BCE; Nadel and Werker 1999). These huts were constructed of perishable organic materials, which, in most instances, were not preserved.

From the Lower Paleolithic period, evidence for hominin exploitation of botanical materials, including collection of wood, is found, albeit rare, indicating the minimal

[1] There were also faunal extinctions during the Holocene that can be related to human activities (overkill, habitat destruction, etc.), with a final wave of extinctions taking place in the late nineteenth and early twentieth centuries with the introduction of firearms (see Yom-Tov and Mendelssohn 1988; Bar-Oz and Weissbrod 2017).

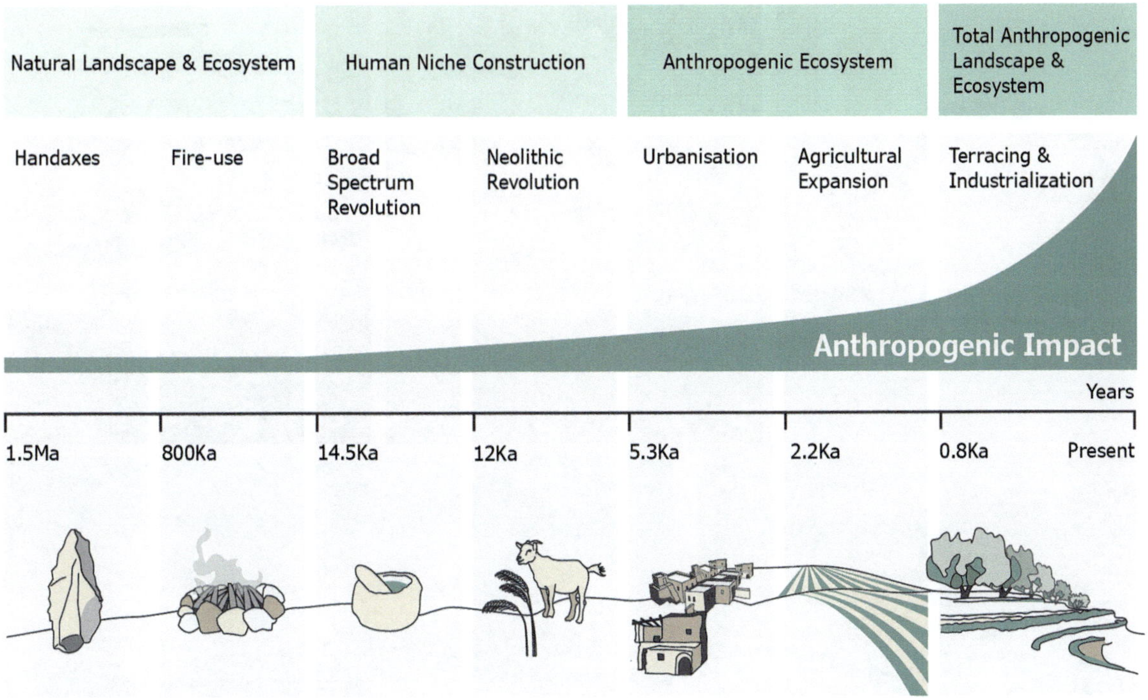

Fig. 5.3 Timeline of the main phases of the Israeli Paleo-Anthropocene (concept adopted after Foley et al. 2013: Fig. 5.1)

impact of early humans on the botanical foundations of the region at this time. The best examples are finds from the waterlogged Acheulean site of Gesher Benot Ya'akov that include the earliest wooden artifact from Israel, a polished plank (Goren-Inbar et al. 2002; Melamed et al. 2016). Wood ash and micro-charcoal derived from a hearth in the late Lower Paleolithic site of Qesem Cave indicate the presence of plant material used as fuel (Shahack-Gross et al. 2014), while there is widespread evidence for the exploitation of a broad spectrum of vegetal resources, both for fuel and as a dietary resource, from several Middle Paleolithic sites such as Amud, Hayonim and Kebara Caves (Madella et al. 2002; Albert et al. 2003; Lev et al. 2005). During the subsequent Epipaleolithic period (Kebaran), waterlogged paleobotanical remains from the site of Ohalo II, reveal the exploitation and consumption of a wide range of annual plants, with a focus on members of the grass family (Poaceae), including wild barley, oats, and emmer wheat (Weiss et al. 2004; Nadel et al. 2012). The morphology of the barley grains attests to a "trial" attempt at cereal cultivation, which may have served as a small-scale prelude to their domestication in the subsequent Neolithic (Snir and Weiss 2014). Remarkably, the Ohalo II botanical finds also include remains of several species of weeds and plants associated with human-disturbed environments. This provides one of the earliest indications of significant anthropogenic intervention in the Paleolithic landscape that may have modified the vegetation and so, the local ecosystem.

A critical innovation in the earliest phase of the Paleo-Anthropocene of Israel was human control of fire, with the first evidence for intentional fire use from the Lower Paleolithic site of Gesher Benot Ya'aqov dated to ~800 ka (Alperson-Afil and Goren-Inbar 2010) corroborated by evidence for fire from the roughly contemporaneous site of Evron Quarry (Stepka et al. 2022). Habitual use of fire occurs only from the late Lower Paleolithic (Acheulo-Yabrudian) onwards and includes evidence for built hearths from several Israeli sites, most notably Qesem Cave (Meignen et al. 2001; Karkanas et al. 2007; Shahack-Gross et al. 2014; Shimelmitz et al. 2014). Fire is considered a vital innovation that facilitated human survival and evolution (Wrangham and Carmody 2010). Although collection of wood, and perhaps tree felling, for fuel took place at this time, there is no evidence that fire was exploited to modify the local Mediterranean vegetation regime, even though bushfires are a critical element in the evolution and sustainability of this natural ecosystem (Naveh 2004).

Ancient human interventions to the physical platform of Israel provide some landmarks that have survived till today (Figs. 5.4a–d and 5.5). The most salient examples are quarries dating to the Lower and especially the Middle Paleolithic, where flint was extracted from primary (including deep, sub-surface) geological formations (Barkai and Gopher 2009; Gopher and Barkai 2011, 2014; Finkel et al. 2018). These locales are characterized by stone tailings, piles of extracted rock waste, some as large as 6 m in diameter,

Fig. 5.4 Anthropogenic geomorphology—ancient human imprint on the current physical platform of the Israeli landscape. **a** Cupmarks in the vicinity of the modern city of Modiin (photo by Oren Ackermann). **b** Limestone Quarry at the area of the modern city of Modiin (photo by Oren Ackermann). **c** Sandstone (known locally as *Kurkar*) Quarry at the site of Nahal Taninim at the coastal plain of Israel (photo by Liora K. Horwitz). **d** Agricultural terraces around Jerusalem (photo by Oren Ackermann). **e** (next page) Abundant slag mounds in Beer-Ora Valley (Southern Arava). The slag deposits represent intensive copper production in the Early Islamic period (From Ben-Yosef et al., 2008, Fig. 5.5e, with Elsevier permission)

Fig. 5.5 Paleolithic flint extraction quarries with stone tailings at Mount Achbara in the Eastern Galilee. (From Finkel et al. 2018, Fig. 5.5b, with Elsevier permission)

amounting to hundreds and over a thousand piles in some localities (Fig. 5.5). The quantity and size of the piles attest to extensive industrial activity, although the duration and/or number of people involved is difficult to assess. These tailings, like later quarries, have modified the physical platform and resulted in the creation of micro-niches for plants and animals (Kaplan 1984). In addition to flint mining for artefact manufacture, from the Middle Paleolithic onwards, there is evidence that ochre was mined locally and was probably used as a symbolic and decorative element (Hovers et al. 2003; Weinstein-Evron and Ilani 1994).

Given that Paleolithic hunter-gatherer communities lived in small, mobile groups and that overall population density in the region at that time was low, despite their extracting natural resources from both plant and animal communities, the impact of their long-term intervention on the natural landscape and ecology of Israel was limited. However, it served as a prelude for things to come as population size increased, people became sedentary, and their socio-economic systems and technological capabilities, became more complex.

5.3 "Human Niche Construction"

This stage of the Paleo-Anthropocene comprises at least three phases, each characterized by a different level of human-landscape relations.

5.3.1 The "Broad Spectrum Revolution"

Undeniably, the most extensive anthropogenic impacts on the Israeli landscape during the initial phase of the Paleo-Anthropocene occurred during the Epipaleolithic Natufian period (~ 12,500–10,500 BCE). Most researchers consider sites of the Early Natufian as the earliest sedentary settlements in the region, although some scholars have argued that this may equally reflect a shift in the scheduling of site occupations, periodic returns to important locations, or changes in social organization rather than sedentism per se (Belfer-Cohen 1991; Boyd 2006; Valla 2018). Nonetheless, the Natufian reflects the introduction of new traditions. As briefly outlined below, many features of this culture reflect greater occupation intensity, and increased size and density of populations than previously.

An important Natufian innovation was stone-built dwellings and installations, including those interpreted as storage facilities (Goring-Morris and Belfer-Cohen 2003; Boyd 2006). The first cemeteries are also attested to in the Natufian, again arguing for some degree of site permanence, as well as extensive anthropogenic disturbance to the sub-soil. For the first time, large numbers of non-portable limestone and basalt stone vessels and implements (mortars, pestles, grinding stones, and basins; Fig. 5.6), some weighing as much as 120 kg, are found on Natufian sites. They attest to some degree of site permanence and to extensive intervention in the physical environment as blocks of stone would have had to be quarried to facilitate their manufacture. Indeed, recent research on provenance of worked basalt objects has demonstrated that they were transported from their place of origin, as finished products or blocks, to sites as far away as 60–120 km (Rosenberg et al. 2021). The increased numbers of ground stone tools are also thought to signify an economic shift, from an extractive to a productive mode of production which is a key attribute of human niche construction (Belfer-Cohen and Hovers 2005).

A further Natufian intervention to the physical platform that has left visible marks in the landscape, up until today, are human-made holes or depressions in the bedrock, termed mortars or cup marks (Fig. 5.4a; Nadel and Lengyel 2009; Filin et al. 2017). It has been argued that these depressions, which vary in depth and shape, were used either for processing food, storing objects, or resulted from quarrying for flint nodules (e.g., Eitam 2009; Nadel and Lengyel 2009). Notably, these depressions have altered the form of the rock surface thereby serving as catchment points for water and soil, and so provide micro-environments for plants and other organisms, as well as locales for erosion and weathering (Allred 2009).

Fig. 5.6 Natufian basalt mortar from the site of En Gev (on the Sea of Galilee) (Courtesy with permission of the collection, the Israel Antiquities Authority Photo, © Israel Museum, Jerusalem, by Laura Lachman)

There is clearly evidence for increased duration and intensity of human settlement occupation at this time (augmented site size, number, and probably occupation span or frequency). This undoubtedly resulted in habitat alteration in close proximity to the sites due to collection of raw materials for construction and fuel (wood, stone), denudation of food resources in the adjacent landscape, disturbance of the physical platform through the erection of structures and installations, human traffic and noise, accumulation of refuse, among many other factors. As first proposed by Tchernov (1984), and since corroborated by others (e.g., Weissbrod et al. 2017), this newly constructed human niche was associated with the first appearance of commensal animals that coexisted with humans and is attested to by finds from Early Natufian sites of, house mouse [*Mus musculus*], black rat [*Rattus rattus*], spiny mouse [*Acomys cahirinus*], house sparrow [*Passer domesticus*], as well as dogs [*Canis domesticus*] and red foxes [*Vulpes vulpes*] (Tchernov 1984; Yeshurun et al. 2009).

The Natufians were complex hunter-gatherers. Their food web, especially of gathered and trapped foods, was significantly broader, and exhibits economic specialization, compared to previous hunter-gatherers in the region. This period has been termed the "Broad Spectrum Revolution" (Flannery 2008), due to evidence for the consumption of so many new species of plants and animals, as well as for the intensification in the exploitation of these resources, with a focus on smaller-sized animal species, including birds and reptiles, in addition to wild cereals (Bar-Yosef 1998; Stiner 2001; Weiss et al. 2004; Weissbrod et al. 2012; Biton et al. 2021). Although climate change has been implicated, the most commonly accepted explanation for this dietary shift is a food crisis brought about by human demographic pressures outstripping the resource capacity of the landscape (see Stiner 2001; Weiss et al. 2004).

In sum, the extensive socio-cultural, technological, and demographic changes identified with the Natufian were noticeably accompanied by far-reaching effects on the local landscape, such that this period may be seen as a turning point in human-environmental interactions in the region.

5.3.2 The Neolithic Revolution

The transition from hunting and gathering to food production took place in the Near East during the Pre-Pottery Neolithic period (~ 10,500–6400 BCE). Following over a million years of simple extractive activities, humans now became the chief modifiers of their environment and a process of human niche construction ensued. As noted by Smith (2007: 188), "Domesticates provided humans with a significant new means of reshaping biotic communities and natural landscapes more to their liking." The new productive economy offered greater stability for human communities, with regular and predictable food production. This was associated with increased sedentism, population size and density. Although the motivation driving domestication remains contentious, with suggestions ranging from demographic pressure, deleterious climatic conditions to scalar stress within communities (e.g., Bar-Yosef and Belfer-Cohen 1989, 2002; Smith 2007; Goring-Morris and Belfer-Cohen 2011), the outcome had a profound impact on all aspects of human lifestyle (e.g., settlement pattern, architecture, subsistence practices, economic systems, technology, rituals, and beliefs—Kuijt and Goring-Morris 2002). The effect on the natural ecosystem and physical landscape was even greater and was probably the most momentous and long lasting of all events in the Paleo-Anthropocene of Israel.

Currently, researchers are still debating whether domestication took place in multiple centers within the Near East or in a single locale. Some of the earliest dates for the appearance of domestic plants (cereals and pulses) and herd animals (sheep, goat, cattle, and pig) are from regions to the north of Israel, specifically southeastern Turkey and Northern Syria, where all the wild progenitors of domestic plants and animals co-occurred (Wilcox 2007; Zohary et al. 2012; Abbo and Gopher 2017; Vigne et al. 2017; Abbo et al. 2022). However, with the possible exception of some species (i.e., wild sheep and wild chickpeas), Israel was also home to many of the wild progenitors of domestic plants and animals (Davis 1982; Horwitz et al. 1999; Zohary et al. 2012; Gourichon and Horwitz 2021), increasing the possibility of local domestication events perhaps following the adoption of the know-how that originated in the Northern Levant. Indeed, aDNA (ancient DNA) analyses have demonstrated that goats and possibly cattle may have undergone autochthonous domestication in the Southern Levant (Daly et al. 2018; Verdugo et al. 2019).

Irrespective of geographic origin, the domestication of founder crops and animals was not a single event but a slow, unidirectional, and long-term process over perhaps some 4000 years that straddled all phases of the Neolithic; from the Pre-Pottery Neolithic A (PPNA; ~ 10,500–8800 BCE), Pre-Pottery Neolithic B (PPNB; ~ 8800–7000 BCE) through Pre-Pottery Neolithic C (PPNC; ~ 6900–6400 BCE). Over this timespan, within the Near East, different species underwent domestication at different times and perhaps in different locations, from where they dispersed to other regions. Thus, the term *Neolithic Revolution* coined by Gordon Childe (1951) which implies a single event is a misnomer. It should be borne in mind that there is evidence from PPNA sites such as Netiv Hagdud and Gilgal in Israel, that economically important plant species (in this case, lentils, barley, and oats), were being manipulated to increase their availability, even before they began to manifest morphometric changes associated with plant domestication (Weiss et al. 2006). Similarly, goats came under human control in this region before they evidenced marked skeletal changes (Zohary et al. 1998; Gourichon and Horwitz 2021).

In Israel, domestic plants preceded animals by some hundreds of years. By the Early PPNB, domestic lentils, faba beans, and barley were known, while einkorn and emmer wheat and chickpeas occurred in Israeli sites only by the Mid-PPNB (Feldman and Kislev 2007; Zohary et al. 2012; Caracuta et al. 2015). Not surprisingly, these finds are all associated with storage facilities. Goats were the first domestic herbivores to appear in the Mid-PPNB, followed by sheep by the Late PPNB and cattle and pigs probably only by the Pre-Pottery Neolithic C/Pottery Neolithic periods (Horwitz et al. 1999; Munro et al. 2018; Gourichon and Horwitz 2021).

It is patent that from their inception, agricultural and animal husbandry activities were among the earliest modes of environmental modification induced by people. They resulted in a changed physical and biotic landscape as areas set aside for pastures and cultivated fields displaced the natural flora and fauna, favoring species of economic interest over wild forms. A core from the Sea of Galilee (Schiebel and Litt 2018) attests to a shift in the ratios of Mediterranean trees and shrubs by the Pottery Neolithic period (PN; ~ 6400–5200 BCE). It could not be conclusively determined whether this was caused by woodland clearance, as had previously been suggested by Rollefson and Köhler-Rollefson (1992), or due to aridification that resulted in the expansion of the Irano-Turanian steppe biome. However, more recent data on sedimentation rates that were extracted from a 297.5 m deep core taken in the Dead Sea (Lu et al. 2017), show a two- to threefold increase in erosion rate in the last 11.6 ka years relative to the period 21.7–11.6 ka. The authors attributed the intensified erosion to vegetation clearance as a result of augmented human intervention in the landscape during the Neolithic, corroborating initial interpretations.

As a result of preparatory deforestation and vegetation clearance in the Neolithic, the agricultural fields would have excluded the original population of wild plants. Fields would have provided a new ecological niche that offered advantages for plant species that could quickly adapt and contend for light, water, and land resources. The archaeobotanical record of the PPNC site of Atlit-Yam provides a good example of such a shift, with both domestic cereals (emmer wheat, small-grained wheat) and pulses (lentil and chickpea) found alongside annual weeds coincident with the introduction of agricultural pests (Hartmann-Shenkman et al. 2015). This may represent the initial establishment of an evolutionary process of adaptation that has survived until today, in which annual plants and weeds underwent adaptations to the growing conditions of the cultivated field and to agricultural processing. For animals, especially the larger-sized wild herbivores, the expansion of human settlements coupled with competition with early domestic herds over grazing areas, undoubtedly contributed to the early extirpation of some wild forms such as the aurochs (the wild progenitor of domestic cattle) whose numbers begin to dwindle in the Neolithic and finally disappeared from Israel by the end of the Early Bronze Age II (~ 3000–2700 BCE, Tsahar et al. 2009). In addition, by the PPNC, pastoralism emerged as a new adaptation in the arid margins of the region—the Negev and Sinai (Goring-Morris 1993; Rosen 2017).

The surplus food generated from domestic food sources facilitated population growth and resulted in greatly expanded settlement numbers, size, and complexity. Cultivation specifically dictated settlement location, with the PPNB villages situated in proximity to alluvial fans to ensure replenishment of soils or else were located in well-watered environments (Goring-Morris and Belfer-Cohen 2011, 2014, 2020). For example, as discussed by Roskin et al. (2022), the geospatial location of the PPNB site of Motza both maximized access to two different ecosystems—facilitating farming and herding, as well as regional cultural connectivity. Overall, the effects of long-term sedentism during the PPNB, appear to have resulted in the decimation of natural resources, which coupled with increased pressure on fields and pasturage adjacent to the settlements caused the collapse of the Late PPNB mega-sites (Köhler-Rollefson 1988; Rollefson 2020). The increased numbers of sedentary people in these villages contributed to their poor health status primarily due to augmented infectious disease rates, many of them zoonotic in origin given the novel association of people and animals (Eshed et al. 2010; Horwitz and Smith 2000; Smith and Horwitz 2007).

Anthroturbation (human-derived bioturbation) was extensive in the Neolithic, including digging foundations for structures, pit graves, and extensive quarrying for flint and ochre, to which was added stone quarrying for architecture (Grosman and Goren-Inbar 2007, 2016).

Water was also first domesticated in the Late Neolithic, entailing the construction of wells to ensure sustainable water sources for the settlements. The earliest water wells in the country, and some of the earliest worldwide, are known from the currently submerged PPNC site of Atlit-Yam (Fig. 5.7a). A fully excavated well at this site was ~ 1.5 m in diameter and 5.7 m deep; its lowest section (1.5 m) was excavated into sandstone with the upper courses (~ 4.2 m) lined with stones (Galili and Sharvit 1998). Another slightly later example is from the Pottery Neolithic (Yarmukian phase, ~ 6400–5800 BCE) at the site of Sha'ar Hagolan and comprises a pit ~ 2 m in diameter at its head and 4.26 m deep, lined with large stones (Garfinkel et al. 2006). Slightly later well heads from the Pottery Neolithic (Wadi Rabah phase, ~ 5800–5200 BCE) submerged site of Kfar Samir, were built of alternating layers of wooden branches and stones (Galili et al. 2018), while those in much later periods were exclusively of stones.

Another instance of Neolithic human intervention is the earliest example of a seawall known to date (Galili et al. 2019). The > 100 m long seawall was constructed 90 m offshore, parallel to the northern Mediterranean Sea shoreline of Israel, on the seaward side of the submerged Pottery Neolithic settlement of Tel Hreiz. It was intended to protect the village from waves and erosion resulting from a rise in post-glacial sea levels. The wall was constructed from extremely large boulders (weighing 200–1000 kg each) that were brought to the site from stream beds located between 1.6 and 3.8 km away, such that its construction was a monumental communal activity. Notably, the seawall would have influenced the topography of the local coast and its hinterland and affected sediment buildup and sea currents. Unfortunately, this endeavor did not succeed as the village was eventually abandoned and is today submerged under ~ 3 m of water.

Fig. 5.7 Water wells and anthropogenic divergence systems, **a** Head of a water well at the submerged Pre-Pottery Neolithic C site of Atlit-Yam (photo courtesy of Ehud Galili), **b** Quarried water channel (Late Byzantine-Ottoman periods) at the site of Nahal Taninim on the coastal plain of Israel (photo by Liora K. Horwitz)

5.3.3 Pyrotechnology: Lime-Plaster, Ceramics, and Metallurgy

Although there is widespread evidence throughout the Neolithic for the use of pisé (a mixture of mud, water, and plant temper) to make floors and walls, the exploitation of lime-plaster is a hallmark of this period. This pyro-technology was practiced locally in the majority of PPNB settlements as lime-plaster was used to cover most floors and walls, and these also show signs of renewal. Despite the easy access to raw material (chalk pebbles and boulders are found in most regions of the country) and ubiquitous finds of plaster in PPNB sites, few lime kilns have been discovered. This may be attributed to the fact that the technology used was relatively simple and would have left negligible traces, while if the kiln was outdoors, erosion and destruction of the installation would have occurred quite rapidly, as described by Goren and Goring-Morris (2008). One recently excavated example is an Early PPNB kiln found at the Nesher-Ramla Quarry site (Toffolo et al. 2017). It would have required the collection of locally occurring raw material and fuel, although the researchers concluded that relatively little lime was needed for the task since it was mixed with other materials (i.e., lime comprised only 25–30% of the total volume of the plastered surfaces). It has been suggested that lime-plaster manufacture resulted in the creation of stone quarries and, more importantly, deforestation as well as pollution (Rollefson and Köhler-Rollefson 1992). However, based on their experimental study, Goren and Goring-Morris (2008: 795) concluded that "this activity would probably have had relatively little impact on the nearby environment of the site. Indeed, we suggest that the environmental impact of the use of larger posts and beams for architectural purposes would likely have been considerably greater than that caused by lime-plaster production."

A second new pyro-technology that was characteristic of the Neolithic was ceramic production. Initially, in the PPNA-PPNB, clay was used for the manufacture of figurines (of animals and humans), beads, and vessels, most of which were unfired (Goren and Biton 2010). However, by the subsequent Pottery Neolithic (~ 6400–5200 BCE), there was extensive exploitation of local clay sources throughout the region for the manufacture of a range of ceramic vessels, as attested to by petrographic analyses (e.g., Goren 1991; Nativ et al. 2012; Shapiro 2013; Ben-Michael 2016), most of which were fired. A related Neolithic industry was the production of mudbricks, the earliest of which date to the PPNA. Mudbricks, either sun-dried and/or heated, were used—albeit rarely, in PPNB and PPNC construction (Galili and Rosen 2011; Ogloblin Ramírez et al. 2020), but increased in use in later periods. An excellent example of this is a rectangular structure and associated silos from the Early Chalcolithic site of Tel Tsaf, dated to 5200–4700 BCE, that were constructed from sun-dried mudbricks (Fig. 5.8a–c; Rosenberg et al. 2020). The researchers noted that the bricks were made from locally sourced raw material and were cemented in place by mortar (either mud-plaster or lime-plaster). Recycling of old decayed mudbricks appears to have been practiced. The raw materials used in the manufacture of ceramics and mudbricks would have been sourced locally at each site and undoubtedly fired on-site as well.

Fig. 5.8 **a** Mudbrick used in construction of a rectangular house and round silos at the Chalcolithic site of Tel Tsaf, Jordan Valley; (photo courtesy of Yosef Garfinkel). **b**, **c** Mudbricks walls, in a close up view. (From: Figs. 5.4a and 5.6b, c, Rosenberg et al. 2020)

A third pyro-technological innovation, namely, metallurgy i.e., copper smelting, first developed in the subsequent Late Chalcolithic period (Ghassulian culture; ~4500–3800 BCE). In this period there is evidence for on-site smelting at numerous sites in Israel, attested to by the remains of metallurgical installations, crucibles, hammer stones, copper ore, and slags. The source of most of the copper ore appears to have been Faynan in Jordan, which is some 100 km away from the main metallurgical production centers in the Northern Negev in Israel. The distance that raw ore was transported, as well as the identification of two different technologies used in artefact production: (i) a two-stage, furnace-based primary smelting of pure ore followed by melting/refining in crucibles, and (ii) lost wax casting, attests to the sophisticated level of craft specialization at this time (e.g., Golden et al. 2001; Ackerfeld et al. 2020; Rose et al. 2021).

Although relatively small in scale, undoubtedly, over time the industrial areas for the production of ceramics, mudbricks, and metallurgy served as sources of pollution, due both to combustion (fuel for the extensive hearths and kilns) as well as the refuse generated at the workshops. Moreover, the plant material and/or sand used as temper as well as clay would have resulted in decimation of vegetated areas as well as grab pits throughout the landscape, while mining for copper ore would have modified the geological platform to access the ore-bearing veins. As for the lime and pottery kilns, the smelting furnaces would have generated smoke and fumes, resulting in extensive air pollution.

Undoubtedly, with their myriad new technologies (e.g., domestication of plants and animals, extensive lithic extraction, plus ceramic, lime plaster and metal production) and revolutionary ways of life (e.g., large sedentary villages, surplus production, pastoralism), the Pre-Pottery Neolithic B and subsequent Chalcolithic periods should be perceived as the tipping points in the Paleo-Anthropocene i.e., a rapid, linear transition toward a new set of relations between people and their natural world.

5.4 "Anthropogenic Ecosystem"

Following the foundation of the first intensively occupied (probably sedentary) villages in the Early Natufian, which can be considered as the precursor of intensive human niche construction in Israel, the anthropogenic impact on the Israeli physical and biotic landscape increased in extent and pace during the Neolithic, as outlined above. The next step in the intensification of human niche construction we identify as coincident with the establishment of large-scale, complex urban centers in the Early Bronze Age II (~ 3000–2700 BCE).

5.4.1 Urbanization

The process of urbanization was associated with architectural monumentality in the form of fortifications, temples, palaces and public buildings, large-scale craft specialization, the creation of surpluses, and the development of market systems (Greenberg 2019). At that time, there is evidence for the first occurrence of fertilized fields dating to the Early Bronze Age II–III (3000–2200 BCE) as

Fig. 5.9 Tells (artificial mounds) in the landscape of Israel. **a** Tel Zayit (photo by Oren Ackermann). **b** Tel Lachish (photo by Oren Ackermann). **c** Cut through a section of Tel Beit Yerach (photo by Oren Ackermann). **d** Aerial view of Tel Burna (photo by Griffin, courtesy of The Tel Burna Archaeological Project). **e** Changes in top soil elemental composition due to anthropogenic activities at Tel Burna, Shephelah. Maps of P, K, Zn and Cu concentrations (From Šmejda et al. 2017a, Fig. 5.3, with Elsevier permission)

discovered adjacent to Tel Yarmouth (Paz et al. 2017). A rural hinterland connected to these urban centers emerged, while in the arid regions of the southern Levant, a mobile pastoralist way of life continued (Horwitz 2022). By the end of the Early Bronze Age, the region experienced the waning of Egyptian colonial control and engaged in broader inter-regional and international networks.

A salient manifestation of early urbanization were large-scale settlements that frequently represent the earliest, extensive occupation layer of many ancient tells (or tels)[2] (Fig. 5.9a–d). These tells represent artificial mounds created from the accumulated, anthropogenically derived debris of superimposed ancient settlements. Tells can reach heights of tens of meters above the immediate surroundings and may even extend the platform of natural hills with massive constructions comprised of fill (Finkelstein 2011). Indeed, tells might be considered the most significant post-Neolithic anthropogenic topographic alternation to the region. Notable human activities associated with tells in general include dense architectural construction and renovation (e.g., Maeir 2012; Greenberg 2019), dedicated ritual areas including public structures, presence (on or adjacent to the tell) of industrial complexes for a range of activities, including production of ceramics (e.g., Eisenberg-Degen 2018), metallurgy (e.g., Eliyahu-Behar et al., 2012, 2013), and specialized work areas and installations for processing of agricultural resources such as olive oil (e.g., Orendi et al. 2017; Maeir et al. 2021), as well as refuse areas for disposal of food waste (Sapir-Hen et al. 2012, 2014). Excavations conducted during the last decades have revealed that human occupation and activities also occurred in the immediate surroundings of tells. For example, at the Early Bronze Age

[2] The spelling is dependent on the name of the site and the translation from Hebrew (tel) or Arabic (tell).

e

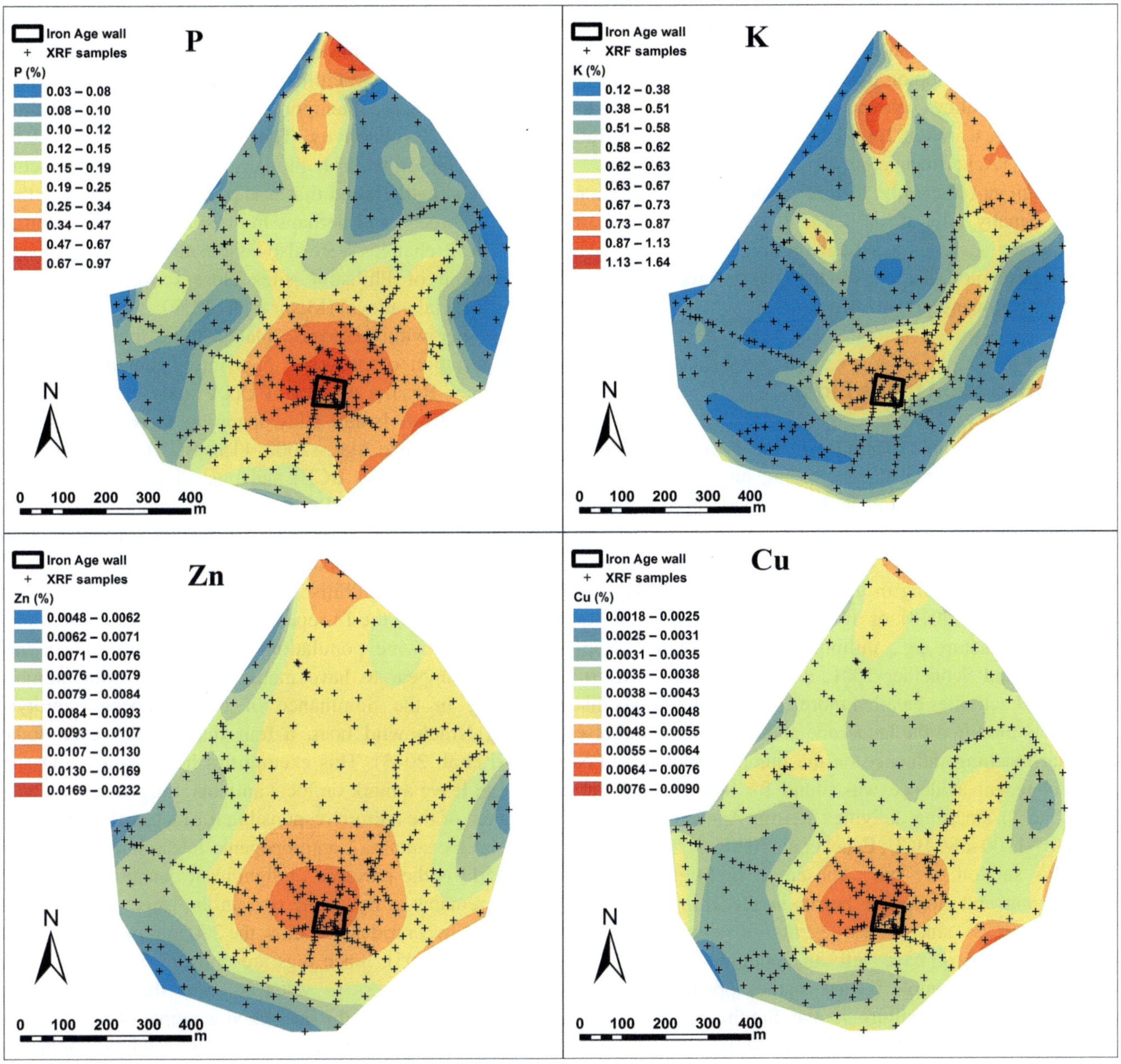

Fig. 5.9 (continued)

site of `En Esur, a large urban complex was unearthed west of the main tell (Elad et al. 2018), while at the base of Tell es Safi/Gath, an Iron Age ritual complex, metal production, and olive oil processing area were revealed (Dagan et al. 2018; Workman et al. 2020).

The destruction or abandonment of tells, either due to warfare, climate change, or population movement, has led to various natural processes, including the collapse of mudbrick structures whose building materials were transformed into sediment that accumulated on the tells' surface. Moreover, the ruins serve as dust traps that increase natural sediment accumulation (Janovský et al. 2022). Notably, the elemental composition of their soils and sediments retain the anthropogenic "imprint" of past human activities. This signature was revealed in the study of the Late Bronze and Iron Age sediments at Tel Burna, located at the Judean foothills (Shephelah), where researchers found a correlation between human activities and the values of a range of elements. Sediments located within the inhabited parts of the tell within the city walls had elevated concentrations of certain elements such as Si, P, K, Cu, Zn, and Ca (Fig. 5.9e) compared to those distant to the occupied (walled) area. This indicates that chemical content can be used to identify anthropogenic processes at archaeological sites and emphasizes the long-lasting impact of human activities on soils (Orendi et al. 2017; Šmejda et al. 2017a, b). Over time, debris derived from the destruction of the tell erodes and is displaced, slipping down the tell slopes to be accumulated in the valleys below, as has been revealed by geoarchaeological research at Tel Lachish (Rosen 1986) and Tell es Safi/Gath (Ackermann et al. 2012, 2014a, b).

Although not the only feature associated with human activities, many tells are still visible in the landscape of Israel and so serve as a salient reminder of the invasive anthropogenic modifications made to the past topography of the region following urbanization.

5.4.2 Alteration to the Biota

There are innumerable examples in the local archaeological record that illustrate the increasing impact of complex urban societies on the abiotic and biotic landscape of ancient Israel over the past 4500 years. Although simple extractive activities of raw materials continued to take place, such as copper and gold mining in the Arava (Avner and Magness 1998; Ben-Yosef 2018) or flint and stone extraction for artifact manufacture e.g., milling stones or tabular scrapers (Rosen and Schneider 2001; Rosen and Goring-Morris 2018), most activities entailed more complex alterations that extensively modified the landscape and ecosystem, and also had a far-reaching influence on the human population. For example, recent evidence was found for heavy metal pollution due to smelting activities in Iron Age city dwellers (dated to ~ 1200–586 BCE; Eshel et al. 2020).

In this section, we have chosen to highlight changes to the fauna and flora that began in the Iron Age II (~ 1000–586 BCE) since these events of significant ecosystem engineering have left an indelible impact until today. The first example is tied to the appearance of the immigrant Philistine culture in the Southern Levant during the Late Bronze II (~ 1400–1200 BCE)/Early Iron Age I; ~ 1200–1000 BCE), a community that was responsible for the introduction of several new plant species not previously documented in the region, including cumin (*Cuminum cyminum*), sycamore (*Ficus sycomorus),* and the opium poppy (*Papaver somniferum),* all species which later became significant cultivars in local cultures (Frumin et al. 2015).

In addition to introducing foreign species into the local agricultural regime, the Philistine immigration was also associated with an extensive change to the composition of local plant species exploitation, with more than 70 new, locally occurring species being incorporated into anthropogenic habitats. It is therefore possible that the Philistine migration heralded a "Green Revolution" that was accompanied by the introduction of new agronomic methods, which differed from those customarily used in Israel until that time (Frumin et al. 2015; Frumin 2017). Indeed, it has been proposed that the convergence of several innovations enabled the Iron Age Philistines to cultivate more extensive areas than other contemporaneous communities, resulting in increased productivity. These innovations included the development of the iron ploughshare and the introduction of the zebu cattle breed that was probably more robust than the local breeds and perhaps more suited for deep ploughing. These changes were also associated with the geospatial location of Philistine sites close to prime cultivated areas (alluvial valleys) (Maeir et al. 2019).

Together with the exotic plants, there is also evidence of importation of domestic pigs into Israel by migrating Philistines. Results of ancient DNA analyses of pig bones recovered from Israeli sites spanning three periods: pre-Philistine (Bronze Age, ~ 2000–1150 BCE), Philistine (Iron Age II, ca1150–586 BCE) and post-Philistine (Persian to Roman periods, ~ 586 BCE–324 CE), indicates a gradual shift from a local, Levantine haplotype to a European one (Meiri et al. 2013). Apparently, the domesticated pigs were translocated by the Philistine population, and from the Iron Age II onwards, the European haplotype became dominant in domestic swine populations in the region. "Philistine" pigs also appear to have mated with local wild animals, resulting in the dominance of the European haplotype among Israeli wild boar, a trait that persists even today (Meiri et al. 2013). This exemplifies how an invading species can leave a mark on local animals, long after the disappearance of the culture responsible for its translocation.

A later example of anthropogenic intervention in the faunal world is the presence of significant quantities of domestic pig remains at Byzantine sites in the Negev, sites that are located in hyperarid areas below the 200 mm isohyet (Horwitz and Studer 2005). Since pigs are obligate drinkers and also require ready access to water to meet their physiological needs (to wallow in), this is a good example of human intervention to facilitate the survival of a desired animal resource.

A later example of human intervention in the natural ecosystem, of a botanical intervention comes from the Persian period (~ 586–332 BCE) is and relates to the introduction of the citron (*Citrus medica*) and walnut (*Juglans regia*). These species are native to the region of modern-day Iran, and it seems likely that they were introduced into the region during the Persian conques (Langgut et al. 2013; Langgut 2017), and were then tended, enabling them to be successfully cultivated in Israel.

The examples given here, although spanning different periods from the Bronze Age through the Persian period, reflect the significant human-engineered alterations to the local biotic regime that transformed the faunal and botanical makeup of the region and laid the foundations for the "Total Anthropogenic Ecosystem and Landscape" we see today.

5.5 "Total Anthropogenic Ecosystem"

The term "Total Human Ecosystem" was first used with reference to Israel in a 1982 article by (Naveh 1982), who may be viewed as the father of landscape ecology in this country. Component parts of this system are the natural bio-ecosystem (natural and semi-natural landscape-natural ecotope), the agricultural bio-ecosystem (agricultural landscape—fields, pasture, lakes, rivers), the rural techno-ecosystem (rural landscape—farms, villages) and the industrial techno-ecosystem (urban industrial landscape with cities, factories). In our view, it is perhaps only from the Early Roman period onwards (~ 63 BCE–132 CE) that we can claim that this stage has been reached.

From the first urban societies in the Early Bronze Age II through to the advent of industrialization during the Ottoman period (eighteenth century CE), the temporal trajectory was for larger as well as increasing numbers of settlements as complex urban societies were formed, developed, collapsed, and were formed again. These events were associated with fluctuations in population size and composition, augmented by immigration, emigration, and conquests. Thus, although the overall trend in the region has been for population increase, peak demographic density in antiquity was attained during the Classical era, in the Late Roman and Byzantine periods (132–324 CE and 324–628 CE respectively) (Tsafrir 1996). This is best evidenced in the southern Shephelah (Judean Foothills) where in an area of 100 km^2, 573 sites and installations dating to the Byzantine period were identified, giving a density of 5.73 sites per km^2 (Dagan 2006). Ongoing sophistication in technological capacity and the creation of surplus and specialized goods, which in turn was coupled with expanded local and international trade (Mazar 1990; Levy 1995; Killebrew and Steiner 2014), resulted in large-scale impacts culminating in the creation of a "Total Human Ecosystem and Landscape." Below we briefly describe four facets of the local ecosystem that began in the Classical era, and which reflect these large-scale and far-reaching changes.

5.5.1 Fertlized Fields

Intensive agrarian activities are attested in this phase by the extensive use of fertilizer which was a homemade compost composed of ash from ovens, animal and human excrement, and cultural material (fragments of pottery, glass, and metal). In the sandy soils around the Byzantine-Islamic settlement of Shivta in the Negev desert, researchers found that the sandy soils had higher than expected phosphate levels, indicating the addition of fertilizer to enhance the soil. In this case, the fertilizer was pigeon dung collected from specially constructed towers that housed thousands of pigeons (Ramsay et al. 2016). Further evidence for manuring of soil was found in ancient fields in the runoff farms of the Negev (Fig. 5.11; Bruins et al. 2020; van Bommel et al. 2021) as well as in the Judean Desert (Hadas 2012). Materials recovered during excavation and micromorphological analyses of thin sections from anthropogenic soils from these sites showed the common presence of small fragments of ceramics as well as very small charcoal and bone fragments, some smaller than 0.01 mm in size, pointing to domestic refuse as the source of the fertilizer in addition to animal dung. Soil fertilization (especially due to the presence of dung, charcoal, and ash) that was carried out several thousand years ago still affects the properties of these sediments today (e.g., van Asperen et al. 2014; Shtienberg et al. 2017; Taxel et al. 2018).

5.5.2 Agricultural Installations

The peak Late Roman through Byzantine (~ 324–628 CE) demographic density is also manifest in extensive agricultural activity, in some instances organized by established political-social institutions such as the Roman state and later the Church during the Byzantine period. For example, an exotic tree species, the Euphrates poplar (*Populus euphratica*), grows near permanent water sources (springs) adjacent to Roman and Byzantine period monasteries in the Judean Desert and the Negev Highlands. They appear to have been planted by the monks from those monasteries. Interestingly, although these monasteries were abandoned more than a thousand years ago, the trees continue to grow and bloom (Fig. 5.10, Ashkenazi et al. 2021).

In addition, increased numbers of agricultural installations are found in the landscape, including the construction of irrigation systems based on springs (Fig. 5.12; Hadas 2012; Yechezkel and Frumkin 2019), aqueducts (Murphy-O'Connor 2003) and wine and olive oil presses, some subterranean, even in the arid zones (Fig. 5.13; Frankel 2009; Chocron et al. 2019; Eisenberg-Degen et al. 2022). Additional evidence pointing to the establishment of a large-scale olive oil industry at this time is that the highest frequency of olive pollen is documented in the palynological record of Israel, attesting to the establishment of industrial olive oil production for export (Neumann et al. 2006, 2007, 2010; Tepper et al. 2022).

5.5.3 Agricultural Terraces, Irrigation Systems, and Runoff Harvesting

Large areas of the present-day landscape of Israel are marked by the carefully constructed stone walls of agricultural terraces that follow the contours of the hillsides

Fig. 5.10 Euphrates poplar (*Populus euphratica*) near Ein Avdat (the spring of Avdat), in the desert of the Negev (photo courtesy of Eli Ashkenazi)

(Fig. 5.4d). In Ramat Bet Shemesh (Shephelah, Judean Foothills), Dagan (2011) showed that 89% of the area is covered by agricultural terraces located in wide valleys, stream beds, and on hillsides. Based on dating of ceramics found on the terraces, he suggested that these were primarily constructed during the Roman (63 BCE–324 CE) and Byzantine (324–628 CE) periods. Ron (1966) demonstrated that 56.4% of the slopes of the Jerusalem hills are covered with ancient agricultural terraces (Fig. 5.4d). However, OSL dating conducted during the last decade has demonstrated that the terraces date back only ~ 800–300 years and that most of them were built during the Ottoman period (~ 500–300 years ago) (Davidovich et al. 2012; Gadot et al. 2016). These constructions might be considered as the most significant geomorphological alternations to the local landscape on the eve of the Industrial Revolution.

Specially constructed irrigation systems (pools, channels) leading to stone-walled agricultural plots containing built agricultural terraces have also been documented in the extreme arid area of the oasis of Ein Gedi in the Judean Desert (Hadas 2012). These installations have been dated to the first–sixth centuries CE. Further to the south, in the Negev Highlands, hundreds of agricultural systems comprising cisterns, built water channels, and pools were surveyed and documented in an area of approximately 300 km^2 and dated, using ceramics and OSL, to the Byzantine and the Early Islamic periods (eighth–eleventh centuries CE) (Haiman 1995a, b; Avni et al. 2013). In this hyper-arid zone, with less than 100 mm of annual rainfall, runoff farming systems were based on innovative technologies for water harvesting (Avni 2022), including constructed lines comprising small heaps of rocks known as "*tuleilat el-'anab*" that enhance runoff generation, in addition to low stone-built channels that were constructed to increase the effectiveness of runoff transportation from the slopes to the agricultural fields in the valleys below (Fig. 5.11a, b; Lavee et al. 1997; Bruins and Ore 2008). When these runoff systems were active, the landscape would have been totally different from the current one and included tens to hundreds of small orchards and vineyards in which were cultivated grapes, figs, olives, carobs, almond/apricot, pomegranates, date palms and also the exotic hazelnut, in addition to Mediterranean tree species for ornamentation (Tepper et al. 2020; Langgut et al. 2021). Thanks to the water that accumulates from old runoff harvesting systems, more than a hundred fruit trees, planted by Bedouin in the Negev Highlands in the early twentieth century, have survived without human maintenance (Ashkenazi et al. 2015).

5.6 Pollen Trends

Pollen records from the Dead Sea have revealed trends in the vegetation of the region over time. For example, evidence shows that cultivated olive trees were present in the Southern Levant as far back as the Early Bronze Age (~ 3800–2000 BCE) and continued to be a significant cultivar through to the end of the Iron Age. Low frequencies

Fig. 5.11 **a** Runoff farm in the Negev Highlands, near the ancient city of Avdat. note: Crops growing in the terrace fields and the slope water conduits with lines of dry annuals. **b** "*Tuleilat el-'anab*", top view (photos by Oren Ackermann)

of arboreal pollen—representing natural/wild tree cover—during the Iron Age (1000–500 BCE), has been interpreted as reflecting deforestation of the natural arboreal cover to make way for olive cultivation (Finkelstein et al. 2022).

Comparison of pollen assemblage composition of subsequent periods, from before and after the Roman-Byzantine period, revealed that human activity caused major changes to the species composition of oak trees. *Quercus ithaburensis* and *Quercus boisseri* were replaced by *Quercus calliprinos*; in an interim stage there was a prevalence of pines (*Pinus* sp.), perhaps as a result of its capacity for speedy reproduction. Another finding arising from the analysis of pollen graphs (Neumann et al. 2007, 2010) is the appearance of thorny burnet, *Sarcopoterium spinosum.* It seems that when there was a decrease in human activity and fields were abandoned (following the Byzantine period), this plant spread and became dominant. This conclusion also arises from observations made in abandoned terraces, where *Sarcopoterium spinosum* reaches coverage levels of up to 80% of the landscape (Fig. 5.14a; Ackermann et al. 2004, 2013). Terrace abandonment leads to the collapse of retaining walls and erosion of soil and sediments from the terrace body (Zgaier and Inbar 2005; Avni 2005; Avni et al. 2013), with the latter occurring at a high rate in the arid Negev highlands (Ackermann et al. 2017). However, in the Judean Hills and the Judean Foothills (Shephelah), areas under sub-humid Mediterranean and semi-arid conditions, the intensity of erosion is lower, since the terraces are covered by dense shrubs such as *Sarcopoterium spinosum.* This plant cover stabilizes and moderates soil erosion (Ackermann et al. 2013, 2017). This regional difference shows that the absence of terrace maintenance is more significant in arid areas, which are more sensitive to erosion, while in more humid areas, dense plant recovery stabilizes the surface and conserves the soil and sediments within the terrace.

Fig. 5.12 The irrigation system at Ein Uzi (the spring of Uzi) in the Jerusalem Hills (photo by Oren Ackermann)

The transition from the Byzantine period to the Early Islamic period (628–1099 CE) was characterized by a significant decrease in population size and abandonment of many sites in Israel. It was also expressed in a significant drop in the frequency of olive pollen while the percentage of arboreal pollen of natural/wild trees increased (Neumann et al. 2007, 2010). The renewal of natural trees was expressed first by an increase in the pollen values of pine (*Pinus*) and later by an increase in the prevalence of oak, *Quercus calliprinos*. The latter peaked during the Mamluk and the beginning of the Ottoman period (~ 1517–1917 CE), reaching an extremely high frequency compared with the previous, pre-Roman periods. In the current Israeli landscape, many archaeological sites are covered by regenerated Mediterranean vegetation, as well as remnants of agricultural vegetation (Fig. 5.14a–c).

As briefly outlined in this section, the region has witnessed an ever-increasing scale of tension between people and their natural surroundings since the first steps toward urbanization. As illustrated in Fig. 5.3, environmental and physical changes track the shift from a natural environment with minimal human interference culminating in a total human ecosystem and landscape.

5.6 From Extractive Production to a Total Anthropogenic Landscape and Ecosystem

Today, people are facing unprecedented and permanent changes to every aspect of their biotic and abiotic environment as a result of human domination over natural systems that have led to immense alterations to the Earth's functioning (Tarolli et al. 2014). For example, industrialization has led to pollution of the air, water, and soil as well as temperature rise, which has been the source of droughts, floods, and sea-level rise (Steffen et al. 2004, 2015; Braje and Erlandson 2013). Urbanization and overdevelopment have led to landscape-scale transformations such as deforestation while hunting, domestication and the translocation of plants and animals has resulted in their extirpation or extinction and the replacement of natural wild forms with homogenized cultivated crops and domestic animals (Vellend et al. 2017; Andermann et al. 2020). Consequently, today, the biomass of domestic animals outweighs that of wild animals, while human mass is an order of magnitude higher than that of all wild mammals combined (Bar-On et al. 2018). Moreover, disproportionate human population growth has resulted in overcrowding, ill health, and starvation in many parts of the globe (Butler 2016). Finally, the weight of global human-made mass far exceeds that of global living mass (Elhacham et al. 2020). These are the defining qualities of the Anthropocene, our current geological epoch (Ellis 2018). To fully understand our present circumstance, we need to examine its roots which lie in the deep history of human actions, the Paleo-Anthropocene. With reference to selected archaeological examples, in this chapter, we have tracked the increasingly complex and intensive niche construction behaviors of peoples who inhabited Israel in the past. Human-driven activities have culminated in extensive ecosystem engineering as well as widespread and dramatic modifications to the physical platform.

The present-day landscape of Israel comprises extensive urban and agricultural areas with natural, open zones in enclaves on their margins or between them. The open areas, which are considered natural, still retain many features that were amassed over close to one and half million years of human impacts, and as such, are essentially also anthropogenic in makeup. Thus today, there are no pristine remnants of the landscape or ecosystems that existed ~ 1.5 Ma years ago.

The first step in the process of natural bio-ecosystem replacement occurred during the Lower and Middle Paleolithic and was the longest period of hominin activity in the Israeli landscape, beginning ~ 1.5 million years ago. Although not devoid of evolution in technology and lifestyle,

Fig. 5.13 Byzantine wine press inside a building in the southern Negev Desert (photo by Davida Eisenberg-Degen and Tali Erickson-Gini, courtesy of the Israel Antiquities Authority)

which reflect the ever-evolving hominin cognitive abilities, this stage was chiefly characterized by simple extractive production of both raw materials and dietary resources. Extractive activities were local and small-scale, having only a negligible impact on the eco-landscape system.

In our view, the first significant milestone in the Paleo-Anthropocene in this region occurred some 14,500 years ago during the Early Natufian (Late Epipaleolithic). This period witnessed the first indications of human niche construction that are associated with sedentism, increased settlement size, population growth, and a shift in the mode of subsistence and diet (the "Broad Spectrum Revolution" including intensive cereal processing). However, over 3000 years passed after this milestone, before there was evidence of further change in the intensity, magnitude, and rate of human niche construction. This second milestone dates to the mid Pre-Pottery Neolithic B (~ 8000–6900 BCE) and follows the advent of domesticated animals and plants during the so-called "Neolithic Revolution". Far reaching changes are evidenced in the landscape at this time associated with the establishment of cultivated fields, grazing regimes for domestic herds, large sedentary villages, and toward the end of the Neolithic, nomadic, or semi-nomadic pastoralism. These new and ever-changing anthropogenic interventions meant that the natural ecosystem had to constantly adapt to human management, leading to co-evolution (Perevolotsky 2005; Perevolotsky et al. 2013). This dynamic laid the basis for the current ecosystem and landscape of Israel.

Subsequent settlement expansion and demographic growth with the establishment of complex urban communities, beginning in the Early Bronze Age II (~3000 BCE), witnessed cycles of colonization, warfare, and conquest, as well as periods of local self-rule. This epoch was characterized by extensive human-mediated landscape modification that included monumental architectural construction, the development of the first large-scale industrial economies, widespread local and international transport and trade networks, as well as the fragmentation of natural habitats and destruction of indigenous vegetation, advancement of large-scale animal husbandry, decimation of wild animals as well as the creation of extensive agro-industries engaged in growing and exporting crops, notably olives for oil, grapes for wine and, in later periods, cotton. These invasive acts

Fig. 5.14 Anthropogenic vegetation establishment and pattern distribution **a** Dense cover of *Sarcopoterium spinosum* L. on an abandoned agricultural terrace in Ramat Beit Shemesh, Shephelah (photo by Oren Ackermann). **b** Horvat Tzunam in the Upper Galilee, covered by dense Mediterranean vegetation known as maquis (photo by Oren Ackermann). **c** A fig tree (*Ficus carica*) grows attached to spring sediments at Ein Kefira (spring of Kefira), near Jerusalem (photo by Oren Ackermann)

culminated in the Islamic period, with the construction of agricultural terraces on more than half of the hill slopes in the Mediterranean phytogeographic zone. This can be considered the last significant, large-scale act marking the Israeli Paleo-Anthropocene.

In the nineteenth and beginning of the twentieth centuries in Israel, the industrialization process of the economy began in earnest. There is evidence for the introduction of modern technological concepts and practice in the fields of construction, agriculture and settlement planning, development of infrastructure for roads, railways, and ports, while the few surviving large wild animals, such as the red deer (*Cervus elaphus*), fallow deer (*Dama mesopotamica mesopotamica*), brown bear (*Ursus syriacus*), and crocodile (*Crocodylus niloticus*), were extirpated following the introduction of firearms (e.g., Yom-Tov and Mendelssohn 1988; Kark 1989, 1995; Bar-Oz and Weissbrod 2017). However, it is the sum total of past activities that began over 1.5 million years ago that have molded and designed the current terrain's physical structure and its biological content. Consequently, today, Israel's natural areas should be considered a total anthropogenic ecosystem and landscape since there is no aspect that has been left untouched by human activities. This process can be observed in the early nineteenth century photograph illustrated at the beginning of the article, showing the first residents of Tel Aviv at the distribution of plots of land for the construction of the first houses, versus that of today. The residents stand in the heart of sand dunes that have since disappeared under the cement, asphalt and glass constructions of the city.

Acknowledgements The authors wish to extend their warm thanks to: Jay Rosenberg for drawing Fig 5.1; Yael Friedman-Levi for drawing Fig 5.3; Kerem Ackermann for providing Fig. 5.2c, Ehud Galili for providing Fig. 5.7a, Yosef Garfinkel for providing Fig. 5.8a, Eli Ashkenazi for providing Fig. 5.10, the Israel Antiquities Authority for providing Figs. 5.6 and 5.13 and the editors -Amos Frumkin, Piotr Migoń, Nurit Shtober-Zisu for their useful comments on an earlier version of the text.

Appendix: General Chronology Table

Period	Phase	Date
Paleolithic	Lower Paleolithic	2,000,000–250,000 BP
	Middle Paleolithic	250,000–48,000 BCE
	Upper Paleolithic	48,000–18,000 BCE
	Epipalaeolithic	18,000–10,500 BCE
Neolithic	Pre-Pottery Neolithic	10,500–6400 BCE
	Pottery Neolithic	6400–5200 BCE
Chalcolithic	Chalcolithic	5200–3800 BCE
Bronze Age	Early Bronze Age	3800–2000 BCE
	Middle Bronze Age	2000–1550 BCE
	Late Bronze Age	1550–1200 BCE
Iron Age		1200–586 BCE
Babylonian & Persian		586–332 BCE
Hellenistic		332–63 BCE
Roman		63 BCE–324 CE
Byzantine		324–628 CE
Islamic		628–1517 CE
Crusader		1099–1291 CE
Ottoman		1517–1917 CE
Modern		1917–current

Source Wikipedia, List of archaeological periods (Levant) Retrieved: May 13, 2023. https://en.wikipedia.org/wiki/List_of_archaeological_periods_(Levant)

References

Abbo S, Gopher A (2017) Near Eastern plant domestication: a history of thought. Trends Plant Sci 22(6):491–511

Abbo S, Gopher A, Kahila Bar-Gal G (2022) Plant Domestication and the Origins of Agriculture in the Ancient Near East. Cambridge University Press, Cambridge

Ackerfeld D, Abadi-Reiss Y, Yagel O, Harlavan Y, Abulafia T, Yegorov D, Ben-Yosef E (2020) Firing up the furnace: new insights on metallurgical practices in the Chalcolithic Southern Levant from a recently discovered copper-smelting workshop at Horvat Beter (Israel). J Archaeol Sci: Rep 33:102578. https://doi.org/10.1016/j.jasrep.2020.102578

Ackermann O (2007) Reading the field: geoarchaeological codes in the Israeli landscape. Isr J Earth Sci 56:87–106

Ackermann O, Maeir AM, Bruins HJ (2004) Unique human-made catenary changes and their effect on soil and vegetation in the semi-arid Mediterranean zone: a case study on *Sarcopoterium spinosum* distribution near Tell es-Safi/Gath, Israel. CATENA 57:309–330

Ackermann O, Greenbaum N, Bruins HJ, Bar-Matthews M, Ayalon A, Almogi-Labin A, Schilman B, Maeir AM (2012) The environmental and paleogeographical history of the Tell es Safi/Gath over the past four millennia. In: Maeir AM (ed) Tell es-Safi/Gath I: report on the 1996–2005 seasons. Ägypten und Alten Testament 69. Harrassowitz, Wiesbaden, pp 531–556

Ackermann O, Zhevelev HM, Svoray T (2013) *Sarcopoterium spinosum* from mosaic structure to matrix structure: impact of calcrete (Nari) on vegetation in a Mediterranean semi-arid landscape. CATENA 101:79–91

Ackermann O, Greenbaum N, Bruins HJ, Porat N, Bar-Matthews M, Almogi-Labin A, Schilman B, Ayalon A, Horwitz LK, Weiss E Maeir AM (2014a) Palaeoenvironment and Anthropogenic activity in the southeastern Mediterranean since the mid-Holocene: the case of Tell es-Safi/Gath, Israel. Quat Int 328–329:226–243

Ackermann O, Greenbaum N, Ayalon A, Bar-Matthews M, Boaretto E, Bruins HJ, Cabanes D, Horwitz LK, Neumann F, Porat N, Weiss E, Maeir AM (2014b) Using palaeo-environmental proxies to reconstruct natural and anthropogenic controls on sedimentation rates, Tell es-Safi/Gath, Eastern Mediterranean. Anthropocene 8:70–82

Ackermann O, Maeir AM, Frumin S, Svoray T, Weiss E, Zhevelev HM, Horwitz LK (2017) The paleo-anthropocene and the genesis of the current landscape of Israel. J Landscape Ecol 10(3):109–140

Albert RM, Bar-Yosef O, Meignen L, Weiner S (2003) Quantitative phytolith study of hearths from the Natufian and Middle Palaeolithic levels of Hayonim Cave (Galilee, Israel). J Archaeoll Sci 30(4):461–480

Allred L (2009) Enchanted Rock: A Natural and Human History. University of Texas Press, Austin

Alperson-Afil N, Goren-Inbar N (2010) The Acheulian site of Gesher Benot Ya'aqov volume II: ancient flames and controlled use of fire. Springer Science & Business Media

Andermann T, Faurby S, Turvey ST, Antonelli A, Silvestro D (2020) The past and future human impact on mammalian diversity. Sci Adv 6(36):eabb2313

Ashkenazi E, Avni Y, Avni G (2012) A comprehensive characterization of ancient desert agricultural systems in the Negev Highlands of Israel. J Arid Environ 86:55–64

Ashkenazi E, Chen Y, Avni Y, Lavee S (2015) Fruit trees' survival ability in an arid desert environment without irrigation in the Negev Highlands of Southern Israel. Israel J Plant Sci 62(1–2):5–16

Ashkenazi E, Tepper Y, Zituni R, Langgut D, Dafni A, Bar-Oz G (2021) Poplar trees in Israel's desert regions: relicts of Roman and Byzantine settlement. J Arid Environ 193:104574

Avner U, Magness J (1998) Early Islamic settlement in the southern Negev. BASOR 310(1):39–57

Avni Y (2005) Gully incision as a key factor in desertification in an arid environment, the Negev highlands, Israel. CATENA 63(2):185–220

Avni Y (2022) The emergence of terrace farming in the arid zone of the Levant—past perspectives and future implications. Land 11(10):1798

Avni G, Porat N, Avn Y (2013) Byzantine-early Islamic agricultural systems in the Negev Highlands: stages of development as interpreted through OSL dating. J Field Archaeol 38(4):332–346

Barash A, Belmaker M, Bastir M, Soudack M, O'Brien HD, Woodward H, Prendergast A, Barzilai O, Been E (2022) The earliest Pleistocene record of a large-bodied hominin from the Levant supports two out-of-Africa dispersal events. Sci Rep 12:1721

Barkai R, Gopher A (2009) Changing the face of the earth: human behavior at Sede Ilan, an extensive Lower–Middle Paleolithic quarry site in Israel. In: Adams B, Blades BS (eds) Lithic Materials and Paleolithic societies. Wiley-Blackwell, pp 174–185

Barnofsky AD, Lindsey E (2010) Timing of Quaternary megafaunal extinction in South America in relation to human arrival and climate change. Quat Int 217(1–2):10–29

Bar-On YM, Phillips R, Milo R (2018) The biomass distribution on Earth. PNAS 115(25):6506–6511

Bar-Oz G, Weissbrod L (2017) The kaleidoscope of mammalian faunas during the Terminal Pleistocene and Holocene in the Southern Levant. In: Enzel Y, Bar-Yosef O (eds) Quaternary of the levant: environments, climate change, and humans. Cambridge University Press, pp 363–368
Bar-Yosef O (1994) The Lower Paleolithic of the Near East. J World Prehist 8(3):211–265
Bar-Yosef O (1998) The Natufian culture in the Levant, threshold to the origins of agriculture. Evol Anthropol 6(5):159–177
Bar-Yosef O, Belfer-Cohen A (1989) The origins of sedentism and farming communities in the Levant. J World Prehist 3(4):447–498
Bar-Yosef O, Belfer-Cohen A (2002) Facing environmental crisis. Societal and cultural changes at the transition from the Younger Dryas to the Holocene in the Levant. In: Cappers RTJ, Bottema S (eds) The dawn of farming in the Near East. Studies in Early Near Eastern Production, Subsistence and Environment 6. Ex oriente, Berlin, pp 55–66
Bar-Yosef O, Belmaker M (2017) Ubeidiya. In Enzel Y, Bar-Yosef O (eds) Quaternary of the Levant: Environments, Climate Change, and Humans. Cambridge University Press, pp 179–185
Belfer-Cohen A (1991) The Natufian in the Levant. Ann Rev Anthropol 20:167–186
Belfer-Cohen A, Hovers E (2005) The ground stone assemblages of the Natufian and Neolithic societies in the Levant—a brief review. J Isr Prehist Soc 35:299–308
Ben-Michael J (2016) The story of the ceramic industry in the Southern Levant. In: Ganor S, Kreimerman I, Streit K, Mumcuoglu M (eds) From Sha'ar Hagolan to Shaaraim Essays in Honor of Prof. Yosef Garijinkel. Israel Exploration Society, Jerusalem, pp 1–18
Ben-Yosef E (ed) (2018) Mining for ancient copper. Essays in memory of Beno Rothenberg. MSSMNIA 37. Emery and Claire Yass. Publications in Archaeology, Winona Lake, IN, Eisenbrauns and Tel Aviv
Ben-Yosef E, Tauxe L, Ron H, Agnon A, Avner U, Najjar M, Levy TE (2008) A new approach for geomagnetic archaeointensity research: insights on ancient metallurgy in the Southern Levant. J Archaeol Sci 35(11):2863–2879
Biton R, Bailon S, Birkenfeld M, Bridault A, Khalaily H, Valla FR, Rabinovich R (2021) The anurans and squamates assemblage from Final Natufian Eynan (Ain Mallaha, Israel) with an emphasis on snake-human interactions. PLoS ONE 16(2):e0247283
Boyd B (2006) On 'sedentism' in the Later Epipalaeolithic (Natufian) Levant. World Archaeol 38(2):164–178
Braje TJ, Erlandson JM (2013) Looking forward, looking back: humans, anthropogenic change, and the Anthropocene. Anthropocene 4:116–121
Bruins HJ, Ore G (2008) Runoff from loess or bedrock? Hillslope geoarchaeology of ancient runoff farming systems at Horvat Haluqim and Har Eldad in the central Negev Desert. Israel J Earth Sci 57(3):231–247
Bruins HJ, Jongmans T, van der Plicht J (2020) Ancient runoff farming and soil aggradation in terraced wadi fields (Negev, Israel): obliteration of sedimentary strata by ants, scorpions and humans. Quat Int 545:87–101
Butler CD (2016) Sounding the alarm: health in the Anthropocene. Int J Environ Res Public Health 13(7):665
Caracuta V, Barzilai O, Khalaily H, Milevski I, Paz Y, Vardi J, Regev L, Boaretto E (2015) The onset of faba bean farming in the Southern Levant. Sci Rep 5(1):1–9
Childe VG (1951) Man makes himself. The New American Library of World Literature, New York
Chocron M, Ackermann O, Zissu B (2019) Byzantine winepresses between the sown and the desert–from The Hebron Mountain to Southern Coastal Plain. Judea and Samaria 28(1):117–137 (in Hebrew with English abstract * 60)
Crutzen PJ (2002a) The "Anthropocene." J Phys 4(12):1–5
Crutzen PJ (2002b) Geology of mankind. Nature 415:23
Crutzen PJ, Stoermer EF (2000) The "Anthropocene." Glob Change Newsl 41:17–18
Dagan A, Eniukhina M, Maeir AM (2018) Excavations in Area D of the lower city: Philistine cultic remains and other finds. Near Eastern Archaeology 81(1):28–33
Dagan Y (2006) Map of Amazya (109). Archaeological Survey of Israel. Israel Antiquities Authority, Jerusalem
Dagan Y (2011) Agricultural activity in the Ramat Bet Shemesh Region. In: Dagan Y (ed) The Ramat Bet Shemesh Regional Project: Landscapes of Settlement from the Paleolithic to the Ottoman Period, IAA Reports No. 47, Jerusalem, Israel Antiquites Authority, Jerusalem, pp 319–340
Daly K, Maisano Delser P, Mullin VE, Scheul A, Mattiangeli V, Teasdale MD, Hare AJ, Burger J, Collins MJ, Verdugo M, Kehati R, Merih Erek C, Bar-Oz G, Pompanon F, Cumer T, Cakirlar C, Mohaseb AF, Decruyenaere D, Davoudi H, Çevik Ö, Vigne J-D, Khazaeli R, Fathi H, Beizaeedoust S, Rahimi Sorkhani R, Vahdati AA, Sauer EW, Maziar HAKS, Pinhasi R, Martin L, Orton D, Arbuckle BS, Benecke N, Manica A, Horwitz LK, Mashkour M, Gasparian B, Bradley DG (2018) Ancient goat genomes reveal mosaic domestication in the Fertile Crescent. Science 361:85–88
Danin A (1988) Flora and vegetation of Israel. In Yom Tov Y, Tchernov E (eds) The Zoogeography of Israel. Dr. W. Junk Publishers. Dordrecht, Netherlands, pp 129–157
Davidovich U, Porat N, Gadot Y, Avni Y, Lipschits O (2012) Archaeological investigations and OSL dating of terraces at Ramat Rahel, Israel. J Field Archaeol 37(3):192–208
Davis SJ (1982) Climatic change and the advent of domestication: the succession of ruminant artiodactyls in the late Pleistocene-Holocene in the Israel region. Paléorient 8:5–15
Eisenberg-Degen D (2018) Tel 'Irit. Hadashot Arkheologiyot: Excavations and Surveys in Israel, 130. Retrieved 5 June, 2921. https://www.jstor.org/stable/26691755
Eisenberg-Degen D, Varga D, Sapir T (2022) Sub/semi-Subterranean complexes in Byzantine Beer Sheva, Negev, Israel, date, use and typology. Athens J Hist 8(3):175–190
Eitam D (2009) Cereals in the Ghassulian culture in Central Israel: grinding installations as a case study. Israel Explor J 59(1):63–79
Elad I, Paz Y, Shalem D (2018) 'En Esur (Asawir), Area M. Hadashot Arkheologiyot: Excavations and Surveys in Israel, 130. Retrieved June 19, 2021, from https://www.jstor.org/stable/26691672
Elhacham E, Ben-Uri L, Grozovski J, Bar-On YM, Milo R (2020) Global human-made mass exceeds all living biomass. Nature 588:442–444
Eliyahu-Behar A, Yahalom-Mack N, Shilstein S, Zukerman A, Shafer-Elliott C, Maeir AM, Boaretto E, Finkelstein I, Weiner S (2012) Iron and bronze production in Iron Age IIA Philistia: new evidence from Tell es-Safi/Gath, Israel. J Archaeol Sci 39(2):255–267
Eliyahu-Behar A, Yahalom-Mack N, Gadot Y, Finkelstein I (2013) Iron smelting and smithing in major urban centers in Israel during the Iron Age. J Archaeol Sci 40(12):4319–4330
Ellis EC (2018) Anthropocene. A very short introduction. Oxford University Press, Oxford
Ellis EC, Gauthier N, Goldewijk KK, Bird RB, Boivin N, Díaz S, Fuller DQ, Gill JL, Kaplan JO, Kingston N, Locke H, McMichael CNH, Ranco D, Torben CR, Shaw MR, Stephens L, Svenning JC, Watson JE (2021) People have shaped most of terrestrial nature for at least 12,000 years. PNAS 118(17):e2023483118
Enzel Y, Bar-Yosef O (eds) (2017) Quaternary of the Levant: environments, climate change, and humans. Cambridge University Press, Cambridge
Eshed V, Gopher A, Pinhasi R, Hershkovitz I (2010) Paleopathology and the origin of agriculture in the Levant. Am J Phys Anthropol 143(1):121–133

Eshel T, Yahalom-Mack N, Tirosh O, Maeir AM, Harlavan Y, Gilboa A, Erel Y (2020) Pollution and human mobility in the southern Levant during the Iron Age using chemical and isotopic analysis of human tooth enamel. J Archaeol Sci 124:105262

Feldman M, Kislev ME (2007) Domestication of emmer wheat and evolution of free-threshing tetraploid wheat. Israel J Plant Sci 55(3–4):207–221

Filin S, Miller V, Rosenberg D, Nadel D (2017) Intra-and inter-site high-resolution geometrical analyses of Natufian bedrock features. Quat Int 439:94–107

Finkel M, Gopher A, Ben-Yosef E, Barkai R (2018) A Middle Paleolithic and Neolithic/Chalcolithic flint extraction and reduction complex at Mt. Achbara, Eastern Galilee, Israel. Archaeol Res Asia 16:14–33

Finkelstein I (2011) Observations on the layout of Iron Age Samaria. Tel Aviv 38(2):194–207

Finkelstein I, Gadot Y, Langgut D (2022) The unique specialised economy of Judah under Assyrian rule and its impact on the material culture of the kingdom. Palest Explor Q 154(4):261–279

Flannery KV (2008) Origins and ecological effects of early domestication in Iran and the Near East. In: Dimbleby GW (ed) The domestication and exploitation of plants and animals. Routledge, pp 73–100

Foley SF, Gronenborn D, Andreae MO, Kadereit JW, Esper J, Scholz D, Pöschl U, Jacob DE, Schöne BR, Schreg R, Vött A, Jordan D, Lelieveld J, Weller CG, Alt KW, Gaudzinski-Windheuser S, Bruhn KC, Tost H, Sirocko F, Crutzen PJ (2013) The Palaeoanthropocene—the beginnings of anthropogenic environmental change. Anthropocene 3:83–88

Frankel R (2009) Introduction. In: Ayalon E, Frankel R, Kloner A (eds) Oil and wine presses in Israel from the Hellenistic, Roman and Byzantine Periods. British Archaeological Reports International Series, Oxford, 1972, pp 1–18

Frumin S, Maeir AM, Horwitz LK, Weiss E (2015) An invasion biology perspective on an ancient human migration: the Philistines as a case study. Sci Rep 5:13308

Gadot Y, Davidovich U, Avni G, Avni Y, Piasetzky M, Faershtein G, Golan D. Porat N (2016) The formation of a Mediterranean terraced landscape: Mount Eitan, Judean Highlands, Israel. J Archaeol Sci: Rep 6:397–417

Galili E, Sharvit J (1998) Submerged Neolithic water-wells from the Carmel Coast of Israel. In: Koschik H (ed) Brunnen der Jungsteinzeit, Internationales Symposium Erkenlenz 27. bis 29. Oktober 1997. Rheinland-Verlag, Köln, pp 31–44

Galili E, Rosen B (2011) Submerged Neolithic settlements off the Mediterranean coast of Israel. In: Benjamin J, Bonsall C, Pickard C, Fischer A (eds) Submerged Prehistory. Oxbow, Oxford, pp 272–286

Galili E, Weinstein-Evron M, Chaim S, Cvikel D, Benjamin J, McCarthy J, Langgut D, Cavanagh D, Sapir Y, Rosen B, Horwitz LK (2018) The archaeology and paleoenvironment of the submerged Pottery Neolithic settlement of Kfar Samir (Israel). Paléorient 44(2):113–132

Galili E, Benjamin J, Eshed V, Rosen B, McCarthy J, Horwitz LK (2019) A submerged 7000-year-old village and seawall demonstrate earliest known coastal defence against sea-level rise. PLoS ONE 14(12):e0222560

Garfinkel Y, Vered A, Bar-Yosef O (2006) The domestication of water: the Neolithic well at Sha'ar Hagolan, Jordan Valley, Israel. Antiquity 80(309):686–696

Gaudzinski S (2004) Subsistence patterns of Early Pleistocene hominids in the Levant—taphonomic evidence from the 'Ubeidiya Formation (Israel). J Archaeol Sci 31:65–75

Gibbard P, Walker M, Bauer A, Edgeworth M, Edwards L, Ellis E, Finney S, Gill JL, Maslin M, Merritts D, Ruddiman W (2022) The Anthropocene as an Event, not an Epoch. J Quat Sci 37(3):395–399

Golden J, Levy TE, Hauptmann A (2001) Recent discoveries concerning Chalcolithic metallurgy at Shiqmim, Israel. J Archaeol Sci 28(9):951–963

Gopher A, Barkai R (2011) Sitting on the tailing piles: creating extraction landscapes in Middle Pleistocene quarry complexes in the Levant. World Archaeol 43(2):211–229

Gopher A, Barkai R (2014) Middle Paleolithic open-air industrial areas in the Galilee, Israel: the challenging study of flint extraction and reduction complexes. Quat Int 331:95–102

Goren Y (1991) The Beginnings of pottery production in Israel: Technology and typology of proto-historic ceramic assemblages in Eretz Israel (6th–4th millennium cal. BCE). Unpublished Ph.D. Dissertation. The Hebrew University (in Hebrew), Jerusalem

Goren Y, Goring-Morris AN (2008) Early pyrotechnology in the Near East: experimental lime-plaster production at the Pre-Pottery Neolithic B site of Kfar HaHoresh, Israel. Geoarchaeology 23(6):779–798

Goren Y, Biton R (2010) Technology of the fired clay objects from Gilgal I. Gilgal: Early Neolithic Occupation in the Lower Jordan Valley. The Excavations of Tamar Noy. ASPR & David Brown/ Oxbow, Oakville, CT, pp 217–221

Goren-Inbar N, Werker E, Feibel CS (2002) The Acheulian Site of Gesher Benot Ya'aqov, Israel: The Wood Assemblage. Oxbow Books, Oxford

Goring-Morris N (1993) From foraging to herding in the Negev and Sinai: the Early to Late Neolithic transition. Paléorient 19(1):65–89

Goring-Morris N, Belfer-Cohen A (2003) Structures and dwellings in the Upper and Epi-Palaeolithic (ca. 42–10k BP) Levant: Profane and symbolic uses. BAR Int Ser 1122:65–82

Goring-Morris AN, Belfer-Cohen A (2011) Neolithization processes in the Levant. Curr Anthropol 52(S4):S195–S208

Goring-Morris AN, Belfer-Cohen A (2014) The Neolithic in the southern Levant: yet another 'unique' phenomenon. La transition Néolithique en Méditerranée (The Neolithic transition in the Mediterranean). Errance, France, pp 59–75

Goring-Morris AN, Belfer-Cohen A (2020) Highlighting the PPNB in the Southern Levant. Neo-Lithics 20:3–22

Gourichon L, Horwitz LK (2021) An inter-regional comparison of animal domestication in the Northern and Southern Levant. Food History 19(1–2):53–78

Greenberg R (2019) The Archaeology of the Bronze Age Levant: from urban origins to the Demise of city-states, 3700–1000 BCE. Cambridge University Press

Grosman L, Goren-Inbar N (2007) "Taming" rocks and changing landscapes. Curr Anthropol 48(5):732–740

Grosman L, Goren-Inbar N (2016) Landscape alteration by Pre-Pottery Neolithic communities in the Southern Levant–the Kaizer hilltop quarry, Israel. PLoS ONE 11(3):e0150395

Hadas G (2012) Ancient agricultural irrigation systems in the oasis of Ein Gedi, Dead Sea, Israel. J Arid Environ 86:75–81

Haiman M (1995a) Agriculture and Nomad-State relations in the Negev Desert in the Byzantine and Islamic periods. BASOR 297:29–53

Haiman M (1995b) An early Islamic period Farm at Nahal Mitnan in the Negev Highlands. Atiqot 36:1–13

Hajdas PK, Leinfelder I, McNeill R, Odada J, Poirier EO, Richter C, Steffen D, Summerhayes W Syvitski C, Vidas JPM, Wagreich D, Wing, SL, Wolfe AP, An Z, Oreskes N (2015) When did the Anthropocene begin? A mid-twentieth century boundary level is stratigraphically optimal. Quat Int 383:196–203

Hartmann-Shenkman A, Kislev ME, Galili E, Melamed Y, Weiss E (2015) Invading a new niche: obligatory weeds at Neolithic Atlit-Yam, Israel. Veg Hist Archaeobotany 24(1):9–18

Horowitz A (1979) The Quaternary of Israel. Academic Press, New York

Horwitz LK (1989) A reassessment of caprovine domestication in the Levantine Neolithic: old questions, new answers. In: Hershkovitz I (ed) People and Culture Change. BAR International Series 508 (i), Oxford, pp 153–181
Horwitz LK (2022) Dynamics of Early Bronze Age Faunal exploitation in the Sinai Peninsula: modes of production, commodity production and trade. In: Adams MJ, Roux V (eds) Transitions During the Early Bronze Age in the Levant: Methodological Problems and Interpretative Perspectives. Aegypten und Altes Testament, vol 109. Zaphon, pp 265–284
Horwitz LK, Smith P (2000) The contribution of animal domestication to the spread of zoonoses: a case study from the southern levant. Anthropozoologica 31:77–84
Horwitz LK, Studer J (2005) Pig production and exploitation during the Classical periods in the Southern Levant. In: Buitenhuis H, Choyke AM, Martin L, Bartosiewicz L, Mashkour M (eds) Archaeozoology Near East volume VI. ARC Publicaties, Groningen, pp 222–239
Horwitz LK, Tchernov E, Ducos P, Becker C, von den Driesch A, Martin L, Garrard A (1999) Animal domestication in the Southern Levant. Paleorient 25(2):63–80
Hovers E, Ilani S, Bar-Yosef O, Vandermeersch B (2003) An early case of color symbolism: ochre use by modern humans in Qafzeh Cave. Curr Anthropol 44(4):491–522
Janovský MP, Horák J, Ackermann O, Tavger A, Cassuto D, Šmejda L, Hejcman M, Anker Y, Shai I (2022) The contribution of POSL and PXRF to the discussion on sedimentary and site formation processes in archaeological contexts of the southern Levant and the interpretation of biblical strata at Tel Burna. Quat Int 618:24–34
Kafle HK, Bruins HJ (2009) Climatic trends in Israel 1970–2002: warmer and increasing aridity inland. Clim Change 96(1–2):63–77
Kaplan Y (1984) The ecosystem of the Yahudia Nature Reserve. Ph.D. Thesis, Agricultural University of Wageningen. Wageningen, The Netherlands
Kark R (1989) The Land that Became Israel: Studies in Historical Geography. Yale University Press
Kark R (1995) The introduction of modern technology into the Holy Land (1800–1914 CE). In: Levy TE (ed) The Archaeology of Society in the Holy Land. Leicester University Press, London, pp 524–541
Karkanas P, Shahack-Gross R, Ayalon A, Bar-Matthews M, Barkai R, Frumkin A, Gopher A, Stiner MC (2007) Evidence for habitual use of fire at the end of the Lower Paleolithic: site-formation processes at Qesem Cave, Israel. J Hum Evol 53(2):197–212
Killebrew A, Steiner M (eds) (2014) The Oxford Handbook of the Archaeology of the Levant. Oxford Handbooks in Archaeology. Oxford University Press, Oxford
Köhler-Rollefson I (1988) The aftermath of the levantine neolithic revolution in the light of ecological and ethnographic evidence. Paléorient 14(1):87–93
Kuijt I, Goring-Morris N (2002) Foraging, farming, and social complexity in the Pre-Pottery Neolithic of the southern Levant: a review and synthesis. J World Prehist 16(4):361–440
Langgut D (2017) The Citrus route revealed: from Southeast Asia into the Mediterranean. HortScience 52(6):814–822
Langgut D, Gadot Y, Porat N, Lipschits O (2013) Fossil pollen reveals the secrets of the Royal Persian Garden at Ramat Rahel, Jerusalem. Palynology 37(1):115–129
Langgut D, Tepper Y, Benzaquen M, Erickson-Gini T, Bar-Oz G (2021) Environment and horticulture in the Byzantine Negev Desert, Israel: sustainability, prosperity and enigmatic decline. Quat Int 593–594:160–177
Lev E, Kislev ME, Bar-Yosef O (2005) Mousterian vegetal food in Kebara cave, Mt. Carmel. J Archaeolog Sci 32(3):475–484
Lavee H, Poesen J, Yair A (1997) Evidence of high efficiency water-harvesting by ancient farmers in the Negev Desert, Israel. J Arid Environ 35(2):341–348
Levy TE (ed) (1995) The Archaeology of Society in the Holy Land. Leicester University Press, London
Lu Y, Waldmann N, Nadel D, Marco S (2017) Increased Sedimentation following the Neolithic Revolution in the Southern Levant. Glob Planet Change 152:199–208
Madella M, Jones MK, Goldberg P, Goren Y, Hovers E (2002) The exploitation of plant resources by Neanderthals in Amud Cave (Israel): the evidence from phytolith studies. J Archaeolog Sci 29(7):703–719
Maeir AM (ed) (2012) Tell es-Safi/Gath I: The 1996–2005 seasons. Ägypten und Altes Testament 69. Harrassowitz Verlag, Wiesbaden
Maeir AM, Ben-Shlomo D, Cassuto D, Chadwick JR, Davis B, Behar AE, Frumin S, Gur-Arieh S, Hitchcock L, Horwitz LK, Manclossi F, Rosen SA, Verduci J, Weiss E, Welch EL, Workman V (2019) Technological insights on Philistine culture: perspectives from Tell es-Safi/Gath. J Eastern Med Archaeol Heritage Stud 7(1):76–118
Maeir AM, Welch EL, Eniukhina M (2021) A note on olive oil production in Iron Age Philistia: pressing the consensus. PEQ 153 (2): 129-144
Mazar A (1990) Archaeology of the Land of the Bible, 10,000–586 B.C.E. Anchor Bible Reference Library, Doubleday
Meignen L, Bar-Yosef O, Goldberg P, Weiner S (2001) Le feu au Paléolithique moyen: recherches sur les structures de combustion et le statut des foyers: l'exemple du Proche-Orient. Paléorient 26:9–22
Meiri M, Huchon D, Bar-Oz G, Boaretto E, Horwitz LK, Maeir AM, Sapir-Hen L, Larson G, Weiner S, Finkelstein I (2013) Ancient DNA and population turnover in Southern Levantine Pigs-Signature of the Sea peoples migration? Nat—Sci Rep 3:3035
Melamed Y, Kislev ME, Geffen E, Lev-Yadun S, Goren-Inbar N (2016) The plant component of an Acheulian diet at Gesher Benot Ya aqov, Israel. PNAS 113(51):14674–14679
Munro ND, Bar-Oz G, Meier JS, Sapir-Hen L, Stiner MC, Yeshurun R (2018) The emergence of animal management in the Southern Levant. Sci Rep 8(1):1–11
Murphy-O'Connor J (2003) The Aqueducts of Israel. J Roman Archaeol Supplementary Ser 46:581–591
Nadel D, Lengyel G (2009) Human-made bedrock holes (mortars and cupmarks) as a Late Natufian social phenomenon. Archaeol Ethnol Anthropol Eurasia 37(2):37–48
Nadel D, Werker E (1999) The oldest ever brush hut plant remains from Ohalo II, Jordan Valley, Israel (19 BP). Antiquity 73(282):755–764
Nadel D, Piperno DR, Holst I, Snir A, Weiss E (2012) New evidence for the processing of wild cereal grains at Ohalo II, a 23,000-year-old campsite on the shore of the Sea of Galilee, Israel. Antiquity 86(334):990–1003
Nativ A, Gopher A, Goren Y (2012) Pottery production at Nahal Zehora II. In: Gopher A (ed) Village communities of the Pottery Neolithic period in the Menashe hills, Israel. Archaeological investigations at the sites of Nahal Zehora. Emery and Claire Yass Publications in Archaeology, Tel Aviv, pp 657–696
Naveh Z (1982) Landscape Ecology as an Emerging Branch of Human Ecosystem Science. Adv Ecol Res (UK) 12:189–237
Naveh Z (2004) Epilogue: toward a transdisciplinary science of ecological and cultural landscape restoration. Restor Ecol 13(1):228–234
Neumann F, Schölzel C, Litt T, Hense A, Stein M (2006) Holocene vegetation and climate history of the northern Golan heights (Near East). Veg Hist Archaeobotany 16(4):329–346
Neumann HF, Kagan EJ, Schwab MJ, Stein M (2007) Palynology, sedimentology and palaeoecology of the late Holocene Dead Sea. Quat Sci Rev 26:1476–1498
Neumann FH, Kagan EJ, Leroy SAG, Baruch U (2010) Vegetation history and climate fluctuations on a transect along the Dead Sea west shore and their impact on past societies over the last 3500 years. J Arid Environ 74(7):756–764

Ogloblin Ramírez I, Galili E, Be'eri R, Golan D, Krakovsky M, Dayan A, Shalem D, Shahack-Gross R (2020) Heated mud bricks in submerged and coastal Southern Levant Pre-Pottery Neolithic C and Late Pottery Neolithic/Early Chalcolithic settlements: diachronic changes in technology and their social implications. J Archaeol Sci: Rep 30:102220

Orendi A, Smejda L, McKinny C, Cassuto D, Sharp C, Shai I (2017) The agricultural landscape of Tel Burna: ecology and economy of a Bronze Age/iron Age settlement in the southern levant. J Landscape Ecol 10(3):165–188

Paz Y, Ackermann O, Avni Y, Ben-Hur M, Birkenfeld M, Langgut D, Mizrahi AS, Weiss E, Porat N (2017) An early Bronze Age fertilized agricultural plot discovered near Tel Yarmouth, Ramat Bet Shemesh, Israel. J Archaeol Sci Rep 15:226–234

Perevolotsky A (2005) Integrating landscape ecology in the conservation of Mediterranean ecosystems: the Israeli experience. Israel J Plant Sci 53(3–4):203–213

Perevolotsky A, Rosen B, Rosenberg D (2013) Man as an ecological super-engineer in the Mediterranean forest. In Perevolotsky A (ed) Conserving and managing Mediterranean ecosystems: the Ramat Hanadiv case study and beyond. Ramat Hanadiv, Zichron Yaakov(in Hebrew)

Prates L, Perez SI (2021) Late Pleistocene South American megafaunal extinctions associated with rise of Fishtail points and human population. Nat Commun 12:2175

Rabinovich R, Gaudzinski-Windheuser S, Kindler L, Goren-Inbar N (2012) The Acheulian Site of Gesher Benot Ya 'aqov. In: The Acheulian Site of Gesher Benot Ya 'aqov, vol III. Springer, Dordrecht, pp 3–14

Ramsay J, Tepper Y, Weinstein-Evron M, Aharonovich S, Liphschitz N, Marom N, Bar-Oz G (2016) For the birds—an environmental archaeological analysis of Byzantine pigeon towers at Shivta (Negev Desert, Israel). J Archaeol Sci: Rep 9:718–727

Rollefson GO, Köhler-Rollefson I (1992) Early Neolithic exploitation patterns in the Levant: cultural impact on the environment. Popul Environ 13:243–254

Rollefson GO (2020) Tumultuous times in the eighth and seventh millennia BC in the Southern Levant. In: Marciniak A (ed) Concluding the neolithic the near East in the second half of the seventh millennium BC. Lockwood Press, Georgia, pp 41–59

Ron Z (1966) Agricultural terraces in the Judean Mountains. IEJ 16:33–49

Rose T, Fabian P, Goren Y (2021) Shedding new light on the pure copper metallurgy of the Chalcolithic Southern Levant through an archaeological experiment. EXARC J 2021/4

Rosen MA (1986) Environmental change and settlement at Tel Lachish, Israel. BASOR 263:55–60

Rosen SA (2017) Revolutions in the Desert. Routledge, New York and London

Rosen SA, Goring-Morris AN (2018) Tabular scraper quarries: a view from Har Qeren in the Western Negev. J Isr Prehist Soc 48:82–96

Rosen SA, Schneider JS (2001) Early Bronze Age milling stone production and exchange in the Negev: preliminary conclusions. J Isr Prehist Soc 31:201–212

Rosenberg D, Love S, Hubbard E, Klimscha F (2020) 7200 years old constructions and mudbrick technology: the evidence from Tel Tsaf, Jordan Valley, Israel. PLoS ONE 15(1):e0227288

Rosenberg D, Gluhak TM, Kaufman D, Yeshurun R, Weinstein-Evron M (2021) Exploring exchange and direct procurement strategies for Natufian food processing tools of el-Wad Terrace, Israel. Sci Rep 11(1):1–8

Roskin J, Asscher Y, Khalaily H, Ackermann O, Vardi J (2022) The palaeoenvironment and the environmental impact of the Pre-Pottery Neolithic Motza megasite and its surrounding Mediterranean landscape in the central Judean Highlands (Israel). Mediterranean Geosci Rev 4(2):215–245

Sapir-Hen L, Bar-Oz G, Sharon I, Gilboa A, Dayan T (2012) Understanding faunal contexts of a complex Tell: Tel Dor, Israel, as a case study. J Archaeol Sci 39(3):590–601

Sapir-Hen L, Bar-Oz G, Sharon I, Gilboa A, Dayan T (2014) Food, economy, and culture at Tel Dor, Israel: a diachronic study of faunal remains from 15 centuries of occupation. BASOR 371(1):83–101

Schiebel V, Litt T (2018) Holocene vegetation history of the southern Levant based on a pollen record from Lake Kinneret (Sea of Galilee), Israel. Veg Hist Archaeobotany 27:577–590

Shahack-Gross R, Berna F, Karkanas P, Lemorini C, Gopher A, Barkai R (2014) Evidence for the repeated use of a central hearth at Middle Pleistocene (300 ky ago) Qesem Cave, Israel. J Archaeoll Sci 44:12–21

Shapiro A (2013) Petrographic examination of selected Pottery Vessels from Horbat Duvshan. Atiqot 73:37–39

Shimelmitz R, Kuhn SL, Jelinek AJ, Ronen A, Clark AE, Weinstein-Evron M (2014) 'Fire at will': the emergence of habitual fire use 350,000 years ago. J Hum Evol 77:196–203

Shtienberg G, Dix JK, Shahack-Gross R, Yasur-Landau A, Roskin J, Bookman R, Waldmann N, Shalev S, Sivan D (2017) Anthropogenic overprints on natural coastal aeolian sediments: a study from the periphery of ancient Caesarea, Israel. Anthropocene 19:22–34

Šmejda L, Hejcman M, Horak J, Shai I (2017a) Ancient settlement activities as important sources of nutrients (P, K, S, Zn and Cu) in Eastern Mediterranean ecosystems–the case of biblical Tel Burna, Israel. CATENA 156:62–73

Šmejda L, Hejcman M, Horak J, Shai I (2017b) Multi-element mapping of anthropogenically modified soils and sediments at the Bronze to Iron Ages site of Tel Burna in the southern Levant. Quatern Int 483:111–123

Smith BD (2007) Niche construction and the behavioral context of plant and animal domestication. Evol Anthropol: Issues, News, and Rev 16(5):188–199

Smith P, Horwitz LK (2007) Ancestors and inheritors: a bio-cultural perspective of the transition to agro-pastoralism in the Southern Levant. In: Cohen MN, Crane- Kramer GMM (eds) Ancient health. Skeletal indicators of agricultural and economic intensification. University Press of Florida, pp 207–222

Smith BD, Zeder MA (2013) The onset of the Anthropocene. Anthropocene 4:8–13

Snir A, Weiss E (2014) A novel morphometric method for differentiating wild and domesticated barley through intra-rachis measurements. J Archaeol Sci 44:69–75

Steffen W, Sanderson A, Tyson PD, Jäger J, Matson PA, Moore B III, Oldfield F (2004) Global change and the earth system: a planet under pressure. The IGBP Book Series. Springer, Berlin, Germany

Steffen W, Broadgate W, Deutsch L, Gaffney O, Ludwig C (2015) The trajectory of the Anthropocene: the great acceleration. The Anthropocene Rev 2(1):81–98

Stepka Z, Azuri I, Horwitz LK, Chazan M, Natalio F (2022) Hidden signatures of early fire at Evron Quarry (1.0–0.8 Mya). PNAS 119 (25):e2123439119

Stiner MC (2001) Thirty years on the "Broad Spectrum Revolution" and Paleolithic demography. PNAS 98(13):6993–6996

Tarolli P, Vanacker V, Middelkoop H, Brown AG (2014) Landscapes in the Anthropocene: state of the art and future directions. Anthropocene 6:1–2

Taxel I, Sivan D, Bookman R, Roskin J (2018) An Early Islamic inter-settlement agroecosystem in the coastal sand of the Yavneh Dunefield, Israel. J Field Archaeol 43(7):551–569

Tchernov E (1975) Rodent Faunas and environmental changes in the Pleistocene of Israel. In: Prakash I, Ghosh, PK (eds) Rodents in desert environments. Monographiae Biologicae, vol 28. Springer, Dordrecht, pp 331–362

Tchernov E (1984) Commensal animals and human sedentism in the Middle East. Ani Archaeol 3:91–115
Tchernov E (ed) (1986) Les Mammiferes du Pleistocene Inferieur de la Vallee du Jourdain a Oueidiyeh. Association Paleorient, Paris
Tchernov E (1988a) The age of 'Ubeidiya Formation (Jordan Valley, Israel) and the earliest hominids in the Levant. Paleorient 14:63–65
Tchernov E (1988b) The biogeographical history of the southern Levant. In: Yom-Tov Y, Tchernov E (eds) The Zoogeography of Israel. Dr. W. Junk Publishers, Dordrecht, pp 159–250
Tepper Y, Porat N, Bar-Oz G (2020) Sustainable farming in the Roman-Byzantine period: dating an advanced agriculture system near the site of Shivta, Negev Desert, Israel. J Arid Environ 177:104134
Tepper Y, Porat N, Langgut D, Barazani O, Bajpai PK, Dag A, Ehrlich Y, Boaretto E, Bar-Oz G (2022) Relict olive trees at runoff agriculture remains in Wadi Zetan, Negev Desert, Israel. J Archaeol Sci Rep 41:103302
Toffolo MB, Ullman M, Caracuta V, Weiner S, Boaretto E (2017) A 10,400-year-old sunken lime kiln from the Early Pre-Pottery Neolithic B at the Nesher-Ramla quarry (el-Khirbe), Israel. J Archaeol Sci Rep 14:353–364
Tsafrir Y (1996) Some notes on the settlement and demography of Palestine in the Byzantine period: the archaeological evidence. In Seger JD (ed) Retrieving the past, essays on the archaeological research and methodology in honor of G. W. Van Beek. Eisenbrauns, Winona Lake, pp 269–283
Tsahar E, Izhaki I, Lev-Yadun S, Bar-Oz G (2009) Distribution and Extinction of Ungulates during the Holocene of the Southern Levant. PLoS ONE 4(4):e5316
Valla FR (2018) Sedentism, the "point of non return", and the Natufian issue. An historical perspective. Paléorient 44(1):19–33
van Asperen HL, Bor AMC, Sonneveld MPW, Bruins HJ, Lazarovitch N (2014) Properties of anthropogenic soils in ancient run-off capturing agricultural terraces in the Central Negev desert (Israel) and related effects of biochar and ash on crop growth. Plant Soil 374(1):779–792
van Bommel D, Bruins HJ, Lazarovitch N, van der Plicht J (2021) Effect of dung, ash and runoff water on wheat and barley grain sizes and stable isotope ratios: experimental studies in ancient desert agriculture (Negev, Israel). J Archaeol Sci Rep 39:103172
Vellend M, Baeten L, Becker-Scarpitta A, Boucher-Lalonde V, Mccune JL, Messier J, Myers-Smith I, Sax D (2017) Plant biodiversity change across scales during the Anthropocene. Annu Rev Plant Biol 68:563–586
Verdugo PM, Mullin VE, Scheu A, Mattiangeli V, Daly KG, Maisano Delser P, Hare AJ, Burger J, Collins MJ, Kehati R, Hesse P, Fulton D, Koukouli-Chrysanthaki C, Adamantios S, Kazantzis G, Kontopoulos I, Bulatovic J, Stojanović I, Benecke N, Linstädter J, Sablin M, Bendrey R, Gourichon L, Arbuckle BS, Mashkour M, Orton D, Horwitz LK, Teasdale MD, Bradley DG (2019) Ancient cattle genomics, origins and climate-linked turnover in the Fertile Crescent. Science 365(6449):173–176
Vigne JD, Gourichon L, Helmer D, Martin L, Peters J (2017) The beginning of animal domestication and husbandry in South west Asia. In Enzel Y, Bar-Yosef O (eds) Quaternary of the Levant: environments, climate change, and humans. Cambridge University Press, pp753–759
Weiberg E, Hughes RE, Finné M, Bonnier A, Kaplan JO (2019) Mediterranean land use systems from prehistory to antiquity: a case study from Peloponnese (Greece). J Land Use Sci 14(1):1–20
Weinstein-Evron M, Ilani S (1994) Provenance of ochre in the Natufian layers of el-Wad cave, Mount Carmel Israel. J Archaeol Sci 21:461–467
Weiss E, Wetterstrom W, Nadel D, Bar-Yosef O (2004) The broad spectrum revisited: evidence from plant remains. PNAS 101(26):9551–9555
Weiss E, Kislev ME, Hartmann A (2006) Autonomous cultivation before domestication. Science 312(5780):1608–1610
Weissbrod L, Bar-Oz G, Yeshurun R, Weinstein-Evron M (2012) Beyond fast and slow: the mole rat *Spalax ehrenbergi* (order Rodentia) as a test case for subsistence intensification of complex Natufian foragers in southwest Asia. Quat Int 264:4–16
Weissbrod L, Marshall FB, Valla FR, Khalaily H, Bar-Oz G, Auffray JC, Cucchi T (2017) Origins of house mice in ecological niches created by settled hunter-gatherers in the Levant 15,000 y ago. PNAS 114(16):4099–4104
Wilcox G (2007) The adoption of farming and the beginnings of the Neolithic in the Euphrates Valley: cereal exploitation between the 12th and 8th millennia cal BCE. In: Colledge S, Conolly J (eds) The origins and spread of domestic plants in Southwest Asia and Europe. Left Coast Press, Walnut Creek, pp 21–36
Workman V, Maeir AM, Dagan A, Regev J, Boaretto E, Eliyahu-Behar A (2020) An Iron IIA Iron and Bronze Workshop in the Lower City of Tell es-Safi/Gath. Tel Aviv 47(2):208–236
Wrangham R, Carmody R (2010) Human adaptation to the control of fire. Evol Anthropol: Issues, News, Rev 19(5):187–199
Yechezkel A, Frumkin A (2019) Spring tunnels in ancient Israel and Jerusalem vicinity: physical, geographical and human aspects. Horiz Geogr 96:154–180 (in Hebrew with English abstract)
Yeshurun R, Bar-Oz G, Weinstein-Evron M (2009) The role of foxes in the Natufian economy: a view from Mount Carmel, Israel. Before Farming 1:1–15
Yom-Tov Y, Mendelssohn H (1988) Changes of the distribution and abundance of vertebrates during the 20th century in Israel. In: Yom-Tov Y, Tchernov E (eds) The Zoogeography of Israel. Dr. W. Junk Publishers, Holland, pp 515–548
Zgaier A, Inbar M (2005) The influence of soil saturation on the stability of abandoned agricultural hillslope terraces under Mediterranean climatic conditions. Dev Earth Surf Process 7:69–86
Zohary M (1973) Geobotanical Foundations of the Middle East. Gustav Fisher Verlag, Stuttgart
Zohary D, Tchernov E, Horwitz LK (1998) The role of unconscious selection in the domestication of sheep and goats. J Zool 245(2):129–135
Zohary D, Hopf M, Weiss E (2012) Domestication of plants in the old world: the origin and spread of domesticated plants in Southwest Asia, Europe, and the Mediterranean Basin. Oxford University Press, New York

Sand Bodies in Israel: Sources, Dynamics, Morphology, Chronology and Relations with Prehistoric to Modern Humans

6

Joel Roskin

Abstract

The main sources of Israel's sand bodies are the Nile River system and the Dead Sea Rift valleys. The main dune fields are along the Mediterranean coast, in the northwestern Negev desert and in the southern Arava valley. Altogether, the sand bodies and sandstone exposures cover ~ 7.5% of the area of Israel. Studies of the dune fields of Israel have contributed significantly to the scientific understanding of aeolian dynamics, aeolian-fluvial interactions, dune field ecology, Quaternary palaeoenvironments and palaeoclimates, and to the changing roles in the lives of prehistoric to modern humans. In particular, the important effect of vegetation, biocrusts and wind power, though not necessarily aridity, on past and present dune mobility and stability has been analyzed and modelled. This chapter reviews the geospatial character, source regions and materials, dynamics, age, and morphology of loose and consolidated sand and dune bodies in Israel, and deals with the forcing factors affecting dune field stability and mobility, and the changing relationship of sand bodies with human activities since prehistory.

Keywords

Dune fields · Sand sheets · Israel · Quaternary · Holocene · Aeolian · Sand sources · Sand-human interactions · Aeolian-fluvial interactions

J. Roskin (✉)
Department of Geography and Environment, Bar-Ilan University, Ramat-Gan, Israel
e-mail: joel.roskin@biu.ac.il

6.1 Introduction

Sand is a ubiquitous sediment in the Israel-Sinai region, mainly in lowlands of its southern and coastal regions (Figs. 6.1 and 6.2a). Sand-size (63–2000 μm) grains dominate in aeolian and fluvial clastic rocks, beaches, sand sheets and dunes. The main sources of the sand and dune bodies are connected with the Nile River system, erosion of sandstone outcrops, mainly along escarpments, and fluvial deposits within structural lows. These sandy environments cover more than ~ 1500 km^2, namely ~ 7.5% of the area of Israel (Fig. 6.2a).

Several inland sand bodies and coastal sand bodies have developed along the Mediterranean Sea (Figs. 6.3, 6.4 and 6.5), in the arid Negev desert (Figs. 6.6, and 6.7) and in the hyper-arid Arava (Dead Sea Rift) valley, between the Dead Sea and the Gulf of Eilat (Fig. 6.8). The main dune fields are the coastal and northwestern Negev dune fields. There are approximately ten different dune morphological types (after Pye and Tsoar 2008; Tsoar 1990) in Israel (Figs. 6.5 and 6.7). Despite being situated along a transitional zone between Mediterranean and hyper-arid climates, the dunes are generally stable today.

Sand grains are also found in several landforms other than dunes. Dune morphology and stratigraphy are unique and usually discontinuous records of past wind properties. Dune field stability and transitions between stable to active modes and vice-versa are highly sensitive to climate and human-induced environmental change. Aeolian movement of sand, mainly of fine sand grains (125–250 μm), usually results in dune formation. Larger sand grain sizes are deposited as sand sheets (Pye and Tsoar 2008), though sand sheet morphologies consisting of fine sand have also been reported (Roskin et al. 2013a, b; 2015).

The coarse component of dust during the Quaternary also consists of very fine sand. These very fine sand grains, along with coarse silt quartz grains, are a component of middle to late Pleistocene loess deposits (Crouvi et al.

A. Frumkin and N. Shtober-Zisu (eds.), *Landscapes and Landforms of Israel*,
World Geomorphological Landscapes, https://doi.org/10.1007/978-3-031-44764-8_6

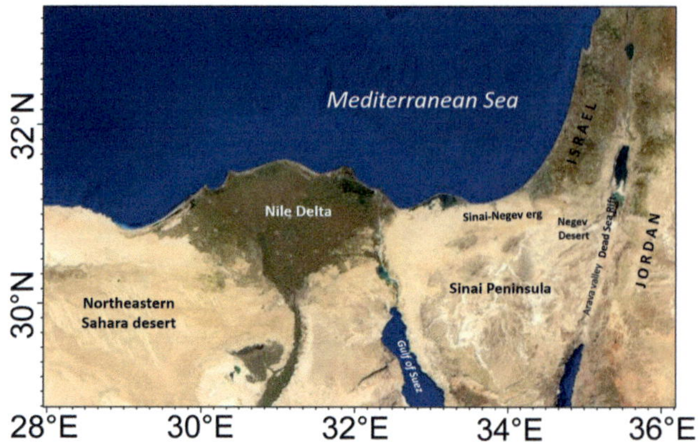

Fig. 6.1 Image of the Nile Delta, Sinai Peninsula and Israel

2008) and often comprise soil-building skeleton material that can be dated by luminescence methods.

Nearly half a century of intensive studies of the Israeli dune fields has resulted in a significant contribution to science. The studies focused on aeolian dynamics and dune hydrology (Breckle et al. 2008; Kidron 2021), ecology (Kutiel 2001; Kutiel et al. 2016), dune damming of wadis (Magaritz and Enzel 1990; Roskin et al. 2022a), Quaternary palaeoenvironments and palaeoclimates (Roskin et al. 2011) and the changing roles of dune dynamics upon the lifestyle of prehistoric to modern humans (Goring-Morris and Goldberg 1990; Vardi et al. 2018). In particular, effects of vegetation (Kutiel and Danin 1987; Kutiel et al. 2004; Tsoar 2005; Dorman 2021), biocrusts (Danin et al. 1989; Kidron et al. 2009; Rozenstein and Karnieli 2015), wind power, and the interactions between them (Yizhaq and Ashkenazy 2022), on past and present dune mobility and stability have been demonstrated and modelled (Tsoar and Møller 1986; Tsoar 2005; Yizhaq et al. 2007).

Based on a review of peer-reviewed papers and reports along with new data and GIS analysis, this chapter presents the geospatial character, source regions and material, dynamics, age and morphology of loose and consolidated sand and dune bodies in Israel. As such, it is the first review of all the sand sources and deposits in Israel. The chapter briefly deals with the forcing factors affecting dune field stability and mobility, as well as the relationship of sand bodies with past and present human activities.

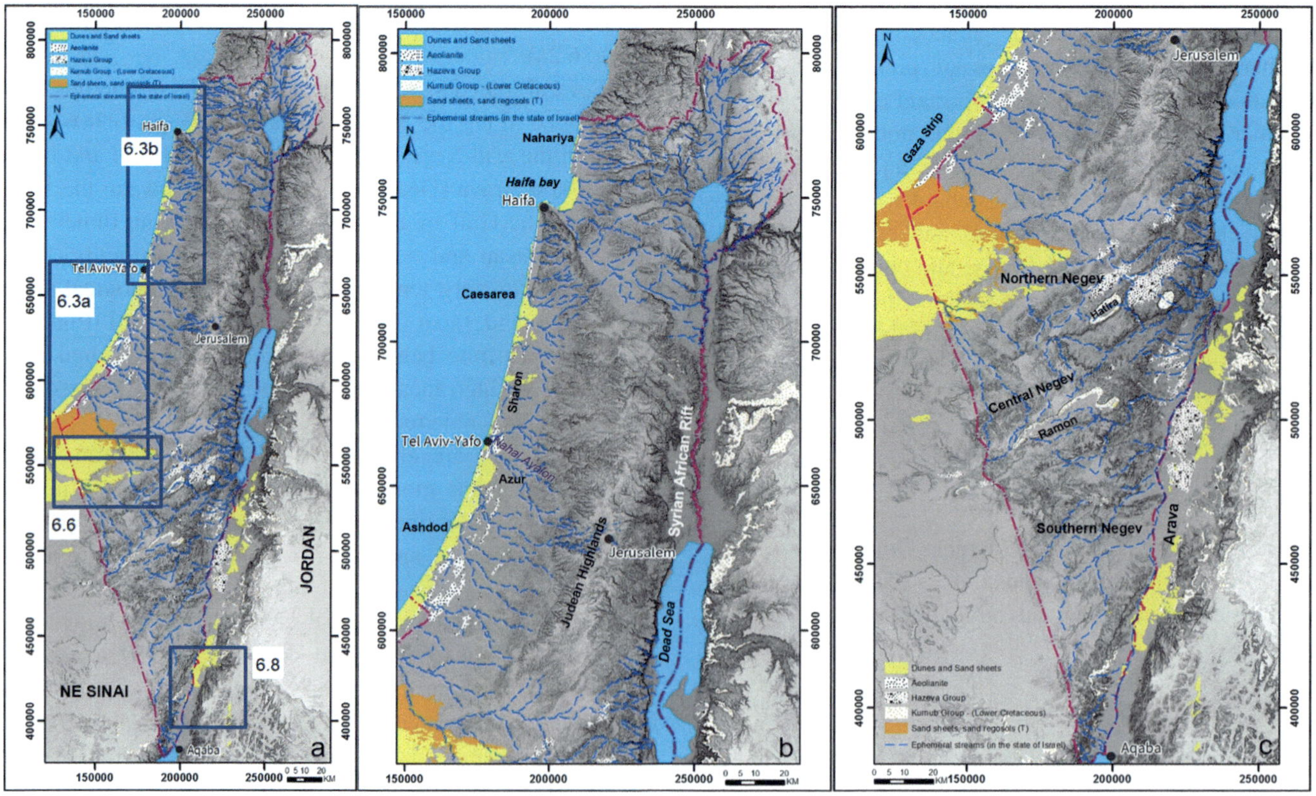

Fig. 6.2 Sand bodies and sandstone units in Israel. Black rectangles delimit other figures. **a** All of Israel. **b** Northern and Central Israel. **c** Southern Israel. GIS geological data from Sneh et al. (1998)

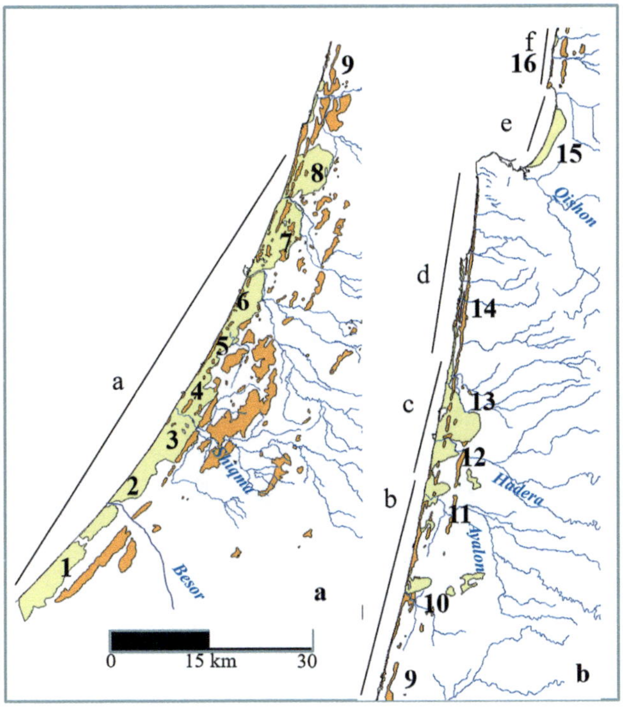

Fig. 6.3 Map of morphological segments of the coast **a–f**, numbered coastal dune fields (yellow) and aeolianite ridges (orange). GIS geology layers from Sneh et al. (1998). See Table 6.1 for details per numbered dune fields

6.2 Geological Sand Providing Systems

6.2.1 The Nile River System

The Nile River drainage system has been a major fluvial sedimentological generator for sediment supply to the eastern Mediterranean region since the Pliocene (Picard 1943; Emery et al. 1960; Goldsmith and Golik 1980; Davis et al. 2012; Garzanti et al. 2015; Bookman et al. 2021) (Fig. 6.1). Longshore currents transporting Nilotic sand along the southeastern Mediterranean continental shelf to the coast of Israel end today in the Haifa Bay (Zevulun valley) that serves as a sedimentological sink (Emery et al. 1960; Goldsmith and Golik 1980; Zviely et al. 2007; Katz and Crouvi 2018; Giaime et al. 2022).

The sand building of the coastal aeolianite dune ridges, Holocene beaches and dunes, and the northwestern Negev dune field (Fig. 6.2a–c) also originates from the Nile system (Roskin et al. 2012; Muhs et al. 2013; Garzanti et al. 2015). Nilotic sand in Israel usually has a sedimentologically mature quartz arenite composition (Muhs et al. 2013). The Nile Delta is built of thick late Pleistocene sand deposits overlaid by fine-grained Nilotic sediments that were deposited during the Holocene (Stanley 1990; Stanley and Warne 1993). From the last Glacial to the Younger Dryas (~ 11

Fig. 6.4 Coastal Quaternary calcified sand. **a**, **b** Olga beach exposure of fine-bedded aeolianite packages. **c** Section of carbonate rich quartz sand at Sha'ari Nikanor excavation in Jaffa. **d** Wall of mansion plot in Jaffa with the lower masonry built of local carbonate rich quartz sand and the upper wall built of aeolianite

Fig. 6.5 Diversity of coastal dune types. **a** Exposed northeast-southwest section of a transverse dune near Ashdod with modern 0.06 ± 0.04 ka (~ 60 yr) OSL age of transverse dune base indicating the time of its stabilization. **b** Aerial photo (1918) of the Palmachim (Yavneh) dune field near Nebi Rubin. Note linear and transverse dunes that generate aeolian-fluvial interactions with the Soreq stream to its north (right). **c** Vertical aerial photograph (DS, 1944) of the Caesarea dune field. Note Plot and Berm agroecosystem on western (left) part and linear and transverse dunes east of the agroecosystem. **d** Coppice dunes in the Olga area exemplifying the hypothesized Aeolian landscape before the construction of the coastal early Islamic agroecosystems and the development of the transverse dunes. **e** Transverse dunes in the central Caesarea dune field

ka), the Nile Delta, which was exposed due to low sea level, delivered sand to the inland Sinai-Negev erg (Roskin et al. 2012, 2014a; Muhs et al. 2013) (Figs. 6.1, 6.2c, and 6.6).

It seems that the Nilotic-sourced dune fields of Israel are one of the most densely and robust-dated sand bodies in the world (Lancaster et al. 2016; Roskin et al. 2017b; Thomas and Bailey 2019). Due to their relatively uniform luminescence qualities, the Nilotic-sourced sand grains found throughout the sediments and soils of Israel are an important subject for Quaternary research and dating (Faershtein et al. 2020).

6.2.2 The Dead Sea Rift Valleys

The Dead Sea Rift (Syrian-African Rift) is the strike-slip contact between the Sinai-Israel sub-plate and the Arabian plate. The Rift borders Israel to the east (Figs. 6.1, and 6.2a). Rift-bordering escarpments and canyons intermittently expose several sand-dominated geological formations and are the source of fluvially deposited sand-size detritus in the Rift valley, mainly in the Arava valley (Fig. 6.2).

Israel's central and southern Arava valley contains two small dune fields, Samar and Yotvata (Fig. 6.8). Several larger dune fields cover low areas near the edges of alluvial fans on the eastern side of the valley, in Jordan. They developed from fluvial detritus originating from the southern Transjordan highlands, which are higher than their Negev counterparts and where numerous canyons deeply incise into basement rocks and sandstones.

In the early Miocene, prior to the rifting and the uplift of the anticlinorium backbone of Israel, large fluvial systems drained the Arabian plate, flowing via southern Israel into the ancient Mediterranean Sea. In the central and northern Negev, and central and northern Arava valley, massive sand and gravel deposits remain from these ancient and disconnected drainage systems, mainly in structural lows. These deposits form the Hazeva Group (Calvo and Bartov 2001).

6.3 Sand in the Geological Record

6.3.1 Cambrian to Cretaceous Sandstones and Associated Dunes

The Lower Cretaceous sandstones are massively exposed in several erosional cirques ("makhtesh", sing.; "makhteshim" plural, in Hebrew) (Finzi et al. 2019) in the northern (Makhtesh Hatira/Hagadol and Makhtesh Hazera/Hakatan) and central anticlinal highlands (Makhtesh Ramon and Makhtesh Arif) of the Negev. They are also exposed in several locations along the Dead Sea Rift escarpments such as the western margins of the southern Arava valley (Weissbrod 2002) (Fig. 6.2c), and in northeastern Israel at the Hula valley escarpment and at the southern flanks of the Hermon mountain range (Fig. 6.2b). The Lower Cretaceous sandstones are mainly of the Kurnub Group (Hatira Fm.). Cambrian sandstone exposures also extend along the western margins of the southern Arava, from the Timna valley to the city of Eilat (Weissbrod 2002). Both are known as the Nubian sandstones. Only in the arid regions of Israel, eroded sand forms sand bodies and dunes.

The Hatira Fm is a 180–430 m thick, friable to moderately consolidated siliciclastic sequence, mainly deposited in a wide range of palaeo-fluvial environments. The makhteshim contain thick deposits of this sand. The Hatira sandstones are pale or vividly coloured by iron oxides. The fine- medium-grained quartz arenites in Makhtesh Hatira are predominately white (Weinberger 1986) and have been reported to have stratigraphies that appear to reflect aeolian environments (Bialik, O. pers comm.). Several exposures are quarried for the glass industry and construction.

The Hatira Fm is usually exposed in the lower parts of escarpments of the cirques, Rift and canyons. Their eroded sand, flushed down the wadis, rarely leads to source-bordering dunes. In the Ardon and Machmal valleys of the eastern Makhtesh Ramon, there are local climbing and coppice dunes such as the climbing red dunes (Blumberg DG, pers comm.; Rozenspat and Avni 2021) that mainly developed from eroded Jurassic Inmar and Hatira Formation sandstones (Fig. 6.2c) (Ward et al. 1993). In the southern Arava valley, poorly sorted sand sheets derived from nearby Lower Cretaceous exposures form megaripples (Yizhaq and Katra 2017) (Fig. 6.2c).

6.3.2 The Hazeva Group

The Hazeva Group, which dates to the early Miocene, is a significant sand source in southeastern Israel. Most of the Hazeva deposits originated from the Arabian crystalline rocks, prior to the subsidence of the Dead Sea Rift. The rifting led to the disruption of the large fluvial sediment transport path of the Hazeva between Arabia and the ancient Mediterranean Sea. The fluvial activity probably also recycled Nubian sandstones. The Hazeva Group is intermittently distributed in the Negev desert highlands, mainly in three synclines and small grabens, and in the central-northern Arava valley (Fig. 6.2c) (Calvo and Bartov 2001). Nowadays, this sand is quarried from several locations for construction use.

Remains of fluvial and lacustrine deposits of the Hazeva Group are divided into several formations (Calvo and Bartov 2001). Several are dominated by pebbles and poorly sorted sand in a wide range of sizes, often cemented by calcite. The Gidron Fm is predominately comprised of

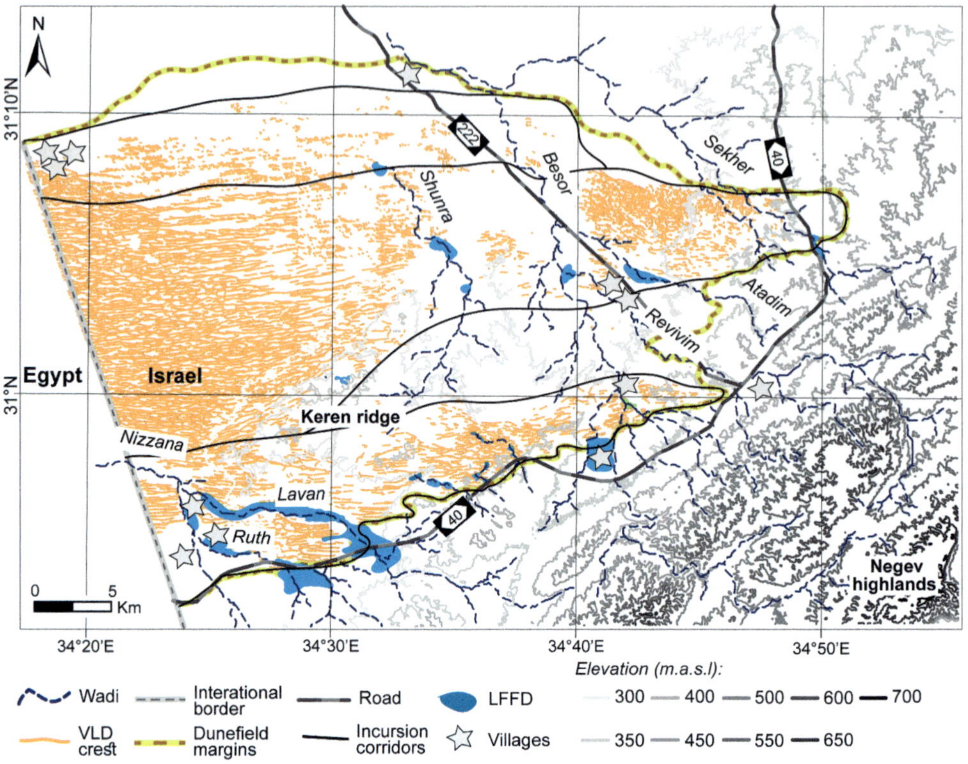

Fig. 6.6 Vegetated linear dunes (VLD) and sand incursion corridors of the northwestern Negev dune field. The blue polygons mark preliminary mapping of low-energy fine-grained sediments (LFFDs) (after Roskin et al. 2014a) deposited in late Quaternary dune-dammed water bodies (see details in Robins et al. 2021; 2022a, b)

Fig. 6.7 Dunes of the northwestern Negev dune field. **a** VLDs with clear interdune corridors in the southwestern part of the dune field (the southern incursion corridor). **b** VLDs with interdune corridors infilled with barchanoid slip faces in the west-central part of the dune field (the central incursion corridor). Both figs. a and b are along the Egypt-Israel border that exemplify the contrast in BSC coating of the dunes. **c** Dune-dammed fine-grained sediment in the foreground and Mitvakh VLD in the background (see Muhs et al. 2013; Robins et al. 2021)

Fig. 6.8 Google Earth image of the dune fields (black polygons) in and above the southern Arava valley of the Dead Sea Rift

subspherical monocrystalline quartz grains (> 75%) with some feldspar sand grains (< 12%), classified as submature quartz wacke to quartz arenite. The Karkom Fm is characterized by alternations of coarse grain sandstone (submature quartz arenite) and conglomerate, with a variable pebble composition. In the Rotem Fm, detrital fragments are mainly sub-rounded monocrystaline quartz and chert (75–100%), feldspar (< 25%), clays and heavy minerals. Pebbles in the Rotem Fm are mainly dark and round "imported chert", indicating the distal Arabian sediment source. The grain-size distribution with medium to coarse sand sizes indicates low-energy fluvial environments (Calvo 2000).

As most of the sand grains are larger than the fine sand-size range that is prone to significant aeolian transport, and since the depositional setting is for the most part wind-protected, the sand deposits often enable for shrub growth and the sand grains appear to have remained close to its often eroded geological depositional location. In the synclines of the northeastern Negev, sand deposits appear on the surface as low and vegetated coppice dunes and sand sheets. There was small-scale aeolian mobilization around 14.5 ka (Dody et al. 2005) in the Rotem dune field, when wind power was at its highest for the late Pleistocene-Holocene transition. This short-term gustiness probably enabled transport of medium to coarse sand fractions (Roskin et al. 2011) (Fig. 6.2c). In the northeastern Arava valley, the Hazeva Group sand deposits may have contributed some of the winnowed fine and mainly very fine sand to the nearby Salmani dune field in Wadi Arabah, Jordan (Saqqa and Atallah 2004) (Fig. 6.2c).

6.4 Dune Fields

6.4.1 Coastal Dunes and Deposits

The Israel coast is located along a Mediterranean climate gradient ranging from ~ 600 mm average annual precipitation in the north to 350 mm in the south (Fig. 6.2b). This gradient appears not to have significantly differentially affected inland sand incursion and deposition. All of the six morphological segments of the coast possess sand bodies (Fig. 6.3), mainly in the form of late Quaternary aeolianite ridges and carbonate rich quartz sand deposits, and Holocene dunes and sand sheets. The coast morphology and sedimentology is dominated by Nilotic sand.

6.4.1.1 Aeolianites and Calcarenite

Lying along the coastal plain and continental shelf of Israel, there are up to eighteen aeolianite ridges of loose to consolidated calcareous sandstone alternating with 2–5 red loam palaeosols (Rhodoxeralfs—Soil Survey Staff 1960; locally recognized as Hamra soils). These Hamra soils are understood to have developed following fine dust accumulation to the sand substrate (Yaalon 1997). The morphology of the 25 km wide southern coastal plain of Israel is comprised of around eight ridges (Fig. 6.3a, b). The number and size of the ridges gradually diminish northward (Sivan et al. 1999) and their morphology is less distinct in the eastern parts of the central coast (Harel et al. 2017).

These aeolianite ridges are ancient lithified dunes of Pleistocene age. They are dominated by quartz grains with the size of 120–180 µm (Neber 2002) and are known locally as "kurkar". The ridges run parallel or semi-parallel to the coastline and control the main and sea-born topographical features of the Israeli coast. The height of the ridges rises from ~ 10 m on the Carmel ridge coast to ~ 50 m in the south. The original dune (ridge) type and morphology reflect a dune whose character is accumulative and not advancing or extending.

Sediment bedding is characterized by planar cross-stratification, with low to high angles of bounding surfaces. Azimuth directions of bedding inclinations are usually to the northeast (Yaalon and Laronne 1971), though recent studies report on a broader range of orientations (Zaineldeen 2010; Mokaya et al. 2022). As such, the overall azimuth of the inclinations of the aeolianitic sand beds coast reflect wind directions that are similar to the current winds which are dominated by westerly and southwesterly sand-transporting winds in winter (Tsoar 1990).

Several hypotheses have been proposed to explain the forcing factors involved in sand accumulation and lithification of the aeolianite ridges (Harel et al. 2017). Sand transported to the beach was blown inland where it accumulated in dunes parallel to the coastline at different periods. However, no correlation or connection between hemispheric paleoclimate events and the establishment of the aeolianite ridges along the coast of Israel has been established (Mauz et al. 2013). Changes in Quaternary Nilotic littoral supply have not been recognized, nor correlated with coastal ridge build-up. The cessation of dune ridge formation has been associated with decreased sand availability (Shtienberg et al. 2017a).

Coastal sand deposition and formation of an overlying red loam and Hamra soils may have occurred nearly synchronously (Sivan and Porat 2004). However, high-resolution portable OSL measurements of the coastal escarpment south of Givat Olga show that the soils have significantly lower luminescence signals than the underlying aeolianite (Mokaya et al. 2022). Only Tsoar (2000) related to the aeolian dynamics of sand accumulation and hypothesized that the aeolianite dune ridges began accumulating as mega-like foredunes. This hypothesis appears probable based on simulations of foredune growth to heights of 25 m (Costas et al. 2022). These ancient foredune-like morphologies were significantly wider and higher than today's modern, tamarisk-topped foredunes that border some of the beaches, and that usually are usually < 12 m high.

In addition to quartz grains, some lithoclasts, feldspars, shell fragments, fragments of red algae, ooids and small quantities of heavy minerals are found in the aeolianite sediments. These along with dust and secretions from plant roots served to enhance calcification (Yaalon 1967; Gvirtzman and Wieder 2001), preserving the internal fine and cross-bedded sand dune structure (Fig. 6.4a, b). Calcium carbonate content is ~ 7–52% (Yaalon 1967) and is indurated by drusy and meniscus cements representing a phreatic and subaerial vadose meteoric environment (Neber 2002).

When carbonate component dominates the aeolianite ridge sediment, it is defined as a calcarenite (Gvirtzman et al. 1984; Sivan et al. 1999; Gvirtzman and Wieder 2001). Calcarenites are found at the top of the eroded aeolianite ridges near the shoreline along the Sharon escarpment within the central coast of Israel and are termed the Tel Aviv calcarenite (Gvirtzman et al. 1984). The shore-bordering calcarenites were dated by IRSL to the middle Holocene (Porat et al. 2004). No distinct climate transition is recognized during the middle Holocene that can explain calcarenite genesis. The rock probably formed prior to the development of the beach escarpment. Then, local biogenic carbonate detritus was more available on the beaches, or the Nilotic sand supply was lower or lagged, and only a short-distance upslope transport of the grains onto the beach-bordering aeolianite ridge was possible.

Calcarenites have been identified inland in the central coastal plain and in the Alexander dune field (Table 6.1), capping aeolianite ridges (Avisar 2022). They probably formed when the coast of the hosting ridge faced an ancient and possibly more easterly shoreline. Here, sand-filled quarries in the calcarenite are evidence of ancient extraction of the relatively hard rock.

Five periods of sand accumulation and soil development in the aeolianite ridges, restricted to the shoreline escarpments and exposures of the central and Carmel coast of Israel, were dated by luminescence methods to 140–12 ka (Frechen et al. 2004; Sivan et al. 2011). Mauz et al. (2013) integrated new and old dating results and stretched the chronology back to ~ 180 ka, stressing several erosional gaps in particular during the last glacial maximum (LGM). Recent analysis of OSL signals suggests that the age ranges of 170–80 ka are probably underestimated and should be treated as minimum ages (Faershtein et al. 2020).

Inland the northern to southern coastal plain, sand sheet deposits and aeolianite dune ridges have been deposited since the early Pleistocene onwards (Davis et al. 2012). Reverse magnetization (< 780 ka) has been found in aeolianites remains at the Evron site on the Galilee coast, dating to 0.99 ± 0.1–0.66 ± 0.11 Ma (Shemer et al. 2019). In the eastern Sharon TT-OSL and OSL ages indicate middle Pleistocene ages (Harel et al. 2017; Shemer et al. 2022). In the far eastern reaches of the southern coastal plain, over 20 km from the shoreline, TT-OSL ages of a Hamra soil/palaeosol, which probably overlies an aeolianite ridge, are 549 ± 36 ka. Here fine-grained sediments, hypothesized as water body remains from an aeolianite dune ridge damming a major drainage basin, date to 490–400 ka (Zaidner et al. 2018). Unconsolidated sand (sheet) deposits and sandy palaeosols at Kerem Shalom on the border with Egypt (Zilberman et al. 2007; Roskin et al. 2013a), dated by TT-OSL and pIRIR250, revealed a minimum age of 715 ka for the base of the section. These ages suggest aeolian sand encroachment and accumulation along the southeastern Mediterranean coastal plain already since the early Pleistocene (Faershtein et al. 2020).

Seen as a whole, the Nilotic coastal sand deposits and aeolianite dune ridge expansion are the longest sequence of aeolian accumulation in Israel. They have significantly shaped the current geomorphology of Israel's coastal plain.

6.4.1.2 Carbonate Rich Quartz Sand Deposits

Deposits of loosely cemented quartz sand sheets and low palaeo-dunes rich with vertical root-like and irregular calcified sand concretions were recently reported. This sediment

Table 6.1 Dune fields of Israel

Dunefield No.	Region + km north of Rafah	Size (km^2)	Maximum inland penetration (km)	Substrate and its upper age (ka)	Beach width (m) + OSL age (ka)	Stream and location	Dune types and stabilization*	Basal—upper sand OSL age range (ka)*	Vegetation association	Dated samples	Main archaeological sites	Sources
1	Katif (South Gaza); 6.7	42	4.0	Aeolianite, clay soils	15–80 m	–	Foredune, transverse, parabolic, linear	–	SA	–		
2	Wadi Gaza (N. Besor); 29	10	3.2		15–90 m	Wadi Gaza (southof duneifield)	Coppice dunes	–	SA	–		
3	Zikim + North Gaza Strip; 39	30	5.95	Calcic loess, Red loam palaeosol 9.4 ± 0.5 ka	15–90 m	Shiqma-	Sand sheet; Transverse/ parabolic	Brown sand unit—1.3 ± 0.1 Loose sand: 0.1 ± 0.03	SA	3		Faershtein et al. (2020)
4	Ashkelon; 45	15	5	Aeolianite, Red loam palaeosol 8.1 ± 0.4	15–90 m	Local blocked drainage in north	Parabolic		SA	1		
5	Nizzanim; 56	12	3.7	Red loam palaeosol, aeolianite, clay 0.73 ± 0.47	15–90 m	Evtakh–in north	Foredune, transverse, parabolic	Sand sheet: Transverse dune base: 0.15 ± 0.60	SA	3		This study
6	Ashdod; 63	36	4.35	Red loam palaeosol—12.0 ± 2.1 aeolianite, clay soils	15–90 m 5.3 ± 3.5 (sand ramp)	Lakhish East and north	Foredune, transverse, parabolic	Brown sand unit—2.1 ± 0.49 Lower loose sand—0.1 ± 0.03 ka Transverse dune base: 0.06 ± 0.04	SA	5	Ashdod Yam Hellenistic wall is upon aeolianite; Foredunes overlay site	A. Yair, pers. comm. This study
7	Yavneh; 77	31	5.2	Hamra, TA calcrenite, clay soils Red loam palaeosol—7.7 ± 1.2 ka;	15–190 m	Sorek East and north	Transverse	0.13 ± 0.04	SA	4	Early Islam inter-settlement agroecosystems overlay sand sheet and pre-date dunes	This study; Roskin et al. (2017a, b, c)
8	Rishon; 86	59	6.3	Red loam palaeosol—8.6 ± 2.5 ka	15–200 m	–	Coppice, nebkha, parabolic	0.07 ± 0.02	SA	2		This study
9	Sharon; 109	13.5	0.45	Hamra, aeolianite, TA calcrenite	5–30	Bordered by escarpment	Coppice, climbing	Base: 2 + 0.2 Upper section: 0.2 + 0.02	SA-MED	17		Porat et al. (2004)
10	Poleg; 117	7.5	3.5	Clay soils	15–200 m	Poleg dunes up mouth of stream	Transverse, coppice	Base: 3.7 ± 2.5 (scattered age) 0.12 ± 0.06	SA-MED	2		

(continued)

Table 6.1 (continued)

Dunefield No.	Region + km north of Rafah	Size (km^2)	Maximum inland penetration (km)	Substrate and its upper age (ka)	Beach width (m) + OSL age (ka)	Stream and location	Dune types and stabilization*	Basal—upper sand OSL age range (ka)*	Vegetation association	Dated samples	Main archaeological sites	Sources
11	Alexander; 133	10	2.2	Hamra, clay soils, aeolianite, TA calcarenite	50–350 6.6 ± 0.9	Alexander: North	Transverse, coppice	Basal: 3.6 ± 0.5	MED	5	Thin sand cover and regosol infills ancient quarry in TA calcarenite	
12	Olga; 139	3.5	2	Hamra, clay soils, aeolianite	10–50	–	Coppice, nebkha, parabolic,	Base: 4.8 ± 0.7 Upper: 0.12 ± 0.05 0.5 ± 0.3	MED	10		
13	Caesarea; 142.5	32	5.8	Beach 3.3 ± 0.5 Hamra 37 ka Clay marsh 21 ± 4	15–200	Hadera: South Ada: Northeast	Transverse, parabolic, linear	Basal regosol: 5.9 Upper: 0.12 ka	MED	20		
14	Dor—Neve Yam; 160	10.2	1.6	Clay 8.8 ± 0.1 (14C)	10–180 5.0 ± 1.1; 4.8 ± 1.1 (IRSL) 5.75 ± 0.58		Foredunes	~1 m above base: 0.05 ± 0.05	MED	1	Includes early Islam sand berm square plots	
15	Zevulun; 193	37	3.8	Sand plain	100–350 4–3.62 ka	Kishon: south; Naaman: North	Foredune ridge sequence; Coppice	3.7–2.4 ka	MED	8	Valley outlined by EB tells	Porat et al. (2008)
16	Akhziv Betzet; 217	4.2	0.35	1–4 m high aeolianite and calcrenite escarpments	5–25	Akhziv: center Betzet: center	Local calcareous sand forms coppice/ foredune like mounds		MED		Calcarenite hosting Phoenician tombs overlaid by coppice dunes	Bowman (1972)

The morphological coastal segments are the: (a) Gaza Strip* and southern coast (b) Sharon (c) Caesarea region (d) Carmel coast (e) Haifa bay-Zevulun valley) (f) Northern coast. + Rafah at the southern edge of the Gaza Strip marks the southern end of the coastal dune fields of Israel directly underlays archaeological sites excavated by the Israel Antiquities Authority (IAA). The sites, located in the central coastal plain, are mainly near the floodplains of major drainage systems. This carbonate rich quartz sand is often capped by calcrete in an irregular epikarst morphology, overlain by a thin, often anthropogenically mixed soil cover. The sediment has been coined as calcic or carbonate rich quartz sand (Roskin et al. 2022a, b) (Fig. 6.4b, c).

The carbonate rich quartz sand underlies archaeological remains from different periods, for example, near the ancient site of Tel Bene Beraq (Be'eri et al. 2019). Here, alongside Nahal (stream) Ayalon (Fig. 6.2b), buried dune-like morphologies with a 3–4 m thick carbonate rich quartz sand section devoid of artefacts are capped with calcrete, whose irregular morphology hosts late Chalcolithic artefacts, while the overlying alluvial soils contain middle to late Bronze finds. At the colluvial base of Tel Azur, south of Nahal Ayalon, the calcrete-cap morphology of the sand is covered with Hamra-like sediments mixed with early Bronze remains.

On the western aeolianite ridge within the city of Jaffa (Yafo) (Fig. 6.2b), Hellenistic graves embedded 1.4–1.9 m deep in carbonate rich quartz sand were found (Fig. 6.4c). The sand had a distinct unimodal particle-size (PSD) distribution with a mode of 210–240 μm (fine sand), which characterizes wind-blown (aeolian) sand. The unimodal PSD of the sand component in the carbonate rich quartz sand unit is very similar to the PSDs of the late coastal Holocene dunes (Roskin et al. 2015, 2017b) (see below). The ~ 22–30% carbonate content of the sand is substantially higher than in the loose dune sand, which is generally < 10% (Roskin et al. 2015) The content resembles the one found in aeolianites (after Yaalon 1967) but has only led to irregular cementation.

The vague remains of horizontal bedding in this carbonate rich quartz sand are different from the distinct cross-bedding found commonly in aeolianites and early modern dune sections (Roskin et al. 2017b). Such bedding is characteristic of gradual and constrained aeolian deposition due to obstacles like shrubs. Calcified and petrified root clasts as found in aeolianites attest to a substantial vegetation cover. Based on the sections measured with a portable, pulsed-photon stimulated luminescence (PPSL/portable OSL), the accumulation patterns of the carbonate rich quartz sand at the aforementioned site suggest gradual aeolian deposition (Roskin et al. 2021). Therefore, this quartz sand is of aeolian origin, but underwent long-term stabilization.

To summarize, the carbonate rich quartz sand unit is sedimentologically distinct. The unit was subject to gradual aeolian deposition followed by or was possibly partly synchronous with precipitation of calcium carbonate dissolved in percolating rainwater as explained for aeolianites (Yaalon 1967). Two OSL ages of its upper section in Jaffa date it to 85–67 ka (Roskin et al. 2022b). These ages position this unit within the lower part of the stratigraphic section of the Sharon coastal escarpment to its north (Porat et al. 2004). It is not fully lithified and also lacks distinct bedding. Due to its shallow subsurface position, the carbonate rich quartz sand probably underwent intense though not necessarily continuous bioturbation, until today, that erased the bedding. Under present-day conditions, it will probably not lithify into a rock that resembles the aeolianites.

The carbonate rich quartz sand appears to have been sporadically quarried in Jaffa for constructing plot walls during the late nineteenth–early twentieth centuries (Fig. 6.4d). This poor masonry material rapidly underwent significant weathering in contrast to their aeolianite counterparts that are abundant in archaeological sites and historic structures (Fig. 6.4d).

6.4.1.3 Mediterranean Coast Dune Fields, Sand Sheets and Soils

Fifteen small (< 60 km^2) dune fields along the Mediterranean coast of Israel and the Gaza Strip covering 394 km^2 are fed by the Nilotic littoral system (Figs. 6.2, and 6.3a, b; Table 6.1). Fourteen are south of the Haifa Bay. Aligning the Haifa Bay, the Zevulun valley dune field was also fed by littoral sand build-up in the sedimentological sink of the Haifa Bay (Zviely et al. 2007; Elyashiv et al. 2016). A unique narrow foredune-like strip comprised of carbonate rich sand borders the rocky beach, north of the city of Nahariya (Bowman 1972; Sivan et al. 1999).

The coastal dune fields stretch from the beach inland wherever the underlying aeolianite ridge topography did not act as a major obstacle to inland sand transport. That being said, in the inter-ridge valleys (troughs), we usually find a greater accumulation of sand and more developed dune morphologies than on the ridge crests (Roskin et al. 2015). We also know that dune fields often develop downwind from wide beaches in coastal lowlands (Shtienberg et al. 2017a), near migrating stream mouths (Lichter and Klein 2011).

Most dune fields with a maximal 5–6 km penetration width and relatively large areas (> 20 km^2) lie south of Tel Aviv. North of Tel Aviv, the Caesarea dune field is the only one with a similar size (Roskin et al. 2015, 2017a, b, c). The sedimentology of the coastal dune field sand is similar from north to south and its character partly mimics the sedimentology of its beach source (Emery et al. 1960).

The calculation of the transport potential of sand is based on the total annual wind power for sand transport (> 5 m/s). This is known as drift potential (DP) (terminology taken from Fryberger and Dean 1979). The DP calculated as 147 vector units for the Ashdod dune field and 207 for the Tel Aviv beach indicate a generally low wind energy (Tsoar 2000; Tsoar and Blumberg 2002). The wind power corresponding to these DP values does not impede vegetation growth and thus, natural stabilization of the dunes is taking place. Inland sand/dune movement only occurs where the sand and dunes are unvegetated (Tsoar 2005). Assuming a similar wind regime since the middle Holocene, the current wind power indicates that the coastal sand bodies were usually vegetated and stable. Exceptions were probably associated with periods when humans destroyed the sand-stabilizing vegetation, mainly by grazing and collection for fuel (Tsoar 2005).

The coastal dune fields today consist of stable and vegetated aeolian landforms that include sand sheets and stable-linear, transverse and parabolic dunes (Fig. 6.5). Prior to modern Israel times, the dunes were made bare by grazing (Fig. 6.5b, c) (Tsoar 1990; Liphschitz and Biger 1997). Since the mid-twentieth century, the dunes gradually stabilized (Levin and Ben-Dor 2004; Levin et al. 2008; Levin, this issue).

Foredunes intermittently border some of the beaches and host tamarisk trees (Tsoar 2000). These accumulating dunes block inland transport of sand. Field relations and preliminary OSL dating suggest that many of these dunes are modern. However, foredunes are hypothesized to have naturally existed in the past when humans did not remove dune field vegetation (Tsoar 2000; Roskin et al. 2015).

While the coastal dune fields lack distinct palaeosols, modern Regosols are found north of Tel Aviv in coppice dunes. Characterized by organic matter at different stages of decomposition (Ravikovitch 1981), these Regosols are associated with modern vegetation cover and generate a feedback mechanism of sand stabilization. In the Alexander dune field (Table 6.1) where sand cover is only 1–3 m thick, organic-rich Regosols developed beneath carob trees, but not beneath eucalyptus trees whose canopy and litter layer seals the surface from infiltration and pedogenesis (Avisar 2022).

The age distribution and inception time of the coastal dune fields have been studied using a database of ~ 100 luminescence ages and several radiocarbon dates drawn from a large body of research (Roskin et al. 2017b). Initially induced by a littoral sand build-up after the decline in the rate of sea level rise at 7–4 ka, the coastal dune fields began to develop in the middle Holocene (Roskin et al. 2015; Shtienberg et al. 2017a, b). Later, their dynamics were controlled by historic land-use intensity and modern land-use/negligence. At ~ 7–5 ka, beach sand build-up (Roskin et al. 2015; Shtienberg et al. 2021, 2022) was followed by a period of inland sand transport forming sand sheets and low coppice-like dunes (Roskin et al. 2015).

A lack of luminescence ages in the coastal dunes between 4–2 ka suggests a gap in sand deposition (Roskin et al. 2017b) and is attributed to sand-stabilizing vegetation cover that probably constrained sand saltation. It is believed that this vegetation was an attractive source of fuel and area for grazing for the growing Hellenistic and Roman-Byzantine (~2.4–1.3 ka) coastal population. The intensive destruction of sand-stabilizing vegetation is also believed to have greatly compromised the stability of coastal sand and dunes. Vegetation destruction also increased sand erodibility and mobilization during winter storms. Gradually, the sand sheet cover expanded to the present-day limits of the dune fields and eventually covered the remains of Byzantine settlements. A gradual but unsteady decline of the post-Byzantine population occurred differentially along the coast due to spatially different governance and land-use practices. This change during early Islamic times is suggested to have gradually led to sand stabilization by natural vegetation growth around 1.1 ka (Roskin et al. 2015).

Based on field evidence and the database Roskin et al. (2017b) suggested that transverse dune construction occurred only in the eighteenth–nineteenth centuries, possibly from existing sand sheets and low coppice dunes. Such a mechanism of dunes emerging from sand sheets has been empirically demonstrated and modelled (Narteau et al. 2022). The upper sand of a transverse dune at the northeastern edge of the Caesarea dune field was OSL-dated to 0.12 ± 0.05 ka (late nineteenth century), showing that by this time, a sand sheet or low dune morphology had reached the current dune field edge. This close-to-maximal spatial expansion of aeolian sand sheets and low dunes occurred prior to the transverse dune mobilization (Fig. 6.5e). Similar, late nineteenth century OSL ages in the western parts of the southern coastal dune fields were calculated, for example, at the base of a ~ 1 m thick coppice dune near Kibbutz Palmachim (0.13 ± 0.04 ka) and at Zikim (0.1 ± 0.03 ka; Faershtein et al. 2020) (Fig. 6.3; Table 6.1).

As with the northeastern Sinai Peninsula (Tsoar 1995), it seems that in approximately the eighteenth–nineteenth century, the rapid Bedouin expansion combined with the growth and influx of Arab peasant village populations (Migdal and Kimmerling 1993) was a decisive factor behind the destruction of sand-covering vegetation. The consequent increasing human activity facilitated erosion of the underlying sand, leading to sand encroachment and dune construction (Roskin et al. 2015, 2017b). Large storm waves may have destroyed foredunes aligning the back beach that were stripped of vegetation, thus increasing the susceptibility of the coastal environment to inland sand transport. For the Pleistocene, when human impact was minute, and shore-bordering aeolianite dune ridges existed, there is limited evidence of similar inland encroaching dune movement (Harel et al. 2017).

Morphological patterns of the late Holocene dunes reflect the orientations of the current sand-transporting winds, whereas the aeolianite ridges display similar wind patterns during the Pleistocene via their internal structure

(Yaalon and Laronne 1971) and not their morphology. Thus, despite similar wind orientations, other environmental conditions behind the formation of aeolianites and Holocene dunes are significantly different.

During the Quaternary, despite similar precipitation and temperature that control the vegetation, similar level of windiness that drives the sand, and a generally similar littoral sand flux from the Nile, the coastal sand deposits have undergone frequent depositional and spatial changes (Bookman et al. 2021). These changes exemplify the environmental sensitivity of the dunes and the connection of the dunes and their morphological types to specific environmental conditions. Altogether, the impact of human activity on coastal sand ecosystems has inversely changed from historic to modern times. The increasing physical human impact and global warming may in time lead to new changes in the dynamics of the remaining dunes and sand bodies.

6.4.1.4 The Northern Coast and Akhziv-Betzet Carbonate Dunes

Local calcareous sand forms vegetated coppice dunes and foredunes with Regosols along ~ 4 km of the backshore of the Akhziv-Betzet coast in northern Israel (Figs. 6.2b, and 6.3; Table 6.1) (Bowman 1972). The narrow dune strip, only < 350 m wide, overlies calcarenite that hosts Phoenician tombs (Sivan et al. 1999) but has not been dated by numerical methods. These small, but nevertheless distinct dunes, currently outside the area of the Nile littoral cell (Almagor et al. 2000), probably formed since the Nilotic littoral system came to terminate in the Haifa bay.

6.4.2 The Northwestern Negev Desert Dune Field

The northwestern (NW) Negev Desert dune field includes dunes and sand sheets covering an area of ~ 1,300 km^2. The dune field lies at the eastern end of the northern Sinai Peninsula dune field (Bubenzer et al. 2020), characterized by elongated and unvegetated linear (seif) dunes (Tsoar 1983; Tsoar et al. 2004). Together, these dune fields constitute the Sinai-Negev erg (Figs. 6.1, 6.2, and 6.6). The mineralogy and geochemistry of the Sinai and Negev sand are generally similar (Roskin et al. 2012; Muhs et al. 2013). The sand throughout the sections of the Negev dune field is generally fine (Roskin et al. 2011; Roskin et al. 2014a), with abundant very fine sand in certain locations within dune sections (Roskin et al. 2014a).

Sandy soils overlying a sequence of sandy and calcareous palaeosols border the NW Negev dune field in the north (Zilberman et al. 2007; Roskin et al. 2013a, b), while middle to late Pleistocene primary (and reworked) loess deposits flank the Negev dune field to the east and south. The average annual precipitation in the dune field is 170 mm in the north to 80 mm in the south (Tsoar and Møller 1986; Tsoar 2008). The current Drift Potential (DP) range, calculated as 21–108 vector units (Tsoar et al. 2008), even lower than DP calculations for the Israeli coast (see above), indicates a natural stable state of the sand bodies.

In the northwestern Negev, the most common dune type is the vegetated linear dune (VLD) (Fig. 6.7a–c). These dunes are currently stabilized and protected from wind erosion by 10–20% perennial shrub cover (Tsoar and Møller 1986; Tsoar et al. 2008; Siegal et al. 2013) and microphytic biological soil crusts (BSCs) (Danin et al. 1989; Karnieli and Tsoar 1995; Tsoar et al. 2008; Roskin et al. 2011; Kidron et al. 2009). The BSCs are mainly concentrated on the dune slopes. Some VLD crests lack crusting and are slightly active (Danin et al. 1989; Tsoar 2008; Kidron et al. 2009; Roskin et al. 2011; Kidron and Zohar 2014). The BSCs and bare sand comprise different ecological environments that support different types of vegetation (Kidron 2015) and faunal activity (Subach 2020), whose roots and burrowing significantly bioturbates the upper parts of the dunes.

The length of vegetated linear dunes is normally limited to less than several kilometres before the adjacent dunes coalesce, often in Y-shape junctions (Tsoar and Møller 1986; Tsoar et al. 2008). This constrains the ability to date single VLD elongation rates accurately. Past VLD elongation rates can only be measured along short increments of the VLD (where OSL dating errors constrain elongation rate calculation) (Roskin and Tsoar 2017; Robins et al. 2021), or along longer general west–east azimuths.

The northwestern Negev dune field has three main east–west dune encroachment corridors with similar chronology ranges (Figs. 6.6, and 6.7a, b). The corridors that follow VLD orientations differ in dune morphology and spacing, width/height ratios, and dune and sand thickness, probably owing to differences in sand supply, vegetation and BSC cover (Roskin et al. 2011). The dunes in the northern incursion corridor are broad and low. The central incursion corridor is characterized by 5–10 m thick interdune sand deposits in the west (Fig. 6.7a). These are overridden by VLDs and barchanoid-like transverse dunes with 5–10-m-high east-facing slip faces, which fill the interdune areas between the VLDs. Here, the cumulative aeolian sand thickness of the dune field reaches a maximum of ~ 25 m. The easternmost lobe of the central incursion corridor has coppice-like dunes that infill the topography of a series of chalk hills (Tsoar and Goodfriend 1994; Roskin et al. 2014b; Barzilai et al. 2015). The dunes of the southern corridor are aligned parallel to the interdune corridors (Fig. 6.7b) and the bulk sand colour is redder than that of the sand of the other corridors (Roskin et al. 2012).

Dune substrates include aeolian-fluvial sediments, calcic palaeosols and Plio-Pleistocene terrace deposits (Zilberman 1991; Roskin et al. 2013a). The calcic, and usually sandy loam palaeosols underlying much of the dune field date to 200–27 ka (Roskin et al. 2013a; Robins et al. 2022a, b). The sand is also probably derived from sand deposits of the Nile Delta (Amit et al. 2011; Roskin et al. 2012; Muhs et al. 2013). Thus, the erg and its underlying palaeosols are part of sand sheets, and a later dune-dominated transport system that starts in the Nile Delta and continues throughout the Sinai-Negev erg (Fig. 6.1) (Roskin et al. 2014a).

The northwestern Negev comprises the most densely and accurately dated linear dune bodies in the INQUA Dunes Atlas chronological database (Lancaster et al. 2016; Thomas and Bailey 2019). Over 250 luminescence ages (TL, IRSL, and mainly OSL) and radiocarbon dates have been retrieved from dune substrates, sand sheets, VLDs, fluvial deposits, and archaeological sites. Despite their being from different deposit types and aeolian landforms, and despite being based on different methodologies, the chronologies of these different deposits generally show good compatibility (Thomas and Bailey 2019).

The distribution of the chronologies points to rapid sand encroachment driven by powerful winds since 23–18 ka (Ben-David 2015; Roskin et al. 2011; Robins et al. 2022b), mainly during the Heinrich 1 event (Roskin et al. 2011; Robins et al. 2021, 2022a) and the Younger Dryas (Enzel et al. 2010; Roskin et al. 2011). The dune encroachments in the southern dune field and its margins (Robins et al. 2021, 2022a, b) caused the damming of fluvial systems (dune damming) and the formation of shallow standing-water bodies that deposited light-coloured loam units that are prominent today in the form of playa-like alluvial plains (Fig. 6.7c). The water bodies supported small and short-term Epipalaeolithic campsites (Roskin et al. 2014b, 2017a; Barzilai et al. 2015; Vardi et al. 2018; Robins et al. 2021, 2022a, b). Preliminary evidence of dune damming was reported near Epipalaeolithic sites in the valleys of the ridge of Gebel Maghara, amidst the northern Sinai dunes (Bar-Yosef and Phillips 1977).

Supervised classification of a Landsat TM (2003) image of exposed surfaces of standing-water deposits show similarities between the flat and light-coloured interdune alluvial deposits in the southwestern Negev dune field and northeastern Sinai near Wadi Al-Arish (Roskin and Rozenstein 2010). It is therefore suggested that Wadi Al-Arish was also blocked by dune encroachment causing extensive upstream and interdune flooding and deposition of fine material. Nahal Besor, the largest wadi dissecting the NW Negev dune field (Greenbaum and Zilberman 2017), may have also been dammed (Robins et al. 2022a).

Chronostratigraphy of dune sections and OSL age clustering at ~ 2–0.8 ka mark a distinct late Holocene dune remobilization episode. This included up to 10 m thick VLD mobilization and also transverse-like dune incursion in VLD interdunes, mainly in the central-western part of the Negev dune field, and surface remobilization and accretion of several metres of sand in its eastern parts (Roskin et al. 2011, 2013b; Robins et al. 2022b). This episode probably developed under short-term windy conditions, albeit less intense than the late Pleistocene encroachment episode, and below the resolution that can be perceived as climate change. Dune mobility was furthered by vegetation consumption and BSC decimation by both nomads and the occupants of the Byzantine towns (Roskin et al. 2013b).

6.4.3 The Rift and Southern Negev

6.4.3.1 Samar and Yotvata Dune Fields in the Southern Arava Valley

In the Arava valley of the Dead Sea Rift lie several dune fields, mostly in Jordan (Figs. 6.2a, and 6.8). The variability of wind direction, wind speed and rates of sand supply led to a variety of simple, compound and complex dune formations. Sand sheets and barchanoid dune types are common. Modern humans have utilized and affected the sand and dunes in the Arava. Quarrying, construction and agriculture have reduced the net areal sand cover. Sand has been applied to enhance agricultural soil in hothouses in the Arava (Hochman et al. 1985).

The small Samar and Yotvata dune fields in the southern Arava valley are composed of the fluvial erosional products of sandstone outcrops. The Samar dune field is the southernmost dune field of Israel. Today it covers 2.6 km^2, though part of the original dune field is now used as an agricultural land (Talbi 2009), while other sections were quarried for sand. The Samar dunes underwent a considerable increase in *Haloxylon persicum* bush density between 1990 and 2009. This change is explained by a 2–6 m deep nitrate-rich elevated groundwater table that developed due to excess water from nearby irrigation (Talbi 2009). This dune field probably formed from sediments washed into the Arava from the Timna valley region along the western Rift escarpment, where Lower Cretaceous and Cambrian sandstones (mainly arkosic and quartz arenite) are widely exposed.

According to Lancaster's (1988) climatic index of dune mobility, based on the understanding that the amount of precipitation is the major control on dune mobility, the hyper-arid Arava climate (P/PET = 0.007) should theoretically cause exceptional dune mobility ($M = 1857$).

However, the immobility of the dunes can be explained by its DP of 60 (Talbi 2009). This DP value demonstrates the major role of wind and not aridity on dune mobility (Tsoar 2005).

The Yotvata dune field is the southwest end of Jordan's Gharandal-Taba dune field. Grain size of dunes is fine to medium in Jordan (Saqqa and Atallah 2004), while the grain size of sand sheets where large ripples develop is mainly coarse (Yizhaq and Katra 2017). The current Yotvata dune field configuration is hypothesized to have developed during the late Pleistocene and Holocene (Saqqa and Atallah 2004). Preliminary OSL dating attempts have failed due to the luminescence properties of the quartz sand grains that originate from nearby Rift escarpment outcrops.

6.4.3.2 Kasuy Carbonate Sand Dune Field and Megaripples

The Nahal Kasuy (Fig. 6.8) dune field is situated in the hyper-arid southern Negev and covers a confined area of ~ 15 km^2. It is bordered by broad, flat-floored wadis between limestone slopes (Ginat 1991) and contains unvegetated seif (linear), falling and climbing dunes. The sand consists of 60% calcite and 35% quartz, originating from the nearby playa-like Uvda valley and Pliocene conglomerate exposures west of the valley.

The dune field surface is partially covered with small megaripples, whose wavelengths are 70 cm and the mean height is about 7 cm. Smaller ripples which are superimposed on the megaripples reflect the most recent wind direction (Yizhaq et al. 2012). The total annual DP for 2007 of 48.67 vector units (Yizhaq et al. 2009) was found capable to seasonally generate megaripples, but is not sufficient to significantly transport sand beyond the dune field limits.

6.5 Ancient Human-Sand Relationships and Effect upon Sand Mobilization and Stabilization

6.5.1 Negev Dunes

Since prehistoric times, humans have significantly and diversely interacted with the NW Negev dune field and the coastal dune field. The coastal dune fields are marginal lands concerning soil productivity in relation to the fine-grained soils to their east (Kutiel et al. 2004; Roskin and Taxel 2021a, b). On the other hand, the Negev dunes have improved water storage capacities in relation to their surrounding areas, and host a thicker shrub cover that has led to rich ecosystems (Danin 1978).

On the fringe of the NW Negev dune field, the density of the early to late Epipalaeolithic (~18–11 ka) sites near dune-dammed water bodies (Goring-Morris and Belfer-Cohen 2017) is anomalous for the arid southern Levant (Sinai and Negev). These sites were probably inhabited in conjunction with times of seasonal water body presence, whose sediments and adjacent dunes and dune remains are dated to similar times (Vardi et al. 2018; Robins et al. 2021, 2022a, b). While increased human presence during a period characterized by greater windiness seems incongruous, the clustering of prehistoric sites and seasonal human movement between the Negev Highlands and dune field lowlands exemplifies short-term human response to a unique time windows of seasonal water availability generated by times of aeolian domination (Robins et al. 2022b).

The 2–0.8 ka sand mobility event (Roskin et al. 2013b) has furnished the debate on climate change and whether increased aridity or storminess caused the collapse of the Northern Negev's Byzantine towns and extensive agricultural infrastructure (Issar et al. 1989; Rubin 1991; Avni et al. 2006; Bar-Oz et al. 2019; Vaiglova et al. 2020). The ruins of the Byzantine city of Halussa (Haluzza) are covered by 1–2 m of apparently very fine to fine sand (Heinzelmann et al. 2020) and a historical text attests to sand mobilization, probably self-inflicted, that only affected its grape vines. This late Holocene episode may have been a result of times of strong winds, probably following short-term droughts such as at a decadal scale (Yizhaq and Ashkenazy 2022), and mainly anthropogenic activity that destroyed stabilizing vegetation and BSCs, but was not associated with climate change (Roskin et al. 2013b).

Intermittent sand activity and stabilization in the NW Negev occurred over the past 150 years. This age range is similar to the age of many transverse coastal dunes. The ages reflect reactivation of crests and slopes of vegetated linear dunes, but not dune elongation (Roskin et al. 2011). These young OSL ages, albeit with relatively large errors, are quite consistent with anthropogenic land-use (surface) changes that decimated biogenic crusts and vegetation. These changes have been captured since the inception of available satellite imagery (Otterman et al. 1975; Karnieli and Tsoar 1995; Tsoar 2008; Rozenstein et al. 2016; Noy et al. 2021). Six OSL ages of ~ 68–40 years (before 2010) probably relate to increased dune activity due to Bedouin presence and grazing, which gradually stopped since the early 1950s, after grazing was banned in this area. The initial VLD re-stabilization and crest activation have even younger OSL ages (Roskin et al. 2011) due to the ban on Bedouin grazing following the re-establishment of the Egyptian-Israeli border in 1982 (Karnieli and Tsoar 1995; Tsoar 2008; Rozenstein et al. 2016; Noy et al. 2021).

In sum, each dune mobility episode in the NW Negev involved a different human-dune relationship. The late Pleistocene dune encroachment was entirely climate-induced and provided intermittent short-term conditions that were favourable to prehistoric humans. The late

Holocene dune mobilization was likely the product of a combination of anthropogenic pressure and short-term climate change whereas modern dune activity can be mainly attributed to human-induced destruction of BSCs and vegetation due to geopolitical changes, with no apparent climate change contribution.

6.5.2 Coastal Dunes

Since prehistory along the coastal plain sand has appeared to not have posed a significant constraint on human activities. Abundant Palaeolithic to Epipalaeolithic until Byzantine sites have been found upon aeolianite ridges and within Hamra soils (Ronen 2007). Several sites are associated with dune-dammed water bodies (Zaidner et al. 2018; Lupu et al. 2021).

Since Hellenistic and Roman times, humans have positively influenced sand incursion in the coastal plain, an effect that gradually declined during the early Islamic period (see Sect. 4.1.3). Then innovative groundwater-harvesting techniques developed in the form of plot and berm (P&B) agroecosystems (coined by Taxel et al. 2018) in the dune fields of Caesarea, Yavneh-Palmachim and apparently Zikim.

P&B agroecosystems are agricultural utilizations of shallow, perched rainfall-fed groundwater tables within loose aeolian sand sheets of agricultural hinterlands. These systems include human-made sand berms that delimit sunken agricultural plots where refuse-enriched sand transformed into distinct anthrosols (Taxel et al. 2018; Roskin and Taxel 2021a, b) (Fig. 6.5c). The plot surfaces are just 1–2 m above the groundwater table (Roskin et al. 2015; Shtienberg et al. 2017b; Bookman et al. 2022) (Fig. 6.5c). The development of the agroecosystem was enabled by the shallow groundwater table that was easily accessible for water extraction by crop roots and/or humans and could also support several crop cycles a year.

The agroecosystems have been dated to the tenth–twelfth century AD, based on ceramics and OSL ages (Roskin and Taxel 2021a). The Yavneh agroecosystem is buried by transverse dunes. These observations strengthen the understanding that at the construction time, only sand sheets and low dunes existed along many parts of the coast.

The P&B agroecosystem exemplifies an innovative adaptation to aeolian and fluvial erosion of sand. Despite the lack of relevant written records, these agroecosystems may have developed in response to religio-administrative calls for a type of mawāt (Arabic: "dead") land reclamation. The term mawāt refers to unowned wastelands, which were an important religion-based matter in Islamic economic history (Roskin and Taxel 2021a, b).

Contemporary Bedouin 'mawasi' agroecosystems are similar to the early Islamic systems. Situated along the northeastern Sinai Peninsula and southern Gaza Strip coast, they are practiced today by sedentary Bedouins using mechanized water drilling and modification of plot and soil surface levels to compensate for fluctuating groundwater levels. Before the mid-twentieth century, Arab peasants cultivated grapes and fruit trees in interdune plantations along the Gaza Strip and southern Israeli coast, where the groundwater table was several metres beneath the surface (Roskin and Taxel 2021a, b).

It seems that the early Islamic P&B agroecosystems along the Israeli coast are the earliest known agroecosystems of their kind and mark an non-linear cultural evolution in agriculture (see Roskin and Taxel 2021a, b). P&B agroecosystems are resourceful rainfall-fed agrotechnologies that are independent of the physical and juridical constraints of surficial water supply systems. They also reflect a tradition which shows an impressive and innovative adaptation to the challenges of aeolian erosion, as well as to local and seasonal groundwater availability without the need for irrigation, channels, canals, overland and fluvial flow. These systems provided an effective form of agriculture which ensured meagre water loss while allowing for several cycles of crops each year.

6.6 Summary

The sand bodies in Israel have developed as a result of the deposition of sand by two main sediment transfer systems along the Sinai-Israel sub-plate margins—the Nile River drainage system, and exposures of sandstones along the Arava valley of the Dead Sea Rift that act as a source of fluvial sand transport and deposition. Here, several local, small dune bodies have formed from sand that originated from washes and proximal alluvial fans.

The two main dune bodies of Israel that are currently stable: the (inland) NW Negev desert dune field and the coastal dune fields along the southeastern Mediterranean Sea, developed from the Quaternary and Holocene Nilotic sand, respectively. Aeolianite ridges shaping the Israeli coast also formed from the Nilotic sand during the Pleistocene.

The Israeli dunes are not large in global standards and are altogether vegetated and hence, stable. In the Holocene, when humans decimated the vegetation cover, sand drift occurred. Dunes became intermittently harnessed for agriculture only since early Islamic times. Today dune fields serve for Bedouin grazing, off-road recreation, local quarrying and military manoeuvres sometimes leading to constrained and local sand mobilization. In contrast to other landforms, these disturbed grounds can rapidly recover, usually in a matter of years, due to BSC growth (Kidron and Drahorad 2022). Sand is also used as a fine substrate for precision agriculture.

The sand deposits and their distinct flora and fauna provide and provided unique ecosystem services (Kutiel 2001) for modern, historic and prehistoric human beings and comprise unique archives of human and environmental history (de Freitas 2020). Understanding the dune dynamics and the utility of these services is vital for the preservation of the sandy environments of Israel and their unique ecosystems that may be found to possess important resources for future needs.

References

Almagor G, Gill D, Perath I (2000) Marine sand resources offshore Israel. Mar Georesour Geotechnol 18(1):1–42

Amit R, Enzel Y, Crouvi O et al (2011) The role of the Nile in initiating a massive dust influx to the Negev late in the middle Pleistocene. Geol Soc Am Bull 123(5–6):873–889

Avisar A (2022) The effect of eucalyptus trees on soil composition and water regime in a coastal sandy habitat: a change in overall ecosystem functioning. Ph.D. thesis (under review), Tel-Aviv University Israel

Avni Y, Porat N, Plakht J, Avni G (2006) Geomorphic changes leading to natural desertification versus anthropogenic land conservation in an arid environment, the Negev Highlands, Israel. Geomorphology 82(3–4):177–200

Bar-Oz G, Weissbrod L, Erickson-Gini T et al (2019) Ancient trash mounds unravel urban collapse a century before the end of Byzantine hegemony in the southern Levant. Proc Natl Acad Sci 116(17):8239–8248

Bar-Yosef O, Phillips JL (1977) Prehistoric investigations in Gebel Maghara, Northern Sinai. Qedem

Barzilai AN, Boaretto E et al (2015) The Natufian site of Nahal Sekher VI: the 2009 excavation season. Mitekufat Haeven/j Israel Prehistoric Soc 45:97–130

Be'eri R, Golan D, Dayan A et al (2019) Bene Beraq, Tel (South). Hadashot Arkheologiyot: excavations and surveys in Israel

Ben-David R (2015) Buildup and breakdown of Sand Dune Barriers in the drainage basins of the NW Negev Desert, Israel, During Late Pleistocene to Holocene climate changes. Eng Geol Soc Territory 3:361–364

Bookman R, Mor-Federman T, Herut B et al (2021) Development of the Nile Littoral Cell during the past 8.2 kyr. Quatern Sci Rev 274:107262.

Bookman R, Shtienberg G, Sivan D et al (2022) From late Quaternary geology to human impact on coastal environments at the northern Sharon and southern Carmel coast. In: IGS annual meeting excursion book 1–33

Bowman D (1972) Ahkziv- Rosh Hanikra. Morphology and sediments of the coastal Strip. In Haifa—in the western part of the Galilee—The regional program of knowing the Land heritage of the Sulam Tzor and Gaaton regional councils, p 53 (in Hebrew)

Breckle SW, Yair A, Veste M (eds) (2008) Arid dune ecosystems: the Nizzana sands in the Negev desert, vol 200. Springer Science & Business Media

Bubenzer O, Embabi NS, Ashour MM (2020) Sand seas and dune fields of Egypt. Geosciences 10(3):101

Calvo R, Bartov Y (2001) Hazeva Group, southern Israel: new observations, and their implications for its stratigraphy, paleogeography, and tectono-sedimentary regime. Israel J Earth Sci 50

Calvo R (2000) Stratigraphy and petrology of the Hazeva Formation in the Arava and the Negev: implications for the development of the sedimentary basins and the morphotectonics of the Dead Sea Rift Valley. Ph.D. thesis, Hebrew Univ., Jerusalem, p 113 (in Hebrew, English abstr.)

Costas, S, Kombiadou, K, Roelvink D (2022) Exploring coastal dune adaptation through a simplified process-based model (No. EGU22-8561). Copernicus Meetings.

Crouvi O, Amit R, Enzel Y, Porat N, Sandler A (2008) Sand dunes as a major proximal dust source for late Pleistocene loess in the Negev Desert, Israel. Quatern Res 70(2):275–282

Danin A (1978) Plant species diversity and plant succession in a sandy area in the Northern Negev. Flora 167(5):409-E1

Danin A, Bar-Or Y, Dor I, Yisraeli T (1989) The role of cyanobacteria in stabilization of sand dunes in southern Israel. Ecol Mediterr 15(1):55–64

Davis M, Matmon A, Rood DH, Avnaim-Katav S (2012) Constant cosmogenic nuclide concentrations in sand supplied from the Nile River over the past 2.5 my. Geology 40(4):359–362

de Freitas JG (2020) Making a case for an environmental history of dunes. Anthropocenes–Hum, Inhuman, Posthuman 1(1)

Dody A, Nahlieli A, Avni Y, Cohen H, Weiner D, Porat N (2005) Late Quaternary deposition and erosion processes along the margins of the Yamin Plain, northeast Negev, Israel. Israel J Earth Sci 57

Dorman M (2021) The importance of annual plants and multi-scalar analysis for understanding Coastal Dune stabilization process in the Mediterranean. Appl Sci 11(6):2821

Elyashiv H, Bookman R, Zviely D, Avnaim-Katav S, Sandler A, Sivan D (2016) The interplay between relative sea-level rise and sediment supply at the distal part of the Nile littoral cell. Holocene 26(2):248–264

Emery KO, Neev D, Bentor YK (1960) Mediterranean beaches of Israel. Jerusalem, Geological Survey of Israel

Enzel Y, Amit R, Crouvi O, Porat N (2010) Abrasion-derived sediments under intensified winds at the latest Pleistocene leading edge of the advancing Sinai-Negev erg. Quatern Res 74(1):121–131

Faershtein G, Porat N, Matmon A (2020) Extended-range luminescence dating of quartz and alkali feldspar from aeolian sediments in the eastern Mediterranean. Geochronology 2(1):101–118

Finzi Y, Avni S, Maroz A, Avriel-Avni N, Ashckenazi-Polivoda S, Ryvkin I (2019) Extraordinary geodiversity and geoheritage value of erosional craters of the Negev Craterland. Geoheritage 11(3):875–896

Frechen M, Neber A, Tsatskin A, Boenigk W, Ronen A (2004) Chronology of Pleistocene sedimentary cycles in the Carmel Coastal Plain of Israel. Quat Int 121(1):41–52

Fryberger SG, Dean G (1979) Dune forms and wind regime. In: A study of global sand seas (1052:137–169). US Government Printing Office Washington

Garzanti E, Andò S, Padoan M, Vezzoli G, El Kammar A (2015) The modern Nile sediment system: processes and products. Quat Sci Rev 130:9–56

Giaime M, Artzy M, Jol HM et al (2022) Refining late-Holocene environmental changes of the Akko coastal plain and its impacts on the settlement and anchorage patterns of Tel Akko (Israel). Mar Geol 447:106778

Ginat H (1991) The geology and geomorphology of Yotvata Region. M.Sc. thesis, Hebrew University, Jerusalem (in Hebrew, English summary)

Goldsmith V, Golik A (1980) Sediment transport model of the southeastern Mediterranean coast. Marine Geol 37(1–2):147–175

Goring-Morris, AN, Belfer-Cohen A (2017) The early and middle Epipalaeolithic of Cisjordan. In: Enzel Y, Bar-Yosef O (eds) Quaternary of the Levant: environments, climate change, and humans. Cambridge University Press, 639–650 (Chap. 47)

Goring-Morris AN, Goldberg P (1990) Late Quaternary dune incursions in the southern Levant: archaeology, chronology and palaeoenvironments. Quatern Int 5:115–137
Greenbaum N, Zilberman E (2017) Palaeogeography and Palaeohydrology of Fluvial Systems in the Levant, Southeastern Mediterranean. In: Enzel Y, Bar-Yosef O, (eds) Quaternary of the Levant: environments, climate change, and humans. Cambridge University Press, pp 401–416 (Chap. 47)
Gvirtzman G, Wieder M (2001) Climate of the last 53,000 years in the eastern Mediterranean, based on soil-sequence stratigraphy in the coastal plain of Israel. Quatern Sci Rev 20(18):1827–1849
Gvirtzman G, Shachnai E, Bakler B, Ilani S (1984) Stratigraphy of the Kurkar Group (Quaternary) of the coastal plain of Israel. Geol Surv Israel, Curr Res 19831984:70–82
Harel M, Amit R, Porat N, Enzel Y (2017) Evolution in of the southeastern Mediterranean coastal plain. In: Enzel Y, Bar-Yosef O (eds) Quaternary of the Levant: environments, climate change, and humans. Cambridge University Press. Cambridge University Press, Cambridge, pp 433–445
Heinzelmann M, Schöne C, Wozniok D, Erickson-Gini (2020) Elusa-ein bislang unbeachtetes Landwirtschaftssystem im Negev? Zwischenbericht zum Elusa-Umlandsurvey (2018 bis 2020). Kölner und Bonner Archaeologica 9:141–160
Hochman E, Vitkon G, Just RE, Zilberman D (1985) The dynamics of agricultural development in sparsely populated areas: the case of the Arava. In: Desert development. Springer, Dordrecht, pp 256–270
Issar A, Tsoar H, Levin D (1989) Climatic changes in Israel during historical times and their impact on hydrological, pedological and socio-economic systems. Paleoclimatology and paleometeorology: modern and past patterns of global atmospheric transport, pp 525–541
Karnieli A, Tsoar H (1995) Spectral reflectance of biogenic crust developed on desert dune sand along the Israel-Egypt border. Remote Sens 16(2):369–374
Katz T, Crouvi O (2018) Sediment flux dynamics over the shallow (25 m depth) shelf of the Mediterranean Sea along the Israeli coast. Mar Geol 406:1–11
Kidron GJ (2015) Dune crests serve as preferential habitats for perennial plants during frequent drought years. J Hydrol 522:295–304
Kidron GJ (2021) The role of biocrust-induced exopolymeric matrix in runoff generation in arid and semiarid zones–a mini review. J Hydrol Hydromechanics 69(4):360–368
Kidron GJ, Zohar M (2014) Wind speed determines the transition from biocrust-stabilized to active dunes. Aeol Res 15:261–267
Kidron GJ, Drahorad SL (2022) Biocrust-induced extracellular polymeric substances are responsible for dune stabilization in the Negev. Land Degrad Dev 33(3):425–438
Kidron GJ, Vonshak A, Abeliovich A (2009) Recovery rates of microbiotic crusts within a dune ecosystem in the Negev Desert. Geomorphology 100(3–4):444–452
Kutiel P (2001) Conservation and management of the Mediterranean coastal sand dunes in Israel. J Coast Conserv 7(2):183–192
Kutiel P, Danin A (1987) Annual-species diversity and aboveground phytomass in relation to some soil properties in the sand dunes of the northern Sharon Plains, Israel. Vegetation 70(1):45–49
Kutiel P, Cohen O, Shoshany M, Shub M (2004) Vegetation establishment on the southern Israeli coastal sand dunes between the years 1965 and 1999. Landsc Urban Plan 67(1–4):141–156
Kutiel PB, Katz O, Ziso-Cohen V, Divinsky I, Katra I (2016) Water availability in sand dunes and its implications for the distribution of Artemisia monosperma. CATENA 137:144–151
Lancaster N (1988) Development of linear dunes in the southwestern Kalahari, South Africa. J Arid Environ 14:233–244
Lancaster N, Wolfe S, Thomas D et al (2016) The INQUA dunes atlas chronologic database. Quatern Int 410:3–10
Levin N, Ben-Dor E (2004) Monitoring sand dune stabilization along the coastal dunes of Ashdod-Nizanim, Israel, 1945–1999. J Arid Environ 58(3):335–355
Levin N, Kidron GJ, Ben-Dor E (2008) A field quantification of coastal dune perennial plants as indicators of surface stability, erosion or deposition. Sedimentology 55(4):751–772
Lichter M, Klein M (2011) The effect of river floods on the morphology of small river mouths in the southeastern Mediterranean. Zeitschrift für Geomorphologie 55(3):317–340
Liphschitz N, Biger G (1997) Sand dunes reclamation by vegetation in Palestine during the British mandate period. Horiz Geogr 46:1–38
Lupu R, Goder-Goldberger M, Ackerman O, Roskin J, Milevski I (2021) Netzer Sereni: a Geometric Kebaran Site in the coastal plain of Israel. J Israel Prehistoric Soc 51:82–105
Magaritz M, Enzel Y (1990) Standing-water deposits as indicators of Late Quaternary dune migration in the northwestern Negev, Israel. Climatic Change 16(3):307–318
Mauz L, Hijma MP, Amorosi A, Porat N. Galili E, Bloemendal J (2013) Aeolian beach ridges and their significance for climate and sea level: concept and insight from the Levant coast (East Mediterranean). Earth-Sci Rev 121:31–54
Migdal J, Kimmerling B (1993) Palestinians: the making of a people
Mokaya L, Bookman R, Roskin J, Filin S, Koren A (2022) High-resolution analysis of late Quaternary aeolianites on the south-eastern Mediterranean coast of Israel European Geoscience Union annual meeting, Vienna, EGU22-10249
Muhs DR, Roskin J, Tsoar H et al (2013) Origin of the Sinai-Negev erg, Egypt and Israel: mineralogical and geochemical evidence for the importance of the Nile and sea level history. Quatern Sci Rev 69:28–48
Narteau C, Lü P, Claudin P et al (2022) Direct validation of dune instability theory. European Geoscience Union annual meeting, Vienna. EGU22-2079
Neber A (2002) Sedimentological properties of Quaternary deposits on the central coastal plain, Israel. PhD thesis, University of Haifa, p 217
Noy K, Ohana-Levi N, Panov N Silver M, Karnieli A (2021) A long-term spatiotemporal analysis of biocrusts across a diverse arid environment: the case of the Israeli-Egyptian sandfield. Sci Total Environ 774:145154
Otterman J, Waisel Y, Rosenberg E (1975) Western Negev and Sinai ecosystems: comparative study of vegetation, albedo, and temperatures. Agro-Ecosyst 2(1):47–59
Picard L (1943) Jaffa-Tel Aviv Sheet 6, 1: 100,000: Israel Geol. Survey Geol Map of Israel Central Part ser B 2–4
Porat N, Wintle AG, Ritte M (2004) Mode and timing of kurkar and hamra formation, central coastal plain, Israel. Israel J Earth Sci 53(1):13–25
Porat N, Sivan D, Zviely D (2008) Late Holocene embayment infill and shoreline migration, Haifa Bay, Eastern Mediterranean. Israel J Earth Sci 57(1)
Pye K, Tsoar H (2008) Aeolian sand and sand dunes. Springer Science & Business Media
Ravikovitch S (1981) The soils of Israel: formation, nature and properties. Tel-Aviv Hakibbutz Hameuchad Publ. House (in Hebrew)
Robins L, Greenbaum N, Yu L, Bookman R, Roskin J (2021) High-resolution portable-OSL analysis of vegetated linear dune construction in the margins of the northwestern Negev dunefield (Israel) during the late Quaternary. Aeol Res 50:100680
Robins L, Roskin J, Yu L, Bookman R, Greenbaum N (2022a) Aeolian-fluvial processes control landscape evolution along dune-field margins of the northwestern Negev (Israel) since the late Quaternary. Quatern Sci Rev 285:107520
Robins L, Roskin J, Yu L, Greenbaum N (2022b) Aeolian-Fluvial response to late-Pleistocene dunefield encroachments along the northwestern Negev dunefield margins (Israel). European Geoscience Union annual meeting, Vienna, EGU22-13303.

Ronen A (2007) Climate, sea level, and culture in the Eastern Mediterranean 20 ky to the present. In: The Black Sea Flood Question: changes in Coastline, climate, and human settlement. Springer, Dordrecht, pp 819–832

Roskin J, Rozenstein O (2010) Evidence for widespread episodic Late Pleistocene and late Holocene dune-dammed lakes in the northeast Sinai (Egypt) and northwest Negev Desert, Israel. Dryland desertification 4th international conference, Sde Boker, Nov 2010

Roskin J, Tsoar H (2017) Late quaternary chronologies of the Northern Sinai/Northwestern Negev dunefield and their palaeoclimatic and palaeoenvironmental implications. In: Quaternary of the Levant. Cambridge University Press, pp 505–520

Roskin J, Taxel I (2021a) Groundwater-fed plot-and-berm agroecosystems in aeolian sand in the Mediterranean Basin. In: Abrunhosa M, Chambel A, Peppoloni S, Chaminé HI (eds) Advances in Geoethics and groundwater management: theory and practice for a sustainable development. Proceedings of the 1st congress on Geoethics and groundwater management (GEOETH&GWM'20), Porto, Portugal 2020, Springer, Cham, pp 473–478

Roskin J, Taxel I (2021b) He who revives dead land": groundwater harvesting agroecosystems in sand along the southeastern Mediterranean coast since early medieval times. Mediterr Geosci Rev 1–26

Roskin J, Porat N, Tsoar H, Blumberg DG, Zander AM (2011) Age, origin and climatic controls on vegetated linear dunes in the northwestern Negev Desert (Israel). Quatern Sci Rev 30(13–14):1649–1674

Roskin J, Blumberg DG, Porat N, Tsoar H, Rozenstein O (2012) Do dune sands redden with age? The case of the northwestern Negev dunefield, Israel. Aeol Res 5:63–75

Roskin J, Katra I, Porat N, Zilberman E (2013a) Evolution of Middle to Late Pleistocene sandy calcareous paleosols underlying the northwestern Negev Desert Dunefield (Israel). Palaeogeogr Palaeoclimatol Palaeoecol 387:134–152

Roskin J, Katra I, Blumberg DG (2013b) Late Holocene dune mobilizations in the northwestern Negev dunefield, Israel: a response to combined anthropogenic activity and short-term intensified windiness. Quatern Int 303:10–23

Roskin J, Katra I, Blumberg DG (2014a) Particle-size fractionation of eolian sand along the Sinai–Negev erg of Egypt and Israel. Bulletin 126(1–2):47–65

Roskin J, Katra I, Agha N, Goring-Morris AN, Porat N, Barzilai O (2014b) Rapid anthropogenic response to short-term aeolian-fluvial palaeoenvironmental changes during the Late Pleistocene–Holocene transition in the northern Negev Desert, Israel. Quat Sci Rev 99:176–192

Roskin J, Sivan D, Shtienberg G et al (2015) Natural and human controls of the Holocene evolution of the beach, aeolian sand and dunes of Caesarea (Israel). Aeol Res 19:65–85

Roskin J, Bookman R, Friesem DE, Vardi J (2017a) A late Pleistocene linear dune dam record of aeolian-fluvial dynamics at the fringes of the northwestern Negev dunefield. Sed Geol 353:76–95

Roskin J, Sivan D, Shtienberg G, Porat N, Bookman R (2017b) Holocene beach buildup and coastal aeolian sand incursions off the Nile littoral cell. In: EGU General Assembly conference abstracts, p 844

Roskin J, Sivan D, Bookman R, Porat N, López GI (2017c) Testing portable luminescence reader signals against late Pleistocene to modern OSL ages of coastal and desert dunefield sand in Israel. In EGU General Assembly conference abstracts, p 11721

Roskin J, Robins L, Yu L, Greenbaum N (2022a) Palaeoclimatic and geomorphic implications of late Quaternary aeolian-fluvial interactions in the northwestern Negev dunefield (Israel)-a review. In EGU General Assembly conference abstracts, p 4344

Roskin J, Rauchberger L, Galit, T, Asscher Y (2022b) Geological character and relative age analysis by portable luminescence (POSL) of carbonate rich quartz sand enable the interpretation of a Hellenistic burial site in Sha'ari Nikanor street (Jaffa). In the center III: time. Annual Research summary meeting of the central district, Israel Antiquities Authority (abstract in Hebrew), 9 June 2022

Rozenstein O, Karnieli A (2015) Identification and characterization of Biological Soil Crusts in a sand dune desert environment across Israel-Egypt border using LWIR emittance spectroscopy. J Arid Environ 112:75–86

Rozenstein O, Siegal Z, Blumber DG, Adamowski J (2016) Investigating the backscatter contrast anomaly in synthetic aperture radar (SAR) imagery of the dunes along the Israel-Egypt border. Int J Appl Earth Obs Geoinf 46:13–21

Rubin R (1991) Settlement and agriculture on an ancient desert frontier. Geograp Rev 197–205

Saqqa W, Atallah M (2004) Characterization of the aeolian terrain facies in Wadi Araba Desert, southwestern Jordan. Geomorphology 62(1–2):63–87

Shemer M, Crouvi O, Shaar R et al (2019) Geochronology, paleogeography, and archaeology of the Acheulian locality of 'Evron Landfill in the western Galilee, Israel. Quatern Res 91(2):729–750

Shemer M, Greenbaum N, Taha N et al (2022) Late Acheulian Jaljulia–Early human occupations in the paleo-landscape of the central coastal plain of Israel. PloS ONE 17(5):e0267672

Shtienberg G, Dix JK, Roskin J et al (2017a) New perspectives on coastal landscape reconstruction during the Late Quaternary: a test case from central Israel. Palaeogeogr Palaeoclimatol Palaeoecol 468:503–519

Shtienberg G, Dix JK, Shahack-Gross R et al (2017b) Anthropogenic overprints on natural coastal aeolian sediments: a study from the periphery of ancient Caesarea, Israel. Anthropocene 19:22–34

Shtienberg G, Gadol O, Levy TE et al (2021) Changing environments and human interaction during the Pleistocene–Early Holocene from the shallow coastal area of Dor, Israel. Quatern Res, 1–18

Shtienberg G, Cantu K, Mischke S, Sivan D, Norris RD, Rittenour TM, Edelman-Furstenberg Y, Yasur-Landau A, Sisma-Ventura G, Levy TE (2022) Holocene sea-level rise and coastal aquifer interactions: Triggering mechanisms for environmental change and impacts on human settlement patterns at Dor, Israel. Quat Sci Rev 294:107740

Siegal Z, Tsoar H, Karnieli A (2013) Effects of prolonged drought on the vegetation cover of sand dunes in the NW Negev Desert: field survey, remote sensing and conceptual modeling. Aeol Res 9:161–173

Sivan D, Gvirtzma G, Sass E (1999) Quaternary stratigraphy and paleogeography of the Galilee coastal plain, Israel. Quatern Res 51(3):280–294

Sivan D, Porat N (2004) Evidence from luminescence for Late Pleistocene formation of calcareous aeolianite (kurkar) and paleosol (hamra) in the Carmel Coast, Israel. Palaeogeo Palaeoclimatol Palaeoecol 211(1–2):95–106

Sivan D, Greenbaum N, Cohen-Seffer R et al (2011) The origin and disappearance of the late Pleistocene–early Holocene short-lived coastal wetlands along the Carmel coast, Israel. Quatern Res 76(1):83–92

Sneh A, Bartov Y, Weissbrod T, Rosensaft M (1998) Geological Map of Israel, 1:200,000, Israel Geological Survey (4 sheets)

Soil Survey Staff (1960) Soil classification, a comprehensive system. U S Department of Agriculture, Washington, DC

Stanley DJ (1990) Recent subsidence and northeast tilting of the Nile delta, Egypt. Mar Geol 94(1–2):147–154

Stanley DJ, Warne AG (1993) Nile Delta: recent geological evolution and human impact. Science 260(5108):628–634

Subach A (2020) Using animal tracks to decipher the foraging mode of species capable of altering between the sit-and-wait and widely foraging modes: a case study of the sand viper Cerastes vipera. Israel J Ecol Evol 66(1–2):94–100

Talbi R (2009) The edge-effect of farming on vegetation density and rodent assemblage of a remnant sandy patch in a hyper-arid environment (Arava Valley, Israel). M.A. thesis, Ben Gurion University in the Negev, p 68
Taxel I, Sivan D, Bookman R, Roskin J (2018) An early Islamic inter-settlement agroecosystem in the coastal sand of the Yavneh Dunefield, Israel. J Field Archaeol 43(7):551–569
Thomas DS, Bailey RM (2019) Analysis of late Quaternary dunefield development in Asia using the accumulation intensity model. Aeol Res 39:33–46
Tsoar H (1983) Dynamic processes acting on a longitudinal (seif) sand dune. Sedimentology 30(4):567–578
Tsoar H (1990) Trends in the development of sand dunes along the southeastern Mediterranean coast. In: Dunes of the European Coasts. Geomorphology-Hydrology-Soils. Catena. Supplement (Giessen) 18:51–60
Tsoar H (1995) Desertification in Northern Sinai in the eighteenth century. Clim Change 29(4):429–438
Tsoar H (2000) Geomorphology and paleogeography of sand dunes that have formed the kurkar ridges in the coastal plain of Israel. Israel J Earth Sci 49(3):189–196
Tsoar H (2005) Sand dunes mobility and stability in relation to climate. Physica A 357(1):50–56
Tsoar H (2008) Land use and its effect on the mobilization and stabilization of the north-western Negev sand dunes. In: Breckle SW, Yair A, Veste M (eds) Arid dune ecosystems. Springer, Berlin, Heidelberg, pp 79–89
Tsoar H, Møller JT (1986) The role of vegetation in the formation of linear sand dunes. In: Aeolian geomorphology. Routledge, pp 75–96
Tsoar H, Goodfriend GA (1994) Chronology and palaeoenvironmental interpretation of Holocene aeolian sands at the inland edge of the Sinai-Negev erg. The Holocene 4(3):244–250
Tsoar H, Blumberg DG (2002) Formation of parabolic dunes from barchan and transverse dunes along Israel's Mediterranean coast. Earth Surf Proc Land 27(11):1147–1161
Tsoar H, Blumberg DG, Stoler Y (2004) Elongation and migration of sand dunes. Geomorphology 57(3–4):293–302
Tsoar H, Blumberg DG, Wenkart R (2008) Formation and geomorphology of the North-Western Negev sand dunes. In: Breckle SW, Yair A, Veste M (eds) Arid dune ecosystems. Springer, Berlin, Heidelberg, pp 25–48
Vaiglova P, Hartman G, Marom N et al (2020) Climate stability and societal decline on the margins of the Byzantine empire in the Negev Desert. Sci Rep 10(1):1–13
Vardi J, Marder O, Bookman R, Friesem DE, Groman-Yeroslavski I, Edeltin L, Porat N, Boaretto E, Roskin J (2018) Middle to late epipaleolithic hunter-gatherer encampments at the Ashalim site, on a linear dune-like morphology, along dunefield margin water bodies. Quat Int 464:187–205
Ward D, Olsvig-Whittaker L, Lawes M (1993) Vegetation-environment relationships in a Negev Desert erosion cirque. J Veg Sci 4(1):83–94
Weinberger G (1986) Sedimentary Environment of the Lower Cretaceous in the Northern Negev. Ph.D. thesis, Ben-Gurion University of the Negev, Beer-Sheva, p 240 (Hebrew with English abstract)
Weissbrod T (2002) Stratigraphy and correlation of the Lower Cretaceous exposures across the Dead Sea Transform with emphasis on tracing the Amir Formation in Jordan. Israel J Earth Sci 51(2):55–78
Yaalon DH (1967) Factors affecting the lithification of eolianite and interpretation of its environmental significance in the coastal plain of Israel. J Sediment Res 37(4):1189–1199
Yaalon DH (1997) Soils in the Mediterranean region: what makes them different?. Catena 28(3–4):157–169
Yaalon DH, Laronne J (1971) Internal structures in eolianites and paleowinds, Mediterranean coast, Israel. J Sediment Res 41(4):1059–1064
Yizhaq H, Katra I (2017) Aeolian Processes of Sand Megaripples in the Southern Negev. Horiz Geogr 93:125–143 (in Hebrew)
Yizhaq H, Ashkenazy Y (2022) Sand dune vegetation-biocrust interaction as a source of spatial heterogeneity. J Hydrol Hydromech 70(2):145–155
Yizhaq H, Ashkenazy Y, Tsoar H (2007) Why do active and stabilized dunes coexist under the same climatic conditions? Phys Rev Lett 98(18)188001
Yizhaq H, Isenberg O, Wenkart R, Tsoar H, Karnieli A (2009) Morphology and dynamics of aeolian mega-ripples in Nahal Kasuy, southern Israel. Israel J Earth Sci 57:149–165
Yizhaq H, Katra I, Isenberg O, Tsoar H (2012) Evolution of megaripples from a flat bed. Aeol Res 6:1–12
Zaidner Y, Porat, N, Zilberman E et al (2018) Geo-chronological context of the open-air Acheulian site at Nahal Hesi, northwestern Negev, Israel. Quatern Int 464:18–31
Zaineldeen U (2010) Palaeowind estimation of cross-bedding within the aeolian Kurkar layers of the Gaza Formation, Gaza Strip, Palestine. Geol Croat 63(1):55–65
Zilberman E (1991) Landscape evolution in the central, northern and northwestern Negev during the Neogene and the Quaternary Geol Surv Isr Rep GSI/45/90
Zilberman E, Porat N, Roskin J (2007) The middle to Late-Pleistocene sand sheet sequence of Kerem Shalom, western Negev-an archive of coastal sand incursion. Ministry of National Infrastructures, Geological Survey of Israel
Zviely D, Kit E, Klein M (2007) Longshore sand transport estimates along the Mediterranean coast of Israel in the Holocene. Mar Geol 238(1–4):61–73

Part II

Northern Israel

7 Lake Kinneret Watershed—Landscape Heterogeneity and Human Impact

Moshe Inbar

Abstract

The 2700 km^2 Lake Kinneret drainage basin exemplifies the diversity and complexity of the Mediterranean climate physiographic region. It is characterized by highly variable geology, geomorphology, soils, climate, vegetation and land use. Human activity proved to be the main factor affecting changes in this environment. This chapter deals with erosion forms and processes, as reflected in the fluvial system morphology. The effects of erosion are of major environmental concern in Mediterranean climate areas, which are characterized by relatively high sediment yields, although rates of erosion variate depending on climatic, physiographic and human factors. The long dry Mediterranean summer is dominated by a dry, high-pressure cell which in winter retreats to allow a series of cold fronts to cross the eastern Mediterranean from the west sea. The Kinneret drainage basin is semiarid (P/ETP = 0.20–0.50). The average rainfall is 700 mm and the evapotranspiration ETP = 1600 mm (P/ETP = 0.41). Only the mountainous areas receive more than 800 mm rainfall and have a subhumid climate. Nine landscape sub-units reflect. The geodiversity and heterogeneity of the Lake Kinneret watershed, characterizing a "mosaic" type of land. This region has a continuous and old history of human presence since early Paleolithic period, but the major effects in the environment occurred in historic times and has been exacerbated in the last decades.

Keywords

Geodiversity · Lake Kinneret · Jordan River · Human activity · Physiographic units

M. Inbar (✉)
School of Environmental Sciences,
University of Haifa, Haifa, Israel
e-mail: inbar@geo.haifa.ac.il

7.1 Introduction

Retaining biodiversity is considered a major goal by the scientific community for humans. However, the abiotic world of mountains, valleys and rocks which are commonly seen as stable and static, appear as endangered systems too. Human activities may degrade geodiversity and their impact will depend on the stability of the geomorphic systems (Schumm 1979; Gray, 2004). For example, geomorphic landforms like glaciers and wetlands can be severely damaged and their extinction is irreversible. A few decades ago, there was no major interest in geodiversity, but in recent years, there is an increasing concern on geodiversity issues, and new research groups on landform assessment and geodiversity are constantly established worldwide.

The Lake Kinneret basin in northern Israel is characterized by rich geodiversity and exemplifies the complexity of physiographic regions located within the Mediterranean climate. The present chapter aims at highlighting its special landforms and landscapes.

7.2 Geographical Setting and Landscape Units

Lake Kinneret drainage basin is part of the Dead Sea Transform that runs from southeastern Turkey to the northern end of the Red Sea Rift. The basin covers an area of 2700 km^2 and is characterized by a wide spectrum of ecosystems, landscapes and processes. The climate is typified by long dry summers and dominated by a dry, high pressure cell which retreats in winter to allow a series of small cyclonic depressions and cold fronts to cross the eastern Mediterranean from the west. In the northern Kinneret basin most of the area is semiarid (P/ETP = 0.20–0.50). The average rainfall is 700 mm and the evapotranspiration ETP = 1600 mm (P/ETP = 0.41). Only the mountainous areas receive more than 800 mm rainfall and have a

A. Frumkin and N. Shtober-Zisu (eds.), *Landscapes and Landforms of Israel*,
World Geomorphological Landscapes, https://doi.org/10.1007/978-3-031-44764-8_7

subhumid climate. In the south, mean annual rainfall is only 350–400 mm.

Geodiversity and heterogeneity of the Lake Kinneret drainage basin is reflected by nine landscape sub-units with highly variable geology, geomorphology, soils, climate, vegetation and land use, characterizing a "mosaic" type of landscapes (Shtober-Zisu and Inbar 2014) (Fig. 7.1).

7.2.1 The Hermon Mountain

The Hermon Mountain is the southern part of the Anti-Lebanon mountain range covering 580 km^2, that forms 21% of the total area of the Kinneret drainage basin and 37% of the Upper Jordan River basin (1500 km^2) (Fig. 7.2). This elongated mountain range is 45 km long, up to 20 km wide, and the summit reaches 2814 m. The Hermon Mountain is built mainly of carbonate rocks from the Jurassic period, with a total thickness of about 2000 m. Several early Cretaceous (146 to 115 Ma) volcanic intrusions are found across the mountain (Wilson et al. 2000). Its complex tectonic history involved the origin of an open NNE-SSW-trending anticline which has been rapidly uplifted in response to the Neogene tectonic activity along the Dead Sea fault system (Shimron 1998). Karst morphology is dominant across the massive limestone layers and the steep climatic gradient (Frumkin et al. 1998). Karstic surface features include lapiez, dolines and sinkholes. The Man Valley (Birket Man) is a large doline at 1500 m asl, with a central pool enlarged by shepherds for cattle drinking water (Fig. 7.2b). Caves are common; however, many are still unknown. The Banyas cave at the foothills of the mountain was the site of the Banyas spring, which feeds one of the three major tributaries of the Jordan River. Today the

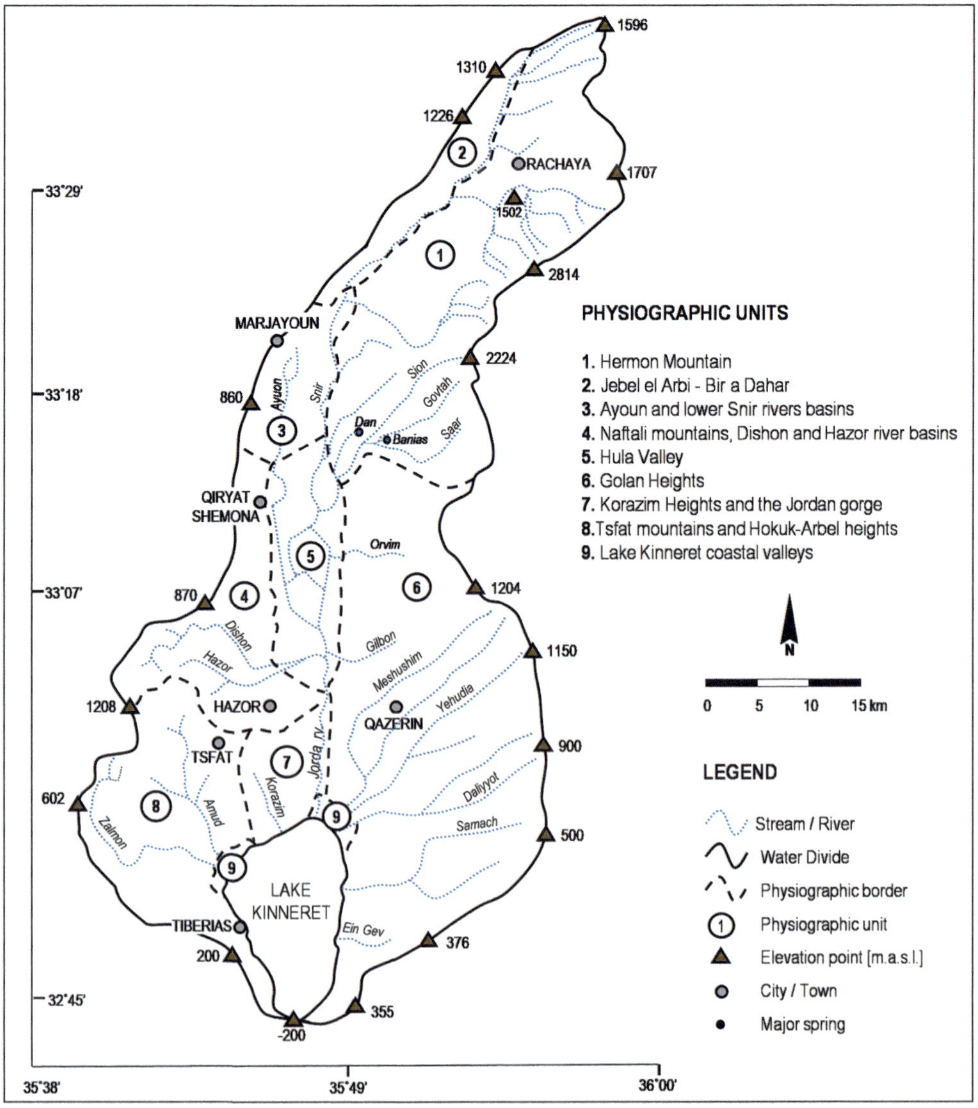

Fig. 7.1 Lake Kinneret watershed physiographic units (Modified after Shtober-Zisu and Inbar 2014)

Fig. 7.2 General view of the Hermon Mountain: **a** Small, parallel, elongated valleys, filled with snow remains, drained towards a karstic valley; **b** Man pond (Birkhet Man) doline at Hermon Mountain at 1500 m asl. Shepherds enlarged the natural depression for animal drinking water

spring emerges several meters below the cave due to natural groundwater decline.

Mount Hermon is eroded at different rates. Using concentrations of in situ cosmogenic ^{36}Cl in bedrock and sediments, it was found that the subalpine region (1700–2200 m, ~1300–1500 mm/yr) erodes at an average hilltop denudation rate of 19 ± 5 mm/ky, while hilltops at intermediate altitudes (1000–1600 m, ca. 900–1200 mm/yr) erode faster, at 42 ± 8 mm/ky (Avni et al. 2019).

Annual precipitation ranges from 1300 mm in the upper part of the mountain range to 600 mm at lower altitudes. Snow is about 50% of total precipitation at altitudes above 1500 m. Patches of snow are distinctive until late spring (Fig. 7.2a). The surface runoff is about 5% of the total water discharge of the upper Jordan basin. The Hermon aquifer feeds the main springs—Dan, Hasbani (Snir), and Banyas—of the Jordan River, as well as the springs of the Damascus basin in Syria.

7.2.2 The Upper Hasbani (Snir) River

This unit covers the western flank of the Hermon Mountain and the eastern part of the Jebel el Arbi and Bir a-Dahar anticlinal ranges in Lebanon. Elevation ranges from 770 to 1200 m asl and steep slopes of carbonate rocks characterize the area. The Hasbani (Snir) River drains the area, flowing southward for 40 km in Lebanon, crossing the border to Israel and joining the Banyas and Dan rivers to form the Upper Jordan River, contributing 26% of its total inflow (Shentsis et al. 2021). Annual precipitation is 900 mm and most of the area drains through the Snir River, mainly after heavy rains, with large amounts of sediments being carried downstream.

7.2.3 The Ayoun Valley and Lower Snir River Basin

This unit is part of the Jordan-Dead Sea Transform extension toward the north. The polje of Ayoun includes the Ayoun valley, which is bounded by faults and filled with alluvial deposits. Perennial flow from springs is mainly diverted for irrigation during the dry period. The eastern part of this unit includes the Lower Snir River, which is deeply incised in Pleistocene basaltic flows (Fig. 7.3).

7.2.4 Naftali Mountains and Dishon-Hazor River Basins

The western margin of the Hula Valley is made of the Naftali mountain range, an 800 m high escarpment formed by regional faulting. It extends from Nahal Kadesh Stream to the south, to kibbutz Kefar Giladi to the north, covering an area of about 50 km^2. The Naftali ridge rises from 80 m asl at its base to 900 m asl at the top and is dissected by over 30 slope-channels. These channels are straight, short, carry only ephemeral runoff, and are developed perpendicular to the piedmont, exhibiting steep gradients (20–60%) towards the valley (Fig. 7.4). The mountain front is exposed to fluvial erosion and intensive gravitational debris movements. A succession of four generations of fan-deposits accumulated during the last 1 million years, facing the channel mouths. The sediments consist of polymictic conglomerates, while carbonate soils and calcrete horizons have developed on top of the sediments. These outcrops extend up to an elevation of 300 m above the base level and their structure implies deposition under pulses of activity, followed by periods of quiescence and calcic-soil development from 1.1 Ma to 3 ky. Ages were obtained by K–Ar

Fig. 7.3 The lower Hazbani (Snir) River incised into the lower Pleistocene basalts, at Rajar (Israel-Lebanon border)

dating and OSL (Shtober-Zisu 2006). Field relationship between the sedimentary units and local faults indicate that no major faulting occurred along the Naftali escarpment during the late Pleistocene. The marginal faults displaced a 0.56 Ma old sedimentary unit during the mid-Pleistocene, and since then tectonic activity has shifted from the western marginal faults eastward, to younger faults deeply buried in the center of the Hula graben (Shtober-Zisu et al. 2008).

The Dishon and Hazor western river basins represent mainly a mountainous area with elevations ranging from 1200 m at the summit of Meron Mountain, with an annual mean rainfall of 1000 mm, to 100 m with 500 mm mean annual rainfall in the Hula Valley. Faults divide this synclinal area into several blocks, built by carbonate rocks from the upper Cenomanian to the Eocene. Caves and concave shaped indentations (Inland notches) are found in the cliffy limestone units (Fig. 7.5). Pleistocene basalt flows cover the Dalton, Alma and Dishon blocks. The western part is characterized by polygenetic, tectonic-karstic depressions. Runoff occurs during heavy rainfall storms only. The annual discharge of the Dishon and Hazor streams is about four millions of cubic meters (MCM). The main springs draining the area, Eynan and Te'o- discharge at the foothills of the mountain front, was 25 MCM before human

Fig. 7.4 Typical steep-gradient slope channels crossing the front of the Naftali Mountains. Four generations of fan deposits have accumulated along the mountain piedmont

Fig. 7.5 Inland notches cut through the carbonate (limestone) slopes of the Nahal Aviv Stream, a major tributary of the Dishon Stream

influence, to 6 MCM at present times. (Human influence will be discussed in the next paragraph).

7.2.5 The Hula Valley

7.2.5.1 General Characteristics

The Hula Valley is the most anthropogenically affected unit in the Kinneret drainage basin. In the last 100 years, human activities completely changed the physical and cultural physiography of the valley.

The Hula Valley is elongated and relatively flat-bottomed, 25 km long and 5–8 km wide. It covers an area of about 150 km^2, gently sloping from 100 m above sea level in the north to 65 m in the south. The general outline of the valley is determined by tectonic features of the transform. The Hula Valley is a deep tectonic depression formed during the last four million years. During the last two million years, the rate of sedimentation has been more or less equivalent to the rate of subsidence (Kafri and Lang 1987), and therefore, the valley floor maintains its relatively stable elevation (Heimann 1990). In a 2.8 km deep core drilled in the valley, several lava flows of 4 Ma to 1 Ma were found, indicating a subsidence rate of 0.5 to 0.7 mm/yr (Heimann 1990) and the coexistence of swamp and lake environment between 250 and 30 ka.

The annual rainfall ranges from 400 mm in the south to 600 mm in the northern part of the basin. The Jordan River tributaries converge in the valley. The Jordan flow is perennial due to the large karstic springs draining the Hermon mountain aquifer. At the valley outlet, the catchment area is 1492 km^2 and the mean annual flow (at the "Obstacle Bridge") is 434 mcm for the 1976–2009 period (Rimmer and Givati 2014). Deep alluvial soils cover the valley and are under intensive agricultural exploitation, producing mainly crops like cotton, peanuts, citrus and avocado (Fig. 7.6a).

7.2.5.2 Drainage of the Hula Lake and Valley

Local tectono-volcanic events led to the formation of a volcanic obstacle (flow) at the outlet of the valley, which caused temporary impoundments of the Jordan River, and the formation of temporary lakes, swamps and wetlands within the valley. The extent of the lake and swamps in the late Pleistocene was determined by climatic factors. During wet periods the lake extended over most of the valley, whereas during dry seasons the extension of the swamps was larger.

Wetlands are rare in the dry subtropical areas of the Levant. The Hula lake and swamp environment is one of the few wetlands in the entire region, and one of the oldest documented lakes and surrounding wetlands in history, with a rich and diverse aquatic biota and dense stands of papyrus. The Hula Valley lakes attracted human settlement as early as the prehistoric times. Freshwater strata of the prehistoric Benot Ya'aqov formation, at the southern outlet of the valley, were dated to 0.78 Ma (Goren-Inbar et al. 2000). Since then, people intervened in the hydrological cycle of the Jordan River through changes in land use, while in modern times major development projects have become further reasons of hydrological and landform alterations. Such changes included forest burning, deforestation, and overgrazing in early times, and the use of irrigation techniques in historic times.

Fig. 7.6 The Hula Valley: **a** a former lake and swamp transformed to fertile agricultural land; **b** thousands of cranes (*Grus grus*) in the restored Agmon Ha'Hula pond. About 500 million birds migrate twice a year through the Hula Valley

The drainage of the last lake/swamp system of the Hula Valley was initiated after the establishment of the State of Israel in 1948. The drainage of the lake and swamps came to an end in 1957, as a total area of 60 km^2 were drained. Further drainage works were conducted after the flood in January 1963, from areas that suffered from high groundwater level and were periodically affected by flooding. The Jordan channel was straightened and enlarged to conduct a maximum flow of 280 m^3/sec, which has a recurrence interval of 100 years.

7.2.5.3 Drainage Effects

The drainage of the Hula Valley caused to various environmental consequences, and these included:

- *Land Reclamation*—About 40 km^2 were gained for agricultural use, and the reclaimed lands were allocated to existing settlements in the region (Fig. 7.5a). An area of 3 km^2 was designated as a nature reserve to preserve the former wetland ecosystem. Most of the former swamp area, covered by peat soils (ca. 20 km^2), was managed by the Hula Authority. However, for more than three decades, most crops were not successful, and at the end of the 1980s, several 100 ha of agricultural land were abandoned.
- *Water table*—The water table that was close to the surface before the drainage, was artificially lowered by 1–2 m after the drainage. Also, the amplitude between winter and summer water levels increased, from 1 to 1.5 m before the drainage to 2–3 m after the drainage (Dimentman et al. 1992). These hydrological changes led to peat soil exposure, decomposition, peat fires and land subsidence.
- *Land subsidence*—Following the drainage, elevation losses at the central Hula Valley were 2.6 m for the period 1958–1988, at a rate of 8 cm/yr. The main causes of subsidence were: (1) compaction by mechanized agricultural machinery; (2) strong easterly winds causing erosion of bare topsoil in spring and autumn, when not covered by crops; (3) consolidation of the subsoil as a result of drainage; (4) dehydration and desiccation of the organic fabric of the soil in the drained and aerated part of the soil profile; (5) oxidation of the organic material; (6) burning of peat by surface and subsurface fires. Some of the underground fires lasted for long periods, being uncontrollable and creating hazardous subsurface caverns; (7) surface erosion by runoff floodwater and washing of soil particles (Levin and Shoham 1986).
- *Nutrient release*—Peat decomposition produced ammonia, which increased the release of nitrates. The built-up nitrates were washed during floods to Lake Kinneret, causing a major ecological problem. Being the main water reservoir of Israel, the supply of 400–3000 tons/year of nitrates increased eutrophication processes, a main concern affecting the lake's freshwater.
- *Peak flood changes and flooding*—The Flood Routing Index is the ratio between the Jordan River peak discharges at the southern outlet of the valley and the northern monitoring station. The mean value for 1959/60–1996/97 was 0.91 against 0.34 for 1935/36–1954/55. Although the catchment area increases by 75%, from 841 km^2 at the northern station to 1492 km^2 at the southern station, the flood peak discharge was less at the southern station. The flood wave movement was affected by the reservoir role of the Hula Valley, whereby the wave was attenuated by storage within the lake and swamps. The drainage diminished the "attenuation" of the flood wave, but there was still an important reservoir function for the valley in the reduction of the Flood Routing Index. The drainage and channeling works increased the capacity of floodwater transmission through the valley and shortened the interval from 2 to 7 days in the previous drainage period to 4–14 h after the drainage. The wave velocity was 4.2 km/h for the valley area and about 5 km/h for the Jordan River canyon area,

at 60 m^3/sec discharge floods after the drainage (Inbar 1982a). The Peakedness Index, which characterizes the flood regime of the river and is measured as the relationship between Qmax (peak discharge of the flood) and the total volume of the flood, rose considerably at the Jordan southern station after the drainage. Bank erosion in the canyon increased due to frequent high-discharge floods (Inbar and Even-Nir 1989).

- *Sedimentological changes*—The mean annual discharge of the suspended sediment of the Jordan River into Lake Kinneret approximates 70,000 tons or 44 tons/km^2/yr (Inbar 1982b). There are very few bedload measurements, but from general equations of the mechanics of bedload transport, it is estimated to be about 5% of the total detrital sediment supply of the river to the lake. The total annual detrital sediment deposited in Lake Kinneret increased about sixfold, from 20,000 tons/year in historic times to 128,000 tons/year right after the Hulla Valley drainage. Before the drainage, most of the sediments carried by the Jordan River were deposited in the swamp area and the Lake Hula. The changes in land use in the catchment area may have increased the sediment production, but not at such a rate.
- *Biotic loss*—The Hula Valley was considered to be the last freshwater wetland outpost facing the Saharo-Arabian desert belt (Dimentman et al. 1992). After the drainage, many animal and plant species were lost from the region, and some endemic species or subspecies were lost forever. As a major "fueling station" for migratory birds, the fishponds were a partial alternative, and flocks had to find other places (Fig. 7.6b).

7.2.5.4 The Restoration Project

The deterioration of the subsiding land caused partial abandonment of cultivation of the affected area. In the 1990s there was a turn to the restoration approach, due to environmental, physical and technological problems. This was not a shift to an ecological approach, but mainly a search for solutions to agricultural technological problems. The rationale was that tourism would be more profitable than agriculture, and the vision was more toward a "safari" site. In environmental issues, the question is whether it is even possible to "manage" a natural ecosystem (Conacher and Conacher 2000). Newson (1997) stated that perhaps the best way for wetland restoration is to return the ecosystem to a close approximation of its condition prior to disturbance, but "setting the clock back" cannot be achieved and restoration projects must adapt to environmental and society needs.

To solve the soil deterioration problems, a committee formulated the Hula Restoration Project aiming (1) to slow down the decomposition and subsidence processes of peat soils; (2) to minimize the flow of pollutants into Lake Kinneret, by diverting all water draining the peat soils to an irrigation reservoir; (3) to develop an ecotourism recreational center (Hambright and Zohary 1998; 1999). A new small lake, called Agmon pond, covering an area of 110 ha, was created in a purposefully excavated depression, and swamp plants like papyrus were artificially planted along the shores and in the pond itself. The new water source attracted thousands of birds in their migration to Africa and back (Figs. 7.6b and 7.7). The project included the drainage and building of a new set of canals, raising the water table to 75 cm below surface, in order to minimize nutrient release from the peat soils and to prevent peat fires. The cost of the project was about 25 million US$, the largest nature restoration investment in Israel, but it was only partially for conservation purposes; the major effort being to allow the agricultural work to continue.

7.2.6 The Golan Heights

The Golan Heights occupy an area of about 1300 km^2 to the east and northeast of the Lake Kinneret, between the Hermon anticline in the north and the Ajlun anticline in the south. They are bounded on the west by the fault system of the Dead Sea Transform, and to the east, they continue into Syria. Elevations in the northern part of the Golan Heights attain ca. 1200 m asl. From there, the plateau descends rapidly westwards to the Hula Valley and gradually to the southern Golan. The Golan is covered by Pliocene–Pleistocene basalts that erupted mainly from volcanic cones (Fig. 7.8a; Mor and Steinitz 1985; Weinstein et al. 2006). The rock column is composed of repeated sequences of basalt flows succeeded by clayey paleosols that formed during periods of quiescence. The thickness of the basaltic sequence exceeds 1000 m in the central part (Meiler et al. 2011), and thins southward and westward to approximately 30–50 m. Several deep canyons, mostly in the southern and western Golan, are carved through the entire basalt succession, exposing the underlying sedimentary units of the Golan syncline. In several canyons, lava flows from the final period of eruptive activity filled the low-lying parts of the existing topography during the Late Pleistocene (Mor 1993; Shtober-Zisu et al. 2018).

The annual rainfall ranges from 500 to 900 mm. Springs are common, but no major steady sources of water are found in the area. About 75% of total runoff is accomplished by floods caused by rainstorms. Rivers are deeply incised in their lower reaches, forming more than 100 m deep canyons and up to 50 m high waterfalls (Fig. 7.8c). Peak discharge and annual water volume were altered by the building of channel water dams for irrigation purposes, changing the natural flow regime (Fig. 7.8b; Shtober-Zisu and Inbar 2005).

Fig. 7.7 The Agmon Ha'Hula pond—a 110 ha pond restored for nature conservation within the Hula Valley. Photo courtesy of Shai Abramson

Fig. 7.8 The Golan Heights; **a** The basaltic plateau is dotted by cinder cones; **b** The Merom Golan reservoir in the central Golan Heights collects runoff during winter floods for irrigation purposes; **c** The 53 m high Gamla waterfall is the highest waterfall in the Golan Heights

7.2.7 The Korazim Heights and the Jordan River Gorge and Delta

The Korazim Heights is a volcanic block to the south of the Hula Valley, incised by the Jordan gorge in its eastern flank. The Tel Ruman volcano was probably the source of the lava in the late Pliocene times, but several other small basaltic cones are also present (Heimann and Ron 1993). Shallow soils cover the area, and land use is mainly pasture.

The Jordan River gorge is 10 km long; the river flows in a narrow and 200–250 m deep canyon entrenched in basalt flows (Fig. 7.9). Along the gorge active landslides provide boulders and most of the coarse sediment is carried by the river to the Betsayda Plain. The river channel is composed of gravel and boulders, and therefore a high shear threshold is needed for bedload transport. Almost all bed load sizes, including boulders of more than 1 m in diameter, move in events with a 1:10 year frequency (Inbar 2002).

From the outlet of the canyon to Lake Kinneret a planimetric view reveals three different fluvial systems: straight channel in the canyon, with an average slope of 2.5%, braided system in the boulder depositional environment, with a channel slope decreasing from 2.5 to 0.5%, and then a meandering system decreasing to 0.1% in the new delta area, entering Lake Kinneret (Fig. 7.10).

The main morphological change due to sediment transport increase was the formation of the Jordan River delta at the Jordan outlet into Lake Kinneret, caused by the rare catastrophic flood of January 1969 (Inbar 1987). Previously, the Jordan River did not develop a delta because of two main reasons: (a) most of the sediments settled in the Hula lake and swamps before its drainage in 1957 and (b) the strong lake current that dispersed the sediments. Since the inception of the delta, there has been a permanent growth in its size by consolidation of sediments, vegetation

Fig. 7.9 The Jordan River canyon. Slopes are eroded by landslides and sediments are transported towards Lake Kinneret. Photo courtesy of Abraham Graicer

development and a continuous sediment supply transported by the river. There is no historic evidence of a delta at the outlet of the Jordan River into Lake Kinneret. Two thousand years ago, Josephus Flavius, a historian and geographer, wrote: "The Jordan River flows directly into the lake", meaning without a delta as most of rivers. The delta covers today an area of 1–1.5 km^2 at low levels of −211–212 m (below sea level) of Lake Kinneret and the channel extends about 1 km into the lake (Fig. 7.10). The delta acts as a sedimentary environment, receiving fine materials, mostly fine sand and silt, transported by the fluvial system. The annual sediment yield from the Jordan basin is relatively low and reaches mean values of 44 tons/km^2/yr (Inbar 1982b). During the 1968/1969 season, large quantities of sediments were transported to the lake. One million tons, or about 20 times the annual average, is the rough estimate of sediment transported by the catastrophic flood. Concentration of suspended sediment during floods reaches values of 1000–7000 ppm. During the 1969 flood, a layer of fine sediments was washed from the alluvial braided area, with an assumed volume of half a million of cubic meters. The present delta itself acts as a morphological factor promoting the accumulation of sediments, it prevents rapid dispersion of sediments by the lake currents and serves as a self-perpetuating sediment trap.

Fig. 7.10 The Jordan delta entering Lake Kinneret. The delta was formed after the exceptional 1969 flood. Airphoto from August 4, 2017, at low level: −213.6 m. Photo courtesy of Abraham Graicer

7.2.8 The Tsfat Monocline and Hokuk-Arbel-Poriya Tilted Blocks

The Tsfat Mountains lie on a monocline structure affected by the main transform system, forming steep slopes from 900 m asl in the upper part to about 300 m at the footslope. Lithology comprises hard, karstified Eocene limestones with small patches of soil exposed mainly between the lapiez. The Hokuk block is covered by basaltic layers and Arbel (Fig. 7.11) is an Eocene limestone block covered partially by a basaltic flow. Runoff occurs during rainstorms. Spring water diverted today for consumption used to maintain a steady base flow throughout the year. The Poriya height is one of the tilted blocks forming part of the eastern lower Galilee, covered with Pliocene basalts, similar to the Arbel height, and bordering the southwestern area of the Kinneret lake.

7.2.9 Lake Kinneret Coastal Valleys

Two major coastal valleys are distinguished: Betsayda and Ginosar valleys. The flat bottomed Betsayda infill was formed by the Jordan and Golan Rivers deposits, several tens of meters thick. The upper layer, some 4 m thick, was formed during historic periods. During high lake water levels, coastal lagoons are formed. Groundwater level fluctuates according to the lake level. The wetland environment is a nature reserve, protected to conserve the richness in flora and fauna. The Ginosar valley is formed by the Amud, Tsalmon and Arbel rivers. The valley serves as a fertile agricultural land for the coastal settlements.

7.3 Environmental Hazards

In the last 100 years several natural disasters affected the Kinneret drainage basin: the 1927 earthquake in the Dead Sea Rift; the 1934 Tiberias floods; and the 1969 Jordan and northern rivers flood, with a recurrence interval of 1:200. Natural disasters were known also earlier, such as the major historic earthquakes of Beit Shean in 757 CE, Benot Yaakov in 1202 CE, and Zefat in 1759 and 1837. Several tsunamis in Lake Kinneret connected to major earthquakes were registered in the last centuries, at 749, 1759, 1837 CE (Ben Menahem 1991). Active landslides occur in the Jordan gorge, in the Golan rivers gorges, and in the Naftali and Zefat mountains escarpments.

7.4 Human Cultural Heritage

The Lake Kinneret (Sea of Galilee) and the upper Jordan River basin have a continuous, very long history of human occupation since early Paleolithic period, about 1.4 Ma, to the present times. Prehistoric archaeological sites were found in the basin: Ubeidiya—early Acheulian, Benot Yaakov site—780 ka, Ma'ayan Baruch—Acheulian,

Fig. 7.11 Nahal Arbel canyon, cuts between the Nitay and Arbel tilted block

Berekhat Ram—Acheulian, Amud cave, Kuneitra—Mousterian, Mahanaym, Ohalo, Ein Gev, Eynan—Neolithic, as well as numerous historic sites.

The region witnessed significant historic settlement like Tel Hazor site that ruled the region, the Israelite settlement at Tel Dan, as well as remarkable events in human history, like Jesus miracles at Lake Kinneret and Saladin battle at Karnei Hittin.

The Kinneret- Jordan basin is probably unique in its wide scenario of human cultures throughout prehistoric and historic periods, as its rich history and heritage warrants its inscription as a 'cultural landscape of universal significance'.

References

Avni S, Joseph-Hai N, Haviv I, Matmon A, Benedetti L, ASTER Team (2019) Patterns and rates of 103–105 yr denudation in carbonate terrains under subhumid to subalpine climatic gradient, Mount Hermon, Israel. GSA Bull 131(5–6):899–912. https://doi.org/10.1130/B31973.1

Ben Menahem A (1991) Four thousand years of seismicity along the Dead Sea Rift. J Geophys Res Sol Ea 96:20196–20216. https://doi.org/10.1029/91JB01936

Conacher A, Conacher J (2000) Environmental planning and management in Australia. Oxford University Press, pp 460

Dimentman C, Bromley HJ, Por FD (1992) Lake Hula, Reconstruction of the Fauna and hydrobiology of a Lost Lake. The Israel Academy of Sciences and Humanities. Jerusalem, pp 170

Frumkin A, Shimron AE, Miron Y (1998) Karst morphology across a steep climatic gradient, southern Mount Hermon, Israel. Z Geomorphol Supp 109:23–40

Goren-Inbar N, Feibel CS, Verosub KL, Melamed Y, Kislev ME, Tchernov E, Saragusti I (2000) Pleistocene milestones on the Out-of-Africa Corridor at Gesher Benot Ya'aqov, Israel. Science 289:944–947. https://doi.org/10.1126/science.289.5481.944

Gray M (2004) Geodiversity: valuing and conserving abiotic nature. John Wiley & Sons, pp 433

Hambright KD, Zohary T (1999) The Hula Valley (Northern Israel) Wetlands Rehabilitation Project. In: Streever W (ed) An international perspective on wetland rehabilitation. Kluwer Academic Pub, pp 173–180

Hambright KD, Zohary T (1998) Lakes Hula and Agmon: destruction and creation of wetland ecosystems in northern Israel. Wetl Ecol Manag 6:83–89

Heimann A, Ron H (1993) Geometric changes of plate boundaries along part of the northern Dead Sea transform: geochronologic and paleomagnetic evidence. Tectonics 12(2):477–491. https://doi.org/10.1029/92TC01789

Heimann A (1990) The development of the Dead Sea Rift and its margins in Northern Israel during the Pliocene and the Pleistocene. Geological Survey of Israel Report 28/90, Jerusalem, pp 87 (in Hebrew, English abstr)

Inbar M (1987) Effects of a high magnitude flood in a Mediterranean climate: A case study in the Jordan River basin. In: Mayer L, Nash D (eds) Catastrophic Flooding. Allen and Unwin, pp 333–353

Inbar M (2002) A geomorphic and environmental evaluation of the hula drainage project Israel. Aust Geog Stud 40(2):155–166. https://doi.org/10.1111/1467-8470.00171

Inbar M (1982a) Spatial and temporal aspects of man-induced changes in the hydrological and sedimentological regime of the upper Jordan River. Isr J Earth Sci 31:53–66

Inbar M (1982b) Measurement of fluvial sediment transport compared with lacustrine sedimentation rates: the flow of the River Jordan into Lake Kinneret. Hydrol Sci J 27(4):439–449. https://doi.org/10.1080/02626668209491125

Inbar M, Even-Nir M (1989) Landslides in the upper Jordan River gorge Pirineos. J Mount Ecol 134:23–40

Kafri U, Lang B (1987) New data on the late quaternary fill of the Hula Basin, Israel. Isr J Earth Sci 36:73–81

Levin I, Shoham D (1986) Ground elevation loss in the reclaimed hula swamp during the period 1958 to 1980 land subsidence. IAHS Pub 151:458–479

Meiler M, Reshef M, Shulman H (2011) Seismic depth-domain stratigraphic classification of the Golan Heights, central Dead Sea Fault. Tectonophysics 510(3):354–369. https://doi.org/10.1016/j.tecto.2011.08.007

Mor D (1993) A time-table for the Levant Volcanic Province, according to K-Ar dating in the Golan Heights, Israel. J Afr Earth Sci 16:223–224. https://doi.org/10.1016/0899-5362(93)90044-Q

Mor D, Steinitz G (1985) K-Ar ages of the Neogene Quaternary basalts around the Yarmuk Valley. Isr Geol Soc Ann Mtg 1985, Yotvata.

Newson MD (1997) Land, water and development, 2nd edn. Routledge, London and New York, p 423

Rimmer A, Givati A (2014) Hydrology. In: Zohary T, Sukenik A, Berman T, Nishri A (eds) Lake Kinneret- ecology and management. Springer Verlag, pp 97–111

Schumm SA (1979) Geomorphic thresholds: the concept and its applications. T I Brit Geogr 485–515

Shentsis I, Rosenthal E, Flexer A, Inbar N (2021) Assessing water withdrawals in scarce-data transboundary areas by use of dynamic precipitation—flow relationships: the case of the Hasbani River Basin. Water 13(11):1440. https://doi.org/10.3390/w13111440

Shimron AE (1998) Tectonic evolution of the southern Mount Hermon. Geological Survey of Israel Report GSI/10/98

Shtober-Zisu N, Inbar M (2005) Downstream effects of river impoundment on hydrological and geomorphological aspects of bedrock rivers (Golan Heights, Israel). IASH Sp Pub 299:219–224

Shtober-Zisu N (2006) Quaternary tectonic geomorphology along the Naftali mountain-front, Ph.D dissertation. Tel Aviv university, pp 161

Shtober-Zisu N, Greenbaum N, Inbar M, Flexer A (2008) Morphometric and geomorphic approaches for assessment of tectonic activity Dead Sea Rift. Geomorphology 102(1):93–104. https://doi.org/10.1016/j.geomorph.2007.06.017

Shtober-Zisu N, Inbar M (2014) Geomorphology. In: Zohary T, Sukenik A, Berman T, Nishri A (eds) Lake Kinneret-ecology and management. Springer Verlag, pp 69–77

Shtober-Zisu N, Inbar M, Mor D, Jicha BR, Singer BS (2018) Drainage development and incision rates in an upper pleistocene basalt-limestone boundary channel: The Sa'ar Stream, Golan Heights, Israel. Geomorphology 303:417–433. https://doi.org/10.1016/j.geomorph.2017.12.002

Weinstein Y, Navon O, Altherr R, Stein M (2006) The role of lithospheric mantle heterogeneity in the generation of Plio-Pleistocene alkali basaltic suites from NW Harrat Ash Shaam (Israel). J Petrol 47(5):1017–1050. https://doi.org/10.1093/petrology/egl003

Wilson M, Shimron AE, Rosenbaum JM, Preston J (2000) Early Cretaceous magmatism of Mount Hermon. Northern Israel. Contrib Mineral Petr 139(1):54–67. https://doi.org/10.1007/s004100050573

8 Volcanic Landscapes in the Golan Heights

Doron Mor, Yishai Weinstein, Ran Calvo and Ram Weinberger

Abstract

The Golan Heights is a basaltic plateau of about 1000 km^2, which declines from 1100 m asl on its northern end to 300–350 m asl on its southern end, above the Sea of Galilee and the Yarmouk River. It is part of the Harrat Ash Shaam volcanic field, which is the largest volcanic field in western Arabia. The volcanics of the Golan Heights cover a syncline, which is located between the anticlines of Ajlun and Mt. Hermon on its southern and northern sides, respectively. The basaltic section is thickest in the central Golan (up to > 600 m in the eastern Golan, next to the water divide), while it decreases to tens of meters to none toward the south and north. Most of the area is covered by basaltic flows, but the landscape of the eastern and northeastern Golan is dotted by more than 60 cinder cones, most of them probably belonging to the last volcanic period, which occurred between 200 and 100 ka. Phreatomagmatic features are relatively scarce, and there are only two cases, which were produced by the interaction of magma with groundwater or with water of a diverted stream. The main water divide is located at the eastern part of the Golan Heights, close to the border between Israel and Syria, dividing between water flowing to the Sea of Galilee and to the Yarmouk River, which re-joins the water from the Sea of Galilee in the Jordan River. A relatively large fraction of the Golan Heights precipitation forms runoff (15%), while another 16% recharges the regional aquifer, which is usually quite deep (several 100's m).

Keywords

Basaltic plateau · Cinder cones · Phreatomagnatism · Sea of Galilee

8.1 Introduction

The Golan Heights is a basaltic highland (plateau), located at the northeastern part of Israel (inset to Fig. 8.1), with altitudes declining from ca. 1100 m at its northern end to 300–350 m at its southern end (Fig. 8.1), above the Sea of Galilee and the Kinnarot Valley. The Golan Heights is located at the northwestern edge of the large volcanic field of Harrat-Ash-Shaam, which covers about 45,000 sq. km in northern Saudi Arabia, central Jordan, southern Syria and northeastern Israel (Fig. 8.2), and is the largest amongst the western Arabia volcanic fields (locally known as "harrats"; inset to Fig. 8.2). The basaltic succession is relatively thin (typically 0–100 m) in the southern Golan (Dafny et al. 2003), as well as in the north, where outcrops of Cretaceous to Eocene sedimentary rocks are often observed (e.g., "Shamir Windows", Sneh and Weinberger 2003; Fig. 8.3). On the other hand, it is quite thick ($\geq$ 250 m, Fig. 8.4a) in the western Golan at the Yehudiyya Plateau (Dafny et al. 2003; Weinstein et al. 2006; Reznik and Bartov 2021), while in the eastern Golan, close to the water divide, it reaches up to >600 m (e.g., Dafny et al. 2003; Fig. 8.4a).

Most of the Golan Heights borders are naturally defined. The western border is mostly tectonically demarcated by

D. Mor (✉)
Kibbutz Merhavia, Israel
e-mail: doronmor35@gmail.com

Y. Weinstein
Department of Geography and Environment, Bar-Ilan University, Ramat Gan, Israel
e-mail: weinsty@biu.ac.il

R. Calvo · R. Weinberger
Geological Survey of Israel, 32 Yesha'ayahu Leibowitz St, 9692100 Jerusalem, Israel
e-mail: rani.calvo@gsi.gov.il

R. Weinberger
e-mail: rami.weinberger@gsi.gov.il

R. Weinberger
Department of Earth and Environmental Sciences, Ben Gurion University of the Negev, Beersheba, Israel

A. Frumkin and N. Shtober-Zisu (eds.), *Landscapes and Landforms of Israel*,
World Geomorphological Landscapes, https://doi.org/10.1007/978-3-031-44764-8_8

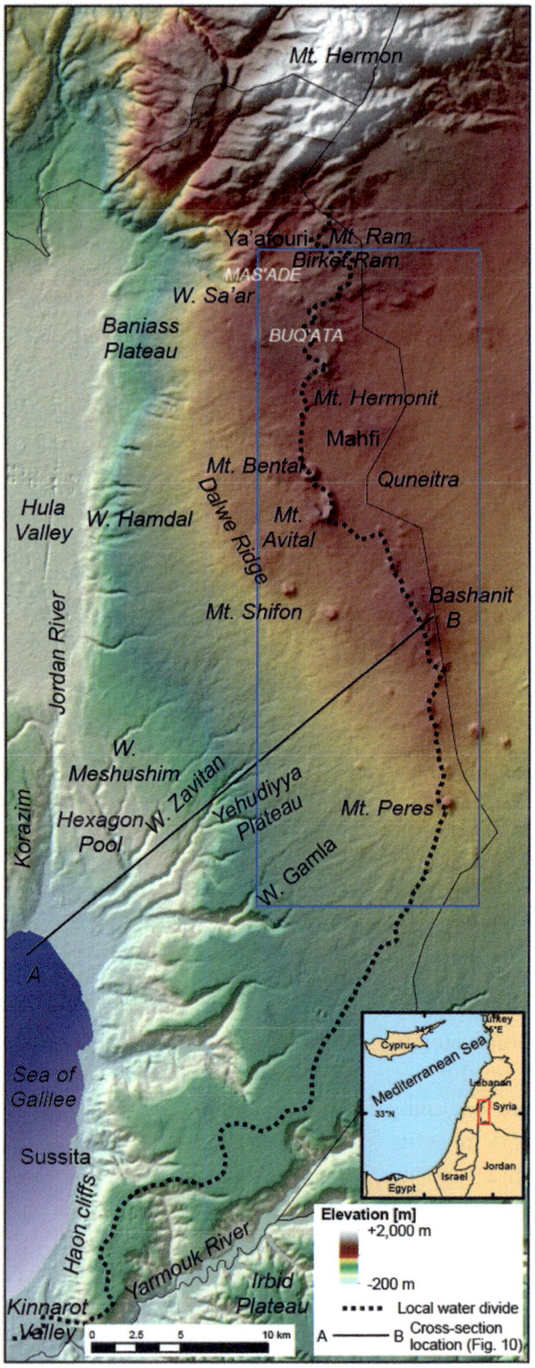

Fig. 8.1 Physical map of the Golan with locations mentioned in the text (blue rectangle denotes the area shown in Fig. 8.6a; for cross section A-B see Fig. 8.20)

the escarpments of the Dead Sea Transform system (DST); on its northern border, the steep southeastern flank of Mt. Hermon anticline defines a sharp, geological boundary of the young volcanic terrain along Wadi Sa'ar; on its southeastern side it is the steep slopes toward the Yarmouk River that delineates the basaltic plateau (Fig. 8.5). On its eastern side, the Golan Heights extend into Syria. This paper focuses on the western Golan, which covers about 1000 km^2 and is politically defined by the Israel-Syria border, cutting slightly east of the N-S-stretching water divide of the Golan (Fig. 8.1).

The plateau is dotted with over 60 cinder cones and some other volcanic features (e.g., basaltic flows, small shield-type volcanoes, tuff rings), which form a typical intraplate volcanic landscape (Schmincke 2004) on its eastern and northern parts (Fig. 8.6a). On its western and southeastern sides, the basaltic plateau is cut by deep canyons and ravines, which commonly drain toward the DST-related rift valley. Due to the differences in the thickness of the basaltic cover (Fig. 8.4a), the stratigraphic section incised by the canyons of the central Golan is wholly basaltic, while in the southern and northern parts of the Golan the canyons also expose sedimentary rocks (Fig. 8.3). The western part of the Golan Heights is divided by two off-DST faults into three subterrains (Fig. 8.6b). The Sheikh-Ali fault cuts and forms a topographic step between the southern Golan plateau and the downfaulted Yehudiyya Plateau of the central Golan, with the latter being the only area in the Golan that gradually declines toward the rift valley (the Sea of Galilee, at -210 m). The Shamir fault, at the northwestern edge of the Golan, separates the downfaulted Banias Plateau from the main Golan Heights to the south (Fig. 8.6b).

The main water divide of the Golan Heights, which separates the drainage basins of the Sea of Galilee and the central Jordan River, mostly stretches on its eastern side, along a high volcanic terrain (1300–700 m asl, Fig. 8.1). The cinder cones and the volcanic features are mostly arranged along a belt made of two parallel rows, aligned NNW-SSE (Fig. 8.6a), while the water divide follows the eastern row (Figs. 8.5 and 8.6a), close to the Israel-Syria border, except for at the Quneitra area, where it shifts to the western row (Fig. 8.5). Basaltic flows stretch from the cinder cone lines both to the east and to the west, producing both low lands (the relatively long-stretching flows) and shoulders (steps) on the flanks or at the cone foot. In the southern part of the Golan, where apparently there are no cones, the water divide runs across the plateau, with very little topographic expression.

Volcanic landscape is an ever-changing scenery that is closely related to the style and activity of volcanism (e.g., Schmincke 2004). The Golan includes a high variety of volcanic features and styles of activity, including strombolian, hawaiian and phreatomagmatic edifices, with times of activity well constrained by dozens of K–Ar and Ar–Ar ages, as well as detailed mapping and subsurface data. As such, the Golan nicely demonstrates a versatile intraplate volcanic landscape in a relatively small area. In this chapter, we will describe the tectonic setting and the volcanic succession, followed by the current volcanic landscape and its

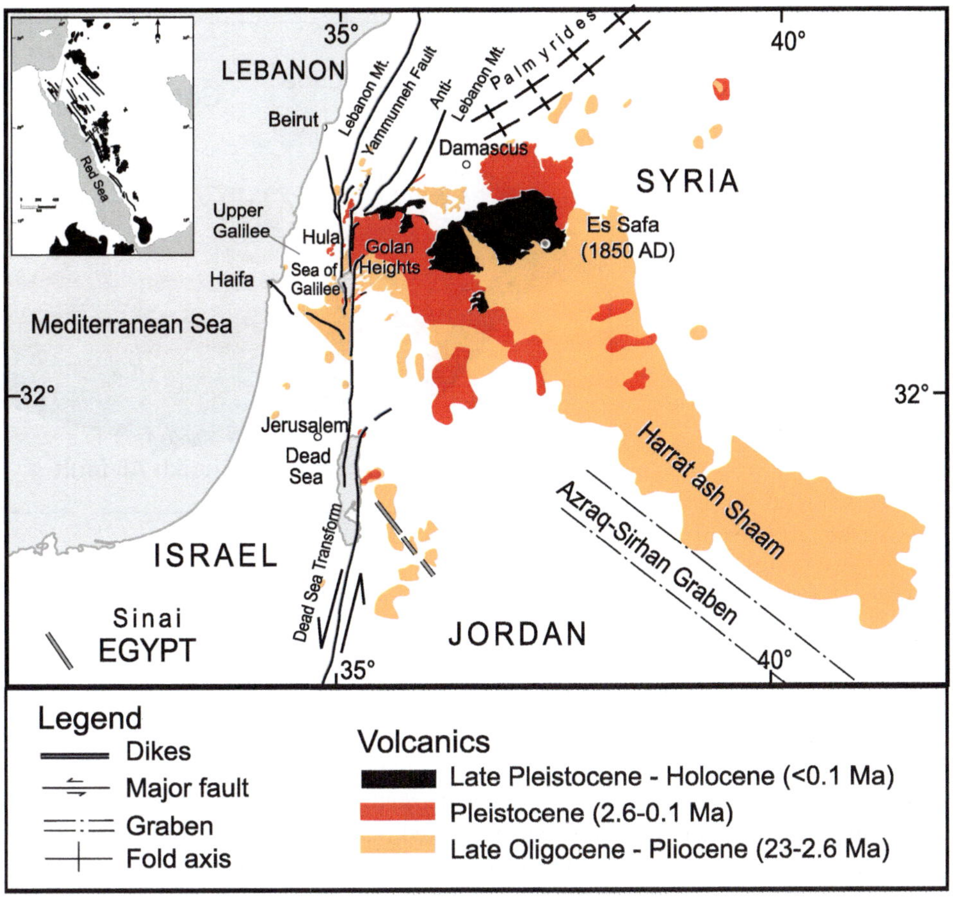

Fig. 8.2 Map of Harrat Ash Shaam, with the Golan Heights located at its northwestern edge (modified after Weinstein and Heimann 2016). Inset shows volcanic fields in the western Arabia

relations to hydrology and fluvio-morphology, which are also affected by the location of the volcanic field next to an active plate boundary, the Dead Sea Transform. Finally, we will shortly discuss the pre-basaltic, Late Miocene landscape.

8.2 Tectonic Setting

The main tectonic feature in the region is that of the ~N-S-trending DST, a left-lateral (sinistral) plate-bounding fault system between Arabia and Africa (Sinai) plates. The DST, which has been active since the Early to Middle Miocene (e.g., Nuriel et al. 2017) delimits the Golan Heights on the west. The DST next to the Golan comprises parallel, mainly left-stepping, fault strands that bound the Hula Valley and the Sea of Galilee pull-apart basins, which are separated by the topographic saddle of the Korazim block (Fig. 8.6b). While on its southern end, the topographic descent from the Golan Heights towards the Sea of Galilee basin is very sharp, with a drop of>500 m over a distance of 1–2 km along one main normal fault (accompanied by bed tilting and abundant slumps), the slopes toward the DST in the central and northern Golan are more moderate. In two areas, along the Sheikh Ali and the Shamir normal faults, the main fault deviates to NE, forming moderately sloping plateaus along the northwestern hanging-walls (Yehudiyya Plateau and Banias Plateau, Fig. 8.6b).

8.3 The Pre-basaltic Geology

Basically, the sub-basaltic structure of the Golan Heights is that of a large syncline, located between the anticline of Ajlun (Jordan) in the south and that of Mt. Hermon (Israel-Syria-Lebanon) in the north (Fig. 8.3). The center of the syncline is located in the central Golan, where the thickest section of overlying basalts was detected (Fig. 8.4a, Dafny et al. 2003). The exposed section includes Lower Eocene to Lower Miocene marine formations (chalks, limestones, chert), covered by Middle Miocene to Pliocene fluvio-lacustrine sediments (chalks, mudstones, sandstones, conglomerates) in the south, and Upper Cretaceous to Middle Eocene carbonate rocks in the north. Jurassic to Lower Cretaceous carbonate rocks, shales and magmatic intrusions are exposed at Mt. Hermon, just north of the Golan Heights. To the south, outcrops are mostly within the deeply entrenched ravines that drain west, to the Sea of Galilee, or to the southeast,

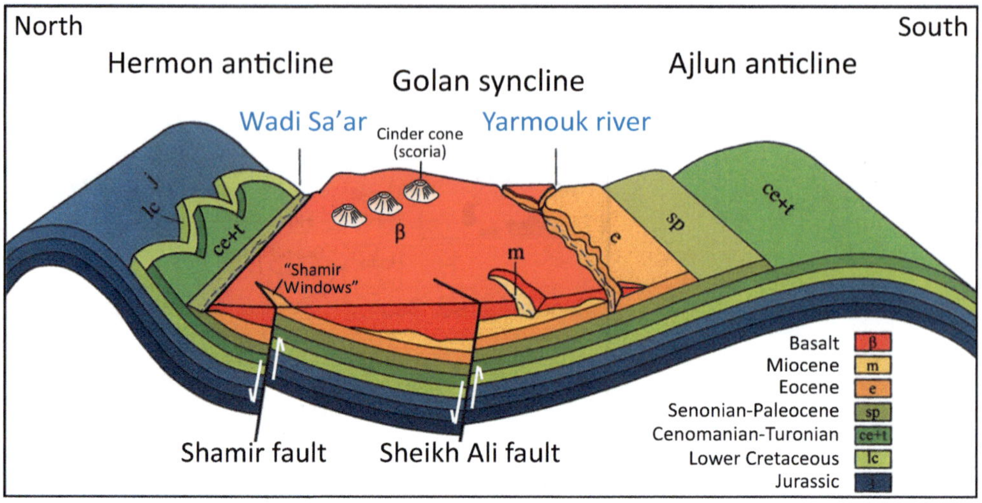

Fig. 8.3 3-D schematic geologic block diagram of the Golan and surroundings (modified after Eidelman and Enmar 2014)

toward the Yarmouk River, which exposes the thicker and deeper parts of the Eocene rocks.

8.4 The Volcanic Succession

The cause for magmatic activity and volcanism in the Harrat Ash Shaam, including the Golan, is greatly debated, with explanations ranging from extension along the Red Sea rift/embryonic ocean (i.e., 'passive magmatism', White and MacKenzie 1989; Regenauer-Lieb et al. 2015), to lateral extension of a hot asthenospheric mantle that propagates northwards from the large Afar Plume (an ascending anomalously hot mantle, that is, "active magmatism") (e.g., Camp and Roobol 1992; Weinstein et al. 2006). Volcanism in the Afar region commenced at about 31 Ma (Oligocene) with the eruption of the Yemen Plateau basalts as part of the Afar Plume activity. In Harrat Ash Shaam, the volcanic activity started as early as 26 Ma (Ilani et al. 2001), and continued with some volcanic paucities until the very recent (e.g., Dubertret and Dunand 1955). In Israel, volcanism can be subdivided into two periods: Miocene and Plio-Pleistocene (e.g., Shaliv 1991). While the Miocene volcanism (17–8 Ma) concentrated on the western side of the DST (Shaliv 1991; Weinstein 2000), it hardly occurred in the Golan Heights. In Ha'on cliffs, east the Sea of Galilee, there are a few basaltic flows within the Miocene sediments, and in Masade village, just next to Wadi Sa'ar, there is an occurrence of ~13 Ma basaltic dike (Weinberger et al. 2003). Except for these rare occurrences, the Golan volcanism mainly fits in the second volcanic period. Pliocene basaltic flows (5.5–3.5 Ma, Mor 1993; Heimann et al. 1996) form the present surface in the southern Golan, where it is commonly covered by a few meters of argillaceous soils. Ages of surface basalts are somewhat younger in the western Golan Heights (Yehudiyya plateau, 2.7–1.7 Ma; Mor 1993; Weinstein et al. 2006), although Pliocene ages were reported at depth (3.5 Ma at 150 m depth, Weinstein et al. 2006). Ages of surface basalts get younger toward the eastern and northeastern parts of the Golan Heights (Fig. 8.6b; Weinstein et al. 2020), with most ages < 1 Ma (Weinstein et al. 2013, 2020; Shtober-Zisu et al. 2018). In places, these young basalts are underlain by older basalts, in particular in the eastern Golan (the area south of Quneitra), where the volcanic section could be as thick as > 600 m (Dafny et al. 2003; Fig. 8.4a). Notably, an age of 3.5 Ma was reported for surface basalts from Mt. Ram at the northeastern end of the Golan (Mor 1993; Heimann et al. 1996; Shtober-Zisu et al. 2018).

While ages of the Pliocene basalts, as well as those of Mid- to Late Pleistocene basalts are well constrained by Ar–Ar ages (Heimann et al. 1996; Weinstein et al. 2013, 2020; Behar et al. 2019); those of the early Pleistocene are less well-defined (only K–Ar ages, Mor 1993; Heimann et al. 1993). Accordingly, the possibility of continuous volcanism between 5.5 and 1.7 Ma cannot be discarded. Nevertheless, clear indications of volcanic paucity between 1.7 and ~1.0 Ma and between 0.6 and 0.2 Ma were detected (Weinstein et al. 2013, 2020). Moreover, while volcanism continued further east (SE Syria) until recent times (e.g., Dubertret and Dunand 1955; Heimann et al. 1993), with a boiling lava lake observed at the Tel Es Safa area (ca. 120 km east of the DST, Fig. 8.2; NASA website) around 1850 AD, no activity took place in the Golan Heights since 100 ka (Weinstein et al. 2020).

Composition-wise, the Pliocene to Early Pleistocene basalts are mainly 'alkali-basalts' (i.e., basalts enriched with alkali elements, mainly Na and K). On the other hand, the

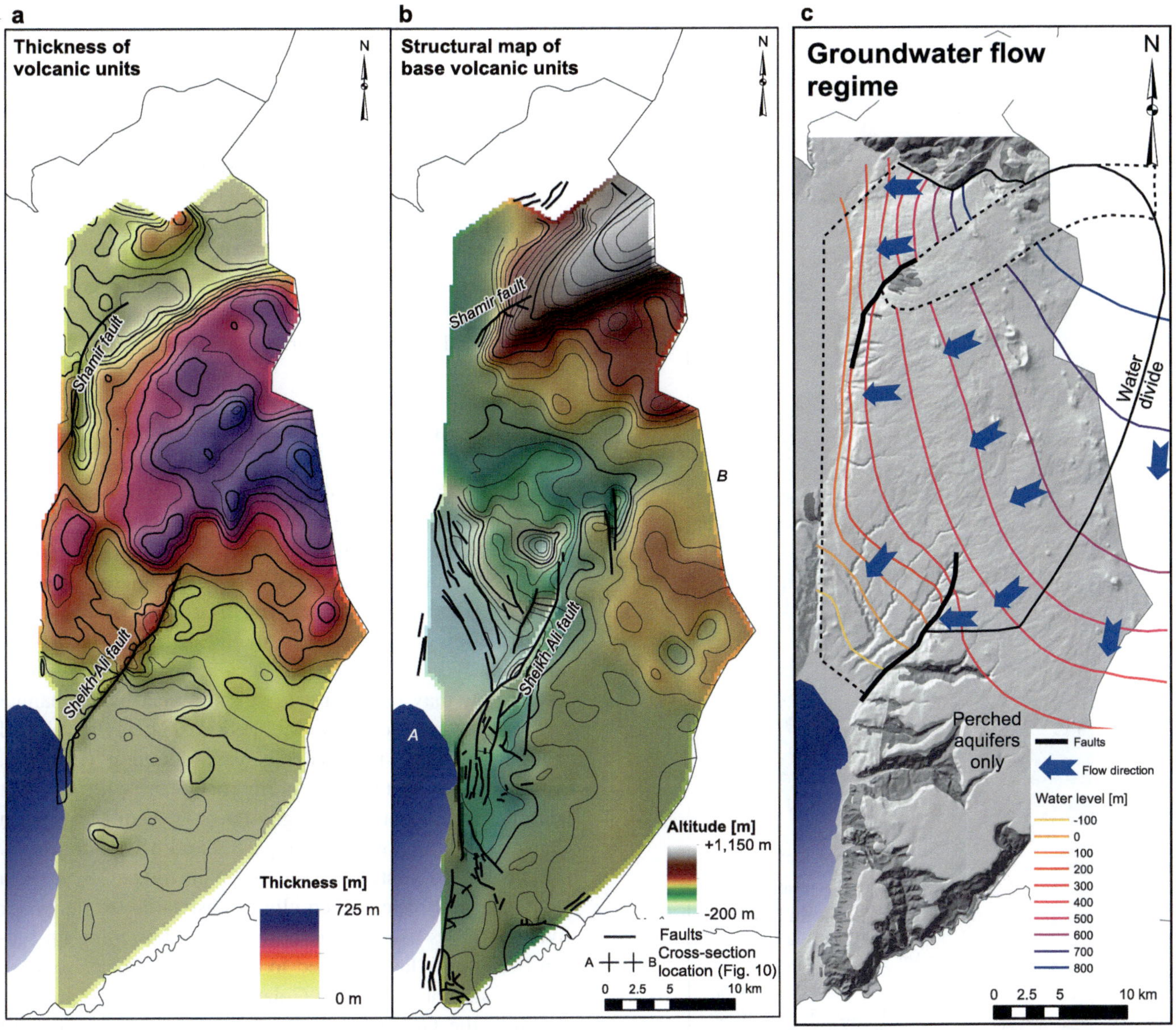

Fig. 8.4 **a** Isopach map of the Plio-Pleistocene basaltic section in the Golan (modified after Dafny et al. 2003); **b** Paleo-relief, structural map of the base of the basaltic section (modified after Dafny et al. 2003); **c** Regional groundwater level map in the Golan (after Dafny et al. 2003).

Mid- to Late Pleistocene basalts (<1 Ma) are mainly basanites, which is the name reserved to basalts enriched with the mineral nepheline (in general, the trend from "regular" basalt to alkali-basalt and to basanites is that of lowering silica and of increasing alkali element content in the rock).

8.5 Volcanic Landscape

Most of the Golan plateau is covered by basaltic flows (Fig. 8.7), whereas the eastern and northern Golan are also dotted by dozens of cinder cones (locally called "Tel"). These are mostly arranged along two N20W-oriented lines (Fig. 8.6a). Many other cones are observed farther east (in Syria), some of them arranged along distinct lines (e.g., prominent N30W cone lineaments ca. 100 km from the Israel-Syria border, north of Tel Es Safa; Garfunkel 1989). Young flows, which are associated with the cinder cones, are mostly short, within 2 km of the associated eruption site, although some of them are relatively thick (10–20 m). However, some of the flows, both of the late phase (<1 Ma, "Mueisse" basalts; Mor 1993), and of the early phase (Pliocene and Early Pleistocene) basalts in the western and southern Golan went much farther, over distances up to>20 km. These differences, which do not necessarily correlate

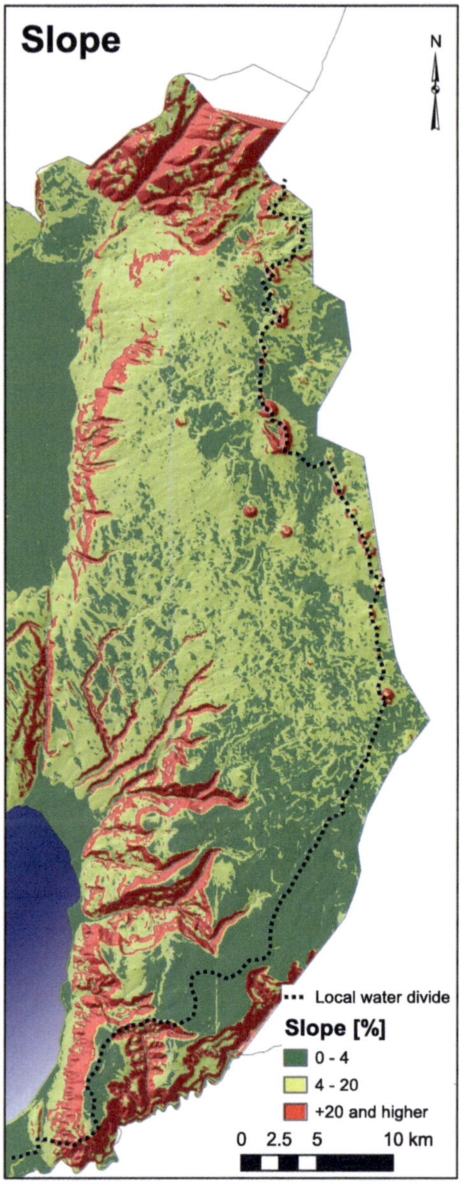

Fig. 8.5 Slope map of the Golan

with flow thickness, probably reflect lava viscosity. This in turn is at least partly related to crystal ('phenocryst') content, which is higher in most young lavas. In some cases, several cinder cones, as well as some basaltic necks, are built on top of locally elevated, small shield-type volcanic structures, such as the Mt. Avital, Dalwe Ridge, Bashanit and Mahfi complexes (Fig. 8.1; Mor 1973; Weinstein 2007). K–Ar and Ar–Ar ages from Mt. Avital and Bashanit indicate that while the shield structures were built during Mid-Pleistocene (~1–0.6 Ma), the cinder cones and associated lavas are mainly from the latest phase (0.2–0.1 Ma; Weinstein et al. 2013, 2020).

While the typical landscape of the northeastern Golan is that of cinder cones, it also hosts several valleys, which are of two types. The first type is located on the Muweisse basalts, which are young lavas that cover low areas between the cones (e.g., between Mt. Avital and the Dalwe Ridge, Fig. 8.1). These valleys are densely cultivated, mainly by apple and other deciduous orchards, and recently by a growing number of vineyards, which allows for the production of the famous Golan Wines. The other type includes two valleys—Quneitra and Ya'afouri, both floored by tuff and fine scoria deposits, produced by the phreatomagmatic eruption of Mt. Avital and Birket Ram, respectively.

8.5.1 Lava Flows

The Golan basalts are of the 'A'ā type (a Hawaiian term for rough rubbly surface composed of broken lava blocks) ranging in thickness from 1–2 to>20 m, showing well-developed bottom "clinkers" (i.e., fragmented lava, Fig. 8.8), when exposed in canyons, road cuts or man-made channels. Top clinkers are usually not observed due to erosion. However, in some cases, local changes to a wholly-clinker flow may be observed (e.g., in a man-made channel at the western slope of Mt. Avital). Nicely-developed columnar jointing in basalts is very common (Fig. 8.8), mainly in the western and northern Golan, with the best examples found in the 'Hexagon Pool' in Wadi Meshushim ("hexagons" in Hebrew; see Figs. 8.9 and 8.1 for location). The central to eastern Golan is characterized by a lava step-style landscape, which climbs eastwards (to the water divide). This is usually due to lava flow shortening up-section (i.e., upper lavas flowed to a shorter distance from the eruption sites at the eastern Golan).

While the Golan basalts are generally younger toward the northeastern Golan (Fig. 8.6b), fine exposures of in-canyon young basalts appear in several locations. This includes (1) the Sa'ar Basalt (0.11 Ma, Shtober-Zisu et al. 2018), which partly fills and creates a stepped landscape down the gorge of Wadi Sa'ar that cuts between the carbonates of Mt. Hermon and the northern Golan young basalts (Figs. 8.1 and 8.3), (2) Wadi Hamdal at the western Golan (next to the Hula Valley, Fig. 8.1) and (3) the Yarmouk River basalts (two flows, 0.8 and 0.4 Ma, Mor 1993; Figs. 8.6b and 8.10), which fill the bottom part of a deep canyon, cutting into Eocene chalks (see below).

8.5.2 Cinder Cones

Most of the Golan cinder cones rise tens of meters to slightly over 100 m above their surroundings (Figs. 8.11).

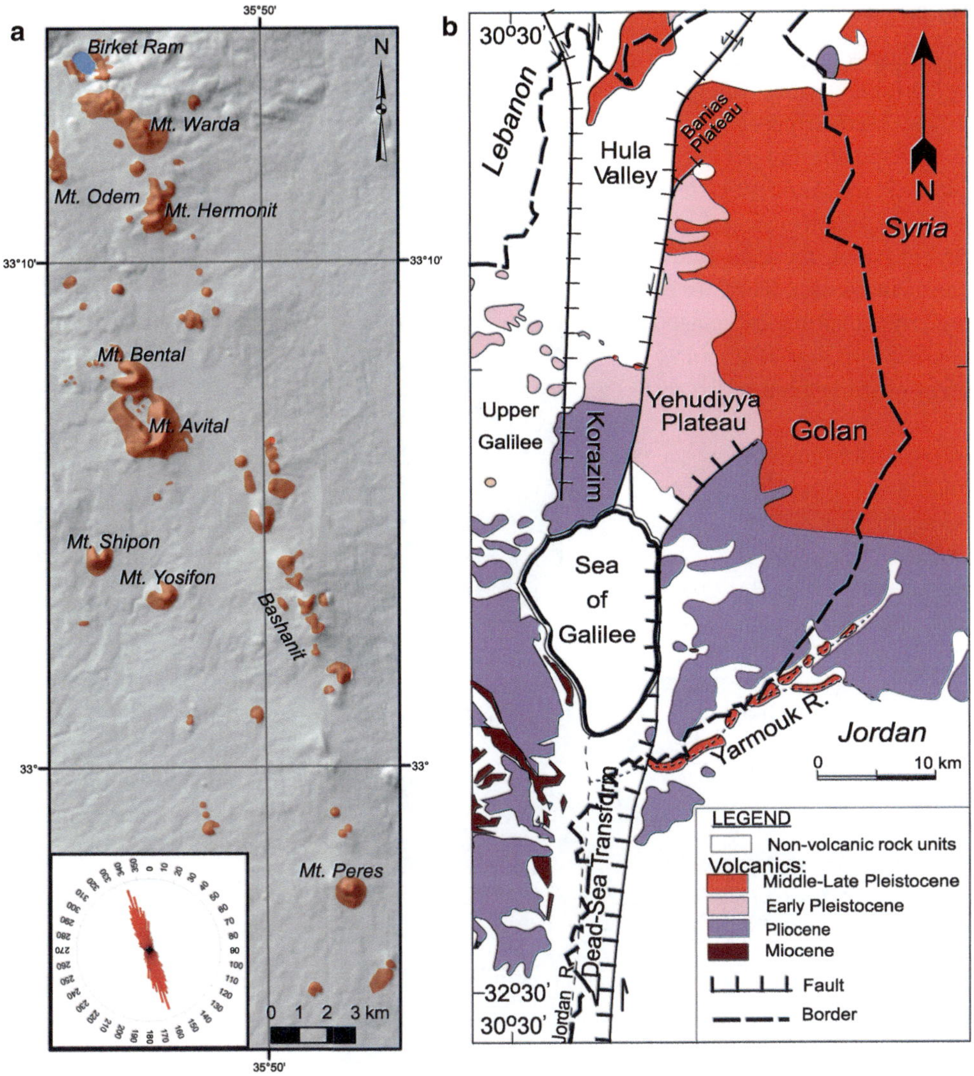

Fig. 8.6 **a** Volcanic cones in the Golan Heights; also shown is a pure cone-to-cone GIS analysis (including direction and length), plotted in a traditional rose diagram; **b** Volcanological map of the Golan (main units determined chronologically, modified after Weinstein et al. 2006).

Exceptional are Mt. Peres, in the south of the cone chain, and Mt. Avital volcanic edifice, west of Quneitra (Fig. 8.1), which rise to more than 200 m above the neighboring lowlands (Figs. 8.12 and 8.13). Craters are rare in the Golan, and the only complete one is preserved at the top of Mt. Peres (Fig. 8.13a). In most cases, craters could hardly be identified or are located at the side (mostly the western to northwestern side) of the cinder cone. A special case is that of Mt. Shifon (Fig. 8.1), which is a double-top hill, connected by a saddle with depressions on its both sides. This was attributed by Mor (1973) to a local fissure eruption, which emplaced more ejecta in its central part (the two tops) than on its both edges.

Cinder cones are mostly composed of coarse scoria (up to tens of cm) and up to 3.5 m aerodynamically-shaped, spindle-style volcanic bombs (Fig. 8.14). Fine scoria (mm to a few cm), erroneously called 'tuff', are also present in some of the cones (mainly, Mt. Peres, Mt. Bental and Mt. Hermonit; Singer et al. 2004; Weinstein and Weinberger 2006), where they are heavily quarried for home and industrial purposes. Tuff and volcanic breccias are found only at two sites—the complex of Mt. Avital and the ma'ar of Birket Ram.

Commonly, the cinder cones are cut on their sides by late lava flows, for example, at Mt. Avital and Mt. Shifon (Fig. 8.1), which apparently represent the culminating volcanic event in these structures (e.g., Weinstein et al. 2013). A special case is that of Mt. Bental, where the erupting flows broke and carried away the whole western flank of the cone, whose detached blocks are still observed as small hills on top of the 2 km-long flow (Mor 1973; Fig. 8.15b).

Ben Asher et al. (2017) showed a large asymmetry in slope erosion (soil creep) of four cones from the Golan, with the south-facing slopes apparently eroding faster than the northern ones, mainly due to the large difference in vegetation coverage.

Fig. 8.7 Basaltic flows in Wadi Gamla, central Golan (photo taken by R. Weinberger)

Fig. 8.8 Columnar basalt on top of bottom clinker, which covers an earlier lava flow in western Golan (photo taken by R. Weinberger)

Another interesting feature is that of several basaltic rings, one inside the other, on top of a small scoria cone (Dalwe ridge, N33°04′39″ E35°45′27″; Fig. 8.13b). These structures were interpreted as repeated lava lake formations in a cinder cone crater, followed by drainage (withdrawal) of the magma, leaving behind concave basaltic layers (Mor 1973).

8.5.3 Phreatomagmatic Structures

The deep groundwater table, down to > 200 m in the eastern Golan (Dafny et al. 2003), results in relatively "dry" volcanism, with hardly any evidence for magma-water interaction (phreatic), except for two structures: Mt. Avital and the ma'ar of Birket Ram (Weinstein 2007; Weinstein and Weinberger 2007; Shaanan et al. 2011; Fig. 8.1). The ma'ar lake (0.9×0.7 km) of Birket Ram (Fig. 8.16a) occurs where the regional groundwater table intersects the surface at the northeastern edge of the Golan, allowing for the development of local phreatomagmatism (Shaanan et al. 2011; see Fig. 8.16b for details). Unlike Birket Ram, the groundwater table at the Mt. Avital area is 250–300 m beneath the surface (Dafny et al. 2003), and the transition to phreatomagmatism is attributed to the diversion of a nearby stream due to local damming by a 115 ka lava flow. This resulted in flooding of the eruption site, which changed its character from "dry" (strombolian) to a "wet" (phreatomagmatic)

Fig. 8.9 The Hexagon Pool in Wadi Meshushim, western Golan (photo taken by D. Rotem)

Fig. 8.10 Cliff of a valley-filling basalt in the Yarmouk River canyon, covering chalks of Eocene age (photo taken by R. Weinberger)

type (Weinstein 2007; Weinstein and Weinberger 2007; Fig. 8.15a,b). The phreatomagmatic eruption, mainly characterized by base surges and volcanic breccia (Fig. 8.17), constructed tuff levees (partial tuff ring), partly welded (e.g., Fig. 8.11), which border the current Avital "depression" on its eastern and western sides (Fig. 8.15a), as well as covered the Quneitra Valley by tuff and fine scoria deposits (Fig. 8.18). The reshaping of the Avital area may have prevented further access of water to the eruption site and restored the dry activity with ~100 ka basaltic flows that breached the structure, and which demonstrate the final activity in the Mt. Avital complex. The diversion of the stream and the flooring of Quneitra Valley by the tuff resulted in the development of a lake, which attracted animals and humans in ancient times (a Musterian site, dated to 54 ka, Ziaei et al. 1990). The dry–wet-dry transitions in the Avital complex are a good example of abrupt landscape changes in the Golan.

8.5.4 Jubas

Jubas are an Arabic term for semi-circular depressions ('wholes' in free translation), tens to a few hundred meters in diameter and up to several tens of meters in depth, which are abundant in the northern Golan (Fig. 8.19). Of the identified 24 jubas, 21 are located on Mid-Pleistocene basalts, while three occur within Eocene carbonates on the western slope of the Golan, above the Hula Valley ("Shamir Windows"; Fig. 8.3). In most cases, their depth/width ratio is < 0.6, with positive slopes (i.e., < 90°) into the depression; however in one case (Jabab Miss, 33°11′32.6″N 35°43′01.2″), the ratio is > 3 (Frumkin and Naor 2019), with partly negative slopes (> 90°). Jubas roughly form a SW-NE lineament (Mor 1973). The formation of these structures is highly debated, with explanations varying between magmatic gas explosions (Mor 1973, 1986) to weathering or collapse features. The latter include basaltic pseudo-karst

Fig. 8.11 The cinder cone of Mt Bental in a photo taken from the south (Mt Avital). On its right side, the photo shows layers of welded tuff, deposited during the phreatomagmatic eruption on the top of Mt Avital (photo taken by Y. Weinstein)

weathering (Inbar 1984), and collapse into either underlying karstic caves (Weinstein 2002) or pit craters (i.e., emptied shallow magma pockets; Frumkin and Naor 2019). The explosion concept is doubted, since no ejecta was found around any of these structures (Weinstein 2002; Frumkin and Naor 2019). On the other hand, the collapse concept was challenged by modelling (Maimon and Heimann 2010), which did not favor the formation of smooth and shallow depressions, although modelling was attributed to a homogeneous medium (rather than to a pile of basaltic flows with abundant discontinuities in the form of columnar joints) and it did not take into account subsequent erosion processes (Frumkin and Naor 2019). These unique features are awaiting further investigation.

Fig. 8.12 The cinder cone of Mt Avital in a photo taken from the southwest. A late stage that erupted from the cone's western flank left its top double-peaked (photo taken by Y. Weinstein)

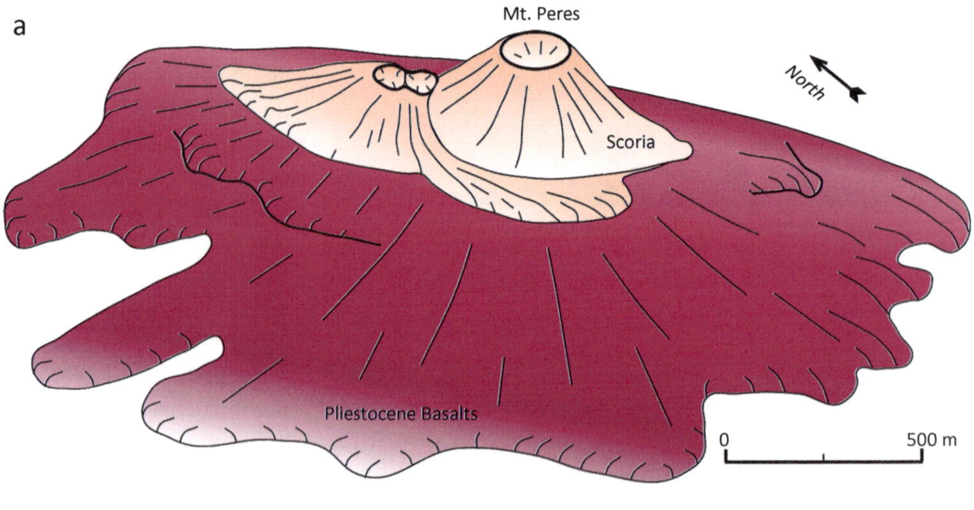

Fig. 8.13 **a** Schematic presentation of Mt. Peres, the only cinder cone with complete round crater in the Golan Heights (Israeli side); **b** Model of the evolution of a basaltic ring feature on top of scoria on the Dalwe Ridge (See Fig. 8.1 for location).

Fig. 8.13 (continued)

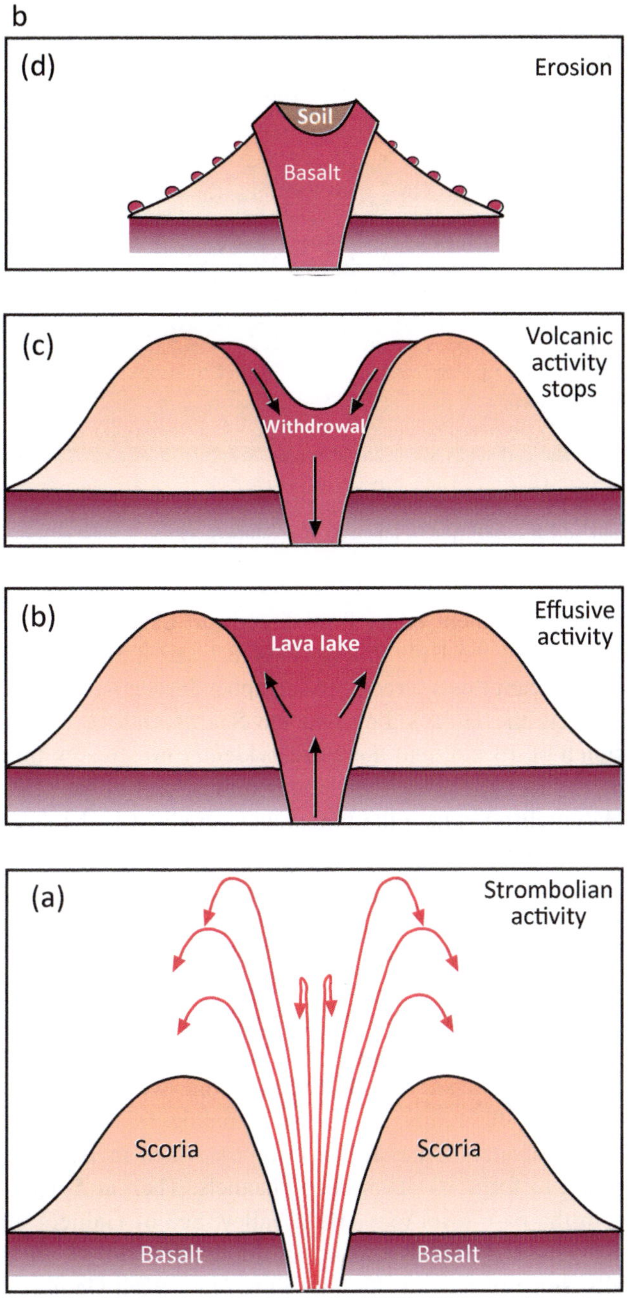

8.5.5 Xenoliths—Visitors from the Mantle

Mantle xenoliths, mostly green to reddish olivine-rich peridotites, are common in both basalts and pyroclastic deposits of the eastern and northern Golan, suggesting a relatively fast ascent from mantle depths. The tuff of Birket Ram, at the northernmost part of the Golan (Fig. 8.1), carries a unique assemblage of xenoliths, which includes upper mantle amphibole-pyroxenites, lower crustal granulites (mainly pyroxene and plagioclase) and amphibole (kaersutite) megacrysts (e.g., Mittlefehldt 1984; Stein et al. 1993), which are shining in black all around the maar lake. The possible genetic relationship between this unique xenolith assemblage and the phreatomagmatism is still poorly understood. No xenoliths were found in the Pliocene and Early Pleistocene basalts of the southern and western Golan.

8.6 Fluvial Morphology

The main water divide of the Golan separates the catchment of Sea of Galilee (Lake Kinneret) and that of the Yarmouk River, which joins the central Jordan River several kilometers south of the lake (Fig. 8.1). Streams in the Sea of

Fig. 8.14 A spindle-shaped bomb in fine scoria at the northeastern flank of Mt Avital (photo taken by R. Weinberger)

Galilee catchment either flow directly to Lake Kinneret or discharge to the Hula Valley (Figs. 8.1 and 8.5). Only one small stream flows directly to the upper Jordan River which connects the Hula Valley and the Sea of Galilee along a stretch of 13 km, and there are evidently no streams flowing to Lake Kinneret in the southernmost part of the Golan (Ha'on cliffs), where the topographic gradient is the steepest (Fig. 8.5).

Streams in both the northern and the southern Golan flow from the east to the west, while in the central Golan (Yehudiyya Plateau, Fig. 8.5) flow pattern is toward the southwest, and streams are typically much longer (25–30 km, compared with 10–20 km in the former). Drainage basins are relatively small, while the largest are in the Yehudiyya Plateau (Wadi Meshsushim, with 160 km^2). Typically, all streams start as very shallow (several meters deep), channels ("massils" in Arabic), which incise downstream to form 10–40 m deep channels. Then at 5–15 km from the base level (e.g., Hula Valley, Sea of Galilee), the channels dramatically transform, through a series of waterfalls, some of them as high as 50 m (e.g., Wadi Gamla), to well-developed steep-sided canyons, with walls as high as 150–200 m. The canyons persist for a few km, and then open into V-shaped valleys, which widen downstream. The transition between shallow channels to deep canyons occurs farther away from the DST in the western and southern Golan (discharging to the Sea of Galilee) than in the northern Golan, where streams drain to the Hula Valley.

Channel incision rates were calculated for streams at both northern and the southern borders of the Golan Heights. Ar–Ar ages of incised and gorge-filling basalts allowed for the determination of incision rates at 20–30 cm kyr^{-1} over the past 760 ka for the Wadi Sa'ar (Shtober-Zisu et al. 2018), where incision was mainly at the expense of the easily-eroded Cretaceous carbonate rocks. These authors also found that back-erosion of the channel followed a higher rate (68 cm $kyr-^{1}$) during the past 100 ka.

On the southern end of the Golan, along the Yarmouk River, incision rates were also assessed based on gorge-filling basalts (Fig. 8.6b). This includes the Yarmouk Basalt, which filled an already-existing 450 m deep gorge (at its outlet to the Jordan Valley) at 0.8 Ma (K–Ar ages, Mor 1993), and which was followed by channel incision of 95–225 m since then, indicating an incision rate of 12–28 cm kyr^{-1} (Zilberman and Begin 1997), very similar to the above Sa'ar stream rates. The top of another gorge-filling basalt, the Raqad Basalt (0.4 Ma, Mor 1993), is located 25–80 m above the current channel, yielding incision rates of 6–20 cm $kyr-^{1}$ (Zilberman and Begin 1997), similar though slightly lower than the above, which could suggest recent slowing of channel deepening. Similar incision rates, also based on basalt K–Ar ages, were found in a gorge east of the Dead Sea (Wadi Zarqa Main, 120 km south of the Yarmouk River; 23–33 cm kyr^{-1} during the last 6 Ma, Steinitz and Bartov 1992; Zilberman and Begin 1997).

The canyons of the central Golan incise 150–200 m into basalts with typical age of 1.8 Ma (Yehudiyya Plateau surface basalts, Weinstein et al. 2006). Assuming incision started soon after basalt emplacement, this suggests an average incision rate of 8–11 cm $kyr-^{1}$, somewhat slower than the above rates. Nevertheless, incision depends on drainage base level subsidence, which occurred later at the northeastern Sea of Galilee (base level for central Golan streams; see Hurwitz et al. 2002), as compared with the Kinnarot Valley and the Hula Valley (base level for central the Yarmouk River and the Sa'ar stream, respectively).

8.7 Hydrology

Precipitation in the Golan typically varies between $>$ 1200 and $<$ 500 mm a^{-1} (Dafny et al. 2003). A relatively large fraction (14–15%) of this develops into surface runoff, which was the reason for the construction of water reservoirs on many streams in the central Golan, mostly just ahead (east) of the waterfall/canyon reaches of the streams. Another 16% of the precipitation recharges the regional aquifer. In most of the Golan, the regional aquifer water table is quite deep, typically 200–300 m below the surface (Dafny et al. 2003), except for its northeastern end, where it is exposed at the surface (the maar lake of Birket Ram, Figs. 8.1 and 8.4c). The aquifer mainly discharges along

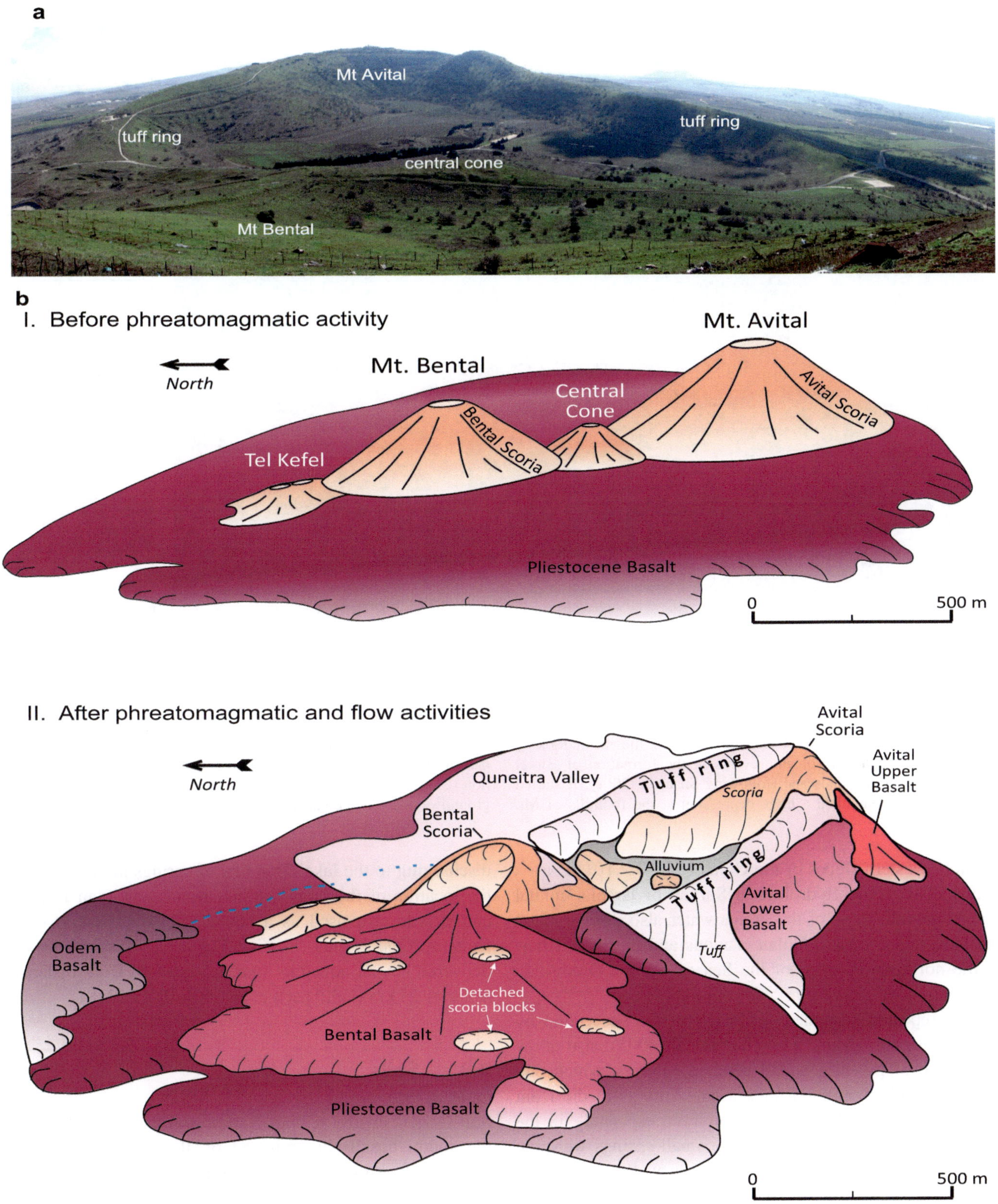

Fig. 8.15 The complex of Mt. Avital. **a** Photo of the tuff ring and the cinder cone of Mt. Avital (photo taken from Mt. Bental to the south); **b** Mt. Avital complex, before and after the phreatomagmatic eruption. At about 120 ka, the complex included four cinder cones, which were emplaced on an older (700–800 ka) small basaltic shield. Then, at about 113–114 ka, a water stream that was blocked by a lava flow changed its course and got into the eruption site, next to the 'central cone'. This resulted in phreatomagmatic blasts of base surges that built levees of tuff and volcanic breccia on the east and west of the complex (forming a central valley), as well as floored the Quneitra Valley (east of the structure) by tuff deposits. Activity in the complex concluded by lava flows that broke from the side of Mt. Bental (107 ka) and close to the top of Mt. Avital (99 ka) (Weinstein et al. 2013)

Fig. 8.16 The maar of Birket Ram. **a** Photo of the maar, looking northwards; **b** Chain of volcanic events in the formation of Birket Ram and the former landscape (following Shaanan, 2009). It started with lava flows that filled a large depression at the foothills of Mt. Hermon, followed by the build-up of a scoria cone at the southwestern part of the current lake, and culminated with a phreatomagmatic eruption (ca. 135 ka, the result of the interaction of magma with shallow groundwater), which created the maar lake (Shaanan et al. 2011)

the northwestern slopes of the Golan (35–50 × 10^6 m^3 a−^{1}, Dafny et al. 2003; Hydrological Survey of Israel's annual reports), while some of the groundwater also discharges at the base of the central Golan canyons and in alluvial fans at the northeastern edge of the Sea of Galilee (Bethsaida Valley).

The repeated sequences of basalt flows, succeeded by clayey paleosols (Mor 1973; Dafny et al. 2003), also allow for the development of small perched aquifers, which discharge as either perennial or seasonal springs at the front of the lava flows. This is very common, especially in the central Golan Heights, where these small springs were the nucleus of small settlements in ancient as well as in modern days.

In the south, the combination of a thin basaltic cover and underlying lacustrine chalk or siltstone allows for the formation of a long line of sub-basaltic springs along the edge of the plateau, above the Sea of Galilee and the ravines discharging to the lake. This line of springs largely affected the location of the large settlements in the Golan Heights, and it was also used as a source of extra water for the Roman Polis, Hypos (Sussita, Fig. 8.1), where water was delivered by an aqueduct (partly tunneled) from distances as far as 25 km (Fahlbusch et al. 1994; Tsuk 2018).

8.8 Soils

Soil cover in the Golan varies significantly. While in the south, basalts are covered by thick montmorillonitic (smectitic) grumusols, the western Golan is mostly rocky with small patches of protogrumusols and basaltic lithosols, and soils of the wetter northern Golan are of the Red Mediterranean type, controlled by kaolinitic clays (Dan and Singer 1973). Fossil, buried soils between lava flows, exposed in gullies and canyons in the western Golan,

Fig. 8.17 Fine-grained surge deposits on top of lithic-rich volcanic breccia on the slopes of Mt Avital (photo taken by Y. Weinstein)

are similar in composition to those in the southern Golan (montmorillonitic, Yaalon et al. 1984), suggesting quite similar climate during the past several million years.

The difference between soil cover of the southern and western Golan (thick grumusols vs. rocky soils) is striking, considering the fact that basalts in both areas are quite similar in composition (alkali basalts, Weinstein et al. 2006) and climate is very similar (dry Mediterranean) with 400–500 mm a^{-1} (e.g., Dafny et al. 2003). Although basalts in the south are older (5–4 vs. ~2 Ma, respectively), it seems unlikely that this is the sole reason for the difference between the two areas. It was suggested by Yaalon et al. (1984) that the main difference is due to surface slope and therefore, soil erosion. Figure 8.5 shows that while slopes in the southern Golan plateau are less than 4% (mostly, no more than 1–2%), those in the central Golan (Yehudiyya Plateau) are mainly 4–20%, which supports the explanation proposed by Yaalon et al. (1984) regarding the differences between the two areas.

A recent study by Za'arur (2021), based on soil chemistry and ^{36}Cl ages, suggests that soils in the south are not older than several thousand years, therefore genetically decoupled from the underlying basalts. This work examined soils from high relief areas, and the ^{36}Cl ages were determined on samples from the steep western Golan. The thick soils of the southern Golan plains are awaiting further research.

8.9 Pre-basaltic Late Miocene Landscape

Figure 8.4b shows a structural map of the base of the Plio-Pleistocene basalts. It appears that the Late Miocene land surface declined from the east to the west-southwest, with altitudes of 1100–350 m asl beneath the current water divide (declining north to south) and 250 m asl in the southern Golan. The NE-SW gradient in the southern Golan is $\ll$ 1%, suggesting a very flat Late Miocene landscape, quite similar to the present. On the other hand, the pre-basaltic gradient in the central to western Golan is significantly higher than in the south (> 1%, see Fig. 8.4b), and the surface gradually declines toward the current Sea of Galilee, similar to the

Fig. 8.18 Inter-layering of tuff and fine scoria in Quneitra Valley, east of Mt Avital. Tuff layers are more massive and are prominent, compared with the more erodible scoria layers (photo taken by D. Mor)

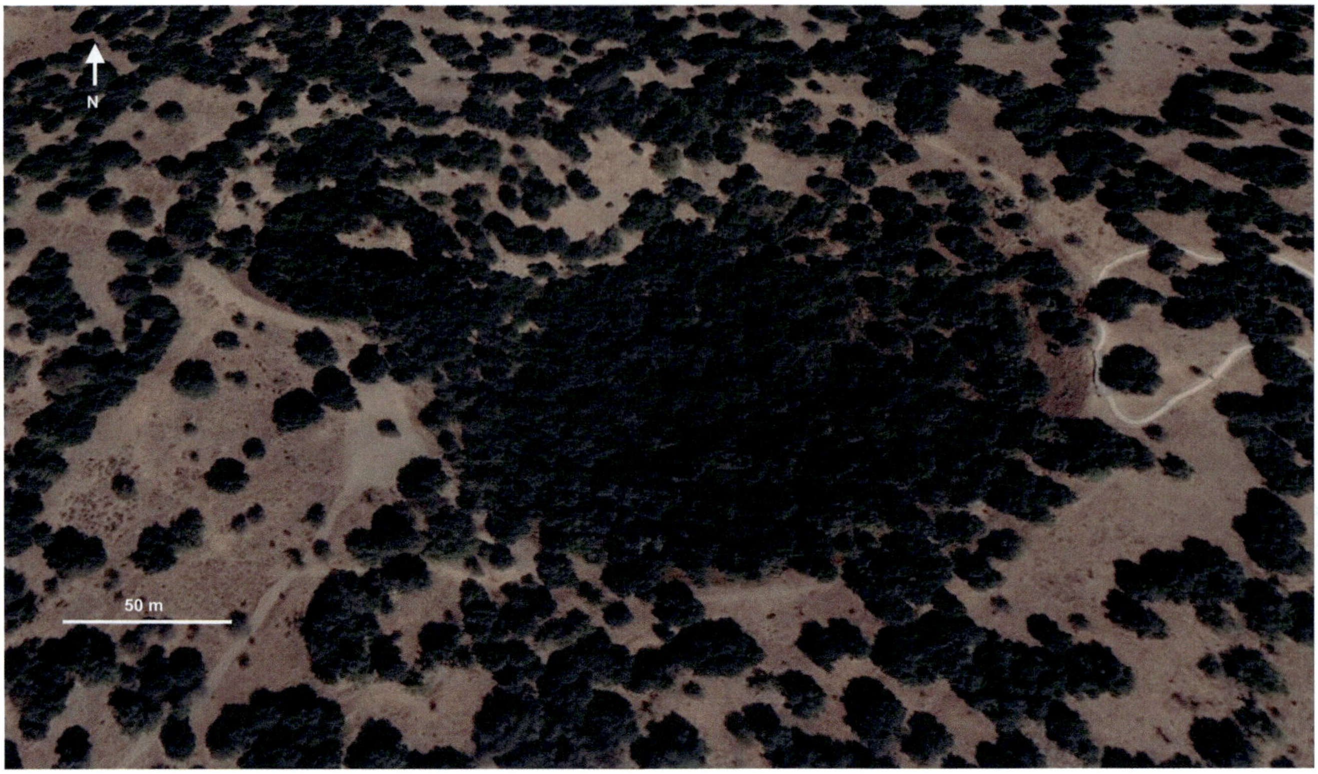

Fig. 8.19 A Google Earth image of the large depression (>200 m wide) of Jubat el Kabire at the northern Golan. Please note the parasitic juba on the western side of the large juba

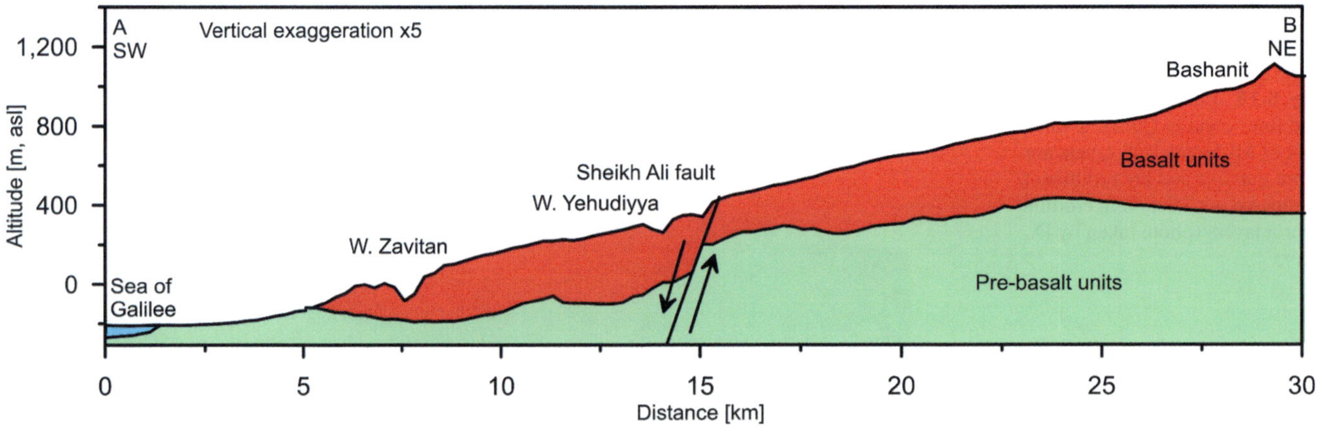

Fig. 8.20 Cross section from Bashanit (the Golan water divide) to Lake Kinneret (A-B in Fig. 8.1)

present landscape. Reznik and Bartov (2021) show a prominent increase in the thickness of the Middle Miocene Hordos Formation (fluviatile to lacustrine sediments) across the Sheikh Ali fault, from 300 m asl on the hanging wall, south of the fault, to 900 m asl on its northern, down-faulted side, which suggests that the fault was already active during the Miocene. The concurrent, although smaller increase in the thickness of basaltic section across the fault (100–300 m, south and north of the fault, respectively; Dafny et al. 2003; see Fig. 8.20) suggests that lavas were preferentially directed toward the downfaulted Yehudiyya block. A local increase in the basaltic section thickness in the area of Katzrin (> 500 m, Fig. 8.4a) suggests the existence of a local depression (Fig. 8.4b; Dafny et al. 2003) with an unclear origin.

References

Behar N, Shaar R, Tauxe L, Asefaw H, Ebert Y, Heimann A et al (2019) Paleomagnetism and paleosecular variations from the Plio-Pleistocene Golan Heights volcanic plateau, Israel. Geochem Geophys Geosys 20:4319–4335

Ben Asher M, Haviv I, Roering JJ, Crouvi O (2017) The influence of climate and microclimate (aspect) on soil creep efficiency: cinder cone morphology and evolution along the eastern Mediterranean Golan Heights. Earth Surf Proc Land 42(15):1649–2662

Camp VE, Roobol MJ (1992) Upwelling asthenosphere beneath western Arabia and its regional implications. J Geophy Res, 97:(B11), 1525–15271

Dafny E, Gvirtzman H, Burg A, Fleischer L (2003) The hydrogeology of the Golan basalt aquifer, Israel. Israel J Earth Sci 52, 139–153

Dan J, Singer A (1973) Soil evolution on basalt and basic pyro-clastic materials in the Golan Heights (Northern Israel). Geo-derma 9:165–192

Dubertret L, Dunand M (1955) The ossiferous seams of Khirbet El Umbachi and of Hebariye (Safa). Ann Arch Syrie IV–V: 59–76

Eidelman A, Enmar L (2014) Geological sites in Israel. Ministry of environmental protection and Israel Nature and Parks Authority, Jerusalem (in Hebrew)

Fahlbusch H, Peleg Y, Tsuk T (1994) Survey of the Sussita-Hippos Aqueducts. Excavat Surv Israel 14:29–30

Frumkin A, Naor R (2019) Formation and modification of pit craters—example from the Golan volcanic plateau, southern Levant. Z Geomorph 62(3):163–181

Garfunkel Z (1989) Tectonic setting of Phanerozoic magmatism in Israel. Isr J Earth Sci 38:51–74

Heimann A, Mor D, Stein M, Foland KA (1993) $^{40}Ar/^{39}Ar$ dating of prehistoric basalts in southwest Syria. Isr Geol Soc Ann Meet Arad Israel (in Hebrew)

Heimann A, Steinitz G, Mor D, Shaliv G (1996) The Cover Basalt Formation, its age and its regional and tectonic setting: Implications from K-Ar and $^{40}Ar/^{39}Ar$ geochronology. Isr J Earth Sci 45:55–71

Hurwitz S, Garfunkel Z, Ben-Gai Y, Reznikov M, Rotstein Y, Gvirtzman H (2002) The tectonic framework of a complex pull-apart basin: seismic reflection observations in the Sea of Galilee Dead Sead transform. Tectonophys 359(3–4):289–330

Inbar M (1984) Lava caves and surface volcano karst features in the basaltic area of the Golan Heights (Israel). Karstologia 4(1):45–49

Ilani S, Harlvan Y, Tarawneh K, Rabba I, Weinsberger R, Ibrahim K et al (2001) New K–Ar ages of basalts from the Harrat Ash Shaam volcanic field in Jordan: Implications for the span and duration of the uppermantle upwelling beneath the western Arabian plate. Geology, 29:171–174

Maimon O, Heimann A (2010) The formation of "Jubah"s (depressions) in the volcanic field of the Golan Heights: collapse into karstic caves or volcanic explosion?. Isr Geol Soc Ann Meet Eilot Israel

Mittlefehldt DW (1984) Genesis of cpx–amphibole xenoliths from Birket Ram: trace element and petrologic constraints. Contributions to Mineralogy and Petrology 88:280–287

Mor D (1973) The volcanism of the central Golan Heights. M.Sc. thesis, The Hebrew University of Jerusalem (in Hebrew)

Mor D (1986) The volcanism of the Golan Heights. Geol Sur Isr Rep GSI/5/86, (Hebrew, English Abstr.), Jerusalem

Mor D (1993) A time-table for the Levant volcanic province, according to K-Ar dating in the Golan Heights. Israel. J Afr Earth Sci 16(3):223–234

Nasa website. Retrieved from: https://earthobservatory.nasa.gov/images/45840/es-safa-volcanic-field-syria

Nuriel P, Weinberger R, Kylander-Clark ARC, Hacker BR, Craddock JP (2017) The onset of the Dead Sea transform based on calcite age-strain analyses. Geology 45(7):587–590

Regenauer-Lieb K, Rosenbaum G, Lyakhovsky V, Liu J, Weinberg R, Segev A, Weinstein Y (2015) Melt instabilities in a cold lithosphere and implications for intraplate volcanism in the Harrat Ash-Shaam volcanic field (NW Arabia). Journal of Geophysical Research, Solid Earth, 120(3):1543–1558

Reznik IJ, Bartov Y (2021) Present heat flow and paleo-geothermal anomalies in the southern Golan Heights, Israel. Earth Space Sci 8:e2020EA001299

Schmincke H-U (2004) Volcanism. Springer, Berlin, 324 pp

Shaanan U (2009) Birket Ram maar. Geol Surv Isr Rep GSI/13/2009 (in Hebrew)

Shaanan U, Porat N, Navon O, Weinberger R, Calvert A, Weinstein Y (2011) OSL dating of a Pleistocene maar: Birket Ram, the Golan Heights. J Volcanol Geotherm Res 201(1–4):397–403

Shaliv G (1991) Stages in the tectonic and volcanic history of the Neogene basin in the Lower Galilee and the Valleys. Geol Surv Isr Rep GSI/11/91 (in Hebrew)

Shtober-Zisu N, Inbar M, Mor S, Jicha BR, Singer BS (2018) Drainage development and incision rates in an Upper Pleistocene Basalt-Limestone Boundary Channel: The Sa'ar Stream, Golan Heights, Israel. Geomorphology 303:417–433

Singer AM, Zarei FM Lange K, Stahr (2004) Halloysite characteristics and formation in the Golan Heights. Geoderma 123:279–295

Sneh A, Weinberger R (2003) Geology of the Metulla quadrangle, northern Israel: implications for the offset along the Dead Sea Rift. Israel J Earth Sci 52:123–138.

Stein M, Garfunkel Z, Jagoutz E (1993) Chronothermometry of peridotitic and pyroxenitic xenoliths: implications for the thermal evolution of the Arabian lithosphere. Geochimica et Cosmochimica Acta 57:1325–1337

Tsuk T (2018) The water supply system. In: Eisenberg M (ed) Hippos-Sussita of the Decapolis, The First Twelve Seasons of Excavations, 2000—2011,Volume II. The Zinman Institute of Archaeology, University of Haifa, Haifa, Israel

Weinberger R, Sneh A, Harlavan Y (2003) The occurrence of Middle-Miocene volcanism at Mt. Hermon, northern Israel. Israel J Earth Sci 52:179–184

Weinstein Y (2000) Spatial and temporal geochemical variability in basin-related volcanism, northern Israel. J Afr Earth Sci 30:865–885

Weinstein Y (2002) Hydro-volcanism and stratigraphy in northern Golan. In" Ken-Tor R (ed) Israel Geological Society annual meeting, geological field trips, pp. 21–38 (in Hebrew)

Weinstein Y (2007) A transition from strombolian to phreatomagmatic activity induced by a lava flow damming water in a valley. J Volcanol Geotherm Res 159:267–284

Weinstein Y, Weinberger R (2006) The geology and volcanological history of Mount Avital. Isr J Earth Sci 55:237–255

Weinstein Y, Navon O, Altherr R, Stein M (2006) The role of fluids and of lithospheric heterogeneity in the generation of alkali basaltic suites from northwestern Arabia. J Petrol 47:1017–1050

Weinstein Y, Weinberger R, Calvert A (2013) High resolution $^{40}Ar/^{39}Ar$ study of Mount Avital, northern Golan: reconstructing the interaction between volcanism and a drainage system and their impact on eruptive styles. Bull Volcanol 75(5):1–12

Weinstein Y, Heimann A (2016) Spatial and temporal patterns in the Levant late Cenozoic volcanism. In Y. Enzel & O. Bar-Yoseph(Eds.), Quaternary of the Levant (pp. 45–51). Cambridge, UK: Cambridge University Press

Weinstein Y, Nuriel P, Inbar M, Weinberger R (2020) Impact of the Dead Sea transform dynamics on adjacent volcanic activity. Tectonics 39:e2019TC005645

White RS, McKenzie D (1989) Magmatism at rift zones: the generation of volcanic continental margins and flood basalts. J Geophy Res, 94:7685–7730

Yaalon DH, Mor D, Netzer M (1984) Basaltic weathering rates and associated soils on the western Golan. Comunicaciones, Universidad de la Laguna, Tenerife, Canary Islands, Congreso Internacional de Suelos Volcanicos

Za'arur S (2021) Age constraints for the Golan Heights volcanic soils, determined from integrating geochemical and geochronological tools. Goldschmidt 2021 Meeting, Lyon

Ziaei M, Schwarcz HP, Hall CM, Grun R (1990) Radiometric dating of the Mousterian site at Quneitra. In: Goren-Inber N (ed) QEDEM, Monographs of the institute of archaeology. The Hebrew University of Jerusalem, vol 31, Chap 14

Zilberman E, Begin ZB (2017) Main stages and rate of the relief development in Israel. Geol Surv Isr Rep GSI/24/97 (in Hebrew)

The Landscapes of Mt. Carmel: A Remarkable Record of Geological and Geomorphological History

9

Nurit Shtober-Zisu

Abstract

Mount Carmel is a notable landmark in the northern region of Israel. It is a precipitous mountain range composed of dolomite, limestone, and chalk, southeast of the Haifa Bay, rising steeply to over 500 m a.s.l. The mountain is rich in its geological and geomorphologic diversity determining a wide range of landforms, a mosaic of settlements, agricultural areas, prehistoric and archaeological sites. The chapter aims to present a macro-level portrait of the landforms of Mt. Carmel, their spatial distribution, origin, evolution and ages, and therefore to emphasize the diversity and beauty of its geomorphological sceneries. The chapter summarizes the lithostratigraphic sequence, the main tectonic uplift phases and ultimately, the predominant landscapes and landforms of Mt. Carmel.

Keywords

Mt. Carmel · Geomorphology · Karst · Abrasion surfaces · Calcrete · Inland notches

9.1 Introduction

Mt. Carmel is a prominent ridge located in northwestern Israel, extending from the Haifa Bay in the north to Ramat Menashe ("Menashe Heights") in the south, and from the Mediterranean Sea in the west to the Jezreel valley in the east. The present landscape of Mt. Carmel reflects the prolonged interactions between lithology, tectonics, and surface processes, resulting in the current topography. The aim of this chapter is to acknowledge the complexity of the regional morphology and to outline the long-term geomorphic history interacting with contemporaneous processes to shape the Carmel landforms and landscape.

Mt. Carmel is characterized by a triangular shape extending over 230 km^2, with a maximum length of 33 km along the Mediterranean coast and maximum width of 12 km. The three vertices of the triangle are located at Rosh Carmel (Stella Maris in Haifa) at the northernmost edge, Ramat Ha'Nadiv (Hotem ha'Carmel) to the south, and Keren Carmel (Muhraqa) to the south-east. The three sides of the triangle form an independent mountainous division, distinguished from their surrounding by prominent topography. Their origin and evolution are reflected in three types of slopes found in nature: the western slopes, which drop precipitously to the coastal plain, were formed by coastal abrasion in the Pliocene; the southern slopes, which bend gradually towards the Menashe syncline, form the Carmel anticlinal flank; and the northeastern edge is marked by the steep Carmel prominent fault escarpment (Fig. 9.1).

9.2 Geological Settings

9.2.1 The Lithostratigraphic Sequence

The paleogeographic configuration of the Levant basin allowed for an extensive carbonate platform to develop within the Tethys sea, from the Aptian (Lower Cretaceous) onward. Most of the exposed rocks of Mt. Carmel are carbonates and constitute part of the Upper Cretaceous Judea Group. The stratigraphy is complicated due to significant facies changes over short distances and due to the existence of multiple faults dissecting the ridge. The various facies changes reflect the location of Mt. Carmel close to the edge of the Cretaceous carbonate platform and the variety

N. Shtober-Zisu (✉)
School of Environmental Sciences, University of Haifa, Haifa, Israel
e-mail: nshtober@research.haifa.ac.il

A. Frumkin and N. Shtober-Zisu (eds.), *Landscapes and Landforms of Israel*, World Geomorphological Landscapes, https://doi.org/10.1007/978-3-031-44764-8_9

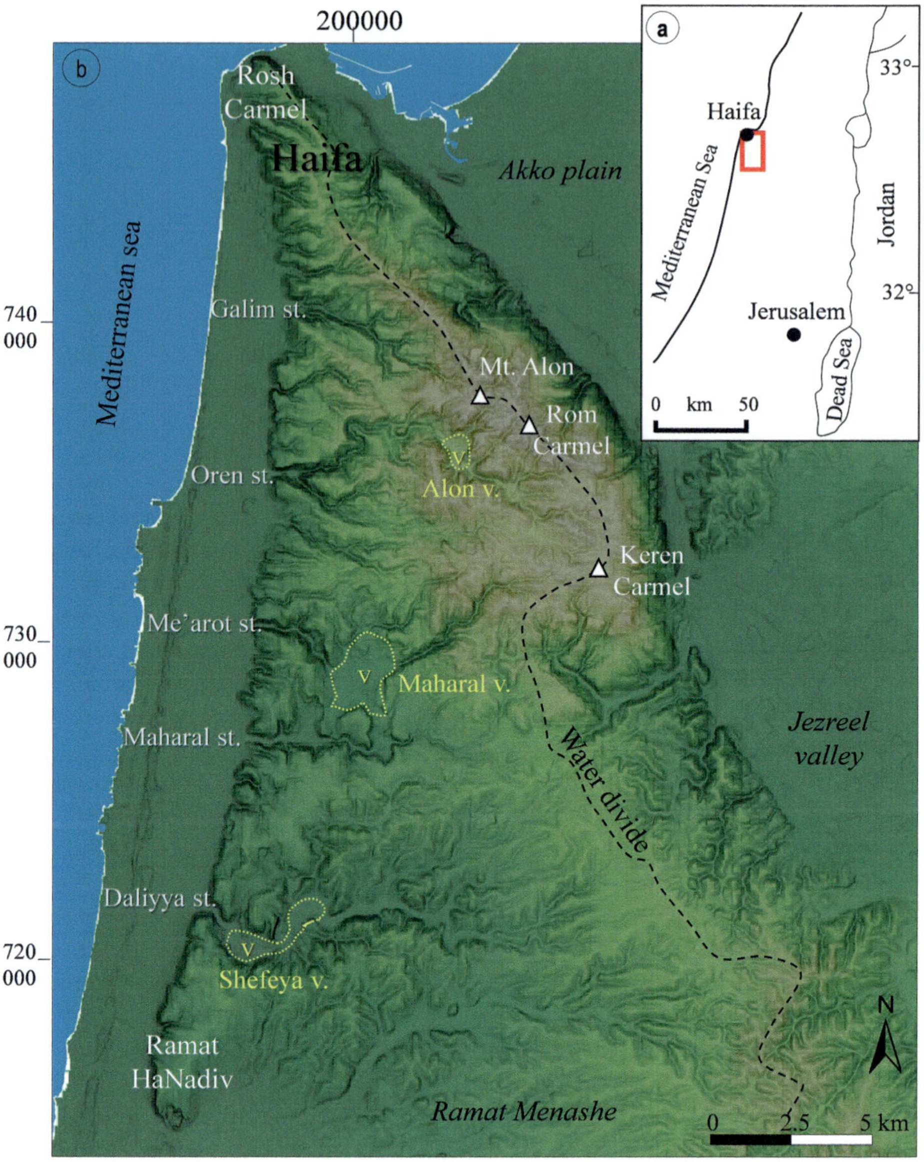

Fig. 9.1 **a** Location of Mt. Carmel; **b** Relief map (courtesy of Rotem Elinson) showing the water divide, major streams, and flat bottom valleys (marked "v")

of depositional environments that characterized this transition zone between the Albian and Turonian stages of the Cretaceous (Segev and Sass 2014).

The sedimentary succession (Fig. 9.2) consists mainly of limestone, dolomite, chalk and chert. The lower part of the sequence consists of stratified dolomites of the Yagur Formation, which accumulated under shallow platform conditions, bordered by a narrow belt of barrier reefs at its western side. The overlying Chalk Complex, composed of Isfye and Arqan formations, was deposited at a depth of a few tens of meters, implying the establishment of open marine environments over the outer platform. The successive Bina Formation from the Turonian stage, consists of both limestone and chalk, with isolated rudist reefs and bioclastic beds suggesting that various depths existed during the deposition (Picard and Kashai 1958; Segev and Sass 2009) (Fig. 9.3).

Reef development was one of the dominant factors controlling sedimentation over the platform. The reefs, indicating long geological history of sea-continent margin, influenced the marine circulation and contributed to much of the coarse and fine skeletal material deposited in the various environments. The reefs are present throughout the whole sequence and are almost entirely composed of rudists with some chondrodonts. Therefore, throughout millions of years the reef experienced significant changes in seawater levels, alternately exposed and submerged due to extreme environmental changes coupled with local tectonics (Sass and Bein 1982).

Four cycles of submarine volcanism, evidenced by volcanic rocks, mostly pyroclastic, locally intercalated with

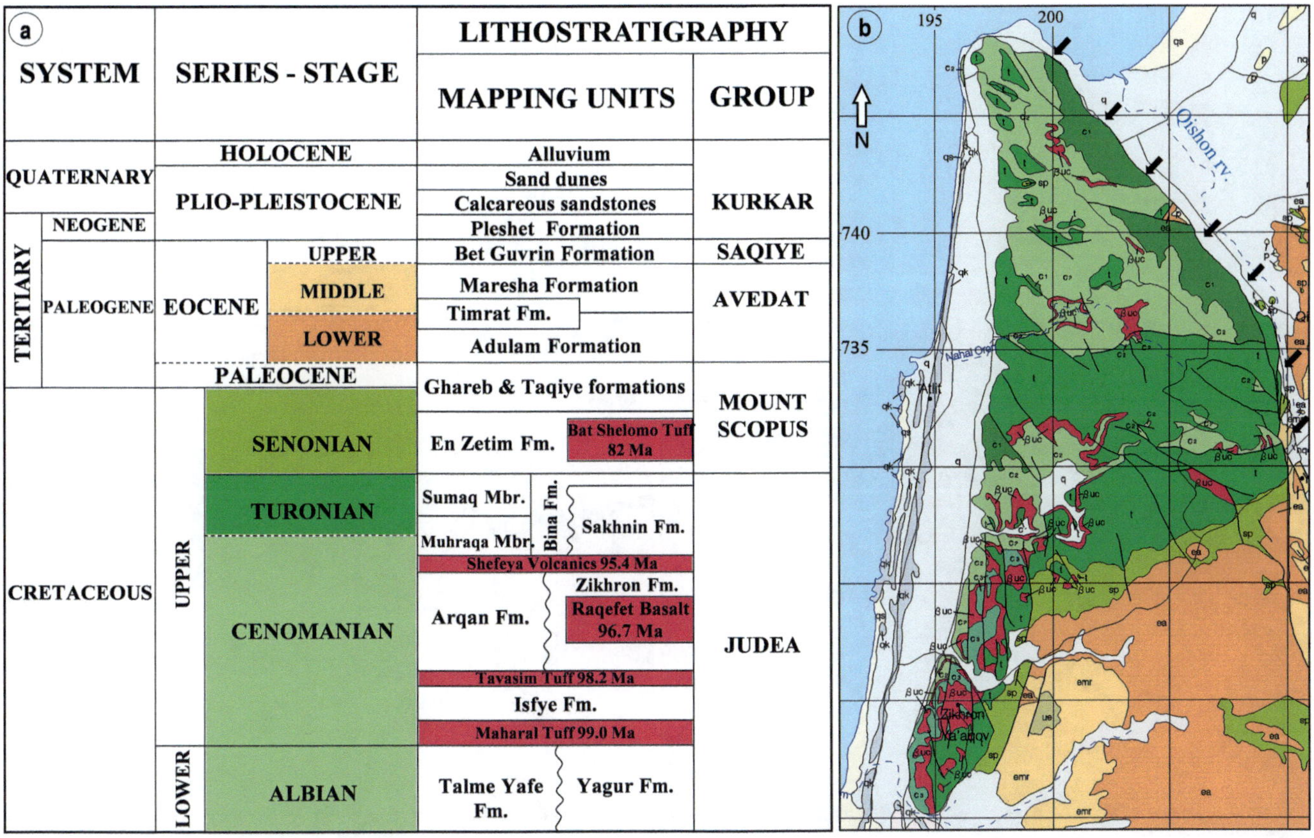

Fig. 9.2 **a** Stratigraphic units of Mt. Carmel, modified after Segev and Sass 2009; **b** Geological map, modified after Sneh et al. 1998, black arrows mark the Carmel-Yagur Fault

the thick (~ 300 m) carbonate sequence, were documented by Segev et al. (2002) (Fig. 9.4). Age constraints for the volcanic phases suggest that deposition occurred within a relatively short time (< 1 Ma) and the sequence represents rising global and local sea-level trends (Bialik et al. 2021). The volcanic cones are basaltic and consist of glass-rich black tuff at the base, variegated colored layers of tuffs within the cones, and yellowish-brown tuffs at the top and at the external parts of the cone (Sass 1980). As the Tethys Sea retreated, the platforms were exposed to tectonic uplift and subaerial erosion, and the Carmel platform became "The Carmel Mountain".

9.2.2 Tectonic Settings

The Carmel structure is probably older than the Levant margin itself and has experienced several phases of activity: (1) Early Mesozoic rifting along the present Levant margin; (2) late Cretaceous until Eocene, folding of the Syrian Arc in a compressional system; (3) late Miocene to early Pliocene NW left-lateral strike-slip faulting and (4) from the Pliocene the prevalence of a tensional tectonic regime, probably still active, forming the well-pronounced Kishon graben (Jezreel valley and Akko plain in Fig. 9.1) (Ben-Gai and Ben-Avraham 1995).

Mt. Carmel is bordered to the northeast by the Carmel Fault (also named Yagur Fault, Fig. 9.5), that is the northwestern continuation of the NW–SE Azraq–Sirhan Graben in Saudi Arabia and Jordan (Schattner et al. 2006). The morphology associated with the fault is expressed by a steep and high (~ 400 m) escarpment, shaping the northeastern edge as a prominent tilted block and separating Mt Carmel from the Galilee and Jezreel valley (Picard and Kashai 1958; Karcz 1959; Fleischer and Gafsou 2003).

The Carmel Fault is considered a major branch of the Dead Sea Transform that was already active during the Cretaceous and was reactivated during the late Eocene. The fault offsets the upper Cretaceous sedimentary rocks on both left-lateral and vertical planes with average displacements of 3–10 km (Freund et al. 1970) and 0.3–1.5 km, respectively (Achmon 1986).

Evidence of significant Quaternary tectonic activity along the fault includes displaced alluvial fans and stream channels, and the formation of shutter ridges and morphological scarps (Zilberman et al. 2008; Ashkar-Halak 2009). Achmon (1986) calculated the Quaternary tectonic uplift to about 200 m, namely an average rate of 100 m/Ma. A paleoseismic

Fig. 9.3 Lithologies of Mt. Carmel; **a** The Yagur dolomitic formation is overlain by the Isfye chalk formation at Nahal Oren canyon; **b** chalk and chert layers of Chreibe Fm. (Arqan Fm.); **c** *Nerinea schiosensis* reef of Arqan Fm. At Ein Qedem spring; **d** Rudists in a boulder at Nahal Mearot stream

Fig. 9.4 Volcanic outcrops in the city of Haifa (dated 98.2 ± 1.1 Ma): **a** The perennial spring of Ein Qedem: the volcanic succession (greyish color, below) is overlain by a reef sequence rich in gastropods; **b** a submarine slump, the cone flank (brownish color) is overlain by soft chalky limestone

Fig. 9.5 The Carmel-Yagur Fault plain exposed at Yoqneam

analysis of the Nesher Fault - a branch of the Carmel-Yagur Fault, showed two periods of tectonic activity during the last 180 ka, accompanied by surface deformation and subsidence. The last stage lasted for almost 50 ka, indicating a minimum subsidence rate of about 60 m/Ma and that the Carmel-Yagur Fault was active during the Late Pleistocene, at least up to 27 ka (Zilberman et al. 2008). The oblique motion along the Carmel Fault was calculated to $\sim$ 0.7 mm/yr left-lateral and $\sim$ 0.6 mm/yr extension rates, resulting in N-S extension across the Carmel Fault (Sadeh et al. 2012).

Along the western side of Mt. Carmel, no fault was documented, and this escarpment was considered an abrasion cliff (Picard and Kashai 1958; Michelson 1968). Therefore, uplift rates are much slower, and the maximum calculated average rates were 0.058–0.067 mm/yr for the last 600 ka (Zviely et al. 2009). According to data from magnetostratigraphy of clastic sedimentary units and calcretes, it appears that much of the latest uplift of Mt. Carmel occurred during 3.58–1.77 Ma, with an uplift rate of between 0.05 and 0.11 mm/yr. Since that time, the uplift of Mt. Carmel has been relatively slow –<0.012 mm/yr (Greenbaum et al. 2012; Mashiah et al. 2009).

9.3 Landscapes and Landforms

Tectonic and erosional factors act together to produce a diverse landscape across the Carmel mountain, including moderate topography, abrasion surfaces, and various geomorphic landforms such as incised canyons and flat-bottomed valleys. Based on this framework, the ridge can be separated into three landscape units: (a) the Upper Carmel, at 450–500 m a.s.l., with moderate elevation differences. The highest peaks are Rom Carmel (546 m), Har Alon (516 m) and Keren Carmel (Muhraqa) (482 m) (Fig. 9.1), which also determine the position of the local water divide; (b) the Intermediate Carmel, located west of the Upper Carmel ridge, characterized by hilly morphology with peaks of between 300 and 400 m a.s.l.; (c) the Lower Carmel, named also Ramat Hanadiv, of low to moderate topography, terminating westwards in steep cliffs bordering the coastal plain and by the major Or-Aqiva fault in the south. This graded landscape reflects periods of severe erosion/abrasion that shaped Mt. Carmel into its present-day morphology.

9.3.1 The Water Divide

As part of the mountainous backbone of Israel, the morphology of Mt. Carmel has been dictated by several tectonic and erosional structures: (a) the Syrian Arc folds; (b) the regional erosion surfaces that formed during the Oligocene by gradual truncation (Quennell 1956; Bar 2009; Avni et al. 2012); and (c) the wide-wavelength (50–60 km) structural arch paralleling the Dead Sea Transform, which has been developing since the Pliocene. Several anticlines of the Syrian Arc Fold system are superimposed on this wide arch and one of them is the Carmel anticline (Matmon and Zilberman 2017).

The mountainous backbone of Mt Carmel is characterized by a dissected plateau rather than a mountainous area, possibly due to Oligocene peneplanization, similar to the Galilee, or other moderate peaks along the Israeli mountainous belt (Wdowinski and Zilberman 1997).

As the Carmel anticline was subsequently faulted by the Carmel-Yagur Fault, its northeastern flank forms a steep escarpment flexed upward toward the axis, resembling an uplifted shoulder. Thus, the Carmel-Yagur Fault dictates the asymmetry of the water divide (Fig. 9.1), and therefore the morphology of the entire drainage system: the northeastern slopes are short and relatively steep (~ 20%), while the westward slopes are characterized by elongated, moderate (~ 5%) crests, terminating in the Carmel coastal plain throughout a steep (~ 30%), 150-m-high cliff parallel to the present-day coastline (Michelson 1968). Structurally, the coastal plain extends offshore to the shelf edge, where it is bounded by a 3 km drop in the continental slope (Steinberg et al. 2010). The present configuration of the mountain relative to the adjacent plains may have started in the Early Miocene (Achmon and Ben-Avraham 1997) or the mid-Oligocene (Schattner et al. 2006).

9.3.2 Abrasion Surfaces Constraining the Morphology of Mt. Carmel

Five abrasion surfaces were recognized at various elevations across Mt. Carmel, providing evidence for fluvial-coastal processes (Mashiah 2011). These surfaces were apparently formed under tectonic stability during equilibrium with stable sea levels: (A) The Isfye surface (515–546 m) extends over the Carmel peak, parallel and in close proximity to the Carmel-Yagur fault; (B) The Daliyat el-Carmel surface (470–497) encircles the Carmel peaks; (C) The Beit Oren surface (350–390 m), which is ~ 110 m lower than surface B; (D) The Ahuza surface (280–390); and (E) The Zichron surface (160–200 m). This "surface staircase" represents the evidence of phase-wise tectonic displacements, with alternating periods of planation and uplift.

Apart from the Zichron surface, the other surfaces are relics of ring-shaped abrasion strips encircling the highest Isfye surface, implying a new stage of elevation, characteristic of mountainous islands surrounded by sea, that were formed during a prolonged period of tectonic stability with low eustatic oscillations. The transition between the Ahuza (D) and the Zichron (E) surfaces is estimated to have happened 6–10 million years ago. The lowest abrasion surface is mostly buried under marine sediments of the coastal plain, and was observed at the Nahal Oren river mouth at 35 m. a.s.l. Calcrete ages show that this abrasion platform was formed after the stabilization of the Carmel block at ~ 1.77 to 3.58 Ma (Zilberman et al. 2008; Mashiah 2011). "This surface staircase represents the evidence of phase-wise tectonic displacements, with alternating periods of planation and uplift and the development of younger surfaces at the expense of older, located higher up".

9.3.3 The Drainage System

The Carmel drainage network comprises ~ 50 drainage basins of various sizes, ~ 30 of which drain the western mountain side towards the Carmel coastal plain (Mashiah 2011), while ~ 20 basins drain the eastern mountain front towards the Akko plain and the Qishon stream (Ashkar Halak 2009). As Mt. Carmel was uplifted out of the Mediterranean Sea, the exposed abrasion surfaces were dissected by fluvial erosion. Thus, a positive correlation between basin size and time of exposure to terrestrial environment occurred, that is, the longer the exposure, the larger the basin size and the higher the stream-order. Five major 4th order streams drain approximately 85% of the western slopes of Mt Carmel: Nahal Galim, Nahal Oren, Nahal Me'arot, Nahal Maharal and Nahal Daliyya (Fig. 9.1). Apparently, the Oren and Galim drainage basins are the oldest systems dissecting Mt Carmel, since they drain the upper abrasion surfaces, starting at the Isfye water divide level, but also cutting through the Daliyat el-Carmel and Beit Oren abrasion surfaces. In the Nahal Oren canyon, a large meander has been incised into the chalk and dolomite thick succession, up to a depth of 140 m and a distance of 1 km from the water divide (Fig. 9.6b). This overfit meander may suggest that Miocene-Pliocene fluvial systems crossed Mt. Carmel in the past, but were completely eroded and therefore became difficult to reconstruct. The Me'arot basin evolved thereafter, draining the middle and lower abrasion surfaces, while the Maharal basin, the last one to develop, drains the lower abrasion surfaces (Mashiah 2011). To the northeast, the drainage basins are dissected by small channels of the 1st–3rd order; most of these systems are deeply incised in the upper parts of the catchments, but are less incised downstream, dispersing over the mountain piedmont through small alluvial fans. Geomorphic activity observed along these channels include debris flows and landslides that may block or displace stream channels, with further interference from active tectonic structures such as shutter ridges and horizontally displaced lineaments and faults (Ashkar-Halak 2009).

The streams of Mt. Carmel are mostly ephemeral, experiencing only sporadic flow events, which occur after the soil is saturated, following high intensity rainstorms, usually from December to March. Lag time is generally a few hours, depending on drainage basin size, soil moisture and rainstorm characteristics. Flows are commonly characterized by a rapid rise in water level, a peak discharge within a few hours, and a swift recession followed by a relatively long "tail", resembling flash floods. Flow event duration ranges between several hours to several days, thereafter the channel returns to its dry condition (Wittenberg et al. 2007) (Fig. 9.6c, d). There are very few perched springs in Mount Carmel due to the lack of extensive marl and clay layers, but they are highly responsive to rainstorms and rainfall amounts (Arbel et al. 2008). Hence, temporary base flows are sustained in limited channel segments which may last between a few days to a couple of weeks (Greenbaum et al. 2021).

Some of the westward-trending valleys display large differences in solar radiation, effective rainfall and evaporation between the south- and north-facing slopes. These form an optimal natural microscale model for unraveling adaptation and speciation of flora and fauna (Nevo 2009). For example, in the Nahal Oren canyon solar radiation is up to 800% more on the south-facing slopes than on the north-facing ones. Solar radiation is therefore considered as a major factor affecting the speciation between the tropical-xeric "African" south-facing slope and the temperate-mesic "European" north-facing slope at a close distance separated by about 250 m (Nevo 2009; Yin et al. 2015) (Fig. 9.6a).

9.3.4 Flat-Bottom Valleys

Along with the rapid chemical weathering (discussed below), the morphological evolution of Mt. Carmel was determined by fluvial activity resulting in streams incision into V-shape canyons. However, contrasting landforms are

Fig. 9.6 Mt Carmel canyons and river channels. **a** The Nahal Oren Canyon shows dramatic differences in solar radiation and therefore speciation divergence between the "African" savannah vegetation developed over the south-facing slope (right) versus the temperate "European" forested north-facing slope (left); **b** The Nahal Oren meander, incised to 120 m at a distance of 1 km from the water divide (a, b-courtesy of Hani Amasha); **c** thin gravel layer accumulated in the Nahal Oren channel; **d** Bedrock channel exposed in Nahal Galim

three flat-bottomed depressions that can be found in the Mt. Carmel area: the Alon, Maharal and Shefeya depressions (Fig. 9.1), which are cut and drained by Nahal Oren, Nahal Maharal and Nahal Dalia, respectively. These flat-bottomed valleys extend over 2–5 km^2, exposing volcanic tuff. The tuff plays an extremely important role in the landscape as the crossing streams expand and widen when passing through the friable and highly erodible soft tuff. The carbonate edges of these valleys form prominent local escarpments encircling the depressions, resembling a bathtub. The Alon and Maharal valleys expose the oldest tuff units (Maharal tuff, 99 Ma), whereas the Shefeya tuff (95.4 Ma) is exposed in the Shefeya valley (Segev 2009; Segev and Sass 2009). Abundant knapped flint, along with Roman to Muslim pottery found in these valleys, proves long-term presence and land use of humans (ʿAd and Sa'id 2021). Currently, these valleys are used for agriculture purposes.

9.3.5 Indurated Alluvial Fans and Colluvial Deposits

The foothills of Mt. Carmel are covered by alluvial and colluvial deposits along both the northeast and western escarpments, in the form of alluvial fans facing channel's mouths, or coarse aprons covering slopes and interfluves (Fig. 9.7). These deposits are mostly indurated by calcium carbonate, typically forming calcretes of several meters to several tens of meters thick. Depending on their origin and environmental setting, the calcretes of Mt. Carmel have a wide range of shapes and forms.

Calcrete accumulation begins as barely perceptible powdery accumulations within the soil or gravel profile. Over time, the carbonate minerals accumulate to form drapes, coatings and bridges, often following original pores or voids created by biological processes. Further accumulation

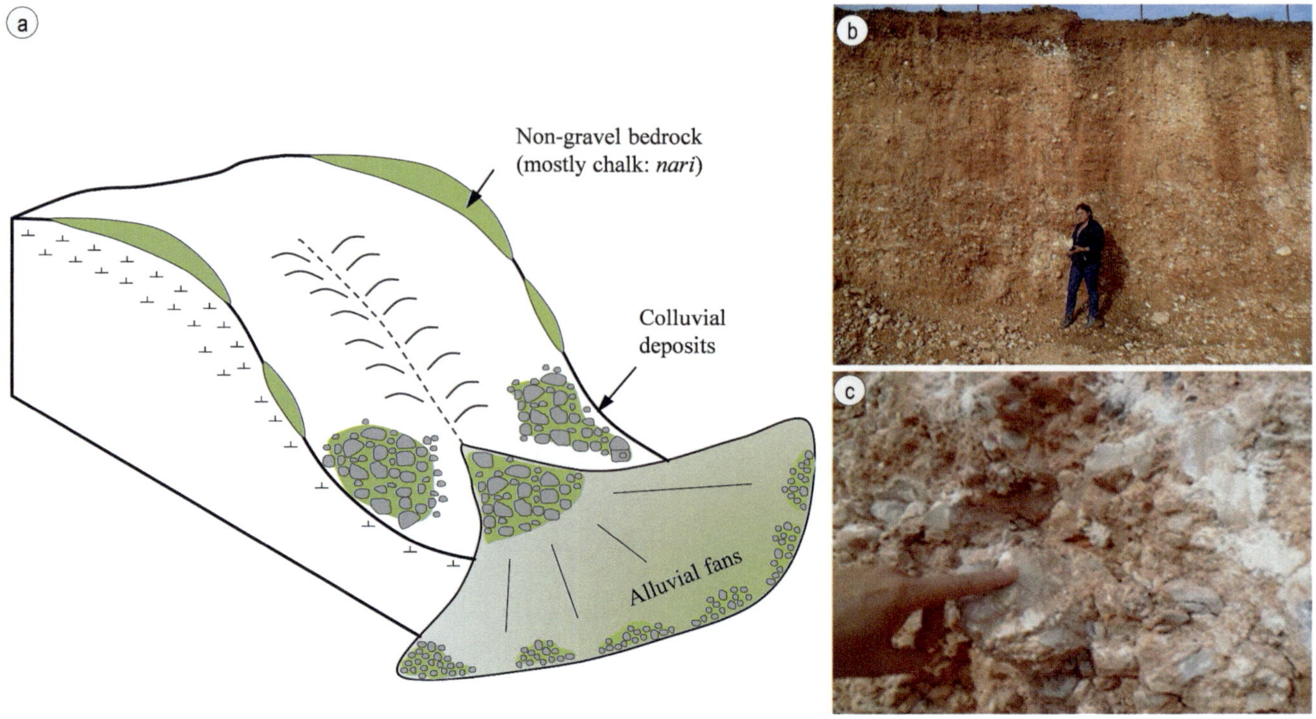

Fig. 9.7 **a** Typical locations of calcrete deposits in Mt. Carmel; **b** Calcretes developed over alluvial fan's distal parts; **c** massive, strongly indurated calcrete layers

creates nodules and coatings that coalesce to form laminar and/or massive horizons (Gile et al. 1966; Chen et al. 2002). At the Mt. Carmel foothills, calcretes are indurated up to the stage IV-V, meaning well-cemented pisoliths and laminae with multiple generations of recementation (Machette 1985).

The calcretization of these multi-phase coarse colluvial aprons and alluvial fans occurred after their accumulation and therefore the age of calcretes corresponds to the end of the uplift stage and landscape stabilization (Mashiah 2011). Paleomagnetic measurements obtained from the indurated calcrete horizons indicates that magnetization of the lower calcrete is at least 1.78 Ma old, implying that the western escarpment of the Carmel block obtained its present morphology during Late Pliocene-Early Pleistocene (Greenbaum 2012).

9.3.6 Stream Terraces and Older Surfaces

Along the stream channels draining the western slopes of Mt. Carmel, a series of strath and fill terraces have been preserved at elevations of 1.5–8 m above the channel. The remains are discontinuous and their lateral exposure usually does not exceed a few meters. The terraces are composed of gravelly alluvial sediment with reddish clay matrix; in places, alternating colluvial and alluvial units appear within the same terrace. Gravels are moderately to poorly rounded implying alternating depositional events, with colluvium reaching the channel. At present, the channels incise into the sedimentary units and bedrock may be exposed at the base of the incision. Along the Nahal Galim stream, a series of terraces was mapped between 60 and 200 m a.s.l. (Fig. 9.8). Most outcrops are located downstream of the confluence of Nahal Galim and its largest tributary—Nahal Neder, and include remnants of terraces and the Nahal Neder alluvial fan entering into Nahal Galim, covering an area of several hundreds of square meters, currently incised by the main Nahal Galim channel to a depth of 15 m (Wittenberg et al. 2002).

9.3.7 Karst Landforms of Mt. Carmel

As previously stated, the sedimentary succession of Mt. Carmel is predominated by carbonate rocks, therefore carbonate solution are the main geomorphic agents active in the weathering processes and the development of karst landforms, especially where limestone and dolomite outcrops prevail. Karst landforms in the region include endokarst features such as caves and mazes (Olami 1984), as well as epikarst features, represented mainly by karren,

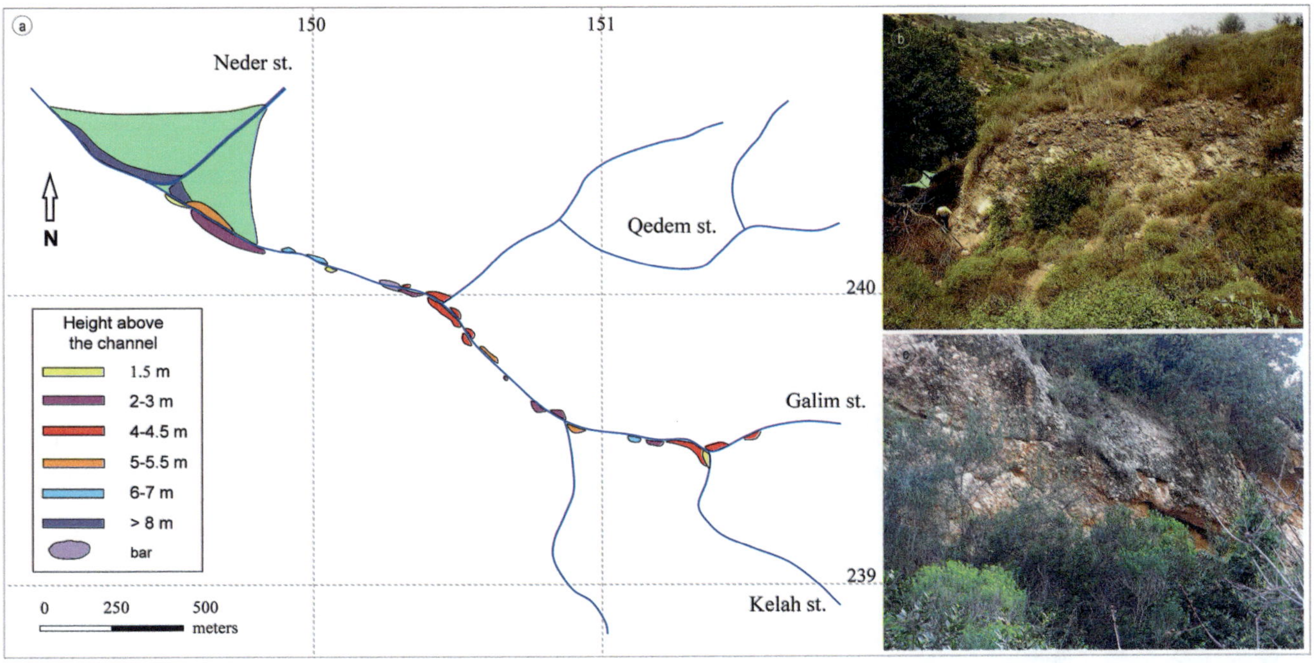

Fig. 9.8 River terraces of Nahal Galim stream, modified after Wittenberg et al. 2002; **a** Location map; **b** the>8m terrace of the Neder tributary; **c** a 5 m high terrace at the Neder-Galim confluence

inland-notches and tufa deposits (Shtober-Zisu et al. 2015) (Fig. 9.9). Although karst processes are very common in the development of Mt. Carmel landforms, the typical karst landforms such as large caves, closed depressions, sinkholes, and travertine are rare or absent.

The western Carmel escarpment has the highest concentration of caves in the ridge and Nahal Me'arot encompasses the best examples of large volume caves along the escarpment, demonstrating bell-shaped chambers and elongated passages with typical phreatic morphologies such as smooth walls, cupolas and elliptic cross-sections (Frumkin and Weinstein-Evron 2021). Most caves of Mt. Carmel belong to old stages of landform development, predating the Plio-Pleistocene uplift and stream entrenchment. Subaerial denudation and slope processes have opened the caves to the surface, enabling long human occupation—among the oldest and longest prehistoric human utilization in the Levant (Nadel et al. 2012; Weinstein-Evron et al. 2012; Ronen 2017) since the lower Paleolithic. The caves of Mt. Carmel are undoubtedly among the most famous prehistoric sites in the world, and as such, are recognized by UNESCO as heritage of humanity and inscribed as a World Heritage Site (Weinstein-Evron 2015).

Mt. Carmel is also rich in epikarst and subaerial carbonate dissolution features. One of the most prominent features visible on the mountain slopes are horizontal "C"-shaped indentations on the carbonate slopes or cliffs of the mountain, forming rock-shelters termed "inland notches" (Fig. 9.9b). They are shaped like half-tubes extending over tens and hundreds of meters along the stream valley slopes. At Mt. Carmel, over 120 inland notches have been mapped, primarily constrained by hard lithology and Mediterranean climate (Shtober-Zisu et al. 2015, 2017; Brook and Shtober-Zisu, 2020). On average, their height and width are 2–2.5 m, but they can reach 6 m in height and 9.5 m in width. These unique landforms originate by dissolution and disintegration of the rock under subaerial environments, by differential weathering of beds with slightly dissimilar petrographic properties. The notches follow specific beds that enable their formation and are destroyed by the collapse of the upper bed. The geomorphic processes that form inland notches combine chemical, mechanical, and biogenic weathering, which generate initial dissolution and subsequent spall weathering (exfoliation) of the bed, forming the notch cavity. The first stage of development is marked by chemical weathering, as water percolates through the contact between two beds of different solubility, creating an initial indentation in the soluble rock. Subsequently, mechanical weathering contributes to the process, as shallow parallel micro- and macro-fissures develop in the back wall of the proto-notch. They expand by dissolution as well as by slaking, and the process is possibly aided by dust (clay) and organisms which accumulate in the fissures. The fissuring process results in spall weathering, exfoliation of the back wall, and accelerated backwards retreat of the cavity. By contrast, the main weathering mechanism that

Fig. 9.9 Typical karstic features of Mt. Carmel: **a** the Yishah cave; **b** inland notches; **c** karren and **d** microkarren

affects the visor which overlies the notch, is breakdown and rockfall (Shtober-Zisu et al. 2015). Thus, their formation and destruction alternate in cyclical episodes and therefore the notches change over time and space (Fig. 9.10).

Occasionally tufa accumulates within notches as stalactites, drapes and crusts, covering the cavity backwalls or floors. The tufa grows during the interval between the backwall erosion and visor collapse, and therefore tufa age is an approximation of the timing of the formation of the surface it developed on, i.e. the age of the notch. The oldest dated tufa sample in Mt. Carmel is 39.0 ± 10.4 ka and the ages of other tufa samples range from 2.1 to 23.3 ka. These ages and the configuration of the tufa on the slopes suggest that the order of magnitude of slope retreat in Mount Carmel ranges from 10^1 to 10^2 mm/ka, corresponding to the rates of tens of meters per million years, similar to the magnitude of denudation found by previous studies in the Mediterranean zone of Israel (Shtober-Zisu et al. 2020).

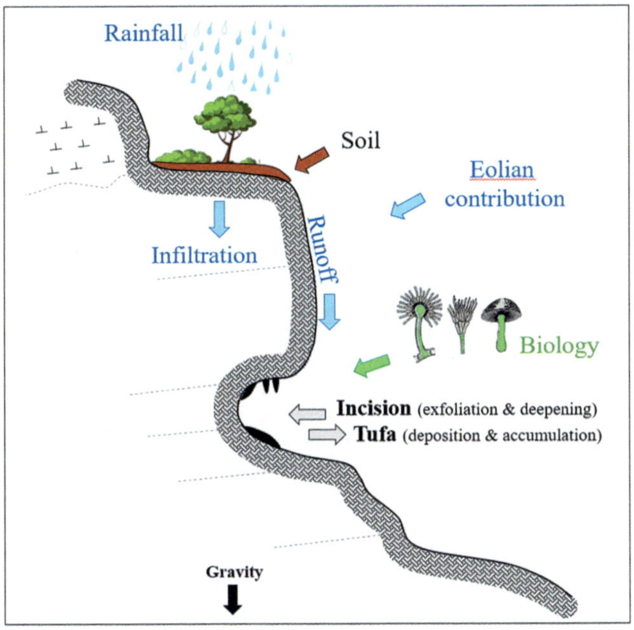

Fig. 9.10 Geomorphic processes that form inland-notches combine chemical, mechanical, and biogenic weathering, which generate dissolution and subsequent spall weathering (exfoliation)

9.4 Summary

Mount Carmel is a notable landmark in the northern region of Israel. It is a precipitous mountain range composed of dolomite, limestone, and chalk, southeast of the Haifa Bay, rising steeply to over 500 m a.s.l. The mountain is rich in its geological and geomorphologic diversity determining a wide range of landforms, a mosaic of settlements, agricultural areas, prehistoric and archaeological sites. The purpose of this chapter was to present a macro-level portrait of landforms of Mt. Carmel, their spatial distribution, origin, evolution and ages, and therefore to emphasize the diversity and beauty of its geomorphological sceneries.

References

Achmon M (1986) The Carmel boundary fault between Yokneam and Nesher (in Hebrew), M.Sc. thesis, Hebrew Univ., Jerusalem

Achmon M, Ben-Avraham Z (1997) The deep structure of the Carmel fault zone, northern Israel, from gravity field analysis. Tectonics 16(3):563–569

'Ad U, Sa'id K (2021) Remains from the Prehistoric to the Late Ottoman Periods at Kerem Maharal. Atiqot 105:69–119 and 168–171

Arbel Y, Greenbaum N, Lange J, Shtober-Zisu N, Grodek T, Wittenberg L, Inbar M (2008) Hydrologic classification of cave drips in a Mediterranean climate, based on hydrograph separation and flow mechanisms. Isr J Earth Sci 57:291–310

Ashkar-Halak L (2009) The morphotectonics of the Carmel Fault. M.Sc. thesis, University of Haifa, Isr Geol Surv Rep GSI/38/2009, 152 pp

Avni Y, Segev A, Ginat H (2012) Oligocene regional denudation of the northern Afar dome: pre and syn-breakup stages of the Afro-Arabian plate. Geol Soc Am Bull 124:1871 https://doi.org/10.1130/B30634.1

Bar O (2009) The shaping of the continental margin of central Israel since the Late Eocene—tectonics, morphology and stratigraphy. Geol Surv Isr Rep GSI/32/2009 [in Hebrew, English abstr]

Ben-Gai Y, Ben-Avraham Z (1995) Tectonic processes in offshore northern Israel and the evolution of the Carmel structure. Mar Petrol Geol 12(5):533–548. https://doi.org/10.1016/0264-8172(95)91507-L

Bialik OM, Samankassou E, Meilijson A, Waldmann ND, Steinberg J, Karcz K. Makovsky Y (2021) Short-lived early Cenomanian volcanic atolls of Mt. Carmel, northern Israel. Sed Geol 411:105805. https://doi.org/10.1016/j.sedgeo.2020.105805

Brook A, Shtober-Zisu N (2020) Rock surface modeling as a tool to assess the morphology of inland notches, Mount Carmel. Israel. Catena 187:104256. https://doi.org/10.1016/j.catena.2019.104256

Chen XY, Lintern MJ, Roach IC (2002) Calcrete: characteristics, distribution and use in mineral exploration. Cooperative Research Centre for Landscape Environments and Mineral Exploration (CRC LEME), Canberra, Australia (170 pp)

Fleischer L, Gafsou R (2003) Top Judea Group Digital Structural Map of Israel: The Geophys Inst Isr Rep GII 753/312/03 scale 1:200,000 (2 sheets)

Freund R, Garfunkel Z, Zak I, Goldberg M, Weissbrod T, Derin B, Bender F, Wellings FE, Girdler RW (1970) The shear along the Dead Sea rift. Philos T R Soc London. Series A Math Phys Sci 107–130

Frumkin A, Weinstein-Evron M (2021) Nahal Me'arot caves: archive of human evolution against the background of prolonged karstic processes. Z Geomorphol Supp 283–300

Gile LH, Peterson FF, Grossman RH (1966) Morphological and genetic sequences of carbonate accumulation in desert soils. Soil Sci 101:347–360

Greenbaum N (2012) Age constrain on the uplift of the Mt. Carmel Block, northwestern Israel-evidence from magnetostratigraphy of clastic sedimentary units and calcretes. Quat Int 279:176. https://doi.org/10.1016/j.quaint.2012.08.259

Greenbaum N, Wittenberg L, Malkinson D, Inbar M (2021) Hydrological and sedimentological changes following the 2010-forest fire in the Nahal Oren Basin, Mt. Carmel Israel–a comparison to pre-fire natural rates. Catena 196:104891. https://doi.org/10.1016/j.catena.2020.104891

Karcz Y (1959) The structure of the northern Carmel. Bull Res Counc Isr 8(2–3):119–130

Machette MN (1985) Calcic soil of the southwestern United States. Geol Soc Am 203:1–21. https://doi.org/10.1130/SPE203-p1

Mashiah M, Greenbaum N, Zilberman E, Ron H, Ronen A (2009) Pleistocene tectonic stability of the western Carmel escarpment–evidence from magnetostratigraphy of calcretes. Geol Surv Isr Rep GSI/19/2009

Mashiah M (2011) Geological, Geomorphologic and Prehistoric along the western front of the Carmel block. Geol Surv Isr Rep GSI/16/2011

Matmon A, Zilberman E (2017) Landscape evolution along the Dead Sea fault and its margins. In: Enzel Y, Bar-Yosef O (eds) Quaternary of the Levant, Environments, Climate Change and Humans, Cambridge University Press, Cambridge, pp 17–30

Michelson H (1968) The Geology of the Carmel Coast, M.Sc. thesis, Jerusalem, Israel Hebrew University, 60 p. (in Hebrew, English abstract)

Nadel D, Shtober-Zisu N, Frumkin A, Yaroshevich A (2012) New Prehistoric Cave Sites in Lower Nahal Oren, Mt. Carmel. Israel. J Prehist Soc 42:1–40

Nevo E (2009) Evolution in action across life at "Evolution Canyons" Israel. Trends Evol. Biol 1(1):3. https://doi.org/10.4081/eb.2009.e3
Olami Y (1984) Prehistoric Carmel. Israel Exploration Society and M. Stekelis Museum of Prehistory, Jerusalem and Haifa
Picard L, Kashai E (1958) On the lithostratigraphy and tectonics of the Carmel. Bull Res Counc Isr. Sect g: Geo-Science 7G:1–19
Quennell AM (1956) The structural and geomorphic evolution of the Dead Sea Rift. Quart J Geol Soc London 114:1–24. https://doi.org/10.1144/gsjgs.114.1.0001
Ronen A (2017) Tabun cave in the Carmel cultural sphere. In: Quaternary of the Levant–environments. In: Enzel Y, Bar-Yosef O (eds) Quaternary of the Levant, Environments, Climate Change and Humans, Cambridge University Press, Cambridge, pp 215–224
Sadeh M, Hamiel Y, Ziv A, Bock Y, Fang P, Wdowinski S (2012) Crustal deformation along the Dead Sea Transform and the Carmel Fault inferred from 12 years of GPS measurements. J Geophys Res: Solid Earth 117(B8). https://doi.org/10.1029/2012JB009241
Sass E (1980) Late Cretaceous volcanism in Mount Carmel. Isr J Earth Sci 29:8–24
Sass E, Bein A (1982) The Cretaceous carbonate platform in Israel. Cret Res 3(1–2):135–144. https://doi.org/10.1016/0195-6671(82)90014-3
Schattner U, Ben-Avraham Z, Reshef M, Bar-Am G, Lazar M (2006) Oligocene-Miocene formation of the Haifa basin: Qishon-Sirhan rifting coeval with the Red Sea-Suez rift system. Tectonophysics 419(1–4):1–12. https://doi.org/10.1016/j.tecto.2006.03.009
Segev A (2009) 40Ar/39Ar and K-Ar geochronology of Berriasian-Hauterivian and Cenomanian tectonomagmatic events in northern Israel: implications for regional stratigraphy. Cret Res 30(3):810–828. https://doi.org/10.1016/j.cretres.2009.01.003
Segev A, Sass E, Ron H, Lang B, Kolodny Y, McWilliams M (2002) Stratigraphic, geochronologic, and paleomagnetic constraints on Late Cretaceous volcanism in northern Israel. Isr J Earth Sci 51
Segev A, Sass E (2009) Geology of Mt. Carmel. Geol Surv Isr Rep GSI/07/2009
Segev A, Sass E (2014) Geology of Mount Carmel—completion of the Haifa region. Geol Surv Rep GSI/18/2014
Shtober-Zisu N, Amasha H, Frumkin A (2015) Inland notches: Implications for subaerial formation of karstic landforms - An example from the carbonate slopes of Mt. Carmel, Israel. Geomorphology 229:85–99. https://doi.org/10.1016/j.geomorph.2014.09.004
Shtober-Zisu N, Amasha H, Frumkin A (2017) Inland notches: lithological characteristics and climatic implications of subaerial cavernous landforms in Israel. Earth Surf Proc Land 42(12):1820–1832. https://doi.org/10.1002/esp.4135
Shtober-Zisu N, Vaks A, Korngreen D, Frumkin A (2020) Slope retreat rates estimated from chronology of tufa deposits sheltered by inland notches on Mt. Carmel, Israel. Geomorphology 367:107319. https://doi.org/10.1016/j.geomorph.2020.107319
Sneh A, Bartov Y, Weissbrod T, Rosensaft M (1998) Geological Map of Israel, 1:200,000. Isr Geol Surv (sheet # 1)
Steinberg J, Gvirtzman Z, Folkman Y (2010) New age constraints on the evolution of the Mt Carmel structure and its implications on a Late Miocene extensional phase of the Levant continental margin. J Geol Soc 167(1):203–216. https://doi.org/10.1144/0016-76492009-089
Wdowinski S, Zilberman E (1997) Systematic analyses of the large-scale topography and structure across the Dead Sea Rift. Tectonics 16(3):409–424. https://doi.org/10.1029/97TC00814
Weinstein-Evron M (2015) The case of Mount Carmel: the Levant and human evolution, future research in the framework of World Heritage. In: Sanz N (ed) Human Origin Sites and the world heritage convention in Eurasia, vol 41. World Heritage Pap, pp 72–92
Weinstein-Evron M, Tsatskin A, Weiner S, Shahack-Gross R, Frumkin A, Yeshurun R, Zaidner Y (2012) A window into early middle Paleolithic human occupational layers: Misliya Cave, Mount Carmel, Israel. PaleoAnthropology 2012:202–228
Wittenberg L, Shtober-Zisu N, Greenbaum N, Inbar M (2002) The paleogeography and stratigraphy of the Nahal Galim stream, Mount Carmel—preliminary results. The 4th Carmel research meeting, University of Haifa (in Hebrew)
Wittenberg L, Kutiel H, Greenbaum N, Inbar M (2007) Short-term changes in the magnitude, frequency and temporal distribution of floods in the Eastern Mediterranean region during the last 45 years—Nahal Oren, Mt. Carmel, Israel. Geomorphology 84(3–4):181–191. https://doi.org/10.1016/j.geomorph.2006.01.046
Yin H, Ben-Abu Y, Wang H, Li A, Nevo E, Kong L (2015) Natural selection causes adaptive genetic resistance in wild emmer wheat against powdery mildew at "evolution canyon" microsite, Mt. Carmel, Israel. PLoS One 10(4):e0122344. https://doi.org/10.1371/journal.pone.0122344
Zilberman E, Greenbaum N, Nahmias Y, Porat N, Ashkar L (2008) Late Pleistocene to Holocene tectonic activity along the Nesher fault, Mount Carmel, Israel. Isr J Earth Sci 57:87–100
Zviely D, Galili E, Ronen A, Salamon A, Ben-Avraham Z (2009) Reevaluating the tectonic uplift of western Mount Carmel, Israel, since the middle Pleistocene. Quat Res 71(2):239–245. https://doi.org/10.1016/j.yqres.2008.11.008

Part III

Coastal Israel

Seascape and Seaforms of the Levant Basin and Margin, Eastern Mediterranean

10

Uri Schattner and Anne Bernhardt

Abstract

The modern seascape (seafloor morphology) of the Levant basin and margin was shaped through an interaction between tectonic, oceanographic, and sediment-transport processes that vary in time and space. These processes include plate and salt tectonics (vertical and lateral motions on land and offshore), oceanographic modifications (circulation, water bodies, and sea-level variations), climatic processes (e.g., aridification), and changes in sedimentation (sediment sources, supply, distribution, and downslope sediment-transport processes). The chapter attempts to disentangle the individual effects of each process and assess the mode of their interaction. We divide the 26,500 km^2 area of the Israeli Exclusive Economic Zone (EEZ) into four distinct morphological domains, examine the seaforms (morphological features) in each domain, and estimate their timing. The evolution of the seafloor in space and time is illustrated for the Levant margin and deep-marine basin.

Keywords

Seascape · Levant · Tectonics · Sedimentation · Erosion · Continental margin · Gravitational processes · Shore-parallel currents · Turbidity flows · Submarine slides

U. Schattner (✉)
School of Environmental Sciences, University of Haifa, Mt. Carmel, 31905 Haifa, Israel
e-mail: schattner@univ.haifa.ac.il

A. Bernhardt
Institute of Geological Sciences, Freie Universität Berlin, Berlin, Germany

10.1 Introduction

10.1.1 Geological Setting

The Levant basin is a marginal sea located in the mid of a convergence zone between the African and Anatolian-Eurasian plates (Faccenna et al. 2013). The basin extends between the tectonically passive margins of northeast Africa, Sinai, Israel, and the Eratosthenes Seamount (Fig. 10.1). The Levant margin is considered passive up to the Carmel structure (LM and CS in Fig. 10.1a, respectively; Ben-Avraham et al. 2006; Schattner and Ben Avraham 2007). However, this passive status is questioned by Plio-Pleistocene tectonic events (Gvirtzman and Steinberg 2012; Lang et al. 2018). The quiescence was interrupted by displacements along several sets of extensional faults relating to submarine slides, and the boundary effect of salt tectonic motion that occurred in the basin (Ryan and Cita 1978; Frey Martinez et al. 2005; Cartwright and Jackson 2008; Katz et al. 2015; Safadi et al. 2017; Ben Zeev and Gvirtzman 2020). From the Carmel structure northward, the margin is tectonically active along the coast of Lebanon, Syria, and the Cyprus arc. The activity represents the plate convergence around the Phoenician basin in the northeastern corner of the Mediterranean (Fig. 10.1; Ben-Avraham et al. 1988, 2006; Schattner and Lazar 2014; Aksu et al. 2021). Despite the expectation of a marginal sea to record the neo-tectonic processes shaping its surrounding landmass (Zhou 2014), the Levant basin, extending from the base of the slope, remained tectonically quiet throughout the Pleistocene (Schattner et al. 2006; Segev et al. 2018).

The Levant is a confined basin. Its primary sediment source is the Nile River (Fig. 10.1), which has supplied terrigenous sediments since the Oligocene, along with additional sources from North Africa (Macgregor 2012; Faccenna et al. 2019). During the Messinian, the flux of sediments decreased to a minimum (Hsü et al. 1973, 1977;

A. Frumkin and N. Shtober-Zisu (eds.), *Landscapes and Landforms of Israel*,
World Geomorphological Landscapes, https://doi.org/10.1007/978-3-031-44764-8_10

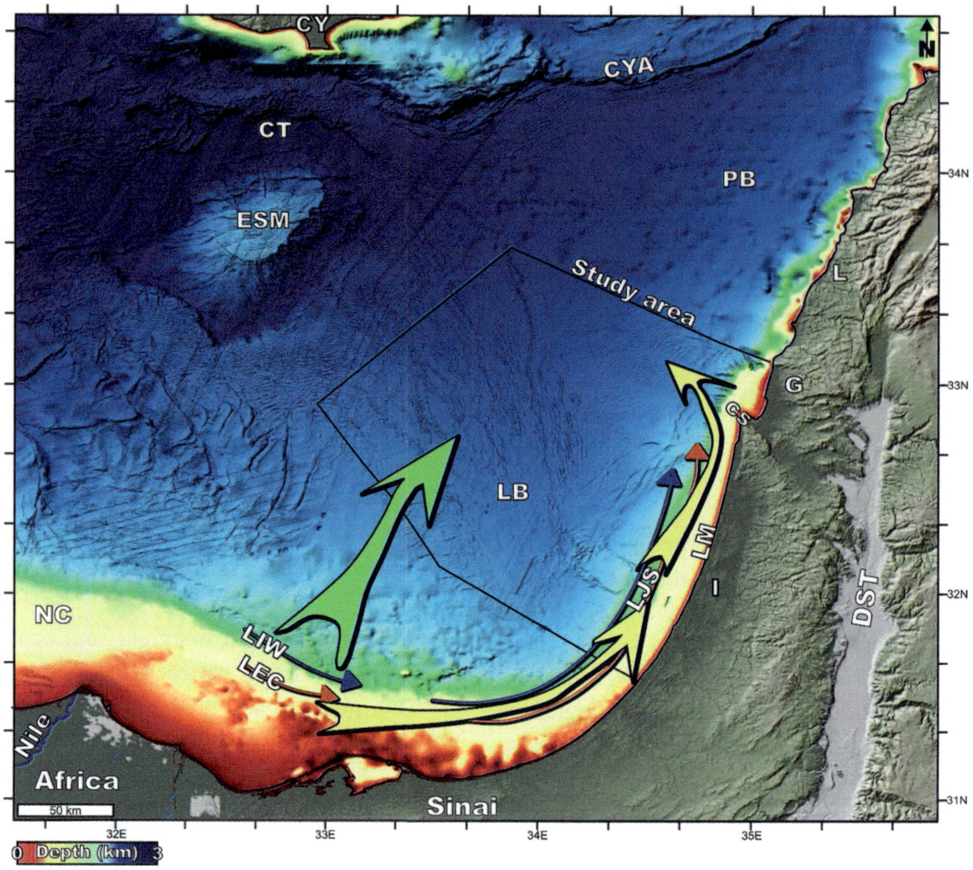

Fig. 10.1 Bathymetry and topography of the eastern Mediterranean (relief from EMODnet Bathymetry Consortium (2020) and GeoMapApp http://www.geomapapp.org; Ryan et al. 2009), featuring the Levant basin (LB), Phoenician Basin (PB), Eratosthenes Seamount (ESM), Cyprus arc (CYA), Cyprus trench (CT), and Cyprus (CY). Black polygon marks the study area, the extent of Fig. 10.2, and the Israeli Exclusive Economic Zone. The Libyan-Egyptian Current (LEC, orange arrow) transports the Atlantic Water mass over the Nile Cone (NC), above the Levant Intermediate Water (LIW, blue arrow). They continue along the Levant Margin (LM) as the Levant Jet System (LJS; after Schattner et al. 2015) which flows along the Israeli coast (I). Thick arrows represent the direct (green) and indirect (yellow) supply of Nile-derived sediments to the deep basin (after Schattner and Lazar 2016; Kanari et al. 2020). Note the NW turn of the yellow arrow immediately south of the Carmel structure (CS). G—Galilee, L—Lebanon. DST—Dead Sea Transform. Coordinates in geographic projection, WGS84 datum

Ryan and Cita 1978; Krijgsman et al. 1999). Evaporites, clastics, and other lithologies were deposited across the desiccated basin, including below the present-day continental rise and slope base (Gvirtzman 1969; Gvirtzman and Buchbinder 1978; Druckman et al. 1995; Feng et al. 2016; Elfassi et al. 2019). The supply of Nile sediments deposited during the Pliocene reached a > 500 m/Ma peak during the Zanclean (mid-Pliocene), and subsequently decreased to ~ 100 m/Ma (Macgregor 2012; Schattner et al. 2017; Li et al. 2021). These sediments built the Nile submarine cone in the southern Levant basin (NC in Fig. 10.1; Ross and Uchupi 1977; Macgregor 2012) and cover the adjacent deep basin extending to the north. Turbidity currents transported the Nile sediments downslope, forming numerous north-trending meandering channels and lobes (Gvirtzman et al. 2015; Schattner and Lazar 2016). These pelagic and terrigenous Pliocene sediments (Carmignani et al. 2009; Paillou et al. 2012; Griffin 2011) accumulated in sections thinning from > 1.5 km over northeast Africa and Sinai continental margin to 400–500 m in the deeper Levant basin (Gardosh et al. 2008; Steinberg et al. 2011; Bar et al. 2013; Lazar et al. 2016; Schattner et al. 2017).

The aridification of North Africa since the late Pliocene (Tiedemann et al. 1989) severely limited fluvial discharge to the Mediterranean coast, excluding the Nile River (Osborne et al. 2008; Paillou et al. 2012). The Nile became almost the exclusive source of Pleistocene sediments to the Levant margin and basin (Schattner et al. 2017). Its submarine cone prograded northward and contributed sediments directly to the deep basin through downslope flows and deep-water channels (Fig. 10.1; Macgregor 2012; Gvirtzman et al. 2015; Schattner et al. 2015; Kanari et al.

2020). However, compared to the widespread sediment supply during the Pliocene and the resulting thick sedimentary succession (hundreds of meters), the deep basin became "starved" during the Pleistocene and accumulated a sedimentary section of only tens of meters in thickness (Ben-Gai et al. 2005; Schattner et al. 2017, 2022).

Meanwhile, during the Pleistocene, the Levant margin accumulated a < 1.8 km thick succession (Lang et al. 2018; Elfassi et al. 2019; Ben Zeev and Gvirtzman 2020; Zucker et al. 2021). Increased tectonic uplift of the Levant landmass by 350–400 m, accompanied by subsidence of the N-S trending Dead Sea Transform, shortened fluvial systems and considerably decreased their supply to the Levant margin (Fig. 10.1; Wdowinski and Zilberman 1997; Sandler and Herut 2000; Ben-Avraham et al. 2005; Bar et al. 2013, 2016; Schattner et al. 2006; Lang et al. 2018). The minor to negligible supply from the Levant landmass emphasizes the dominance of shore-parallel transport of terrigenous sediment from the Nile to the Levant margin (Yellow arrows in Fig. 10.1). Most of this supply to the Levant shelf, slope, and deep basin was delivered by shore-parallel marine counter-clockwise circum-Mediterranean currents (Gallego-Torres et al. 2011; Malanotte-Rizzoli et al. 2014). The currents mobilized sediments along the North African shelves and upper slopes toward the Levant margin (Fig. 10.1; Zvieli et al. 2007; Schattner et al. 2015; Schattner and Lazar 2016; Zucker et al. 2021).

Seafloor and subsurface data indicate that the shore-parallel Levant Jet System (LJS in Fig. 10.1) transports the sediments between the coastline and 350 m depth, up to 15 km offshore the coastline (Schattner et al. 2015; Schattner and Lazar 2016; Kanari et al. 2020; Schattner 2021). The LJS flow is primarily controlled by periodic/seasonal fluctuations, while the location of its core is controlled by eustatic sea-level cycles (Schattner 2021). Modeling calculated a broader depth range for the northward along-margin flow, down to the ~ 1000 m isobath (Zucker et al. 2021). These authors suggest that the flow transports fine-grained sediment along the entire slope, down to the slope-rise transition. However, no indications for margin-parallel seafloor contour currents below a water depth of 350 m are documented to date.

10.1.2 Oceanographic Settings

Seafloor currents (also called bottom currents) are among the most immediate factors involved in reworking and reshaping seabed morphology through erosion, sediment transport, and deposition. These currents are driven by thermo-haline circulation. They mobilize the deepest water mass over the seafloor, forming an interface of mutual influence between the flow and the seafloor morphology (Pinet and Popenoe 1982; Rebesco et al. 2014). The seafloor composition, grain-size distribution, and relief dictate local friction and, consequently, shore-parallel flow pattern (laminar, helical, Langmuir), velocity, direction, and erosion/deposition pattern (Jobe et al. 2012). Hence, the morphological imprint of the bottom currents on the seafloor and the formation of indicative seaforms are frequently used as indicators of modern and paleo-flow paths of the currents and lateral and vertical shifts of the flow core over time (Brand et al. 2015). Therefore, the seascape and seaforms may serve as reconstruction proxies for the oceanic conditions involved in their formation.

The water stratigraphy consists of four masses circulating the Mediterranean (Hecht et al. 1988). Their vertical extent and lateral location over the seafloor vary seasonally/periodically. The high temperature and saline Levant Surface Water (LSW) comprises the top tens of meters (~ 80 m). The bottom currents of LSW flow over the inner shelf. Below, the Atlantic Water (AW) extends down to ~ 120 m. This water mass arrives across the Mediterranean, carried by a jet stream until it courses northward along the outer Levant shelf and shelf edge as the LJS (Millot 1985; Ruiz et al. 2009; Malanotte-Rizzoli et al. 2014; Schattner et al. 2015). The Levantine Intermediate Water (LIW) underlying the AW originates southeast of Crete, circulates throughout the Levant, and exits toward the Atlantic (Millot 2009). The LIW flows along the upper continental slope, down to ~ 350 m (Schattner et al. 2015).

The shore-parallel Levant Jet System (LJS) transports the LSW, AW, and LIW water masses along the Levant continental margin (Schattner et al. 2015). Their flow along isobaths forms four distinct morphological belts (Fig. 10.2b). Paleo-current indicators and grain-size distributions show that over the last ~ 6 ka of highstand conditions, the LJS had a consistent NNE orientation (Almogi-Labin et al. 2012; Schattner et al. 2015; Katz and Cruvi 2018; Schattner 2021). Subsurface data suggests the LJS flowed along the Levant margin at least since ~ 200 ka (Schattner and Lazar 2016). Similar to the modern flow (Rosentraub and Brenner 2007), the paleo-flow probably underwent seasonal/periodic modifications (Schattner 2021). The flow core also shifted laterally and vertically due to sea-level variations, similar to the Gulf Stream axis shifts during the Miocene (Pinet and Popenoe 1982). During lowstand conditions and shelf exposure, the LJS shifted to flow over the reduced area of the submerged shelf. Transgression enabled a landward and upward shift of the flow core until reaching the maximum landward position during highstands, when the entire shelf and upper slope were subjected to the currents at the LJS bottom (modern situation; Fig. 10.1).

The lowermost water mass is the Eastern Mediterranean Deep Water (EMDW; Hecht et al. 1988; Gertman 2014). The EMDW originated in the Adriatic Sea until 1995,

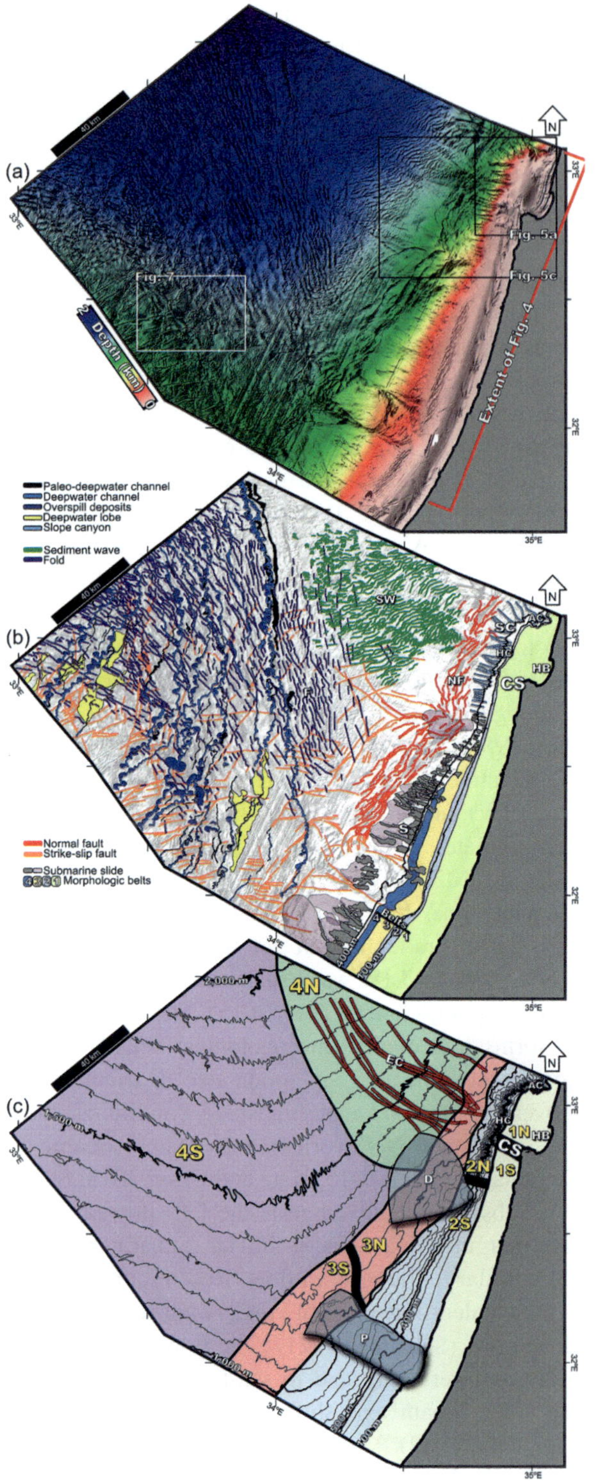

Fig. 10.2 **a** Relief of the study area based on a 100 m cell multibeam bathymetry grid (after Kanari et al. 2020). **b** Shaded relief bathymetry overlaid by the seaforms discussed in the text. Light green, light blue, yellow, and blue morphological belts are numbered 1–4 (respectively, after Schattner et al. 2015). Isobath contours of 100 m and 400 m are marked. **c** Morphological domains of the study area overlying isobaths. SC—slope canyons, SW—sediment waves, NF—normal faults, S—submarine slides (after Gadol 2019), P—Palmahim slide, D—Dor slide, EC—erosional channels, CS—Carmel structure, HB—Haifa Bay, HC—Hilazon canyon, AC—Achziv slope canyons. Coordinates in geographic projection, WGS84 datum

when it started to flow from the Aegean Sea into the deep eastern Levant basin (Theocharis et al. 2002). It is unclear (but not unlikely) if similar alterations occurred in the geological past. In the Levant basin, the EMDW occupies a depth range below ~ 400 m and hence has the most extensive interface with the seascape relative to the other water masses. Its flow parameters over the seafloor are unknown due to the lack of direct oceanographic measurements. Numerical modeling places the core of the shore-parallel flow along the lower continental slope (Zucker et al. 2021). However, the modeling results are not supported by morphological evidence, since the seascape below ~ 400 m water depth lacks along-margin current-indicating morphologies (Schattner and Lazar 2016), while featuring numerous submarine slides (detailed in Sect. 3.2.1; Frey Martinez et al. 2005; Katz et al. 2015; Gadol et al. 2019) and slope canyons (detailed in Sect. 3.2.2).

10.2 Scope of the Study

This chapter focuses on the 26,500 km^2-wide Israeli Exclusive Economic Zone (EEZ), covering most of the Levant basin and margin (Fig. 10.1). The present-day seascape of the study area has developed over geological history through alternating tectonic, climatic, oceanic, sedimentary, and structural geology processes. These processes acted in various combinations as part of the regional, circum-Mediterranean conditions. Furthermore, throughout the development of the area, the influence of old tectonic and sedimentary structures on later seafloor processes is apparent in each morphological domain. The present study attempts to evaluate the role of individual controlling processes and suggests a unifying scenario for their interactions that shaped the modern seascape.

Disentangling the processes will aid in understanding the age of the seabed, the direct and indirect effect of tectonics, and the role of sea-level change on the morphology of the margin and deep basin. Understanding the sediment-distribution patterns and rates (erosion/deposition/transport budget) is crucial for conducting more accurate provenance and source-to-sink studies (Gong et al. 2021) that will also assist in addressing the growing problem of microplastic in marine systems (Kane and Clare 2019; Sharma 2021). The Mediterranean was shown to be one of the major sinks of plastic litter, with ~ 7% of the world's microplastic, most of which concentrates on the nearshore seafloor (Pham et al. 2014; Suaria et al. 2016) and large accumulations of macro-litter (Pierdomienico et al. 2019). The present study will highlight the transport pathways along the Levant margin and into the basin, and the effect of existing seafloor relief on sediment transport.

10.3 Seafloor Mapping

During the last two centuries, hydrographic mapping of the Levant seafloor gradually revealed its seascape. The first surveys were conducted by the British Admiralty in 1861–62 and 1930–33 (British Admiralty, 1863a, b, 1930; Hall, 1996; Sade et al. 2007), followed by the Survey of Palestine Fisheries Service (four 1:100,000 charts by Rosenau 1938). Emery and Bentor (1960) surveyed the continental shelf and slope, down to 200 m, and produced a 10 m contour map. Israel Oceanographic and Limnological Research and the Geological Survey of Israel conducted the first methodologic survey that used tide-corrected single-beam spot soundings (IOLR 1971) that yielded a series of 1:100,000 scale charts. Hall (1994, 1998) compiled all data sources to formulate the first bathymetric map of the Levant basin and margin, and produced 1:50,000 scale Mercator charts in a 25 m grid offshore. These pioneering hydrographic studies provided the first glance at the seafloor shape. However, due to technological limitations, and the availability of data points only along ship tracks, only large-scale geomorphologic structures were detected. For example, slides crossing the slope were identified as 'disturbances' (Almagor 1980; Garfunkel 1984).

The main breakthroughs in understanding seascape geomorphology are attributed to multibeam mapping and three-dimensional seismic data interpretation. These data sources provide continuous coverage of the seafloor. The multibeam mapping covered the entire Israeli Mediterranean EEZ (26,500 km^2) in three main phases between 2001 and 2016 (summarized in Kanari et al. 2020). These surveys were carried out as part of the National Bathymetric Survey and by the Israel Ministry of Energy, through collaborative work of the IOLR and the Geological Survey of Israel (Sade et al. 2007; Tibor et al. 2013; Hall 2015; Hall et al. 2017). Additional studies extracted the seafloor relief from 3D seismic reflection surveys (Gvirtzman et al. 2015).

10.4 Seascape Domains

The Levant seascape is divided into four main domains: (1) continental shelf, (2) slope, (3) slope-rise transition, (4) deep basin (Figs. 10.2c and 10.3). When needed, the domains are sub-divided into a southern and northern segment (S and N, respectively). The seaforms occurring in these domains are described below.

10.4.1 Domain 1—The Continental Shelf and Its Edge

The Levant continental shelf narrows gradually northward from ~ 20 km at 31.8° N to around 10 km at 32.7° N, and is almost non-existent offshore southern Lebanon (Figs. 10.1 and 10.4). The shelf gradient varies only subtly between 0.45 and 0.5° (Fig. 10.4). The margin is subdivided into two morphologic segments by the basinward protrusion of the Carmel structure (CS in Figs. 10.1, 10.2, 10.3, 10.4, 10.5 and 10.6; Ben-Avraham et al. 2006). This tectonic division, which includes the deep subsurface up to the modern seafloor, has a major effect on the structure, stratigraphy, and sediment transport along this segment of the margin (Garfunkel and Almagor 1985; Schattner et al.

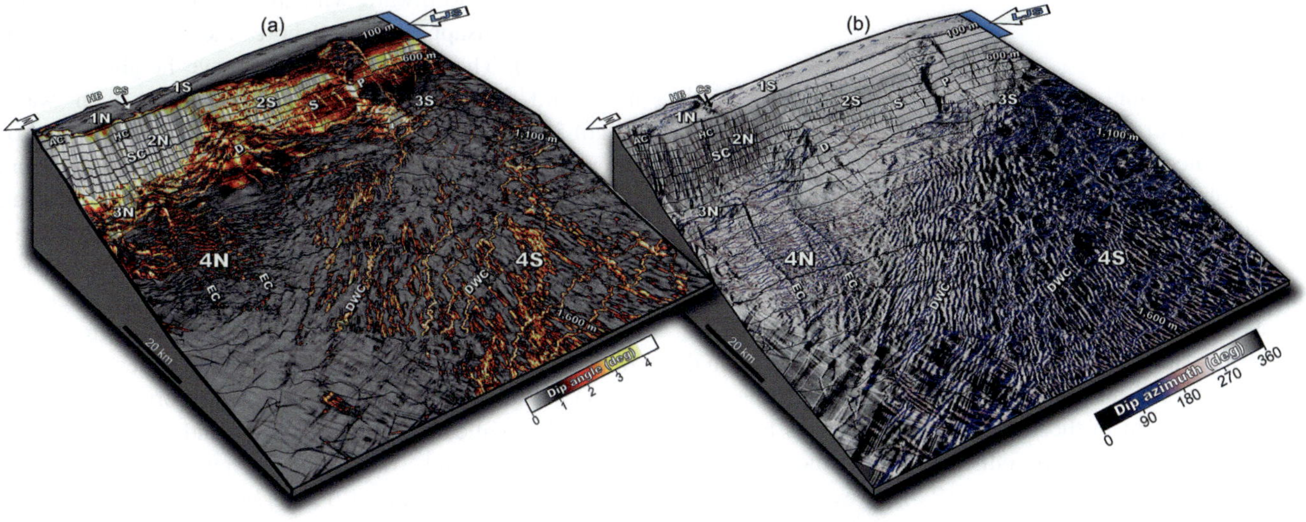

Fig. 10.3 Three-dimensional representation of the bathymetric relief of the study area colored by dip angle (**a**) and dip azimuth (**b**). Isobath contours are marked every 100 m. Domain names and abbreviations as on Fig. 10.2. DWC—deep-marine channel

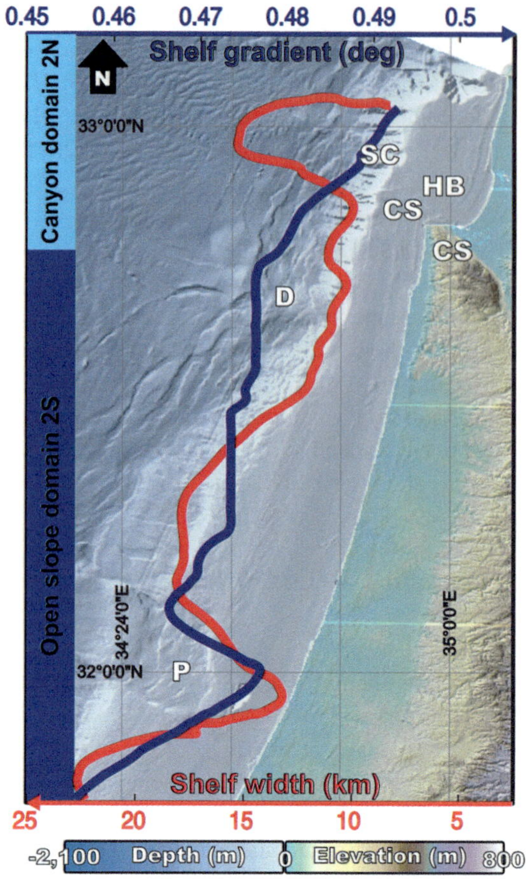

Fig. 10.4 Graphs representing the width (red, bottom horizontal axis) and gradient (blue, upper horizontal axis) of the continental shelf vs latitude. The graphs are draped over their corresponding latitude in the relief map. Note the shelf narrowing and steepening next to the Palmahim slide (P); widening and moderating between the P and D (Dor) slides; narrowing and moderating next to the D slide; and widening and steepening north of the Carmel structure (CS). SC—slope canyons, HB—Haifa Bay. Coordinates in geographic projection, WGS84 datum

2006). The Carmel nose is a shallow, ~ 25 m water depth, protrusion of Cretaceous bedrock that extends from a prominent elevated onshore structure (Garfunkel and Almagor 1985; Cohen et al. 1990; Ben-Gai and Ben-Avraham 1995; Achmon and Ben-Avraham 1997; Schattner et al. 2006; Segev and Sass 2009). Its rugged relief extends across the entire ~ 10 km of the continental shelf. South of the Carmel, the shelf narrows northward from ~ 25 km to ~ 10 km over the Carmel structure, along with overall steepening (domain 1S; Figs. 10.2c and 10.4). North of the Carmel, the shelf widens to ~ 15 km in the Haifa Bay and narrows to almost zero north of the western Galilee until it is incised by the Achziv slope canyon (Ac in Fig. 10.5; see Sect. 3.2.2; Sade et al. 2006).

The shelf consists of four distinct shore-parallel morphological belts that follow isobath contours (Belts 1–4 in Fig. 10.2b; Schattner et al. 2015). Belt B1 extends between the 0–80 m isobaths underneath the Levant Surface Water. B1 is the only belt that appears both south and north of the Carmel structure. South of the Carmel B1 appears as a shore-parallel erosional truncation along the inner shelf. In places, it shows elongated fossilized aeolianite ridges (Figs. 10.2a and 10.3; Sivan and Porat 2004; Katz and Mushkin 2013) surrounded by younger soft Nile-derived sediments (Eytam and Ben-Avraham 1992; Almogi-Labin et al. 2012; Kanari et al. 2020; Schattner 2021). The ridges also appear on the Carmel structure and the inner shelf north of it (Figs. 10.5c and 10.6; Sade et al. 2006; Kanari et al. 2020). They comprise part of the underlying erosional surface exposed during the Last Glacial Maximum (LGM; Surface A in Neev et al. 1966; Almagor 2005; Schattner et al. 2010, 2012, 2015; Lazar et al. 2016; Schattner and Lazar 2016).

Along the shelf edge, the relief of B1 is concave and smooth (Figs. 10.2, 10.3 and 10.4). It comprises the top surface of shore-parallel draped clinoform contourites (Fig. 10.2a). Belt B1 continues over the Carmel structure northward (Figs. 10.2c and 10.5c). However, its underlying post-LGM unit is absent over the Carmel structure and re-appears north of it (Figs. 10.5d and 10.6; Schattner et al. 2010). The relatively smooth relief of Haifa Bay and Western Galilee is interrupted by numerous pockmarks and low-cohesion zones, both related to ongoing gas/fluid emission (Schattner et al. 2012).

The deeper shore-parallel morphological belts appear only south of the Carmel structure (domain 1S in Fig. 10.2c). Belt B2 forms a series of NNE-trending elongated sediment waves along the shelf break, between 80 and 130 m. This belt, which underlays the Atlantic Water mass, narrows northward with the steepening of the shelf edge, until disappearing offshore the Carmel structure (Fig. 10.2c). The average sediment-wave height decreases northward from 3.52 m to 2.67 m, whereas their wavelength increases from 500 to 800 m. The smooth belt B3 extends between 130 and 200 m, under the transition zone between the Atlantic Water and the Levant Intermediate Water. The latter flows along the upper slope and forms belt B4 between 200 and 350 m. This belt comprises sediment waves similar to B2 (Fig. 10.2b). The boundaries between the belts consistently follow isobaths (Schattner et al. 2015). Two shore-perpendicular rotational slides, Palmahim and Dor slides, erode the shore-parallel morphology of the shelf (also called 'disturbances'; P and D, respectively, in Figs. 10.2c and 10.3; Almagor and Garfunkel 1979; Almagor 1984; Garfunkel 1984; Mart 1984; Gadol et al. 2019; Lifshiz and Makovsky 2019).

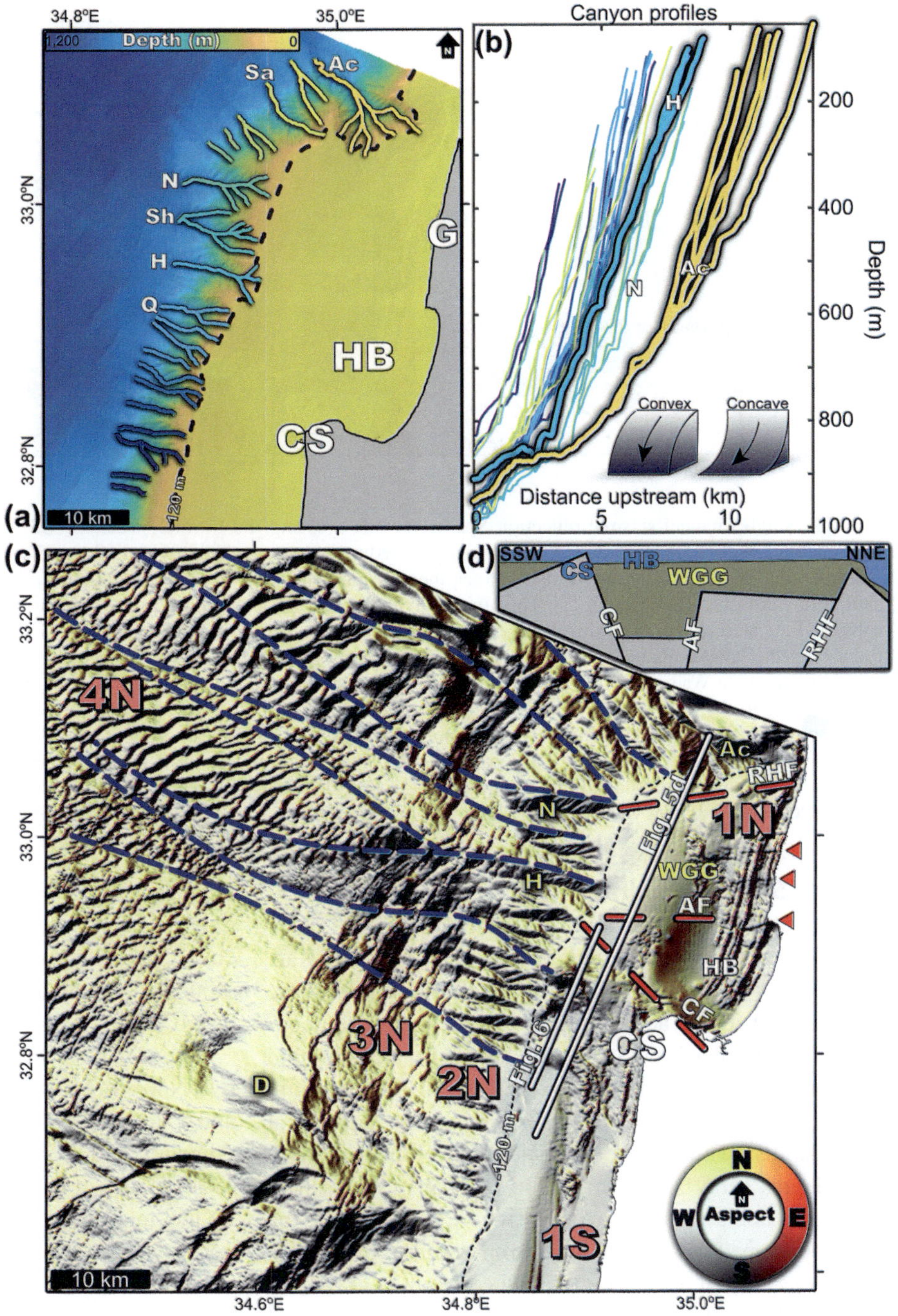

Fig. 10.5 **a** Amplified bathymetry of the slope canyon domain 2N (location in Fig. 10.1). Slope canyons are marked in colors corresponding to their latitude and the thalweg profiles in (b). Names of major canyons are Qishon (Q), Hilazon (H), Shomrat (Sh), Nahariya (N), Sa'ar (Sa), and Achziv (Ac). CS—Carmel structure, HB—Haifa Bay, and G—Galilee. **b** Depth profiles of the slope canyon thalwegs. Note the concave profile of the Achziv canyon (Ac, yellow) compared to the convex one of Hilazon canyon (H, light blue). All analyses were carried out using the functions of TopoToolbox (Schwanghart and Scherler 2014). **c** Shaded aspect map (dip azimuth) emphasizing the possible extension of the slope canyons across the slope-rise transition and continental rise (domain 4N, dashed blue lines). Dashed red lines mark the marine extensions of the Carmel and Rosh Hanikra faults (CF and RHF, respectively) that bound the Western Galilee Graben (WGG), and the Ahihud fault (AF), which bounds the Qishon graben (after Garfunkel and Almagor 1985; Schattner et al. 2006). Red triangles mark E-W faults crossing the inner shelf offshore the Galilee coast. Coordinates in geographic projection, WGS84 datum. **d** Schematic NNE-trending profile representing the structural transition across the Carmel structure into Haifa Bay. Location marked on panel (c)

10.4.2 Domain 2–The Continental Slope

The continental slope below ~ 400 m water depth shows two distinct morphologies, divided here into domains 2S and 2N (Figs. 10.2 and 10.3). The southern slope (2S in Figs. 10.2c and 10.3) has a gradient of less than 5°, featuring numerous kilometer-scale slumps (~ 10 to 50 km^2), and the rotational Palmahim slide, which is an order of magnitude larger (Almagor and Wiseman 1977; Mart et al. 1978; Almagor and Garfunkel 1979; Garfunkel et al. 1979; Almagor 1984; Mart 1984; Almagor and Hall 1980, 1984; Katz 2012; Katz et al. 2015; Gadol et al. 2019). The northern termination of domain 2S is marked by the narrowing and steepening of the continental slope, and the reduction of slump abundance and scale (Schattner et al. 2015; Gadol et al. 2019). To the north, domain 2N exhibits 20 slope canyons that begin to appear across the head scar of the Dor slide and continue to characterize northern Israel, Lebanon, and Syria slopes (Figs. 10.2, 10.3, 10.5; Ben-Avraham et al. 2006; Carton et al. 2009).

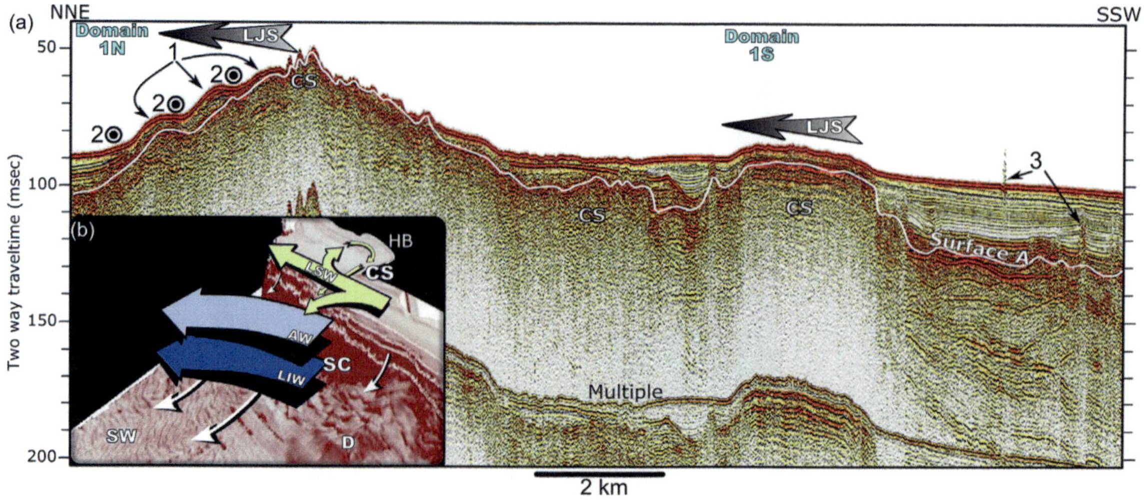

Fig. 10.6 Single-channel seismic reflection Sparker profile crossing the Carmel structure (CS) in a NNE orientation (location in Fig. 10.5c; after Schattner et al. 2015). South of CS (domain 1S), sediments accumulated over the Last Glacial Maximum unconformity (Surface A; after Schattner et al. 2010). Hardly any post-LGM sediment accumulates on the rugged southern flank of CS. The northern CS flank is covered by a thin northward progradation unit (1). This unit is divided into three bathymetric steps, by currents exiting the southern Haifa Bay toward the west (2), illustrated in (b). Occasional gas escape features appear across the profile (3). (b) Schematic illustration of the effect of the Carmel structure on the course of the Levant Jet System. The LJS mobilizes the Levant Surface Water (LSW), above the Atlantic Water (AW), and Levant Intermediate Water (LIW). The LJS advances along the margin, until the Carmel structure obstructs its flow northward. The main flow continues over to the WNW, above domains 2N, 3N, and 4N (Fig. 10.2c). Meanwhile, the LSW follows morphological belt B1 (Fig. 10.2b), crosses above the CS, deposits sediments over its northern flank, and continues northward to fill Haifa Bay. Excess sediments fall along the slope canyons (Fig. 10.5). In the Haifa Bay (HB) part of LSW branches eastward, turns clockwise to the inner part of the bay, and injects sediments westward along the northern flank of CS (marked as 2 in Fig. 10.6a). The injected sediments arriving with the flow fall down the southern slope canyons (SC). SW-sediment waves, D-Dor slide

10.4.2.1 Slides Domain 2S

Most of the relief of domain 2S situated between 31.8°N and 32.7°S comprises hundreds of open-slope slides (Katz 2012; Katz et al. 2015; Gadol et al. 2019). While these slides are distributed across the variable depths of the continental slope, their head scars are clustered mainly around 400 m water depth (Gadol et al. 2019). Over the Palmahim slide, the smaller slides of the 400 m cluster appear at grater depths, while additional slides developed at shallower depths (gray slides in Fig. 10.2b). Their scars show steeper gradients from 5° to more than 20°. Some of the slides developed secondary and even tertiary piggyback generations, which appear on the seabed (Katz et al. 2015). Some of the mobilized sediments rest on the slope and form the slump bodies, while others cover the upper few meters of the slope-rise transition (domain 3S in Figs. 10.2 and 10.3; Gadol et al. 2019). Gadol et al. (2019) identified several short, narrow, and localized slope-gullies on top of few slide bodies. Their comprehensive quantitative analysis showed a general increase in the number of slide scars from south to north and a decrease in the individual slide area, depth, and total transported volume.

The shape, dimensions, and mechanism of the Palmahim and Dor slides differ from the open-slope surface slumps (Almagor and Garfunkel 1979; Garfunkel et al. 1979; Garfunkel 1984; Katz et al. 2015; Gadol et al. 2019). Palmahim and Dor are rotational slides extending from the outer shelf at water depths shallower than 100 m across the entire slope to the slope-rise transition. Their head scars eroded the shelf morphological belt up to B1 (Figs. 10.2c and 10.3).

The ~ 21 km wide Palmahim slide is bounded by ~ N60°W trending normal faults that define its head and extensional domain (Figs. 10.2 and 10.3; Mart 1984; Almagor and Garfunkel 1979; Almagor 1984; Garfunkel 1984; Gadol et al. 2019). The slide is located at a 14° turn in the margin strike, from ~ N32° E to N18° E (south to north). Hence, the slide dip is at an angle of ~ 85° to the strike of the slope. The slide-surface gradient of ~ 2° resembles the lower slope to the north and south. However, it is shifted westward into the basin.

The Dor slide is partially responsible for the shelf narrowing to ~ 10 km at the northern part of domain 1S, south of the Carmel structure (Figs. 10.2 and 10.3). The slide widens across the slope, from ~ 8 km at the shelf break to ~ 15 km at the slope base. This slide is not confined by apparent boundary faulting. Its steep head scar continues to fan-like radial slide with a central bathymetric bulge located over the slope-rise transition (domain 3 in Fig. 10.2c). Both the Palmahim and Dor slides were suggested to result from the ongoing motion of the Plio-Pleistocene section over the Messinian mobile unit (Gadol et al. 2019).

10.4.2.2 Slope Canyons Domain 2N

The transition between domains 2S and 2N occurs at the head scar of the Dor slide, offshore the Carmel structure (Figs. 10.2, 10.3 and 10.5). From this location northward, the continental slope steepens, and its morphology is dominated by slope canyons (Almagor 2005; Ben-Avraham et al. 2006; Schattner et al. 2006; Gvirtzman et al. 2015). We traced 20 slope canyons along this ~ 40 km long slope domain (Fig. 10.5a) and analyzed their morphological expression using TopoToolbox functions (Schwanghart and Scherler 2014).

Our analysis shows that most canyons are confined to the continental slope, while their heads are located close to the -120 m water depth contour. All canyons show a low sinuosity (1.1 ± 0.05 mean/std), with an exception reaching 1.4 in the lower Achziv canyon. Their typically concave profiles show gradients from ~ 35° close to their heads toward 0° at the toe of the continental slope (Fig. 10.5b). The length of the slope canyons varies between 2 and 13 km, with the Achziv, Hilazon, and Nahariya canyons having the longest thalwegs. The slope canyons show a general increase in length from south to north.

North of the Nahariya canyon, the shelf is incised by the Achziv Canyon and its tributaries which stands out with the longest profile, largest area, most concave-up profile shape, and pronounced dendricity. It is the most evolved slope canyon in the study area, with at least five canyon heads incising deeply into the shelf (Fig. 10.5; Almagor 1993; Paldor et al. 2020a,b). The Hilazon Canyon also stands out from the rest (H in Fig. 10.5c). This slope canyon is located along the marine extension of the Carmel fault. Its canyon head slightly indents the shelf edge (Fig. 10.5a); its thalweg is characterized by a pronounced convexity in mid-slope profile (Fig. 10.5b); and it is the second-longest slope canyon mapped. The extent of the slope canyons into the deep-marine basin is obscured by normal faulting along the slope-rise transition (domain 3N in Figs. 10.2c and 10.3). Nevertheless, the dip azimuth values suggest that several slope canyons continue to the rise at depths greater than 1 km (dashed blue lines on aspect map, Fig. 10.5c).

10.4.3 Domain 3–Slope-Rise Transition

The ~ 10 km-wide transition between the slope base and the continental rise is divided into two segments. From the base of the Palmahim slide northward, the slope-rise transition is extensively faulted by NE-trending normal faults (domain 3N in Figs. 10.2c and 10.3; Ben-Avraham 1978; Neev et al. 1985; Tibor and Ben-Avraham 1992; Ben-Avraham et al. 2006; Schattner et al. 2006; Gvirtzman et al. 2015; Kanari et al. 2020). In this region, Kanari et al. (2020) mapped 430 faults with a mean N48°E orientation, which comprise 47% of the total number of faults in the Israeli EEZ studied (Fig. 10.2). These faults follow the morphology of the slope base: they commence at the toe of the Palmahim slide, bound the toes of the slide-dominated southern slope domain (2S in Figs. 10.2c and 10.3), cross the bathymetric bulge of the Dor slide in a semi-radial pattern, and continue along the base of the northern slope domain (2N in Figs. 10.2c and 10.3). This extensional zone was previously defined as a 'deformation belt' and 'disturbed zone' which continues along the base of the Lebanese slope (Ben-Avraham 1978; Neev et al. 1985; Tibor and Ben-Avraham 1992; Ben-Avraham et al. 2006; Schattner et al. 2006). Its development was attributed to NW-trending salt tectonics related to the subsurface Messinian mobile unit (Gradmann et al. 2005; Cartwright and Jackson 2008; Gvirtzman et al. 2015; Kanari et al. 2020; Zucker et al. 2020; Hamdani et al. 2021). From the Palmahim slide southward, the slope-rise transition domain (3S in Fig. 10.2c) is covered by lobes of slides and does not show traces of the extensive extension of its northern counterpart (Gvirtzman et al. 2015; Gadol et al. 2019; Kanari et al. 2020).

10.4.4 Domain 4–The Deep Basin

The deep basin is divided into two domains, based on the dominant seaforms and grain-size distribution. The southern domain 4S (Figs. 10.2c and 10.3) descends from 1000 m in the east to ~ 2000 m water depth at the northwesternmost edge of the EEZ. The grain size of the surface sediment is smaller than of the northern domain 4N (Fig. 10.2) as measured and calculated by Kanari et al. (2020; Fig. 10.2c there).

10.4.4.1 Domain 4S

Domain 4S comprises four morphological seaforms: seafloor folds, faults, deep-marine channels, sedimentary lobes, and erosional channels (Figs. 10.2, 10.3; Gvirtzman et al. 2015; Schattner and Lazar 2016; Kanari et al. 2020). Although these morphologies appear on the modern seabed, they may have formed in the geological past and remained undeformed (see discussion).

Seafloor Folds

Kanari et al. (2020) mapped 971 cylindrical folds across ~ 12,500 km^2 of the deep basin, between water depths of 1300–2110 m (Fig. 10.2b). In the study area, their median orientation is N34° W ± 16°. However, this orientation is part of a larger concentric fold belt that arches around the periphery of the deep Nile cone, beyond the study area (Zucker et al. 2020). Similar folds were also detected further northeast, in the subsurface of domain 4N (see below; Schattner and Lazar 2016).

The length of the folds ranges between 0.2 and 17.2 km, with a median of 2.4 km. Their average relief above the surrounding seafloor is 5.2 m, and in some cases, reaches up to 26 m. Their intermediate synclines form semi-confined basins that continue into the low-lying seabed formed between down-plunging neighboring folds (Figs. 10.2b, 10.3b and 10.7). These localized basins allow post-folding sediment to accumulate and smooth the bathymetric relief (Kanari et al. 2020; see "Lobes" in Sect. 4.4.1.3 below).

Faults

From a total of 918 faults mapped by Kanari et al. (2020) across the Israeli EEZ, 488 faults (53%) appear across domain 4S (orange lines in Figs. 10.2b and 10.7b). These faults displace folds, deep-marine channels, and flat seafloor (Fig. 10.7). Two hundred and ten (210) of the faults are NNE-trending strike-slip faults with an average N14°E orientation. Some of these faults form local push-up and pull-apart structures up to 3 km wide (Kanari et al. 2020). The remaining 278 strike-slip faults show a concentric arrangement with an average N85°E orientation (orange

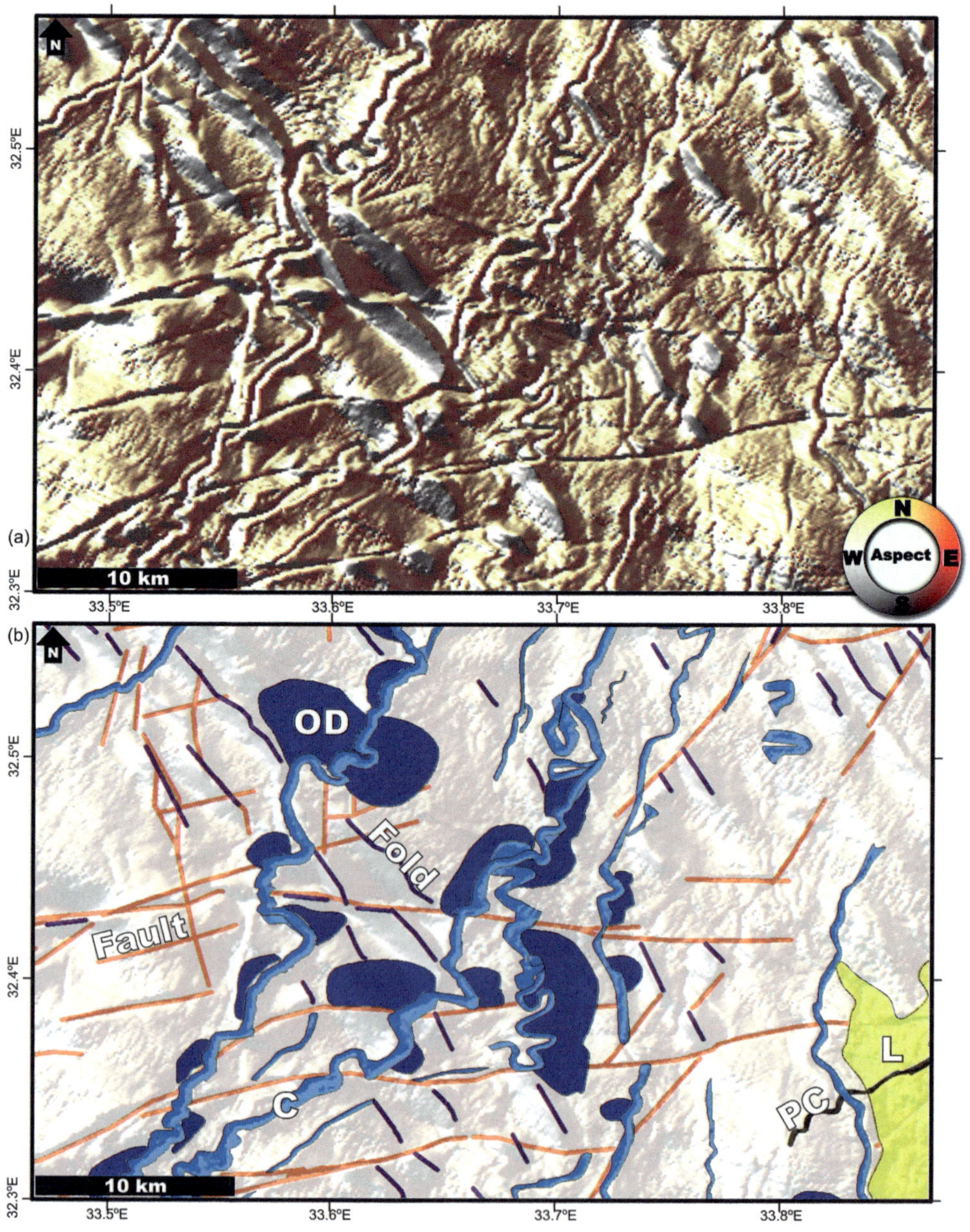

Fig. 10.7 a Shaded aspect map (dip azimuth) across the southern part of the study area (location in Fig. 10.1b), and its interpretation (b), showing morphological evidence for the NNE transport of sediment along paleo-channels (PC, black), deep-marine channels (C, blue), their associated overspill deposits (OD, dark blue), and lobes (L, yellow). These sedimentary features cross faults (orange) and folds (purple) related to the late Pliocene salt-tectonic phase. The low sedimentation rate in the deep basin (domain 4S) during the Pleistocene–enabled preservation of the late Pliocene salt-tectonic-derived seaforms (faults and folds), and the fixation of the deep-marine channels to a general northward direction. Coordinates in geographic projection, WGS84 datum

lines in Fig. 10.7b). Some of the strike-slip and concentric faults are rooted within the Messinian mobile salt layer, displacing the entire overlying Plio-Pleistocene unit up to the surface (Anagnostoudi et al. 2018; Lazar et al. 2018). In places, deep-marine channels cover the surface expression of the faults and obscure their trend (Fig. 10.7). None of these faults were detected within domain 4N (Figs. 10.2c and 10.3).

Deep-Marine Channels

Three patterns of seafloor channel-related deposition are present in domain 4S of the deep basin: modern channels, paleo-channels, and deep-marine fan lobes (Light blue, black, and yellow areas in Figs. 10.2b and 10.7b; Gvirtzman et al. 2015; Schattner and Lazar 2016; Zucker et al. 2017; Kanari et al. 2020). While these patterns vary in size, extent, timing, and formation mechanism, their morphological characteristics indicate that all the channels formed by gravity-driven flow from south to north, and none developed from east to west.

The paleo-channels are shorter than the modern ones (which erode through the paleo-channels; Fig. 10.2b), more segmented/truncated, and partly or entirely filled with sediments, so that their relief is much shallower than the modern ones. Kanari et al. (2020) mapped 20 modern channels, which are arranged in three separated belts (Figs. 10.2b; Fig. 10 in Kanari et al. 2020). The channels cross domain 4S from south to north, continue beyond the study area, and terminate in the Cyprus trench between the Eratosthenes Seamount and Cyprus (CT in Fig. 10.1; Arvidsson et al. 1998; Robertson 1998; Loncke et al. 2006; Klimke and Ehrhardt 2014; Zucker et al. 2017). The lengths of the individual channels range between 30 and 300 km. The average spacing between the channels decreases from 8–11 km in the western belt to 4–5 km, and 4–8 km in the central and eastern belts, respectively. The channels' sinuosity shows similar values: 1.4 in the western belt, 1.4 ± 0.3 and 1.2 ± 0.1 in the southern and northern central belt, whereas the eastern belt shows a northward increase in sinuosity from 1.4 to 1.75. The channel width spans around 400 m, with exceptions such as the 300 km long easternmost channel, which is 500–1000 m wide. The formation timing of the modern channels is unknown. However, their morphological appearance suggests they are continuously eroded by sediment-gravity flows and not passively filled by hemipelagic sediment. In places, they captured some of the paleo-channels (Kanari et al. 2020).

Most of the meandering sections of the modern deep-marine channels are accompanied by zones of overbank deposits (dark blue areas in Figs. 10.2b and 10.7b). Kanari et al. (2020), who identified these seaforms, noted that 125 out of the total 143 overspill zones occur along the western levee of the channels. These zones comprise concentric rings of sediment waves that bound the circumference of meander bends. Their relief decreases away from the meander bend (Kanari et al. 2020).

A set of triangular submarine lobes (yellow areas in Figs. 10.2b and 10.7b) cover only the southern part of domain 4S, between the deep-marine channels. The lobes rise to ~ 1 m above the surrounding seafloor. They cover and fill consecutive synclines between the folds described in Sect. 4.4.1.1. Narrow ~ 500 m-wide channels connect lobes across folds (Figs. 10.2b; Fig. 11c in Kanari et al. 2020). Equivalent lobes were not detected along the eastern and northern parts of the deep basin (Kanari et al. 2020).

10.4.4.2 Domain 4N

Sediment Waves

Domain 4N comprises two main seaforms: NE-trending sediment waves and WNW-trending erosional channels. Previous studies identified 668 sediment waves at a water depth range of 1150–1900 m (Kanari et al. 2020) and described them as elongated dunes with an average N65° E orientation (Figs. 10.2b, 10.3 and 10.5c). Similar sediment waves were also identified offshore southern Lebanon, but not across domain 4S (Gvirtzman et al. 2015; Schattner and Lazar 2016; Kanari et al. 2020; Schattner et al. 2022).

Most of the sediment waves exhibit an asymmetric profile with a steeper land-facing flank. Their consistent asymmetry was interpreted as upslope-migrating cyclic steps created by unconfined turbidity currents that undergo hydraulic jumps (Schattner and Lazar 2016; Slootman and Cartigny 2020). Their relative height reaches up to 6 m above the surrounding seafloor, and their length ranges between 3 and 14 km (Kanari et al. 2020). The longest and highest waves are next to the outlet of the slope canyons, particularly the Hilazon canyon, which faces the eastern center of the sediment wave field (Figs. 10.2b, 10.3 and 10.5).

The modern sediment waves are the uppermost unit of a wedge-shaped Pleistocene succession that thickens toward the base of the continental slope. The wedge includes several consecutive parasequences that accumulated on top of the folded Pliocene unit (Schattner and Lazar 2016). Hence, the morphology of the sediment wave field represents contemporary erosional and depositional seafloor landforms that are not folded nor faulted, unlike the underlying Pliocene unit.

Erosional Channels

The continental rise of domain 4N is incised by ~ 15 channels, which extend from the slope-rise transition toward the deep basin (Fig. 10.2c; Schattner and Lazar 2016; Schattner et al. 2022). These ~ 2 km wide, N56°W-trending channels form almost linear depressions representing pathways of repetitive erosion and negligible passive infill by hemipelagic sediments (Fig. 10.5c). Their thalweg relief decreases downstream from ~ 30 m in the east to almost zero in the northwest. Their banks are not accompanied by levee formation. Past studies may have misinterpreted the channels as the NW-trending marine continuation of the Carmel fault from Levant landmass, or troughs between diapiric ridges (Kafri and Folkman 1981; Garfunkel and Almagor 1985). However, modern seismic data do not show any deep-rooted faulting or diapirism underneath the erosional channels (Schattner and Lazar 2016).

10.5 Discussion—Controlling Processes

The Levant seascape developed through tectonic, structural geology, sedimentary, and oceanographic processes that formed the mosaic of seaforms described above. The following paragraphs describe the individual and combined effects of these processes, using the following logic: the first to be described are ongoing processes that set the conditions through which the seascape had evolved. Shorter-term processes that modify the conditions are described later.

10.5.1 Sediment Distribution

Following the aridification of north Africa, the Nile River became almost the exclusive sediment supplier to the eastern Mediterranean during the Pleistocene. Its direct supply to the basin built the Nile shelf and cone (Fig. 10.1). This ongoing fine-grained gravity-driven supply formed the lobes in the southern part of domain 4S, filled (but not covered) the synclines, and fed generations of deep-marine channels that crossed the basin northward (Figs. 10.1, 10.2 and 10.3; Gvirtzman et al. 2015; Schattner et al. 2015; Kanari et al. 2020).

Meanwhile, in the upper 350 m of the water column, the thermo-haline-driven shore-parallel Levant Jet System of currents (LJS) transported sediments from the Nile outlet and adjacent shelf eastward. The LJS was the exclusive supplier to the Levant shelf and slope throughout the Pleistocene (Schattner et al. 2015; Schattner and Lazar 2016; Kanari et al. 2020). It was subjected to seasonal and periodic alteration in pace and transport capacity. In addition, the location of its core flow shifted basinward (and downward) during regression, and landward (and up) during transgression (Schattner 2021). Hence, sea-level variations directly affected the course of flow, its sediment-transport capacity, and the contribution to the slope and basin floor sediment deposition.

The supply to the deep Levant basin differs between the domains 4S and 4N (Figs. 10.2 and 10.3). Based on an analysis of the morphological development of the deep-marine channels through time in domain 4S, Sagy et al. (2020) coupled the cessation and activation of channels to discharge flooding events of the Nile River. These authors deduced that sea-level fluctuations had a small effect on sediment accumulation in the deep basin. These deductions agree with the grain-size distribution, seafloor morphology, and subsurface data presented by Schattner and Lazar (2016) and Kanari et al. (2020). On the other hand, Niyazi et al. (2018), who studied the same channels and region, suggested that their development is controlled by relative sea-level variations and the Nile River discharge.

Unlike the discrepancy regarding domain 4S, sedimentation across domain 4N (Figs. 10.2 and 10.3) does seem to be affected by the sea-level variations during the Pleistocene (Schattner and Lazar 2016; Schattner et al. 2022). The asymmetric profile of the sediment waves formed across this domain fits the characteristics of upslope-migrating cyclic steps. Their steeper upslope face and the lack of bottom currents below the Eastern Mediterranean Deep Water mass (Schattner and Lazar 2016) indicate that the sediment waves of domain 4N were formed by unconfined turbidity currents arriving from the nearby canyon-dominated continental slope (domain 2N in Figs. 10.2, 10.3 and 10.5). Such upslope-migrating sediment waves are typical for many deep-marine settings and migrate upslope due to hydraulic jumping of turbidity currents (Slootman and Cartigny 2020). This interpretation is supported by recent 18-months-long direct measurements by the DEEP-MED station, which showed that the supply from the margin could travel over 100 km into the basin (Katz et al. 2020). Therefore, the sediment waves of domain 4N are exclusively fed by the shelf and the slope (not from the basin). The processes controlling the sediment-wave formation combine oceanographic and sedimentary aspects such as sea-level variations, thermo-haline flow, and sediment delivery along the course of the LJS (i.e., along the shelf and upper slope). The sediment delivery to domain 4N passed through the slope canyons that developed on the basin-facing margin of the Western Galilee Graben (Fig. 10.5c). Similar oceanographic-sedimentary connection between the shelf-slope-rise is absent in the southern Levant domains (1S, 2S, 3S, 4S in Figs. 10.2c and 10.3).

10.5.2 Tectonic Processes

Tectonic processes defined the basin structure and extent during its Mesozoic opening (Garfunkel 1998; Schattner and Ben-Avraham 2007; Segev et al. 2018). They acted again during the post-rift inversion stage between the

Paleocene and Lower Miocene (Ben-Gai, 2018; Segev et al. 2018). This compressional phase coincides with the formation of the Dead Sea Transform plate boundary between the Arabian and Sinai plates (~ 20 Ma; DST in Fig. 10.1; Avni et al. 2012; Nuriel et al. 2017), which may also correlate with the early stages of the Eastern Mediterranean closure. Segev et al. (2018) showed that after the post-rift inversion ended, the uppermost upper Miocene to recent succession in the basin did not experience tectonic deformation. Schattner et al. (2006) showed that despite the ongoing displacement along the Dead Sea Transform, its branches, and the vertical tectonic accentuation on land, the Levant basin remained tectonically isolated. Nevertheless, seismicity was recorded mainly across the northern Levant basin, between the Camel Structure, Cyprus, and the Phoenician basin (Fig. 10.1; Salamon et al. 1996; Schattner et al. 2006; Mohamed et al. 2015; Wetzler and Kurzon 2016; Sharon et al. 2020).

The morphology of the margin shows both direct and indirect influence of tectonics. The displacement along the Dead Sea Transform was accompanied during the Pleistocene by an increasing transpression component (Butler et al. 1998; Schattner and Weinberger 2008). The resulting uplift of the shore-parallel mountain range had four central influences:

(1) The uplift, that was sub-parallel to subsidence of the Dead Sea Transform valley (DST in Fig. 10.1), drastically shortened fluvial systems that previously reached the Levant margin. The topographic accentuation severely decreased river supply to the Levant margin, which became almost exclusively dependent on the Nile River supply.

(2) The uplift was also induced by excess weight of the Nile sediment cone in the southern Levant that tilted the basin floor downward while the land area gained elevation (Tibor and Ben-Avraham 1992; Gvirtzman and Steinberg 2012). The tilt axis was located along the southern Levant coast (domain 1S; hinge line in Gvirtzman and Klang 1972; Bein and Gvirtzman 1977; Ginzburg and Gvirtzman 1979), and its eastern end did not extend beyond the Dead Sea Transform. Its associated uplift, which took place through isostatic equilibrium (Segev et al. 2006), exhumed the eastern part of the margin that became the coastal area (Gvirtzman and Steinberg 2012). As a result, the width of the submerged part of the margin narrowed, the shore-parallel LJS was progressively pushed westward, and the sediments it carried along the margin fed the prograding clinoforms. For these reasons, the uppermost clinoform, which comprises the modern seafloor, shows an exclusive shore-parallel component without any cones of local rivers. These processes were enhanced by the Nile-induced tilting.

(3) The third major tectonic process was the uplift and protrusion of the Carmel structure on land and offshore (Cohen et al. 1990; Achmon and Ben Avraham 1997; Rotstein et al. 2004; Schattner et al. 2006). The marine extension of the Carmel structure placed a barrier to the northward shore-parallel sediment-bearing flow, and hence regulated the accumulation patterns along the southern Levant margin (Goldsmith and Golik 1980; Zvieli et al. 2007). During low sea levels, the Carmel structure was exposed across the entire margin. It blocked the northward flow of the LJS and diverted it to the northwest to the continental slope (southern part of domain 2N; Figs. 10.1, 10.5 and 10.6). The transported sediments were routed into the southern slope canyons (Fig. 10.5a) and continued toward the sediment wave field of domain 4N (Fig. 10.5c). Back at the shelf, vorticity channeled some of the diverted shore-parallel flow around the Carmel structure into the partially submerged Haifa Bay (light green arrows in Fig. 10.6a; Schattner et al. 2015). These flows circled the bay clockwise and exited westward along the northern flank of the Carmel structure (left side of Fig. 10.6). This current directed the suspended sediments arriving at the shelf edge to the southern slope canyons (Fig. 10.5a).

Low sea levels shifted the coastline, the LJS, and its resulting morphological belts westward (Schattner et al. 2015; Schattner and Lazar 2016; Schattner 2021). The shift enabled a more direct supply to the upper slope. In the southern slope domain 2S, this extra supply may have increased the load over feeble slides and triggered their failure. Hence, low sea levels might be related to excessive sliding over the southern slope that lacks tectonic faulting (domain 2S in Figs. 10.2c and 10.3). This connection indicates that the slope of domain 2S is out of stability balance (Almagor and Garfunkel 1979; Almagor and Wiseman 1977; Frey Martinez et al. 2005). The slides covering most of the slope and the slope-rise transition indicate that the repetitive mass movement reworked the overall slope to moderate its profile (Fig. 10.2b). Various studies related slide triggers with regional seismicity, variations in sea levels and base level, and an excessive load of shelf-derived material over susceptible localities (Katz et al. 2015; Hübscher et al. 2016; Gadol et al. 2019). This excessive, coarse-grained load was sourced by the ongoing supply from the Nile via the LJS (Almagor and Schilman 1995; Schattner et al. 2015; Schattner 2021). The load triggered localized failures that slid downslope and even across the slope-rise transition (Gadol et al. 2019).

In the northern slope domain 2N, the location of the canyon heads along the − 120 m contour indicates that they were connected to the coast during lowstands, such as the Last Glacial Maximum (Fig. 10.5). Previous studies showed that during the glacial maximum sea level was at − 120 m below present since the vertical tectonic and isostatic variations are negligible for this time frame (Fig. 10.2a; Sivan et al. 2001, 2004, 2010; Lambeck et al. 2002). Most of the shelf was exposed, and the coastline was ~ 10 km west of its present location (Enzel et al. 2008). The slope canyons funneled sediment-gravity flows toward the continental rise and incised the deeper 4N domain and formed the WNW-trending erosional channels (EC in Fig. 10.2c; Schattner and Lazar 2016; Kanari et al. 2020; Schattner et al. 2022; Jaijel et al. 2023).

During high sea-level conditions, the shore-parallel LJS flows carried sediments into Haifa Bay and progressively filled it (Pomerancblum 1966; Nir 1980; Inman and Jenkins 1984; Almagor 2005; Zviely et al. 2006, 2007; Avnaim-Katav et al. 2012; Schattner et al. 2015). However, the extent of the fill was dictated by the boundaries of the Western Galilee Graben (WGG in Fig. 10.5c, d). The western bound of the graben determined the location and steepness of the slope and its associated canyons (from Hilazon canyon (H) northward; Fig. 10.5a). These canyons conveyed sediments to the continental rise (domain 4N in Figs. 10.2c, 10.3 and 10.5c; Jaijel et al. 2023), mainly during highstand conditions. The narrow horst defining the northern bound of the Western Galilee Graben induced the formation of the Achziv canyon, which acts as the northern limit of the Nile littoral cell (Zviely et al. 2007).

Canyon occurrence has been previously associated with offshore faulting (Almagor 1993). The Rosh Hanikra fault coincides spatially with prominent canyon heads of the Achziv canyon (Fig. 10.5c). The Hilazon canyon appears to be a particularly efficient sediment pathway based on its exceptional width, indented head scar into the shelf edge, convex thalweg profile, and the concentration of high and long sediment waves at its outlet to the continental rise (Fig. 10.5; Schattner and Lazar 2016; Schattner et al. 2022). The location of the Hilazon canyon coincides with the intersection between the marine extension of the Ahihud and Carmel faults and possible additional slope faulting, close to the elevated marine extension of the Carmel structure (Fig. 10.5c, d; Schattner et al. 2006). It is also located along the course of an enhanced water flow that periodically injected surface water from southern Haifa Bay across the shelf into the basin (Schattner et al., 2015). Hence, the overall northward increase in tectonism of the continental margin (uplift, steepness, and seismicity) controlled the slope angle of domain 2N, which led to the formation of slope canyons that channeled excess sedimentary supply to the continental rise (paths marked by dashed blue lines in Fig. 10.5c).

In summary, tectonic processes which control the width of the margin, the transition from passive to active margins, and the Carmel structure divider define the settings for oceanic flow, shore-parallel sediment transport, the degree of slope stability, and mode of downslope sediment transport. The slope type (slide vs. canyon-dominated slope) determines the mode of sediment distribution across the slope-rise transition: slumps in the south and turbidity currents in the north arriving from domains 2S and 2N, respectively.

10.5.3 Salt Tectonics

The formation of many of the morphological features was induced by salt tectonics. These features include folds, faults in the basin and the slope-rise transition, and the Palmahim and Dor slides. However, the co-appearance of these features on the modern seascape does not suggest they were formed during the same time frame or through a sub-recent process. Elfassi et al. (2019) suggested the following triggers and timing of the salt-tectonic phases: (1) minor motion during the Pliocene, which (2) peaked during the Gelasian (2.6–1.8 Ma) along with slumping from the margin, and (3) decreased until the present.

The accumulation of the Nile cone of sediments in the deep Levant basin offshore northern Egypt (Fig. 10.1) reached an overburden threshold during the Plio-Pleistocene transition. The load induced a lateral motion of the underlying Messinian unit radially away from the Nile cone across its entire circumference as part of thin-skin salt tectonics (Gvirtzman et al. 2015; Schattner and Lazar 2016; Elfassi et al. 2019; Kanari et al. 2020; Zucker et al. 2020). In the Levant Basin, the Messinian mobile unit advanced in a general NNE direction (Netzband et al. 2006), and formed the concentric folds and NNE-trending strike-slip faults across domain 4S (Figs. 10.2, 10.3 and 10.7). The spatial variations in fold symmetry indicate that the strain distribution was uneven across the basin, with no dominant trend. Kanari et al. (2020) explained this as a result of internal volume zonation within the Messinian mobile unit (Anagnostoudi et al. 2018) expressed in its compressibility and hence in the seafloor folds. The spatial variations could also result from sub-phases of folding driven by salt tectonics.

Seismic data indicate that the folding event occurred during a short time window during the Plio-Pleistocene transition. Across the entire basin (domains 4S and subsurface of 4N; Fig. 10.2), the entire Pliocene section was folded with the same wavelength, while the top Pliocene formed the seafloor (Schattner and Lazar 2016; Feng et al.

2017; Schattner et al. 2017). Hence, the folding event began at the end of the Pliocene. The appearance of the folds on the modern seafloor of domain 4S results from low sedimentation rates during the Pleistocene (Ben-Gai et al. 2005) that managed to fill the synclines only partially, coverage of the southern ones by lobes, and crossing of the entire folded area by deep-marine channels that continue until they reach their base level in the Cyprus trench (Figs. 10.1, 10.2 and 10.3; Schattner and Lazar 2016). However, the burial of the salt-tectonic driven folds underneath the Pleistocene sediment-wave field indicates a higher sedimentation rate over domain 4N (Figs. 10.2, 10.3 and 10.5c). The sediment-wave field received indirect supply from the Nile source, via the shore-parallel LJS that arrived through the slope canyons. Since this Pleistocene succession is not folded nor faulted, the salt-tectonic folding phase is restricted to the Plio-Pleistocene transition (Schattner and Lazar 2016; Schattner et al. 2017, 2022). This transition was accompanied by intensified tectonic and probably seismic activity across the entire eastern Mediterranean basin and margin, resulting from the incipient collision of the Eratosthenes Seamount with the Cyprus arc (Schattner 2010). Hence, the combination of sediment load, salt tectonics, and regional tectonic turmoil induced the relatively short folding phase that did not repeat during the Pleistocene, and therefore, most of the seascape of domain 4S was formed during the Plio-Pleistocene transition.

Several clues suggest that the folding phase was accompanied by the formation of the NNE-trending faulting and incision of the deep-marine channels of domain 4S (Figs. 10.2, 10.3 and 10.7). The orientation of the faults coincides with the radial salt-tectonic motion away from the Nile cone. They are detectable on the seafloor due to low sedimentation rates during the Pleistocene. The interactions between the faults and channels indicate co-development. When the folding-faulting phase ended sometime during the early Pleistocene, sediments continued to travel through the deep-marine channels. These occasional flows developed the generations of channels detected on the seabed along with their meanders and overspill deposits. However, the directionality of the channels remained stationary, toward the base level in the Cyprus trench. Therefore, the channels and faults developed synchronously, both became fixed to their locations, and the low sedimentation rates assured their preservation until present at the seafloor.

The normal faulting along the slope-rise transition (Figs. 10.2, 10.3 and 10.5c) indicates a different phase of salt tectonics. Cartwright and Jackson (2008) suggested that this faulting resulted from tilting of the basin that mainly began in the Pliocene. They linked the tilting with the loading of the Nile cone and uplift of the Levant landmass (Tibor and Ben-Avraham 1992; Gvirtzman and Buchbinder 1978; Gvirtzman and Steinberg 2012) that occurred through isostatic equilibrium (Segev et al. 2006). The resulting basinward salt flow thinned the Messinian unit below the slope-rise transition, which induced extensional faulting that occurred while the Plio-Pleistocene sediments accumulated. The growth nature of the faults and the fact they were not buried under the sediments arriving from the slope indicate that the extensional faulting continued throughout the Pleistocene until present (Tibor and Ben-Avraham 1992; Ben-Avraham et al. 2006; Gradmann et al. 2005; Netzeband et al. 2006; Schattner et al. 2006; Elfassi et al. 2019).

10.6 Concluding Remarks

The seascape of the Levant basin and margin developed through an interaction between tectonic, oceanographic, and sedimentary processes acting at different scales, depth ranges, and paces. The 26,500 km^2 area of the Israeli Exclusive Economic Zone (EEZ) can be divided into four distinct morphological domains, with different seaforms, formative processes, and timing. The dominance of terrigenous sedimentation and the limitation of its source to the Nile simplifies the analysis.

Our results indicate that the seascape of the study area comprises a mosaic of relief features of different ages. The morphology of the shelf domain is controlled mainly by thermo-haline marine circulation, sea-level variations, and their interaction with regional vertical tectonic motion. The most prominent tectonic feature along the margin is the Carmel structure that obstructed the northward shore-parallel sediment transport and partially diverted it over the deeper basin. This crucial diversion connects the sediment deposition in the deep domain 4N with shelf processes. In contrast, the lack of similar shelf-basin connections in the southern Levant does not promote the formation of slope canyons.

In general, the slope morphology developed through downslope gravity-driven processes. The southern, slide-dominated domain 2S is gradually moderating basinward of the shelf break, whereas the northern canyon-dominated domain 2N gradually steepens. This difference is enhanced by the excess sediment flow through the slope canyons, and the general northward increase in tectonic and seismic activity, from the Carmel structure northward.

The morphology of the deep basin was shaped during two main phases of salt-tectonic motion, relating to the Messinian mobile unit in the subsurface. The latest Pliocene radial motion away from the Nile cone formed folds and faults across most of the deep Levant basin. In the southern domain 4S, the folds and faults appear on the seafloor due to low sediment supply from the Nile, negligible pelagic sedimentation, and no contribution from the Levant margin, specifically from the slide-dominated domain (2S).

In contrast, the ongoing sediment supply to the northern deep domain 4N covered the late Pliocene folds and faults by shelf-derived sediments due to the NW diversion of the LJS by the Carmel structure. The normal faulting along the slope-rise transition represents an ongoing extension driven by salt tectonics. These faults are partially covered by young slides in the southern domain (3S), and sediment flow paths extending from the slope canyon outlets (domain 3N) toward the sediment wave field of domain 4N.

References

Achmon M, Ben-Avraham Z (1997) The deep structure of the Carmel fault zone, northern Israel, from gravity field analysis. Tectonics 16:563–569

Aksu A, Hall J, Yaltırak C (2021) Miocene–Quaternary tectonic, kinematic and sedimentary evolution of the eastern Mediterranean Sea: a regional synthesis. Earth Sci Rev 103719

Almagor G, Hall JK (1980) Morphology of the continental margin off northern Israel and southern Lebanon. Isr J of Earth Sci 29:245–252

Almagor G (1984) Salt-controlled slumping on the Mediterranean slope of central Israel. Mar Geophys Res 6:227–243

Almagor G (1993) Continental slope processes off northern Israel and southernmost Lebanon and their relation to onshore tectonics. Mar Geol 112:151–169

Almagor G (2005) The Mediterranean coast of Israel. Geol Surv Isr Rep: GSI-13-2005:50

Almagor G, Wiseman G (1977) Analysis of submarine slumping in the continental slope off the southern coast of Israel. Mar Geor Geot 2:349–388

Almagor G, Garfunkel Z (1979) Submarine slumping in continental margin of Israel and northern Sinai. AAPG Bull 63:324–340

Almagor G (1980) Halokinetic deep-seated slumping on the Mediterranean slope of northern Sinai and southern Israel. Mar Geotech 4:83–105

Almagor G, Hall JK (1984) Morphology of the Mediterranean Continental Margin of Israel: (a compilative summary and a bathymetric chart). Geol Surv Isr 77

Almagor G, Schilman B (1995) Sedimentary structures and sediment transport across the continental slope of Israel from piston core studies. Sedimentology 42:575–592

Almogi-Labin A, Calvo R, Elyashiv H, Amit R, Harlavan Y, Herut B (2012) Sediment characterization of the Israeli Mediterranean shelf (10–100 m). Geol Sur Isr: Rep GSI/27/2012:43

Anagnostoudi T, Venderville BC, Gaullier V, Lazar M, Schattner U (2018) Salt-related gravity-driven tectonics in the Levant Basin: insights from physical experiments. In GeoMod2018, (Barcelona, Spain)

Arvidsson R, Ben-Avraham Z, Ekstrom G, Wdowinski S (1998) Plate tectonic framework for the October 9, 1996, Cyprus earthquake. Geoph Res Let 25:2241–2244

Avnaim-Katav S, Almogi-Labin A, Sandler A, Sivan D, Porat N, Matmon A (2012) The chronostratigraphy of a Quaternary sequence at the distal part of the Nile littoral cell Haifa Bay. Isr J Quat Sci 27:675–686

Avni Y, Segev A, Ginat H (2012) Oligocene regional denudation of the northern Afar dome: Pre- and syn-breakup stages of the Afro-Arabian plate. Geo Soc Am Bull 124:1871–1897

Bar O, Gvirtzman Z, Feinstein S, Zilberman E (2013) Accelerated subsidence and sedimentation in the Levant Basin during the Late Tertiary and concurrent uplift of the Arabian platform: tectonic versus counteracting sedimentary loading effects. Tectonics 32:334–350

Bar O, Zilberman E, Feinstein S, Calvo R, Gvirtzman Z (2016) The uplift history of the Arabian Plateau as inferred from geomorphologic analysis of its northwestern edge. Tectonoph 671:9–23

Bein A, Gvirtzman G (1977) A Mesozoic fossil edge of the Arabian plate along the Levant coastline and its bearing on the evolution of the Eastern Mediterranean. In: Structural history of the Mediterranean basins, (Editions Technip Paris), pp 95–110

Ben Zeev Y, Gvirtzman Z (2020) When two salt tectonics systems meet: gliding downslope the Levant Margin and salt out-squeezing from under the Nile Delta. Tectonics 39:e2019TC005715

Ben-Avraham Z (1978) The structure and tectonic setting of the levant continental margin, eastern Mediterranean. Tectonoph 46:313–331

Ben-Avraham Z, Kempler D, Ginzburg A (1988) Plate convergence in the Cyprean Arc. Tectonoph 146:231–240

Ben-Avraham Z, Lazar M, Schattner U, Marco S (2005) The Dead Sea fault and its effect on civilization. In: Wenzel F (ed) Lecture notes in earth sciences: perspectives in modern seismology. Springer, Heidelberg, pp 147–170

Ben-Avraham Z, Schattner U, Lazar M, Hall JK, Ben-Gai Y, Neev D, Reshef M (2006) Segmentation of the Levant continental margin, eastern Mediterranean. Tectonics 25:TC5002

Ben-Gai Y, Ben-Avraham Z (1995) Tectonic processes in offshore northern Israel and the evolution of the Carmel structure. Mar Petr Geol 12:533–548

Ben-Gai Y, Ben-Avraham Z, Buchbinder B, Kendall CGSC (2005) Post-Messinian evolution of the Southeastern Levant Basin based on two-dimensional stratigraphic simulation. Mar Geol 221:359–379

Ben-Gai Y (2018) The Leviathan Miocene Diapir–A novel insight into the tectonic evolution of the Southern Levant Basin, Eastern Mediterranean. Mar and Petr Geol 89:701–713

Brand L, Wang M, Chadwick A (2015) Global database of paleocurrent trends through the Phanerozoic and Precambrian. Scientific Data 2:150025

British Admiralty Hydrographic Office (1863a) Chart of the Bay of Haifa or Khaifa, surveyed by Lieut FGD Bedford RN under the direction of Comm AJ Mansell RN, 1862

British Admiralty Hydrographic Office (1863b) Chart of the Bay of Acre or Akka, surveyed by Lieut FGD Bedford RN under the direction of Comm AJ Mansell RN, 1862

British Admiralty Hydrographic Department (1930) Chart E3261—Bay of Akka. Hydrographic fairsheet for bathymetric survey carried out by H. M. Surveying ship "Endeavour" under the command of Capt. JA Edgell RN in 1930, Mercator projection at a scale of 1:24,980, depths in feet, 1 sheet

Butler RWH, Spencer S, Griffiths H (1998) The structural response to evolving plate kinematics during transpression: evolution of the Lebanese restraining bend of the Dead Sea transform. Geol Soc Lon Sp Publ 135:81–106

Carmignani L, Salvini R, Bonciani F (2009) Did the Nile River flow to the Gulf of Sirt during the late Miocene? Boll Soc Geol Ital 128:403–408

Carton H, Singh SC, Tapponier P, Elias A, Briais A, Sursock A, Jomma R, King GCP, Daëron M, Jacques E, Barrier L (2009) Seismic evidence for Neogene and active shortening offshore of Lebanon (Shalimar cruise). J Geoph Res 114:B07407

Cartwright J, Jackson M (2008) Initiation of gravitational collapse of an evaporite basin margin: the Messinian saline giant, Levant Basin, eastern Mediterranean. Geol Soc of Am Bull 120:399–413

Cohen Z, Kapstan V, Flexer A (1990) The tectonic mosaic of the southern Levant: implications for hydrocarbon prospects. J of Petr Geol 13:437–462

Druckman Y, Buchbinder B, Martinotti GM, Simon Tov R, Aharon P (1995) The buried Afiq Canyon (eastern Mediterranean, Israel): a case study of a Tertiary submarine canyon in Late Messinian times. Mar Geol 123:167–185

Elfassi Y, Gvirtzman Z, Katz O, Aharonov E (2019) Chronology of post-Messinian faulting along the Levant continental margin and its implications for salt tectonics. Mar and Petr Geol 109:574–588

EMODnet Bathymetry Consortium (2020) EMODnet Digital Bathymetry (DTM 2020) EMODnet Bathymetry Consortium. https://doi.org/10.12770/bb6a87dd-e579-4036-abe1-e649cea9881a

Enzel Y, Amit R, Dayan U, Crouvi O, Kahana R, Ziv B, Sharon D (2008) The climatic and physiographic controls of the eastern Mediterranean over the late Pleistocene climates in the southern Levant and its neighboring deserts. Glob Plan Ch 60:165–192

Eytam Y, Ben-Avraham Z (1992) Morphology and sediments of the inner shelf off northern Israel. Isr J of Ear Sci 41:27–44

Faccenna C, Becker TW, Jolivet L, Keskin M (2013) Mantle convection in the Middle East: reconciling Afar upwelling, Arabia indentation and Aegean trench rollback. Ear Plan Sci Let 375:254–269

Faccenna C, Glišović P, Forte A, Becker TW, Garzanti E, Sembroni A, Gvirtzman Z (2019) Role of dynamic topography in sustaining the Nile River over 30 million years. Nat Geosci 12:1012–1017

Feng YE, Steinberg J, Reshef M (2017) Intra-salt deformation: implications for the evolution of the Messinian evaporites in the Levant Basin, eastern Mediterranean. Mar Petr Geol 88:251–267

Feng YE, Yankelzon A, Steinberg J, Reshef M (2016) Lithology and characteristics of the Messinian evaporite sequence of the deep Levant Basin, eastern Mediterranean. Mar Geol 376:118–131

Frey Martinez J, Cartwright J, Hall B (2005) 3D seismic interpretation of slump complexes: examples from the continental margin of Israel. Basin Res 17:83–108

Gadol O, Kanari M, Katz O, Makovsky Y (2019) High-resolution seismic imaging and preliminary geohazard estimation across the three bathymetric archetypes present along the Israeli continental slope. Paper presented at: AAPG Geoscience Technology Workshop

Gallego-Torres D, Martinez-Ruiz F, Meyers PA, Paytan A, Jimenez-Espejo FJ, Ortega-Huertas M (2011) Productivity patterns and N-fixation associated with Pliocene-Holocene sapropels: paleoceanographic and paleoecological significance. Biogeosci 8:415–431

Gardosh M, Druckman Y, Buchbinder B, Calvo R (2008) The Oligo-Miocene deep-water system of the Levant Basin. Geol Surv Isr: Rep GSI/33/2008:80

Garfunkel Z (1984) Large-scale submarine rotational slumps and growth faults in the eastern Mediterranean. Mar Geol 55:305–324

Garfunkel Z (1998) Constrains on the origin and history of the Eastern Mediterranean basin. Tectonoph 298:5–35

Garfunkel Z, Arad A, Almagor G (1979) The Palmahim disturbance and its regional setting. Isr Geol Surv Bull 72:56

Garfunkel Z, Almagor G (1985) Geology and structure of the continental margin off northern Israel and the adjacent part of the Levantine Basin. Mar Geol 62:105–131

Gertman I (2014) Salinity and temperature trends in the southeastern Mediterranean. Paper presented at: EastMed Symposium Agenda (Limassol, Cyprus)

Ginzburg A, Gvirtzman G (1979) Changes in the crust and in the sedimentary cover across the transition from the Arabian Platform to the Mediterranean Basin: evidence from seismic refraction and sedimentary studies in Israel and in Sinai. Sed Geol 23:19–36

Goldsmith V, Golik A (1980) Sediment transport model of the southeastern Mediterranean coast. Mar Geol 37:147–175

Gong Y, Pease V, Wang H, Gan H, Liu E, Ma Q, Zhao S, He J (2021) Insights into the evolution of a rift basin: Provenance of the middle Eocene-lower Oligocene strata of the Beibuwan Basin, South China Sea from detrital zircon. Sed Geol 419:105908

Gradmann S, Huebscher C, Ben-Abraham Z, Gajewski D, Netzeband GL (2005) Salt tectonics off northern Israel. Mar and Petr Geol 22:597–611

Griffin DL (2011) The late Neogene Sahabi rivers of the Sahara and the hamadas of the eastern Libya-Chad border area. Palaeog, Palaeoc, Palaeo 309:176–185

Gvirtzman G (1969) Subsurface data on the Saqiye Group in the Coastal Plain and Hashephela regions, Israel. Geol Surv Isr Rep OD/1/69

Gvirtzman G, Klang A (1972) A structural and depositional hinge-line along the coastal plain of Israel, evidenced by magnetotellurics. Geol Surv of Isr Rep GSI/ 55:18

Gvirtzman G, Buchbinder B (1978) The late Tertiary of the coastal plain and continental shelf of Israel and its bearing on the history of the eastern Mediterranean. In Initial reports of the Deep Sea Drilling Project. Geol Surv Isr 42

Gvirtzman Z, Steinberg J (2012) Inland jump of the Arabian northwest plate boundary from the Levant continental margin to the Dead Sea Transform. Tectonics 31(4)

Gvirtzman Z, Reshef M, Buch-Leviatan O, Groves-Gidney G, Karcz Z, Makovsky Y, Ben-Avraham Z (2015) Bathymetry of the Levant basin: interaction of salt-tectonics and surficial mass movements. Mar Geol 360:25–39

Hall JK (1994) Bathymetric chart of the Eastern Mediterranean Sea, Scale 1:625,000 at lat. 46 N; Mercator projection. Geol Surv of Isr map

Hall JK (1996) Digital topography and bathymetry of the area of the Dead Sea depression. Tectonoph 266:177–185

Hall JK (1998) Israel Mediterranean 1:50,000 scale coastal chart series: explanatory notes. Geol Surv of Isr Rep GSI/1/98, 38 p

Hall JK, Lippman S, Tibor G, Gardosh M, Sade AR, Sade H, Golan A, Amit G, Gur-Arie L, Nissim I (2015) A new bathymetric map for the Israeli EEZ: preliminary results. Min of Enand Wat. Natural Resources Administration in Jerusalem, Israel

Hall JK, Tibor G, Kanari M, Sade AR, Sade H, Golan A, Amit G, Gur-Arie L, Ketter T (2017) A new multibeam bathymetric map of the Israeli exclusive economic zone. IOLR Rep IOLR-H24/2017

Hamdani I, Aharonov E, Olive JA, Parez S, Gvirtzman Z (2021) Initiating salt tectonics by tilting: viscous coupling between a tilted salt layer and overlying brittle sediment. J Geophys Res Sol Ear e2020JB021503

Hecht A, Pinardi N, Robinson AR (1988) Currents, water masses, eddies and jets in the Mediterranean Levantine Basin. J Phys Ocean 18:1320–1353

Hsu KJ, Cita MB, Ryan WBF (1973) The origin of the Mediterranean evaporites. DSDP Init Rept 13:1203–1231

Hsu KJ, Montadert L, Bernoulli D, Bianca Cita M, Erickson A, Garrison RE, Kidd RB, Melieres F, Muller M, Wright R (1977) History of the Mediterranean salinity crisis. Nature 267:399–403

Hübscher C, Betzler C, Reiche S (2016) Seismo-stratigraphic evidences for deep base level control on middle to late Pleistocene drift evolution and mass wasting along southern Levant continental slope (Eastern Mediterranean). Mar Petr Geol 77:526–534

Inman DL, Jenkins SA (1984) The Nile littoral cell and man's impact on the coastal zone of the southeastern Mediterranean. In: Scripps Institution of Oceanography. Reference Series, University of California, La Jolla, vol 43, pp 84–31

Jaijel R, Biton E, Weinstein Y, Ozer T, Katz T (2023) Observations of turbidity currents in a small, slope-confined submarine canyon in the Eastern Mediterranean Sea. Ear Plan Sci Let 604:118008

Jobe ZR, Lowe DR, Morris WR (2012) Climbing-ripple successions in turbidite systems: depositional environments, sedimentation rates and accumulation times. Sedimentology 59:867–898

Kafri U, Folkman Y (1981) Multiphase reverse vertical tectonic displacement across major faults in northern Israel. Ear Plan Sci Let 53:343–348

Kanari M, Tibor G, Hall JK, Ketter T, Lang G, Schattner U (2020) Sediment transport mechanisms revealed by quantitative analyses of seafloor morphology: new evidence from multibeam bathymetry of the Israel exclusive economic zone. Mar Petr Geol 104224

Kane IA, Clare MA (2019) Dispersion, accumulation, the ultimate fate of microplastics in deep-marine environments: a review and future directions. Front Ear Sci 7

Katz O (2012) Mapping of submarine landslides along the Mediterranean continental slope, off the coast of Israel. Geol Surv Isr 13

Katz O, Mushkin A (2013) Characteristics of sea-cliff erosion induced by a strong winter storm in the eastern Mediterranean. Quat Res 80:20–32

Katz O, Reuven E, Aharonov E (2015) Submarine landslides and fault scarps along the eastern Mediterranean Israeli continental-slope. Mar Geol 369:100–115

Katz T, Crouvi O (2018) Sediment flux dynamics over the shallow (25 m depth) shelf of the Mediterranean Sea along the Israeli coast. Mar Geol 406:1–11

Katz T, Weinstein Y, Alkalay R, Biton E, Toledo Y, Lazar A, Zlatkin O, Soffer R, Rahav E, Sisma-Ventura G (2020) The first deep-sea mooring station in the eastern Levantine basin (DeepLev), outline and insights into regional sedimentological processes. Deep Sea Research Part II. Top Stu Ocean 171:104663

Klimke J, Ehrhardt A (2014) Impact and implications of the Afro-Eurasian collision south of Cyprus from reflection seismic data. Tectonoph 626:105–119

Krijgsman W, Hilgen F, Raffi I, Sierro F, Wilson D (1999) Chronology, causes and progression of the Messinian salinity crisis. Nature 400:652–655

Lambeck K, Yokoyama Y, Purcell T (2002) Into and out of the Last Glacial Maximum: sea-level change during oxygen isotope stages 3 and 2. Quat Sci Rev 21:343–360

Lang G, Lazar M, Schattner U (2018) Structural implications of strain localization towards a continental transform fault: the example of the shift between the Levant margin and the Dead Sea Fault plate boundary. Mar Petr Geol 89:402–414

Lazar M, Lang G, Schattner U (2016) Coincidence or not? Interconnected gas/fluid migration and ocean–atmosphere oscillations in the Levant Basin. Geo-Marine Let 36(4):293–306

Lazar M, Schattner U, Reshef M (2018) Chapter 13—The Levant Basin. In: Lofi J, Déverchère J, Gaullier V, Gillet H, Gorini C, Guennoc P, Loncke L, Maillard A, Sage F, Thinon F (eds) Seismic atlas of the "messinian salinity crisis" markers in the mediterranean and black seas—volume 2. Commission for the Geological Map of the World and Société Géologie de France

Li P, Kneller B, Hansen L (2021) Anatomy of a gas-bearing submarine channel-lobe system on a topographically complex slope (offshore Nile Delta, Egypt). Mag Geol 106496

Lifshiz I, Makovsky Y (2019) The kinematics of the dor disturbance offshore Israel: the controls of Pre-MSC morphology on the thin skin deformation patterns and styles. Paper presented at: AAPG Geosc Tech Workshop

Loncke L, Gaullier V, Mascle J, Vendeville B, Camera L (2006) The Nile deep-sea fan: an example of interacting sedimentation, salt tectonics, inherited subsalt paleotopographic features. Mar Petr Geol 23:297–315

Macgregor DS (2012) The development of the Nile drainage system: integration of onshore and offshore evidence. Petr Geosc 18:417–431

Malanotte-Rizzoli P, Font J, García-Ladona E, Pascual A, Tintoré J, Triantafyllou G (2014) Physical forcing and physical/biochemical variability of the Mediterranean sea: a review of unresolved issues and directions for future research. Ocean Sci 10:281–322

Mart Y (1984) The tectonic regime of the southeastern Mediterranean continental margin. Mar Geol 55:365–386

Mart Y, Eisin B, Folkman Y (1978) The Palmahim structure—a model of continuous tectonic activity since the upper Miocene in the southeastern Mediterranean off Israel. Earth Plan Sci Let 39:328–334

Millot C (1985) Some features of the Algerian current. J Geoph Res 90:7169–7176

Millot C (2009) Another description of the Mediterranean sea outflow. Prog Ocean 82:101–124

Mohamed EK, Hassoup A, Abou Elenean K, Othman AA, Hamed DEM (2015) Earthquakes focal mechanism and stress field pattern in the northeastern part of Egypt. NRIAG J Astr Geoph 4:205–221

Neev D, Edgerton HE, Almagor G, Bakler N (1966) Preliminary results of some continuous seismic profiles in the Mediterranean shelf of Israel. Isr J Earth Sci 15:170–178

Neev D, Greenfield L, Hall JK (1985) Slice testonics in the eastern Meditteranean basin. In: Stanley DJ, Wezel FC (eds) Geological evolution of the eastern Mediterranean basin. Springer, New York, pp 249–269

Netzeband GL, Hübscher C, Gajewski D (2006) The structural evolution of the Messinian evaporites in the Levantine Basin. Mar Geol 230:249–273

Nir Y (1980) Recent sediments of Haifa bay. Geol Surv Isr: RepMG/11/80, 8

Niyazi Y, Eruteya OE, Omosanya KO, Harishidayat D, Johansen SE, Waldmann N (2018) Seismic geomorphology of submarine channel-belt complexes in the Pliocene of the Levant Basin, offshore central Israel. Mar Geol 403:123–138

Nuriel P, Weinberger R, Kylander-Clark A, Hacker B, Craddock J (2017) The onset of the Dead sea transform based on calcite age-strain analyses. Geology 45:587–590

Osborne AH, Vance D, Rohling EJ, Barton N, Rogerson M, Fello N (2008) A humid corridor across the Sahara for the migration of early modern humans out of Africa 120,000 years ago. Proc Nat Acad Sci 105:16444–16447

Paillou P, Tooth S, Lopez S (2012) The Kufrah paleodrainage system in Libya: a past connection to the Mediterranean sea? Compt Rend Geosc 344:406–414

Paldor A, Katz O, Aharonov E, Weinstein Y, Roditi-Elasar M, Lazar A, Lazar B (2020a) Deep submarine groundwater discharge - evidence from Achziv submarine canyon at the exposure of the Judea group confined aquifer, Eastern Mediterranean. J Geoph Res Oceans 125:e2019JC015435

Paldor A, Aharonov E, Katz O (2020b) Thermo-haline circulations in subsea confined aquifers produce saline, steady-state deep submarine groundwater discharge. J Hydr 580:124276

Pham CK, Ramirez-Llodra E, Alt CH, Amaro T, Bergmann M, Canals M, Davies J, Duineveld G, Galgani F, Howell KL (2014) Marine litter distribution and density in European seas, from the shelves to deep basins. PLoS ONE 9:e95839

Pierdomenico M, Casalbore D, Chiocci FL (2019) Massive benthic litter funneled to deep sea by flash-flood generated hyperpycnal flows. Sci Rep 9:1–10

Pinet PR, Popenoe P (1982) Blake Plateau: Control of Miocene sedimentation patterns by large-scale shifts of the Gulf Stream axis. Geology 10:257–259

Pomerancblum M (1966) The distribution of heavy minerals and their hydraulic equivalents in sediments of the Mediterranean continental shelf of Israel. J Sed Res 36

Rebesco M, Hernández-Molina FJ, Van Rooij D, Wåhlin A (2014) Contourites and associated sediments controlled by deep-water circulation processes: state-of-the-art and future considerations. Mar Geol 352:111–154

Robertson AHF (1998) Tectonic significance of the Eratosthenes Seamount: a continental fragment in the process of collision with a subduction zone in the eastern Mediterranean (Ocean Drilling Program Leg 160). Tectonoph 298:63–82

Rosentraub Z, Brenner S (2007) Circulation over the southeastern continental shelf and slope of the Mediterranean Sea: direct current measurements, winds, numerical model simulations. J Geoph Res: Oceans 1978–2012:112

Ross AD, Uchupi E (1977) Structure and sedimentry history of the south-eastern Mediterranean sea—Nile cone area. AAPG 61:872

Rotstein Y, Shaliv G, Rybakov M (2004) Active tectonics of the Yizre'el valley, Israel, using high-resolution seismic reflection data. Tectonoph 382:31–50

Ruiz S, Pascual A, Garau B, Faugère Y, Alvarez A, Tintoré J (2009) Mesoscale dynamics of the Balearic Front, integrating glider, ship and satellite data. J of Mar Sys: 78, S3–S16.Ryan WBF (1978) Messinian badlands on the southeastern margin of the Mediterranean Sea. Mar Geol 27:349–363

Ryan WBF, Cita MB (1978) The nature and distribution of Messinian erosional surfaces—indicators of a several-kilometer-deep Mediterranean in the Miocene. Mar Geol 27:193–230

Ryan WBF, Carbotte SM, Coplan JO, O'Hara S, Melkonian A, Arko R, Weissel RA, Ferrini V, Goodwillie A, Nitsche F (2009) Global multi-resolution topography synthesis. Geoch Geoph Geosys 10(3)

Sade A, Hall JK, Golan A, Amit G, Gur-Arie L, Tibor G, Ben-Avraham Z, Ben-Dor E, Fonseca L, Calder BR, Mayer LA (2006) Acoustic backscatter at 95 kHz from the Mediterranean seafloor off Northern Israel. Laminated poster at 1:50,000 scale, 1 sheet. Geol Surv Isr Rep GSI/20/2006 and IOLR Rep H44/2006.

Sade AR, Hall JK, Amit G, Golan A, Gur-Arieh L, Tibor G (2007) The Israel national bathymetric survey—a new look at the seafloor off Israel. Isr J of Ear Sci 55:185–187

Safadi M, Meilijson A, Makovsky Y (2017) Internal deformation of the southeast Levant margin through continued activity of buried mass transport deposits. Tectonics 36:559–581

Sagy Y, Dror O, Gardosh M, Reshef M (2020) The origin of the Pliocene to recent succession in the Levant basin and its depositional pattern, new insight on source to sink system. Mar Petr Geol 120:104540

Salamon A, Hofstetter A, Garfunkel Z, Ron H (1996) Seismicity of the eastern Mediterranean region: Perspective from the Sinai Subplate. Tectonoph 263:293–305

Sandler A, Herut B (2000) Composition of clays along the continental shelf off Israel: contribution of the Nile versus local sources. Mar Geol 167:339–354

Schattner U, Ben-Avraham Z, Lazar M, Hübscher C (2006) Tectonic isolation of the Levant basin offshore Galilee-Lebanon–effects of the Dead Sea fault plate boundary on the Levant continental margin, eastern Mediterranean. J Str Geol 28:2049–2066

Schattner U, Ben-Avraham Z (2007) Transform margin of the northern Levant, eastern Mediterranean: from formation to reactivation. Tectonics 26:1–17

Schattner U, Weinberger R (2008) A mid-Pleistocene deformation transition in the Hula basin, northern Israel: Implications for the tectonic evolution of the Dead Sea Fault. Geoch Geoph Geosys 9:Q07009

Schattner U, Lazar M, Tibor G, Ben-Avraham Z, Makovsky Y (2010) Filling up the shelf—a sedimentary response to the last post-glacial sea rise. Mar Geol 278:165–176

Schattner U, Lazar M, Harari D, Waldmann N (2012) Active gas migration systems offshore northern Israel, first evidence from seafloor and subsurface data. Cont Shelf Res 48:167–172

Schattner U, Lazar M (2014) Flip convergence across the Phoenician basin through nucleation of subduction. Gond Res 25:729–735

Schattner U, Gurevich M, Kanari M, Lazar M (2015) Levant jet system—effect of post-LGM seafloor currents on Nile sediment transport in the eastern Mediterranean. Sed Geol 329:28–39

Schattner U, Lazar M (2016) Hierarchy of source-to-sink systems—example from the Nile distribution across the eastern Mediterranean. Sed Geol 343:119–131

Schattner U, Lang G, Lazar M (2017) Pliocene or Pleistocene, that is the question—new constraints from the eastern Mediterranean. In: Enzel Y, Bar-Yosef O (eds) Quaternary environments, climate change, humans in the levant. Cambridge University Press, pp 63–74, 12 p

Schattner U (2010) What triggered the early-to-mid Pleistocene tectonic transition across the entire eastern Mediterranean? Earth Plan Sci Let 289:539–548

Schattner U (2021) Isolating the role of shore-parallel sediment transport in continental shelf build-up. Cont Shelf Res 104480

Schattner U, Kanari M, Goodman-Tchernov BN, de Mahiques MM, Bernhardt A (2022) Sedimentary response of the deep eastern Mediterranean basin to the north African desertification, sea level variation and regional tectonics. Basin Res 34(2):662–687

Schwanghart W, Scherler D (2014) TopoToolbox 2–MATLAB-based software for topographic analysis and modeling in Earth surface sciences. Earth Surf Dyn 2:1–7

Segev A, Rybakov M, Lyakhovsky V, Hofstetter A, Tibor G, Goldshmidt V, Ben Avraham Z (2006) The structure, isostasy and gravity field of the Levant continental margin and the southeast Mediterranean area. Tectonoph 425:137–157

Segev A, Sass E (2009) The geology of the Carmel region. Albian-Turonian volcano-sedimentary cycles on the northwestern edge of the Arabian platform. Geol Surv Isr: Rep GSI/7/2009:77

Segev A, Sass E, Schattner U (2018) Age and structure of the Levant basin, Eastern Mediterranean. Earth Sci Rev 182:233–250

Sharma S, Sharma V, Chatterjee S (2021) Microplastics in the Mediterranean Sea: sources, pollution intensity, Sea health. Regulatory Policies. Fron in Mar Sci 8:634934

Sharon M, Sagy A, Kurzon I, Marco S, Rosensaft M (2020) Assessment of seismic sources and capable faults through hierarchic tectonic criteria: implications for seismic hazard in the Levant. Nat Haz Earth Sys Sci 20:125–148

Sivan D, Lambeck K, Toueg R, Raban A, Porath Y, Shirman B (2004) Ancient coastal wells of Caesarea Maritima, Israel, an indicator for relative sea level changes during the last 2000 years. Earth Plan Sci Let 222:315–330

Sivan D, Porat N (2004) Evidence from luminescence for Late Pleistocene formation of calcareous aeolianite (kurkar) and paleosol (hamra) in the carmel coast, Israel. Palaeog Palaeoc Palaeo 211:95–106

Sivan D, Schattner U, Morhange C, Boaretto E (2010) What can a sessile mollusk tell about neotectonics? Earth Plan Sci Let 296:451–458

Sivan D, Wdowinski S, Lambeck K, Galili E, Raban A (2001) Holocene sea-level changes along the Mediterranean coast of Israel, based on archaeological observations and numerical model. Palaeog Palaeoc Palaeo 167:101–117

Slootman A, Cartigny MJ (2020) Cyclic steps: review and aggradation-based classification. Earth Sci Rev 201:102949

Steinberg J, Gvirtzman Z, Folkman Y, Garfunkel Z (2011) Origin and nature of the rapid late Tertiary filling of the Levant Basin. Geology 39:355–358

Suaria G, Avio CG, Mineo A, Lattin GL, Magaldi MG, Belmonte G, Moore CJ, Regoli F, Aliani S (2016) The Mediterranean Plastic Soup: synthetic polymers in Mediterranean surface waters. Sci Rep 6:1–10

Theocharis A, Klein B, Nittis K, Roether W (2002) Evolution and status of the Eastern Mediterranean Transient (1997–1999) J Mar Sys 33:91–116

Tibor G, Ben-Avraham Z (1992) Late Tertiary seismic facies and structures of the Levant passive margin off central Israel, eastern Mediterranean. Mar Geol 105:253–273

Tibor G, Sade R, Hall JK (2013). Data collection and processing of multibeam data from the deepwater offshore Israel. IOLR Rep: H-31/2013

Tiedemann R, Sarnthein M, Stein R (1989) Climatic changes in the western Sahara: Aeolo-marine sediment record of the last 8 million years (Sites 657–661) Paper presented at: Ruddiman W, Sarnthein M. Proc ODP Sci Res 108:241–277

Wdowinski S, Zilberman E (1997) Systematic analyses of the large-scale topography and structures across the Dead Sea Rift. Tectonics 16:40–424

Wetzler N, Kurzon I (2016) The earthquake activity of Israel: revisiting 30 years of local and regional seismic records along the Dead Sea transform. Seis Res Let 87:47–58

Zhou D (2014) Marginal Seas. Ency Mar Geosc 10

Zucker E, Gvirtzman Z, Steinberg J, Enzel Y (2017) Diversion and morphology of submarine channels in response to regional slopes and localized salt tectonics, Levant Basin. Mar Petr Geol 81:98–111

Zucker E, Gvirtzman Z, Steinberg J, Enzel Y (2020) Salt tectonics in the Eastern Mediterranean Sea: where a giant delta meets a salt giant. Geology 48:134–138

Zucker E, Gvirtzman Z, Granjeon D, Garcia-Castellanos D, Enzel Y (2021) The accretion of the Levant continental shelf alongside the Nile Delta by immense margin-parallel sediment transport. Mar and Petr Geol 126:104876

Zviely D, Kit E, Klein M (2007) Longshore sand transport estimates along the Mediterranean coast of Israel in the Holocene. Mar Geol 237:61–73

Zviely D, Sivan D, Ecker A, Bakler N, Rohrlich V, Galili E, Boaretto E, Klein M, Kit E (2006) Holocene evolution of the Haifa Bay area, Israel, and its influence on ancient tell settlements. Holocene 16:849–861

The Coastal Dunes of Israel and their Transformation in the Past 200 years

11

Noam Levin

Abstract

The coastal dunes of Israel represent a highly dynamic landscape, which can serve as an indicator of both climatic and anthropogenic changes. The current phase of coastal dunes is relatively young, representing an incursion which mostly started in the post-Byzantine and early Moslem periods, probably due to changes in anthropogenic land use. Until the first half of the twentieth century, the coastal dunes were mostly active, and the low interdune areas were often used for traditional Mawasi agriculture, relying on accessible underground water. Both live and dead plants of the coastal dunes were used by the local Arab and Bedouins people for their herds, building materials, heating, and medicines. Following the establishment of the State of Israel, centuries' old practices of overgrazing and cutting of the dune vegetation came to an end, and given the low wind power along the coast, these dunes underwent a process of spontaneous stabilization, associated with an increase in vegetation cover and a decrease in dune movement rates. In the past 140 years, Israel's coastal dunes were put under great development pressures, and large parts of them have been transformed to other land uses, including agricultural land, built-up areas, and other human uses. Since the 1980s, however, the importance of the dunes as a habitat for psammophile species and for recreation has been recognized, and conservation efforts have been put forward, both by designating protected areas and via studies examining the possibility of reactivating the dunes to preserve some of their ecological and cultural values.

N. Levin (✉)
Department of Geography, The Hebrew University of Jerusalem, Mt Scopus, 9190501 Jerusalem, Israel
e-mail: noamlevin@mail.huji.ac.il

Remote Sensing Research Centre, School of the Environment, The University of Queensland, Brisbane, Australia

Keywords

Coastal dunes · Stabilization · Grazing · Bedouins · Parabolic dunes

11.1 Introduction

11.1.1 Israel as Part of the Littoral Cell of the Nile

The majority of the coastal area of Israel is part of the littoral cell of the Nile River, which extends from Alexandria, Egypt, to Acre on the Bay of Haifa, Israel (Inman and Jenkins 1985; Fig. 11.1). The submarine continental terrace of Israel and the northern Sinai Peninsula becomes narrower and steeper from south to north, as the distance from the Nile River increases (Almagor and Perath 2012; Fig. 11.1). The coastal plain of Israel and northern Sinai also narrows from south to north, from a width of about 40 km in northern Sinai, to less than 5 km in the north of Israel, at Rosh ha Nikra (Almagor and Perath 2012). Sand from the Nile River is transported by wave-induced longshore current along the coast of the Sinai Peninsula and Israel and is deposited on beaches fringing the coastline (Emery and Neev 1960; Goldsmith and Golik 1980; Stanley 1989; Almagor and Perath 2012).

The coastal plain is characterized by widespread aeolian landforms, comprising three distinct varieties: (i) coastal dunes, involving both active and stabilized dunes; (ii) lithified dunes which formed parallel calcareous aeolianite (kurkar) ridges both on land and under the sea, thus indicating paleo-dune formation episodes when the sea level was different than today (Mauz et al. 2013; Harel et al. 2017); (iii) paleosols, locally known as hamra soils (Rhodoxeralfs), a mature, non-calcareous red Mediterranean soil of sandy clay loam (Yaalon 1959; Fig. 11.2). The kurkar ridges of Israel, which are composed

A. Frumkin and N. Shtober-Zisu (eds.), *Landscapes and Landforms of Israel*, World Geomorphological Landscapes, https://doi.org/10.1007/978-3-031-44764-8_11

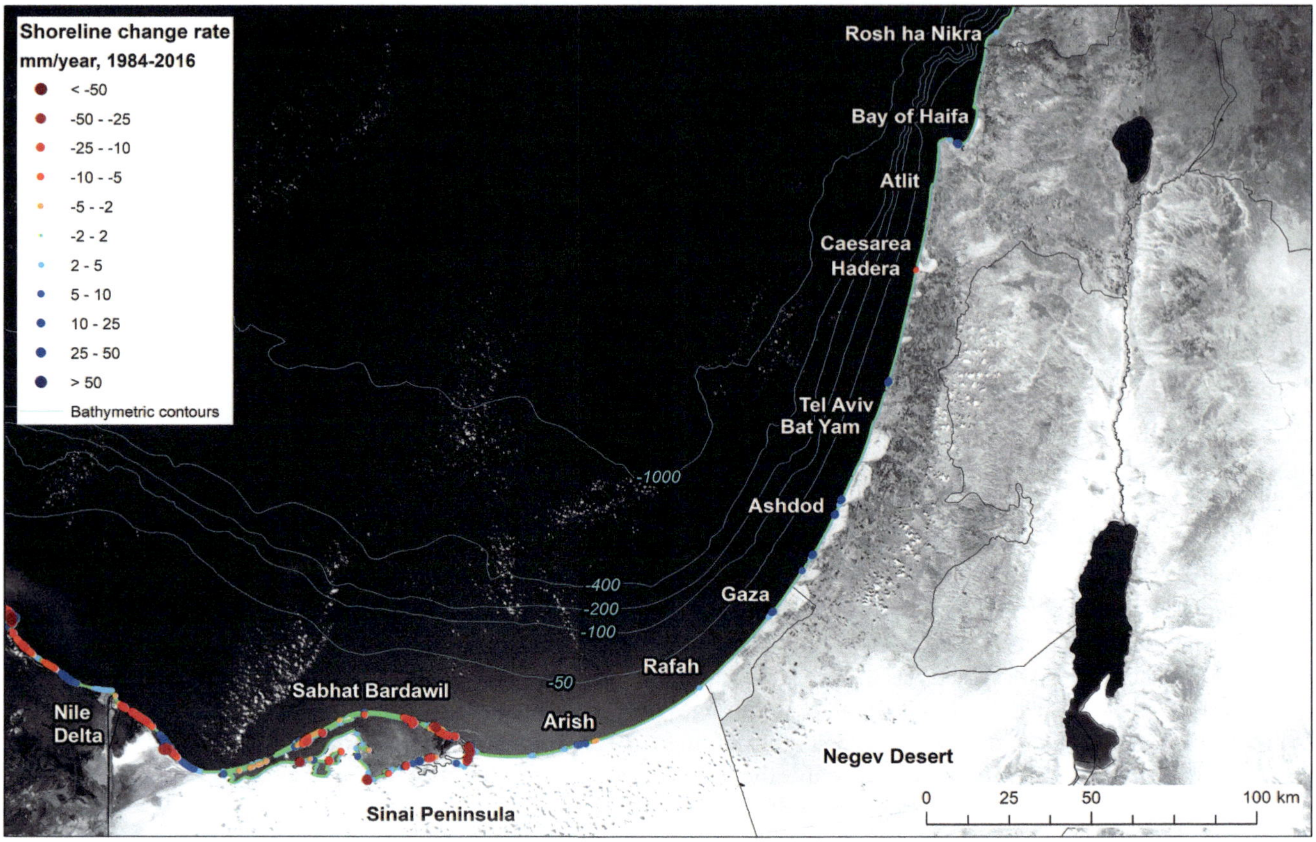

Fig. 11.1 The littoral cell of the Nile. The background satellite image is of Corona, 21 August 1964, and is overlaid by shoreline change rates based on Luijendijk et al. (2018). Note the bright patches of coastal dunes along the coast of Israel, and that most of the shoreline of Israel is relatively stable, in comparison with that of the Nile Delta. The bathymetric contours are based on the General Bathymetric Chart of the Oceans, presented here to visualize the underwater extent of the cone of the Nile Delta

of alternating kurkar and hamra units, are formed of sediments deposited during the late Pleistocene. Whereas in the past this alteration between kurkar and hamra was attributed to paleoclimatic events or to changes in sea level, Sivan and Porat (2004) found this alteration to be unrelated with climatic events. Rather, kurkar formation was found to be associated with high $CaCO_3$ content and high deposition rates (> 1 m/1000 years), whereas the formation of hamra soils is associated with low $CaCO_3$ content and slower deposition rates (< 0.3 m/1000 years) (Porat et al. 2004). Some kurkar units have lithified quite recently (e.g., 5000 years ago; Porat et al. 2004), indicating coastal dune activity during that period. In this chapter, I focus on the present (Holocene) phase of evolution of coastal dunes along the coastal plain of Israel.

11.1.2 Sections of the Coastal Dunes of Israel

Almagor and Perath (2012) have divided the coastal plain of Israel into six sections (Fig. 11.2b): (i) Pleshet (from Gaza Strip to the south up to Tel Yona, south of Bat Yam, about 51 km): this section does not have significant parallel kurkar ridges, nor significant coastal cliffs, and has wide coastal dune fields with sand blowing inland from the beach; (ii) HaSharon (from Tel Yona up to the Hadera River, about 50 km): this section is defined by the existence of a steep coastal cliff, which has blocked the formation of extensive coastal dunes inland, except for small coastal dune fields near river mouths; (iii) Caesarea (from the Hadera River to the Taninim River, about 9 km): this section is mostly sandy, with an extensive coastal dune

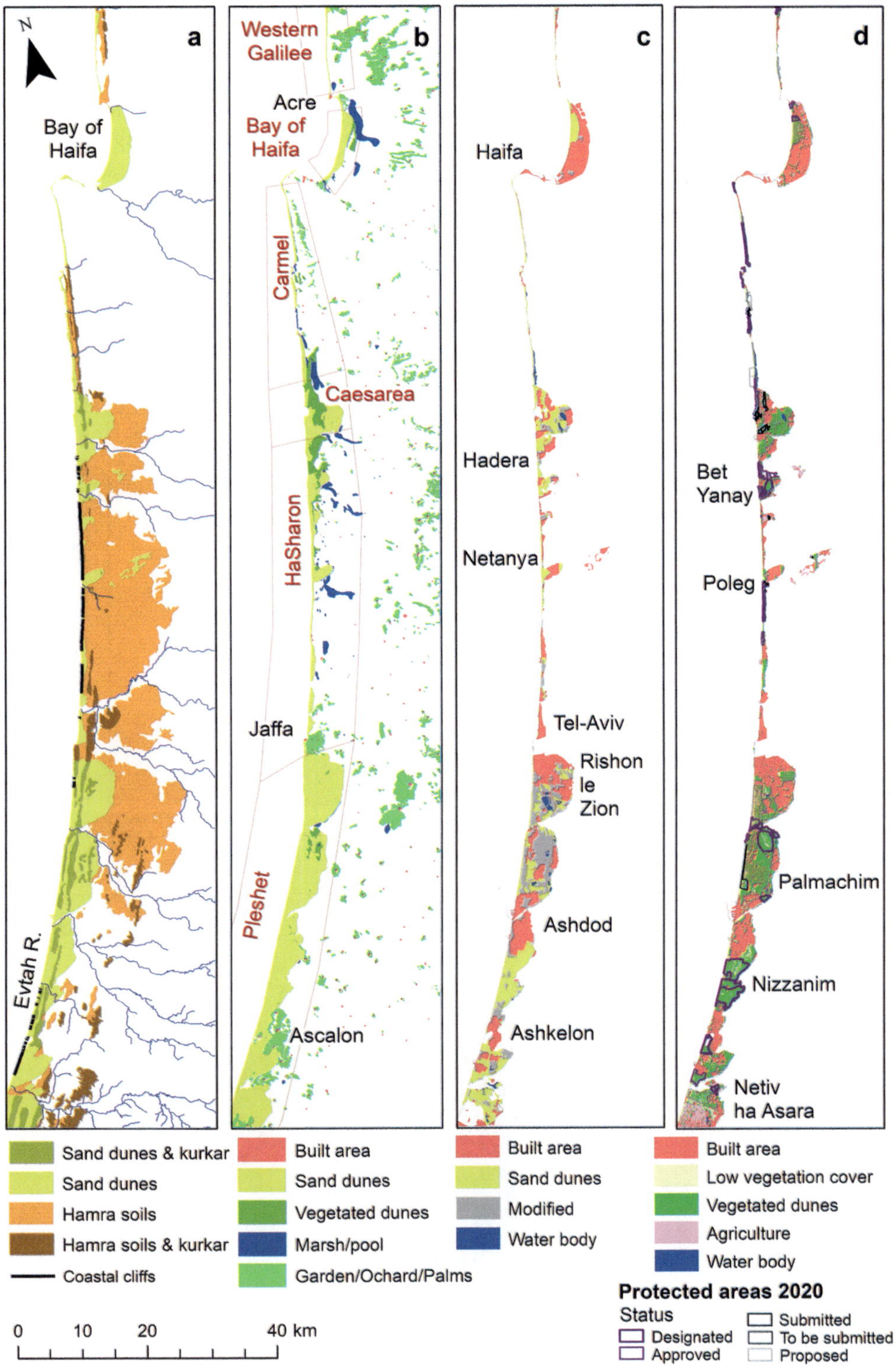

Fig. 11.2 The coastal dunes of Israel, from left to right: **a** Aeolian soils following Dan (1972) and coastal cliffs; **b** Land use in the late 19th century based on the Survey of Western Palestine of the Palestine Exploration Fund, published in 1880; **c** Land use within coastal dunes in 2000 following Frumkin-Ahiron et al. (2003); **d** Land cover within coastal dunes in 2018 based on HaMaarag—Israel's National Ecosystem Assessment Program, and Israel's protected areas as of 2020. The six sections of the coastal dunes are schematically shown in **b** as brown polygons

field; (iv) the Carmel coast (from the Taninim River to Tel Shikmona, about 33 km): this section is narrow, with a limited extent of coastal dunes; (v) the Bay of Haifa (between Haifa and Acre, about 13 km): this is the termination of the littoral cell of the Nile, and coastal dunes have been accumulating north of the Kishon River; (vi) the coast of the western Galilee (from Acre to Rosh HaNikra, about 20 km): the beach is more rocky, and about 80–90% of the sand is derived from local biogenic sources. A detailed description of the coastal dune fields (divided into 12 sections), their changes during the twentieth century, and their conservation status is given by Frumkin-Ahiron et al. (2003).

11.2 Age of Coastal Dunes in Israel

11.2.1 When Did the Current Coastal Dunes of Israel Form?

Based on the model of Sivan et al. (2001), the sea level along the coast of Israel was about − 13.5 m at 8000 years ago and reached its present level about 2000–3000 years ago, hence providing the maximum age of the current (Holocene) phase of evolution of Israel's coastal dunes. Various towns and agricultural areas along the coastal plain of Israel, dating back to the Roman period, have since been covered by coastal dunes (Rim 1950; Issar 1968), also seen from aerial photos (Fig. 11.3), including the towns of Caesarea (Reifenberg 1950a, b) and Ashkelon (Sasson

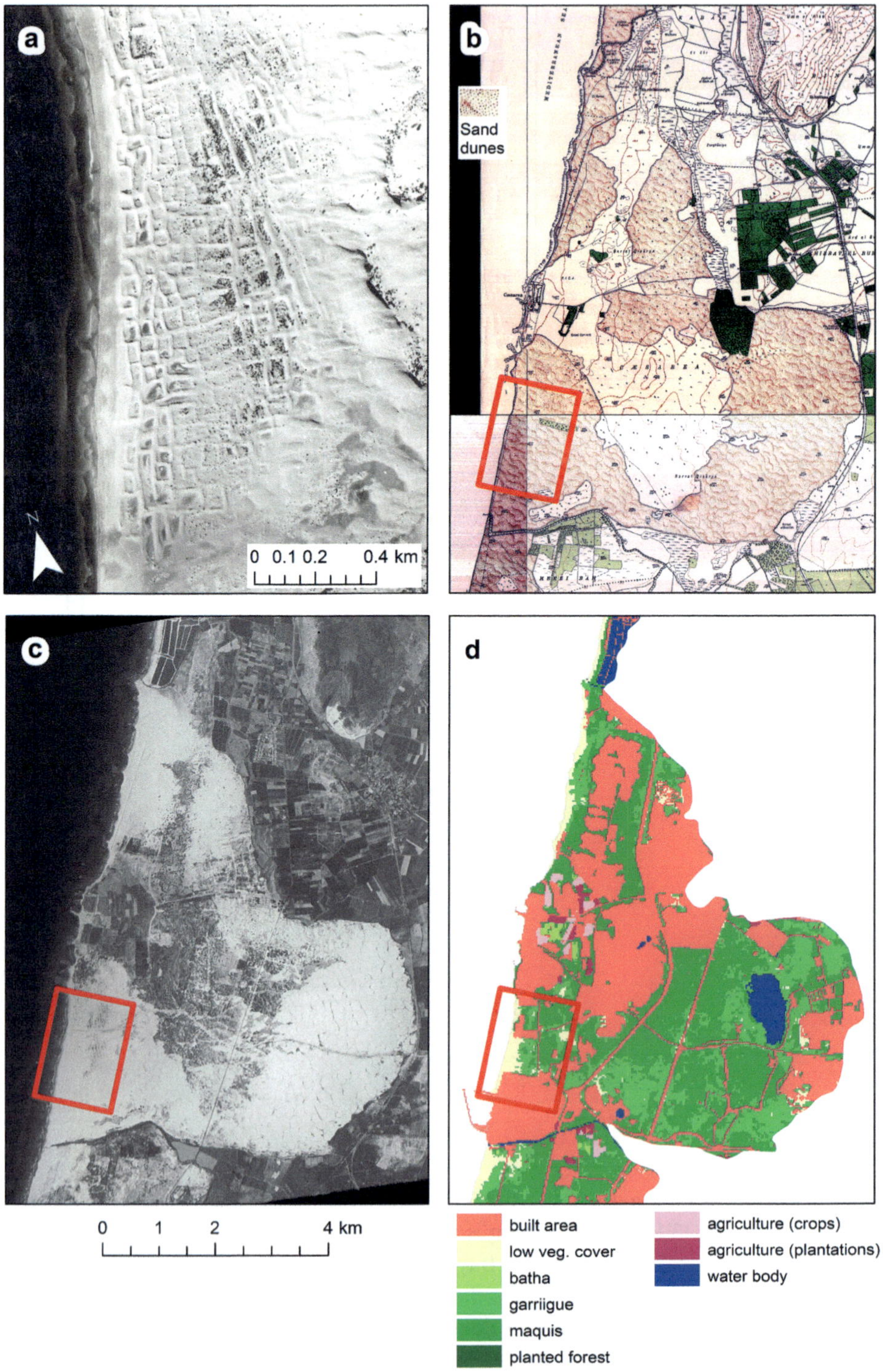

Fig. 11.3 The coastal dunes of Caesarea: **a** aerial photo from December 1945, showing the plot and berm agroecosystem south of Caesarea, covered by dune sand; the location and extent of **a** is shown by a red rectangle in **b**–**d**; **b** British Mandate historical maps from the 1930s at a scale of 1:20,000, showing two distinct areas of sand dunes, separated by a lower vegetated area; **c** a Corona satellite image from June 1963, showing two distinct areas of sand dunes, separated by a lower vegetated area; **d** land cover within the coastal dunes in 2018 based on HaMaarag—Israel's National Ecosystem Assessment Program

2019). The town of Caesarea was founded by King Herod in 22 B.C., and Reifenberg (1950a, b) considered the construction of the mole for the great port of Caesarea as a factor which enhanced the deposition of sand south of it, leading to the widening of the beach and enabling sand to be entrained and transported inland. While Caesarea is described in the Babylonian Talmud (written between the third and fifth centuries AD) as located within dunes (Talmud Bavli, Megilla 6a: "*This is Caesarea, daughter of Edom, which is seated between the dunes*"), it is probable that when it was founded it was not surrounded by dunes. Reifenberg stated that the town of Caesarea continued to flourish for centuries, until its destruction (along with other coastal towns) by the Egyptian Mameluks in the thirteenth century, after which sand dunes encroached inland everywhere. Consequently, the current phase of coastal dune development in Israel is usually described as being post-Roman or post-Byzantine (Tsoar 1990; Roskin et al. 2013; Fischer and Taxel 2021). Levin (2006) estimated the average rate of dune movement between 1874 and 1930 (based on historical maps) to have been about 5.1 m/year. Given that the maximum inland penetration of coastal dunes south of Jaffa was 6 km, he suggested that coastal dunes started encroaching inland about 1200 years ago (assuming no major climatic changes). Confirming this widely held hypothesis, OSL dating of sand dunes in the Caesarea area by Roskin et al. (2015) revealed the existence of both Holocene dunes dated to about ~ 5.9–3.3 ka BP and a large cluster of dune dates going back to the early Moslem period, ~ 1.1–1.2 ka BP (800–900 A.D.). As of the late nineteenth century, before extensive building and land cover transformation (Schaffer and Levin 2014), most of the coastal dunes of Israel were active and with little vegetation cover. Within the coastal dunes of Caesarea, two distinct areas of active dunes can be observed, separated by a lower area of old stabilized sands (Kutiel et al. 1979; Fig. 11.3). This separation into two areas of active dunes may indicate that in the vicinity of Caesarea there were two phases of dune activity, both taking place probably during the past 2000 years.

11.2.2 What Caused the Incursion of the Present Phase of Coastal Dunes?

Explanations given for the subrecent inland incursion of coastal dunes in Israel, often also associated with the desertification of the Negev and the collapse of the Byzantine towns there, include both anthropogenic factors (Huntington 1911; Reifenberg 1950b) and climatic hypotheses of desertification (Issar and Tsoar 1987). However, considering recent literature and ongoing stabilization of coastal dunes in Israel in the twentieth century (as will be described later on in this chapter), it is more likely that the present phase of coastal dune development in Israel can be attributed to changes in land use. Whereas in the past scholars often attributed the fluctuation of the border between the wilderness and the sown to climatic variability, geo-political factors may have been more influential in this transition area, impacting land use, e.g., in the case of the Negev desert (Finkelstein et al. 2018; Avni et al. 2019). Roskin et al. (2013) suggested that the late Holocene dune mobilization both in the NW Negev and in the coastal belt may be attributed to anthropogenic activities of grazing and shrub uprooting (for fuel) by nomadic population. For example, the deliberate destruction of Palestine's coastal towns by the Mamluks in the late thirteenth century, and their subsequent abandonment (Amiran 1953; Fuess 2017), may have enabled nomadic people (Bedouins) to settle along the coastal plain. However, according to the present assessment the incursion of coastal dunes occurred a few centuries before the destruction of the coastal towns by the Mamluks and is yet to be explained (see Roskin et al. 2015).

11.3 Morphology of Coastal Dunes

Various morphological varieties of dunes can be found along the coastal dune belt of Israel, some related to active dunes, and others to dunes in various stages of stabilization (Fig. 11.4). These different types are shaped by several controlling factors, such as distance from the shoreline (wave action), underlying topography that the dunes transgress, wind regime (directionality), vegetation cover, and human disturbance.

11.3.1 Nebkhas

Nebkhas, also known as coppice dunes or hummock dunes (Pye and Tsoar 1990; Hesp and McLachlan 2000), are small mounds of sand formed around shrubs which lower wind speed, thus trapping sand (Tengberg and Chen 1998; Fig. 11.4b). The nebkhas tend to increase in size and height with time and may grow to a significant size, if they happen to form around a tree (Fig. 11.5e). Eventually, as time passes by and the shrub which induced the formation of the nebkha dies, the nebkha itself will erode (Tengberg and Chen 1998). The formation of nebkhas is an essential stage in the formation of vegetated dunes, whether they are vegetated foredunes or parabolic dunes. In a detailed study using erosion pins on 30 nebkhas located on a stabilizing dune in the Nizzanim coastal dune field, Ardon et al. (2009)

Fig. 11.4 Selected photos of sand dune areas in Israel: **a** the beach and foredunes south of Ashdod on a windy day; **b** semi stabilizing dunes with small nebkhas south of Ashdod; **c** an exposed windward dune area with stabilizing dunes in the background, south of Ashdod; **d** an active slipface and a fully vegetated interdune area, south of Ashdod; **e** fully stabilized sand hills, south of Ashdod; **f** fully stabilized area in the former dune field of Caesarea

found that deposition rates were highest for nebkhas located on the dune crest (compared with those located on the windward and leeward sides of the dunes), and that within and around the nebkhas, deposition rates exceeded erosion rates, except for the windward side of the nebkha.

11.3.2 Foredunes

Foredunes are formed behind the beach and beyond the area reached by the waves, when the conditions for vegetation establishment are met, usually by pioneer plant species

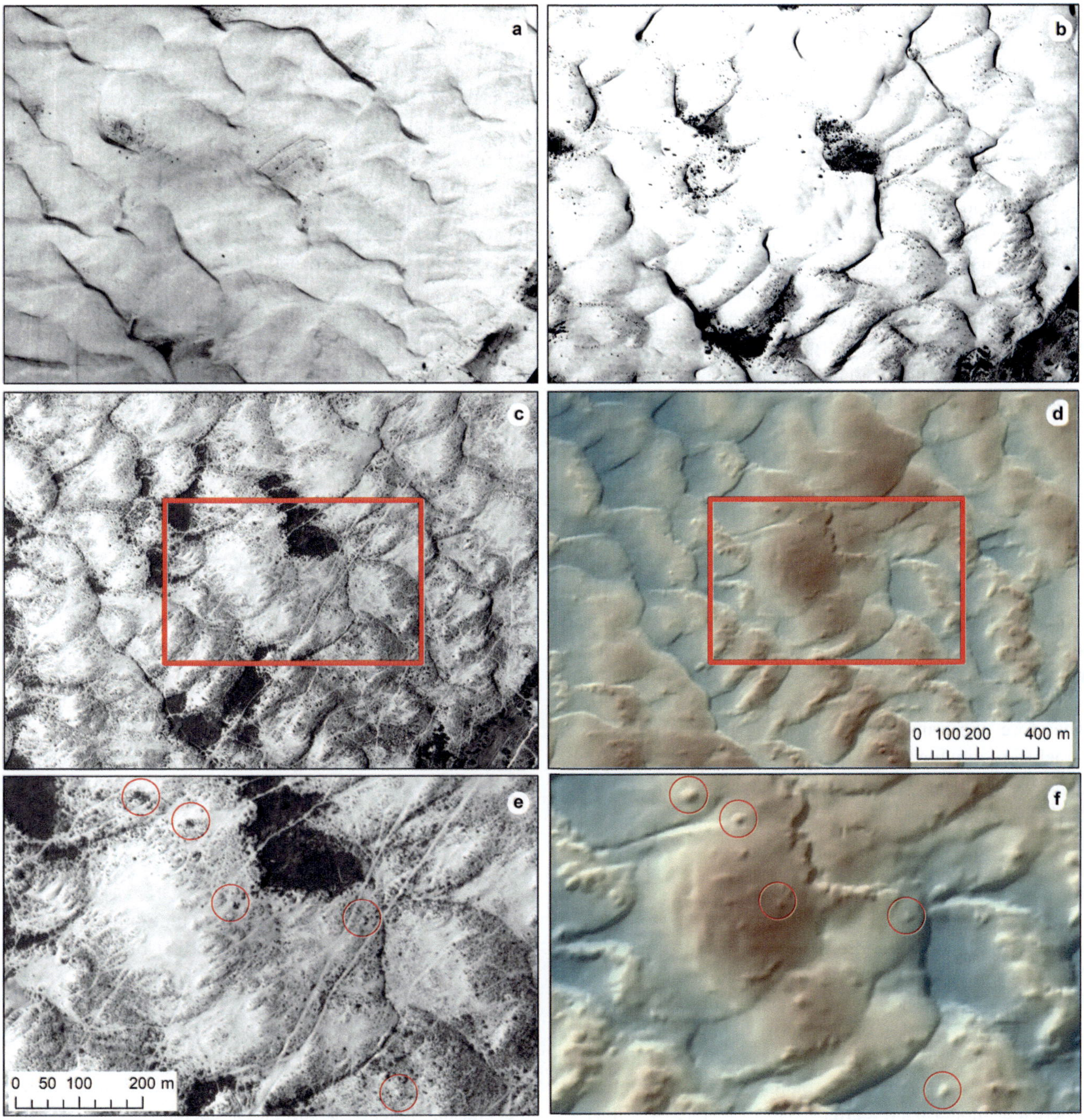

Fig. 11.5 The transformation of coastal dunes from transverse barchanoid dunes to parabolic dunes, as the dune stabilize, as seen on the dunes of Nizzanim: **a** aerial photo from December 1944; **b** aerial photo from 1974; **c** aerial photo from June 2002; **d** digital elevation model of the dunes in the early 2000s (Source: Levin et al. 2008b) with hill shading; **e** aerial photo from June 2002, with five large nebkhas highlighted by red circles; **f** digital elevation model with five large nebkhas highlighted by red circles

that can withstand higher salt concentration and high rates of sand erosion and deposition (Hesp 2002). While foredunes can indicate processes of sediment transport to the beach and often form parallel ridges along the coast (Tsoar 2000; Levin et al. 2017; Fig. 11.4a), long stretches of foredunes are not common along the Israeli coastline, probably because of the presence of coastal cliffs, and human disturbance by off-road vehicles and trampling. At present, one of the best locations to observe active foredunes along the coast of Israel is at Bet Yanay (Fig. 11.2d), where ample wind power and sand supply are driving the foredunes inland, often covering parts of a local road.

11.3.3 Barchanoids

Barchanoid dunes form in areas with a unidirectional wind regime and can be separated into two landforms: individual, well-separated barchan dunes are commonly found in dune fields with relatively low sand thickness, whereas transverse barchanoid dunes tend to form where the equivalent sand thickness is relatively high (Fryberger and Dean 1979; Wasson and Hyde 1983). Along the coastal plain of Israel, the most common dune type before the establishment of the State of Israel and consequent dune stabilization were transverse barchanoids (Fig. 11.5a). Wind directionality for sand drift is commonly quantified using the ratio between the resultant drift potential (RDP) and the drift potential (DP); the values of RDP/DP range between 0 (multidirectional wind) and 1 (unidirectional wind) (Fryberger and Dean 1979). The wind regime along the coastal plain of Israel is unidirectional, and wind directional variability is relatively low, with annual RDP/DP values usually above 0.5 (for Gaza and Tel-Aviv; Levin and Ben-Dor 2004) and an average annual RDP/DP of 0.71 for the coastal dunes south of Tel-Aviv (Tsoar 2005). Given that strong winds which are powerful enough to entrain and transport sand grains mostly occur in Israel during winter months, these strong winds mostly blow from the west to south-west direction (Tsoar and Blumberg 2002). Therefore, the transverse dunes along the coastal plain of Israel were moving inland in a north-east direction, with the majority of sand transport taking place in the winter months (Levin et al. 2006, 2007a).

11.3.4 Seif Dunes

While Seif dunes are more a feature of the desert of the Negev and the Sinai Desert, the formation of Seif dunes from one of the horns of barchan dunes was already described by Bagnold (1941) and was shown to take place in the coastal dune field of El-Arish, i.e., along the coastal plain of northern Sinai, to the south of Israel's coastal plain, where a bi-modal wind regime exists (Tsoar 1974, 1984),

11.3.5 *Parabolic Dunes*

Since the establishment of the State of Israel, the transverse barchanoid dunes along the coast of Israel have been undergoing stabilization by vegetation (the reasons for this process will be described later in this chapter) (Levin and Ben-Dor 2004). As first described by Anthonsen et al. (1996), under the influence of vegetation, transverse dunes undergo a morphological transformation into parabolic dunes, whose horns are facing backwards, in the windward direction.

Given that the transformation of these dunes to parabolic dunes is a relatively recent process along the coast of Israel, they do not have elongated shapes as described by Pye (1982) for dunes in Queensland, Australia. This transformation process was first described for the coastal dunes of Nizzanim by Tsoar and Blumberg (2002) about 50 years after the stabilization of these coastal dunes started (Fig. 11.5). They attributed the transition from transverse to parabolic dunes to vegetation growth on the crest of the dune, where eroded sand from the windward slope is trapped by vegetation (on nebkhas), which succeeds in establishing there because on the crest there is neither erosion nor deposition. Reitz et al. (2010) stated that the transition to parabolic dunes starts as vegetation initially colonizes the horns of the dune, where erosion and deposition rates are lower, whereas Barchyn and Hugenholtz (2012) attributed the transition to a parabolic form to the colonization of vegetation on the slipface of the dune, when the deposition rates are below the threshold that vegetation can withstand. Using numerical models, it has been suggested that the process of transformation of transverse dunes to parabolic dunes is reversible (Durán et al. 2005).

11.4 Controls on Coastal Dune Formation

11.4.1 Sources of Sand

Whereas the beaches and dunes along the southern and central coast of Israel are mostly composed of quartz grains, those to the north of the Bay of Haifa are almost entirely composed of calcium carbonate. This indicates the Nilotic source of quartz sand along most of the coast of Israel and the local source of sand in the northern beaches of the western Galilee (Emery and Neev 1960). Indeed, the width of the coastal dune fields of Israel decreases from south to north, with the exception of the coastal dunes of Caesarea, which are wider than those to the south of them. The northward longshore transport of sand along the coast of Israel terminates in the Bay of Haifa, where some sand is trapped, and which was partly filled in by aeolian dunes in the past 4000 years (Porat et al. 2008), whereas the rest of the sand is moved to deeper waters (Almagor et al. 2000; Zviely et al. 2007).

11.4.2 Coastal Cliffs

Coastal cliffs, comprised of alternating aeolianite rock and hamra soils, are found along 69 out of 173 km of the shoreline between Rafah in the south and Atlit (to the south of Haifa; Almagor and Perath 2012) (Fig. 11.2). South of Bat Yam, these coastal cliffs are usually less than 10 m high, whereas between Bat Yam and Hadera they reach heights up to 50 m (Almagor and Perath 2012). While cliff erosion was found to contribute about 50% of the sand transported along the beaches of Israel (Mushkin et al. 2016), these cliffs have another important role in controlling the spatial distribution of Israel's coastal dunes. Tsoar and Blumberg (1991) found that even moderate cliff slopes of 10–15° prevented sand grains from climbing the cliffs, at a height of 22–25 m in the study area to the south of Ashdod. Therefore, the abundance of extensive coastal dune fields to the south of Tel-Aviv, and their relatively smaller extent to the north of Tel-Aviv, can be attributed to a large extent to the higher coastal cliffs between Bat Yam and Hadera.

11.4.3 Rivers

Both to the south and north of Tel-Aviv, river mouths are an important avenue via which sand can encroach inland, with relatively few impediments. Indeed, along the coastal plain of Israel various coastal dune fields emerge at river mouths and extend from there to the north-east (Fig. 11.2). River outlets also preclude the formation of foredunes, thus enhancing the encroachment of sand inland (Tsoar 2000). Following the classification scheme of Liu and Coulthard (2015), the interaction between fluvial and aeolian processes along most of the Israeli coast may be characterized as "fully fluvial dominant", with the rivers acting as a barrier to dune movement. However, examination of the courses of the rivers along the coastal plain reveals that before entering the sea, many of the rivers bend toward the north, a few km east of the shoreline (Fig. 11.2a). This results from inland migration of the coastal dunes to the north-east, forming a barrier to the flow of the rivers (Almagor and Perath 2012). The only river in Israel which flows through a coastal dune field, with semi-active dunes to the south-west of the river and stabilized dunes to the north-east of the river, is the Evtah River (Kutiel et al. 2016), which has a small watershed, and flows intermittently through the Nizzanim coastal dune field (Fig. 11.2a).

11.4.4 Wetlands

Wetlands and coastal dunes are closely associated and are often found in the same place, due to the combination of two factors: (i) coastal dunes form an impediment for the rivers flowing to the sea, and thus wetlands (either seasonal or year-round) can often be found on the edge of the coastal dunes; (ii) given that within coastal dunes almost all rainfall infiltrates and usually there are no drainage networks, seasonal lagoons will form with excess rainfall, when the local level of the groundwater aquifer will rise above the elevation of the interdune areas (wet dune slacks; Grootjans et al. 1998; Levin et al. 2007b). During the late Pleistocene-Holocene, there were several marsh episodes along the coastal plain of Israel (Cohen-Seffer et al. 2005), the most recent one associated with the neglect of the coastal plain during the Middle Ages (Kark and Levin 2012; note the extent of wetlands along the coastal plain during the nineteenth century as shown in Fig. 11.2b). Most of the marshes and winter ponds which existed along the coastal plain of Israel downwind of the coastal dunes have been drained during the twentieth century, and only a small remnant of these still exist (Levin et al. 2009). While the interdune slacks within Israel's coastal dunes are usually not inundated, they do form a unique habitat, which was used in the past for the "Mawasi agriculture" (see in Sect. 5.1), and these slacks have since become fully vegetated, mostly by *Retama raetam* (Levin et al. 2008a; Almog 2011; Fig. 11.4d).

11.4.5 The Aswan Dam

Until 3000 years ago the Nile Delta was in its expansion phase (Marriner et al. 2012), and over the past 3000 years the delta of the Nile prograded by 1–12 km into the Mediterranean Sea (Sneh et al. 1986). While the construction of the Low and the High Aswan Dam (completed in 1964) is often stated as the cause for the accelerated erosion of the Nile Delta coast during the twentieth century, Stanley (1996) attributed this erosion to the extremely dense network of irrigation and drainage channels (more than 10,000 km long) within the delta, where sediment is retained. During the present destruction phase of the Nile Delta (Stanley and Warne 1998), sediment is eroded from the delta and deposited eastwards, with areas of severe erosion mostly confined to the promontories of the present-day mouths of the Nile at Damietta and Rosetta (White and Asmar 1999; Fig. 11.1). Given the great uncertainties in longshore transport rates of sand along the Mediterranean coast (Almagor et al. 2000) and the lack of erosion along the northern beaches of the Sinai Peninsula (Luijendijk et al. 2018), it is probable that there will be no lack of sediments of Nilotic origin reaching Israel's beaches in the next few centuries (Rohrlicht and Goldsmith 1984).

11.4.6 Artificial Structures Along the Coast

Artificial structures along Israel's coastline locally interfere with longshore transport along the beaches, with sand accretion to the south of port facilities, and sand erosion to the north (Almagor et al. 2000; Zviely and Klein 2003). Nonetheless, the impact of these structures on longshore transport is mostly local, impacting the shoreline and the beaches up to 2–3 km downdrift from the obstacles (Luijendijk et al. 2018; Fig. 11.1). Any possible impacts of artificial structures along the coast on sand encroachment to the dunes inland should be evaluated at the scale of several centuries, and not decades, given the relatively low rates of sand erosion and dune movement in Israel.

11.5 Traditional Human Uses of the Coastal Dunes Before 1948

11.5.1 Traditional Agriculture

A traditional method of agriculture has been practiced by Arab peasants for centuries in the coastal dunes of Israel and northern Sinai, known by the local term of "Mawasi" (meaning suction in Arabic; Tsoar and Zohar 1985; Tsoar 1990; Fischer and Taxel 2021; Roskin and Taxel 2021; Fig. 11.5). Given that rainwater infiltrates through sand dunes and the proximity of the underground water table to the surface near the coast, the low interdune areas were often used for agriculture by clearing the sand dune to the water table and using animal manure and compost to fertilize the sandy soil. These agricultural plots were used for growing palm trees, grapevines, and a variety of fruit trees (e.g., figs), some of which have survived for decades (Almog 2011; Sasson 2019) after the abandonment of the Mawasi agriculture within the coastal dunes of Israel after 1948. Historical aerial photos from World War II, as well as historical maps going back to Napoleon's expedition to Egypt and Palestine in 1799, as well as British Mandate maps of Palestine, provide evidence for the extent of Mawasi agriculture, and even to some of the types of plants which were grown there (Fig. 11.6). Extensive agricultural plots covering about 3 km^2 also existed within the coastal dunes of Caesarea and can probably be dated to the 9th–10th century A.D. (Porat 1975) or maybe even before (Ad 2009); the remains of raised plots can be clearly seen on aerial photos (Shtienberg et al. 2017; Fig. 11.3a). A similar plot and berm agroecosystem, which was active between the tenth and early twelfth century in the Yavneh coastal dune field, was described by Taxel et al. (2018). Charles Warren provided a lively description of Mawasi agriculture after his visit to the area of Ascalon (Ashkelon) in 1867 (pp. 463–464, Warren 1876): "*It is curious, in traversing these sand-hills, to note what the industrious hands of individual men can do, to come upon the site of some orchard which has, perhaps, been covered up, more or less, for hundreds of years. Suddenly the travelers come upon a crater in the sandy desert, some forty feet deep, at the bottom of which flourishes one apple tree, then, after a few paces, perhaps a fig tree growing under the same conditions, and so on, until lastly the cotter's house is reached, with its little patch of ground, lying far below the level of the sand, and looking as though any tempest would engulf it, and the world knows it no more. The husbandman's chief care consists in dragging up the sand in baskets from the bottom of the crater to the surface. If individuals can thus keep off this mighty rolling monster, what may not a good Government do?*". An additional type of planted fruit tree worth noticing along the coastal dunes is that of the Sycamore fig (*Ficus sycomorus*). These trees were planted in a long row between the edge of the coastal dunes and the alluvial plains and fields to the east of the dunes; remnants of the sycamore row exist to this day between Gaza and Tel-Aviv (Fischer and Taxel 2021). Originally the cultivation of the tropical Sycamore figs was an Egyptian specialty, and from there it spread to southern Palestine (Galil and Eisikowitch 1968; Zohary and Spiegel-Roy 1975). Sycamore figs were important culturally, their fruits were sold in markets (Sasson 2019), and they were also mentioned in various places in the Bible as a tree signifying the coastal plain.

11.5.2 Grazing, Cutting, and Uses of Native Vegetation

Dune plants have been utilized by the local population both in desert areas and in coastal dune areas for a wide variety of uses. The Bedouins made use of both live and dead plants (Levin and Ben-Dor 2004). This included food for their herds as well as for humans, provision of building materials, fuel for heating, and medicines (Bailey and Danin 1981; Briggs et al. 1999). *Artemisia monosperma*, one of the most common plants found on Israel's coastal dunes, is used both for grazing animals of the Bedouins

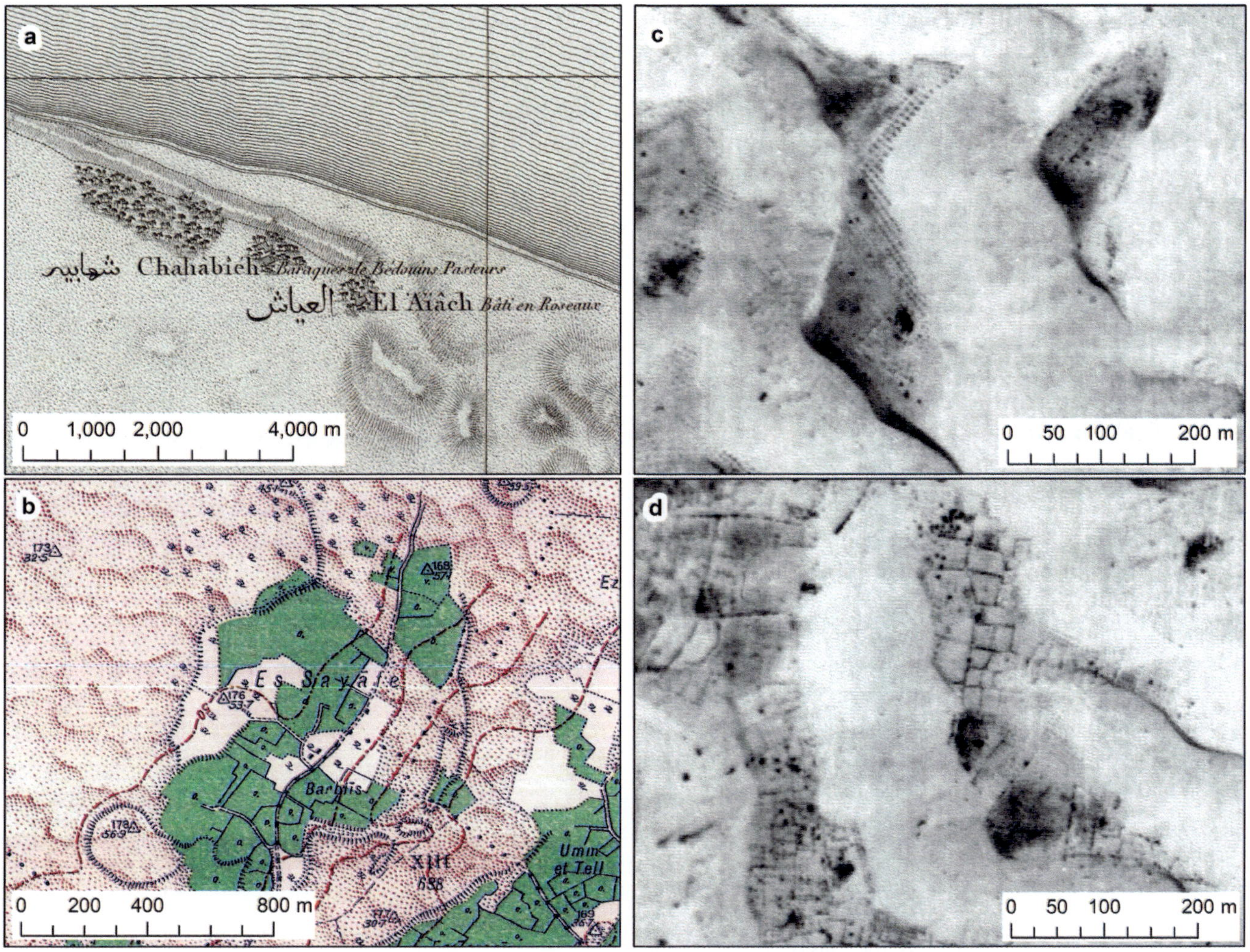

Fig. 11.6 Traditional Mawasi agriculture as shown on historical maps and historical aerial photos: **a** an extract from the Damietta sheet of Jacotin's map (published in 1818), showing palm trees and shacks of Bedouin farmers between the coastal dunes; **b** an extract of the British 1:20,000 map, sheet of Beit Hanun (published in 1931), showing agricultural areas (orchards and vineyards) within the coastal dunes; **c**, **d** agricultural plots in the interdune areas, as shown on an aerial photo from January 1945, within the Ashdod-Nizzanim dunes

(Meir and Tsoar 1996), as well as for constructing windbreaks around tents and construction of booth dwelling (Bailey and Danin 1981). *Retama raetam*, another common perennial dune plant, is used by the Bedouins as a medicinal plant for both people and their animals (treating aches, open sores, and cuts), for pins of tents or camel saddle, traps, and for charcoal (Bailey and Danin 1981). Jarvis, who was the British governor of Sinai in the 1920s and 1930s, infamously stated that the Bedouin should be seen not as the son of the desert but as its father, and that accompanied with his goats and camels, the Bedouins led to overgrazing and desertification (Epstein 1939; Jarvis 1943). While there was certainly overgrazing and cutting of vegetation throughout the land of Israel for centuries, modifying its landscape and reducing vegetation cover substantially (Danin 1988; Schaffer and Levin 2014) through direct grazing as well as via trampling (Kidron 2016), Bailey (1975) claimed that the view of Jarvis regarding the Bedouins' overuse of vegetation was exaggerated.

11.5.3 Distribution of Villages and Towns

While during several historical periods (Roman, Byzantine, and early Islamic to name a few) there were quite a few towns and villages along the coastal plain of Israel, by the nineteenth century, out of eight old towns, two ceased to exist (Apollonia and Caesarea), and two have been reduced to the size of villages (Ashkelon and Dor-Tantoura) (Amiran 1953). This was part of the abandonment of the coastal plain since the intentional destruction of the coastal towns by the Mamluks, which probably led to the settlement of Bedouin tribes later on along the coastal plain. Examining the Palestine Exploration Fund (PEF) Survey of Western Palestine map (published in 1880; Schaffer and Levin 2016) it can be observed that most of the permanent agricultural areas (orchards and vineyards) were located in the mountain areas, that coastal dunes, marshes, and winter ponds were abundant along the coastal plain, and that within the coastal dune areas, the highest density of gardens and orchards was in the vicinity of Ashkelon and Gaza (Fig. 11.2).

11.6 Coastal Dune Stabilization

11.6.1 The Causes of Spontaneous Dune Stabilization

In the past century, coastal dune fields around the world were found to be mostly stabilizing, with human intervention playing a major role in this trend (Gao et al. 2020). Given the low wind power along the coastal plain of Israel, climatically these coastal dunes can be expected to become stabilized, however, given constant human pressures in the past centuries, involving grazing and cutting of vegetation, the coastal dunes of Israel were active until 1948—the year of establishment of the State of Israel (Tsoar 2005; Fig. 11.7). After 1948, there was a dramatic change in the population along the coastal plain of Israel. The Bedouin population in the Negev fled (some cases of expulsion are also known; Falah et al. 1985) and their numbers decreased from about 70,000 to 11,000, with those who were left being confined to the Sayig zone around Beer Sheva (Levin and Ben-Dor 2004). Given the decrease in human pressure of traditional land uses over the coastal dunes of Israel, a process of spontaneous stabilization started, in which the dune slacks became fully vegetated in just about a decade, and the dunes themselves have been also undergoing a gradual process of stabilization ever since (Fig. 11.5). This process was first noted by Danin and Nukrian (1991) in the coastal dunes of Netiv HaAsara (north of Gaza) and was later studied mostly in the coastal dune field of Nizzanim, between Ashkelon and Ashdod (Kutiel et al. 2004; Levin and Ben-Dor 2004), where Tsoar and Blumberg (2002) also recognized the transformation of dune morphology from transverse to parabolic as part of this process. The transition between active and stabilized dunes may be reversible and cyclic and is controlled by the frequency, intensity, and duration of both climatic and anthropogenic perturbations (Hugenholtz and Wolfe 2005; Yizhaq et al. 2009). Along the coastal dune belt of Israel, and especially south of Tel-Aviv, it was noted that the coastal dunes were reactivated between the late 1960s and late 1970s, and that they began stabilizing again since the late 1970s (Tsoar and Blumberg 2002; Levin and Ben-Dor 2004) (Fig. 11.9). This temporary reactivation of coastal dunes was found to result from the return of the Bedouin population to the area after 1967, following the end of the military rule over the Arab population in Israel, and results of the Six Days War which abolished the borders between Israel, the West Bank, Gaza Strip and the Sinai Peninsula. A drought in the area of Beer Sheva in the 1970s led to the migration of Bedouins and their herds northwards, and an estimated amount of 10,000 goats, 5000 sheep, and 200 camels was grazing on the coastal dunes between Gaza and Tel-Aviv (Levin and Ben-Dor 2004). This period of reactivation ended in the late 1970s, when the State of Israel started enforcing laws for protecting vegetation from uncontrolled grazing, followed from 1978 by operations of actively confiscating goats and removing herds from the coastal dunes (Levin and Ben-Dor 2004). The results of the removal of the grazing pressures were noted by local rangers of the Israel Nature Reserves Authority, with a quick

Fig. 11.7 German aerial photos of the fully active dunes: (top) between Ashdod and Ashkelon, view from the north southwards, taken on January 1918; (bottom) the dunes over which Tel-Aviv was built, view from the east, taken on November 1917 (Jaffa is located at the top left the photo)

recovery of native vegetation (Berliner 1979), also documented by a renewed stabilization process of the dunes (Levin and Ben-Dor 2004).

11.6.2 The Rate of Dune Stabilization

The rate of coastal dune movement in Israel was estimated to be about 5.1 m/year prior to 1948 (based on historical maps; Levin 2006) and about 3.5 m/year (based on historical aerial photos) in the early 1950s (Tsoar and Blumberg 2002; Levin and Ben-Dor 2004; Figs. 11.8a and 11.9). The rate of coastal dune movement along most of the Israeli coast has since been greatly reduced (apart from the brief period of dune reactivation in the 1970s), and at present most of the coastal dunes of Israel are either stable or semi-stable. Present dune movement, sand erosion, and deposition are mostly limited to some foredunes (as in the dunes of Bet Yanay; Fig. 11.8b), to specific locations within the dune fields where surface roughness is lower (Levin et al. 2008b; Fig. 11.8c), or to individual dunes where human pressures of trampling and off-road driving are intensive enough (as in the case of the "Big Dune", which in easily accessible to people in Ashdod; Tsoar and Cohen-Zada 2020). The abundance and percentage of the cover of biological soil crust is lower in the coastal dunes in comparison with the desert dunes of the Negev (Danin et al. 1989), and thus, its role in dune stabilization for the coastal dunes is smaller (Levin et al. 2007a). Given the competition over water in the open matrix between the semi-shrubs of

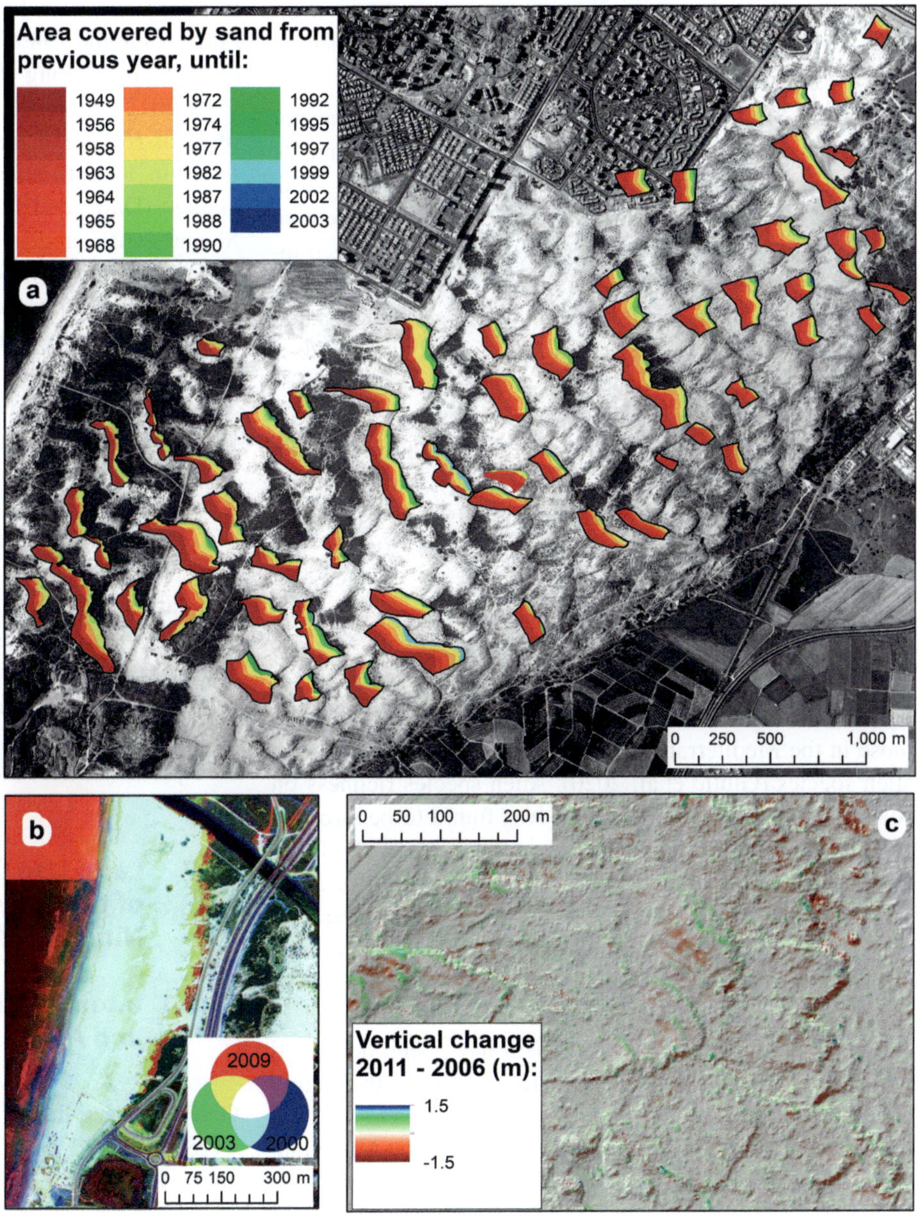

Fig. 11.8 Visualizing dune activity: **a** areas covered by sand as a consequence of dune movement in the Ashdod dune field, as extracted from a temporal series of aerial photographs; **b** false color composite of aerial photos of the Bet Yanay fore dunes, showing the inland advance of the dunes in red and yellow; **c** vertical changes in dune surface based on two LiDAR scans of the Caesarea dunes: notice that surface changes are very localized within these parabolic dunes

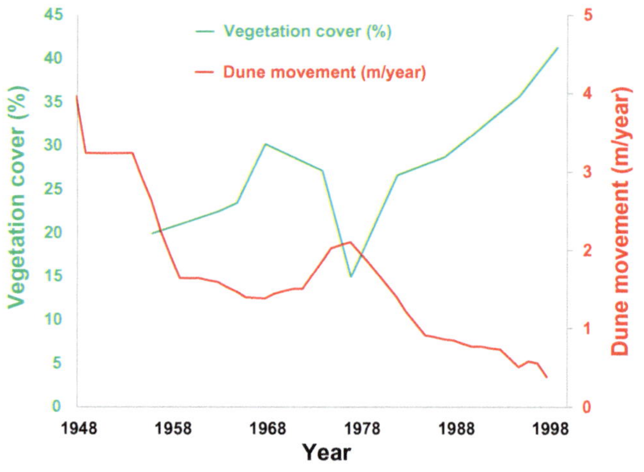

Fig. 11.9 Temporal changes in the stabilization process of the Ashdod dunes, as monitored by dune movement rate and vegetation cover. Note the stabilization until the late 1960s, the reactivation of the dunes until the late 1970s, and the renewed stabilization since then

Artemisia monosperma, it is probable that the room for expansion of these shrubs is limited, and full coverage of the dunes by vegetation will not take place (Kutiel et al. 2016). As part of the stabilization process, dune movement decreases, vegetation cover increases, leading to the trapping of fine grained materials and to an increase in silt and clay content in the top soil of the dunes (Danin and Yaalon 1982; Meir and Tsoar 1996). In addition to the gradual increase in vegetation cover over the dunes, its composition has also changed, and *Ammophila arenaria*, a pioneer species which is indicative of active coastal dunes (Kutiel 1979; Danin and Nukria 1991; Danin 2005), is now mostly found only in foredune areas and hardly in any of the coastal dunes beyond the foredune (Levin et al. 2008a).

11.6.3 The Role of Invasive Species

About 7% of the flora of European coastal dunes (including those in the Mediterranean) is estimated to be composed of alien species (Giulio et al. 2020). Alien species richness on coastal dunes at the country level was found to be strongly associated with anthropogenic factors (Tordoni et al. 2021). In various coastal dune systems around the world, invasive species have been a key factor leading to their stabilization, e.g., *Ammophila arenaria* in South Africa (Hertling and Lubke 1999) and north America (Buell et al. 1995). Five of the worst invasive plant species in Israel are found in coastal dune areas, and three of them have been described by Dufour-Dror (2012) as able to transform the sandy habitat and to compete with native species: *Acacia saligna*, *Carpobrtus edulis*, and *Heterotheca subaxillaris*. Australian acacia species are widely used to stabilize dunes (Wilson et al. 2011), and the *Acacia saligna* was introduced in Israel for dune stabilization during the British Mandate period. Within the Nizzanim coastal dune field, *Acacia saligna* was planted by the Jewish National Fund (JNF) in 1962 for dune stabilization and protection of orange orchards, and between its planting and 1999, the area covered by the *Acacia* trees grew by 166% (Bar et al. 2004). However, it was found that the Acacia trees only invaded disturbed habitats (e.g., sand (extraction) pits and depressions between the dunes) and did not spread onto active or semi-stabilized dunes (Bar et al. 2004). Given the negative impacts of Acacia trees on native biodiversity of Israel's coastal dunes (Cohen and Bar 2017; Anglister et al. 2019), active efforts were undertaken to remove Acacia trees, and their planted grove within the Nizzanim coastal dune field was treated in the early 2010s. The *Carpobrtus edulis* was introduced to Israel from South Africa for gardening purposes in the 1940s and has since spread to foredunes and coastal cliffs along the Mediterranean shoreline (Dufour-Dror 2012). Of most concern, probably, is the herbaceous plant of *Heterotheca subaxillaris*, which was introduced in Israel in 1975 to stabilize the coastal dunes in the coastal Galilee (Dafni and Heller 1980) and in Giv'at Olga (Ben-Shahar 2010), and in just 25 years has spread across all of the coastal dunes, including semi-stabilized dunes (Dufour-Dror 2012). Within the dunes of Palmachim, almost 30% of the area is impacted by the invasive plants of *Heterotheca subaxillaris* and *Acacia saligna* (Gal et al. 2008). Ben-Shahar (2010) found that while in its native habitat in the USA, *Heterotheca subaxillaris* is an annual plant, in Israel it has sometimes become a bi-annual and even a perennial herb. Sternberg (2016) found that the introduced population of *Heterotheca subaxillaris* developed roots exceeding 5 m in length, more than four times the length of native populations of *Heterotheca*, thus demonstrating a fast adaptation to the ecosystem of coastal dunes. Kozhoridze et al. (2022) found the mean expansion rate of Heterotheca subaxillaris, 107% per decade (between 1990 and 2020), to be much higher than that of Acacia saligna, between 48 and 54% per decade for the same period.

11.7 Land Use Changes and Management of the Coastal Dunes Since the Late Nineteenth Century

11.7.1 Intentional Dune Stabilization and Afforestation

Given the activity of the coastal dunes in the past, they were perceived as a threat to agricultural areas, villages, and infrastructure. Afforestation practices in Israel mostly started in the 1920s, and during the British Mandate period, most of the afforestation was done by the Palestine Forestry Department (PFD), including the planting of trees over 1271 hectares of sand dunes (Biger and Liphschitz 1995). Experiments were conducted by the PFD in planting Eucalypt trees on the coastal dunes of Acre and Gaza; however, these were mostly unsuccessful (Biger and Liphschitz 1995). Two Australian acacia species, *Acacia cyanophylla* and *Acacia longifolia*, were found effective for sand dune stabilization and windbreaks (Biger and Liphschitz 1995). Within the northern part of the coastal dune area of Acre, successful results in dune stabilization were achieved by the PFD in the early 1920s using *Ammophila arenaria* and *Tamarix aphylla* (Liphschitz and Biger 1997). The British also invested efforts in stabilizing coastal dune areas between the shoreline and the city of Gaza, altogether stabilizing about 260 hectares by 1947 (Liphschitz and Biger 1997). The British chose areas to stabilize dunes based on their threats to infrastructure, e.g., their work for stabilizing a small section of dunes south of Ashdod to protect the railway line (Liphschitz and Biger 1997). Additional local efforts were conducted by various Jewish agencies during the British Mandate period, near Rishon le Zion, Netanya and Caesarea; however, they were not very successful (Liphschitz and Biger 1997). Following the establishment of the State of Israel, many experiments regarding dune stabilization were conducted by the Soil Conservation Service within the Ministry of Agriculture, using various plant species, as well as mechanical methods (spraying of chemicals and erection of fences), aiming to create artificial foredunes, as attempted for the coastal dune field south of Acre (Tsuriell 1974, 1983). Intentional dune stabilization efforts receded with time, mostly because of the spontaneous stabilization of Israel's coastal dunes since the 1950s, as reviewed above.

11.7.2 Sand Mining

Mining of beach sand (the coarse-grained fraction consisting of skeletal debris and wave-milled kurkar grains, locally known as Zifzif) for the building industry was common practice along Israel's beaches since the early twentieth century (Almagor and Perath 2012). It is estimated that at least 10 million m^3 of sand (equivalent to about 40 years of sand reaching Israel's beaches via longshore transport) were quarried, out of which 4 million m^3 just between the beaches of Nizzanim and Palmachim (Almagor and Perath 2012). This uncontrolled sand mining led to enhanced erosion of many of the beaches and coastal cliffs, with the highest magnitudes of beach retreat in the northern part of Israel, where many coastlines retreated landward by more than 20 m between 1948 and 1989 (Almagor et al. 2000). Beach sand mining came to its end following the "Zifzif law" of 1964, which prohibited the removal of sand from the beaches (150 m inland and seaward of the shoreline). After 1964, sand mining shifted eastwards, and it is estimated that since the 1960s more than 60 million m^3 of sand were quarried from Israel's coastal dunes, damaging many of them (Barzilay 2017).

11.7.3 Land Cover and Land Use Changes Over the Coastal Dunes

Coastal dunes around the Mediterranean Sea are under various pressures from anthropogenic activities (Van der Muelen and Salman 1996). Over the twentieth century, Israel's landscape underwent great transformations, with a dramatic increase in population size, associated with transformations of many of the open spaces and natural habitats into agricultural land, built-up areas, and other human uses (Schaffer and Levin 2014). It is estimated that out of a total area of about 360 km^2 of coastal dunes (as depicted on the 1880 PEF Survey of Western Palestine), only about 185 km^2 have remained (Table S3 in Schaffer and Levin 2014), out of which 80 km^2 consist of former coastal dunes which were disturbed (e.g., by sand mining for the building industry from the coastal dunes; Milgrom 2008) or are under development (Frumkin-Ahiron et al. 2003). Coastal dune areas were mostly transformed into the following land uses: built-up areas (40%), agricultural fields (5%), orchards (4.4%), artificial water bodies (waste facilities and reservoirs for recharging the coastal aquifer; 2.6%), and planted forests (1.8%) (Schaffer and Levin 2014; the percentages refer to the dune area as mapped in 1880; Fig. 11.10). Frumkin-Ahiron et al. (2003) estimated that building and development plans are threatening the existence of about 86 km^2 of coastal dunes, and if that comes into effect, an area of only 99 km^2 of coastal dunes will remain, with only 58 km^2 being coastal dunes in a semi-natural "non-disturbed" state. The decrease in habitat area of the coastal dunes and their increased fragmentation both interfere with free inland flow of sand from the beach and create additional edge effect pressures, with domestic pets and urban species changing the composition of the fauna within the dunes (Perry and Dmi'el 1995; Manor and Saltz 2004).

11.7.4 Conservation

In the past, coastal dune areas were perceived as prime areas for development, whereas alluvial soils were reserved for agriculture. Given the abovementioned development pressures, it seems that the coastal dune habitat

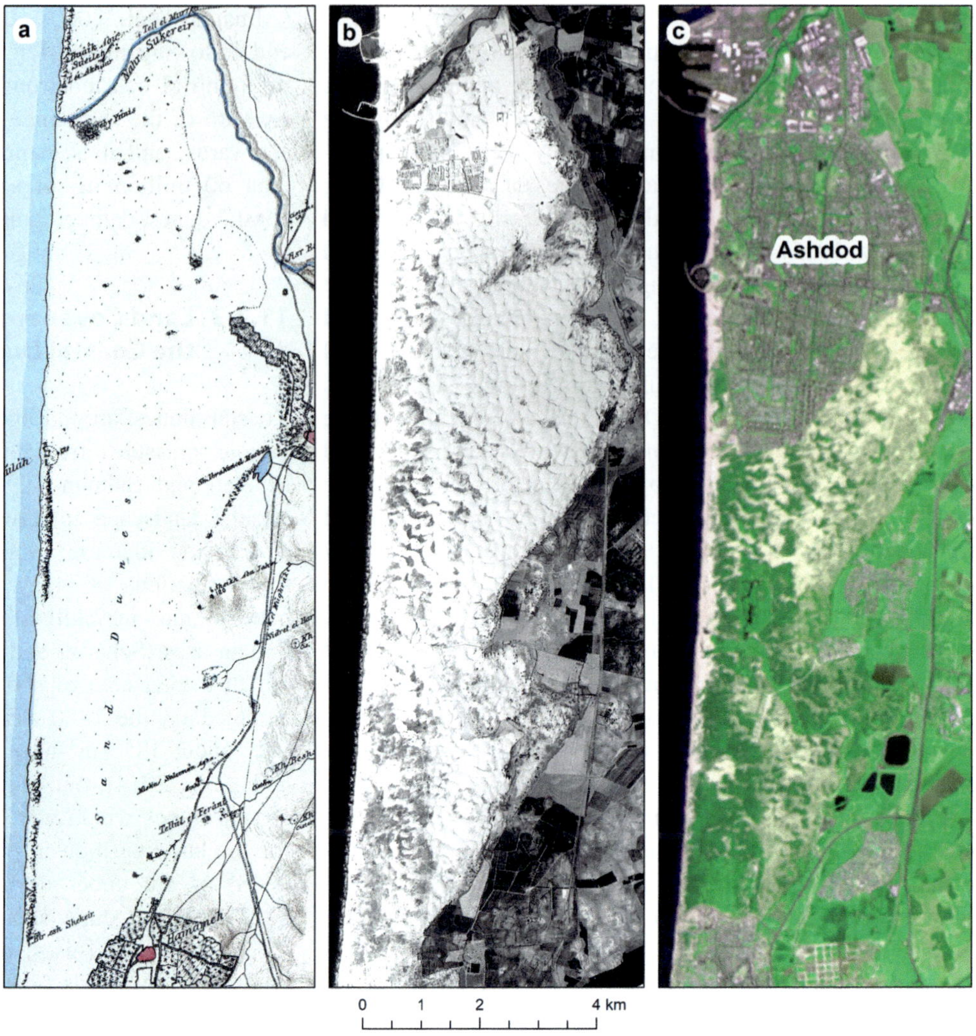

Fig. 11.10 Landscape changes in the Ashdod-Nizzanim dune field: **a** the dunes as shown by dotted areas on the Survey of Western Palestine map of the Palestine Exploration Fund, published in 1880; **b** a 1963 Corona satellite image of the dunes with many of the interdune areas already stabilized, also showing the emergence of the new city of Ashdod and its port, founded in 1961, in the north; **c** a Landsat satellite image from April 2021, with most of the dunes stabilized or semi-stabilized, and much of the dunes converted to built-up areas

is threatened, under-represented within Israel's network of protected areas (Rotem and Weil 2014), and requires further conservation (Levin and Shmida 2007; Weil and Levin 2015). Israel's coastal dunes are important as a unique habitat for psammophile species: 26% of Israel's endemic plants (31 species) are located within the coastal dunes, and they even host an endemic rodent, the *Gerbillus allenbyi* (Kutiel 2001). As of 2000, about 75 km^2 of the coastal dunes were included within protected areas, but only 48 km^2 of these dunes were coastal dunes with low degree of disturbance (Frumkin-Ahiron et al. 2003). The largest stretch of protected coastal dunes is located between the cities of Ashdod and Ashkelon, known as Ashdod-Nizzanim coastal dune field, altogether covering an area of 19.5 km^2. The conservation values of this dune field were first recognized in the 1980s, with the establishment of the "Dunes Park" in 1985 and the Shikmin Field School of the Society for Protection of Nature in Israel, and later on, the inclusion of much of this area within National Master Plan of Forestry (Tama 22) under the jurisdiction of the Jewish National Fund in 1995. However, the final designation of the Ashdod-Nizzanim as a protected area in 2005 and 2008 was the result of advocating by environmental organizations, as well as an outcome of the unique political opportunity created following the Israeli disengagement from Gaza in 2005 (Paz 2009).

11.7.5 Reactivation of Coastal Dunes

Until the 1980s, the common paradigm in the management of protected areas in Israel was that human interference should be minimized, and therefore, grazing was mostly forbidden within protected areas, to allow vegetation to recover from previous periods of overgrazing which has stripped much of the landscape from dense vegetation cover (Perevolotsky and Seligman 1998). In the case of Israel's coastal dunes, the abandonment of grazing and cutting of vegetation led to their spontaneous stabilization and to the transformation from mobile dunes to semi-stabilized and stabilized dunes. This transformation threatens psammophile species, some of which are endemic to these coastal dunes, hence, since the 1990s, it has been recognized that vegetation removal can be beneficial to preserve such species, and various studies have examined the beneficial effects of removing shrub cover (Kutiel et al. 2000a; Bar 2013, Pua 2022). Most experiments on dune reactivation were conducted within the Ashdod-Nizzanim coastal dune field, which has since been designated as one of Israel's Long Term Ecological Research (LTER) sites (Lehrer et al. 2013). When comparing the effectiveness of shrub removal by hand or by shovel, it was found that mechanical removal of shrubs was more effective, as it is able to also remove the roots of the shrubs and to level nebkha mounds (Tsoar et al. 2005). However, without further disturbances, and given the low wind power along Israel's coast, dune vegetation recovers quickly. Various spatial patterns of shrub removal were compared in these studies: vegetation removal in a gridded pattern on fixed dunes, as well as the entire removal of vegetation cover from the wind-facing slope of semi-stable dunes (Bar 2013; Bird et al. 2020). While the experiments in which full removal of vegetation from the windward slope were more successful, as can be learned from the recolonization of those areas by indicator species common to active dunes, constant disturbances are needed to prevent the dune from stabilizing again. Israel's coastal dunes seem to be well adapted to recover from various disturbances, including recreational impacts of trampling by pedestrians and off-road vehicles (Kutiel et al. 2000b), unless these disturbances are continuous over time (compare with Tsoar and Cohen-Zada 2020). While reintroducing grazing regimes to the dunes may be more sustainable, an experiment conducted with camels in the Nizzanim dunes pointed out that camels are selective feeders and on their own would not damage the vegetation enough so as to reactivate the dunes (Katz et al. 2016). In addition, vegetation removal experiments in Israel were very local and did not aim at recreating the connection between the beach, foredunes, and the more inner dunes, in order to re-establish past corridors of wind and sand movement inland (Bar 2013).

11.8 Concluding Remarks

The coastal dunes of Israel cover a relatively small area and are young in age. Most of these dunes have been transformed to other land uses or have become stabilized following land use changes. Thus, the relatively undisturbed coastal dunes which are included within protected areas cover less than 20% of the original area of coastal dunes. However small, the coastal dunes of Israel are an important landscape feature of the country, both for heritage values, endemic biodiversity, replenishment of the coastal aquifer as well as for tourism and recreation (Bar et al. 2016). In addition, Israel's coastal dunes proved to be a fertile study area, where significant studies were conducted, such as those focused on the transformation of barchanoid dunes to parabolic dunes as vegetation cover increases (Tsoar and Blumberg 2002), the hysteretic nature of these coastal dunes and their sensitivity to changes in land use practices of cutting of vegetation, grazing and trampling (Levin and Ben-Dor 2004; Tsoar 2005), and intensive experiments in dune reactivation using various methods (Bar 2013, Bar 2022). Coastal dunes present one of the most dynamic landscapes of Israel, quickly responding to climatic and anthropogenic pressures at various spatial and temporal scales, and therefore deserve an important place in our hearts and minds.

References

Ad U (2009) Or'Aqiva: remains of a farming complex and irrigation system from the end of the byzantine-beginning of the early Islamic periods in the agricultural hinterland of Caesarea. Atiqot 133–134:49–60 (in Hebrew)

Almagor G, Perath I (2012) The Mediterranean sea coast of Israel. Geol Surv Israel Rep GSI/28/2012

Almagor G, Gill D, Perath I (2000) Marine sand resources offshore Israel. Marine Georesour Geotec 18(1):1–42

Almog R (2011) Soil moisture regime in the sandy depressions of the southern Mediterranean coastal plain, PhD Dissertation. Hebrew University of Jerusalem, Department of Geography

Amiran DH (1953) The pattern of settlement in Palestine. Israel Explor J 3(3):192–209

Anglister N, Yom-Tov Y, Motro U (2019) The impact of *Acacia saligna* and the loss of mobile dunes on rodent populations: a case study in the Ashdod-Nizzanim sands in Israel. Isr J Plant Sci 66(3–4):162–169

Anthonsen KL, Clemmensen LB, Jensen JH (1996) Evolution of a dune from crescentic to parabolic form in response to short-term climatic changes: Råbjerg Mile, Skagen Odde, Denmark. Geomorphology 17(1):63–77

Ardon K, Tsoar H, Blumberg DG (2009) Dynamics of nebkhas superimposed on a parabolic dune and their effect on the dune dynamics. J Arid Env 73(11):1014–1022

Avni Y, Avni G, Porat N (2019) A review of the rise and fall of ancient desert runoff agriculture in the Negev Highlands—a model for the southern Levant deserts. J Arid Env 163:127–137

Bagnold RA (1941) The physics of blown sand and desert dunes: New York. William Morrow & Company

Bailey C (1975) Fundamentals of nature conservation in the Bedouin culture. Annual conference in memory of Yizhaki Netzer 7. http://beduin.snunit.org.il/arti/0750.html

Bailey C, Danin A (1981) Bedouin plant utilization in Sinai and the Negev. Econ Bot 35(2):145–162

Bar P (2013) Restoration of coastal sand dunes for conservation of biodiversity: the Israeli experience. In: Restoration of coastal dunes, pp 173–185. Springer, Berlin

Bar P (2022) Geo-ecology of dunes in the coastal plain of Israel. Dan Perry Publishing (in Hebrew)

Bar P, Cohen O, Shoshany M (2004) Invasion rate of the alien species Acacia saligna within coastal sand dune habitats in Israel. Isr J Plant Sci 52(2):115–124

Bar P, Becker N, Segev M (2016) Sand dunes management: a comparative analysis of ecological versus economic valuations applied to the Coastal region in Israel. Reg Environ Change 16(4):941–950

Barchyn TE, Hugenholtz CH (2012) A process-based hypothesis for the barchan–parabolic transformation and implications for dune activity modelling. Earth Surf Proc Land 37(13):1456–1462

Barzilay B (2017) Quarrying of marine sand in Israel—crime and punishment. Karka 74:75–98 (in Hebrew)

Berliner R (1979) Proposal for a nature reserve in the southern Shfela sand dunes near Zikim, 12.7.1979. Israel Nature Reserves Authority archives, Tel Aviv (in Hebrew)

Ben-Shahar S (2010) Causes, processes and attributes which make a plant specie more invasive: the case of *Heterotheca subaxillaris* (*Asteracea*) in Coastal Sands of Israel. MA thesis, Ben Gurion University of the Negev

Biger G, Liphschitz N (1995) Australian trees in the land of Israel 1865–1950. J Isr Hist 16(3):235–244

Bird TLF, Bouskila A, Groner E, Bar Kutiel P (2020) Can vegetation removal successfully restore coastal dune biodiversity? App Sci 10(7):2310

Briggs J, Badri M, Mekki AM (1999) Indigenous knowledges and vegetation use among bedouin in the Eastern Desert of Egypt. App Geogr 19(2):87–103

Buell AC, Pickart AJ, Stuart JD (1995) Introduction history and invasion patterns of Ammophila arenaria on the north coast of California. Conserv Biol 9(6):1587–1593

Cohen O, Bar P (2017) The impact of *Acacia saligna* invasion on the indigenous vegetation in various coastal habitats in Israel and its implication for nature conservation. Isr J Plant Sci 64(1–2):111–121

Cohen-Seffer R, Greenbaum N, Sivan D, Jull T, Barmeir E, Croitoru S, Inbar M (2005) Late Pleistocene-Holocene marsh episodes along the Carmel coast, Israel. Quat Int 140:103–120

Dafni A, Heller D (1980) The threat posed by alien weeds in Israel. Weed Res 20(5):277–283

Dan J, Yaalon DH, Koyumdjisky H, Raz Z (1972) The soil association map of Israel. Isr J Earth Sci 21:29–49

Danin A (1988) Flora and vegetation of Israel and adjacent areas. Zoogeogr Israel 30:251–276

Danin A (2005) The sandy areas of Caesarea, a rare situation of alpha and beta diversity linked by plant succession. Isr J Plant Sci 53(3–4):247–252

Danin A, Yaalon DH (1982) Silt plus clay sedimentation and decalcification during plant succession in sands of the Mediterranean coastal plain of Israel. Isr J Earth Sci 31:101–109

Danin A, Nukrian R (1991) Dynamics of dune vegetation in the southern coastal area of Israel since 1945. Doc Phytosociol 13:281–296

Danin A, Bar-Or Y, Dor I, Yisraeli T (1989) The role of cyanobacteria in stabilization of sand dunes in southern Israel. Ecologia Mediterranea 15(1):55–64

Dufour-Dror JM (2012) Alien invasive plants in Israel. Jerusalem

Durán O, Schatz V, Herrmann HJ, Tsoar H (2005) Transformation of barchans into parabolic dunes under the influence of vegetation. In: The proceedings of powders and grains, vol 2, pp 951–954

Emery KO, Neev D (1960) Mediterranean beaches of Israel. State of Israel, Ministry of Agriculture, Division of Fisheries, The Sea Fisheries Research Station

Epstein E (1939) Bedouin of the Negeb. Palest Explor Q 71(2):59–73

Falah G (1985) The spatial pattern of Bedouin sedentarization in Israel. GeoJournal 11(4):361–368

Finkelstein I, Adams MJ, Dunseth ZC, Shahack-Gross R (2018) The archaeology and history of the Negev and neighbouring areas in the third millennium BCE: a new paradigm. Tel Aviv 45(1):63–88

Fischer M, Taxel I (2021) Life in the dunes: the Western Hinterland of Yubna/Yavneh in Late Ottoman and British Mandate Times. J East Med Archaeol Heritage St 9(1–2):29–63

Frumkin-Ahiron T, Frumkin R, Roudich R, Melloul A, Levin N, Papay N (2003) Conservation of the Coastal Sand Dunes—a policy report, The Surveys Unit—Open Landscape Institute—The Society for the Protection of Nature in Israel, The Ministry of Environment, The Nature and Parks Authority, The Jewish National Fund, The Water Commission and the Jerusalem Institute for Israel Studies, 126 p (in Hebrew)

Fryberger SG, Dean G (1979) Dune forms and wind regime. In: A study of global sand seas. US Government Printing Office Washington, vol 1052, pp 137–169

Fuess A (2017) Rotting ships and razed harbors: the naval policy of the Mamluks. Mamluk St Rev 5:45–71

Gal A, Lahav H, Ramon U (2008) Palmachim survey. The Society for Protection of Nature in Israel (in Hebrew)

Galil J, Eisikowitch D (1968) Flowering cycles and fruit types of Ficus sycomorus in Israel. New Phytol 67(3):745–758

Gao J, Kennedy DM, Konlechner TM (2020) Coastal dune mobility over the past century: a global review. Prog Phys Geog 44(6):814–836

Giulio S, Acosta ATR, Carboni M, Campos JA, Chytrý M, Loidi J et al (2020) Alien flora across European coastal dunes. App Veg Sc 23(3):317–327

Goldsmith V, Golik A (1980) Sediment transport model of the southeastern Mediterranean coast. MarGeol 37(1–2):147–175

Grootjans AP, Ernst WHO, Stuyfzand PJ (1998) European dune slacks: strong interactions of biology, pedogenesis and hydrology. Trends Ecol Evol 13(3):96–100

Harel M, Amit R, Porat N, Enzel Y (2017) Evolution of the Southeastern Mediterranean coastal plain. In: Enzel Y, Bar-Yosef O (eds) Quaternary of the levant environments, climate change, and humans. Cambridge University Press, Chapter 50, pp 433–445

Hertling UM, Lubke RA (1999) Use of *Ammophila arenaria* for dune stabilization in South Africa and its current distribution—perceptions and problems. Environ Man 24(4):467–482

Hesp P (2002) Foredunes and blowouts: initiation, geomorphology and dynamics. Geomorphology 48(1–3):245–268

Hesp P, McLachlan A (2000) Morphology, dynamics, ecology and fauna of Arctotheca populifolia and Gazania rigens nabkha dunes. J Arid Environ 44(2):155–172

Hugenholtz CH, Wolfe SA (2005) Biogeomorphic model of dunefield activation and stabilization on the northern Great Plains. Geomorphology 70(1–2):53–70

Huntington E (1911) Palestine and its transformation. Houghton Mifflin

Inman DL, Jenkins SA (1985) The Nile littoral cell and man's impact on the coastal zone of the southeastern Mediterranean. In: Coastal engineering 1984, pp 1600–1617

Issar A (1968) Geology of the central coastal plain of Israel. Isr J Earth Sci 17:16–29

Issar A, Tsoar H (1987) Who is to blame for the desertification of the Negev, Israel. In: Proceedings, IAHS Symposium, pp 577–583

Jarvis CS (1943) Tribes of Arabia. Jerusalem (translated into Hebrew)

Kark R, Levin N (2012) The environment in Palestine in the late Ottoman period, 1798–1918. In: Miller C Tal A, Orenstein D (eds) Between ruin and restoration: an environmental history of Israel, University of Pittsburgh Press, pp 1–28

Katz O, Kam M, Carmi A, Degen AA, Henkin Z, Bar P (2016) Activity and short-term impacts of dromedary camels (*Camelus dromedarius*) foraging on perennial coastal sand dune vegetation. J Arid Environ 133:47–53

Kidron GJ (2016) Goat trampling affects plant establishment, runoff and sediment yields over crusted dunes. Hydrol Proc 30(13):2237–2246

Kozhoridze G, Ben-Dor E, Sternberg M (2022) Assessing the dynamics of plant species invasion in Eastern-mediterranean coastal dunes using cellular automata modeling and satellite time-series analyses. Remote Sens 14(4):1014

Kutiel P (2001) Conservation and management of the Mediterranean coastal sand dunes in Israel. J Coast Cons 7(2):183–192
Kutiel P, Danin A, Orshan G (1979) Vegetation of the sandy soils near Caesarea, Israel. I. Plant communities, environment and succession. Isr J Plant Sci 28(1):20–35
Kutiel P, Peled Y, Geffen E (2000a) The effect of removing shrub cover on annual plants and small mammals in a coastal sand dune ecosystem. Biol Cons 94(2):235–242
Kutiel P, Eden E, Zhevelev Y (2000b) Effect of experimental trampling and off-road motorcycle traffic on soil and vegetation of stabilized coastal dunes, Israel. Environ Cons 27(1):14–23
Kutiel P, Cohen O, Shoshany M, Shub M (2004) Vegetation establishment on the southern Israeli coastal sand dunes between the years 1965 and 1999. Landscape Urban Plan 67(1–4):141–156
Kutiel P, Katz O, Ziso-Cohen V, Divinsky I, Katra I (2016) Water availability in sand dunes and its implications for the distribution of *Artemisia monosperma*. CATENA 137:144–151
Lehrer D, Becker N, Bar P (2013) The value of coastal sand dunes as a measure to plan an optimal policy for invasive plant species: the case of the *Acacia saligna* at the Nizzanim LTER coastal sand dune nature reserve, Israel. In: Restoration of coastal dunes. Springer, Berlin, pp 273–288
Levin N (2006) The Palestine exploration fund map (1871–1877) of the holy land as a tool for analysing landscape changes: the coastal dunes of Israel as a case study. Cartogr J 43(1):45–67
Levin N, Ben-Dor E (2004) Monitoring sand dune stabilization along the coastal dunes of Ashdod-Nizanim, Israel, 1945–1999. J Arid Environ 58(3):335–355
Levin N, Shmida A (2007) Determining conservation hotspots across biogeographic regions using rainfall belts: Israel as a case study. Isr J Ecol Evol 53(1):33–58
Levin N, Kidron GJ, Ben-Dor E (2006) The spatial and temporal variability of sand erosion across a stabilizing coastal dune field. Sedimentology 53(4):697–715
Levin N, Kidron GJ, Ben-Dor E (2007a) Surface properties of stabilizing coastal dunes: combining spectral and field analyses. Sedimentology 54(4):771–788
Levin N, Tsoar H, Maia LP, Sales VC, Herrmann H (2007b) Dune whitening and inter-dune freshwater ponds in NE Brazil. CATENA 70(1):1–15
Levin N, Kidron GJ, Ben-Dor E (2008) A field quantification of coastal dune perennial plants as indicators of surface stability, erosion or deposition. Sedimentology 55(4):751–772
Levin N, Ben-Dor E, Kidron GJ, Yaakov Y (2008b) Estimation of surface roughness (z0) over a stabilizing coastal dune field based on vegetation and topography. Earth Surf Proc Land 33(10):1520–1541
Levin N, Elron E, Gasith A (2009) Decline of wetland ecosystems in the coastal plain of Israel during the 20th century: implications for wetland conservation and management. Lands Urb Plan 92(3–4):220–232
Levin N, Jablon PE, Phinn S, Collins K (2017) Coastal dune activity and foredune formation on Moreton Island, Australia, 1944–2015. Aeolian Res 25:107–121
Liphschitz N, Biger G (1997) Sand dunes reclamation by vegetation in Palestine during the British mandate period. Horiz Geogr 46:21–38 (in Hebrew)
Liu B, Coulthard TJ (2015) Mapping the interactions between rivers and sand dunes: implications for fluvial and aeolian geomorphology. Geomorphology 231:246–257
Luijendijk A, Hagenaars G, Ranasinghe R, Baart F, Donchyts G, Aarninkhof S (2018) The state of the world's beaches. Sci Rep 8(1):1–11
Manor R, Saltz D (2004) The impact of free-roaming dogs on gazelle kid/female ratio in a fragmented area. Biol Cons 119(2):231–236
Marriner N, Flaux C, Morhange C, Kaniewski D (2012) Nile Delta's sinking past: Quantifiable links with Holocene compaction and climate-driven changes in sediment supply? Geology 40(12):1083–1086
Mauz B, Hijma MP, Amorosi A, Porat N, Galili E, Bloemendal J (2013) Aeolian beach ridges and their significance for climate and sea level: Concept and insight from the Levant coast (East Mediterranean). Earth-Sci Rev 121:31–54
Meir A, Tsoar H (1996) International borders and range ecology: the case of Bedouin transborder grazing. Human Ecol 24(1):39–64
Milgrom T (2008) Environmental aspects of rehabilitating abandoned quarries: Israel as a case study. Landscape Urban Plan 87(3):172–179
Mushkin A, Katz O, Crouvi O, Alter SR, Shemesh R (2016) Sediment contribution from Israel's coastal cliffs into the Nile's littoral cell and its significance to cliff-retreat mitigation efforts. Eng Geol 215:91–94
Paz T (2009) Environmental conflicts and open space protection in the Nitzanim-Ashdod sand dunes, Israel, 1948–2008. MA thesis, Ben Gurion University of the Negev
Perevolotsky A, Seligman NG (1998) Role of grazing in Mediterranean rangeland ecosystems. Bioscience 48(12):1007–1017
Perry G, Dmi'el R (1995) Urbanization and sand dunes in Israel: direct and indirect effects. Isr J Ecol Evol 41(1):33–41
Porat Y (1975) The gardens of Caesarea. Qadmoniot: J Antiq Eretz-Israel Bible Lands 30/31:90–93 (in Hebrew)
Porat N, Wintle AG, Ritte M (2004) Mode and timing of Kurkar and Hamra formation, central coastal plain, Israel. Isr J Earth Sci 53(1):13–25
Porat N, Sivan D, Zviely D (2008) Late Holocene embayment infill and shoreline migration, Haifa Bay, Eastern Mediterranean. Isr J Earth Sci 57(1):21–31
Pye K (1982) Morphological development of coastal dunes in a humid tropical environment, Cape Bedford and Cape Flattery, North Queensland. Geografiska Annaler: Ser Phys Geogr 64(3–4):213–227
Pye K, Tsoar H (1990) Aeolian sand and sand dunes. Springer Science & Business Media
Reifenberg A (1950a) Caesarea: a study in the decline of a town. Isr Explor J 1(1):20–32
Reifenberg A (1950b) The struggle between the desert and the sown. Jerusalem, 131 pp (in Hebrew)
Reitz MD, Jerolmack DJ, Ewing RC, Martin RL (2010) Barchan-parabolic dune pattern transition from vegetation stability threshold. Geophys Res Lett 37(19):L19402
Rim M (1950) Sand and soil in the coastal plain of Israel: a study of the rate of accumulation. Isr Explor J 1(1):33–48
Rohrlicht V, Goldsmith V (1984) Sediment transport along the southeast Mediterranean: a geological perspective. Geo-Marine Lett 4(2):99–103
Roskin J, Taxel I (2021) "He who revives dead land": groundwater harvesting agroecosystems in sand along the southeastern Mediterranean coast since early medieval times. Med Geosci Rev 3(3):293–318
Roskin J, Katra I, Blumberg DG (2013) Late Holocene dune mobilizations in the northwestern Negev dunefield, Israel: a response to combined anthropogenic activity and short-term intensified windiness. Quat Int 303:10–23
Roskin J, Sivan D, Shtienberg G, Roskin E, Porat N, Bookman R (2015) Natural and human controls of the Holocene evolution of the beach, aeolian sand and dunes of Caesarea (Israel). Aeolian Res 19:65–85
Rotem D, Weil G (2014) Natural ecosystem-units in Israel and the Palestinian authority-representativeness in protected areas and suggested solutions for biodiversity conservation. J Landscape Ecol 7(1):91–109

Sasson A (2019) Historical geography of the Palestine southern coastal plain in the late Ottoman period–the Ashkelon region as a case study. Middle Eastern Stud 55(6):974–1004
Schaffer G, Levin N (2014) Mapping human induced landscape changes in Israel between the end of the 19th century and the beginning of the 21th century. J Landscape Ecol 7(1):110–145
Schaffer G, Levin N (2016) Reconstructing nineteenth century landscapes from historical maps—the Survey of Western Palestine as a case study. Landscape Res 41(3):360–379
Shtienberg G, Dix JK, Shahack-Gross R, Yasur-Landau A, Roskin J, Bookman R et al (2017) Anthropogenic overprints on natural coastal aeolian sediments: a study from the periphery of ancient Caesarea, Israel. Anthropocene 19:22–34
Sivan D, Porat N (2004) Evidence from luminescence for Late Pleistocene formation of calcareous aeolianite (kurkar) and paleosol (hamra) in the Carmel Coast, Israel. Palaeogeog Palaeoecol 211(1–2):95–106
Sivan D, Wdowinski S, Lambeck K, Galili E, Raban A (2001) Holocene sea-level changes along the Mediterranean coast of Israel, based on archaeological observations and numerical model. Palaeogeogr Palaeoecol 167(1–2):101–117
Sneh A, Weissbrod T, Ehrlich A, Horowitz A, Moshkovitz S, Rosenfeld A (1986) Holocene evolution of the northeastern corner of the Nile Delta. Quat Res 26(2):194–206
Stanley DJ (1989) Sediment transport on the coast and shelf between the Nile Delta and Israeli margin as determined by heavy minerals. J Coastal Res 5(4):813–828
Stanley DJ (1996) Nile Delta: extreme case of sediment entrapment on a delta plain and consequent coastal land loss. Mar Geol 129(3):189–195
Stanley DJ, Warne AG (1998) Nile Delta in its destruction phase. J Coastal Res 14(3):795–825
Sternberg M (2016) From America to the Holy Land: disentangling plant traits of the invasive *Heterotheca subaxillaris* (Lam.) Britton Rusby. Plant Ecol 217(11):1307–1314
Taxel I, Sivan D, Bookman R, Roskin J (2018) An Early Islamic inter-settlement agroecosystem in the coastal sand of the Yavneh Dunefield, Israel. J Field Archaeol 43(7):551–569
Tengberg A, Chen D (1998) A comparative analysis of Nebkhas in central Tunisia and northern Burkina Faso. Geomorphology 22(2):181–192
Tordoni E, Bacaro G, Weigelt P, Cameletti M, Janssen JA, Acosta AT et al (2021) Disentangling native and alien plant diversity in coastal sand dune ecosystems worldwide. J Veg Sci 32(1):e12861
Tsoar H (1974) Desert dunes morphology and dynamics, El Arish, northern Sinai. Z Geomorph 20:41–61
Tsoar H (1984) The formation of Seif dunes from barchans-a discussion. Z Geomorph 28(1):99–103
Tsoar H (1990) Trends in the development of sand dunes along the southeastern Mediterranean coast in Dunes of the European coasts. Geomorphology-hydrology-soils. Catena Supp (giessen) 18:51–60
Tsoar H (2000) Geomorphology and paleogeography of sand dunes that have formed the Kurkar ridges in the coastal plain of Israel. Isr J Earth Sci 49(3):189–196
Tsoar H (2005) Sand dunes mobility and stability in relation to climate. Physica A 357(1):50–56
Tsoar H, Zohar Y (1985) Desert dune sand and its potential for modern agricultural development. In: Desert development. Springer, Dordrecht, pp. 184–200
Tsoar H, Blumberg D (1991) The effect of sea cliffs on inland encroachment of aeolian sand. In: Aeolian grain transport. Springer, Vienna, pp 131–146
Tsoar H, Blumberg D (2002) Formation of parabolic dunes from barchan and transverse dunes along Israel's Mediterranean coast. Earth Surf Proc Land 27(11):1147–1161
Tsoar H, Cohen-Zada AL (2020) The conflict between urban planning and the preservation of sand dunes—the case of the big dune in the city of Ashdod, Israel. City Environ Interactions 5:100034
Tsoar H, Shachak M, Blumberg DG (2005) Ecological aspects of vegetation removal from the coastal sand dunes of Israel. Proceedings 'dunes and estuaries, pp 487–493
Tsuriell DE (1974) Sand dune stabilization in Israel. Int J Biomet 18(2):89–93
Tsuriell DE (1983) The Sandy Haifa acre plain as a typical phytoecosystem of Israel's Coast. In: Sandy beaches as ecosystems. Springer, Dordrecht, pp 756–757
Van der Meulen F, Salman AHPM (1996) Management of Mediterranean coastal dunes. Ocean Coast Manage 30(2–3):177–195
Warren C (1876) Underground Jerusalem: an account of some of the principal difficulties encountered in its exploration and the results obtained. With a narrative of an expedition through the Jordan Valley and a visit to the Samaritans. R. Bentley and Son
Wasson RJ, Hyde R (1983) Factors determining desert dune type. Nature 304(5924):337–339
Weil G, Levin N (2015) Can siting algorithms assist in prioritizing for conservation in a densely populated and land use allocated country?–Israel as a case study. Isr J Ecol Evol 61(1):50–60
White K, El Asmar HM (1999) Monitoring changing position of coastlines using Thematic Mapper imagery, an example from the Nile Delta. Geomorphology 29(1–2):93–105
Wilson JR, Gairifo C, Gibson MR, Arianoutsou M, Bakar BB, Baret S et al (2011) Risk assessment, eradication, and biological control: global efforts to limit Australian acacia invasions. Divers Distrib 17(5):1030–1046
Yaalon DH (1959) Classification and nomenclature of soils in Israel. Bull Res Counc Israel G 8:91–118
Yizhaq H, Ashkenazy Y, Tsoar H (2009) Sand dune dynamics and climate change: a modeling approach. J Geophys Res 114(F1)
Zohary D, Spiegel-Roy P (1975) Beginnings of fruit growing in the old world. Science 187(4174):319–327
Zviely D, Klein M (2003) The environmental impact of the Gaza strip coastal constructions. J Coastal Res 19(4):1122–1127
Zviely D, Kit E, Klein M (2007) Longshore sand transport estimates along the Mediterranean coast of Israel in the Holocene. Mar Geol 238(1–4):61–73

Part IV

Central Israel, Dead Sea and Judean Desert

A Climatic Eco-geomorphological Transect Across the Major Climatic Belts of Israel—Processes, Patterns, and Desertification

12

Hanoch Lavee

Abstract

In this chapter, the typical landforms and processes along the climatic eco-geomorphological transect running from the arid region near the Dead Sea, through Mildly-arid and Semi-arid regions, to the Mediterranean climate region in the Judean Mountains, are described and analyzed. Based on field measurements of the spatial and temporal variability of rainfall and temperature, as well as soil properties and hydro-geomorphological processes, the trends at the regional scale and the spatial patterns at the plot scale are discussed. Regarding the regional scale, a high correlation was found between the average climatic conditions and the relevant variables and processes such as soil aggregate stability and organic matter content, infiltration rate, and runoff coefficient. However, a step-like threshold exists at the semi-arid area which sharply separates between the Mediterranean eco-geomorphic system and the arid systems. This sharp change indicates that these systems are controlled by different processes. While the Mediterranean eco-geomorphic system is controlled by biotic processes, the arid eco-geomorphic systems are controlled by abiotic factors. Regarding the plot scale, the spatial response to rainfall in both the Mediterranean climate system and the arid system is homogeneous, but in opposite directions. While in the Mediterranean climate zone infiltration is the dominant process all over the hillslopes, in the arid zone overland flow is generated over the entire hillslopes. In contrast, the semi-arid zone is characterized by a mosaic-like pattern, containing source patches that contribute overland flow and sink patches in which the direct rain and the water inflow infiltrate. The sharp difference in the geomorphic processes between the Mediterranean climate region and the arid region and the expected climate change in Israel means that Israel is threatened by desertification even in the case of relatively small climatic change. On the basis of the findings in this research, the expected spatial distribution changes of the surface cover components in the Mediterranean climate region and the semi-arid region are described.

12.1 Introduction

The purpose of this chapter is to describe the landscapes and relevant processes along a climatic eco-geomorphological transect running from an arid region near the Dead Sea in the east, through mildly-arid and semi-arid regions, to the Mediterranean climate region in the Judean Mountains near Jerusalem in the west.

The chapter is based on the results of a multidisciplinary project carried out for nearly a decade, during the 1990s, aimed at exploring the possible impact of climate change on desertification. The project yielded dozens of articles related to various environmental aspects and scientific questions such as soil properties dynamics (soluble salts (Tzvikel et al. 2007), carbohydrates (Steinberger et al. 1999), organic carbon (Jarmer et al. 2010) and aggregates stability (Lavee et al. 1996)), soil moisture spatial patterns (Katra et al. 2007), soil development (Wieder and Lavee 1990), overland flow generation and soil erosion (Lavee et al. 1991), the effect of hillslope aspect on soil and vegetation properties (Kutiel and Lavee 1999), vegetation cover dynamics (Kutiel et al. 1995; Shoshany et al. 1995), vegetation cover spatial patterns (Shnerb et al. 2003), and vegetation response to climate change (Kutiel et al. 2000). In these and other articles diverse techniques were employed such as field and laboratory measurements, rainfall simulation experiments in the field and in the laboratory, remote sensing, and thermal infrared image interpretation and

H. Lavee (✉)
Department of Geography and Environment, Laboratory of Soil and Geomorphology, Bar Ilan University, Ramat Gan, Israel
e-mail: hanoch.lavee@gmail.com

A. Frumkin and N. Shtober-Zisu (eds.), *Landscapes and Landforms of Israel*,
World Geomorphological Landscapes, https://doi.org/10.1007/978-3-031-44764-8_12

modeling. Most of the findings that are presented in this chapter were discussed by Lavee et al. (1998).

Following, the processes at each of the climatic regions along the abovementioned transect and the elements that differentiate between the arid zones and the Mediterranean climate zone are analyzed. Based on the correlation between eco-geomorphic processes and climatic conditions, the option of desertification due to the expected climate change is also discussed.

12.2 Study Area

The study area includes the eastern flank of the Judean Mountains in the west and the Judean Desert that stretches eastward of the Judean Mountains to the Dead Sea in the east (Figs. 12.1 and 12.2). It is about 35 km wide and 85 km long from north to south.

12.2.1 Topography, Geology and Drainage System

The Judean Mountains that reach elevations of 800–1000 m asl consist of carbonate rock formations of the Albian-Turonian Judea Group, mainly limestone and dolomite with some marl and chalk (Fig. 12.3a). They are part of the SSW–NNE-trending Hebron and Ramallah anticlinorium. East of the mountain crest, at the western border of the Judean Desert, there is a steep topographic descent, associated with a structural flexure, whose maximum elevation ranges from 800 to 500 m. The rocks in this area are mainly limestones of the Turonian Judea Group (Fig. 12.3b). In its northern part, chalk and chert of the Senonian Mount Scopus Group are also exposed. A dense channel system descends down this flexure. East of the flexure the Judean Desert forms a plateau characterized by rounded hills whose average elevation is about 600 m in the west and

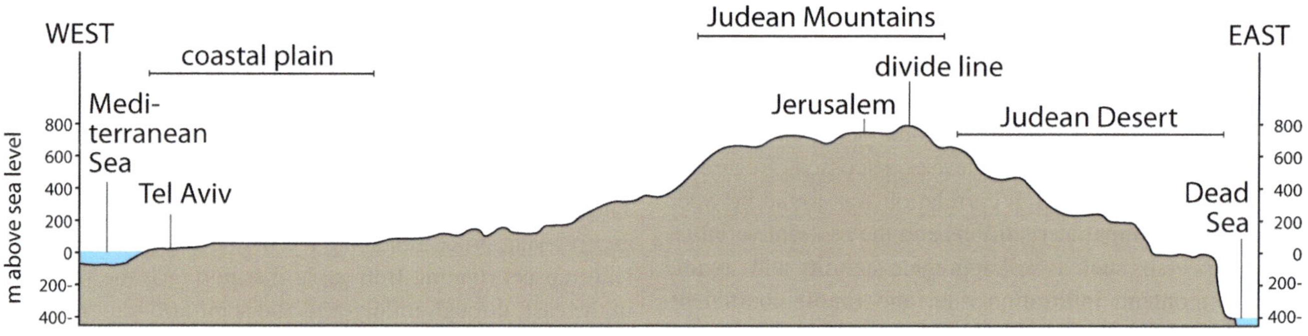

Fig. 12.1 A schematic topographic cross-section from the Mediterranean Sea to the Dead Sea. The distance is 80 km

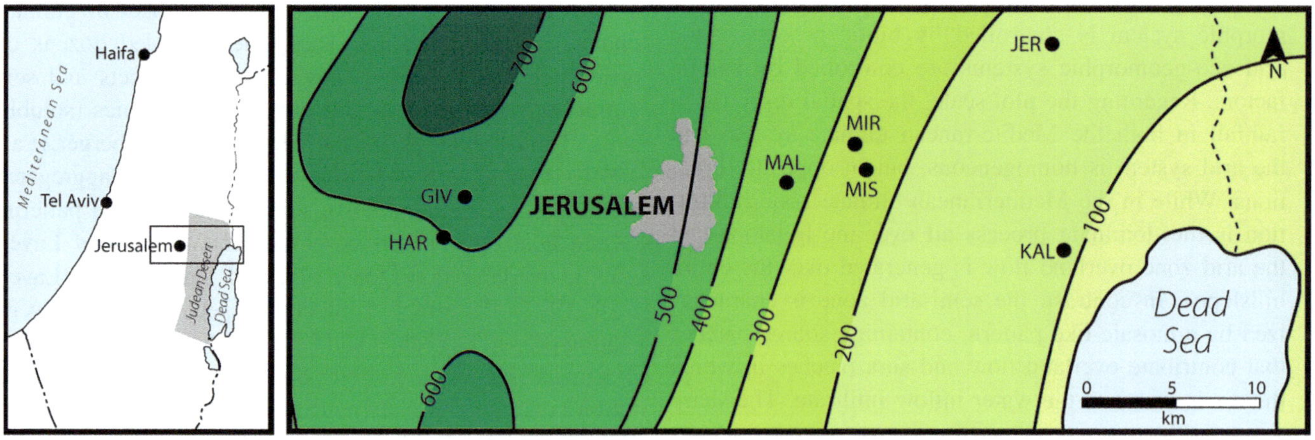

Fig. 12.2 Location of research sites (GIV = Giv'at Ye'arim, HAR = Har Tayasim, MAL = Ma'ale Adumim, MIS = Mishor Adumim, MIR = Mishor Adumim B, JER = Jericho, KAL = Kalia) along the climatic eco-geomorphological transect (mean annual rainfall, in millimeters, is indicated by isohyets). Note the area of the Judean Desert on the map to the left

Fig. 12.3 Typical landscapes of the Mediterranean (**a**), semi-arid (**b**), mildly-arid (**c**), and arid (**d**) climatic belts along the transect across the Judean Desert. Pay attention to 1. The sharp change in vegetation cover between the Mediterranean belt and the arid belts is explained in Sect. 12.3.2. 2. Canyons as shown in 3b cross the Judean Desert. See Sect. 12.2.1

about 150 m asl in the east. The general structure of the plateau is a shallow syncline, but there are a number of small anticlines too. The anticlines are mostly composed of Turonian limestone, and most of the synclines expose chalk of the Senonian age (Fig. 12.3c). Sub-parallel canyons, in which flash flood events may occur a few times a year, are starting on the eastern flank of the Judean Mountains (Fig. 12.3b) and crossing this plateau eastward. In the plateau, at the bottom of these canyons Turonian limestone is usually exposed (Fig. 12.4). At the eastern edge of the plateau, a narrow flat strip with a slight inclination to the east exists, where Turonian limestone is exposed (Fig. 12.3d). Here, all the sub-channels gather into the main stream channels, the drainage system is sparse, but the main streams widen and deepen. From there eastwards, the desert drops as a steep cliff toward the Dead Sea Valley (Fig. 12.1). At the cliff the main streams descend in giant canyons, where Cenomanian limestone and dolomite are exposed, and in addition, there are young channels that descend from the top of the cliff in huge waterfalls. This cliff is part of the western transform fault line of the Dead Sea Rift valley. The elevation difference across the cliff is 450 m on average.

12.2.2 Climate, Soil, and Vegetation

The Judean Mountains are dominated by Mediterranean climate, that is hot and dry summers and cool, rainy winters. Brown-reddish Terra Rossa soils overlie the area's hard limestone and dolomite bedrocks. These soils are rich in clay minerals and are carbonate-poor. The natural vegetation on these soils is usually a Mediterranean forest, dominated by kermes Oak (*Quercus coccifera*) and Palestine Pistachio (*Pistacia terebinthus*). Due to damages caused by man and livestock over a long time throughout history, in

Fig. 12.4 A typical view of the plateau area. Note the Turonian limestone exposure in the canyon at the center of the picture

most of the area the trees were naturally been replaced by Mediterranean semi-shrubs and herbaceous or have been reforested with the Aleppo Pine (*Pinus haleppensis*).

The Judean Desert is not part of the global desert belt. Due to the prevailing westerly winds and the topographic barrier of the Judean Mountains, most of the rain, produced by clouds moving in winter from the Mediterranean Sea eastward, falls on the coastal plain and the Judean Mountains. The leeward, eastern side of the mountains remains under rain shadow effect. As the clouds glide eastward over the flexure of the Judean Desert, their temperature rises and humidity decreases. The result is that rains are relatively rare east of the water divide line and the mean annual rain decreases from around 700 mm in the Judean Mountains west of Jerusalem to less than 100 mm near the Dead Sea (Fig. 12.2). In addition, the Judean Desert is characterized by high temperatures and strong sunlight during the day. Mean annual temperature rises in an easterly direction, from 17 °C in the Judean Mountains near Jerusalem to 23 °C near the Dead Sea. The soil and vegetation types in the Judean Desert vary depending on the bedrock and the climatic conditions. In the western part of the Judean Desert, in an area with over 300 mm annual rainfall, brown Rendzina soil having clay texture covers limestone and hard chalk, while pale Rendzina with silty clay texture is found on soft chalk. Annual plants, gramineae such as Bulbous Meadow (*Poa bulbosa*) and wild wheat, barley and oats, and semi-shrubs like Thorny Burnet (*Sarcopoterium spinosum*) are growing on these shallow soils. Rock fragments are spread at the surface. In the eastern part of the Judean Desert, under more arid conditions, shallow Lithosols are the main soil type, with a dense cover of rock fragments. The vegetation in this area includes semi-shrubs like Herba Alba (*Artemisia sieberi*) and Wooly Sage (*Salvia lanigera*), geophytes such as Sea Squill (*Drimia aphylla*) and annuals such as Pink Sorrell (*Rumex cyprius*) and Damascus Oxtongue (*Picris longirostris*). The plants are sparsely spread over the hillslopes and concentrate in the channels. The area near the Dead Sea, eastward of the Rift valley cliff, is characterized by saline Alluvial-arid-brown-soil with sparse vegetation cover, mainly the trees Umbrella Thorn Acacia (*Acacia tortilis*) and Christ's Thorn Jujube (*Ziziphus spina-christi*), the shrubs White Weaping Broom (*Retama raetam*) and Arabian Boxthorn (*Lycium shawii*), and the semi-shrubs Jointed Anabasis (*Anabasis articulata*), Bushy Bean Caper (*Zygophillum dumosum*), and Common Reaumuria (*Reaumuria hirtella*).

12.3 Chemical, Mechanical, and Biological Processes

Chemical, mechanical, and biological processes operate the eco-geomorphological systems in the arid, mildly-arid, semi-arid, and Mediterranean climate zones. Among the factors influencing these processes, rainfall exerts a dominant control.

During the last decade of the twentieth century, rainfall and temperature, as well as soil properties and geomorphological processes, have been monitored along an eco-geomorphological transect running from the Judean Mountains near Jerusalem to the Dead Sea area, over a distance of 35 km.

12.3.1 Experimental Sites—Setting and Methods

12.3.1.1 Instrumentation and Measurements

Seven research sites were established on hillslopes along this transect (Fig. 12.2), having a similar lithology (limestone or hard chalk), similar gradient, and similar aspect. The sites differ from each other in their elevation above sea level and climate conditions (Table 12.1). Of the seven research sites, GIV, MAL, MIS, and KAL sites best characterize the Mediterranean, semi-arid, mildly-arid, and arid climatic zones, respectively. Therefore, most of the research was conducted at these sites.

Typical general view of the Mediterranean, semi-arid, mildly-arid, and arid climatic zones is shown in Fig. 12.3, and typical view of a hillslope at each of these climatic zones is shown in Fig. 12.5. At sites GIV, MIS, and KAL instrumented stations were established, containing rain gauges, temperature probes at 30 cm above the ground surface and 5 and 20 cm beneath the surface, 18 gypsum blocks for measuring soil moisture—9 blocks at 5 cm depth and 9 blocks at 20 cm depth, 3 runoff plots of 3 m width having different lengths of 7, 14, and 21 m, and runoff and sediment collectors. These instruments were connected to data-loggers.

6–12 soil samples, taken from each site four times a year to represent different seasons, were brought to the laboratory to determine physical and chemical soil properties such as aggregate size and stability, organic matter content, and soluble salt ions concentration. In addition, in order to study the spatial distribution of these soil properties, 16 soil

Table 12.1 Basic characteristics of the research sites

Site	Mean annual Rainfall (mm)	Mean Annual Temp. (°C)	Elevation a.s.l. (m)	Slope Grad. (deg.)	Orientation (azim.)
Giv'at Ye'arim (GIV)	620	17	650	13	142
Har Tayasim (HAR)	600	17	740	12	150
Ma'ale Adumim (MAL)	330	19	330	13	145
Mishor Adumim A (MIS)	260	20	230	11	140
Mishor Adumim B (MIR)	260	20	270	11	135
Jerico (JER)	180	23	– 70	14	140
Kalia (KAL)	120	23	– 60	12	150

Fig. 12.5 Typical view of hillslopes in the Mediterranean (**a**), semi-arid (**b**), mildly-arid (**c**), and arid (**d**) climatic belts

samples were taken randomly, but not under shrubs, from the upper 2 cm of the soil, from plots 10 m × 10 m in size, at sites GIV, MAL, MIS, and KAL.

Rainfall simulation experiments were conducted in the field and in the laboratory in order to study the response of the soil of each site to rainfall.

12.3.1.2 Data Interpretation

For each soil property, the overall average value at each of the four representative research sites (average value of all the soil samples of each season in all the seasons all over the measurement years) reflects the long-term influence of average climatic conditions in that climatic zone on this soil property. The spatial variability of this average value of soil property along the climatic transect represents the long-term effect of climate conditions at the regional scale. [The short-term effect of climate conditions can be studied, at each site, using the temporal variability, i.e., the differences between average values of seasons or years]. The differences between values obtained at different points at each research site (climate zone) at each sampling date (season) represent the plot scale spatial variability that controls the functioning of the systems in terms of the spatial patterns of water redistribution.

12.3.1.3 Lithology

The lithology of site GIV is limestone and dolomite (Aminadav formation of the Cenomanian age). MAL site is composed of limestone (Nezer formation of the Turonian age). Site MIS consists of hard chalk with a small amount of chert of the Mishash formation (Senonian age), and KAL site is composed of limestone (Nezer formation of the Turonian age).

12.3.1.4 Rainfall

Figures 12.6 and 12.7 show the daily rainfall distribution at the Mediterranean, mildly-arid, and arid climate zones, at Giv'at Ye'arim (GIV), Mishor Adumin (MIS), and Kalia (KAL) research sites, respectively. Figure 12.6 presents an example of a rainy winter, 1991–92, in which rainfall amounts were almost twice the annual mean, while Fig. 12.7 presents an example of the relatively dry winter of 1993–94.

Analysis of the distribution of rainfall events, on the basis of the definition of a 'rainfall event' that includes the requirement of having an interval of at least one hour with no rain between two consecutive rain events, showed that:

1. An average rainfall event in the arid zone lasts for about 35 min, and the duration increases with increasing mean annual rainfall and reaches about 65 min in the mildly-arid zone and more than two hours in the Mediterranean climate zone.
2. The intervals between rainfall events increase with aridity. The median interval lasts about three days in the Mediterranean climate zone, five days in the mildly-arid zone, and seven days in the arid zone. A similar trend was found by Kutiel (1985).

12.3.2 Trends at the Regional Scale

Figures 12.8, 12.9, 12.10, 12.11, 12.12, and 12.13 are based on field and laboratory experiments and show the change of the average values of several variables and processes along the climatic transect. The results lead to two main conclusions.

First, there is a high correlation between the average climatic conditions and the relevant variables and processes. Average soil organic matter content, average aggregate stability, and infiltration rate decrease with aridity (Figs. 12.8, 12.9, and 12.10). Runoff increases from the Mediterranean site toward the arid sites (Fig. 12.11), and the runoff coefficient (=the ratio runoff amount/rainfall amount) is very small in the Mediterranean site and increases dramatically with aridity (Figs. 12.12 and 12.13). These trends result from different average climatic conditions that exist, for a relatively long period of time, in the different climatic zones along the climatic eco-geomorphological transect, which led to the development of different soil types and vegetation communities.

Second, the rate of change of most of these variables and processes along the climatic transect is not linear. A step-like threshold exists at the semi-arid area (about 300 mm annual rainfall), which sharply separates the Mediterranean and arid eco-geomorphic systems. This sharp change can be clearly seen especially regarding the aggregate stability (Fig. 12.9), the infiltration rate (Fig. 12.10), the laboratory experiments of overland flow generation (Fig. 12.11), and the runoff coefficients at the overland flow plots in the field (Figs. 12.12 and 12.13). It also can be seen in the sharp change in vegetation cover between the Mediterranean area and the arid areas in Fig. 12.3. The fact that the transition between the Mediterranean climate region and the arid region is acute and not gradual indicates that when the soil properties pass a certain threshold value, the system ceases to be controlled by certain processes and begins to be controlled by other processes. While the arid eco-geomorphic systems are controlled by abiotic factors such as soluble salts content and mechanical crust formation, the Mediterranean eco-geomorphic system is controlled by biotic processes such as plant growth, microbial activity, and organic matter production and decomposition.

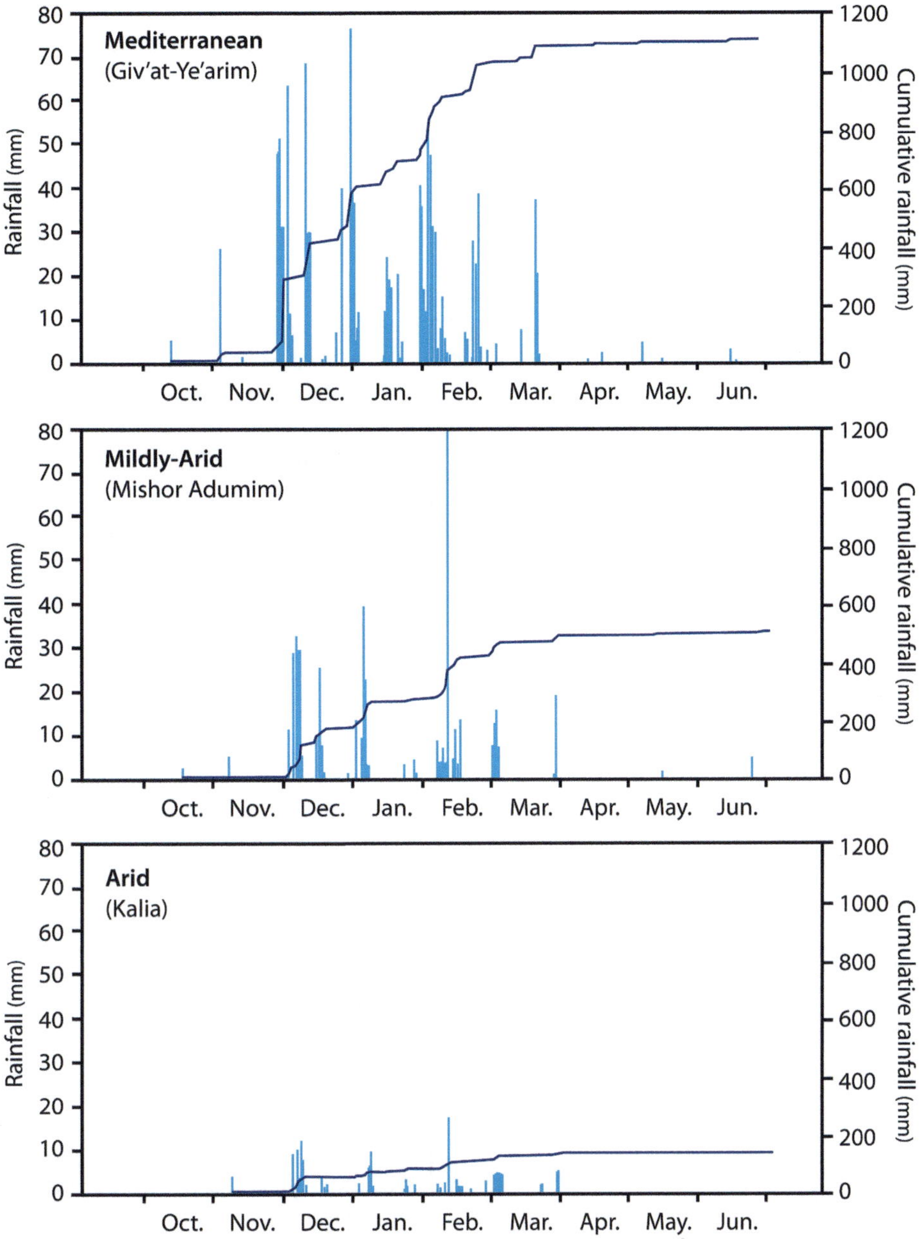

Fig. 12.6 Daily rainfall distribution of a very wet year, 1991–2, at the research sites GIV, MIS, and MAL

12.3.3 Spatial Patterns at the Plot Scale

The spatial redistribution of rainfall at the plot scale, in each of the climatic zones, depends on the spatial variability and the magnitude of the values of the factors affecting rainfall-infiltration relationships, mainly the soil organic matter content and the aggregate stability.

12.3.3.1 The Mediterranean Climate Zone

At the Giv'at Ye'arim (GIV) site, the temporal variability of soil organic matter content (Fig. 12.8) and aggregate stability (Fig. 12.9) is relatively high. The spatial variability of these two soil properties (Figs. 12.14 and 12.15, respectively) is also relatively high. This relatively high heterogeneity is a result of degradation of this area due to long history of intensive grazing—until the beginning of the twentieth century—that caused the disappearance of most of the trees and left behind mainly shrubs and some bare soil patches (Fig. 12.5a). Nevertheless, despite this heterogeneity, due to dense vegetation cover and the fact that even the lowest values of organic matter content and aggregate stability are high enough to cause high permeability in the soil and to prevent crust formation, most of the raindrops infiltrate at the points where they reach the soil surface. This explains the very small amount of overland flow in the lab experiments (Fig. 12.11) and the very low values of

Fig. 12.7 Daily rainfall distribution of a relatively dry year, 1993–4, at the research sites GIV, MIS, and MAL

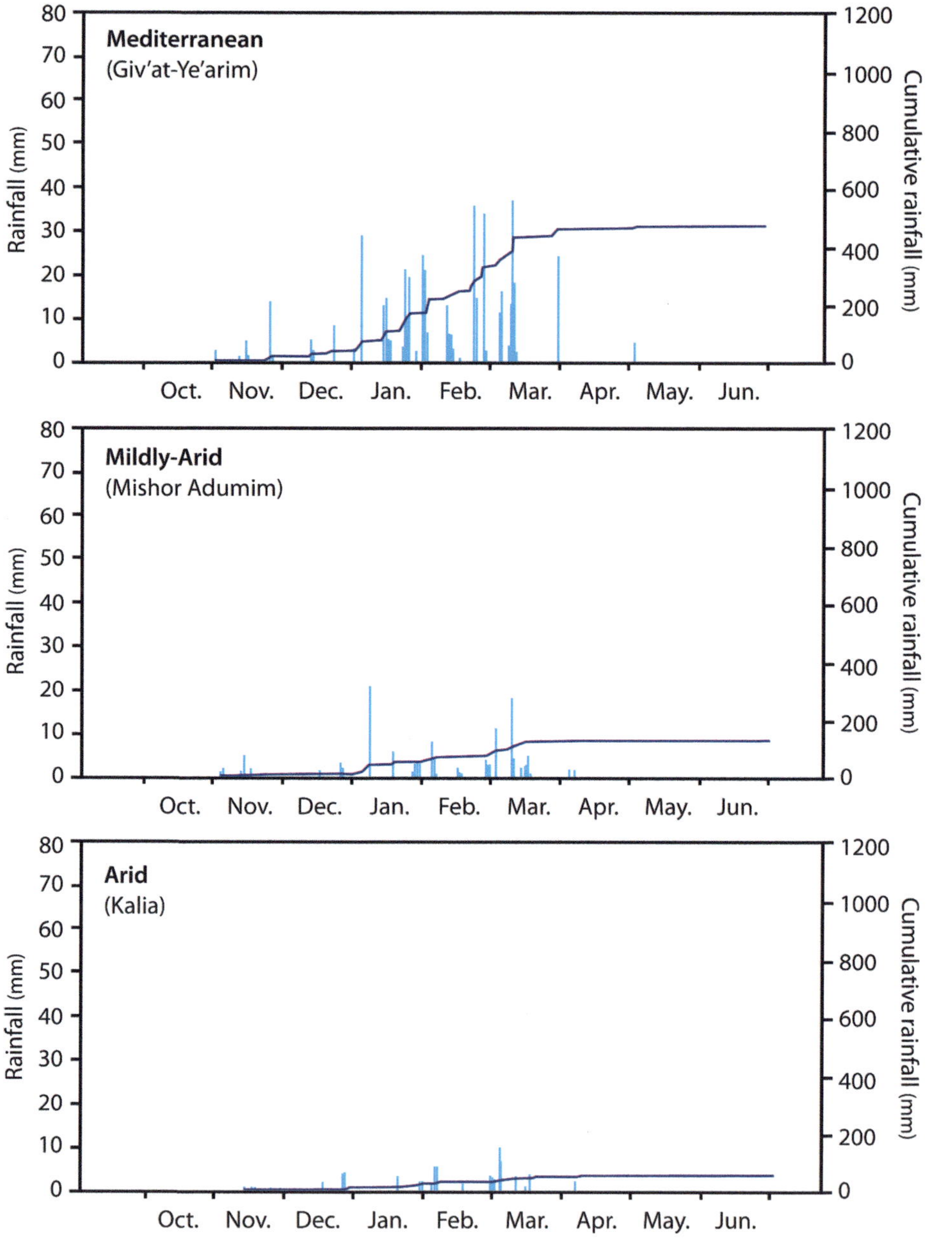

runoff coefficient (Figs. 12.12 and 12.13) at the runoff plots of this site. Over similar but non-degraded Mediterranean climate soils, such as at Mount Meron in the Upper Galilee in northern Israel, where average annual rainfall is 900 mm, almost no overland flow was generated under laboratory rainfall simulation (Fig. 12.11). This means that in the Mediterranean eco-geomorphic system, in spite of containing different ecological patches, from the geomorphological viewpoint of probability for overland flow generation the entire area behaves uniformly as infiltration is the dominant process all over the hillslope. In addition, under the climatic conditions in the Judean Mountains, where the annual rainfall amount is around 600 mm, the probability of the soil to reach saturation is very low, and overland flow events are very rare. This is in agreement with previous conclusions (Whipkey 1965; Kirkby and Chorley 1967; Dunne and Black 1970) that in humid areas overland flow is generated only when and where saturation conditions have been reached.

12.3.3.2 The Arid Zone

At the mildly-arid (Mishor Adumim, MIS) and arid (Kalia, KAL) sites, the temporal variability of organic matter content (Fig. 12.8) and aggregate stability (Fig. 12.9), as well as their spatial variability (Figs. 12.14 and 12.15, respectively) are very low. Due to the low values of organic matter content and aggregate stability, and the almost complete absence of vegetation cover (Figs. 12.3c, d and 12.5c, d),

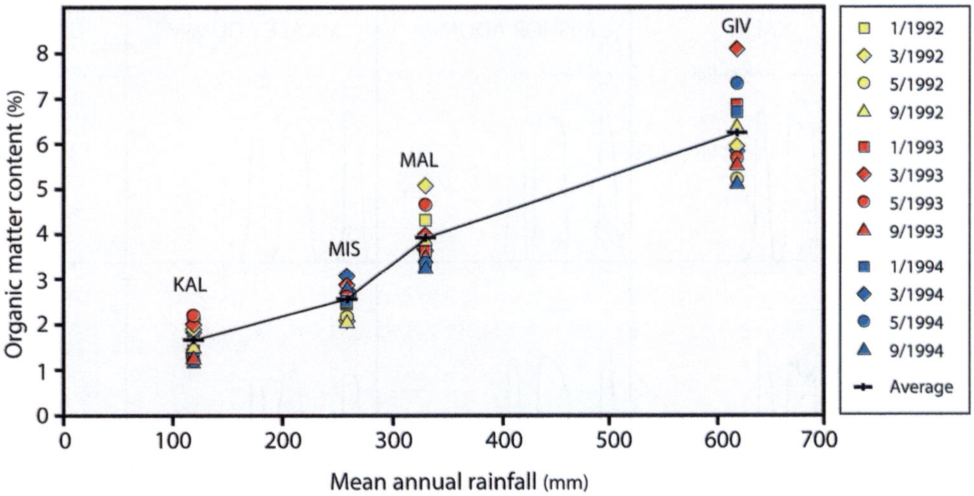

Fig. 12.8 Organic matter content in the soil (at 0–2 cm depth) along the climatic transect, at sites GIV, MAL, MIS, and KAL. Each point is the average of 6–12 samples that were taken at each site at each sampling date

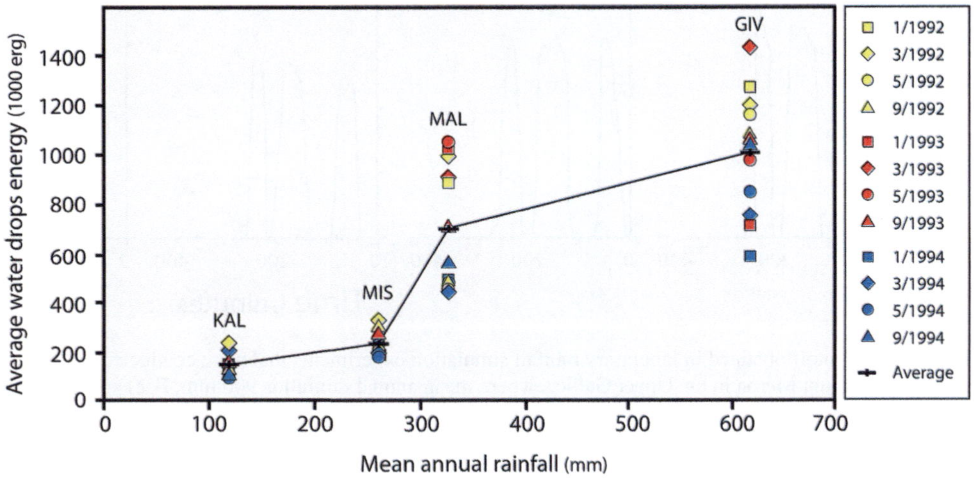

Fig. 12.9 Aggregate stability variations along the climatic transect, at sites GIV, MAL, MIS, and MAL. The aggregate stability is represented by the average energy needed to decompose an aggregate. From each soil sample 30 aggregate, 4–5 mm in size, were tested, using the single drop test. The average of the 30 aggregates was averaged for the 6–12 soil samples that were taken at each site at each sampling date. This average of averages is represented by points in this figure

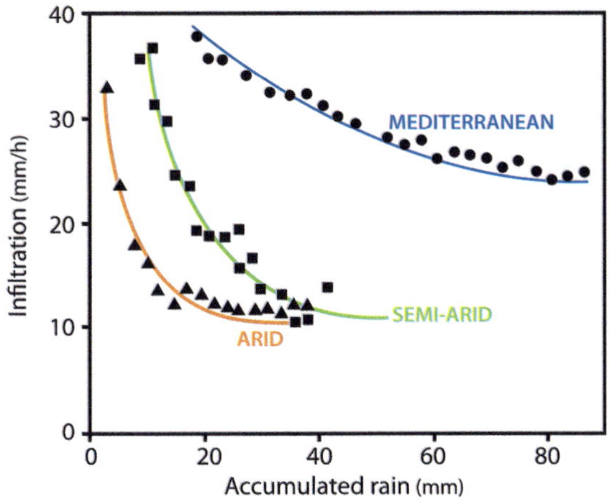

Fig. 12.10 Typical infiltration curves for the Mediterranean, semi-arid, and arid areas along the climatic eco-geomorphological transect. It can be seen, for example, that after 10 mm of rainfall the infiltration rates in the Mediterranean, semi-arid, and arid areas were ≫40, 35, and 15 mm/h, respectively

mechanical crust is easily formed at the soil surface and the infiltration rate reaches low values a few minutes after the rain started (Fig. 12.10, Arid site). Under these conditions overland flow is widespread over the entire hillslopes, the overland flow yield is very high (Fig. 12.11), and the runoff coefficients are high at the mildly-arid site (MIS) and very high at the arid site (KAL) (Figs. 12.12 and 12.13).

An interim conclusion is that the responses to rainfall in both the Mediterranean climate and the arid systems are spatially homogeneous, but in opposite directions. While in the Mediterranean climate zone infiltration is the dominant process all over the hillslopes, in the arid zone overland flow is generated over the entire hillslope. However, in the arid areas the overland flow is discontinuous on hillslopes longer than 10–30 m, mainly because the typical rain events (spells) are very short, a few minutes on average. The soil is usually dry before the rain starts and even at the end of a short rain it is still not fully compacted and the final infiltration rate was not reached yet. Under this condition and

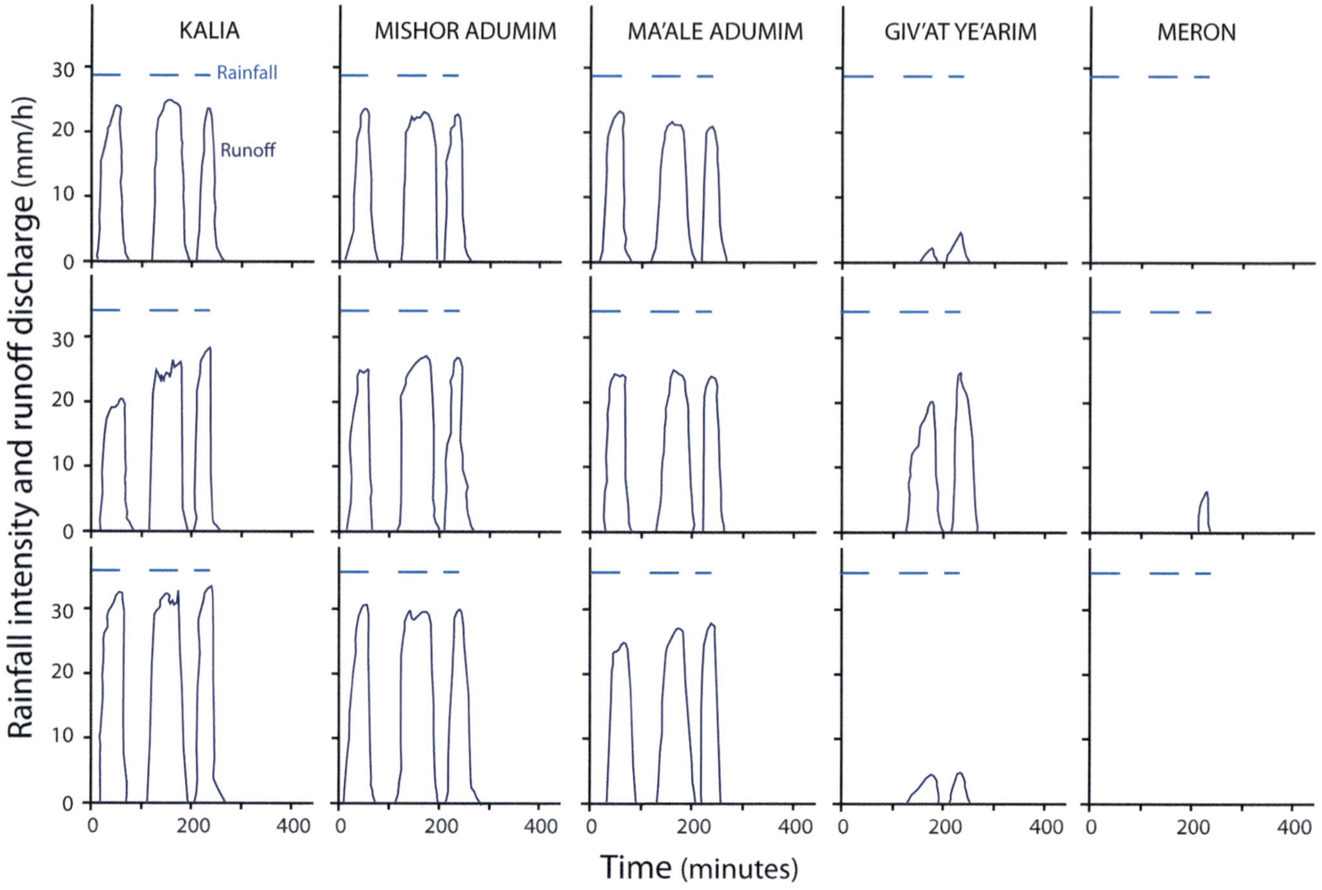

Fig. 12.11 Runoff obtained in laboratory rainfall simulation experiments that were conducted on soils collected between shrubs in the research sites and in Mount Meron in the Upper Galilee, where mean annual rainfall is 900 mm. For each of the soils three experiments, with rain intensities of 29, 33, and 35 mm/h were conducted, each consisting of a 60 min rain shower, break of 60 min, second shower of 60 min, break of 30 min, and last shower of 30 min

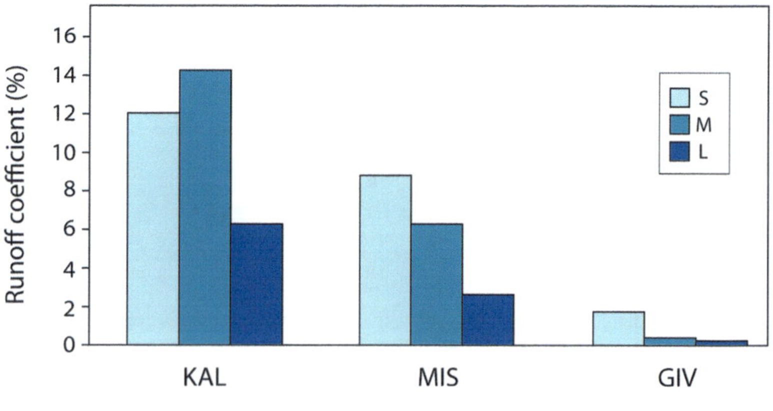

Fig. 12.12 Average runoff coefficients, for the years 1992–94, calculated on the basis of overland flow yields at the runoff plots in the arid (KAL), mildly-arid (MIS), and Mediterranean (GIV) areas. S, M, and L represent the shortest (7 m long), medium (14 m long), and longest (21 m long) plots, respectively

when taking into account that overland flow velocity is in the order of 0.5–3.0 m/min, the overland flow runs for only a short distance, usually not longer than 10–30 m before infiltrating into the shallow soil. Thus, overland flow that has been generated in the upper part of a hillslope will hardly reach the channel. The fact that the contributing area of the hillslope is only its lower part explains the relatively similar values of runoff coefficients at the short and medium-length plots at the arid sites of KAL and MIS and the lower runoff coefficient at the long plot (Fig. 12.12). This phenomenon is even more pronounced if the lower part of the hillslope is covered by a colluvium having a relatively high infiltration rate (Yair and Lavee 1985; Lavee and Yair 1990).

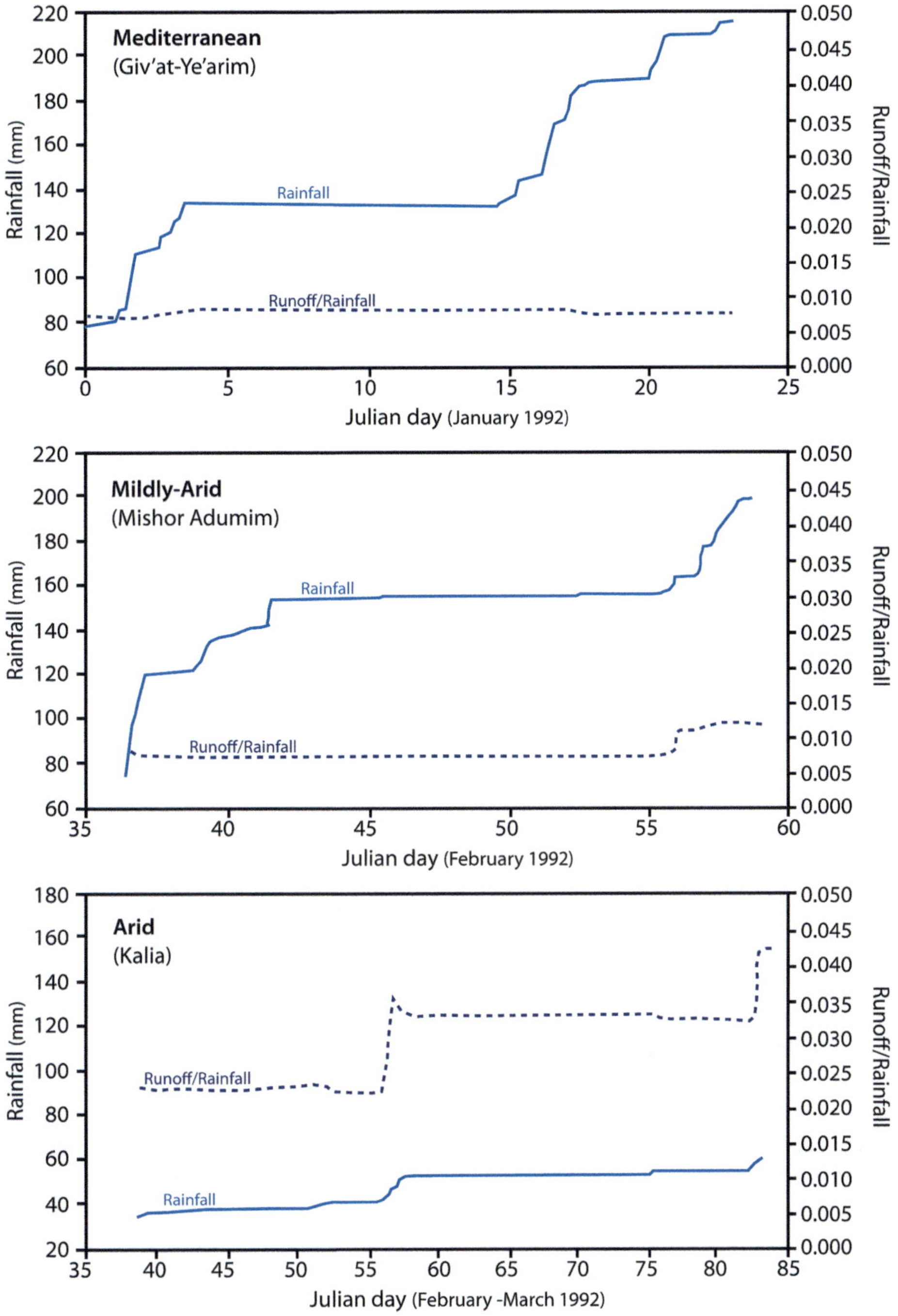

Fig. 12.13 Accumulated rainfall and accumulated runoff coefficient (accumulated runoff/accumulated rainfall) al the Mediterranean (GIV), mildly-arid (MIS), and arid (KAL) research sites. The accumulated runoff coefficient was calculated on the basis of measurements at the long (21 m) runoff plots. The low values of runoff coefficients at these relatively long plots are due to the fact that in addition to the effect of soil properties and vegetation cover on the infiltration rate, overland flow that was generated at the upper part of these plots infiltrated on its way downslope and did not contribute to the overland flow yield of the plots

12.3.3.3 The Semi-Arid Zone

The boundary area between the arid zone and the Mediterranean climate zone, i.e., the semi-arid area, represented by site MAL (Ma'ale Adumim), is characterized by high temporal and spatial variability of organic matter content and aggregate stability (Figs. 12.8, 12.9, 12.14, and 12.15). This high variability reflects the spatial heterogeneity of surface components and the existence of various patches which are dominated by shrubs, stones, some types of crusts and bare soil, respectively (Fig. 12.5b). These different patches function differently in terms of their response to rainfall and their effect on water conservation and redistribution. At some patches, such as those dominated by bare soil, a mechanical crust or big stones, especially when partly embedded in the topsoil (Lavee and Poesen 1991), overland flow generation is the dominant process. At other patches, like those covered by shrubs, mosses, or small stones that lie on top of the soil surface, the direct rainfall

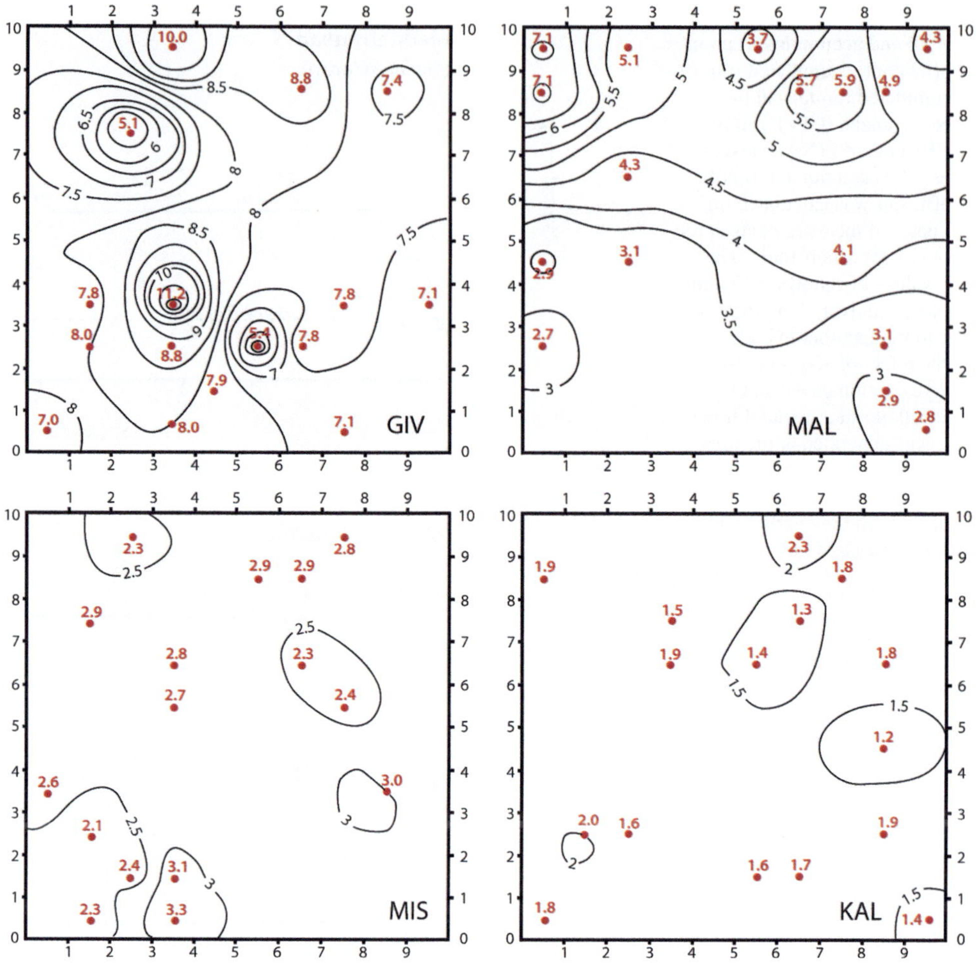

Fig. 12.14 Spatial distribution of organic matter content (%) in the upper 2 cm of the soil at sites GIV, MAL, MIS, and KAL. The maps are based on soil samples that were taken from 16 points within a plot, 10 m × 10 m in size, at each site in March 1993

plus the overland flow from adjacent patches infiltrate into the soil. In other words, the semi-arid area is characterized by a mosaic-like pattern, containing 'arid'/'dry' patches that contribute overland flow and 'humid'/'wet' patches that accept overland flow. Within such a mosaic pattern, the different patches might be distributed regularly, in clusters, or randomly. The development of such mosaic pattern enables most of the rainfall to be retained in the hillslopes. Similar spatial patterns and functions have been reported in several semi-arid areas, e.g., by Schlesinger et al. (1990), Tongway and Ludwig (1990), Montana (1992), Sanchez and Puigdefabregas (1994).

12.3.3.4 Summary—Functional Patterns of Conserving Water

Figure 12.16 presents different water redistribution processes and spatial functional patterns under different eco-climatic conditions, from arid through semi-arid, to humid zones.

The area having between 450 and 550 mm of rainfall per year may belong to the semi-arid or to the Mediterranean type eco-geomorphic system, depending on the degree of human interference, through introducing fires, grazing, or deforestation. A sharp threshold separates the arid eco-geomorphic systems, controlled by abiotic factors such as soluble salts content and mechanical crust formation, from the humid eco-geomorphic systems, controlled by biotic processes such as plant growth, microbial activity, and organic matter production and decomposition.

From the functional point of view of conserving water resources for germination and vegetation growth, the humid system has enough rainfall for maintaining 100% vegetation cover based on the direct rainfall at any point. Accordingly, the entire hillslope can absorb the rainfall through infiltration. Spatial differences in water uptake can be observed only at a scale smaller than a patch. Hortonian overland flow usually does not develop, except for the rare events in which soil saturation has been reached.

The Semi-arid area has not enough rainfall to maintain 100% vegetation cover. Thus, overland flow contributing patches and water-accepting patches develop next to one another. The rainfall amount is sufficient to support vegetation growth in accepting patches if adjacent contributing patches have a size between 0.25 and 3 m^2. Temporal

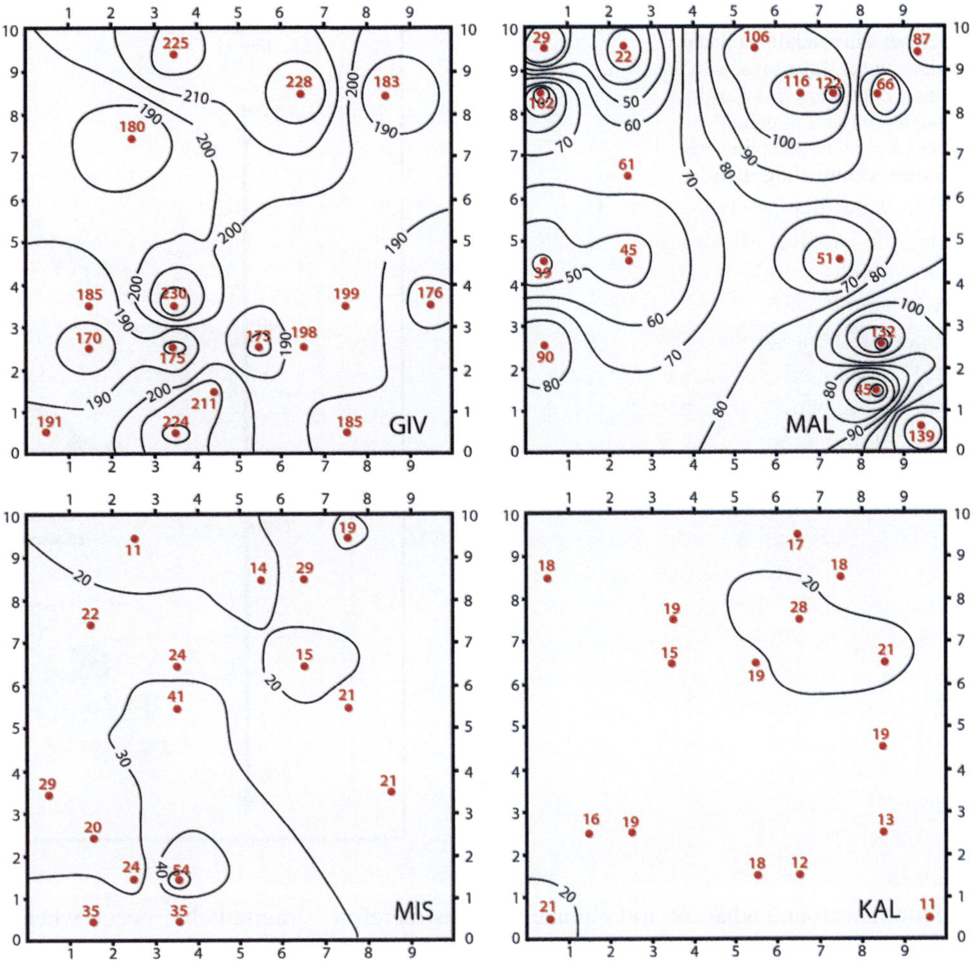

Fig. 12.15 Spatial distribution of aggregate stability (average number of drops needed to decompose an aggregate) in the upper 2 cm of the soil at sites GIV, MAL, MIS, and KAL. The maps are based on soil samples that were taken from 16 points within a plot, 10 m × 10 m in size, at each site in March 1993. From each soil sample 30 aggregate, 4–5 mm in size, were tested, using the single drop test

variability of the annual rainfall may cause changes in the size and distribution of the patches in a way that most rainfall will be retained on the hillslope.

Under arid conditions, the redistribution of water can be observed only at the hillslope scale. A water contributing area having the size of a patch is not enough to support perennial vegetation growth due to the low amount of rainfall. The upper part of the hillslope contributes overland flow, sediments, and nutrients to the lower colluvial section, and in some cases, the entire hillslope contributes overland flow to the channel, so that the perennial vegetation is limited to the lower part of the hillslope or to the channel area only.

12.4 Desertification

Keeping in mind the huge effect of climatic conditions on the eco-geomorphic processes, as shown above, one should think about the impact of the expected climate change on such a vulnerable area as the semi-arid area is and the areas on both sides of it.

12.4.1 Climate Change Features Relevant to Geomorphology and Desertification

Köppen (1936) defined the Mediterranean climate (Csa) as one in which:

1. The mean temperature of the coldest month is between – 3 and 18 °C.
2. The summer is dry, and the rainfall amount of the wettest month is at least three times higher than that of the driest month.
3. The mean temperature of the warmest month is above 22 °C.
4. The mean annual rainfall amount (in mm) is higher than 20 times the mean annual temperature (in Celsius degrees).

The first three conditions (1, 2, and 3 above) also hold for semi-arid and arid regions adjacent to Mediterranean climate zones. The crucial difference between the

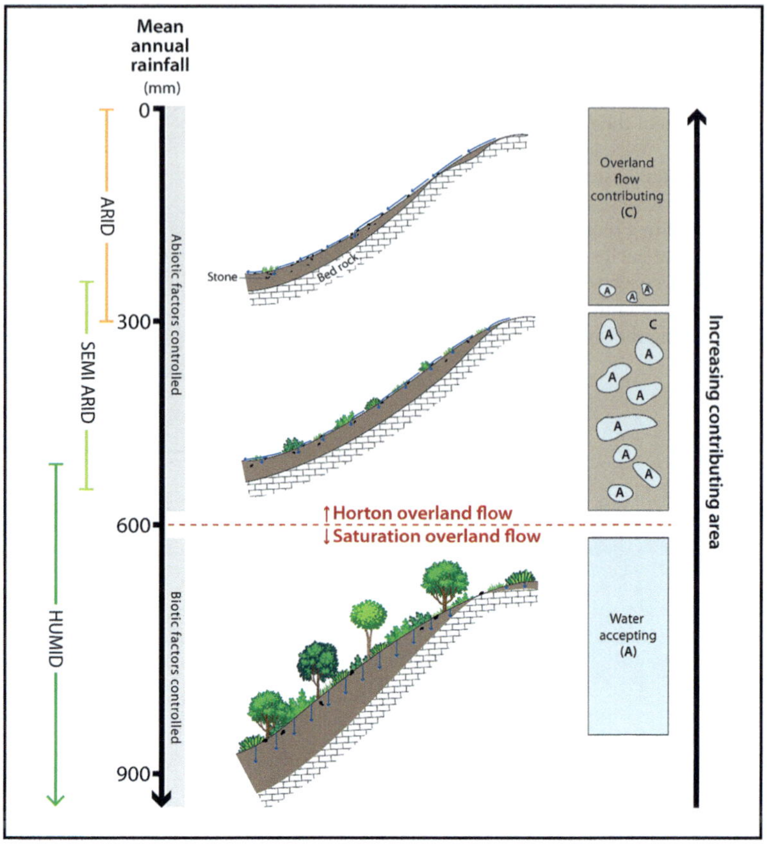

Fig. 12.16 Water redistribution processes and spatial functional patterns under different eco-climatic conditions, from arid through semi-arid to humid zones (A = water-accepting area; C = water contributing area)

Mediterranean and adjacent arid climate zones is therefore the mean annual rainfall.

Regarding the expected climate change, there is a general agreement that higher temperatures are expected, especially over most continental areas which are now already hot and dry. However, the scenarios for changes in precipitation vary from one model to another and are less reliable (Wigley 1992; Palutikof et al. 1996; Palutikof 2003). Nevertheless, Palutikof et al. (1999) claimed that some of the existing drier parts of Spain, France, Israel, and North Africa will see a decline in winter precipitation, with the most extreme cases occurring in the areas that are already more arid. Philandras et al. (2011) found, within the east Mediterranean regions, trends of significant decrease in total annual rainfall amount and in the annual number of rainy days during the period 1901–2009.

To forecast the potential impact of climate change on geomorphological processes and desertification, it is not enough to assume that the annual rainfall amount and annual number of rain days will decrease but also to consider changes in other rainfall characteristics such as rainfall intensities and extreme events. A logical future scenario is that along with increasing temperature and decreasing annual rainfall amount, the rainfall characteristics—duration and temporal distribution of rain events and of rain intensity—will change similarly to the present changes in rainfall that occur when shifting from the Mediterranean climate zone, which is characterized by relatively low temperatures and high rainfall amount, to the semi-arid and arid zones that have higher temperatures and lower rainfall amount.

Figure 12.17 shows the effect of decreasing annual rainfall, especially when combined with increasing intervals between rain events and increasing temperature, on eco-geomorphic processes and spatial patterns of water redistribution. Such a change will lead to reduced availability of water for vegetation germination and growth and for microbial activity (Thornes 1985); the organic matter content in the soil will decrease, and soluble salt concentration will increase, mainly due to an increase in sodium content (Imeson et al. 1982). These are rapid responses. In the long term, the clay content in the soil will decrease. That will affect aggregation processes so that aggregates size and stability will decrease (Reid and Goss 1981; Tisdall and Oades 1982; Rengasamy and Olsson 1991). All of this will lead to lower water-holding capacity, lower permeability, higher probability of crust formation (Farres 1978), and thus to a dramatic decrease in infiltration rates (Dunne et al. 1991; Lavee et al. 1991). The result will be less water in the root zone and more water moving as overland flow and eroding the most fertile topsoil layer (Bryan et al. 1984; Imeson 1986; Kirkby 1987; Thornes 1990). Consequently, the

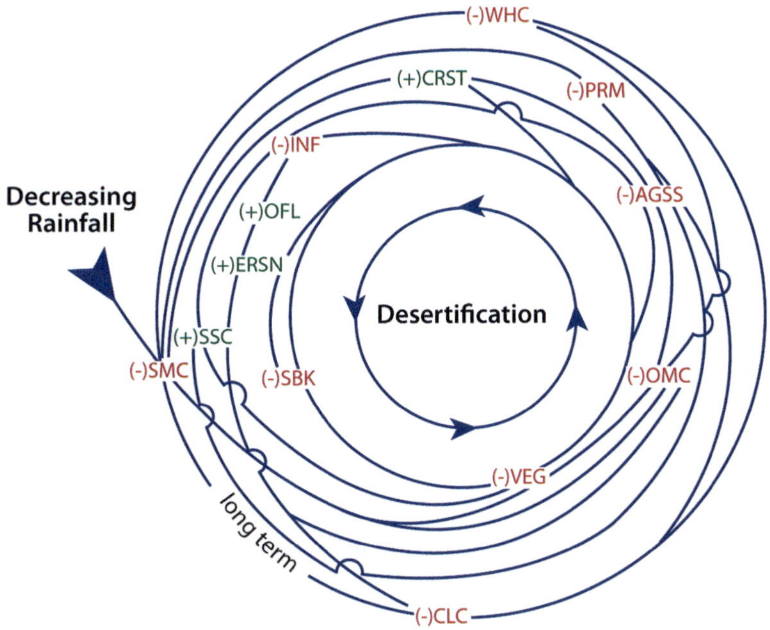

Fig. 12.17 Inter-relationships within an eco-geomorphological system in response to climate change. The lines connect between variables/processes that have direct relationships (AGSS = aggregate size and stability; CLC = clay content; CRST = crusting; ERSN = erosion; INF = infiltration; OFL = overland flow; OMC = organic matter content; PRM = permeability; SBK = seed bank; SMC = soil moisture content; SSC = soluble salt concentration; VEG = vegetation; WHC = water-holding capacity)

seedbank and nutrients content will diminish, and a second cycle of decreasing soil organic matter content will start. This positive feedback leads to desertification.

12.4.2 Possible Landscape Changes in the Judean Desert

At present, along with the temperature increase eastward of the Judean Mountains, the annual rainfall amount and the duration of rainfall events decrease, while the intervals between rainfall events increase. Sharon and Kutiel (1986) concluded that, in Israel, the percentage of the total annual rainfall, falling during peaks > 30 mm h^{-1}, also increases with aridity. Under these conditions and assuming a climate change in the direction mentioned above, the area between the Judean Mountains and the Dead Sea is an ideal field laboratory for evaluating the effect of such climate change on geomorphological processes, landforms, and desertification.

It was shown that regarding the soil properties and the geomorphic processes a threshold exists at the semi-arid zone which sharply separates the Mediterranean climate zone from the arid zones. This means that even a relatively small climatic change is enough for shifting the border between the Mediterranean and the arid eco-geomorphic systems. As many Mediterranean climate regions and semi-arid areas lie adjacent to arid ones, they are threatened by desertification.

As climate change leads to a decrease in available water in the eco-geomorphic system, the spatial distribution of surface cover components might be changed in the following ways.

The Mediterranean climate systems will lose their trees and some shrubs, and only the most resilient shrubs will survive. Mechanical and biogenic crusts will appear at the developing open bare soil patches. These patches will function as water and sediment contributors, and the spatial pattern will become similar to that of the semi-arid area. Increasing human pressure in response to resources degradation will accelerate this process.

In semi-arid areas, the size of the accepting patches will decrease due to both less direct rainfall on these patches and less water supply from the adjacent contributing patches. The plants at the lower part of the accepting patches will die. Such a mechanism will cause a gradual increase in the size of the contributing patches and the disappearance of all shrubs from the drier upper part of the hillslopes. As the erosion rate increases, more stones and bedrock outcrops will appear at the surface. Most, if not all, of the hillslope area will become a contributing area. In such a way, the semi-arid systems will become true arid ones.

In conclusion, the absolute size of each contributing unit as well as the ratio of contributing/accepting area is predicted to increase. Changes in these parameters together with a change in the spatial patterns of contributing versus accepting water areas can be used as indicators of desertification. Moreover, the results are of great relevance for rehabilitation strategies as they illustrate how valuable is the vegetated-bare soil mosaic in trapping soil moisture at the hillslope. Triggering the regeneration of appropriate patterns in the semi-arid zone should be preferred as a more sustainable option than treating areas as homogeneous landscape units.

Acknowledgements The author thanks John Wiley & Sons for their permission to use parts of the article 'The impact of climate change on geomorphology and desertification along a Mediterranean-Arid transect' published in the Journal of Land Degradation and Development (1998) 9: 407-422.

References

Bryan RB, Imeson AC, Campbell IA (1984) Solute release and sediment entrainment on micro catchments in the Dinosaur Park Badlands, Alberta. Can J Hydrol 71:79–106

Dunne T, Black RD (1970) Partial area contribution to storm runoff in a small New-England watershed. Water Resour Res 6:1296–1311

Dunne T, Zhang W, Aubury BF (1991) Effects of rainfall, vegetation and microtopography on infiltration and runoff. Water Resour Res 27:2271–2285

Farres P (1978) The role of time and aggregate size in the crusting process. Earth Surf Proc 3:243–254

Imeson AC (1986) An eco-geomorphological approach to the soil degradation and erosion problem. In: Fantechi R, Magaris NS (eds) Desertification in Europe. Reidel, Dordrecht, pp 110–125

Imeson AC, Kwaad FJPM, Verstraten JM (1982) The relationship of soil physical and chemical properties to the development of Badlands in Morocco. In: Bryan RB, Yair A (eds) Badlands geomorphology. GeoBooks, Elsevier, Norwich, pp 44–70

Jarmer T, Hill J, Lavee H, Sarah P (2010) Mapping topsoil organic carbon in non-agricultural semi-arid and arid ecosystems of Israel. Photogramm Eng Remote Sens 75:85–94

Katra I, Blumberg DG, Lavee H, Sarah P (2007) Topsoil moisture patterns on arid hillsides—Micro-scale mapping by thermal infrared images. J Hydrol 334:359–367

Kirkby MJ (1987) Modelling some influences of soil erosion, landslides and valley gradient drainage density and hollow development. Catena Suppl 10:1–14

Kirkby MJ, Chorley RJ (1967) Throughflow, overland flow and erosion. Bull Int Assoc Sci Hydrol 12:5–21

Köppen W (1936) Das geographisca System der Klimate. In: Köppen W, Geiger R (eds) Handbuch der Klimatologie. Berlin, Band I, Teil C, pp 1–44

Kutiel H (1985) The multimodality of the rainfall course in Israel as reflected by the distribution of dry spell. Archive Für Meteorologie Geophysik and Bioklimatologie 36B:15–27

Kutiel P, Lavee H (1999) Effect of slope aspect on soil and vegetation properties along an aridity transect. Israel J Plant Sci 47:169–178

Kutiel P, Kutiel H, Lavee H (2000) Vegetation response to possible scenarios of rainfall variations along a Mediterranean—extreme arid climatic transect. J Arid Environ 44:277–290

Kutiel P, Lavee H, Shoshany M (1995) Influence of a climatic gradient upon vegetation dynamics along a Mediterranean-arid transect. J Biogeogr 22:1065–1071

Lavee H, Yair A (1990) Spatial variability of overland flow in a small arid basin. IAHS 189:105–120

Lavee H, Poesen JWA (1991) Overland flow generation and continuity on stone-covered soil surfaces. Hydrol Process 5:345–360

Lavee H, Imeson AC, Pariente S, Benyamini Y (1991) The response of soils to simulated rainfall along a climatological gradient in an arid and semi-arid region. Catena Suppl 19:19–37

Lavee H, Sarah P, Imeson AC (1996) Aggregate stability dynamics as affected by soil temperature and moisture regimes, Geografiska Annaler 78A(1):73–82

Lavee H, Imeson AC, Sarah P (1998) The impact of climate change on geomorphology and desertification along a Mediterranean-Arid transect. J Land Degrad Dev 9:407–422

Montana C (1992) The colonization of bare areas in two-phase mosaics of an arid ecosystem. J Ecol 80:315–327

Palutikof JP (2003) Analysis of mediterranean climate data: measured and modelled. In: Bolle HJ (ed) Mediterranean climate, regional climate studies. Springer, Berlin, pp 125–132

Palutikof JP, Conte M, Casimiro Mendes J, Goodess CM, Espirito Santo F (1996) Climate and climatic change. In: Brandt CJ, Thornes JB (eds) Mediterranean desertification and land use. Wiley, Chichester, pp 43–86

Palutikof JP, Trigo RM, Adcock ST (1999) Scenarios of future rainfall over the Mediterranean: Is the region drying? In: Balabanis P, Peter D, Ghazi A, Tsogas M (eds) Mediterranean desertification—research results and policy implications. Proceedings of the international conference, vol 1, 29 October to 1 November, Crete, Greece

Philandras CM, Nastos PT, Kapsomenakis J, Douvis KC, Tselioudis G, Zerefos CS (2011) Long term precipitation trends and variability within the Mediterranean region. Nat Hazard 11:3235–3250

Reid JB, Goss MJ (1981) Effect of living roots of different plant species on the aggregate stability of two arable soils. J Soil Sci 32:521–541

Rengasamy P, Olsson KA (1991) Sodicity and soil structure. Aust J Soil Res 29:935–952

Sanchez G, Puigdefabregas J (1994) Interactions of plant growth and sediment movement on slopes in a semi-arid environment. Geomorphology 9:243–260

Schlesinger WH, Reynolds JF, Cunningham GL, Huenneke LF, Jarrell WM, Virginia RA, Whitford WG (1990) Biological feedbacks in global desertification. Science 247:1043–1048

Sharon D, Kutiel H (1986) The distribution of rainfall intensity in Israel, its regional and seasonal variations and its climatological evaluation. J Climatol 6:277–291

Shnerb NM, Sarah P, Lavee H, Solomon S (2003) Reactive glass and vegetation pattern. Phys Rev Lett 90(3):038101

Shoshany M, Lavee H, Kutiel P (1995) Seasonal vegetation cover changes as an indicator of soil types along a climatological gradient: a mutual study of environmental patterns and controls using remote sensing. Int J Remote Sens 16(12):2137–2151

Steinberger Y, Lavee H, Barness G, Davidor M (1999) Soil Carbohydrates along a topo climatic gradient in a Judean desert ecosystem. Land Degrad Dev 10(6):523–530

Thornes JB (1985) The ecology of erosion. Geography 70:222–235

Thornes JB (1990) The interaction of erosional and vegetation dynamics in land degradation: spatial outcomes. In: Thornes JB (ed) Vegetation and erosion. Wiley, New York, pp 41–54

Tisdall JM, Oades JM (1982) Organic matter and water stable aggregates in soils. J Soil Sci 33:141–163

Tongway DJ, Ludwig JA (1990) Vegetation and soil patterning in semi-arid mulga lands of Eastern Australia. Aust J Ecol 15:23–34

Tzvikel S, Lavee H, Sarah P (2007) Temporal evolution of salts in Mediterranean soils transect under different climatic conditions. CATENA 70:282–295

Whipkey RZ (1965) Subsurface stormflow on forested slopes. Hydrol Sci Bull 10:74–85

Wieder M, Lavee H (1990) Micromorphological characteristics induced by subsurface water flow in the Judean Desert. Dev Soil Sci 19:235–243

Wigley TML (1992) Future climate of the Mediterranean basin with particular emphasis on changes in precipitation. In: Jeftic L, Milliman JD, Sentini G (eds) Climatic change and the mediterranean. Arnold, London, pp 15–44

Yair A, Lavee H (1985) Runoff generation in arid and semi-arid zones. In: Anderson MG, Burt TP (eds) Hydrological forecasting. Wiley, Chichester, pp 183–220

Karst Landforms Along the Backbone Hills of Central Israel

13

Amos Frumkin

Abstract

Selected examples of karst features and landscapes are discussed, demonstrating the karst landforms origin and development of the mountainous backbone of Israel. These range across various scales, from mega-landscape features, such as polje and epigenic karst basin, through rounded hills, caves, and dolines, to rock-pillar karren, mostly under fluviokarst conditions. Considering the varied features, it is deduced that karst dissolution has played a major role in landscape development since the late Eocene regression.

Keywords

Epigenic karst · Cave · Karst drainage · Denudation · Fluviokarst

13.1 Introduction

The backbone hills of central Israel comprise a thick (~ 1500 m) sequence of carbonate rocks of late Cretaceous to Eocene age (Sass and Bein 1982; Segev et al. 2011). Non-carbonate intercalations such as clays constrain karstification by blocking the hydrologically conductive system of fractures and bedding planes. The present water-divide region, lying between the Mediterranean and Jordan valley base levels, has been exposed above the sea level since the late Eocene or Oligocene, when the Neotethys–Mediterranean Sea started retreating (e.g., Gvirtzman et al. 2011).

Rising today up to ~ 1000 m.a.s.l., the region displays various karst landforms, of which only a small sample can be presented in the current chapter. The selected karst landforms (Fig. 13.1a) highlight the region's long history of dynamic karstification, combined with other geomorphic and tectonic processes over diversified spatial conditions.

13.2 Long-Term Landform Evolution Context

The backbone hills encompass the Anabta and Ramallah-Faria anticlines (outcrops of mostly Cretaceous rocks) and Shechem syncline (outcrops of Eocene carbonates) (Fig. 13.1b). These Syrian-Arc folds were truncated by a planation surface during the Oligocene–Early Miocene times, followed by uplift and arching during the Plio-Pleistocene. Consequently, a broad structural arch (40–60 km wide) developed between the Mediterranean and the Dead Sea–Jordan valley with an amplitude of ~ 400 m in central Israel (Wdowinski and Zilberman 1997; Segev et al. 2006). The structural arching was induced by the subsidence of the Dead Sea rift on the east and the Nile River sediment loading in the Mediterranean Sea on the west that has occurred since the Miocene (Steinberg et al. 2014). Some old faults and flexures, previously participating in the Syrian-Arc folding, were rejuvenated during the arching, such as the eastern and western flexures of Faria (Mimran 1984) and Hevron (Reches et al. 1981) anticlines, respectively. The relief was further increased by faulting and subsidence of the Dead Sea rift valley and associated Samaria grabens (Shaliv et al. 1992). These enhanced the modern fluvial entrenchment and surface flow, limiting infiltration and reducing the fully-karstified terrain (i.e., area with complete drainage to the subsurface).

The water divide is irregular in plan-view and elevation (Fig. 13.1a). Closed karst basins still occur in some places (rectangles 2 and 6 in Fig. 13.1a), recalling the dominance

A. Frumkin (✉)
Institute of Earth Sciences, The Hebrew University of Jerusalem, 9190401 Jerusalem, Israel
e-mail: amos.frumkin@mail.huji.ac.il

A. Frumkin and N. Shtober-Zisu (eds.), *Landscapes and Landforms of Israel*,
World Geomorphological Landscapes, https://doi.org/10.1007/978-3-031-44764-8_13

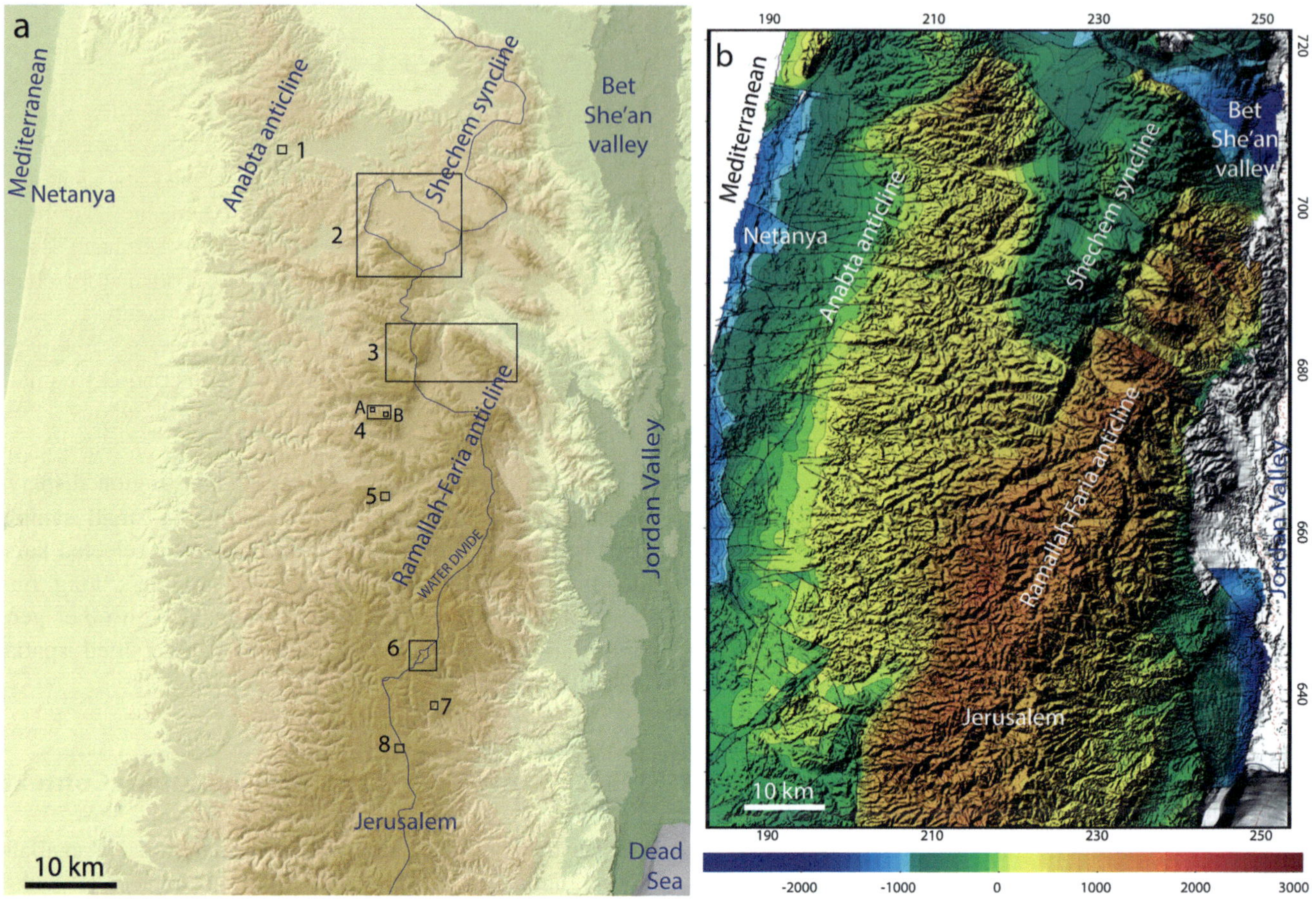

Fig. 13.1 The discussed part of the backbone hills of central Israel. **a** Topographic map showing the location of the discussed karst features (numbered rectangles) in the vicinity of the water divide (gray line); **b** structural map of top Judea Group with shaded topography, showing that the features are located across the folded Syrian Arc: anticlines (Anabta and Ramallah-Faria) and Shechem syncline. Israel Grid ITM coordinates (modified after maps by the Geophysical Institute and Geological Survey of Israel)

of karst processes under Miocene conditions of a low and flat carbonate plateau. This was transformed subsequently by tectonics and erosion which increased slope gradients and associated fluvial erosion.

While the extent of chemical weathering is widely recognized in carbonate landscapes, its relative magnitude and the role of physical erosion is not well-constrained. Ryb et al. (2014a) demonstrated the dominance of karstification relative to physical processes over the backbone hills of Israel by comparing spring solute loads and discharge measurements with denudation rates obtained from cosmogenic ^{36}Cl. They showed that dissolution rates on the carbonate hillslopes are dominant, while physical erosion is negligible. Ryb et al. (2014b) measured ^{36}Cl denudation rates along a climatic transect from the arid Dead Sea to the Mediterranean climate area of Judea and suggested a transition between chemically-dominated denudation and mechanically-dominated denudation around the isohyets of 100–200 mm/yr. The denudation rate close to the water divide was 12–23 mm/ky, and the measured denudation rates rise with the amount of precipitation. On average, the Mediterranean climate interfluves denude at a rate of 21 ± 7 mm/ky. Ben-Asher et al. (2021) measured similar denudation rates of 13–29 mm/ky in soil-covered bedrock and 14–21 mm/ky on exposed rock surfaces in the Judea hills (SW edge of the Ramallah anticline) and Mt. Gilboa (NE edge of the Shechem syncline). Similar values were found at Mt Carmel (northern Israel), where slope retreat rates range from 10^1 to 10^2 mm/ky (and 20–35 mm/ky in one particular case (Shtober-Zisu et al. 2020).

Bar et al. (2016) argued that extrapolating such recent denudation rates to the period since the exposure of the region, particularly during the planation surface development, contradicts the geological field data. Therefore, they suggested that the amount of accumulated chemical weathering is much lower than suggested by Ryb et al. (2015).

Two phases of denudation occurred in the region. The first was dominated by a low relief regime, during the Oligocene-early Miocene. Karst denudation dominated, like in present interfluves and karst plateaus. Most of the regional denudation on the structurally-elevated areas occurred during this stage. This denudation had truncated the entire region without association to the present-day precipitation. No remains of this ancient (e.g., early Miocene) alluvium have been found on the backbone planation surface in the vicinity of the water divide, unlike in the Negev Desert. As shown below, parts of the backbone hills comprise karst depressions covered by thick alluvium. Such areas must have been more common during the early development of the exposed planation surface. The denudation rate under such alluvium-covered valleys in unknown, but must have been limited due to the low permeability of the clay-rich cover. This may explain the lower denudation rates suggested by Bar et al. (2016).

The second phase of denudation occurred during the late Neogene-Quaternary, following tectonic uplift, faulting, and arching. Karst denudation remained dominant at interfluves and karst plateaus, but their area decreased dramatically due to fluvial entrenchment. The significance of fluvial and slope erosion processes increased especially in fluvial channels and along steep slopes.

13.3 Rounded Convex Hills

A typical landscape in the entire discussed region comprises rounded convex hills, whose diameter may range from hundreds of metres to kilometers, and height may range from

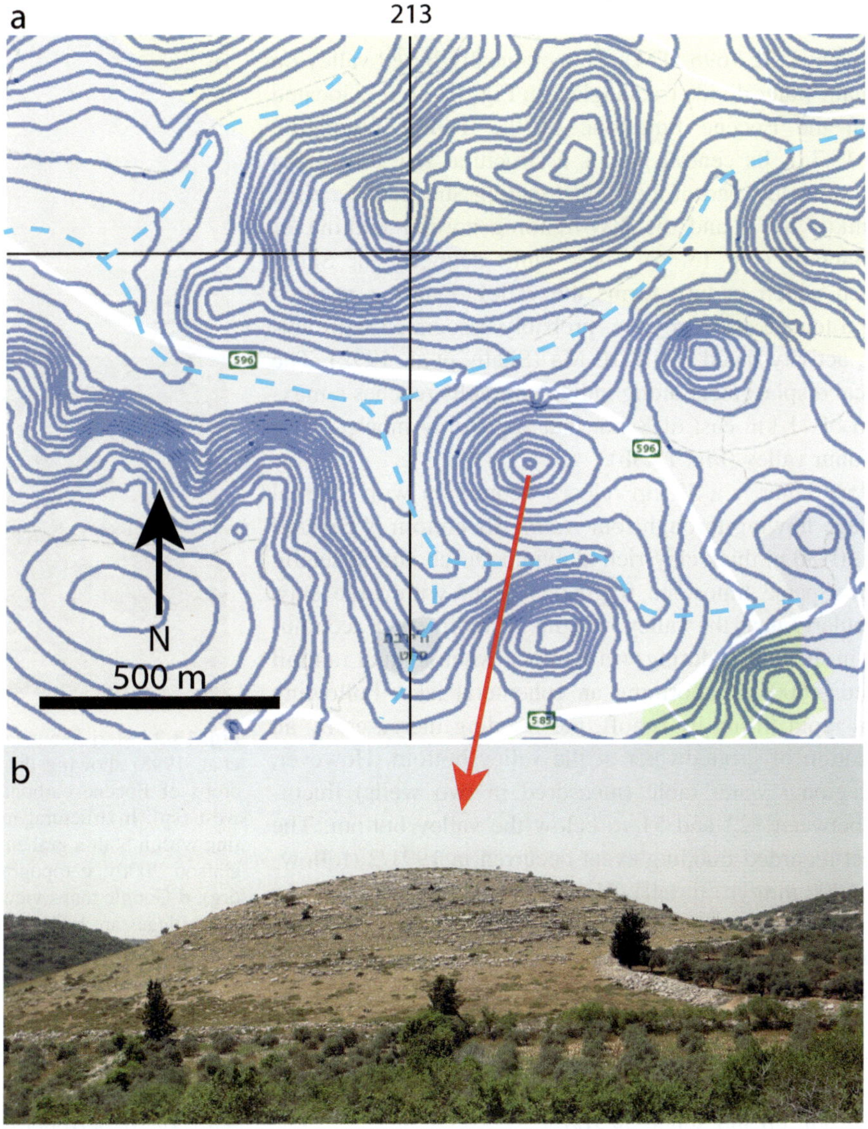

Fig. 13.2 Examples of rounded convex hills on late Cretaceous carbonates, northern Samaria (site 1 in Fig. 13.1a). **a** topographic map (5 m interval contours); **b** profile of a typical hill (view westward)

tens to few hundred metres. These are commonly separated from one another by entrenched fluvial courses (*wadis*) and in some cases by tectonic scarps. Their form is age-dependent, reaching rounded profiles after several million years of karstification and slope erosion. The equilibrium rounded shape (Fig. 13.2, 213,280/704,530 Israel Transverse Mercator grid [ITM]) is associated with karst and slope processes, depending on the slope angle. Carbonate dissolution dominates the denudation at the hilltops (Ryb et al. 2014a, b), while the slopes are also affected by fluvial entrenchment, tectonics, and mass movement. A complex relationship between hilltop curvature, soil production, and climate was observed by Ben-Asher et al. (2021). The detailed slope profiles depend on various agents, e.g., the lower slope can be planar or convex if controlled by a fault or rapidly entrenching stream, whereas vertical cliffs develop on hard, massive layers.

13.4 Marj Sanur Polje

Marj Sanur (227/696 ITM) is a tectonically-based valley on the main water divide (rectangle 2 in Fig. 13.1). It is located within the Eocene limestone of the Timrat Formation (Fig. 13.3a). Its central part is a structural low associated with both a low part of the NNE-trending Shechem syncline (Frumkin 2017) and the NW-trending normal faulting of NE Samaria (Fig. 13.3b). These faults delineate the Sanur Graben which in turn forms the valley. Faulting resulted from a tensile stress regime, probably associated with volcanic activity dated to 16–10 Ma (Shaliv et al. 1992). The vertical displacement along individual faults reaches a maximum of ~ 1 km east of Sanur, but apparently much less in the Sanur valley (Fig. 13.3b).

Marj Sanur is a 55 km^2 closed depression with internal drainage flowing through four *wadis*. Its bottom is covered with 10–30 m thick clay-rich alluvium (Ju'ub and Schetelig 2002). Sparse limestone fragments occur within the clay, particularly near the valley margins. Runoff water accumulates in the valley during wet winters, when annual rainfall exceeds ~ 600 mm, forming an ephemeral lake. Following initial rapid filling by runoff, the flooding increases by an emanation of groundwater at the valley bottom. However, the regional water table (measured in two wells) fluctuates between 123 and 51 m below the valley bottom. The largest recorded flooding event occurred in 1991/2 (following > 1000 mm/yr rainfall). Water depth in the mid-lake was estimated as ~ 6 m, and its area exceeded 10 km^2. Unlike in average years when the valley dries in summer, in 1991/2 the valley remained flooded until winter 1992/3, when the water level rose again. During this event the farmers lost not only their winter crops, but also their summer crops. Roads crossing the valley margins were inundated for considerable periods (Ju'ub and Schetelig 2002).

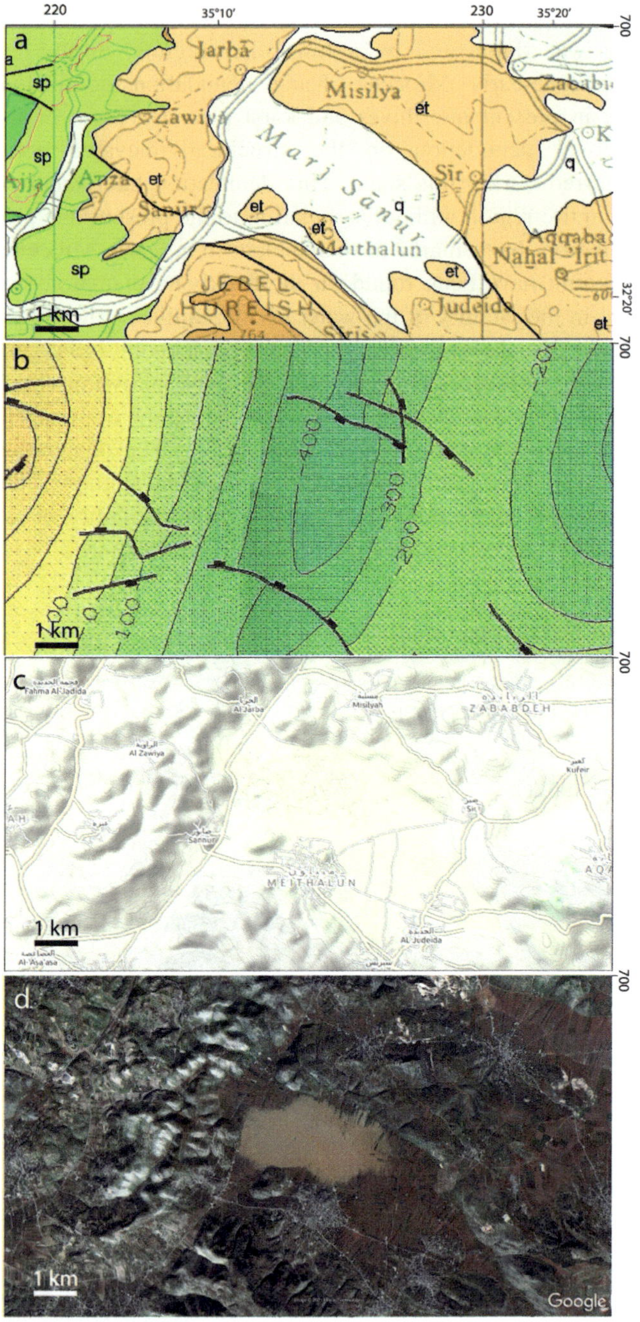

Fig. 13.3 Marj Sanur Polje. **a** excerpt from a geological map (Sneh et al. 1998) showing the clay-covered bottom (q) valley within outcrops of Eocene carbonates (et). Chalks and marls crop out at the west (sp); **b** structural map of the same area showing the polje setting within both a graben and a syncline (modified after Fleischer and Gafsou 2000); **c** topographic map (from http://www.opencyclemap.org). **d** Google maps view of the inundated polje (January 2013). Note that villages are built on higher grounds, avoiding the inundation hazard, thus the water affects mainly the agriculture

The geomorphic term 'karst polje', the largest type of karst depression, refers to large (> 1 km) closed depressions with conspicuously flat bottoms developed on karst rocks with steep peripheral slopes. The angle between the side slope and the polje bottom is clearly demarcated, and the flat floor meets the surrounding slopes with a sharp break. The flat floor may have an intermittent or a perennial stream which may be liable to flood and becomes a lake (Bonacci 2013; Kranjc 2013). All these attributes apply to Marj Sanur (Fig. 13.3), which is thus classified as a karst polje.

13.5 Regional Planation Surface Manifested in Central Samaria

Central Samaria, comprising Kabir (231/683 ITM) and Ebal (226/682 ITM) mountains (rectangle 3 in Fig. 13.1), presents a good view of the regional planation surface (Avni et al. 2012; Bar et al. 2016) which developed following the emergence from the Eocene sea. The planation surface cuts both synclinal and anticlinal areas (Fig. 13.4a), demonstrating the consequence of a long-term ancient denudation, accomplished mainly by karstification (Frumkin 1993). The following uplift, faulting and arching (Fig. 13.4a) triggered incision along the erodible rocks exposed between the present-day mountains. Thus, the planation surface is expressed today by the skyline of the regional backbone hills (Fig. 13.4).

13.6 Surface Collapse Associated with Hypogenic Caves

Hypogenic caves are abundant along Israel's central hills (Frumkin and Fischhendler 2005; Frumkin et al. 2017a). Here we present two features affecting subaerial collapse landforms (rectangle 4 in Fig. 13.1). The subaerial landforms appear in outcrops of non-karstic Senonian-Paleocene chalk and marl, which prevent infiltration of meteoric water. The deep collapsed voids indicate deep dissolution by rising plumes of undersaturated water within the underlying karstic limestones.

Hasif Shaft (rectangle 4A in Fig. 13.1, 221,500/675,960 ITM) demonstrates the connection of the surface to a deep, partly collapsed, chamber cave, which is accessed via ropes (Fig. 13.5c–d). The 40 m × 30 m-size chamber cave at the bottom of the shaft formed within Turonian karstic limestone (Fig. 13.5a) and developed within groundwater under confined conditions, prior to the regional uplift and the associated fall of the water table. Recently (in the last century or two according to local people), the overlying low-cohesivity Senonian chalk has collapsed, forming a vertical shaft with ~ 10 m diameter opening within a steep

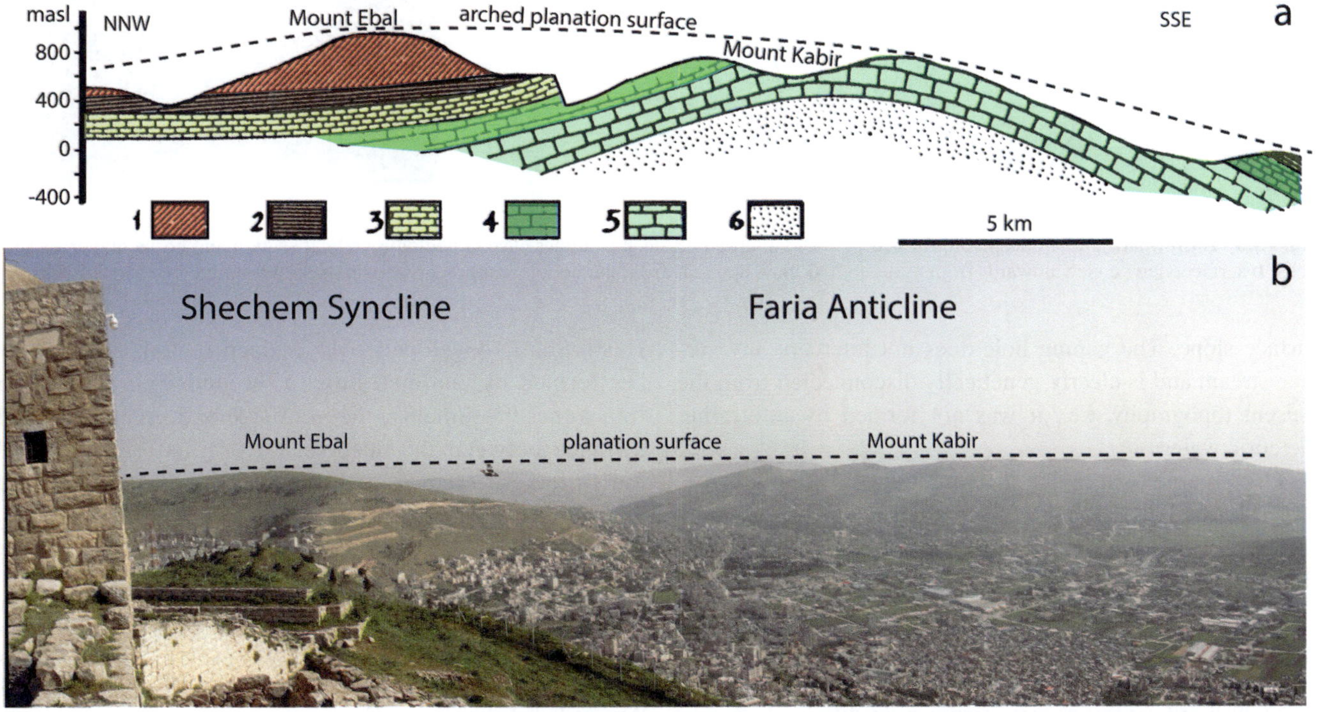

Fig. 13.4 The regional planation surface formed mainly by karstification following the late Eocene regional marine regression. Note that the planation surface truncated the then-existing Syrian-Arc structures, and that at this location deeper truncation occurred in the presently drier eastern part, as dictated by the anticlinal structure. The planation surface was subsequently arched by younger tectonics and dissected by associated erosion. Note also deep young incision into the chalk and limestone between Mt. Ebal and Kabir. a. Schematic geological section: 1—Eocene limestone, 2—Senonian-Paleocene chalk and marl, 3—Turonian limestone, 4—Cenomanian dolostone, 5—Albian carbonates, 6—Early Cretaceous, mainly sandstone. b. View northward of the planation surface from Mt. Gerizim

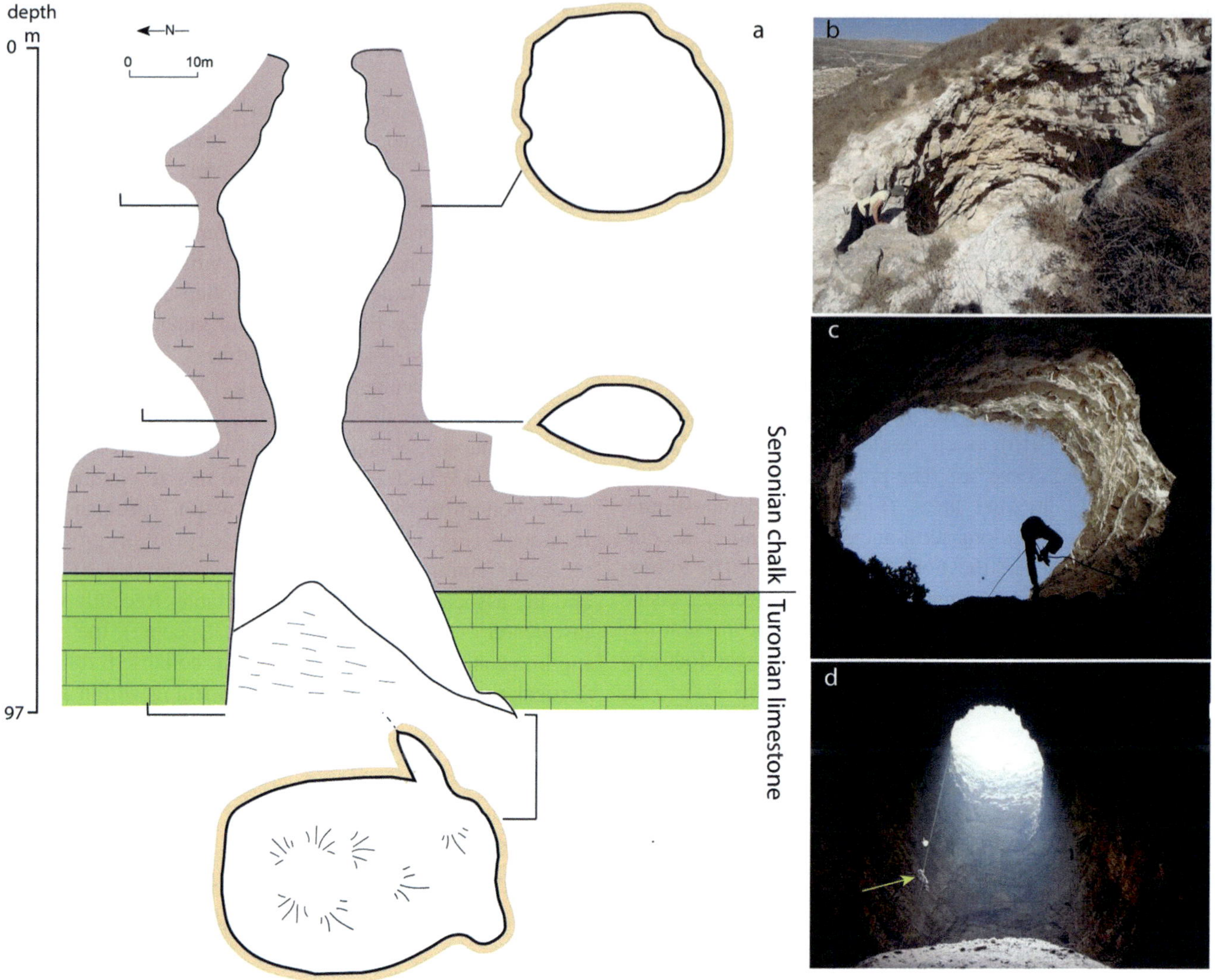

Fig. 13.5 Hasif Shaft: a 97 m deep void formed by roof collapse of a deep chamber cave. **a** geologic vertical section and horizontal cross sections; **b** surface view; **c** view upwards from mid-depth. **d** view upwards from the bottom, rappelling person (arrow) for scale

surface slope. The gaping hole does not intercept any surface stream and is clearly genetically disconnected from the present topography, i.e., it was not formed by infiltrating meteoric water.

The Jneine sinkhole (rectangle 4B in Fig. 13.1, 223,190/674,970 ITM) is a shallower and wider surface collapse depression, formed in Paleocene marls (Fig. 13.6). It is located at the confluence of two first-order streams, draining ~ 0.1 km^2 (Fig. 13.6a). The flow infiltrates at the sinkhole bottom and cannot drain subaerially. Being easily accessible, the sinkhole is cultivated as an orchard, enjoying preferable humidity and protection (Fig. 13.6b) in a generally unfavorable environment. The surface outcrops are not karstified, but some flowstone calcite deposits are observed, possibly reaching the sinkhole by mass movement from above (Fig. 13.6c). It is presumed that this sinkhole, similar to the nearby Hasif Shaft, developed over a deep-seated cave chamber formed by confined flow in the underlying Turonian limestone. Presumably, the void size and greater vertical distance between the deep-seated original chamber and the surface resulted in a collapse doline rather than in a deep shaft.

13.7 Nehemya Stone Pillars

Stone pillars in the massive Cenomanian dolostone of Aminadav Formation (Fig. 13.8c) reach a height of six metres (rectangle 5 in Fig. 13.1, 221,633/666,706 ITM), rendering them one of the finest examples of this karren type in central Israel (Fig. 13.7). They demonstrate a long-term vertical dissection and differential denudation of a dolomitized and partly silicified fossil rudist reef.

Fig. 13.6 Jneine sinkhole. **a** the collapsed sinkhole is seen as a dark-green rectangular-like area in the center of an orthophoto-topographic map (5 m contour interval). Blue arrows indicate first-order streams captured by the sinkhole; **b** side view of the sinkhole; **c** speleothem overgrowth on a limestone block found within the sinkhole

Fracture control is important for stone pillars, which may have initially developed under a soil cover. Indeed, many types of karren fields including grikes, clints, cutters, and pinnacles form initially under a soil cover (Lundberg 2013). The Nehemya stone pillars seem to be influenced by sets of fractures that supported the accumulation of dust, terra rossa soil and vegetation, which allowed for increased water aggressiveness and rapid dissolution along the discontinuity planes (Ford and Williams 2007). Positive feedback mechanism promotes the deepening of the fracture-controlled depressions, while the intervening pillars undergo slower dissolution. The sloping surface allowed for soil erosion when vegetation was eliminated following climatic deterioration.

Stone pillars are generally arranged in networks of various densities and shapes. Their walls are steep, vertical, or overhanging, and their surfaces reflect the manner of their formation and development (Knez and Slabe 2013). Their formation process can vary, but their very existence implies long-term moderate climate, because under freezing or arid climates mechanical weathering precludes their development by disintegrating the rock. Nevertheless, bedding planes induce instability and collapse of the higher pillars (Fig. 13.7b), and thus, the stratal thickness controls the pillar's maximal height.

13.8 Ofra Karst Plateau

The Ofra karst plateau (rectangle 6 in Fig. 13.1, 225/651 ITM) is a 5 km^2 endorheic basin (Frumkin et al. 2017b). The exposed Cenomanian carbonate Aminadav Formation

Fig. 13.7 Nehemya stone pillars, with a person for scale. **a** general view of the pillar cluster. **b** a pillar in close-up, showing porous dolostone rock with a top part disintegrating along a bedding plane, suggesting imminent collapse

(Fig. 13.8) is massive in the upper part and well-bedded to laminar in the lower part of the unit. The overlying and underlying formations are rich in chalk and marls, lacking karst features (Fig. 13.8c). The Aminadav epikarst comprises a few metres of disintegrated rocks, partly covered by terra rossa soil which also fills deep pockets (Fig. 13.9a, b).

The Ofra plateau demonstrates the entire set of features of active epigenic karstification, including karren fields (Fig. 13.9a), stone pillars (Fig. 13.9e), solution and collapse dolines (Fig. 13.9c), and sinking ephemeral streams, capturing rare overland flow events. Vadose caves are common, having mainly vertical shafts (Figs. 13.8d, and 13.9f) and some short sub-horizontal passages decorated with speleothems (Fig. 13.9d). The deeper dolines are associated with collapse or subsidence of epikarstic disintegrated bedrock above vadose shafts. This karst plateau is still actively evolving under the dry Holocene climate, as indicated by unsaturated water with increasing solute content along the flow route measured on pitted shaft walls (Fig. 13.9g).

Fifty-five vadose caves are known in the plateau as of 2023. They typically consist of a series of vertical shafts with less common short sub-horizontal, canyon-like sections (Fig. 13.9f), reaching a maximal depth of 103 m. The caves include active dissolution features such as scallops (Fig. 13.9g), as well as speleothems (Fig. 13.9d), indicating a wide range of solute content in the water, associated with changes in aggressivity. In one case a vadose cave clearly dissects a maze cave, probably of older hypogenic origin. Other cases of ancient phreatic or hypogenic cavities are less distinctive in the Ofra plateau.

13.9 Migron Cliff Inland Notches

The inland notches at Migron (rectangle 7 in Fig. 13.1, 225,330/644,180 ITM) are some of the best examples of inland cliff notches in Israel. Inland notches are sub-horizontal indentations shaped like half tubes, developed on the carbonate cliffs of the Mediterranean to semi-arid zones (Shtober-Zisu et al. 2017). The Migron notches extend over tens of metres along the eastern bank of a stream valley in the Cenomanian carbonate reef dolostone of Weradim Formation (Fig. 13.8c). They form due to slight differences (1–15%) in the porosity of the visor and the cavity bedrock layers. The cavity layer is more porous and therefore more prone to dissolution, exfoliation, and faster retreat (Shtober-Zisu et al. 2015). The presence of bacteria, algae, and other lithobionts suggests that biologic weathering is active too, facilitated by the shaded and relatively moist notch environment. Slower retreat of the visor bed occurs up to a critical point where the visor collapses. The Migron notches are clearly associated with bedrock properties. For example, at Migron the bedrock was faulted, and the notches layer was cut vertically. The notches follow the porous layer across the fault, indicating the control of bedrock properties (Fig. 13.10a). Where the bedrock layer diminishes in thickness within a fore-reef inclined and truncated setting, the notch size diminishes according to the lateral thinning of the layer (Fig. 13.10b). Similar inland notches have been dated by Shtober-Zisu et al. (2020) to the late Quaternary. Young notches are still beginning to form today within appropriate layers, following the slope denudation (Fig. 13.10c).

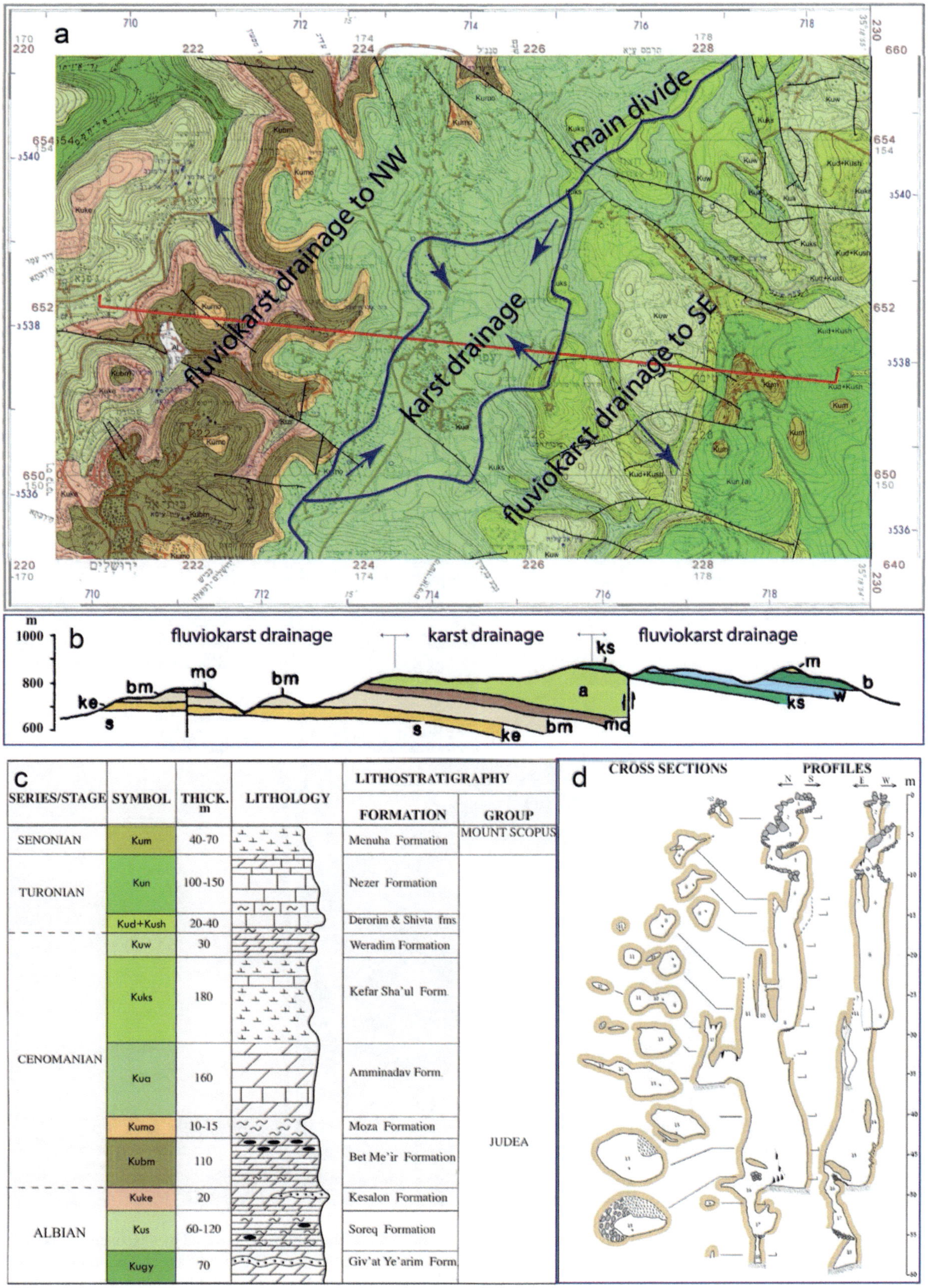

Fig. 13.8 Ofra karst plateau. **a** geological map showing the area with internal karst drainage on the main water divide, between the Mediterranean (left) and Dead Sea (right) fluviokarst catchments (modified after Shachnai 2000). **b** geological section (along the red line indicated in **a**). The Ku prefix of the formation names is omitted. **c** geological columnar section of the region, showing the exposed late Cretaceous formations (Shachnai 2000). **d** cross sections and profiles of Ha-sini Cave, a typical vadose shaft system in the Ofra karst plateau

Fig. 13.9 Landscapes and landforms of Ofra karst plateau. **a** vertical drone view of a karren field, partly covered by vegetated soil pockets. Person for scale (arrow). **b** one-metre-wide section through soil pockets with subsurface karren. **c** surface view of the karren field with a small collapse doline overlying Ha-sini Cave. Person for scale (arrow). **d** both saturated and unsaturated runoff water reach 50 m below surface, forming speleothems and dissolution microkarren side by side, Nof-Amona Cave. **e** stone pillar perforated by a horizontal passage with a three years old girl for scale. **f** Vadose shaft, presently dry and pitted by corrosion, accommodating an aggressive waterfall during the winter. **g** micro-dissolution scallops and flutes, 25 m below surface, Ha-sini Cave. Aggressive super-critical film flow was documented here in the winter

Fig. 13.10 Inland notches in late Cretaceous reef dolostone (Weradim Formation, Fig. 13.8c). Persons for scale. **a** double notch developed in a previously faulted layer. The two notches follow the offset erosion-prone layers. **b** truncated fore-reef dipping strata etched by a notch, decreasing in size to the right as dictated by the decreasing thickness of the layer. **c** young notch at the earliest stage of development in the foreground, and in a more mature stage in the background, within the same layer. The difference is attributed to the younger age of the foreground slope

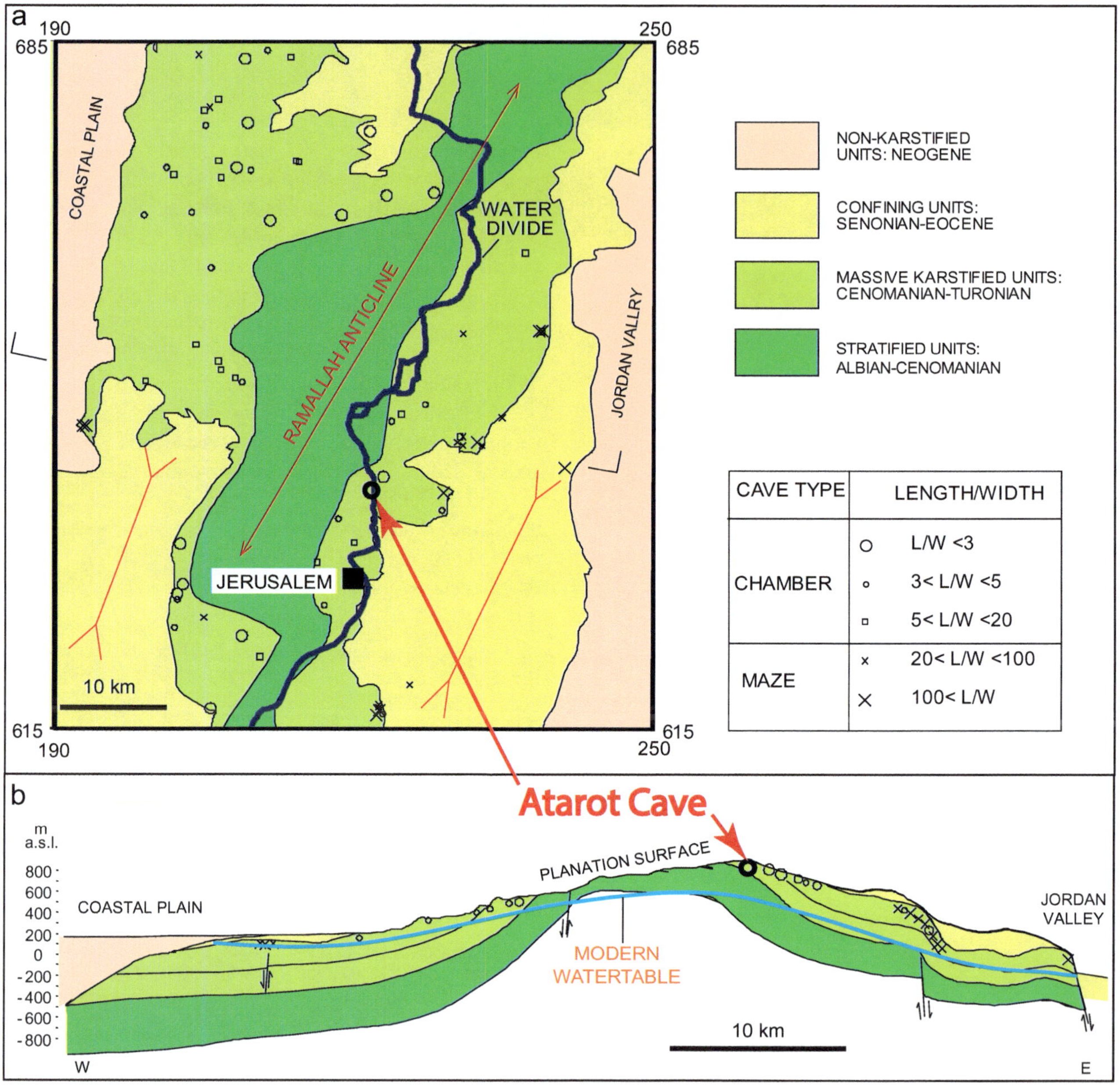

Fig. 13.11 Chamber and maze caves distribution in the southern Samaria and Jerusalem hills. Blue line (in a) is the water divide. ITM coordinates. **a** generalized geological map. **b** generalized geological section (location indicated on **a**). Atarot Cave location is projected

13.10 Atarot Cave

Chamber caves are common at the backbone and in the western Samaria hills, while maze caves are common in the confined lower areas (Frumkin and Fischhendler 2005). Their geometry is indicated by small and large length/width ratio, respectively (Fig. 13.11). The Atarot Cave (rectangle 8 in Fig. 13.1, 220,855/639,827) is the largest chamber cave in Israel, measuring 230 m × 140 m in area (Fig. 13.12). The cave is formed within the Cenomanian Weradim (dolostone) Formation (Fig. 13.8c). Its location at the present water divide, coupled with the large size and the need for a long duration of groundwater flow for speleogenesis, indicates that it formed under an ancient setting, probably during the Oligocene to early Miocene, as suggested for hypogenic caves further to the east (Frumkin et al. 2017a).

Following regional uplift, arching, and water-table fall, the cave shifted to the unsaturated (vadose) zone at the modern water divide. A continuous dripping of infiltration water supports rich speleothem deposition, rendering

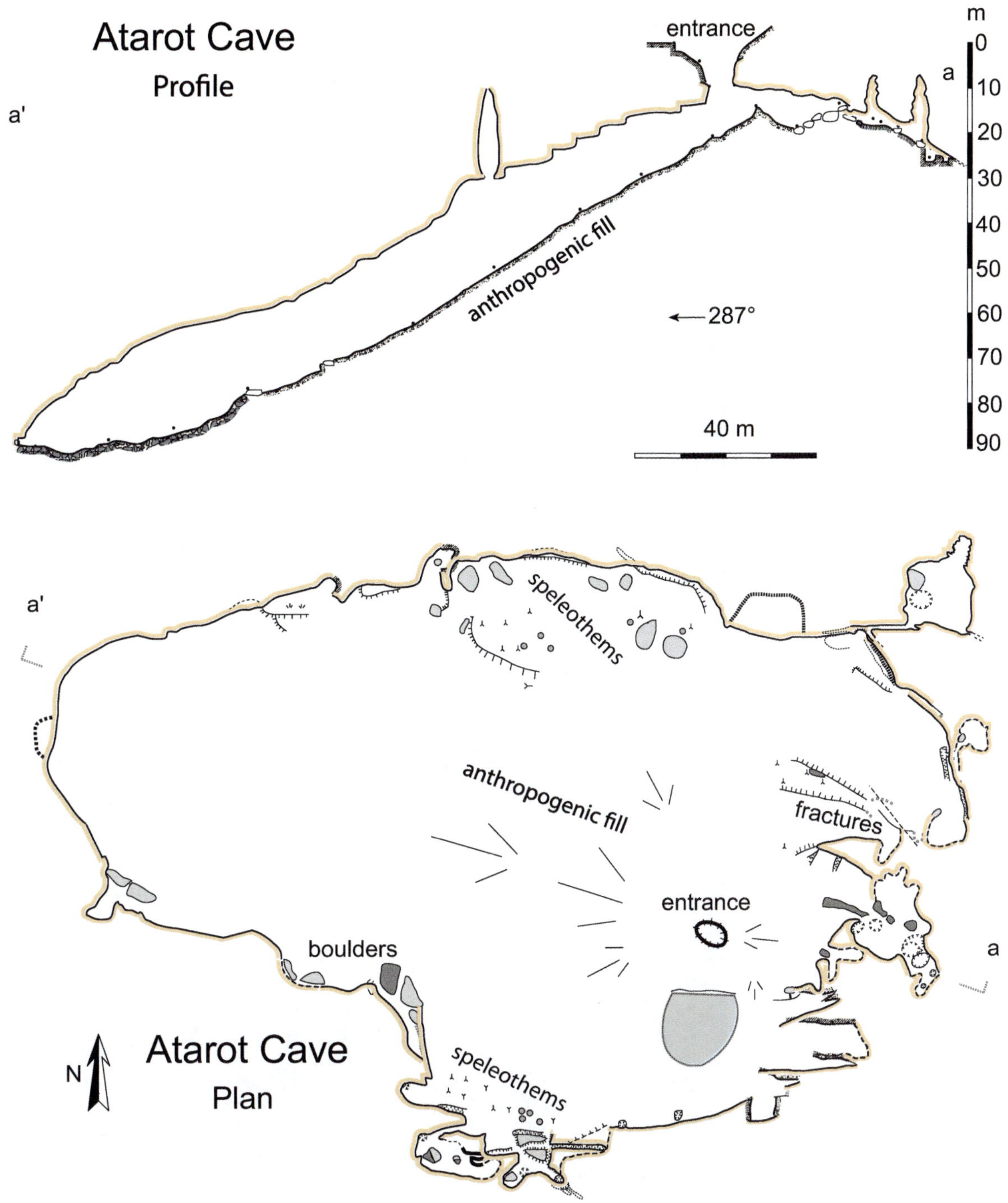

Fig. 13.12 Profile and plan of the Atarot Cave, the largest chamber cave in Israel (surveyed by the Israel Cave Research Center). Speleothems are well-preserved at the northern and southern edges, while the center is covered by anthropogenic fill

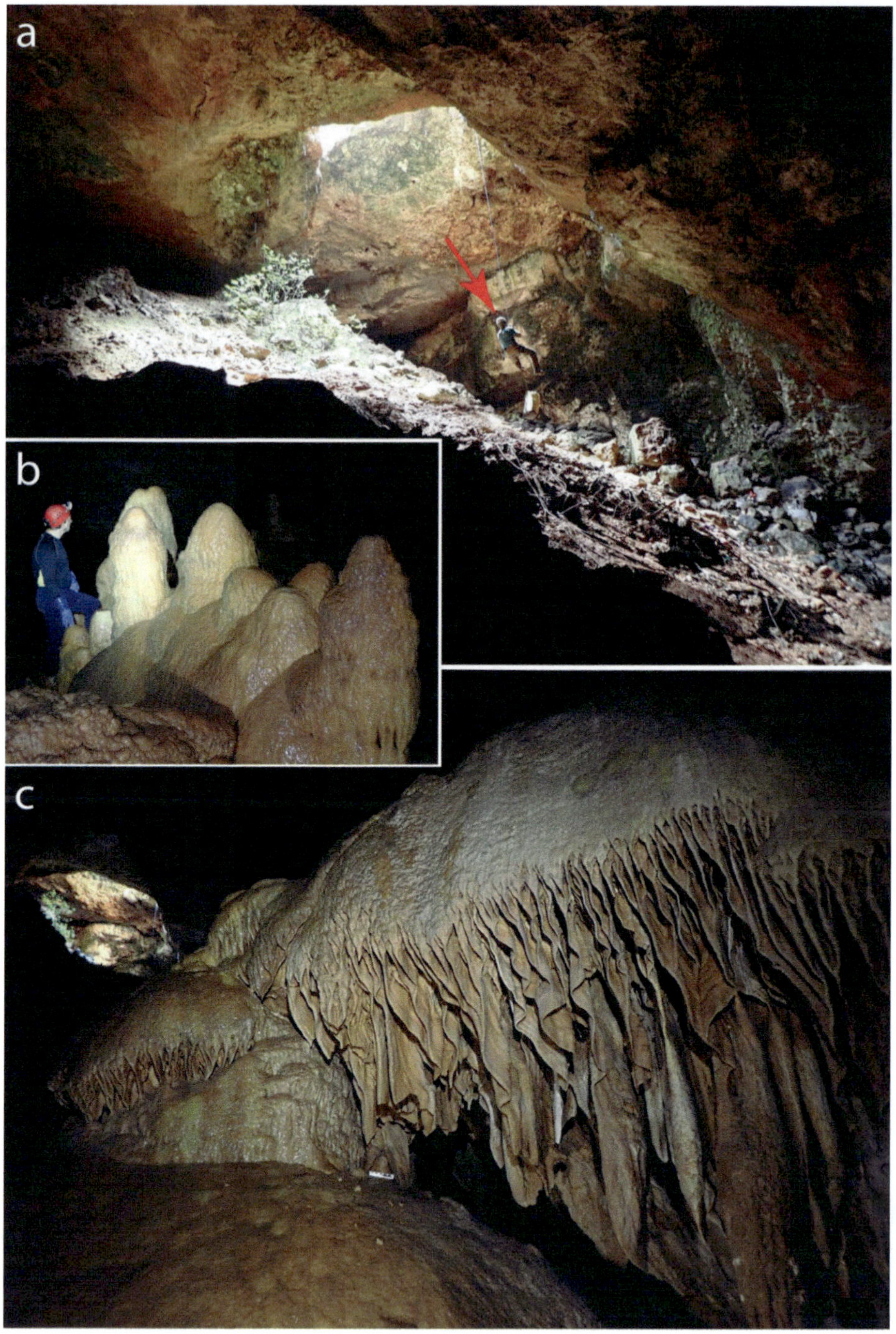

Fig. 13.13 Interior of the Atarot Cave. **a** entrance at the top of the chamber's ceiling. Person for scale (arrow). **b** a cluster of stalagmites, northern area. **c** 2-m-high speleothems forming draperies, northern area. The entrance area is seen on the left in the background

the cave one of the most decorated in Israel (Fig. 13.13). Unfortunately, the cave suffered from anthropogenic debris dumping in the twentieth century, after being exposed by quarrying, and before governmental protection was established.

13.11 Conclusions

The presently active and relict examples of karst landforms can be generally classified as dry-Mediterranean fluviokarst, comprising both subaerial solution landforms (exokarst) and underground landforms (endokarst) (White and White 2013). Together, these make up a well-developed epigenic and hypogenic karst landscape. The discussed landforms include a wide range of positive and negative landforms, as well as subsurface speleological features, all indicating long-term (millions to tens millions of years) dissolution and underground drainage. Such features are expected in karst regions (Frumkin and Shroder 2013), even where the surface is not indicative of well-developed karstification. The compilation of such features allows for the integration of a 'karst feature complex', typical of a dry-Mediterranean karst landscape, which is spatially modified by long-term tectonic and climatic variability (Daoxian 2013).

Furthermore, various relict features, such as abundant hypogenic caves, and remains of a karstified planation surface indicate that prior to late Cenozoic uplift and arching the region was generally flat and karst processes dominated (Frumkin 1993). The remaining karstified areas with internal drainage, such as the Ofra plateau and Sanur Polje, can thus be the remains of a much larger full-karst terrain. Many other depressions along the Samaria backbone hills (Fig. 13.1) have subsequently developed partial surface drainage but still retain features of karst plateaus.

References

Avni Y, Segev A, Ginat H (2012) Oligocene regional denudation of the northern Afar dome: pre- and synbreakup stages of the Afro-Arabian plate. Geol Soc Am Bull 124:1871–1897. https://doi.org/10.1130/B30634.1

Bar O, Zilberman E, Feinstein S, Calvo R, Gvirtzman Z (2016) The uplift history of the Arabian Plateau as inferred from geomorphologic analysis of its northwestern edge. Tectonophysics 671:9–23. https://doi.org/10.1016/j.tecto.2016.01.004

Ben-Asher M, Haviv I, Crouvi O, Roering JJ, Matmon A (2021) The convexity of carbonate hilltops: ^{36}Cl constraints on denudation and chemical weathering rates and implications for hillslope curvature. GSA Bull. https://doi.org/10.1130/B35658.1

Bonacci O (2013). Poljes, ponors and their catchments. In: Shroder JF, Frumkin A (eds) Treatise on geomorphology. Academic Press, San Diego, vol 6, pp 112–120. https://doi.org/10.1016/B978-0-12-374739-6.00112-3

Fleischer L, Gafsou R (2000) Structural map of Israel. Geophysical Institute of Israel, Israel

Ford D, Williams PD (2007) Karst hydrogeology and geomorphology. Wiley, Chichester, 562p. https://doi.org/10.1002/9781118684986

Frumkin A (1993) Karst origin of the upper erosion surface in the Northern Judean Mountains, Israel. Isr J Earth Sci 41:169–176

Frumkin A (2017) The underground water systems of Ma'abarta—Flavia Neapolis, Israel. Geoarchaeology 33:127–140. https://doi.org/10.1002/gea.21637

Frumkin A, Fischhendler I (2005) Morphometry and distribution of isolated caves as a guide for phreatic and confined paleohydrological conditions. Geomorphology 67:457–471. https://doi.org/10.1016/j.geomorph.2004.11.009

Frumkin A (ed), Shroder JF (ed in chief) (2013) Karst geomorphology. In: Treatise on geomorphology 6, San Diego, Elsevier, Academic Press, 483p

Frumkin A, Langford B, Lisker S, Amrani A (2017a) Hypogenic Karst at the Arabian platform margins: implications for far-field groundwater systems. Bull Geol Soc Am 129(11–12):1636–1659. https://doi.org/10.1130/B31694.1

Frumkin A, Valdman E, Langford B, Yaaran S, Buslov V (2017b) The southernmost authigenic karst plateau of the Levant. In: Proceedings of the 17th international congress of speleology, Sydney, vol 2, pp 13–16

Gvirtzman Z, Steinberg J, Bar O, Buchbinder B, Zilberman E, Siman-Tov R, Calvo R, Grossowicz L, Almogi-Labin A, Rosensaft M (2011) Retreating late tertiary shorelines in Israel: implications for the exposure of north Arabia and Levant during Neotethys closure. Lithosphere 3(2):95–109. https://doi.org/10.1130/L124.1

Ju'ub GA, Schetelig K (2002) Using the concept of artificial recharge to solve the flooding problem of Marj Sanur in Jenin Governorate-Palestine. In: Management of aquifer recharge for sustainability, proceedings of the 4th international symposium on artificial recharge of groundwater, Swets & Zeitlinger, Lisse, pp 419–423. https://doi.org/10.1201/9781003078838

Knez M, Slabe T (2013) Stone forests and their rock relief. In: Shroder JF, Frumkin A (eds) Treatise on geomorphology. Academic Press, San Diego, vol 6, pp 139–156

Kranjc A, Sazu P (2013) Classification of closed depressions in carbonate karst. In: Shroder JF, Frumkin A (eds) Treatise on geomorphology. Academic Press, San Diego, CA (2013), vol 6, pp 104–111

Lundberg J (2013) Microsculpturing of solutional rocky landforms. In: Shroder JF, Frumkin A (eds) Treatise on geomorphology. Academic Press, San Diego, vol 6, pp 121–138

Matmon A, Zilberman E, Enzel Y (1998) Morphometric analysis for determining the age of escarpments: an example from the Galilee, northern Israel. Geol Surv Isr Rep GSI/31/98, Jerusalem

Mimran Y (1984) Unconformities on the eastern Flank of the Fari'a anticline, and their implications on the structural evolution of Samaria (Central Israel). Isr J Earth Sci 33(1–2):1–11

Reches ZE, Hoexter DF, Hirsch F (1981) The structure of a monocline in the Syrian Arc system, Middle East-Surface and subsurface analysis. J Petrol Geol 3(4):413–426. https://doi.org/10.1111/j.1747-5457.1981.tb00939.x

Ryb U, Matmon A, Erel Y, Haviv I, Benedetti L, Hidy AJ (2014) Styles and rates of long-term denudation in carbonate terrains under a mediterranean to hyper-arid climatic gradient. Earth Planet Sc Lett 406:142–152. https://doi.org/10.1016/j.epsl.2014.09.008

Ryb U, Matmon A, Erel Y, Haviv I, Katz A, Starinsky A, Angert A, Aster Team (2014a) Controls on denudation rates in tectonically stable Mediterranean carbonate terrain. Geol Soc Am Bull 126: 553–568. https://doi.org/10.1130/B30886.1

Ryb U, Matmon A, Haviv I, Benedetti L (2015) Exhumation and uplift coupled with precipitation along the western Dead Sea Rift margin. Geol 43(6):483–486

Sass E, Bein A (1982) The Cretaceous carbonate platform in Israel. Cret Res 3:135–144. https://doi.org/10.1016/0195-6671(82)90014-3

Segev A, Rybakov M, Lyakhovsky V, Hofstetter A, Tibor G, Goldshmidt V, Ben Avraham Z (2006) The structure, isostasy and gravity field of the Levant continental margin and the southeast Mediterranean area. Tectonophysics 425(1–4):137–157. https://doi.org/10.1016/j.tecto.2006.07.010

Segev A, Schattner U, Lyakhovsky V (2011) Middle-Late Eocene structure of the southern Levant continental margin—Tectonic motion versus global sea-level change. Tectonophysics 499:165–177. https://doi.org/10.1016/j.tecto

Shachnai E (2000) Geological map of Ramallah 1:50,000. Geol Surv Israel
Sneh A, Bartov Y, Rosensaft M (1998) Geological map of Israel 1:200,000 sheet 2. Geol Surv Israel, Jerusalem
Shaliv G, Mimran Y, Hatzor Y (1992) The sedimentary and structural history of the Bet She'an area and its regional implications. Isr J Earth Sci 40:161–179
Shtober-Zisu N, Amasha H, Frumkin A (2015) Inland notches: Implications for subaerial formation of karstic landforms—an example from the carbonate slopes of Mt. Carmel, Israel. Geomorphology 229:85–99. https://doi.org/10.1016/j.geomorph.2014.09.004
Shtober-Zisu N, Amasha H, Frumkin A (2017) Inland notches: lithological characteristics and climatic implications of subaerial cavernous landforms in Israel. Earth Surf Proc Land 42:1820–1832. https://doi.org/10.1002/esp.4135
Shtober-Zisu N, Vaks A, Korngreen D, Frumkin A (2020) Slope retreat rates estimated from chronology of tufa deposits sheltered by inland notches on Mt. Carmel, Israel. Geomorphology 367:107319. https://doi.org/10.1016/j.geomorph.2020.107319
Steinberg J, Gvirtzman Z, Garfunkel Z (2014) Flexural response of a continental margin to sedimentary loading and lithospheric rupturing: The mountain ridge between the Levant Basin and the Dead Sea Transform. Tectonics 33:166–186. https://doi.org/10.1002/2013TC003330
Wdowinski S, Zilberman E (1997) Systematic analyses of the large scale topography and structure across the Dead Sea Rift. Tectonics 16(3):409–424. https://doi.org/10.1029/97TC00814
White WB, White EL (2013) Karst landforms: scope and processes in the early twenty-first century. In: Shroder JF, Frumkin A (eds) Treatise on geomorphology. Academic Press, San Diego, CA, vol 6, pp 319–326
Yuan D (2013) Variations of Karst geomorphology over geoclimatic gradients. In: Shroder JF, Frumkin A (eds) Treatise on geomorphology. Academic Press, San Diego, vol 6, pp 319–326. https://doi.org/10.1016/B978-0-12-374739-6.00107-X

14 Surface Landforms of Mount Sedom Diapir, Dead Sea Basin, Israel

Amos Frumkin and Israel Zak

Abstract

Mount Sedom diapir is one of the few places on Earth where rock-salt is exposed, due to extreme aridity. The relief and surface features of the diapir basically reflect its parent-rock geological structure, stratigraphy, and lithology on one hand, and recent erosion and dissolution on the other one. Major landforms include lines of sliding faults, dissolution furrows, dolines of dissolution and collapse origin, karstic shafts, and an irregular drainage system dominated by many blind valleys. The diapir rock-salt is covered by residual caprock, in turn partly overlain by less consolidated insoluble sediments. Gravels and terraces of abrasion of the regressing lake shore appear in places. Exposed salt outcrops are relatively rare and undergo rapid dissolution, demonstrated by karst features, from sharp rillenkarren to the largest salt caves known globally. The extreme solubility of the underlying salt influences the surface landscape by inducing high permeability, which promotes runoff to be swallowed into the underlying salt. The young relief and erodible sediments allow for various rock towers and inselbergs which remain standing after surrounding erosion took place.

Keywords

Salt · Caprock · Evaporite karst · Karren · Doline

Israel Zak—Deceased author.

A. Frumkin (✉)
Institute of Earth Sciences, The Hebrew University of Jerusalem, 9190401 Jerusalem, Israel
e-mail: amos.frumkin@mail.huji.ac.il

14.1 Introduction

Salt diapirs, also termed domes, are rarely exposed on the Earth surface, due to the extreme solubility of salt, which promotes subsurface dissolution by groundwater (Frumkin 2013). Their landforms seem to resemble features on Mars, such as 'honeycomb' terrains, which may indicate diapirs on Mars (Diot et al. 2014; Bernhardt et al. 2016; Weiss et al. 2017; Frumkin et al. 2021), and possibly other extraterrestrial sites. Understanding such landscapes on Earth is a pre-requirement for analysis of the extraterrestrial evaporite terrains (Baioni and Wezel 2010).

Most of the early studied salt bodies (e.g., Barton 1925) are highly deformed, overlain by a thick sedimentary cover, and never penetrated the surroundings. Their surface features reveal, therefore, only little of the details of the lithology, structure, and geological evolution of the salt bodies. Powers (1926) concluded that the physiographic expression of interior salt domes may take one of several types, depending largely upon the proximity of the salt to the surface and to the amount of uplift of the formation. Bevier (1925) described the physiography of the Damon Mound salt body (Texas, US), where radial drainage lines extend in all directions down the sides of the mound. He stated that the present topography, while corresponding in outline with the core of the dome, has apparent relationship to the detail of the subsurface structure.

Recent studies of diapirs, which reach the topographic surface, present only partial descriptions of some landforms, e.g., in Iran (Bruthans et al. 2000, 2010; Abirifard et al. 2017), Chile (De Waele et al. 2009), and Spain (Mottershead et al. 2008). However, the top of actively uplifting diapirs under semi-arid conditions is commonly flowing rapidly sideways, similar to glaciers, and therefore, many diapirs extrude onto the surface as sheets of allochthonous and highly deformed salt plugs (Talbot and Pohjola 2009). This tendency obscures the unique diapiric morphology of most diapirs reaching the surface.

A. Frumkin and N. Shtober-Zisu (eds.), *Landscapes and Landforms of Israel*,
World Geomorphological Landscapes, https://doi.org/10.1007/978-3-031-44764-8_14

Mount Sedom is the surface expression of a deep rooted salt body extending to a drilled depth of a few km. It displays a large variety of morphological features characteristic for the still rising salt diapir underneath. The salt was deposited in the Sedom lagoon within the Dead Sea rift during the late Miocene.

During the Holocene, there has been no fresh water aquifer within Mount Sedom (Frumkin 1994b). Therefore, the main body of salt bedrock is continuously extruded without undergoing total dissolution. Holocene landforms develop under the influence of rare rainfall events in the mostly unvegetated, erodible surface.

The diapir is located on the SW shore of the Dead Sea, within the strike-slip dominated Dead Sea rift system (e.g., Garfunkel 1981). It is a N-S trending asymmetric salt wall (Fig. 14.1), parallel to the main faults of the Dead Sea rift. It forms a long and narrow hill (10 km by 2 km, Fig. 14.2), rising 260 m above the Dead Sea shore (currently evaporation ponds). The Hebrew name Sedom reverberates the biblical name Sodom (Usdum in Arabic). The diapir has a simple structure (relative to other diapirs), exposing steep-dipping to locally overturned, but internally only slightly deformed beds of rock-salt with some non-salt layers (mainly anhydrite, dolomite, and clastics) of the Sedom Formation, a late Miocene sequence, over 2000 m thick.

The Sedom diapir provides superb rare exposures of well-bedded rock-salt, overlain by salt-depleted residual caprock (Figs. 14.2b and 14.3b), locally covered by Lisan 'marls' and gravels. The Lisan Formation was deposited in the Late Pleistocene Lake Lisan whose level was higher than other mid- to late Pleistocene rift lakes. The salt is visible mainly along the walls of the diapir, in particular where the Dead Sea coastal erosion caused notching, collapse, and associated retreat of a marginal vertical cliff (Fig. 14.3) and in the multitude of caves and shafts of the highly developed Holocene karst system.

The diapir borders are delimited mainly by salt-intrusion faults, with further contribution from the Dead Sea coastal erosion. In the north-eastern side, this erosion probably consumed a salt sheet which appears to have extended up to few km away from the present mountain (Alsop et al. 2015).

Here, the surface features of the diapir are discussed, attempting a comprehensive analysis of landforms typical for an actively rising salt diapir. Its well-developed diapiric landscape is associated with the hyper-arid climate: mean annual precipitation is ~ 46 mm, and annual potential evaporation ~ 2200 mm. Rainfall events whose intensity is > 4 mm/10 min produce high discharge flow in the channels, causing flash floods with a recurrence interval > 1/year. Flash floods last up to a few hours and are commonly followed by long dry periods.

14.2 Landscape Development and Major Features

14.2.1 Caprock Relief

The caprock of Mount Sedom is a few cm to 50 m thick, consisting of the hardly-soluble, residual components of the Sedom Formation parent rock sequence (mostly anhydrite). A pre-Holocene, fossil dissolution surface, termed salt table (also called salt mirror), separates the caprock from the underlying parent salt (Figs. 14.3b and 14.4). The surface topography roughly correlates with the salt table elevation, indicating a mean caprock thickness of 27 m (Zucker et al. 2019).

The caprock was formed when the rising top of the salt diapir entered an aggressive (capable of dissolving salt) aquifer associated with the Pleistocene water bodies of the Dead Sea basin, such as Lake Lisan and its predecessor Lake Samra (Weinberger et al. 2007). Consequently, the diapir head dissolved, and the horizontal salt table formed at the top of the rock-salt. The hardly-soluble minerals remained as an overlying residual caprock, which developed from the top downward as the dissolution process developed, with the newer residual material accumulating underneath the protecting cover of the older freed residue.

On the surface, an 'etched-looking' caprock relief (Figs. 14.1a, 14.3a and 14.4) traces the facies and strike of the parent underlying strata. This surface landscape is unique in appearance and extension and contributes to the understanding of the underlying primary evaporite facies. The surface caprock relief is mainly affected by the amount and stratification of the hardly-soluble components of the Sedom Formation parent rock, and the inclination of its layers (Fig. 14.4). The sedimentary features of the Sedom diapir, on its various scales, are thus more or less preserved in the caprock. Hardly-soluble layers, freed from the salt matrix, continue un-interrupted across the salt table, into the overlying caprock and up to the surface. Horizontal parent strata (rare in Mount Sedom) produce plain relief (Fig. 14.4a), while vertical parent strata with alternating salt and insolubles produce ridges and valleys oriented along the bedding strike (Figs. 14.4b and 14.5a). Such valley and ridge landscape is also observed over inclined bedrock strata (Figs. 14.3b and 14.4c).

More complex deformational features within the parent rock are also transferred into and reflected in the caprock. Relict structures of folded salt beds appear in the caprock as folded ghosts, occasionally with the rest of the folded salt beds visible under the separating salt table (Fig. 14.5b, c). Thus, the caprock landforms help with the interpretation of the structure and deformation processes occurring in the course of diapirism.

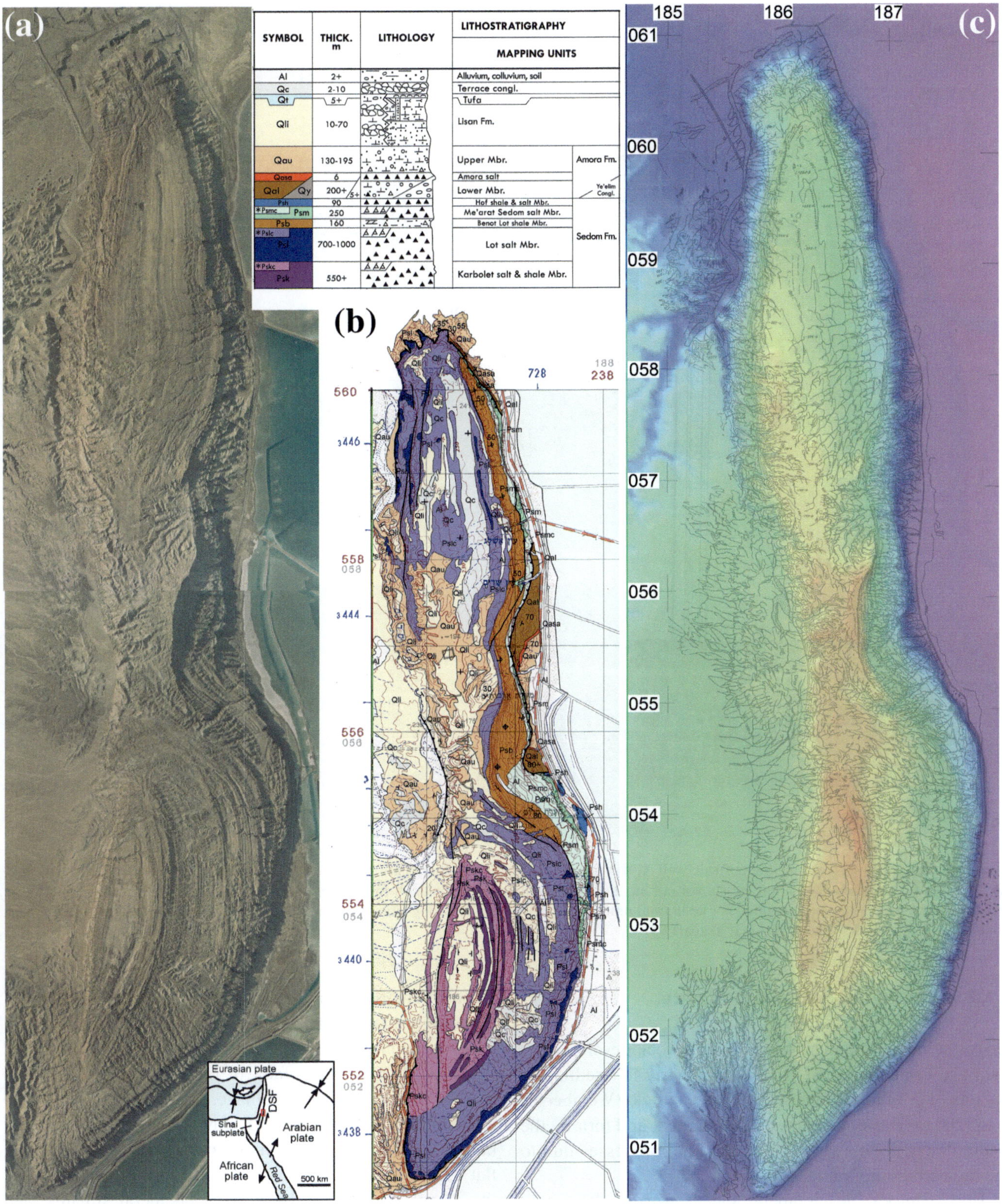

Fig. 14.1 General presentations of Mount Sedom surface. **a** air photograph and tectonic location of Mount Sedom along the Dead Sea Fault (DSF); **b** simplified geological map (modified after Zak 1967; Agnon et al. 2006); **c** topographic map (10 m contours, stream beds shown) (modified after Frumkin et al. 2021)

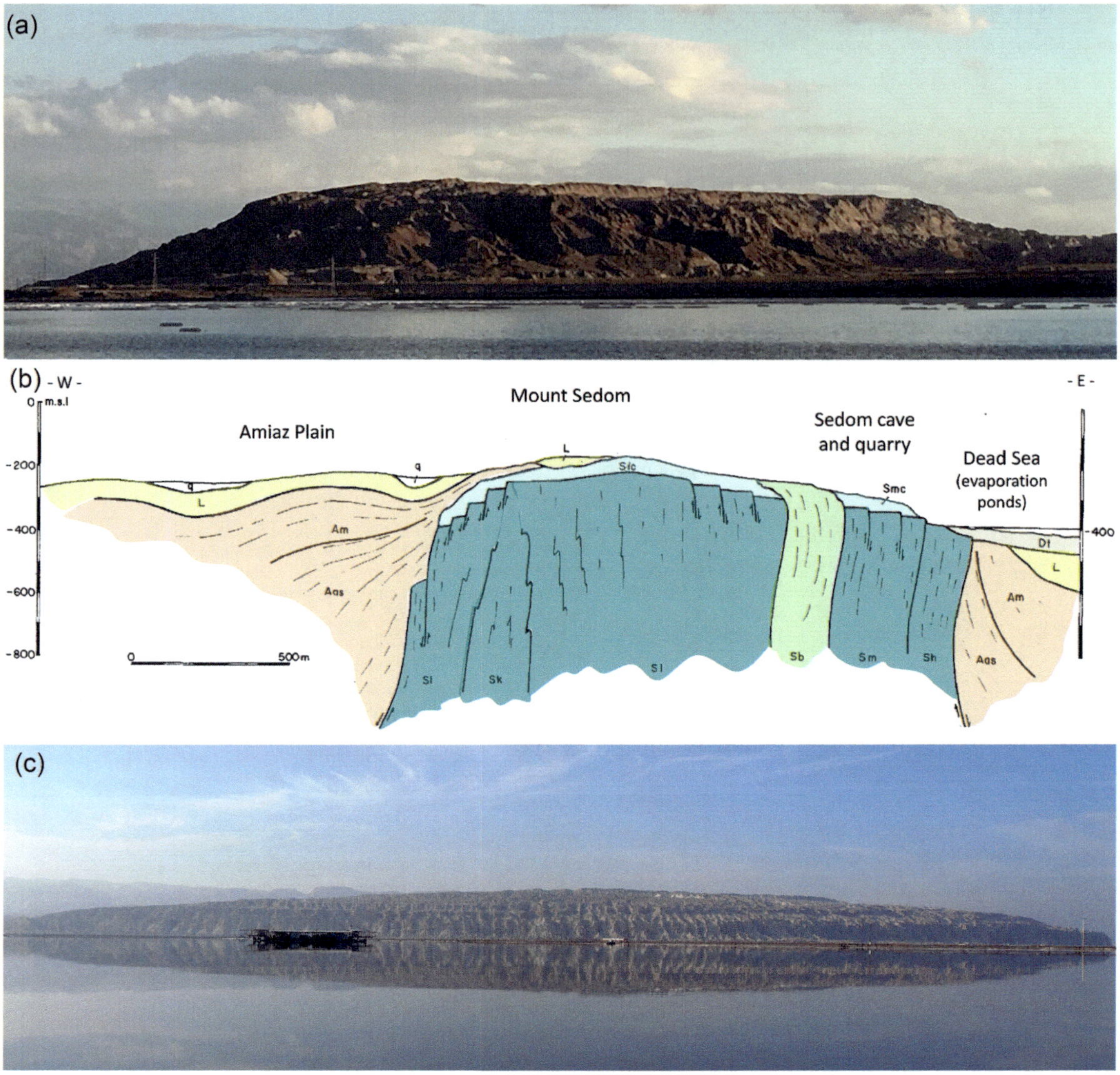

Fig. 14.2 Side views of Mount Sedom salt wall. **a** the northern edge, view southward (photo by Amos Frumkin); **b** geological cross section. Salt units of Sedom Formation are shown in blue and green, intercalated shales in light green, caprock in light blue, Amora Formation in orange, and Lisan Formation in yellow. For additional units see Fig. 14.1b; **c** The long side of Mount Sedom salt wall, with reflection in a potash pond, seen from the east (photo by Assaf Tsabar)

14.2.2 Overlying Surficial Deposits

Coeval with the late Pleistocene final caprock development, the Lake Lisan deposited its lacustrine marly-like sediments on top of Mount Sedom, above the caprock (Weinberger et al. 2007). These were uplifted, deformed, and torn by the rising diapir, followed by rapid erosional processes due to the high erodibility of the sediments.

Both caprock development and lacustrine deposition ceased at the terminal Pleistocene (~14–11 ka), when the Lake Lisan receded abruptly and its associated aquifer became saline (Weinberger et al. 2006). The surface was partly covered by gravels, such as pebbles and cobbles of braided streams and fan deltas deposited by ephemeral stream valleys (wadis) flowing from the Cretaceous outcrops in the west on the sub-horizontal top of the diapir. This situation persisted for no more than a few thousand years, until the western fault of the rising diapir detached it from the east-flowing drainage. The thin (few cm up to a few meters thick) veneer of gravels started to be eroded, but it is still preserved on high ground such as interfluves, flat topped hills, and plateaus (Fig. 14.6a). In most regions, the gravel is dissected by young ephemeral streams, and the pebbles are transported down active slopes and along entrenched valleys. The present-day catchments of Mount Sedom are local and small, restricted to the diapir surface. As a result, large cobbles and rare boulders which were originally (terminal Pleistocene) deposited by major flash

Fig. 14.3 The walls of Mount Sedom diapir. **a** air photograph of Sedom Cave area, scale bar is ~ 50 m. White arrows indicate the escarpment formed by the marginal fault and Dead Sea coastal erosion. Other features indicated by colored arrows (see details below): red—strike of layers dictating surface morphology and small subsequent tributaries, blue—direction of major consequent streams, green—active shaft-ponors swallowing sinking streams; note traces of superficial entrenchment bypassing the shafts toward abandoned shafts and diapir edge, black—residual white hill (inselberg) of Lisan Formation; **b** Northern Mount Sedom escarpment viewed from the east, showing how diapir lithology is reflected in the caprock: white arrow indicates the salt table, which is crossed by insoluble residual layers (red arrows) that collapsed in the caprock due to salt dissolution, but retained their original volume (compare Fig. 14.4c) (photo by Amos Frumkin); **c** outlet of Qolnoa Cave within the east escarpment of southern Mount Sedom. The structure was used as a cinema for potash workers utilizing the cool cave air flow as a natural air-condition (photo by Amos Frumkin); **d** pool of flash-flood water at the foot of southernmost Mount Sedom (photo by Boaz Langford 2020)

floods flowing from large regional catchments are not moved by runoff during the Holocene (Fig. 14.6b).

14.2.3 Structural and Tectonic Features

The diapir is still rising at an average rate of ~ 1 cm/year (Frumkin 1996a; Pe'eri et al. 2004; Weinberger et al. 2006; Zucker et al. 2019). The sub-horizontal surface of the latest Pleistocene is presently faulted tens of meters above the Dead Sea level and tilted toward it. Thus, more intensive erosion occurred on the eastern side, where the relief of the rising diapir is amplified by the falling Dead Sea base level, as well as by coastal erosion. In contrast, the relief is smaller on the western side of the mountain, although the western marginal fault is young, steep, and smeared with the overlying Lisan marls (Fig. 14.7a). Between the major piercing faults, the water divide area of the diapir is relatively flat in places (Figs. 14.2 and 14.6b). In addition, higher density fault-scarp steps, located mainly along the eastern and western rims, are related to the marginal shear zone of the diapir, but they appear also within the diapir body. Where the beds are almost vertical, as in the southern part of Mount Sedom, bedding plains act as sliding faults (Fig. 14.7b) which separate the internal salt units from one another. Between these structures, longitudinal collapse synclines and grabens are observed, with corresponding residual anticlines and horsts. The cross section through the southern peak of Mount Sedom (Fig. 14.2b) demonstrates that the surface topography of the southern region is shaped like an asymmetric telescopic antenna with step faults (Zucker et al. 2019). The continuous deformation of the

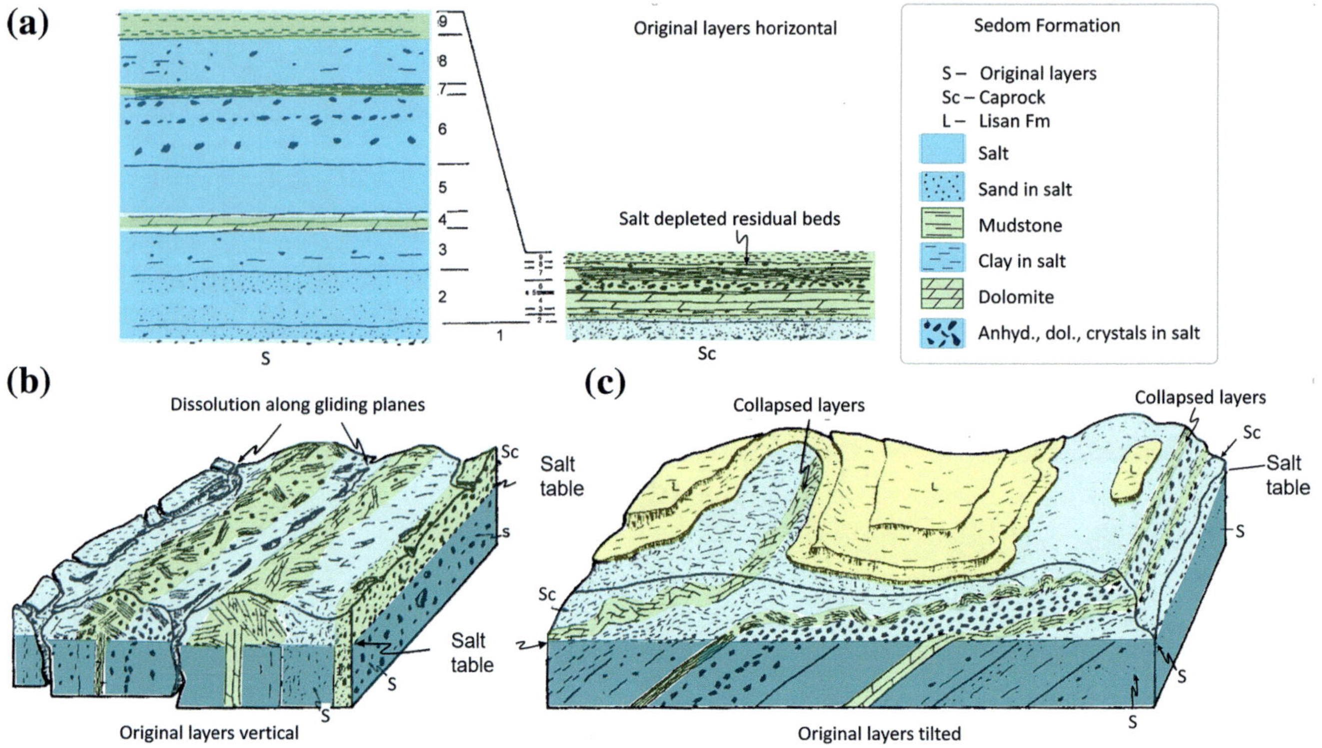

Fig. 14.4 Diapir lithology reflected by surface relief: salt beds (blue) which dissolved in the aquifer above the salt table are reflected by surface valleys; insoluble residual layers (green) collapsed during aquifer dissolution due to salt depletion, but retain their original volume, forming ridges. **a** horizontal bedding, rare at Mount Sedom; **b** vertical beds, common in southern Mount Sedom; **c** inclined beds common in northern Mount Sedom (compare Fig. 14.3b). Overlying Lisan sediments (yellow) are deformed into synforms and antiforms (modified after Frumkin et al. 2021)

diapir results in many open fractures (Fig. 14.7c, d), which are particularly abundant at the highly deformed north and south edges of the diapir. These commonly attract runoff, acting as dissolution lines. Where water flow is concentrated from a catchment stream, they become a cave inlet, where the stream sinks into a cave (Fig. 14.7e).

The main landform features described above are presented collectively in a map (Fig. 14.8) demonstrating the major impact of structural lineaments on the landscape (Frumkin et al. 2021), as observed also from the air (Fig. 14.1a). Variations of this general landscape are associated mainly with lithology (Fig. 14.1b), stream entrenchment (Fig. 14.1c), and karstification. These will be further discussed below.

14.2.4 Holocene Landscape Development

During the terminal Pleistocene–early Holocene, the lake level fell dramatically to ~ −400 m. The receding lake shaped a subaerial planation surface slightly tilted toward the Dead Sea basin. The remnants of this plateau-like erosional surface are still observed mainly along the drainage divide of Mount Sedom (Fig. 14.2), but also at lower regions where the erosional surface remnants are dissected by later erosion, tilting, and deformation. The receding lake continued abrading terraces and notches at the eastern escarpment, where lower cliff-face lake waves erosion promoted steepening and rock-falls. Following lake retreat, the planation surface was covered by pebbles and cobbles of braided ephemeral streams and fan deltas flowing from the Cretaceous outcrops at the west (Fig. 14.6), on the sub-horizontal planation surface at the top of the diapir (Zak 1967; Frumkin et al. 2021).

During the Pleistocene–Holocene transition, regional large streams were diverted by the uplifting diapir northward and southward of the salt wall. During the Holocene, new local stream catchments developed on top of the diapir, mostly draining eastward and partly westward, according to the local slope. Therefore, the larger Holocene streams are consequent, flowing along the early-Holocene surface slope. Their smaller tributaries are commonly subsequent, following paths determined by the 'etched' valleys and mid-Holocene erosion of weaker rocks (Fig. 14.3a).

In spite of the extremely arid climate, the Holocene geomorphic processes, such as stream erosion and dissolution, are vigorous. The entrenchment rate of streams is high both subaerially (Gerson 1977) and in salt caves (Frumkin and Ford 1995). The correlation of width and drainage area of its ephemeral streams is similar to that in humid environments (Wolman and Gerson 1978). The reason may be the lithology (imperviousness and soft caprock) and associated dense drainage network, combined with its high uplift rate.

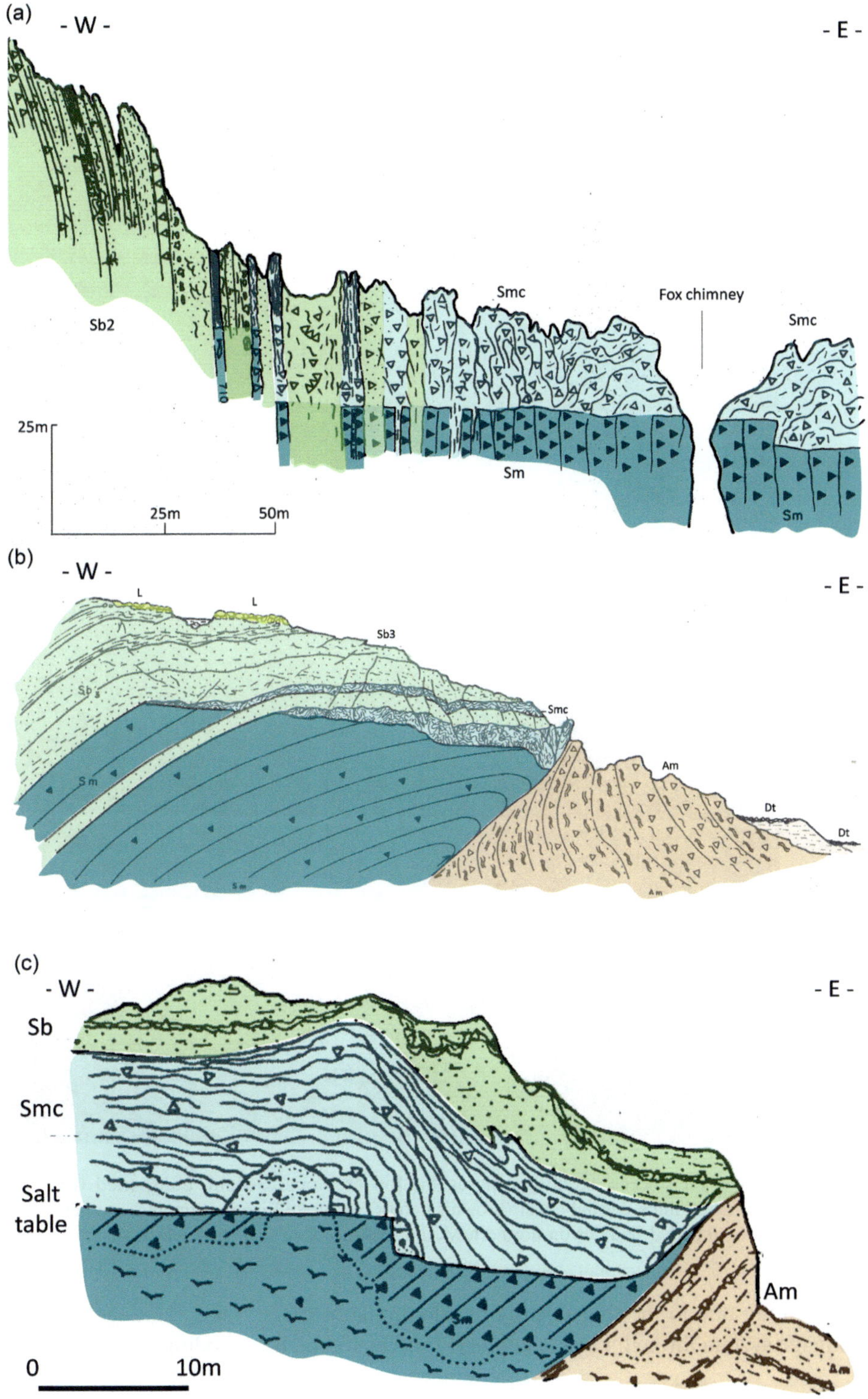

Fig. 14.5 Geological sections of sites at the eastern margin of Mount Sedom, showing relations between bedrock, caprock, and surface features. **a** vertical beds with lithology ranging from solution-resistant bedrock (green), well-preserved at the surface, to Sedom salt, transformed to anhydrite caprock (blue) above the salt table, with the Fox chimney capturing runoff into the Sedom Cave. Insoluble clastics formed upstanding linear ridges. During aquifer dissolution of the surrounding salt, the ridge eventually undergoes gravity-driven collapse; **b** overturned shales (light green), collapsed over and within desalted caprock. Amora shales (orange) were intruded by the salt along a thrust fault, subsequently eroded forming residual pendants (Fig. 14.14b, c); **c** overturned shales (green) and a 'ghost' fold in caprock, reflecting a paleo-fold in the salt bedrock

Most streams have not reached an equilibrium profile (Frumkin and Ford 1995). Their highest and steepest knick-points are usually close to the marginal faults or at the point of stream entry into rock-salt, where a deep (10s meters) shaft is common.

The soft slopes of Mount Sedom yield large amounts of debris during rainfall events; mean annual denudation is at least 1 mm (Wolman and Gerson 1978). However, severe storms may erode as much as 10 times the annual average (Gerson and Inbar 1974). Flash flood water has high concentrations of suspended clastics and bedload, ranging from 10 to 80% by weight. During intense rain events, runoff collects upon the relatively impermeable caprock and flows partially into fissures and stream caves. This process is

Fig. 14.6 Latest Pleistocene exotic gravels on the top of Mount Sedom, transported from the west by large streams (photographs Amos Frumkin). **a** a few m thick detrital cover (brownish color) on white Lisan sediments, southern Mount Sedom; **b** thin veneer of pebbles, cobbles, and a limestone rounded boulder on northern Mount Sedom. Note that these clasts originate from the Cretaceous outcrops at the background skyline, in the west

accelerating due to the dynamically developing subsurface drainage system through karst conduits (Frumkin 1994a, b).

14.3 Holocene Karst Features

14.3.1 Blind Valleys and Stream Sinks

During the early Holocene, fluvial drainage was subaerial. Gradual capture of surface streams into the subsurface salt promoted the formation of blind valleys, which are dominant today. However, thick caprock and Lisan sediments coupled with alluvium deposited within such valleys may still choke underdeveloped fractures, forming alluviated surface depressions. In such choked depressions, flash-flood runoff accumulates for up to a few days before infiltrating, evaporating, and drying (Fig. 14.9a, b). Such depressions retain soil humidity for long periods, locally supporting hotspots of vegetation, which include various shrubs as well as *Tamarix* trees (Frumkin 2009a). Radiocarbon-dated twigs deposited in caves demonstrate a mid-Holocene (~5000 years ago) wetter climate, supporting

Fig. 14.7 Tectonic lineaments of Mount Sedom. **a** the western main fault of Sedom diapir (arrow) demonstrating caprock smeared with overlying Lisan marls (photo by Amos Frumkin); **b** step faults sliding along vertical bedding plains (arrows), southern Mount Sedom (photo by Amos Frumkin); **c** air photo by of open tension cracks in caprock, southern Mount Sedom (image width ~ 100 m, photo by Ariel Frumkin); **d** surface view of the same fractures (photo by Amos Frumkin); **e** an open fracture capturing a stream into the Malham Cave, southern Mount Sedom (photo by Vladimir Buslov)

oaks (*Quercus calliprinos* Webb.) under such conditions (Frumkin et al. 1991).

During the Holocene, the infiltrating water dissolved the rock-salt below the caprock, producing solution dolines (Fig. 14.9b). Where the hydraulic conductivity of the fractures is large enough to accommodate the flash flood discharge and detrital load, a stream sink (also termed swallet, ponor) develops, directing the water into the caves. The top of the stream sink can occur in Lisan sediments (Fig. 14.9c), caprock (Fig. 14.9d), or rock-salt (Fig. 14.9e), depending on the local outcrop.

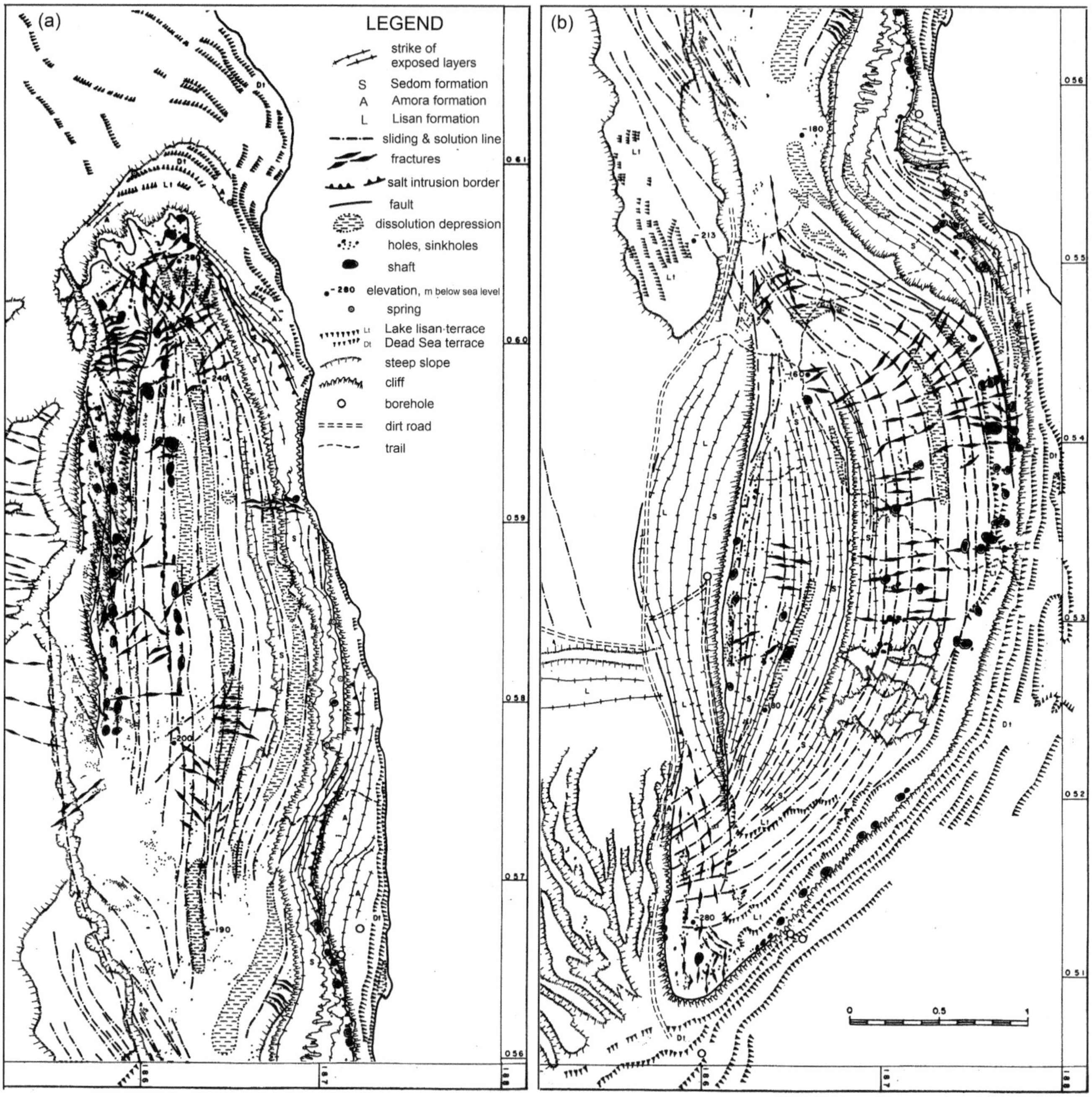

Fig. 14.8 Landform map of northern (**a**) and southern (**b**) Mount Sedom, excluding streams (which are shown in Fig. 14.1c, and their catchments in Fig. 14.10) (modified after Frumkin et al. 2021)

The rock-salt sinks develop rapidly into major shafts which can reach ~ 100 m in depth (Fig. 14.9e). The morphology of a shaft depends on its lithology, and the shaft cross section and wall features change as it dissects various rock types. Ultimately, the runoff water reaches the salt whose high solubility promotes intensive karstification and cave streams. Inlet caves, drained by diffuse infiltration through the alluviated bottom of the cave, occupy the central part of the diapir, while integrated cave systems, directing flash floods all the way down to the base level, commonly drain its peripheral south-eastern margins, and less commonly the other margins (Fig. 14.10). The integrated caves include the globally longest salt cave, the 10.2 km long Malham Cave, as well as other multi-km long

Fig. 14.9 Closed depressions and stream sinks on Mount Sedom. **a** shallow dissolution depression with remains of a flash-flood mud and vegetation using soil moisture (photo by Amos Frumkin); **b** alluviated and vegetated solution doline, northern Mount Sedom; **c** stream sink in the Lisan sediments, directing flash floods into the salt through caprock, southern Mount Sedom (photo by Amos Frumkin); **d** stream sinking vertically into a caprock shaft, southern Mount Sedom (photo by Vladimir Buslov); **e** rock-salt shaft capturing a stream, northern Mount Sedom (photo by Vladimir Buslov)

caves (Frumkin 1994a). The overall cave length surveyed in Mount Sedom until May 2021 is over 38 km.

Radiocarbon dates of twigs deposited by flash floods in the caves show that they formed during the last 8000 years (Frumkin 1996b). The caves started forming immediately after the exposure of the rock-salt above the receding lake, indicating that the entire Mount Sedom landscape is indeed of Holocene age.

While the salt caves gradually capture additional catchments, much of Mount Sedom (43%) is still drained subaerially by surface streams toward the Dead Sea base level or to streams reaching it (Fig. 14.10).

14.3.2 Collapse Sinkholes Demonstrating Wide-Scale Karstification

The older cave chambers or stream passages which captured large runoff discharge for thousands of years have in some places developed laterally. Sub-horizontal flow and karstification within the salt commonly undermines the cave roof which ultimately collapses. Upon reaching the surface, large sinkholes (also termed collapse dolines) are formed (Fig. 14.11a). Most commonly, these sinkholes have vertical caprock walls (Fig. 14.11c), while the underlying rock-salt is exposed in some cases (Fig. 14.11d), giving rise

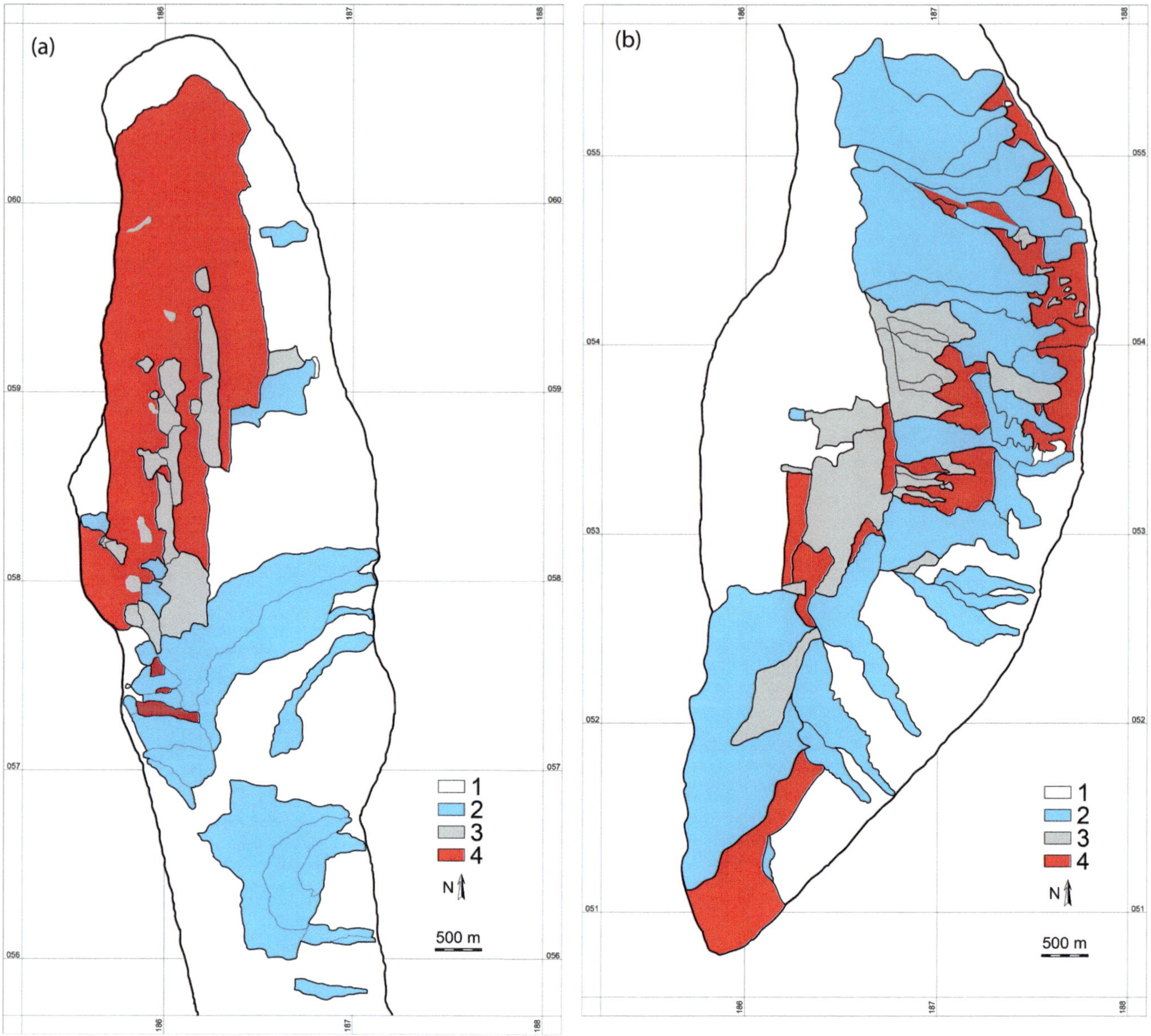

Fig. 14.10 Drainage type map of northern (a) and southern (b) Mount Sedom. Legend: 1—subaerial, non-karstic runoff; 2—rapid flow to base level through integrated cave systems; 3—drainage through inlet caves—non-integrated caves with diffuse infiltration through their alluviated bottom; 4—drainage into small cavities, mostly too small to permit human entry

to salt karren (below). The sinkholes commonly provide entries into salt caves through openings at their bottom.

Smaller sinkholes are developed also within Lisan sediments, associated with deep karstification in the underlying salt, and piping within the less soluble caprock and Lisan rocks (Fig. 14.11e).

14.3.3 Solutional Micro-sculpturing of Rock-Salt

Salt rillenkarren are micro-sculpturing suites of packed-together rainfall-controlled Hortonian channels and ridges that form on a bare and pure rock-salt surfaces. They are controlled by direct rainfall which flows gravitationally

Fig. 14.11 Collapse features. **a** air photograph of a major sliding, solution, and collapse line (image width ~ 230 m). A stream (top) captured into the main fracture formed a 60 m diameter sinkhole upon partial collapse of a larger cave (north is toward the left); **b** surface view of the southern, well-preserved part of the tension fracture line in caprock (shown in the right side of **a**) (photo by Amos Frumkin); **c** surface view of a large sinkhole (shown in the lower left of **a**, formed by combined salt tectonics, dissolution, and collapse. Walls are in caprock, persons at the right for scale (photo by Boaz Langford); **d** major sinkhole formed at the sinkpoint of the southern stream, Malham Cave. Note the gray-colored salt outcrop below the salt table and the overlying caprock (photo by Amos Frumkin); **e** collapse of Lisan sediments into deeper piping and dissolution voids (photo by Amos Frumkin) (Frumkin et al. 2021)

along the deepest available routes, deepening them further with positive feedback. Salt rillenkarren are generally sharper and have higher relief than the limestone ones. Their form depends on the steepness of the slope. On semi-horizontal salt surfaces, they appear as small pinnacles, resembling a bed of sharp nails (Fig. 14.12a). On moderate slopes, they tend to form sharp knife-edged crests separated by 1–2 cm wide channels with u-shaped cross section (Fig. 14.12b). The sides of the ridges diverge irregularly sideways. On vertical walls, the rillenkarren are vertically straight, elongated, and parallel to one another (Fig. 14.12c).

Where the salt outcrop is covered by a few mm-thick residual insolubles, rillenkarren are subdued. A cover of a few cm-thick insolubles results in an irregular step-like karren morphology, with individual step size ranging from a few cm to 1 m size (Fig. 14.12d). Dissolution is relatively intensive on the unprotected vertical walls where salt is exposed. Re-evaporation of salty infiltration rainwater leaves patches of white salt on the flat surfaces. Irregularly-sloped surfaces with variable amounts of insoluble coverage produce combinations of rillenkarren and step-like features (Fig. 14.12e). Rock-salt, which was exposed to slightly aggressive Dead Sea water (at the eastern margin of Mount Sedom), has developed convoluted irregular dissolution micro-voids (Fig. 14.12f).

14.4 Non-salt Karst and Pseudokarst

14.4.1 Micro-sculpturing

Anhydrite caprock presents various macromorphologies depending on its lithologic and structural features (Figs. 14.4 and 14.5). Where the caprock comprises fine-grained cohesive anhydrite, rillenkarren may develop by the rare rainfall under the present-day hyper-arid climate (Fig. 14.13a). These are commonly longer, shallower, narrower, and less pronounced, compared with salt rillenkarren. They form in clusters, commonly along the side-slopes of stream channels whose flash floods had washed the unconsolidated surficial caprock, exposing inclined surfaces of compact anhydrite.

The Lisan sediments are softer, producing various micro-sculpturing pseudokarst features, associated with badland landscape. In some cases, the micro-landforms follow the laminar bedding, and in others, vertical flutes are formed (Fig. 14.13b). These are commonly associated with piping cavities formed along permeable fractures and layers.

The most resistant bedrock layers in Mount Sedom are dolomitic, which are rarely interbedded within the Sedom Formation. Such dolostone layers were not dissolved in the aquifer which formed the caprock. On these layers, one can observe original features such as ripples and animal tracks formed on the bottom of Sedom lagoon, prior to the consolidation of the Sedom Formation bedrock (Fig. 14.13c). Anhydrite layers may also preserve original micro-landforms such as ripples.

14.4.2 Inselbergs and Pinnacles

Isolated rock columns and hills are some of the most prominent landforms in Mount Sedom. The high erodibility of Lisan sediments tends to remove chunks of its layers, leaving behind isolated vertical-sided hills (Fig. 14.14a).

Pinnacles, or natural rock towers, appear in the Amora Formation overburden comprising limy marls which are commonly steeply inclined following their piercing by the diapir (Fig. 14.14b, c). The pinnacles are relics of a harder layer surrounded by preferentially-eroded soft marls.

One of the earliest written documentations of a prominent landform is a 'pillar of salt' (Genesis 19:26) associated with Sodom (Sedom). This intriguing ancient text describes both the morphology and the lithology of the feature. If this verse indeed refers to an ancient observation of a pillar of salt, it must have been at the only salt outcrop in the Dead Sea region, namely Mount Sedom. The most prominent salt pinnacle (roofed by a protecting caprock 'hood') in Mount Sedom, resembling a human statue in profile, carries the name 'Lot's wife', commemorating the legendary salt pillar of Genesis (Fig. 14.14d). This pinnacle is actually a relic of the wall of the collapsed upper level of Sedom Cave (Frumkin 2009b). Interestingly, radiocarbon dating indicates that the cave level associated with the pillar is ~ 4000 years old, the same period to which the biblical story is commonly attributed.

Fig. 14.12 Dissolutional micro-sculpturing of rock-salt outcrops. **a** rillenkarren on sub-horizontal surface, resembling sharp nails. Foreground is 1 m wide (photo by Amos Frumkin); **b** rillenkarren on moderate slope outcrop, forming sharp knife-edged crests and u-shaped channels (photo by Amos Frumkin); **c** karren on vertical walls of a shaft (viewed downwards); the karren are straight, elongated, and sub-parallel. Foreground is ~ 2 m wide, person in the bottom for scale (photo by Vladimir Buslov); **d** irregular step-like karren morphology develop under a few cm-thick insoluble. Foreground is ~ 4 m wide (photo by Amos Frumkin); **e** mixed karren types under irregularly-sloping surfaces with variable amounts of insolubles (photo by Boaz Langford 2020); **f** exposed to slightly aggressive Dead Sea water during mid-Holocene, this outcrop developed convoluted irregular dissolutional micro-voids. Outcrop is ~ 6 m wide (photo by Amos Frumkin)

Fig. 14.13 Micro-sculpturing of non-salt outcrops (photographs Amos Frumkin). **a** solutional rillenkarren on caprock anhydrite; **b** pseudokarst vertical flutes on the walls of a piping sinkhole in the Lisan sediments; **c** preserved ripples on inclined bedding plane surfaces of dolomitic layers, formed originally on the bottom of Sedom lagoon

14.5 Conclusions

Salt outcrops are globally rare and occur at Mount Sedom due to the extreme aridity of the region. While the Mount Sedom diapir is mainly composed of rock-salt below the surface, the salt is observed mainly along its margins and in shafts. The remaining surface of the diapir comprises residual caprock, derived from impurities which escaped total dissolution, capped by Lisan fine sediments and gravels, deposited in the Lake Lisan during the latest Pleistocene. The caprock morphology reflects the underlying structural and stratigraphic features of the salt. These are preserved in Mount Sedom better than in other exposed diapirs, which undergo horizontal flow. The rock-salt is protected from direct rainfall in most areas, but where salt is directly affected by rainwater or runoff, it develops prominent highly-focused karst features, comparable to limestone under humid climates. These include knife-sharp rillenkarren on bare salt outcrops, as well as shafts and sinkholes which divert runoff to the underground courses and cave streams. Dolines are common, ranging from small (a few meters diameter) alluviated and sometimes vegetated solutional depressions to large collapse sinkholes, tens of meters in diameter and depth. Most of the diapir surface is drained to the subsurface via swallets (ponors), dolines, or fractures. The relief of Mount Sedom increases with time due to its fast diapiric evolution and limited precipitation. The continuous uplift promotes deep erosional incision and excellent exposure of the diapiric features.

Fig. 14.14 Inselbergs, pillars, pinnacles, and rock towers form as erosional vertical-sided inselbergs within the erodible Lisan (**a**) and Amora formations (**b**, **c**). The salt and caprock pillar (**d**), dubbed 'Lot's Wife' is a remnant of the wall of a collapsed cave (Frumkin 2009a, b), associated with an ancient tradition (Genesis 19:26) (photographs Amos Frumkin)

Acknowledgements The Israel Cave Research Center team assisted in cave surveying. Nurit Shtober-Zisu has re-painted Figs. 14.2b, 14.4, and 14.5. Assaf Tsabar kindly provided his photos (Fig. 14.2c). Ariel Frumkin provided drone photographs.

References

Abirifard M, Raeisi E, Zarei M, Zare M, Filippi M, Bruthans J Talbot CJ (2017) Jahani salt diapir, Iran: hydrogeology, karst features and effect on surroundings environment. Int J Speleol 46(3):445–457

Agnon A, Weinberger R, Zak I, Sneh A (2006) Geological map of Israel 1: 50000, Sedom sheet 20-I II. Geological Survey of Israel

Alsop GI, Weinberger R, Levi T, Marco S (2015) Deformation within an exposed salt wall: recumbent folding and extrusion of evaporites in the Dead Sea Basin. J Struct Geol 70:95–118

Baioni D, Wezel FC (2010) Morphology and origin of an evaporitic dome in the eastern Tithonium Chasma, Mars. Planet Space Sci 58(5):847–857

Barton DC (1925) The salt domes of south Texas. AAPG Bull 9(3):536–589

Bernhardt H, Reiss D, Hiesinger H, Ivanov MA (2016) The honeycomb terrain on the Hellas basin floor, Mars: a case for salt or ice diapirism. J Geophys Res Planets 121(4):714–738

Bevier GM (1925) The Damon mound oil field, Texas. AAPG Bull 9(3):505–535

Bruthans J, Šmíd J, Filippi M, Zeman O (2000) Thickness of cap rock and other important factors affecting the morphogenesis of salt karst. Acta carsologica 29(2):51–64

Bruthans J, Filippi M, Zare M, Churáčková Z, Asadi N, Fuchs M, Adamovič J (2010) Evolution of salt diapir and karst morphology during the last glacial cycle: effects of sea-level oscillation, diapir and regional uplift, and erosion (Persian Gulf, Iran). Geomorphology 121(3–4):291–304

De Waele J, Picotti V, Zini L, Cucchi F, Forti P, Galli E, Rossi A (2009) Karst phenomena in the Cordillera de la Sal (Atacama, Chile). Geoacta Spec Publ 2:113–127
Diot X, El Maarry MR, Schlunegger F, Norton KP, Thomas NH, Grindrod PM (2014) The geomorphology and morphometry of the banded terrain in Hellas basin, Mars. Planet Space Sci 101:118–134. https://doi.org/10.1016/j.pss.2014.06.013
Frumkin A (1994a) Morphology and development of salt caves. J Caves Karst Stud (NSS Bull) 56:82–95
Frumkin A (1994b) Hydrology and denudation rates of halite karst. J Hydrol 162(1–2):171–189
Frumkin A (1996a) Uplift rate relative to base level of a salt diapir (Dead Sea, Israel), as indicated by cave levels. In: Alsop I, Blundell D, Davison I (eds) Salt Tectonics, Geological Society of London, SP, vol 100, pp 41–47
Frumkin A (1996b) Determining the exposure age of a karst landscape. Quatern Res 46:99–106
Frumkin A (2009a) Stable isotopes of a subfossil Tamarix tree from the Dead Sea region, Israel, and their implications for the Intermediate Bronze Age environmental crisis. Quatern Res 71:319–328
Frumkin A (2009b) Formation and dating of a salt pillar in Mount Sedom diapir, Israel. Geol Soc Am Bull 121(1/2):286–293
Frumkin A (2013) Salt Karst. In: Frumkin A (volume ed), Shroder J (ed. in chief) Treatise in geomorphology. Elsevier, Academic Press, San Diego, vol 6, pp 208–424
Frumkin A, Ford DC (1995) Rapid entrenchment of stream profiles in the salt caves of Mount Sedom, Israel. Earth Surf Proces Landforms 20(2):139–152
Frumkin A, Magaritz M, Carmi I, Zak I (1991) The Holocene climatic record of the salt caves of Mount Sedom, Israel. Holocene 1(3):191–200
Frumkin A, Pe'eri S, Zak I (2021) Development of banded terrain in an active salt diapir: potential analog to Mars. Geomorphology 389:107824. https://doi.org/10.1016/j.geomorph.2021.107824
Garfunkel Z (1981) Internal structure of the Dead Sea leaky transform (rift) in relation to plate kinematics. Tectonophysics 80:81–108
Gerson R (1977) Sediment transport for desert watersheds in erodible materials. Earth Surf Proces 2(4):343–361
Gerson R, Inbar M (1974) The field study program of the Jerusalem-Elat symposium, 1974. Reviews and summaries of Israeli research projects. Z Geomorphol Suppl 20:7–11
Mottershead DN, Duane WJ, Inkpen RJ, Wright JS (2008) An investigation of the geometric controls on the morphological evolution of small-scale salt terrains, Cardona, Spain. Environ Geol 53(5):1091–1098
Negev Y, Cohen E, Vlaykova A, Langford B (2021) The new survey of Malham Cave—the longest salt cave in the world. Niqrot Zurim 21:157–168 (Hebrew, English abstract)
Pe'eri S, Zebker HA, Ben-Avraham Z, Frumkin A, Hall JK (2004) Spatially-resolved uplift rate of the Mount Sedom (Dead Sea) salt diapir from InSAR observations. Israel J Earth Sci 53(2):99–106
Powers S (1926) Interior salt domes of Texas. AAPG Bull 10(1):1–60
Talbot CJ, Pohjola V (2009) Subaerial salt extrusions in Iran as analogues of ice sheets, streams and glaciers. Earth Sci Rev 97(1–4):155–183
Weinberger R, Begin ZB, Waldmann N, Gardosh M, Baer G, Frumkin A, Wdowinski S (2006) Quaternary rise of the Sedom Diapir, Dead Sea basin. In: Enzel Y, Agnon A, Stein M (eds) New frontiers in Dead Sea paleoenvironmental research, GSA Special Paper, Boulder, GSA, vol 401, pp 33–51
Weinberger R, Bar-Matthews M, Levi T, Begin ZB (2007) Late-Pleistocene rise of the Sedom diapir on the backdrop of water-level fluctuations of Lake Lisan, Dead Sea Basin. Quat Int 175(1):53–61
Weiss DK, Head JW (2017) Salt or ice diapirism origin for the honeycomb terrain in Hellas basin, Mars?: implications for the early martian climate. Icarus 284:249–263
Wolman MG, Gerson R (1978) Relative scales of time and effectiveness of climate in watershed geomorphology. Earth Surf Proces 3(2):189–208
Zak I (1967) The geology of Mount Sedom, PhD thesis (Hebrew, English Summary). The Hebrew University of Jerusalem
Zucker E, Frumkin A, Agnon A, Weinberger R (2019) Internal deformation and uplift-rate of Salt walls detected by a displaced dissolution surface, Dead Sea basin. J Struct Geol 127:103870

Landscape Response to the Dead Sea Level Fall in Recent Decades

15

Elad Dente

Abstract

The geomorphic response of channels to base-level fall is a major factor in landscape evolution and a basic question in geomorphology. The response of the tributaries of the Dead Sea to its progressively level fall in recent decades (> 40 m, > 1 m per year) forms a new and evolving landscape; the tributaries provide a rare opportunity to observe, directly measure, and analyze active field-scale fluvial processes in response to base-level fall. These channels are evolving within diverse settings, which enable isolation of specific factors and processes that control their evolution. This rapid ongoing geomorphic evolution is being studied in recent years using high spatial and temporal resolution datasets. The studies yielded both practical and theoretical insights on various geomorphic subjects, including channel meandering, waterfall migration, sediment redistribution, flood effects, karst and channel interactions, and more. As the Dead Sea level continues to fall, the potential of the lake region as a unique field-scale fluvial geomorphology laboratory is far from being exploited.

Keywords

Dead Sea · Base-level lowering · Incision · Meandering · Landscape evolution · Channel morphology

E. Dente (✉)
School of Environmental Sciences, University of Haifa, 3498838 Haifa, Israel
e-mail: edente@geo.haifa.ac.il

15.1 Introduction

Bounded by normal faults of the Dead Sea Rift, the saline ~ 300-m-deep Dead Sea is the lowest waterbody on Earth. The Dead Sea is also a terminal lake located in the center of an international drainage basin (~ 42,000 km^2), which crosses several climate zones and is characterized by mean annual precipitation ranging from < 50 to > 1000 mm per year (Greenbaum et al. 2006; Lensky et al. 2005). Since the middle of the twentieth century, the Dead Sea is experiencing a hydrological deficit due to increasing water demand in and around its drainage basin, and industrial harvest of its brine (Lensky and Dente 2015). As a result, the Dead Sea level dropped by > 40 m in recent decades, from − 395 m above sea level (masl) in 1980 to − 437.5 masl in 2023 triggering the evolution of > 40 different tributaries around the lake (Figs. 15.1 and 15.2). The variability in the characteristics of these channels, for example, channel-bed lithology, hydrologic regime, and topographic settings, enables identifying and isolating specific factors and geomorphic effects. Therefore, the Dead Sea level fall provides a rare opportunity to directly observe and analyze geomorphic processes in response to base-level fall at the field scale.

The geomorphic research in the Dead Sea is based on diverse datasets. In the last century, the Dead Sea level fall and the channel responses were documented in numerous aerial photographs with high spatial resolution and increasing temporal resolution (the Geological Survey of Israel and the Survey of Israel). High-resolution Digital Elevation Models (DEMs) are fundamental to field-scale geomorphic study. Most of the DEMs (0.5 m per pixel) of the western shore of the Dead Sea were acquired by the Geological Survey of Israel from nearly annual surveys of airborne Laser Detection and Ranging (LiDAR). In recent years, DEMs of local study sites along the Dead Sea shore have reached much higher spatial and temporal resolutions using drone-based photogrammetry. The

A. Frumkin and N. Shtober-Zisu (eds.), *Landscapes and Landforms of Israel*,
World Geomorphological Landscapes, https://doi.org/10.1007/978-3-031-44764-8_15

Fig. 15.1 **a** Incised spring-fed channels within the lacustrine sediments that were exposed in Ein Feshkha nature reserve during the Dead Sea level fall in recent decades (photographed by Iyad Swaed; see location in Fig. 15.2). **b** Coastal staircase morphology on the Dead Sea retreating coastline (photographed by Piotr Migoń). Each step is created during one winter, and its size and shape are determined by the topographic slope, the rate of the Dead Sea level changes, and the magnitude and frequency of waves (Enzel et al. 2022)

bathymetry of the Dead Sea was measured in a 5-m per pixel multibeam survey by John Hall (Sade et al. 2014), enabling the evaluation of future topographic constraints on the responding channels. Most of the Dead Sea tributaries are quite accessible and allow field work for detailed documentation. For less accessible segments, researchers used helicopters, boats, and time-lapse cameras. Remotely sensed imagery from satellites, such as Landsat (the U.S. Geological Survey) and Sentinel (European Space Agency), provide high temporal resolution of days to weeks for the study of the large main tributaries of the Dead Sea, Nahal HaArava and the Jordan River. These datasets enable one, for example, to detect the effect of individual floods on rapid morphological changes such as channel cutoffs.

15.2 Fluvial Response to Base-Level Fall

Streams and drainage basins are first-order agents of landscape evolution and sediment transport (e.g. Leopold and Wolman 1957; Schumm 1977); they respond to controls such as climate, hydrologic regime, lithology, and regional slope. Therefore, they serve in identifying and constructing such controls in the past on Earth and even on other planets

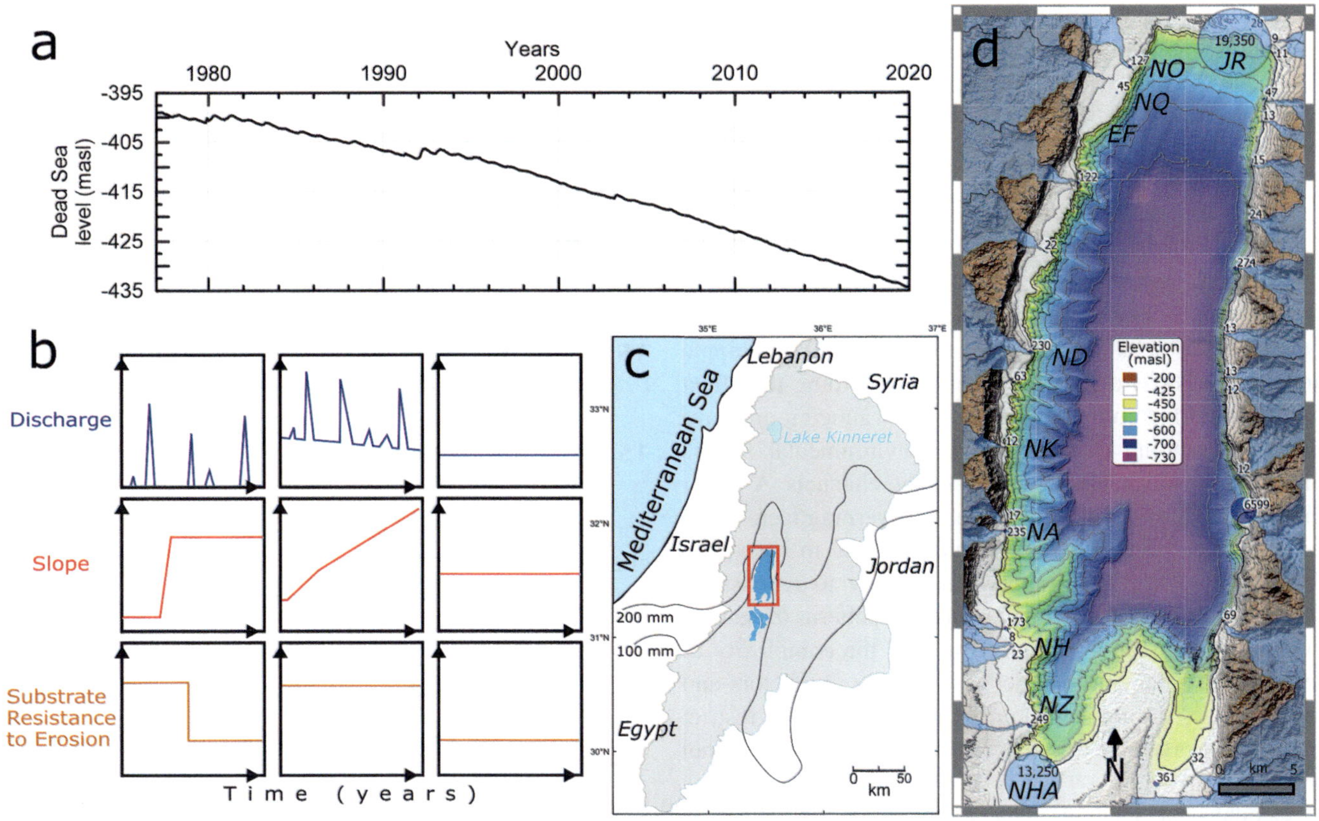

Fig. 15.2 **a** The Dead Sea level fall measured by the Hydrological Survey of Israel. **b** Schematic demonstration of the variability of the Dead Sea tributaries settings in the hydrologic regime (ephemeral, perennial, and spring-fed channel), emerging slope (sharp transition from moderate to steep, steepening, and stable slope), and resistance of the substrate to erosion (transient, resistant, and erodible substrate). **c** The Dead Sea drainage basin (gray) with 100 and 200 mm mean annual isohyets. **d** Topographic and bathymetric map of the Dead Sea, modified from Lensky and Dente (2015) and based on data from Sade et al. (2014). Most of the white area was exposed during the Dead Sea level fall in recent decades, and the yellow area will be completely exposed earlier than 2035. Blue polygons denote the downstream part of the lake's tributary watersheds (their area in square km is noted in the adjacent labels). Location of the channels mentioned in this chapter are (from north to south): JR—the Jordan River, NO—Nahal Og, NQ—Nahal Qumeran, EF—Ein Feshkha channels, ND—Nahal Darga, NK—Nahal Kedem, NA—Nahal Arugot, NH—Nahal Hever, NZ—Nahal Ze'elim, NHA—Nahal HaArava

and moons (Aharonson et al. 2002; Blum and Aslan 2006; Baker et al. 2015; Wang et al. 2019). Furthermore, fluvial adjustments and deposits are sources of information on water, fossil fuels, and mineral resources. Additional motivation for fluvial research stems from their destructive potential through floods, erosion, and transport of debris and contaminants along streams (Ben Moshe et al. 2008; Bennett et al. 2013).

A base level of a channel is the lowest point to which the channel can erode (e.g. an ocean, a lake, and other channels) (Powell 1875; Davis 1902). Base-level fall is a major trigger for changes in the fluvial system. When the level of the channel mouth is lowered, the potential energy increases, resulting in cascading effects upstream the drainage basin, with channel geometry modifications through lateral and vertical erosion (incision; Fig. 15.3). Therefore, a base-level fall changes the potential of the channel to transport and deposit sediment (Leopold and Bull 1979; Begin et al. 1981; Schumm 1993).

Researchers have studied the evolution of fluvial systems in response to base-level fall in many places, scales, and intervals, from the response of the largest rivers to eustatic global level fall of the ocean during glacial periods (up to 130 m sea-level fall in the last glacial maxima) (e.g. Leeder and Stewart 1996; Woolfe et al. 1998; Muto and Steel 2002; Tornqvist et al. 2006; Wang et al. 2019), through the response of the fluvial system to the level fall of the Mediterranean Sea during the Messinian crisis (102–103 m) (Loget and Van Den Driessche 2009; Paillou et al. 2009), to the response of a channel to a level fall of lakes on Earth and on Mars (101–102 m) (e.g. Davis et al. 2009; Adams 2012; Skorko et al. 2012).

Geomorphologists and sedimentologists (e.g. Davis 1893; Wolman and Miller 1960; Leopold and Bull 1979;

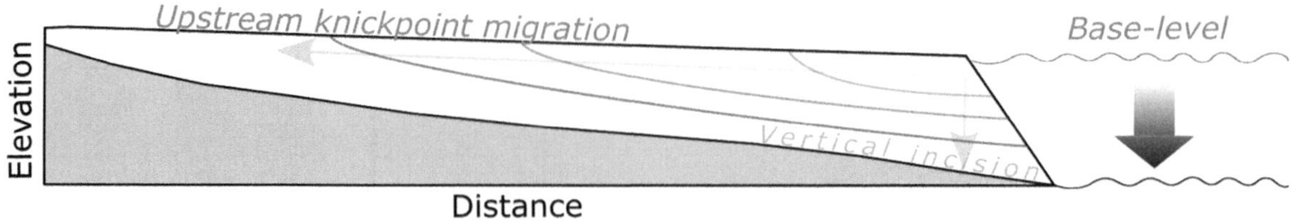

Fig. 15.3 Channel response to base-level fall through a vertical incision and upstream knickpoint migration. The rate, magnitude, form of the response are subjected to factors such as the lithological setting and the hydrologic regime of the channel

Dunne 1980; Schumm 1993; Posamentier 2001; Whipple and Tucker 2002) dealt with questions regarding the effect of base-level fall on fluvial systems. They aimed to improve our understanding of the effect of environmental factors on the morphology and dynamics of the channels. A major question in this field is what are these governing factors and how do they affect the geomorphic response to base-level fall? The answers can improve the ability to reconstruct past conditions and sediment transport patterns based on evidence for fluvial evolution. However, the complexity of the fluvial system and interactions between the factors and the processes during and after the response greatly restrict the ability to identify causality between factors and channel responses.

Geophysical methods enable to locate and analyze remnants of ancient fluvial systems such as channel deposits, meanders, and deltas, which are commonly subaqueous and/or buried (Weimer and Posamentier 1994; Woolfe et al. 1998; Posamentier 2001; Sylvester et al. 2012). Although these methods provide field-scale evidence for ancient fluvial evolution, their spatial and temporal resolutions are relatively low, regarding the studied channel and the environmental factors under which these fluvial systems evolved. To overcome this disadvantage, researchers have used physical experiments (flumes) and numerical models, in which they control the conditions and measure the geomorphic effect in very high spatial and temporal resolutions (e.g. Schumm and Khan 1971; Shepherd and Schumm 1974; Begin et al. 1981; Koss et al. 1994; Tucker and Whipple 1999; Haviv et al. 2010; Tal and Paola 2010). These studies have yielded significant insights regarding the mechanism of fluvial channel evolution. However, because of the different spatial and temporal scaling, identification of the governing mechanisms controlling the fluvial responses remains a challenge and it is hard to upscale conclusions from these simulations to the field scale (e.g. van Heijst and Postma 2001; Kleinhans et al. 2014). Some of the abovementioned knowledge gaps were filled by researchers using the unique setting on the receding Dead Sea.

15.3 Rapid Fluvial Landscape Evolution in Response to the Dead Sea Level Fall

15.3.1 Ephemeral Channels

An extensive study on the response of the ephemeral alluvial channels to the Dead Sea level drop was conducted by Ben-Moshe et al. (2008) and Bowman et al. (2007), emphasizing the great potential of these settings to gain insights into basic questions in fluvial geomorphology. In addition, the motivation for these studies stemmed from ongoing damage to structures around the Dead Sea. In 2001–2002, the incision of the ephemeral channels was already damaging main roads and bridges (Fig. 15.4).

By fitting a linear diffusion incision model to the main alluvial channels on the western Dead Sea coast, Ben-Moshe (2008) investigated the controlling factors over the recent evolution of these channels. The linear diffusion model was found to be suitable for simulating stream incision by Begin et al. (1981), based on Begin's classic flume experiments conducted in his Ph.D. Almost three decades later, Ben–Moshe (2008) utilized the Dead Sea setting to validate this fundamental model at a field-scale and showed that it can describe the evolution of channels with an occasional flow. The validation yielded a good fit between the modeled and reconstructed longitudinal profiles of the channels, and enabled predictions of future channel-bed elevations for better engineering planning under the progressive Dead Sea level fall.

Ben Moshe (2008) showed that the bathymetric slope emerging during the level drop is a first-order control over the magnitude of the vertical erosion, as suggested by Bowman (1988) for the response of Nahal Ze'elim to the level fall. In addition, Ben Moshe (2008) pointed that the potential discharge (calculated by the mean annual rainfall over a drainage basin) in the channels strongly impacts the incision rate.

Using a physically-based model, Salomon (2016) further studied the impact of the hydrological regime on the longitudinal evolution of relatively steep alluvial

Fig. 15.4 An example of progressive fluvial incision and its destructive consequences, Nahal Arugot, the Dead Sea (Fig. 15.2 for location). **a** New knickpoint (marked with white arrow) at the mouth of Wadi Arugot in 2017. The next flood will translate the potential energy to a vertical incision and upstream knickpoint retreat (see Fig. 15.3 for schematic illustration). **b** The collapse of the main road bridge over Nahal Arugot during a flood in 2001 (photos by Yuval Bartov; a video documentation of the collapse: https://youtu.be/s9GySqNj0cU), as a result of vertical incision and widening of the channel in response to the Dead Sea level fall. **c** Nahal Arugot in 2015 after massive structural interventions, including the construction of a new bridge with deep foundations (on the background), and channel-bed stabilization with large boulders and concrete pavement (photo by Liran Ben Moshe; Ben Moshe and Lensky 2020). While in the short term, the incision in the paved segment is inhibited, the progressive incision in the downstream segment of the channel emphasizes the need for long-term and continuous maintenance in response to Dead Sea level fall

channels responding to base-level fall. Based on available data and additional observations from Wadi Darga at the western coast of the Dead Sea (Fig. 15.2) and the model, Salomon (2016) determined that the shape of the flash-flood hydrographs, the frequency and order of the floods, and their mean peak discharge control the resulted channel profile. For example, the model results showed that for the same discharge volume and duration, a steady flow would result in a lower incision in comparison with a characteristic flood hydrograph with higher peak discharge (Salomon 2016).

The impact of flood frequency and magnitude on channel evolution can be even more substantial, as Eyal et al. (2019) showed based on field evidence from two adjacent channels in the northwest section of the Dead Sea (Fig. 15.5). While the channel characterized by larger and more frequent floods, Nahal Og, followed the receding Dead Sea shore and incised to form a deep gorge, the smaller and dryer channel, Nahal Qumeran, remained (partially) hydrologically disconnected from the lake (Eyal et al. 2019). The exposure of a relatively wide and low-sloped bathymetry in front of Nahal Qumeran mouth

15.3.2 Perennial Channels

Most of the drainage basin of the Dead Sea is characterized by a semi-arid to hyper-arid climate, however, some of its tributaries are flowing continuously. Among others, the perennial streams include the channels of Nahal HaArava (Fig. 15.7), the Jordan River (Fig. 15.8), and several spring-fed streams, e.g. in the Ein Feshkha natural reserve (Fig. 15.1). These tributaries present diverse characteristics in terms of hydrologic regime, the substrate into which they incise, and slopes that were exposed at their mouths during the lake level fall. Nahal HaArava and the Jordan River channels experienced extensive geomorphic responses that reached segments over 10 km upstream of their respective channel mouths. This is a much longer channel response in comparison to other tributaries around the Dead Sea, where the response is limited to<3 km by human-made structures such as roads that artificially inhibit the free upstream migration of the incision signal (Hassan and Klein 2002; Ben Moshe et al. 2008; Dente et al. 2017, 2018) (Fig. 15.3).

All these studies point out that the emerging slope is the major factor controlling incision and other channel

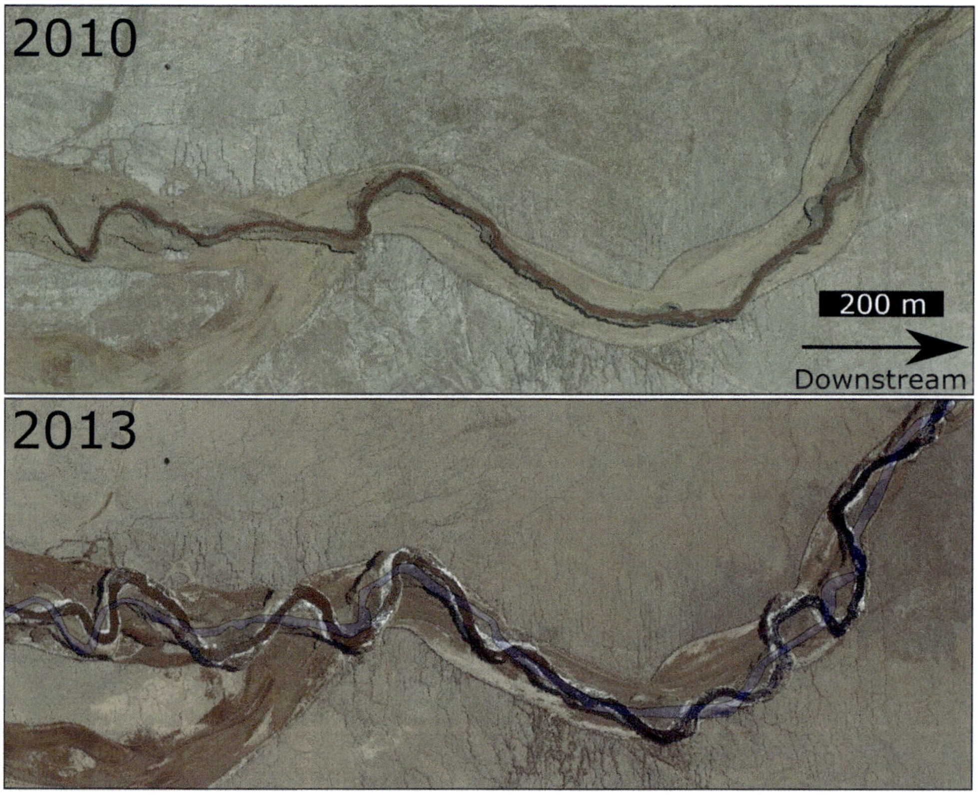

Fig. 15.7 The meandering and incising Nahal HaArava channel in response to the Dead Sea level fall, 2010–2013. A transparent blue polygon of the 2010 channel is shown for comparison on the 2013 aerial photograph (orthophotos: The Geological Survey of Israel)

Fig. 15.8 A completely new erosional landscape on the northern shore of the Dead Sea, created in the recent 50 years. During recent decades, the downstream segment of the Jordan River has evolved from a relatively shallow channel into a>30m deep meandering gorge in response to the Dead Sea level fall (photo: Terrascan). Note the remnants of the erosional terraces that exhibit lateral and vertical changes of the riverbed throughout the evolution. For example, the terraces of 1991 and 2002 are marked above the active channel in 2015. See Fig. 15.10 for the longitudinal profile evolution of the river since 1970

adjustments during base-level fall (Fig. 15.9) (Dente et al. 2017, 2018, 2021). When the emerging bathymetric slope (α in Fig. 15.9) is similar to the channel bed slope, the channel will mainly elongate toward and with the receding basin shoreline, rather than incise (Dente et al. 2017). This holds for perennial channels, in which the continuous flow can keep pace with the water-body regression. The Nahal HaArava evolution showed that this lengthening stage could be extended by large enough deposition near the channel mouth and rapid delta progradation, even in a steepening bathymetric setting (https://youtu.be/qmblBaTt-bs). Sufficient deposition is dependent on the sediment flux of the channel, but also on the depositional characteristics of the delta. For example, the Jordan River carries relatively high sediment flux. However, buoyant sediment plumes, steep bathymetry, and sediment reworking by waves and currents in the Dead Sea prevent delta depositions near the channel mouth (Dente et al. 2018).

The crucial role of the mouth slope in the response of different channels to the same base-level fall is further emphasized by comparing the evolutions of the channels of the Jordan River and Nahal HaArava (Fig. 15.10). In the Jordan River, the exposure of high mouth slopes (> 1%) began in the 1960s, when its delta front was exposed. Then, river incision was initiated, transferring the emerging high slopes signal of the delta front to upstream segments (Dente et al. 2018). In contrast, the response of Nahal HaArava during the 1960s–1980s was limited to lengthening on top of the desiccating, wide, and low-gradient lake floor of the southern Dead Sea basin. Only during the late 1980s, Nahal HaArava channel encountered a steep mouth slope, and its incision has begun (Fig. 15.10) (Dente et al. 2017).

Demonstrating these differences in the causes and consequences of adjacent tributaries draining to the same lake is important for analyzing the evolution of fluvial systems during ancient level-fall, for which data is limited. The entire response of some perennial channels can be restricted to elongation, without sufficient incision that enables sediment transport to the deep basin (Leeder and Stewart 1996; Woolfe et al. 1998; van Heijst and Postmal 2001; Tornqvist et al. 2006; Dente et al. 2017; Eyal et al. 2019). After the emergence of the break in the continental slope, a steeper slope (β in Fig. 9d) will trigger a deeper incised channel (Wood et al. 1993b; Dente et al. 2021).

Within a homogeneous substrate, the incision will continuously propagate landward and would trigger changes in the channel morphology. In such a case, the longitudinal profile of the channel can be described in terms of continuous diffusion, i.e. the incision is practically dependent on the local changes in the gradient (Begin et al. 1981) (Fig. 9d). This was the case of the Jordan River evolution (Figs. 15.8 and 15.10). However, if the channel, while incising, encounters a resistant layer, a spatially discontinuous response will be forced. In this case, the incision signal initiated in response to an emerging steep slope will be restricted to the downstream reach by a slowly migrating vertical knickpoint, without major modifications in upstream segments (Fig. 9e). Such resistant layer, a consolidated rock salt, restricted the evolution of Nahal HaArava until 2001 when the channel and the knickpoint incised entirely to a level below the hard layer. Then, a spatially continuous incision along the entire channel length was gained (Fig. 15.10) (Dente et al. 2017).

The studied perennial tributaries of the Dead Sea, regardless of their discharge, sediments, or width, exhibited a positive association between a steepening slope and channel sinuosity (Fig. 9f). Sinuosity increased in the segment downstream of the knickpoint of Nahal HaArava (Dente et al. 2017) (Fig. 15.7), in the confined downstream section of the Jordan River (Dente et al. 2018) (Fig. 15.8), and along the Ein Feshkha channels (Figs. 15.1 and 15.11) during exposure of steep mouth slopes (Dente et al. 2021). These observations of sinuosity development that accompanies channel incision shed light on the long-studied relationship between slope and meandering rivers. Previous field-scale studies of this correlation were mainly based on channels in equilibrium, across various substrate erodibilities, and using channel slope rather than the regional slope (e.g. Davis 1893; Harden 1990; van den Berg 1995; Kleinhans and van den Berg 2011). Therefore, these studies yielded contradicting conclusions and impressions. The Ein Feshkha channel setting enabled to isolate the relationship between slope and sinuosity and provided the clearest evidence that during steepening of the regional slope in a homogenous substrate and under steady discharge, sinuosity will increase (Dente et al. 2021).

Overall, the observations and insights from Ein Feshkha agree with the results of previous flume simulation studies (Koss et al. 1994; Schumm and Khan 1971; Shepherd 1972). However, the hydrologic regime in Ein Feshkha channels and in these simulations, a relatively stable discharge without episodic overbank floods (Vachtman and Laronne 2011, 2013), represents only a portion of river hydraulic regimes. The study of the Jordan River evolution revealed that the interaction of ongoing incision, increasing bed slope, and hydrologic regime that is characterized by occasional overbank flow, generates a dichotomy in the evolving channel pattern (Dente et al. 2018). In the downstream, deeply incised section, where high-magnitude floods ares confined within banks and do not reach the relatively high floodplain, sinuosity increases with the steepening slope, as observed in the smaller Ein Feshkha channels (Fig. 15.8). In the upstream section, where the channel incises but floods can still reach the floodplains, avulsions and chute cutoffs prevail, and the channel straightens (Fig. 15.9h, i). Observations from the Nahal HaArava channel support this insight; high-magnitude floods (mainly generated by tropical plume storms) yielded major planform changes and exerted a large avulsion around the retreating vertical knickpoint (Armon et al. 2018; Dente et al. 2017, 2018).

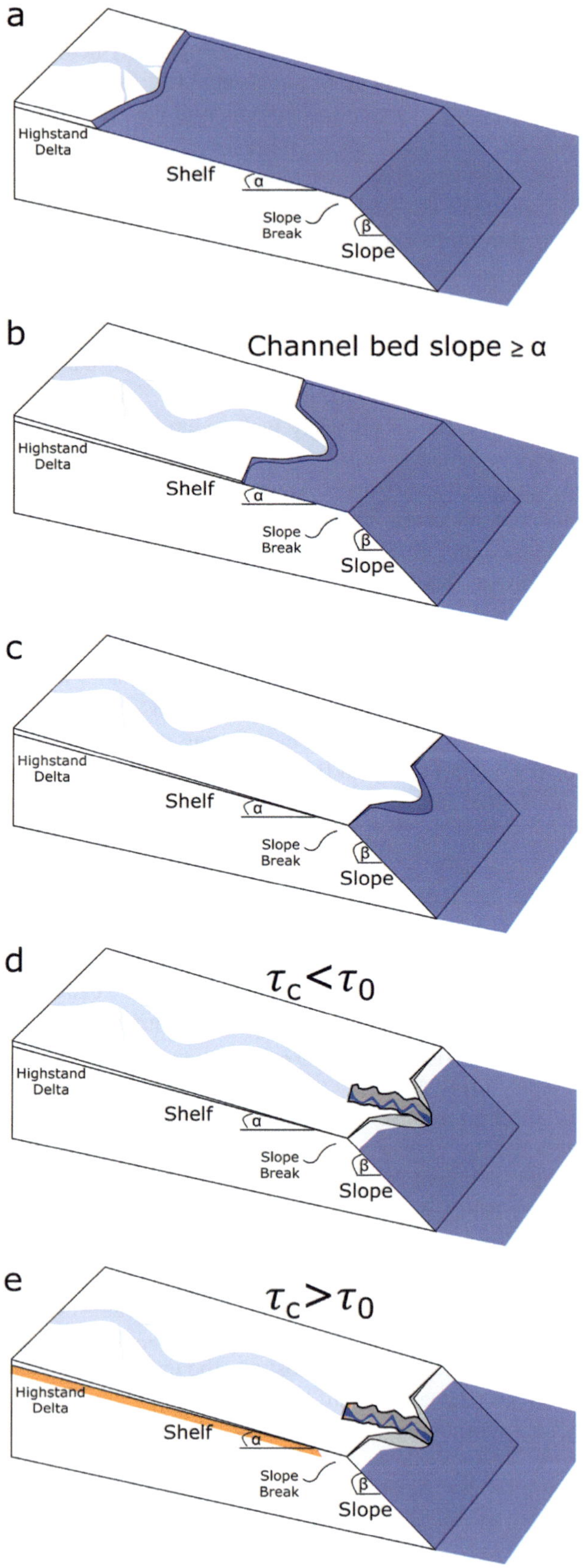

Fig. 15.9 Channel evolution in response to base-level fall (based on insights from the fluvial response to the Dead Sea level fall). **a** the initial fluvial and topographic settings, where α is the gradient of the shelf and β is the gradient of the continental slope. **b** channel elongation and delta progradation on top of the shelf without incision, due to channel slope ≥ α. **c** channel mouth slope moderation by delta progradation. **d** initiation of incision after channel mouth slope ≤ β; the flow shear stress, τ0, exceeds the critical shear stress, τc. **e** if present, resistant substrate (orange) inhibits the incision and the knickpoint upstream migration where τ0 < τc. **f** incision, confinement, and meandering in response to the exposer of steep mouth slope, ~ β, during the base-level fall. **g** overbank floods on the original floodplain can still occur upstream to the confined channel. **h** overbank floods can react with steep banks of the incised segment to create headcut retreat. **i** chute cutoffs in the upstream segment yields channel straightening, while the sinuosity of the downstream segment increases

Fig. 15.9 (continued)

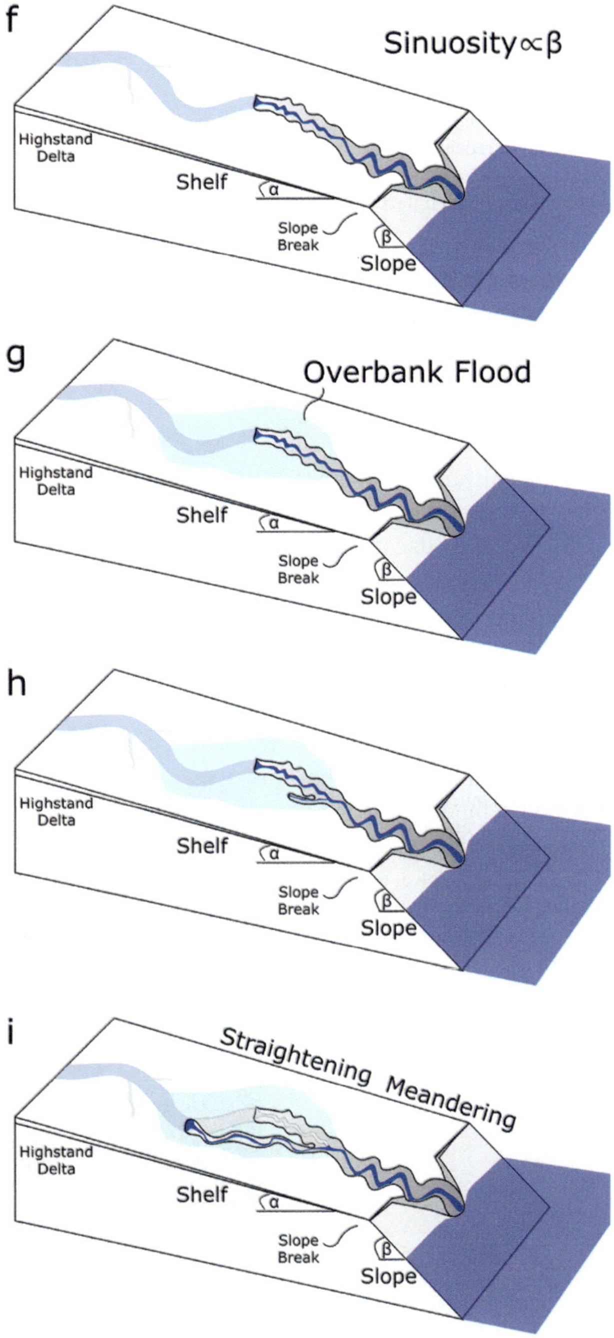

The potential for chute cutoff is a function of the flood magnitude and duration and the erodibility of the floodplain; such cutoffs can be triggered by enhanced elevation differences between the upstream and downstream limbs of an individual curve along the channel (Constantine et al. 2010). The results from the studied perennial tributaries of the Dead Sea imply that such elevation differences prevail along incising channels. Knickpoints and natural dams, induced by failures of the high banks, are frequently observed (Dente et al. 2017, 2018, 2021),

The integration of the insights from Ein Feshkha channels and the Jordan River can promote better interpretations of ancient planform evolution. For example, abundant evidence for chute cutoffs along an incised channel can imply that frequent overbank floods prevailed in the channel during incision, whereas absence of or rare evidence for chute cutoffs can be the result of infrequent high-magnitude floods during the evolution. Although, as noted, the erodibility of the floodplain is a major control on the likelihood for chute cutoff and can mask these hydrological interpretations.

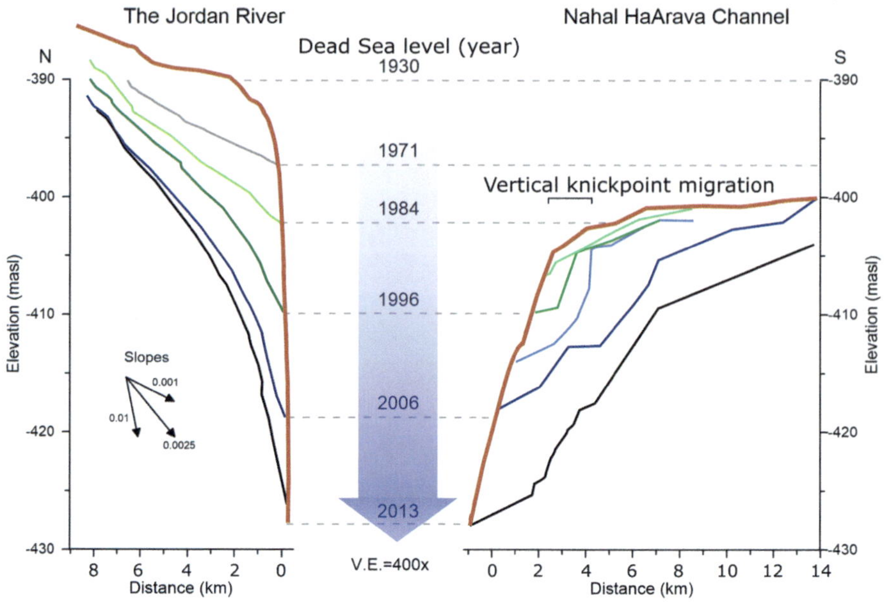

Fig. 15.10 A comparison between the longitudinal evolution of the Jordan River and Nahal HaArava channels in response to the Dead Sea level drop in recent decades. Brown lines denote the exposed bathymetry in which the channels incised

Fig. 15.11 A 10-m deep incised meander within one of the channels in Ein Feshkha nature reserve (flow is to the left). These spring-fed channels lengthened, incised, and meandered at different rates under the control of the emerging slope at the receding Dead Sea shore

15.4 Summary Remarks

The Dead Sea potential as a unique field-scale fluvial geomorphology laboratory is far from being fully exploited. The conditions under which the tributaries of the lake actively evolve are highly varied. Many perennial and ephemeral channels on the eastern and steeper coast of the Dead Sea (Kingdom of Jordan) were not studied in detail for their fluvial response in recent decades (Fig. 15.2c, d). As the lake continues to fall at a rate of > 1 m per year, we will be able to observe, measure, and model more changes, controls, and processes in the responding channels. Further monitoring of changes should be encouraged as this area provides unique real-time field observations that can largely enhance our understanding of geomorphic processes under base-level fall and their general and practical consequences.

Acknowledgements I wish to thank Yehouda Enzel, Efrat Morin, Nadav Lensky, Amos Frumkin, Liran Ben Moshe, Reut Solomon, and Haggai Eyal for their constructive discussion and comments. I am also grateful to Iyad Swaed, Haggai Eyal, Liran Ben Moshe, Piotr Migoń, and Yuval Bartov for contributing some of the photographs in this chapter.

References

Abelson M, Yechieli Y, Baer G, Lapid G, Behar N, Calvo R, Rosensaft M (2017) Natural versus human control on subsurface salt dissolution and development of thousands of sinkholes along the Dead Sea coast. J Geophys Res Earth Surf 122:1262–1277. https://doi.org/10.1002/2017JF004219

Adams K (2012) Response of the Truckee River to lowering base level at Pyramid Lake, Nevada, based on historical air photos and LiDAR data. Geosphere 8:607. https://doi.org/10.1130/GES00698.1

Aharonson O, Zuber MT, Rothman DH, Schorghofer N, Whipple KX (2002) Drainage basins and channel incision on Mars. Proc Nat Ac Sci USA 99:1780–1783. https://doi.org/10.1073/pnas.261704198

Armon M, Dente E, Smith JA, Enzel Y, Morin E (2018) Synoptic-scale control over modern rainfall and flood patterns in the levant drylands with implications for past climates. J Hydromet 19:1077–1096. https://doi.org/10.1175/JHM-D-18-0013.1

Avni Y, Lensky N, Dente E, Shviro M, Arav R, Gavrieli I, Yechieli Y, Abelson M, Lutzky H, Filin S, Haviv I (2015) Self-accelerated development of salt karst during flash floods along the Dead Sea Coast, Israel. J Geophys Res Earth Surf 1–22. https://doi.org/10.1002/2015JF003738.Received

Baker VR, Hamilton CW, Burr DM, Gulick VC, Komatsu G, Luo W, Rice JW, Rodriguez JAP (2015) Fluvial geomorphology on Earth-like planetary surfaces: a review. Geomorphology 245:149–182. https://doi.org/10.1016/j.geomorph.2015.05.002

Begin ZB, Meyer DF, Schumm SA (1981) Development of longitudinal profiles of alluvial channels in response to base-level lowering. Earth Surf Proc Land 6:49–68. https://doi.org/10.1002/esp.3290060106

Ben Moshe L, Lensky NG (2020) Increased deviation of Dead Sea tributaries from steady state and recommendations for infrastructure planning in changing conditions. Geol Surv Isr Rep GSI/27/202

Ben Moshe L, Haviv I, Enzel Y, Zilberman E, Matmon A (2008) Incision of alluvial channels in response to a continuous base level fall: field characterization, modeling, and validation along the Dead Sea. Geomorphology 93:524–536. https://doi.org/10.1016/j.geomorph.2007.03.014

Bennett GL, Molnar P, Mcardell BW, Schlunegger F, Burlando P (2013) Patterns and controls of sediment production, transfer and yield in the Illgraben. Geomorphology 188:68–82. https://doi.org/10.1016/j.geomorph.2012.11.029

Blum MD, Aslan A (2006) Signatures of climate vs. sea-level change within incised valley-fill successions: quaternary examples from the Texas GULF Coast. Sed Geol 190:177–211. https://doi.org/10.1016/j.sedgeo.2006.05.024

Bowman D (1988) The declining but non-rejuvenating base level—The Lisan lake, the Dead Sea area, Israel. Earth Surf Proc Land 13:239–249. https://doi.org/10.1002/esp.3290130305

Bowman D, Shachnovich-Firtel Y, Devora S (2007) Stream channel convexity induced by continuous base level lowering, the Dead Sea, Israel. Geomorphology 92:60–75. https://doi.org/10.1016/j.geomorph.2007.02.009

Bowman D, Svoray T, Devora S, Shapira I, Laronne JB (2010) Extreme rates of channel incision and shape evolution in response to a continuous, rapid base-level fall, the Dead Sea, Israel. Geomorphology 114:227–237. https://doi.org/10.1016/j.geomorph.2009.07.004

Bowman D, Devora S, Svoray T (2011) Drainage organization on the newly emerged Dead Sea bed, Israel. Quat Int 233:53–60. https://doi.org/10.1016/j.quaint.2010.05.022

Constantine JA, McLean SR, Dunne T (2010) A mechanism of chute cutoff along large meandering rivers with uniform floodplain topography. Bull Geol Soc Am 122:855–869. https://doi.org/10.1130/B26560.1

Davis WM (1893) Comment on The Osage river and its meanders. Scince XXII

Davis WM (1902) Baselevel, grade and peneplain. J Geol 10:77–111

Davis M, Matmon A, Zilberman E, Porat N, Gluck D, Enzel Y (2009) Bathymetry of Lake Lisan controls late Pleistocene and Holocene stream incision in response to base level fall. Geomorphology 106:352–362. https://doi.org/10.1016/j.geomorph.2008.11.014

Dente E, Lensky NG, Morin E, Grodek T, Sheffer NA, Enzel Y (2017) Geomorphic response of a low-gradient channel to modern, progressive base-level lowering: Nahal HaArava, the Dead Sea. J Geophys Res Earth Surf 122:2468–2487. https://doi.org/10.1002/2016JF004081

Dente E, Lensky NG, Morin E, Dunne T, Enzel Y (2018) Sinuosity evolution along an incising channel: new insights from the Jordan River response to the Dead Sea level fall. Earth Surf Proc Land 44(3):781–795. https://doi.org/10.1002/esp.4530

Dente E, Lensky NG, Morin E, Enzel Y (2021) From straight to deeply incised meandering channels: Slope impact on sinuosity of confined streams. Earth Surf Proc Land 46(5):1041–1054. https://doi.org/10.1002/esp.5085

Dunne T (1980) Formation and controls of channel networks. Progr Phys Geogr Earth Environ 4:211–239. https://doi.org/10.1177/030913338000400204

Enzel Y, Mushkin A, Groisman M, Calvo R, Eyal H, Lensky N (2022) The modern wave-induced coastal staircase morphology along the western shores of the Dead Sea. Geomorphology. https://doi.org/10.1016/j.geomorph.2022.108237

Eyal H, Dente E, Haviv I, Enzel Y, Dunne T, Lensky NG (2019) Fluvial incision and coarse gravel redistribution across the modern Dead Sea shelf as a result of base-level fall. Earth Surface Proces Landforms: esp.4640. https://doi.org/10.1002/esp.4640

Eyal H, Enzel Y, Meiburg E, Vowinckel B, Lensky NG (2021). How does coastal gravel get sorted under stormy longshore transport? Geophys Res Lett 48:e2021GL095082. https://doi.org/10.1029/2021GL095082

Fernández R, Parker G, Stark CP (2019) Experiments on patterns of alluvial cover and bedrock erosion in a meandering channel. Earth Surf Dynam Disc: 1–40. https://doi.org/10.5194/esurf-2019-8

Gardner TW (1975) The history of part of the Colorado River and its tributaries: an experimental study. In: Four Corners Geological Society Guidebook: Field Conference, 8th, Canyonlands, Utah, pp 87–95

Gardner TW (1983) Experimental study of knickpoint and longitudinal profile evolution in cohesive, homogeneous material. Geol Soc Am Bull 94:664–672. https://doi.org/10.1130/0016-7606(1983)94<664

Greenbaum N, Ben-Zvi A, Haviv I, Enzel Y (2006) The hydrology and paleohydrology of the Dead Sea tributaries. Geol Soc Am Sp Pap 401:63–93. https://doi.org/10.1130/2006.2401(05)

Grimaud J-L, Paola C, Voller V (2015) Experimental migration of knickpoints: influence of style of base-level fall and bed lithology. Earth Surf Dynam Disc 3:773–805. https://doi.org/10.5194/esurfd-3-773-2015

Harden DR (1990) Controlling factors in the distribution and development of incised meanders in the central Colorado Plateau. Bull Geol Soc Am 102:233–242. https://doi.org/10.1130/0016-7606(1990)102<0233:CFITDA>2.3.CO;2

Hassan MA, Klein M (2002) Fluvial adjustment of the Lower Jordan River to a drop in the Dead Sea level. Geomorphology 45:21–33. https://doi.org/10.1016/S0169-555X(01)00187-8

Haviv I, Enzel Y, Whipple KX, Zilberman E, Matmon A, Stone J, Fifield KL (2010) Evolution of vertical knickpoints (waterfalls) with resistant caprock: insights from numerical modeling. J Geophys Res 115:F03028. https://doi.org/10.1029/2008JF001187

Istanbulluoglu E, Bras RL, Flores-Cervantes H, Tucker GE (2005) Implications of bank failures and fluvial erosion for gully development: field observations and modeling. J Geophys Res 110:F01014. https://doi.org/10.1029/2004JF000145
Kleinhans MG, van den Berg JH (2011) River channel and bar patterns explained and predicted by an empirical and a physics-based method. Earth Surf Proc Land 36:721–738. https://doi.org/10.1002/esp.2090
Kleinhans MG, van Dijk WM, van de Lageweg WI, Hoyal DC, Markies H, van Maarseveen M, Roosendaal C, van Weesep W, van Breemen D, Hoendervoogt R, Cheshier N (2014) Quantifiable effectiveness of experimental scaling of river- and delta morphodynamics and stratigraphy. Earth-Sci Rev 133:43–61. https://doi.org/10.1016/j.earscirev.2014.03.001
Koss JE, Ethridge FG, Schumm SA (1994) An experimental study of the effects of base-level change on fluvial, coastal plain and shelf systems. J Sed Res B64:90–98. https://doi.org/10.1306/d4267f64-2b26-11d7-8648000102c1865d
Leeder MR, Stewart MD (1996) Fluvial incision and sequence stratigraphy: alluvial responses to relative sea-level fall and their detection in the geological record. Geol Soc London Sp Pub 103:25–39. https://doi.org/10.1144/GSL.SP.1996.103.01.03
Lensky NG, Dente E (2015) The hydrological processes driving the accelerated Dead Sea level decline in the past decades. Geol Surv Isr Rep GSI/16/2015
Lensky NG, Dvorkin Y, Lyakhovsky V, Gertman I, Gavrieli I (2005) Water, salt, and energy balances of the Dead Sea. Water Resour Res 41:1–13. https://doi.org/10.1029/2005WR004084
Leopold B, Bull B (1979) Base level, aggradation, and grade. P Am Philos Soc 123(3):168–202
Leopold BL, Wolman MG (1957) River channel patterns: braided meandering and straight. Geol Surv Prof Paper 282 [online]. Available from: https://pubs.usgs.gov/pp/0282b/report.pdf
Loget N, Van Den Driessche J (2009) Wave train model for knickpoint migration. Geomorphology 106:376–382. https://doi.org/10.1016/j.geomorph.2008.10.017
Muto T, Steel RJ (2002) In defense of shelf-edge delta development during falling and lowstand of relative sea level. J Geol 110:421–436. https://doi.org/10.1086/340631
Paillou P, Schuster M, Tooth S, Farr T, Rosenqvist A, Lopez S, Malezieux JM (2009) Mapping of a major paleodrainage system in eastern Libya using orbital imaging radar: the Kufrah River. Earth Planet Sc Lett 277:327–333. https://doi.org/10.1016/j.epsl.2008.10.029
Posamentier HW (2001) Lowstand alluvial bypass systems: incised vs. unincised. AAPG Bull 85:1771–1793
Powell J (1875) Exploration of the Colorado river of the west and its tributaries. Government Printing Office, Washington
Sade A, Hall JK, Amit G, Tibor G, Schulze B, Gur-Arieh I, Ten-Brink U, Ben Avraham Z (2014) Multibeam bathymetric map of the Dead Sea. Isr Geol Soc Ann Meet
Salomon R (2016) Evolution of gravel-bed channels in response to non-steady flow regime. The Hebrew University of Jerusalem
Schumm SA (1977) The fluvial system. Wiley, New York, p 1977
Schumm SA (1993) River response to baselevel change: implications for sequence stratigraphy. J Geol 101:279–294
Schumm SA, Khan HR (1971) Experimental study of channel patterns. Nature 233:407–409. https://doi.org/10.1038/233407a0
Shepherd RG (1972) Incised river meanders: evolution in simulated bedrock. Science 178:409–411. https://doi.org/10.1126/science.178.4059.409
Shepherd RG, Schumm SA (1974) Experimental study of river incision. Geol Soc Am Bull. https://doi.org/10.1130/0016-7606(1974)85<257
Skorko K, Jewell PW, Nicoll K (2012) Fluvial response to an historic lowstand of the Great Salt Lake, Utah. Earth Surf Proc Land 37:143–156. https://doi.org/10.1002/esp.2226
Storz-Peretz Y, Bowman D, Laronne JB, Svoray T (2011) Rapid incision of a small, coarse and steep fan-delta in response to base-level fall: the case of Nahal Qedem, the Dead Sea, Israel. Earth Surf Proc Land 36:467–480. https://doi.org/10.1002/esp.2066
Sylvester Z, Deptuck ME, Prather BE, Pirmez C, O'byrne C (2012) Seismic stratigraphy of a shelf-edge delta and linked submarine channels in the northeastern Gulf of Mexico. Application of the principles of seismic geomorphology to continental-slope and base-of-slope systems: case studies from seafloor and near-seafloor analogues. SEPM Special Publication, pp 31–59
Tal M, Paola C (2010) Effects of vegetation on channel morphodynamics: results and insights from laboratory experiments. Earth Surf Proc Land 35:1014–1028. https://doi.org/10.1002/esp.1908
Tornqvist TE, Wortman SR, Mateo ZRP, Milne GA, Swenson JB (2006) Did the last sea level lowstand always lead to cross-shelf valley formation and source-to-sink sediment flux? J Geophys Res Earth Surf 111:1–13. https://doi.org/10.1029/2005JF000425
Tucker GE, Whipple KX (1999) Dynamics of the stream-power river incision model: implications for height limits of mountain ranges, landscape response timescales, and research needs. J Geophys Res 104:661–674
Vachtman D, Laronne JB (2011) Flow structure over bed irregularities in a straight cohesive open channel. J Hydraul Eng 137:1335–1346. https://doi.org/10.1061/(ASCE)HY.1943-7900.0000447
Vachtman D, Laronne JB (2013) Hydraulic geometry of cohesive channels undergoing base level drop. Geomorphology 197:76–84. https://doi.org/10.1016/j.geomorph.2013.04.039
van den Berg JH (1995) Prediction of alluvial channel pattern of perennial rivers. Geomorphology 12:259–279. https://doi.org/10.1016/0169-555X(95)00014-V
van Heijst MWIM, Postma G (2001) Fluvial response to sea-level changes: a quantitative analogue, experimental approach. Basin Res 13:269–292. https://doi.org/10.1046/j.1365-2117.2001.00149.x
Wang R, Colombera L, Mountney NP (2019) Geological controls on the geometry of incised-valley fills: insights from a global dataset of late-quaternary examples. Sedimentology. https://doi.org/10.1111/sed.12596
Watson RA, Holohan EP, Al-Halbouni D, Saberi L, Sawarieh A, Closson D, Alrshdan H, Abou Karaki N, Siebert C, Walter TR, Dahm T (2018) Sinkholes, stream channels and base-level fall: a 50-year record of spatio-temporal development on the eastern shore of the Dead Sea. Solid Earth Disc 1–43
Weimer P, Posamentier HW (1994) Siliciclastic sequence stratigraphy: recent developments and applications. Am Assoc Petrol Geol 7:210
Whipple KX, Tucker GE (2002) Implications of sediment-flux-dependent river incision models for landscape evolution. J Geophys Res 107(B2):ETG-3
Wolman GM, Miller JP (1960) Magnitude and frequency of forces in geomorphic processes. J Geol 68:54–74
Wood L, Ethridge F, Schumm S (1993a) An experimental study of the influence of subaqueous shelf angles on coastal plain and shelf deposits. In: Recent developments in siliciclastic sequence, pp 381–393. https://doi.org/10.1306/M58581C15
Wood L, Ethridge F, Schumm S (1993b) The effects of rate of base-level fluctuation on coastal-plain, shelf and slope depositional systems: an experimental approach. Sequence Stratigraphy and Facies Associations, pp 381–393. https://doi.org/10.1002/9781444304015
Woolfe KJ, Larcombe P, Naish T, Purdon RG (1998) Lowstand rivers need not incise the shelf: an example from the Great Barrier Reef, Australia, with implications for sequence stratigraphic models. Geology 26:75–78. https://doi.org/10.1130/0091-7613(1998)026<0075:LRNNIT>2.3.CO;2
Yechieli Y, Abelson M, Bein A, Crouvi O, Shtivelman V (2006) Sinkhole "swarms" along the Dead Sea cost: Reflection of disturbance of lake and adjacent groundwater systems. Bull Geol Soc Am 118:1075–1087. https://doi.org/10.1130/B25880.1

Part V

Southern Israel

Landscape Evolution of the Central Negev

16

Yoav Avni

Abstract

This chapter presents the main features composing the Central Negev landscape and discusses their origin and evolution. The present landscape contains a mosaic of morphotectonic elements developed during the last 90 Ma through several geologic events. The Syrian Arc fold structures, expressed by the Ramon and Arif-Badad monoclines and their adjacent synclines, Avedat and Loz-Neqarot, developed at 90–50 Ma and still dominate the present-day landscape. The regional truncation surface (RTS) that originated during the Oligocene and Early Miocene (30–20 Ma) controls the contemporary skyline, while large fault lines, tectonic depressions and dome structures that locally deformed the general flat relief testify to the Early Miocene to Early Pleistocene (18–1.5 Ma) tectonic activity. This activity terminated the evolution of the Sinai-Israel microplate and generated the development of the cliff systems and the deeply incised canyons and valleys. Although the present-day landscape is under ongoing development, in most cases, it is possible to link the young geomorphic features to older structures that were rejuvenated. This pattern is best demonstrated by the deeply incised valleys developed within the core of the Syrian Arc anticlines (Makhteshim). Although their development was initiated in the Late Cretaceous, their near-present dimensions were controlled by the Oligocene truncation event, while their development towards the oval-incised valleys was caused by younger tectonic events occurring during the Miocene. Most of the time, the landscape evolutionary processes were active under arid climates, contributing to good preservation of morphotectonic and geomorphologic features. These environmental conditions promoted the human occupation of this desert region, located at the crossroad between Africa and Asia, over the past 1.5 Ma.

Keywords

Morphotectonics · Landscape evolution · Desert environments · Drainage systems · Cliff formation · Drainage reorganization

16.1 Introduction to the Central Negev

The Central Negev occupies the central part of the Negev Desert, which comprises the southern half of Israel (Fig. 16.1). The region is forming a large high-elevated dome (Fig. 16.2) while its borders are controlled by main tectonic and morphologic structures. The northern border of the Central Negev follows the E-W Zin Fault located near Sede Boker, marking the transition from the Avedat syncline to the anticline structures of the northern Negev, while its southern border is located along the E-W Paran Fault and Paran stream channel (Fig. 16.3). In the east, the Central Negev region borders the deep depression of the Arava Valley, developed along the Dead Sea Fault system, while along its western border, the Central Negev forms a high-elevation terrain that sharply descends toward the northern Sinai lowlands, drained by the numerous tributaries of Wadi el Arish (Figs. 16.1 and 16.2).

Within these boundaries, the Central Negev occupies approximately 4000 km^2. This large region is one of the most diverse geologic and morphotectonic regions in Israel, having the highest altitude in southern Israel and comprising several summits reaching elevations of more than 1000 m (the highest peak is Mount Ramon: 1037 m; Fig. 16.3).

The region contains the largest structure of the Syrian Arc fold belt (Krenkel 1924) developed in the Negev: the

Y. Avni (✉)
Geological Survey of Israel, 32 Yesha'ayahu Leibowitz st., 9692100 Jerusalem, Israel
e-mail: yavni@gsi.gov.il

A. Frumkin and N. Shtober-Zisu (eds.), *Landscapes and Landforms of Israel*, World Geomorphological Landscapes, https://doi.org/10.1007/978-3-031-44764-8_16

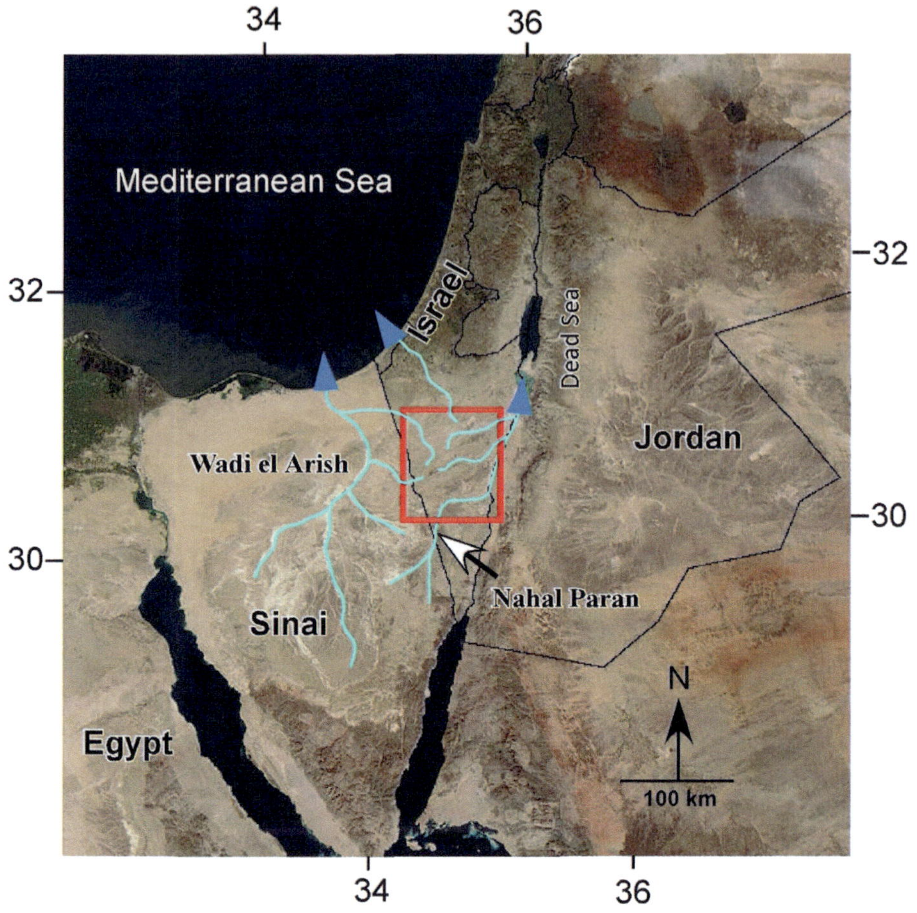

Fig. 16.1 General setting of the Central Negev (indicated by red rectangle). Main drainage basins are marked in blue

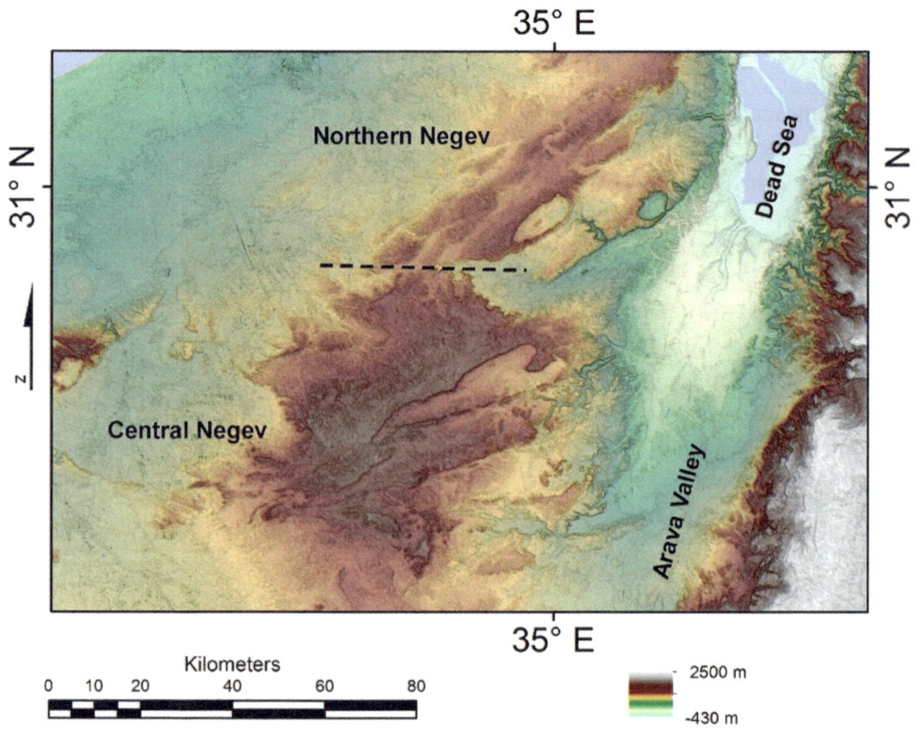

Fig. 16.2 The Central and Northern Negev regions in DTM imagery. Note that the Central Negev is shaped as a large dome while the Northern Negev Highlands is shaped as an elongated arch. The dashed line is marking the border between these regions. *Source* Shuttle Radar Topography Mission, SRTM data

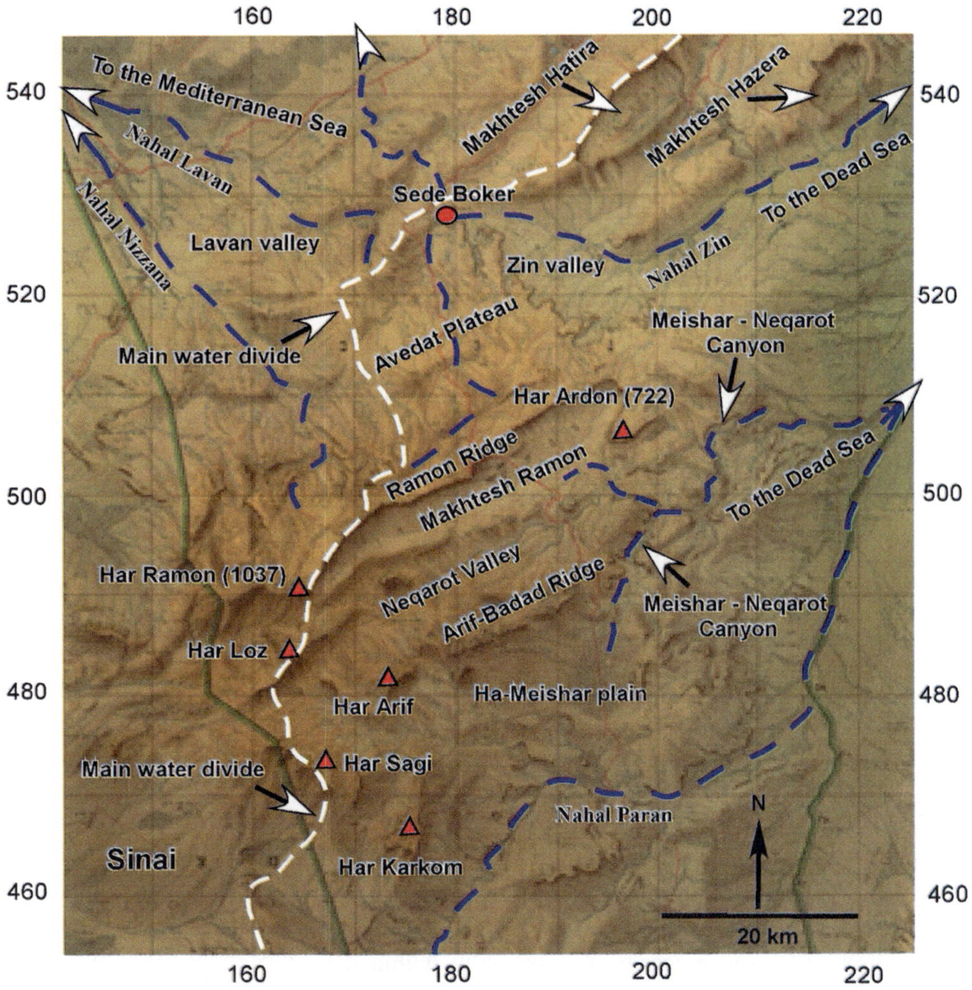

Fig. 16.3 The main physiographic features in the Central Negev. The dashed white line marks the location of the main water divide, whereas dashed blue lines indicate watercourses (Nahal = Stream). International borders are marked in green

Ramon Monocline, and five of the six major east–west fault lines in the Negev-Sinai Shear Belt (Bartov 1974; Fig. 16.5). In addition, the region comprises the largest erosive valley that developed within the core of an anticlinal structure: Makhtesh Ramon (Figs. 16.2, 16.3 and 16.5). In contrast to its high elevation, most of its relief is shaped as a large plateau, dissected by numerous valleys, drained toward the Mediterranean Sea in the northwest and the Dead Sea in the northeast (Figs. 16.1 and 16.3). The region presents abundant geological and geomorphological elements that developed during the most important geological events that shaped the Negev region in the past 90 Ma. The Syrian Arc structures, although developed some 90–50 Ma ago, still dominate the present-day landscape. However, younger features were developed interacting with the main geological events that controlled the plate separation between Africa and Arabia, triggering the creation and evolution of the Sinai-Israel microplate during the last 25 Ma. Since the past 20–18 Ma, the region has been affected by tectonic activity localized along the Dead Sea Fault system, accommodating a plate separation of 105 km between the Arabian plate and Sinai-Israel microplate, named the Dead Sea Transform.

The ongoing tectonism involved several stages of uplift that triggered the formation of several large drainage basins, with a few phases of drainage reorganization, including large-scale drainage capture events. During the Pliocene and Pleistocene the climate deteriorated from semi-arid to arid and landscape evolutionary processes were active under drier climate, contributing to the special conservation of old landscape elements. As a result, the region is rich in preservation (and documentation) of long-lasting human occupation, recorded in the region during the last 1.5 Ma.

The aim of this chapter is to present the long-term development of the landscape in the Central Negev and to evaluate the link between landscape evolutionary process and major morphotectonic events.

16.2 Geologic and Tectonic Composition

16.2.1 Exposed Geological Strata: Lithological Units Composing the Central Negev

The geological strata exposed in the region accumulated as a thick stratigraphic column ranging from 2000 to 2300 m, covering a time span from the Middle Triassic to the recent (Zilberman and Avni 2004; Avni et al. 2017b; Fig. 16.4). During most of this long period, the region was influenced by shallow marine ingressions over a relatively flat continental margin located at the northeastern edge of the African continent, facing the Tethys Ocean. The section contains marine sediments such as limestones and dolostones that interfinger with continental sediments such as sandstones, clays and conglomerates. While the exposed Triassic to Lower Cretaceous units, composed mainly of sandstone and clay, are confined to the eroded core of the Ramon and Arif-Badad Monoclines, the Upper Cretaceous Judea Group and the Eocene Avedat Group, composed of massive beds of dolostone, limestone and chalk, occupy extensive areas in the Central Negev. The most important stratigraphic elements for landscape evolutionary processes are the major transitions from resistant to nonresistant units, such as the transition from friable sandstone in the Lower Cretaceous Hatira Formation to hard carbonates in the Judea Group deposited above (Fig. 16.4). This transition forms cliffs and escarpments, especially within the eroded and dissected anticlines. Another important stratigraphic transition is located between the chalky-marly units composing the Ghareb and Taqiye formations that belong to the Mount Scopus Group of Late Cretaceous–Paleocene age, and the hard carbonates units in the Avedat Group of Eocene age (Fig. 16.4). This transition is forming large-scale cliffs and escarpments such as the Loz cliffs that developed on the southeastern edge of Har (mount) Loz presented on Fig. 16.3. In places where the hard carbonates layers in the Eocene strata were removed by erosion, prominent areas of badlands were developed in the nonresistant marly units, especially in the Zin Valley and the Nahal Lavan Basin located in the northern sector of the region (Fig. 16.3). In the same way, extensive areas of badlands developed in the south, in areas bordering the table mountains of Har Sagi and Har Karkom (Figs. 16.2 and 16.3). The youngest rock units exposed in the region are composed of continental deposits of conglomerates, sandstones and lacustrine marls deposited during the Neogene to the recent.

16.2.2 Syrian Arc Structures—Ramon and Arif-Badad Monoclines

First described by Krenkel (1924), the Syrian Arc fold belt extends from the Palmyrides in Syria across northern and central Israel to the Central Negev and Sinai Peninsula. The Central Negev encompasses two large monoclinal structures—Ramon and Arif-Badad—located to its south (Fig. 16.5). In the vicinity to these uplifted structures, three synclines were developed: the Avedat syncline, developed to the north of the Ramon Monocline; the Loz-Neqarot syncline, developed between the two monoclines; and the southern Negev broad syncline, developed to the south of the Arif-Badad Monocline. The latter syncline is connected to the large syncline occupying most of central Sinai (Bartov 1974; Fig. 16.6).

These fold structures generally strike from ENE-WSW to NNE-SSW have developed along deep-seated faults originally established at the end of the Precambrian orogeny and were later reactivated as normal faults during the Triassic (Garfunkel 1978). From the Late Turonian to the Early Eocene (90–50 Ma), compression associated with intraplate shortening reactivated these faults as reverse faults, creating asymmetric folds near the surface (Guiraud and Bosworth 1997; Walley 1998; Segev and Rybakov 2010). The evolution of these structures in several stages from the Santonian to Early Eocene, when the region was located near the southern edge of the Tethys Ocean, facilitated the formation of several angular unconformities within the sedimentary strata. Along the crests of the larger structures, such as the Ramon Monocline, terrestrial conglomerates are associated with these unconformities, indicating that the most uplifted structures emerged from the sea, forming elongated islands (Avni 1993).

At the end of the Late Cretaceous, the Ramon Monocline was uplifted and deeply truncated, exposing its sandstone core composed of the Lower Cretaceous Hatira Formation (Avni 1993). This conclusion is based on the identification of quarzitic sands forming lenses below the Eocene section deposited in the western sector of the Ramon Monocline (Avni 1991), indicating that a section of more than 500 m of carbonate rocks was removed from the crest of the Ramon Monocline before the Eocene. Simultaneously, less uplifted structures such as the Arif-Badad Monocline remained below sea level and accumulated thin sections of the Mount Scopus Group, condensed to approximately 10–50 m, while much thicker sections, up to 200–400 m, were deposited in the synclines bordering these monoclines. A second event of gradual uplift and more regional truncation occurred in the Oligocene (Avni et al. 2012b) that

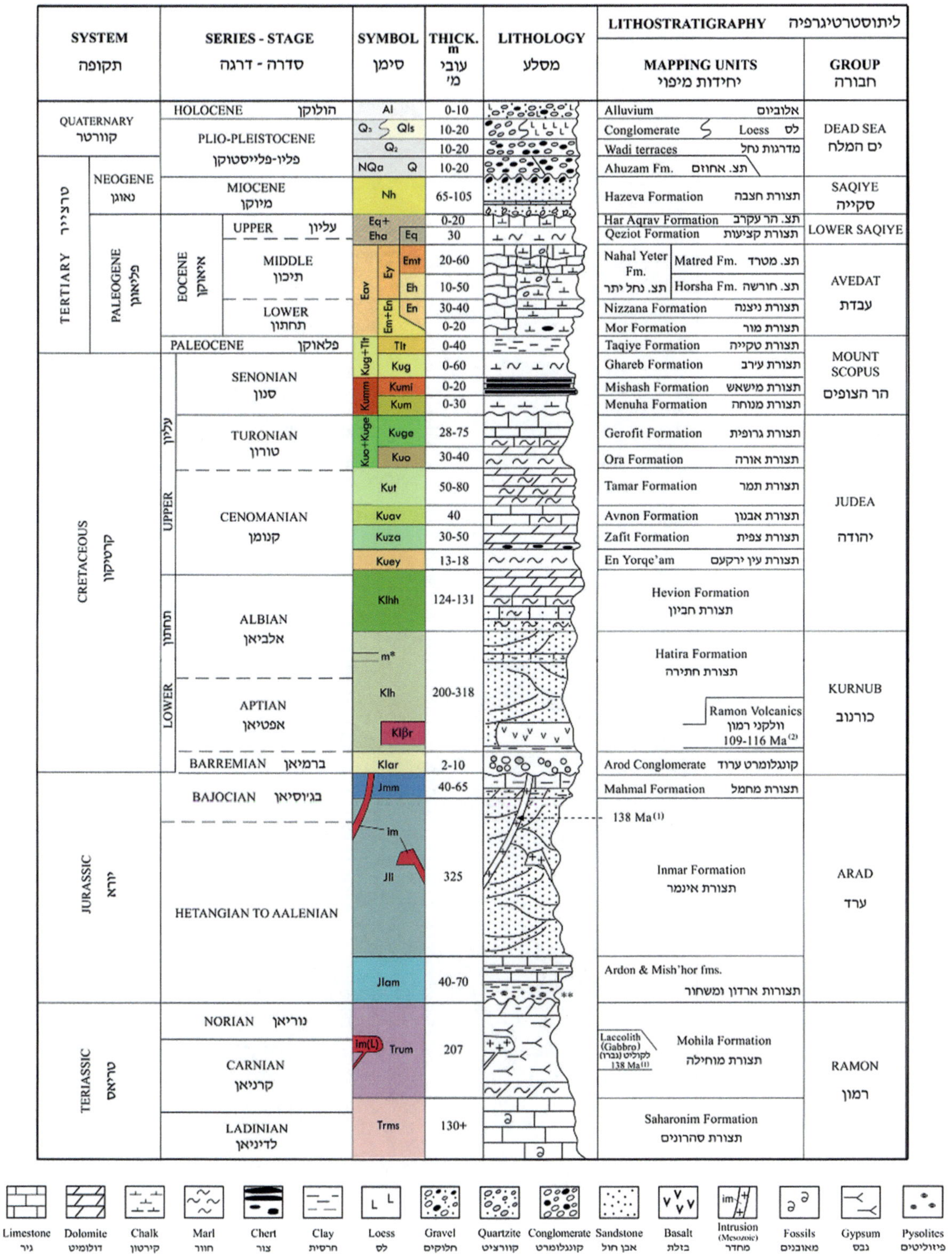

Fig. 16.4 Stratigraphic section of the Central Negev (after Zilberman and Avni 2004)

removed an additional ~ 500 m of rock column from the structural summit of the Ramon Monocline, developing a regional truncation surface (see below). However, due to a Miocene deformation phase, the Syrian Arc structures were tectonically rejuvenated and still dominate the present-day landscape of the Central Negev.

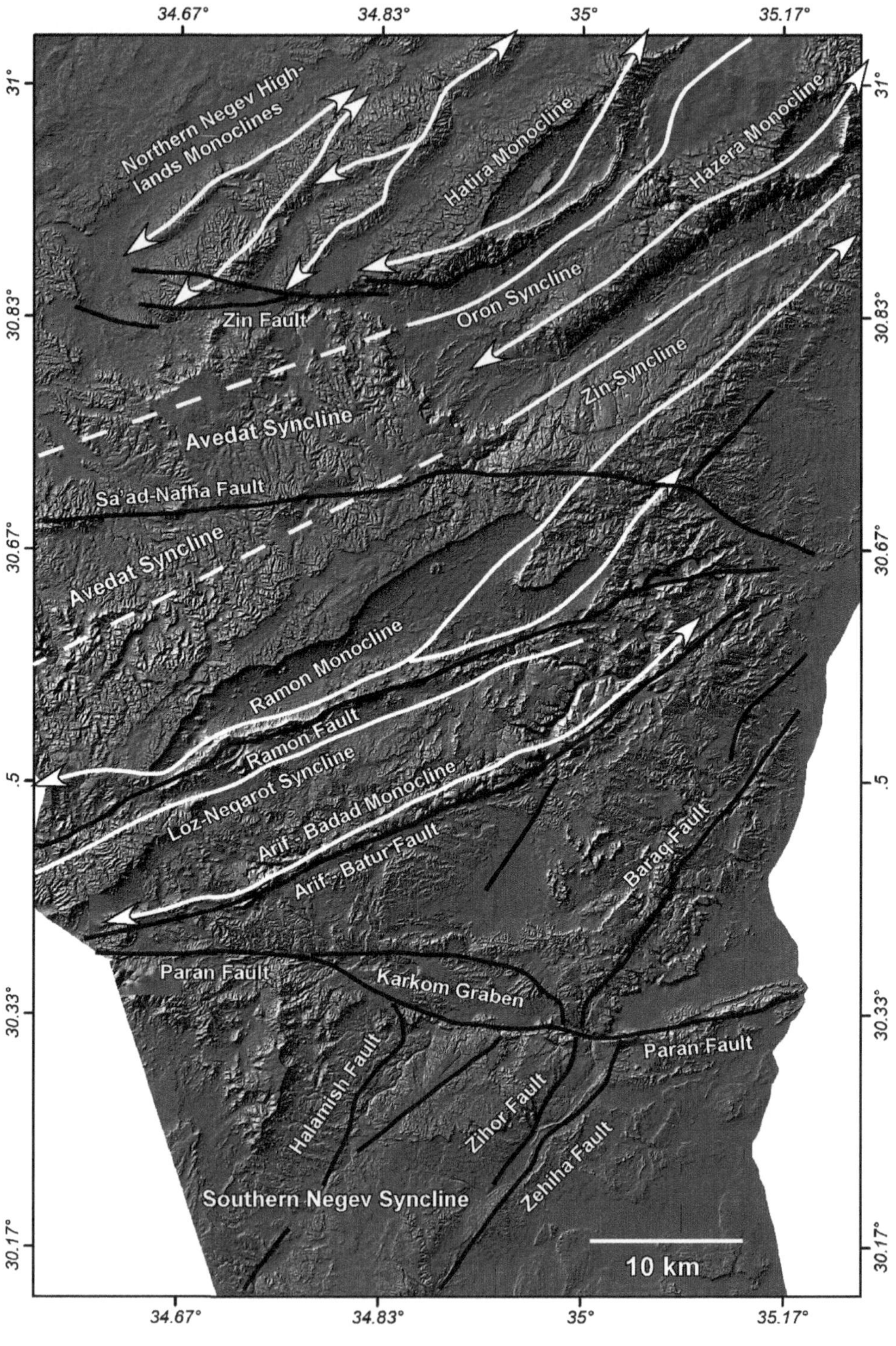

Fig. 16.5 Main structural elements in the Central Negev. In white: fold structures of the Syrian Arc. In black: major faults (shaded relief map: Hall 1997; GSI)

16.2.3 East–West Faults in the Sinai-Negev Shear Zone

These approximately east–west trending faults (Fig. 16.5) include six major faults extending hundreds of kilometers across the Sinai-Israel microplate, while the extensions of these faults traverse the Eastern Desert of Egypt and the Jordanian Plateau east of the Dead Sea Transform.

These faults include, from north to south, the Zin, Sa'ad-Nafha, Ramon, Arif-Bator, and Paran faults—all of which traverse the Central Negev (Fig. 16.5)—and the Themed Fault located in the southern Negev (Bartov 1974). Almost all faults present evidence of vertical and right-lateral displacements ranging from a few hundred meters to a few kilometers that occurred in several phases from the Triassic to

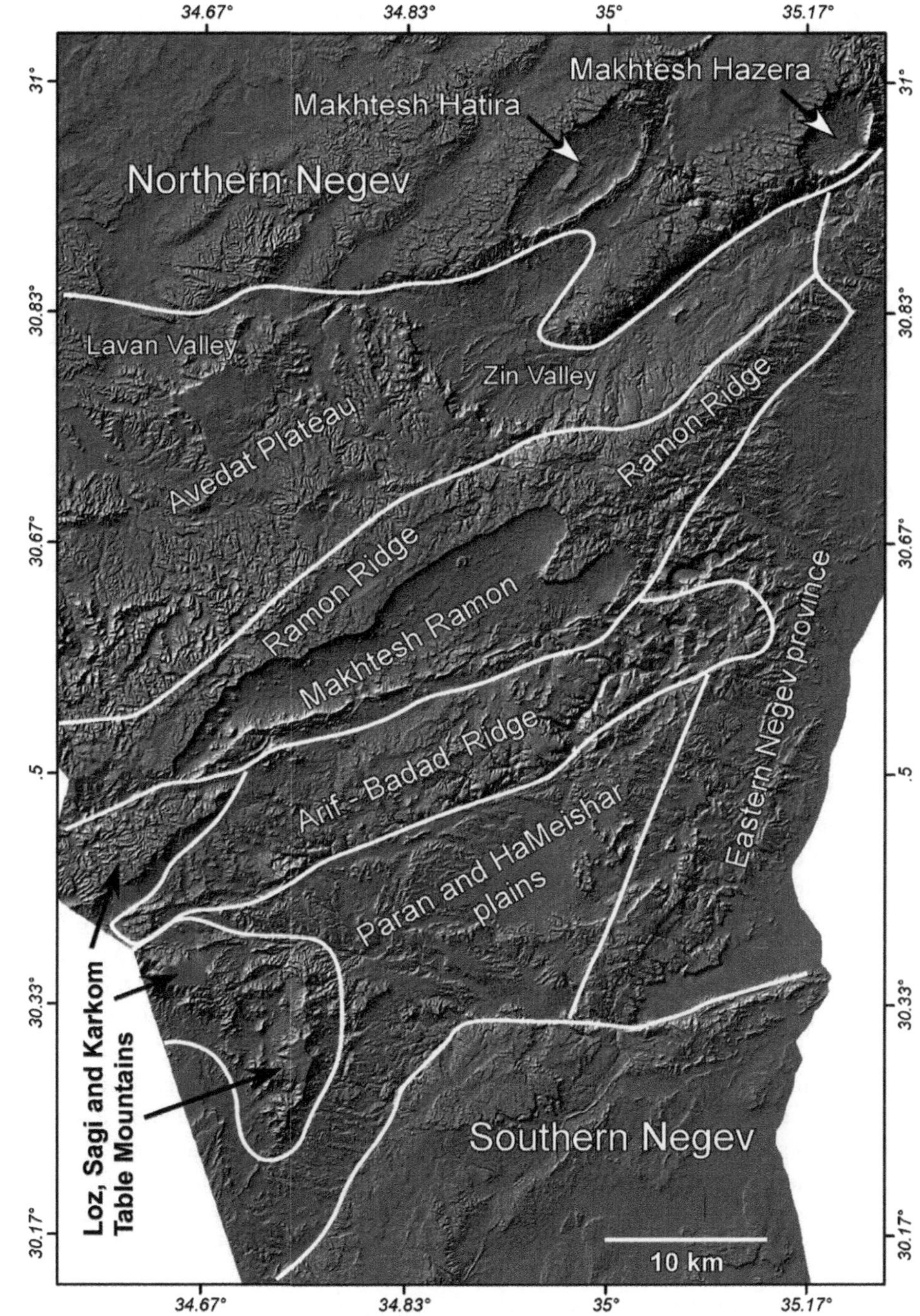

Fig. 16.6 Subdivision of the Central Negev into geographic provinces (shaded relief map: Hall 1997; GSI)

the Early Pleistocene (Garfunkel 1964; Bartov 1974; Avni 2017a, b; Wienberger et al. 2020).

In the case of the Central Negev, the Ramon and Arif-Bator Fault lines delineate the southern flanks of the Ramon and Arif-Badad monoclinal structures. However, in the eastern sector of the Central Negev, these fault lines diverge toward the east, while the monoclinal structures continue toward the northeast, expressing their diverse origins (Fig. 16.5).

The Paran Fault splits and develop a large graben between two parallel segments of the fault line, known as the Karkom Graben (Fig. 16.5). The activation of the graben in the early Miocene caused the syntectonic deposition of the Hazeva Formation within the graben, accumulating more than 1000 m of sediments (Calvo 2002; Wienberger et al. 2020). In the middle Miocene, a large tectonic event occurred along the Paran Fault and caused the collapse of the Hazeva Formation section into the graben. In addition, this tectonic phase activated the Sa'ad Nafha, Ramon and Arif-Batur Fault lines, creating grabens and locally compressed domes along their traces, locally preserving sections of the Hazeva Formation. While very little tectonism was recorded along the Paran Fault after the middle Miocene, several tectonic events were recorded along the Arif-Bator Fault line during the late Miocene and Early Pleistocene. These events caused a major uplift of the Arif-Badad Monocline that, concurrently with the incision of the paleo-Paran-Neqarot River (Avni et al. 2000), caused the development of the largest canyon in the Negev—the Meishar-Neqarot Canyon—that deeply crosses the uplifted structures of the Arif-Badad and Ramon ridges (Fig. 16.3).

16.2.4 Northeastern Longitudinal Faults

This set of faults, each extending 30–70 km, is oriented N10°–40° E and crosses the Central Negev west of the Arava Depression, comprising several en-echelon faults and flexures (Fig. 16.5). The most important faults of this group are the Halamish and the Zehiha-Baraq faults (Fig. 16.5), while other minor faults cross the syncline south of the Arif-Badad Monocline, the Arif-Badad Monocline and the eastern sector of the Ramon Monocline. South of the Paran Fault, the Zhiha and Zihor faults continue toward the south, dissecting the eastern sector of the southern Negev Province and forming a narrow graben in which Nahal Zihor is located (Avni et al. 2000; Fig. 16.5).

The fault planes of this group are almost vertical, and the kinematic indicators along the faults show mainly normal offset, combined with minor components of left-lateral motion (Avni 1998; Avni et al. 2000). They displace the Arava Formation of Early Pleistocene age by ~ 100 m, forming several tilted blocks superimposed on top of an asymmetric arch structure that is 60–80 km wide and 700–200 m high, developed between the central Sinai Basin and Dead Sea Fault zone (Avni et al. 2001). In the Central Negev, this arch, which developed in the Early Pleistocene, caused the uplift of a large asymmetric dome located in the western sector of the Central Negev region, and formed the highest present mountain peaks (Figs. 16.2 and 16.3). The uplift of the dome reorganized all previous geological structures, including anticlines and synclines of the Syrian Arc and fault lines, reshaping the present morphotectonic structure of the Central Negev. The long eastern flank of the dome caused tilting of the region toward the east, enhancing earlier events of uplift and northeastward tilting that developed in the middle Miocene.

Beside clear evidence of the Early Pleistocene tectonic movements, most of these faults show earlier phases of activity, which predated the early Miocene activation of the Dead Sea Transform (Avni et al. 2000). A possible late Precambrian age for the entire fault system was inferred from geophysical surveys performed in the southern Negev (Bartov et al. 2004). Minor tectonic deformation and microseismicity are still recorded along these faults. The faults, especially the general eastward tilt that accompanied their activity, had a large impact on landscape evolution and drainage reorganization that occurred in the region during the Pleistocene (see below, Sect. 16.4.5).

16.3 Subdivision of the Central Negev into Geographic Provinces (Fig. 16.6)

Based on its morphotectonic configuration, the Central Negev is divided into several sub-regions:

16.3.1 The Ramon Ridge

The Ramon Ridge forms the largest morphological structure in the heart of the Central Negev. It is approximately 70 km long, and its axis is oriented southwest-northeast, extending between the northern Sinai lowland in the west and the western margin of the Arava Valley in the east, while its width ranges between 5 km in the west and 14 km in its central part. The southeastern flank of the ridge is sharply inclined toward the Neqarot Valley, whereas the northwestern flank gently descends below the Avedat Plateau (Figs. 16.3, 16.6 and 16.7). This morphologic configuration is very much in accordance with the structure of the Ramon Monocline of the Syrian Arc fold belt (Krenkel 1924; Garfunkel 1964). Although the structural reconstruction of the Ramon Monocline indicates the development of an elongated dome, oriented parallel to the Ramon Fault and reaching a structural elevation of approximately 2000 m (Garfunkel 1964; Avni 1991), the present-day elevation of the Ramon Ridge summits slightly exceeds 1000 m. Moreover, all present summits are located in the western part of the structure (Fig. 16.3), approximately 20 km to the southwest relative to the original structural summit of the Ramon Monocline (Garfunkel 1964; Avni 1991).

The contemporary Ramon Ridge forms a large plateau, developed as a product of deep truncation of the Ramon Monocline and its neighboring regions, which occurred mainly during the Oligocene (Avni et al. 2012b). This plateau is slightly tilted eastward and gradually descends toward the Arava Valley. This tilt is most visible while comparing the present elevation of Mount Ramon (1037 m) with the summit of Har Ardon (722 m); both once shared the same, almost horizontal surface (Fig. 16.3). As Har Ardon is located in the eastern sector of the Ramon Ridge, the approximately 300 m decrease in elevation along the 30-km-long transect connecting the western Ramon

Fig. 16.7 The Ramon Ridge and Makhtesh Ramon. Note the flat plateau shaping the present-day landscape above the deep valley of Makhtesh Ramon. The flat plateau is a remnant of the Regional Truncation Surface (RTS) developed during the Oligocene

summits and the eastern peaks of the ridge expresses a 1% eastward tilt. This tilt occurred after the creation of the plateau in the late Oligocene and after the deposition of the early Miocene Hazeva Formation (see below, Sect. 16.4.3).

Several drainage systems drain the Ramon Ridge. Its northwestern and western sectors are drained toward the Mediterranean Sea via the tributaries of Wadi el Arish (Nahal Nizzana, Nahal Loz; Fig. 16.3), while its central and eastern parts are drained toward the Arava Valley and Dead Sea via Nahal Zin and Nahal Neqarot (Fig. 16.3). In the central part of the ridge, a unique geomorphic phenomenon—the erosive valley of Makhtesh Ramon—has developed due to deep incision into thick layers of friable sandstone units of Mesozoic age exposed in the core of the truncated structure (Avni 1991, 1993; Zilberman 2000; Fig. 16.7).

16.3.2 The Avedat Plateau and Zin Valley

The Avedat Plateau occupies the area extending to the north of the Ramon Ridge until the Zin Fault zone, which defines the northern border of the Central Negev (Figs. 16.5 and 16.6). This almost tabular region, 70–90 km long and 20–30 km wide, is composed mainly of limestone and chalk rocks of the Avedat Group of Eocene age, deposited in a large syncline developed to the north of the Ramon Monocline (Figs. 16.5 and 16.8). The elevation of this terrain decreases from approximately 1000 m in the southwestern sector of the Avedat Plateau to approximately 500 m in the north near the Zin Fault. Toward the west, north and northeast, the Avedat Plateau is dissected by numerous streams forming deep canyons and valleys that incised the relatively soft rocks—chalk, marl and clay—composing a section of the Mount Scopus Group and the lower part of the Avedat Group (Figs. 16.4 and 16.8). The amalgamation of several streams on both sides of the main water divide developed two wide erosional valleys located in the northern sector of this terrain. These are the wide valleys of Nahal Lavan and Nahal Nizzana open to the Mediterranean Sea and the wide valley of the lower Zin Basin draining to the Dead Sea (Figs. 16.3 and 16.6). Deep incision of the drainage system into the soft units in the Mount Scopus Group, whereas an extensive area is still covered by highly resistant layers of the Avedat Group, created large cliff systems. Among the largest are the Ha-Zinim cliffs to the south of the Zin Valley and the cliffs bordering the Matred Plateau south of the Lavan Valley (Figs. 16.3 and 16.6). Along a transition from marls to limestone beds, several springs emerge, such as Ein Qudeirat in the west and Ein Avedat and Ein Ziq in the east, each of them is forming small oasis in the desert environment.

16.3.3 The Arif-Badad Ridge

The ridge, 60 km long and 6–10 km wide, is oriented parallel to the Ramon Ridge, following the trace of the Arif-Bator

Fig. 16.11 The Paran stream entering the Karkom Graben. Outcrops of the Early Miocene Hazeva Formation are in the foreground while the hills in the background are composed of the Eocene Avedat Group

16.3.5 The Paran and Ha'Meishar Plains

East of the table mountain province and south of the Arif-Badad Ridge, a large lowland region developed. As the main trunk of the Paran stream delineates the border between the Central Negev and southern Negev (Figs. 16.3 and 16.6), we refer here only to the region located to the west and north of the Paran stream, although this lowland province extends well toward the east and south into the southern Negev region.

Within these borders, the area is approximately 40 km long (in the north–south direction) and 20–30 km wide (in the west–east direction). The elevation is 500 m in the south and gradually decreases toward the north, reaching 400 m near the southern flank of the Arif-Badad Ridge. This relatively flat terrain is composed mainly of massive chert layers of the Mishash Formation, forming the middle part of the Mount Scopus Group (Fig. 16.4). The exposed geological composition is the outcome of widespread truncation of the massive layers in the Avedat Group in this province during past erosional events, causing denudation of the thick clay and marl units composing the upper part of the Mount Scopus Group. A major tectonic element—the Paran Fault—crosses this terrain at its middle in the east–west direction (Fig. 16.5). Along this fault, a large tectonic basin—the Karkom Graben—has developed, marked by a thick section of the early Miocene Hazeva Formation that was preserved within it (Fig. 16.11). However, this graben has little morphotectonic impression in the present-day relief, as it was totally buried below younger continental sedimentary units, mainly by the Arava Formation of Pliocene to Early Pleistocene age (Avni et al. 2000, 2001).

At present, the region is drained toward the Dead Sea by two drainage basins: the Paran, which drains the southern portion of the province (Fig. 16.11), whereas the northern part is drained by the Ha' Meishar-Neqarot stream system. This is contrary to the Pliocene–Early Pleistocene drainage configuration in which only one stream—the Paran-Neqarot major stream—drained the entire region (Avni et al. 2000, 2001). The separation of the basins was the result of Early Pleistocene deformation, as the Paran stream and its tributaries deeply incised the eastern Negev Province, up to 160 m, on their way to the central Arava Valley, while the Ha'Meishar Plain to the north remains intact as a fossil landscape of Pliocene age.

16.3.6 The Eastern Negev Province

This terrain occupies a long strip that is 70 km long and 3–20 km wide, forming the eastern sector of the Central Negev bordering the Arava Valley, from the Zin Valley in the north to the Paran stream in the south. Its geological composition expresses the diverse structure of the eastern Negev, in which the Ramon and Arif-Badad ridges and their neighbouring synclinal structures descend toward the Arava Depression (Fig. 16.12). This morphotectonic configuration is well reflected by the exposed rock formations composed of diverse units of the Judea, Mount Scopus and Avedat Groups. The elevation of this terrain gradually decreases from 500–400 m in the southwest to below sea level in the northeast. The whole area drains toward the Dead Sea via several deeply incised valley systems in which the major ones are occupied by the Zin, Neqarot, Ashosh and Paran

Fig. 16.12 The Marzeva stream and eastern Negev province, looking east toward the Arava Valley

streams (Figs. 16.3 and 16.12). Most of these drainage basins take the form of deep gorges and canyons, expressing the relatively young process of drainage reorganization that has occurred since the Early Pleistocene (Avni et al. 2000; Avni, 2017a).

16.4 Morphotectonics and Landscape Evolution of the Central Negev

Based on the geological and geomorphological features described in the previous sections, the landscape evolutionary processes and morphotectonic stages that shaped the present landscape of the Central Negev include:

16.4.1 Stage 1: The Oligocene Truncation Surface

The regression of the sea from the Central Negev in the late Eocene–early Oligocene (35–30 Ma; Avni et al. 2012b; Avni 2017a, b) marked the final transition from marine deposition to the continental regime, which occurs until now. This transition was the consequence of gradual tectonic uplift at the regional scale related to the thermal activation of the northeastern sector of the African–Arabian plate as an initial stage preceding the rifting of the Red Sea (Cloos 1953; White and McKenzie 1989; Pik et al. 2003). This morphotectonic process led to the formation of a ~ 3000 × 1500 km crustal doming centered on both sides of the present Red Sea, including the Sinai Peninsula and Negev (Cloos 1953; Avni et al. 2012b). In the Central Negev, this slow-rate uplift activated old tectonic features such as the Syrian Arc structures and the Negev-Sinai shear belt, forming differential structural blocks (Fig. 16.13). Erosive forces acting simultaneously with the slow and continuous uplift developed a widespread erosion surface known as the Oligocene Regional Truncation Surface (RTS). The truncation was caused by the long (10–15 Myr) activity of non-localized erosion described by Avni et al. (2012b). During this stage, each of the structural blocks composing the region was truncated differently, according to its structural elevation, exposing different rock units on its upper surface (Figs. 16.13 and 16.14).

In the region surrounding the Ramon Monocline, the regional truncation surface is well preserved and consists of a large high-elevation plateau extending over ~ 500 km^2 in the present-day landscape (Fig. 16.7). Diverse geological units ranging from the Triassic to late Eocene were truncated by this nearly flat surface. The deepest truncation is observed along the axis of the monocline, as the Oligocene truncation was superimposed on a series of earlier local erosional surfaces that developed during the Syrian Arc folding stages in the Late Cretaceous. Avni (1991, 1993) estimated the net truncation related to the Oligocene erosional event in this terrain to ~ 500 m, which was 50% of the total truncation, estimated to ~ 1000 m. In the synclines bordering the Ramon Monocline, the truncation ranged ~ 100–300 m, exposing Upper Eocene to Upper Cretaceous rocks, as the synclines were less elevated and therefore their truncation was moderate (Fig. 16.14).

It is important to note that although occurring some 30–20 Ma, the regional truncation surface still dominates the present-day landscape (Figs. 16.7, 16.8 and 16.12).

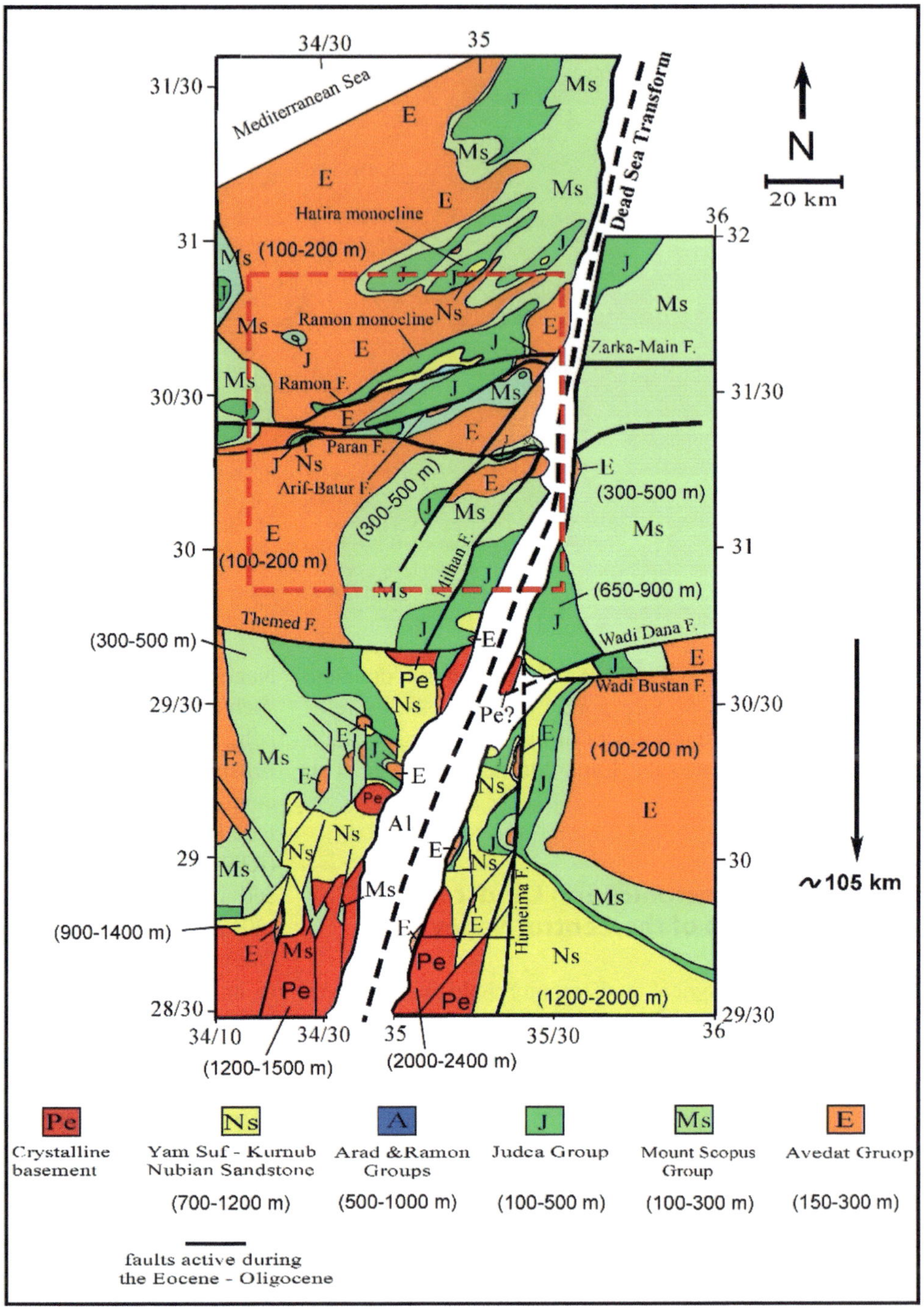

Fig. 16.13 Post-Oligocene reconstructed geological map showing the truncation of the Central Negev (marked by a red square) during the Oligocene by the RTS. Note the diverse stratigraphic units exposed on the surface. Numbers at the bottom: the thickness of the stratigraphic units. Numbers within the figure: the amount of rock truncation in each geological terrain. Modified after Avni et al. (2012b)

This surprising conclusion is consistent with the fact that the flat summits and the top of the cliff systems are composed of highly resistant carbonate units. The denudation rate calculated by Matmon (2017) on these rocks under the current arid climate ranges from 1 to 3 m/million years, implying that these geomorphic elements still preserve the near-original position of the RTS. On the other hand, most contemporary valleys developed within nonresistant rocks, such as chalk and clay units exposed by the RTS (Fig. 16.14b). In some cases, as in the Ramon Monocline, the truncation exposed the Early Cretaceous to Jurassic sandstone core, promoting the future development of the deeply incised valley known as Makhtesh Ramon (Figs. 16.3, 16.6 and 16.7).

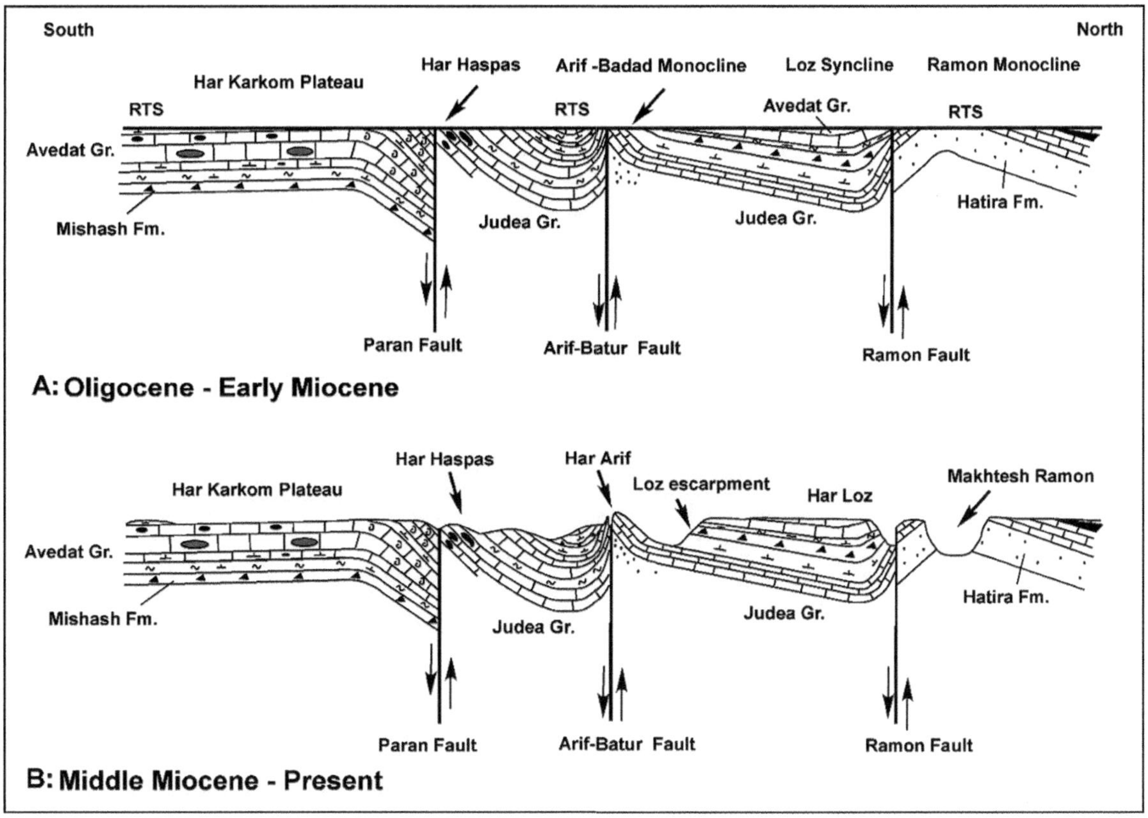

Fig. 16.14 Differential truncation of the Central Negev during the Oligocene (A) and the present configuration of the same region (B). Note that only minor erosion occurred in the region since the truncation event

16.4.2 Stage 2: Early Miocene Tectonism and Deposition of the Hazeva Formation

The transition from the nonlocalized erosional regime forming the RTS to the deposition of thick fluvial sediments of the Hazeva Formation (Calvo 2002) on the RTS was predated by a tectonic event that occurred in the early Miocene. This tectonic event caused uplift of the northern Negev Highlands and early incision of Makhtesh Hatira located in the core of the Hatira Monocline (Zilberman 1992; Figs. 16.5 and 16.6). This tectonic phase, which occurred approximately 24–20 Ma, was related to the tectonic uplift of the entire region that occurred concurrently with the downfaulting of the Red Sea and Suez Rift (Avni 2021).

Following the uplift of the eastern margin of the Red Sea in the early Miocene, a phase of massive erosion occurred, and the eroded sediments were transported to the entire Negev region by a large river system. These sediments are known as continental deposits of the Hazeva Formation (Calvo 2002). In the Central Negev, remnants of these deposits are found on top of the highest summits of the Central Negev, such as on Har Harif, Har Ardon and Har Karkom (Fig. 16.3). They were deposited on top of the RTS or within gently incised, wide valleys eroded below the original RTS. These sediments accumulated in tectonic depressions and grabens developed along the Sinai- Negev shear zone, such as in the Karkom and Nahal Teref grabens, in thicknesses ranging from several hundreds to more than 1000 m (Fig. 16.15; Calvo 2002; Zilberman 1992; Zilberman and Calvo 2013). However, in other locations in the region, the thickness of the Hazeva Formation is much smaller, ranging from several tens to a few hundreds of meters. Of special importance is the Har Karkom outcrop presenting only 4–5 m of the uppermost unit of the Hazeva Formation and containing well-rounded chert pebbles. These pebbles, originating from silicified Eocene beds exposed in the Central Jordan Plateau, were preserved within karstic shafts developed on the very top of the Har Karkom composed of the Matred Formation of the Avedat Group (Fig. 16.4). These field relationships hint at the elevated position of Har Karkom relative to the nearby Karkom Graben (Figs. 16.5 and 16.15), in which a section of over 1000 m of Hazeva Formation was deposited, containing all five members of the formation (Calvo 2002; Zilberman and Calvo 2013). Direct evidence for synchronous tectonic activity along most of the Central Negev faults during deposition of the Hazeva Formation provided

Fig. 16.15 The upper part of the Hazeva Formation in the Karkom Graben. Note the westward tilt of the strata, indicating on post early Miocene tectonic deformation within the Karkom Graben

by Weinberger et al. (2020) indicates that the Hazeva Formation syn-tectonically accumulated over the Negev (Weinberger et al. 2020).

Karstic shafts, a few caves with speleothems and wide zones of massive rock alteration from original dolostone to coarse crystalline calcite have been found in diverse locations in the Central Negev, especially along the Ramon Monocline. These features were probably produced by massive percolation of river water into the subsurface, originating from Oligocene to early Miocene fluvial systems, the later of those were the river system deposited the Hazeva Formation (Avni 2017b). In addition, groundwater flowing from the uplifted east side of the Red Sea and the Dead Sea fault system formed major karst caves in the region, such as the Marzeva Cave (Frumkin et al. 2017).

Sediments of the Hazeva Formation were found on the top of the cliff system bordering the present deep valley of the Makhtesh Ramon. Therefore, it is concluded that the Hazeva drainage system managed to cross the truncated Ramon structure from southeast to northwest and that the Makhtesh Ramon was not yet existent at this stage (early Miocene) but developed afterward (Avni 1993, 2021).

16.4.3 Stage 3: Middle Miocene Uplift and Erosional Response

The transition from a relatively flat landscape on which the Hazeva Formation was deposited to the present-day uplifted and dissected terrain was related to major morphotectonic events. The first event followed the gradual activation of the eastern boundary of the Sinai-Israel microplate, and the creation of the Dead Sea Transform initiated 20.8–18.5 Ma (Nuriel et al. 2017). The development of the Dead Sea Transform along the pre-existing Dead Sea Fault System (Avni et al. 2012b) initiated left-lateral plate separation between the Sinai-Israel microplate and Arabian Plate, accumulating approximately 105 km of sinistral displacement since the early Miocene.

The first sign of uplift that accompanied the activity of the Dead Sea Transform was evident approximately 16–15 Ma by the deposition of the Ein Yahav Formation (formerly the Hufeira Member of the Hazeva Formation, Calvo 2002). This formation consists of a combination of redeposited sediments from the Hazeva Formation and fresh rock fragments eroded from large variety of rock formations exposed below the Hazeva Formation sediments. The Ein Yahav Formation was deposited and preserved within tectonic and topographic depressions, such as the Arava Depression. At this stage, the Negev-Sinai shear belt was activated, creating several push-up structures and tectonic depressions, such as the uplifted structures of Har Arif and Har Nafha and the grabens of Karkom and Nahal Teref, which preserved sections of the Hazeva and Ein Yahav formations (Zilberman 1992; Avni 2017a, b). In addition, the Central Negev was uplifted, forming a large asymmetric dome that resembled its present-day configuration. As any considerable uplift is accompanied by synchronous incision, we can trace the timing and dimension of this uplift event using the geomorphic evidence of landscape evolution, especially the incision of new drainage systems and deposition of post-Hazeva Formation units.

The uplift of the Central Negev and its neighboring regions to the north, as well as the creation of a new depression along the Arava Valley and Dead Sea Basin, forced the post-Hazeva Formation drainage system to split into western and eastern provinces. The northwestern sector of the Central Negev was drained to the Mediterranean Sea, while its eastern and southern sectors consisted of catchments open to the evolving Arava—Dead Sea Basin. In the initial stage of the Middle Miocene, a substantial part of the Southern and Central Negev were drained by the Paran-Neqarot trunk river that crossed the eastern part of the Central Negev dome toward the Dead Sea Basin (Avni et al. 2000). This river developed the largest and deepest (400–350 m) canyon in the Negev, deeply incised in the eastern sector of the Central Negev dome, creating the present Mishar-Neqarot Canyon (Fig. 16.3). As the incision was probably contemporaneous with the uplift, it is reasonable to assume that the course of the Paran-Neqarot mainstream was established prior to the uplift and that this large river maintained its route during the development of the dome as an antecedent river. Similar drainage lines crossing prominent uplifted terrains and anticlinal structures in the northern Negev and Sinai are denoted by the local term "Dika". The most famous are the "Dika" of Nahal Ha'Besor crossing the Boker anticline and the "Dika" of Wadi El-Arish in northern Sinai crossing the Gabel Hallal-Gabel Dalafa anticlinorium.

As part of the uplift and drainage reorganization, the deep valley of Makhtesh Ramon started to evolve, followed by the first development of its cliff system (Fig. 16.16). This large and deep valley was drained by a branch of the Paran-Neqarot river toward the Dead Sea Basin. It is important to note that although the creation of the Makhtesh Ramon cliff system was initiated some 16–14 Ma (Avni 2021), these cliff systems and local water divides situated above them have showed long-term stability since then. The initial location of nearly all cliff systems in the Central Negev can be explained by the structural and geometrical relationships presented in Fig. 16.17. After their first initiation and stabilization, the cliffs were subjected to slope processes, but the slow rate of weathering and exhumation described by Matmon (2017) prevented massive retreat. Additional stabilization of the cliff slopes was related to massive landslides that covered much of the slopes, composed of friable sandstone and shale units, and by thick carbonate rock slabs originating from the upper cliff units, protecting the underlying weak lithologies against surface erosion. The outcome of this stabilization process was the occurrence of nearly stable cliffs over geological time, and the conservation of water divides at their near-original position.

Fig. 16.16 The major cliff system of Makhtesh Ramon in DTM imagery. *Source* SRTM data

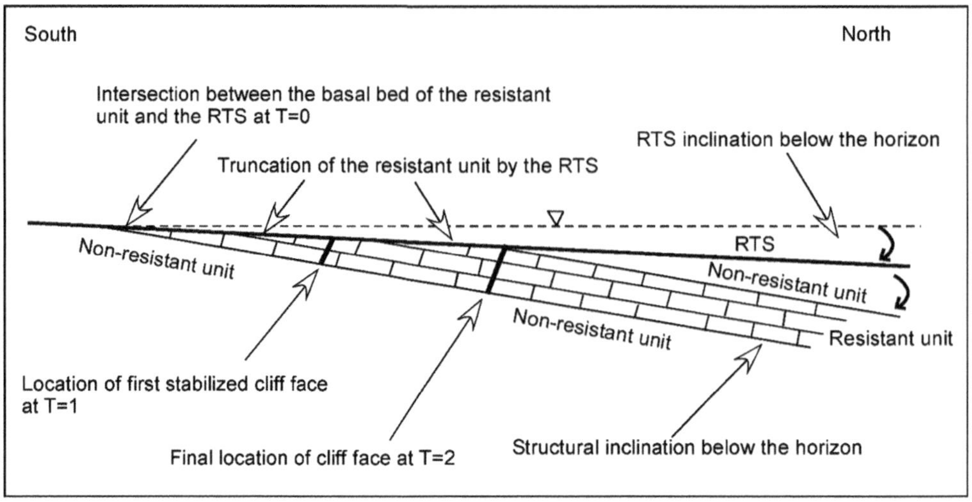

Fig. 16.17 A conceptual model for the formation of major cliffs on the Sinai-Israel microplate. Dash line marks the horizon, black arrows mark the angle between the horizon and the Oligocene Regional Truncation Surface (RTS) and the angle between the RTS and the strata inclination. The cliffs were formed following the structural and geometrical relationships between the inclined strata and the RTS. After initiation and first stabilization (T = 0 and T = 1), slow slope processes and landslides contributed to the near stabilization of the cliffs (T = 2) (see Fig. 16.13 and explanations in text)

However, during later stages of the morphotectonic evolution of the region, significant breaching of the original water divides occurred in some critical locations, causing considerable drainage reorganization, as will be discussed below.

The tectonic events in the late Early and Middle Miocene were synchronous with the Miocene deformation phase of the Syrian Arc structures, located mainly in the Palmyride structures of Syria and in the eastern Mediterranean Basin (Walley 1998; Sagy et al. 2018). As this Miocene deformation phase was superimposed on the original Syrian Arc structures developed since the late Turonian, it caused rejuvenation of these structures, controlling their present-day dominant morphotectonic position in the landscape of the Central Negev.

16.4.4 Stage 4: The Deposition of the Arava and Ahuzam Formations at the Pliocene/Pleistocene Transition

Following the long period of erosion accompanying the Middle Miocene uplift of the Central Negev, deposition of fluvial units within the continuously incised drainage basins started only in the late stage of their gradual development and only after the erosive forces resulting from the initial steep topographic gradients gradually decreased. In the western province drained to the Mediterranean Sea, the Pliocene–Lower Pleistocene fluvial conglomerates of the Ahuzam Formation were deposited and intercalated in the northwest with shallow marine to estuarine deposits of the Pleshet Formation (Zilberman 1991; Bar 2009). At the same time, the fluvial-lacustrine Arava Formation was deposited in the eastern province, draining toward the Arava Depression and Dead Sea (Avni et al. 2001). As part of the general fluvial environment, several fresh to brackish lakes developed along the main drainage system of the Paran-Neqarot river, mainly in the Kuntila Basin, Karkom Graben and next to the outlet of this major river to the northern Arava Depression near Ein Yahav (Avni et al. 2001). The age of the lower part of the Arava Formation (Saggi Member) ranges from 4 to 3 Ma (Avni et al. 2000), while the age of the top Arava Formation (Kuntila Member) was confirmed using cosmogenic isotopes to range between 2.2 and 1.7 Ma (Matmon et al. 2009; Guralnik et al. 2010). Within the lacustrine units deposited in Kuntila Lake, a magnetic shift was identified, best correlated to the Oldovai excursion (1.945–1.778 Ma) (Larrasoaña et al. 2020).

The occurrence of lacustrine sediments in the Arava Formation hints at relatively wet periods that developed in the region during the Pliocene–Early Pleistocene. The manifestation of relatively wet intervals is also indicated by the appearance of extensive landslides developed along cliff systems and regional escarpments in the Central Negev. These landslides fed the uppermost outcrops of the Arava and Ahuzam formations, deposited near the base of these cliffs (Fig. 16.18). As these landslides are still very close to the present-day position of the cliff face, they serve as evidence for the minor retreat of the cliff system, demonstrating considerable stability of the cliffs and their associated water divides during the last 4–2 Ma (Avni 2021).

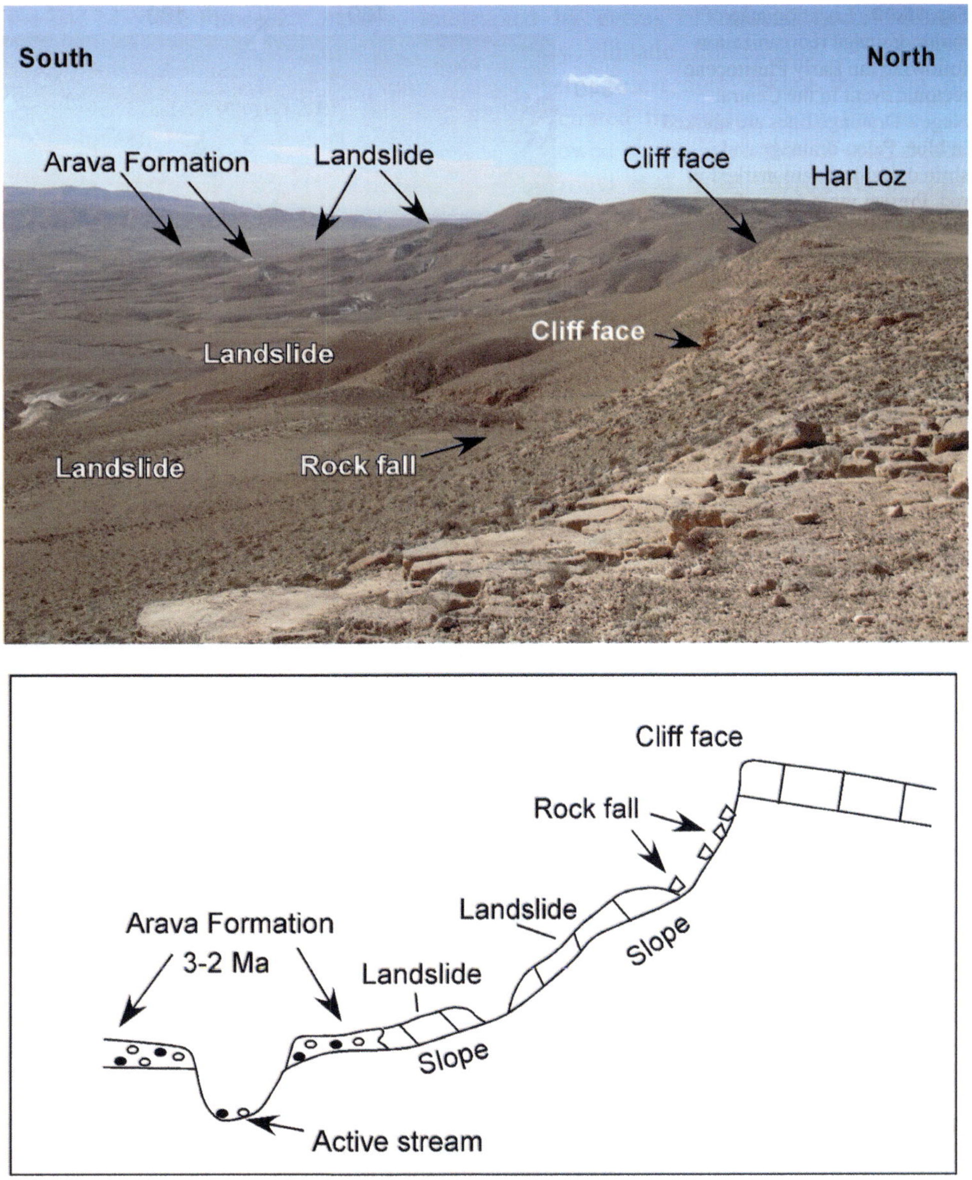

Fig. 16.18 Cliffs, slopes and landslides in the Central Negev. Note the interfingering relationships between the landslides and the Pliocene—Early Pleistocene Arava Formation outcrops

16.4.5 Stage 5: Post-Arava Formation Tectonic Deformation

Major regional tectonic deformation occurred in the Central Negev during the Early Pleistocene, dated by cosmogenic isotopes to 1.5–1 Ma (Matmon et al. 2009; Guralnik et al. 2010). At this stage, a broad asymmetric dome developed in the Central Negev, enhancing an older Miocene relief and forming a 60–80 km wide deformation belt, whose long eastern limb was tilted toward the Arava Depression in the east. The rise of this gently tilted dome was accompanied by the activation of the northeastern longitudinal faults that displaced the Arava Formation, forming several tilted blocks. During the early stage of tectonic deformation, a syntectonic sequence of conglomerates, lacustrine sediments and red palaeosols was deposited. These sediments were integrated into the Lower Pleistocene Sede Zin and Zehiha formations on both sides of the regional water divide (Issar et al. 1984; Ginat 1997; Avni 1998).

The combined effects of uplift, eastward tilting and faulting shaped the present-day landscape configuration of the Central Negev. The western part of this region was uplifted by an additional 200 m relative to the northern Sinai lowland, achieving its current topographic elevation of over 1000 m. The uplift was accompanied by subsidence of the eastern sector of the Central Negev toward the Arava Depression that was deepened by an additional ~ 700–400 m, achieving its present-day elevation. These topographic changes triggered substantial changes in the landscape and drainage systems. One of the cardinal manifestations of this landscape was the regional truncation surface (RTS), which formed in the Oligocene as a geomorphic response to a regional event of uplift and planar erosion resulting from the tectonic and thermal activation of the region preceding the continental breakup. This long event, dated between 34 and 24 Ma (Avni et al. 2012b) controls the contemporary distribution of rock formations on the surface, with minor modification by later erosion, especially in deeply incised valleys, while the summits and plateaus consisting of highly resistant rocks remain almost intact (Figs. 16.7 and. 16.14).

The second event was related to the initial faulting and localized uplift along the Red Sea-Suez Rift initiated some 24–20 Ma. In this stage, the western border of the Sinai-Israel microplate evolved, including the evolution of major cliff systems in central Sinai and the uplift of the northern Negev Highlands, triggering the development of Makhtesh Hatira and its cliff system. The major erosion caused by the uplift of the Red Sea margins caused the deposition of the early Miocene Hazeva Formation over wide areas in the Central Negev.

The third event that shaped the Central Negev was associated with the development of the Dead Sea Transform, which was accommodated within the older Dead Sea Fault system (Avni et al. 2012b). This event occurred 20–18 Ma and formed the eastern boundary of the Sinai-Israel microplate. During the major uplift that followed (16–10 Ma), the Makhtesh Ramon and its cliff system were formed, together with the establishment of a regional drainage system draining the uplifted terrains toward the Mediterranean Sea and Dead Sea basins. In later stages, and after achieving relative geomorphologic stability, fluviatile-lacustrine units in the Arava and Ahuzam formations were deposited along the drainage courses. A prominent tectonic phase occurred approximately 1.5–1 Ma, probably as the region bordering the active Dead Sea Transform was making morphotectonic adjustments to the evolving new plate boundary. This event triggered new uplift and faulting in the Central Negev dome (Fig. 16.2), which was associated with a larger regional arching toward the Dead Sea rift documented in Judea, Samaria and Galilee, leading to drainage reorganization, such as in the Paran and Zin drainage basins. This tectonic event is still shaping the present-day landscape. The landscape evolutionary process was accompanied by an intermittent presence of humans inhabiting the Central Negev Desert during the last 1.5 Ma.

Acknowledgements Omry Barzilai and Maya Oron from the Israel Antiquity Authority are acknowledged for assisting in elaborating the paragraph entitled "Hominids in the Central Negev".

References

Amit R, Enzel Y, Sharon D (2006) Permanent quaternary hyperaridity in the Negev, Israel, resulting from regional tectonics blocking Mediterranean frontal systems. Geology 34:509–512

Avni Y (1991) The geology, paleogeography and landscape evolution in the central Negev highlands and the western Ramon structure. Isr Geol Surv Rep GSI/6/91. Jerusalem

Avni Y (1993) The structural and landscape evolution of the western Ramon structure: Israel J. Earth Sci 42:177–187

Avni Y (1998) Paleogeography and tectonics of the central Negev and the Dead Sea Rift western margin during the late Neogene and Quaternary. Geol Surv Isr Rep GSI/24/98, 231 pp. (in Hebrew, English abstr.)

Avni Y (2017a) Tectonic setting and physiography setting of the Levant. In: Enzel Y, Bar-Yosef O (eds) Quaternary of the Levant (Environment, Climate change and Humans). Cambridge University Press, pp 3–16

Avni Y (2017b) The Early Pleistocene tectonic deformation in the Central Negev and its influence on the landscape evolution. Field Excursion Guide Book. In: Elisha B, Wieler N (eds) pp 8–40. Isr Geol Soc Annu Meet. Mitzpe Ramon (in Hebrew).

Avni Y (2021) The tectonic significance of the main cliff systems in the Sinai—Israel micro plate—example from the Makhtesh Hatira cliffs and the Ramat Beqa—Mizpe Shivta escarpment. In: In: Kenigsberg H, Eshed Y (eds) Field excursion guide book, Isr Geol Soc Annu Meet. Yeruham (in Hebrew), pp 5–42

Avni Y, Zilberman E (2007) Landscape evolution triggered by neotectonics in the Sede Zin region, Central Negev, Israel. Israel J Earth Sci 55:189–208

Avni Y, Bartov Y, Garfunkel Z, Ginat H (2000) Evolution of the Paran drainage basin and its relations to the Plio-Pleistocene history of the Arava Rift western margin, Israel. Israel J Earth Sci 49:215–238

Avni Y, Bartov Y, Garfunkel Z, Ginat H (2001) The Arava formation—a pliocene sequence in the Arava Valley and its western margin, southern Israel. Israel J Earth Sci 50:101–120

Avni Y, Porat N, Plakht J, Avni G (2006) Geomorphologic changes leading to natural desertification processes versus anthropogenic land conservation in an arid environment, the Negev Highlands, Israel. Geomorphology 82:177–200

Avni Y, Porat N, Avni G (2012a) Pre-farming environment and OSL chronology in the Negev Highlands, Israel. J Arid Environ 86:12–27

Avni Y, Faershtein G, Porat N (2017a) Studies of stream terraces in the Negev Highlands and their relationship with the Levant alluvial chronologist. In: Enzel Y, Bar-Yosef O (eds) Quaternary of the Levant (Environment, Climate Change and Humans). Cambridge University Press, pp 457–469

Avni Y, Segev A, Ginat H (2012b) Oligocene regional denudation of the northern Afar dome: pre and syn breakup stages of the Afro-Arabian plate. Geol Soc Am Bull 124:1971–1897

Avni Y, Baeker A, Zilberman E (2017b) The Geological Map of Israel (1:50,000) Be'erot Oded Sheet (21-IV). Geol Surv Israel

Avni Y, Oron M, Cohen-Sasson E, Porat N, Barzilai O (2021) Chronosequences of alluvial terraces and fossilized water bodies as a predictive model for detecting Lower and Middle Palaeolithic sites in the Negev desert Israel. Quat Sci Rev 268:107114

Bar O (2009) The shaping of the continental margin of the central Israel since the Late Eocene-tectonics, morphology and stratigraphy. Isr Geol Surv Rep GSI/32/2009, 207pp (in Hebrew, English abst)

Bartov Y (1974) A structural and paleogeographical study of the Central Sinai faults and domes. Ph.D thesis, Hebrew University of Jerusalem, 143pp (in Hebrew, English abst.)
Bartov Y, Frieslander Y, Avni Y, Goldman M, Shamir G, Ginat H (2004) Possible Precambrian origin for the Plio-Quaternary NNE trending fault system in the southern Negev—evidence from field mapping and geophysical surveys: Isr Geol Soc ann meet. Abstract vol p 13
Bar-Yosef O, Belfer-Cohen A (2013) Following Pleistocene road signs of human dispersals across Eurasia. Quat Int 285:30–43
Calvo R (2002) Stratigraphy and petrology of the Hazeva Formation in the Arava and Negev: Implications for the development of sedimentary basins and morphotectonics of the Dead Sea Rift valley. Geol Surv Israel rep GSI/22/02, 264p
Cloos H (1953) Conversation with the Earth: New York, Knopf. 413p
Crouvi O, Amit R, Enzel Y, Porat N, Sandler A (2008) Sand dunes as a major proximal dust source for late Pleistocene loess in the Negev Desert, Israel. Quat Res 70:275–282
Crouvi O, Amit R, Enzel Y, Gillespie AR (2010) Active sand seas and the formation of desert loess. Quat Sci Rev 29:2087–2098
Enzel Y, Amit R, Dayan U, Crouvi O, Kahana R, Ziv B, Sharon D (2008) The climatic and physiographic controls of the eastern Mediterranean over the Late Pleistocene climates in the southern Levant and its neighboring deserts. Global Planet Change 60:165–192
Frumkin A, Langford B, Lisker S, Amrani A (2017) Hypogenic Karst at the Arabian platform margins: implications for far-field groundwater systems. Bull Geol Soc Am 129:1636–1659
Garfunkel Z (1964) Tectonic problems along the Ramon [M.Sc. thesis]: Jerusalem, Israel, Hebrew University of Jerusalem, 68p (In Hebrew)
Garfunkel Z (1978) The Negev—regional synthesis of sedimentary basins: International Congress Sedimentological, 10th, Jerusalem, Guide book, v 1, pp 35–101
Ginat H (1997) Paleogeography and landscape evolution of the Nahal Hiyyon and Nahal Zihor basins (sedimentology, climatic and tectonic aspects). Geol Surv Isr Rep GS1I19/97 (in Hebrew, English abstr)
Guralnik B, Matmon A, Avni Y, Fink D (2010) ^{10}Be exposure ages of ancient desert pavements reveal Quaternary evolution of the Dead Sea drainage basin and rift margin tilting. Earth Planet Sc Lett 290:132–141
Guiraud R, Bosworth W (1997) Senonian basin inversion and rejuvenation of rifting in Africa and Arabia: synthesis and implications to plate-scale tectonics. Tectonophysics 282:39–82
Issar A, Kamieli A, Bruins HJ, Gilead I (1984) The Quaternary geology and hydrology of Sede Zin. Negev. Israel. Isr J Earth Sci 33:34–42
Krenkel E (1924) Der Syrische Bogen: Zentralblatt für Mineralogie. Geologie Und Palaeontologie 9:274–281
Larrasoaña JC, Waldmann N, Mischke S, Avni Y, Ginat H (2020) Magnetostratigraphy and paleoenvironments of the Kuntila Lake sediments, southern Israel: Implications for late Cenozoic climate variability at the northern fringe of the Saharo-Arabian desert belt. Front Earth Sci 8:173
Marks AE (1976) Prehistory and Paleoenvironments of the Central Negev, Israel, Vol. I, The Avdat/Aqev Area. Part 1. Dallas, Texas: Southern Methodist University (SMU) Press
Marks AE (1977) Prehistory and Paleoenvironments of the Central Negev, Israel, Vol. II, The Avdat/Aqev Area, Part 2 and the Har Harif. Dallas, Texas: Southern Methodist University (SMU) Press
Marks AE (1983) Prehistory and Paleoenvironments of the Central Negev, Israel, Vol III. Dallas, Texas: Southern Methodist University (SMU) Press
Matmon A (2017) Cosmogenic-Isotope Based Erosion Rates along the Western Margin of the Dead Sea Fault. In: Y. Enzel and O. Bar-Yosef (eds.): Quaternary of the levant (Environment, Climate change and Humans). Cambridge University Press, pp 391–400
Matmon A, Simhai O, Amit R, Haviv I, Porat N, McDonald E, Benedetti L, Finkel R (2009) Where erosion ceases: desert pavement coated surfaces in extreme deserts present the longest-lived landforms on Earth. Geol Soc Am Bull 121:688–697
Nuriel P, Weinberger R, Kylander-Clark ARC, Hacker BR, Craddock JP (2017) The onset of the Dead Sea Transform based on calcite age-strain analyses. Geology 45:587–590
Pik R, Marty B, Carignan J, Lave J (2003) Stability of the Upper Nile drainage network (Ethiopia) deduced from (U-Th)/He thermochronometry: implications for uplift and erosion of the Afar plume dome. Earth Plan Sci Lett 215:73–88
Sagy Y, Gvirtzman Z, Reshef M (2018) 80 m.y. of folding migration: new perspective on the Syrian arc from Levant Basin analysis. Geology 46:175–178
Segev A, Rybakov M (2010) Effects of cretaceous plume and convergence, and early tertiary tectonomagmatic quiescence on the central and southern Levant continental margin. J Geol Soc London 167:731–749
Walley CD (1998) Some outstanding issues in the geology of Lebanon and their importance in the tectonic evolution of the Levantine region. Tectonophysics 298:37–62
White RS, McKenzie D (1989) Magmatism at rift zones: the generation of volcanic continental margins and flood basalts. J Geophys Res 94:7685–7729
Weinberger R, Nuriel P, Kylander-Clark A, Craddock J (2020) Temporal and spatial relations between large-scale fault systems: Evidence from the Sinai-Negev shear zone and the Dead Sea Fault. Earth Sci Rev 211:103377
Zilberman E (1991) Landscape evolution in the Central, Northern and Northwestern Negev during the Neogene and the Quaternary. Ph.D. Thesis. The Hebrew University; and Israel Geol Surv Rep GSI 45/90 Jerusalem (in Hebrew, English abstract)
Zilberman E (1992) Remnants of Miocene landscape in the central and northern Negev and their paleogeographic implications. Isr Geol Surv Bull 83: 54 p
Zilberman E (2000) The formation of the "Makhteshim"—Unique erosion cirques in the Negev, southern Israel. Isr J Earth Sci 49:127–142
Zilberman E, Avni Y (2004) The geological map of Israel (1:50,000) Mizpe Ramon Sheet (21-II). Geol Surv Isr
Zilberman E, Calvo R (2013) Remnants of Miocene fluvial sediments in the Negev Desert, Israel, and the Jordanian Plateau: Evidence for extensive subsiding basin in the north-western margins of the Arabian plate. J African Earth Sci 82:33–53

"Makhteshim"—Unique Arid Land Erosion Cirques in the Negev

17

Ezra Zilberman

Abstract

Several peculiar erosion cirques, termed "*makhteshim*" in Hebrew (sing. "*Makhtesh*"), occur in the monoclinal ridges of the Syrian Arc Fold Belt (SAFB) in southern Israel and eastern Sinai. The *makhteshim* are closed valleys (kind of erosion cirques) surrounded by steep cliffs 200–400 m high, with a single outlet incised in the steep south eastern flanks of some of the largest monoclines. The present study examines the initial conditions, which seem to be required for the development of the *makhteshim* in the Negev. These are: (1) a stratigraphic sequence of hard carbonate rocks overlying an easily eroded sandstone unit; (2) an asymmetric fold, which was also a prominent topographic ridge; (3) an erosion surface that truncated the hard carbonate rocks of the fold's crest and exposed the underlying friable sandstone; (4) a deep transverse valley, which incised in the erosion surface and the structural ridge. This valley is the precursor of the present outlet of the *Makhtesh.* In the Negev, all the above initial conditions already existed in the Early Miocene. However, the final opening stage of each *Makhtesh* was triggered by the establishment of a deep erosion base level southeast of each anticlinal ridge. For *Makhtesh* Hatira, such an erosion base level formed locally and for a limited period in the Early Miocene. The opening of the other *makhteshim* started in the Late Miocene or Early Pliocene triggered by uplift and eastward tilting of the eastern part of the Negev toward the subsiding Dead Sea Rift. A rapid erosion of the inner parts of the truncated anticlines through the eastward-tilted Miocene transverse valleys, established the present morphology of the *makhteshim.*

Keywords

Makhtesh · Erosion cirque · Landscape development · Negev desert

17.1 Introduction

In the Negev desert (southern Israel) and northern Sinai desert (Egypt), there are several peculiar erosion cirques, which are termed "*makhteshim*" (plural of *Makhtesh*, which is a bowl used for grinding in Hebrew). A *Makhtesh* is a deep, oval valley surrounded by high cliffs nearly all along its perimeter, except a narrow outlet, and situated at the crest of an asymmetric fold (monocline). Three *makhteshim* are known from the Negev desert (southern Israel) (Fig. 17.1).

Hazera, Hatira and Ramon; and one from the Sinai desert (eastern Egypt): Halal. Their dimensions range from 5 km × 7 km to 12 km × 42 km and they are up to 400 m deep.

The *makhteshim* are characterized by the following geological and morphological features: (a) Each *makhtesh* is located along an anticlinal crest, which forms a prominent topographic ridge; (b) The anticlines are asymmetric (monoclines), with generally a south eastern steeper flanks (60°–90°) than the north western ones (6°–15°); (c) The cliffs surrounding the *makhteshim* are built of hard carbonate rocks at the top, and friable sandstone at the bottom (Fig. 17.2); (d) Each *Makhtesh* is drained by a single drainage system, with one outlet through a deep gorge incised in the south eastern, steeply inclined flank of the anticline (Fig. 17.3).

Although fold structures are common in continental platforms, oval erosion "cirques" similar to those in southern Israel and Sinai are rare. Elongated valleys along anticlinal ridges (termed "breached valleys") are known from other

E. Zilberman (✉)
Geological Survey of Israel, 32 Yesha'ayahu Leibowitz St., 9692100 Jerusalem, Israel
e-mail: ezra.zilberman@gsi.gov.il

A. Frumkin and N. Shtober-Zisu (eds.), *Landscapes and Landforms of Israel*,
World Geomorphological Landscapes, https://doi.org/10.1007/978-3-031-44764-8_17

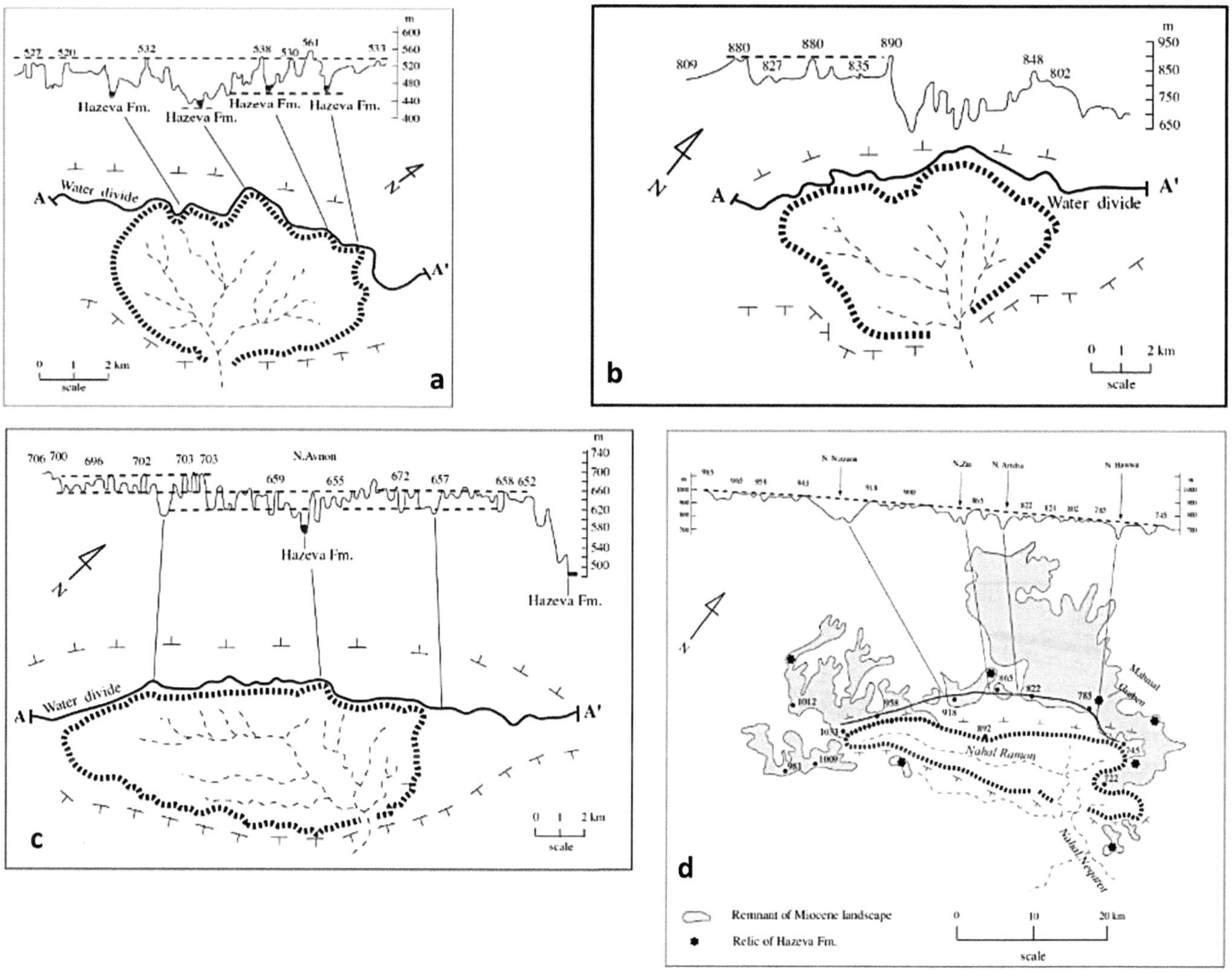

Fig. 17.3 Topographic profiles of the northwestern water divides of **a** Hazera, **b** Halal, **c** Hatira, and **d** Ramon anticlines, with maps of the associated *makhteshim*. The abandoned valleys in the northwestern cliffs of *Makhtesh* Hatira, *Makhtesh* Hazera and *Makhtesh* Halal are well demonstrated in the profiles drawn along the water divides of the anticlinal ridges. *Makhtesh* Ramon is developed in the middle of a plateau, capped by remnants of the Miocene Hazeva Formation (Modified after Zilberman 2000)

17.3 Landscape Development in Southern Israel

The present landscape in the Negev started to develop after the Late Eocene regression of the sea. Since then, continental environment has been present in the Negev and deposition of two main sedimentary units occurred: the Early Miocene fluvial-lacustrine Hazeva Formation, which is of regional extent, and the Plio-Pleistocene fluvial-lacustrine sediments of the Arava Formation, which is mostly confined to the Dead Sea Rift and its drainage basin. The Hazeva Formation was deposited in an extensive subsiding basin (Zilberman 1992; Calvo 2002; Zilberman and Calvo 2013), located south of the Judea Mts. This basin has developed before the breakup of the African plate, when both sides of the contemporary Dead Sea Rift were still connected and formed a continuous low-relief terrain drained toward the Mediterranean Sea. The distribution of the Arava Formation was highly influenced by the subsiding Dead Sea Rift along the trace of the Dead Sea Transform (Garfunkel 1981), and the newly developing local drainage systems.

In this study, the development of the *makhteshim* is linked to the main stages of landscape evolution of the Negev. Therefore, a brief review of these stages is presented below.

17.3.1 Landscape Evolution of Southern Israel

Four main stages can be distinguished in the evolution of the Negev landscape, after the final regression of the Eocene sea:

(1) Oligocene—a long stable period of subaerial erosion, during which an extensive low-relief terrain (peneplain) formed in the area. (Picard 1951; Quennell 1958; Garfunkel and Horowitz 1966; Avni et al. 2012) (Figs. 17.4 and 17.5c).

(2) Latest Oligocene to earliest Miocene—first tectonic activity along the Dead Sea Transform (Nuriel et al. 2017) and limited reactivation of the SAFB, leading to the formation of moderate relief (70–120 m) along anticlinal ridges in the northern Negev (Garfunkel 1988; Zilberman 1992) (Fig. 17.5d). Local fluvial systems transported coarse gravel from the anticlinal ridges to the adjacent synclinal valleys (Bentor and Vroman 1951, 1954, 1960; Garfunkel and Horowitz 1966). Later on, during the early Miocene, the uplift movement inverted and an extensive sedimentary basin has developed throughout the Negev, western Jordan and eastern Sinai. This subsiding basin was filled by a thick sequence (up to 1000 m) of the Hazeva Formation, mostly fluvial-lacustrine sandstone with interbeds of well-rounded chert pebbles and cobbles, transported from remote sources in the Arabian plate by a regional fluvial system (Zilberman and Calvo 2013). This fluvial-lacustrine sequence gradually covered most of the previous relief of southern Israel, including the ridges of the SAFB (Zilberman 1984, 1992) (Fig. 5e).

(3) During the late early Miocene, the Arabian plate broke up along the sinistral Dead Sea Transform, which separated it from the Israel-Sinai micro-plate (Garfunkel 1981, 1988; Nuriel et al. 2017). During the middle and late Miocene and the Pliocene, this reorganization of the regional plates was associated with the northward movement of the Arabian plate in relation to Israel-Sinai micro-plate, the subsidence of the Dead Sea Rift and uplift of the Negev region (Garfunkel 1981; Bar et al. 2016a). The uplift of the Negev was associated with wide arching (Picard 1951; Wdowinski and

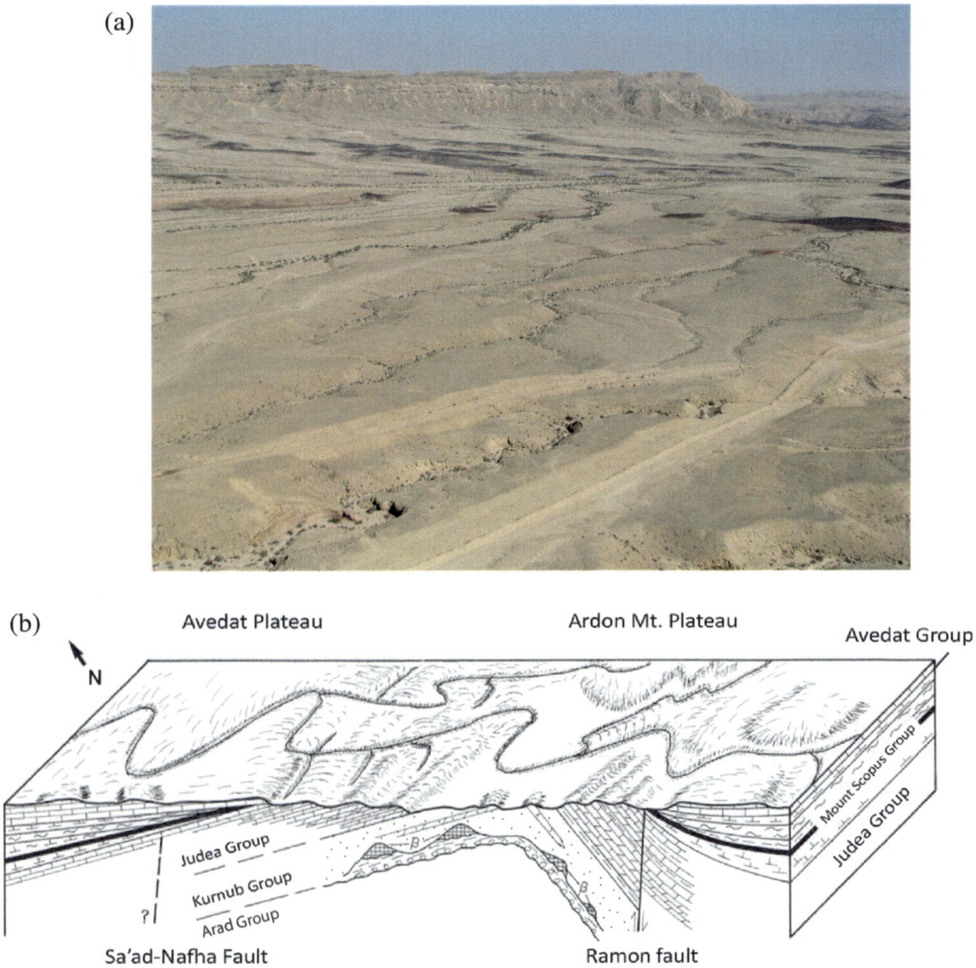

Fig. 17.4 **a** Ardon Mountain; a view to the north east. The slopes of this high plateau build the north eastern cliff of *Maktesh* Ramon. Remnants of the Miocene Hazeva Formation, which preserved on its top enable to reconstruct the Early Miocene landscape and the westward-flowing streams that crossed the truncated anticline. **b** Reconstructed Early Miocene erosion surface that truncated the north eastern part of the Ramon anticline (Ardon Mountain, Fig. 17.4a), based on preserved remnants of the Early Miocene fluvial Hazeva Formation. The northwest-flowing streams transported clastic sediments of the Hazeva Formation from remote sources in the Arabian plate. Jurassic and probably Triassic sequences were exposed along the south western part of the truncated anticline (Modified after Zilberman 1992). (For lithological patterns see Fig. 17.2; β—Early Cretaceous Basalt)

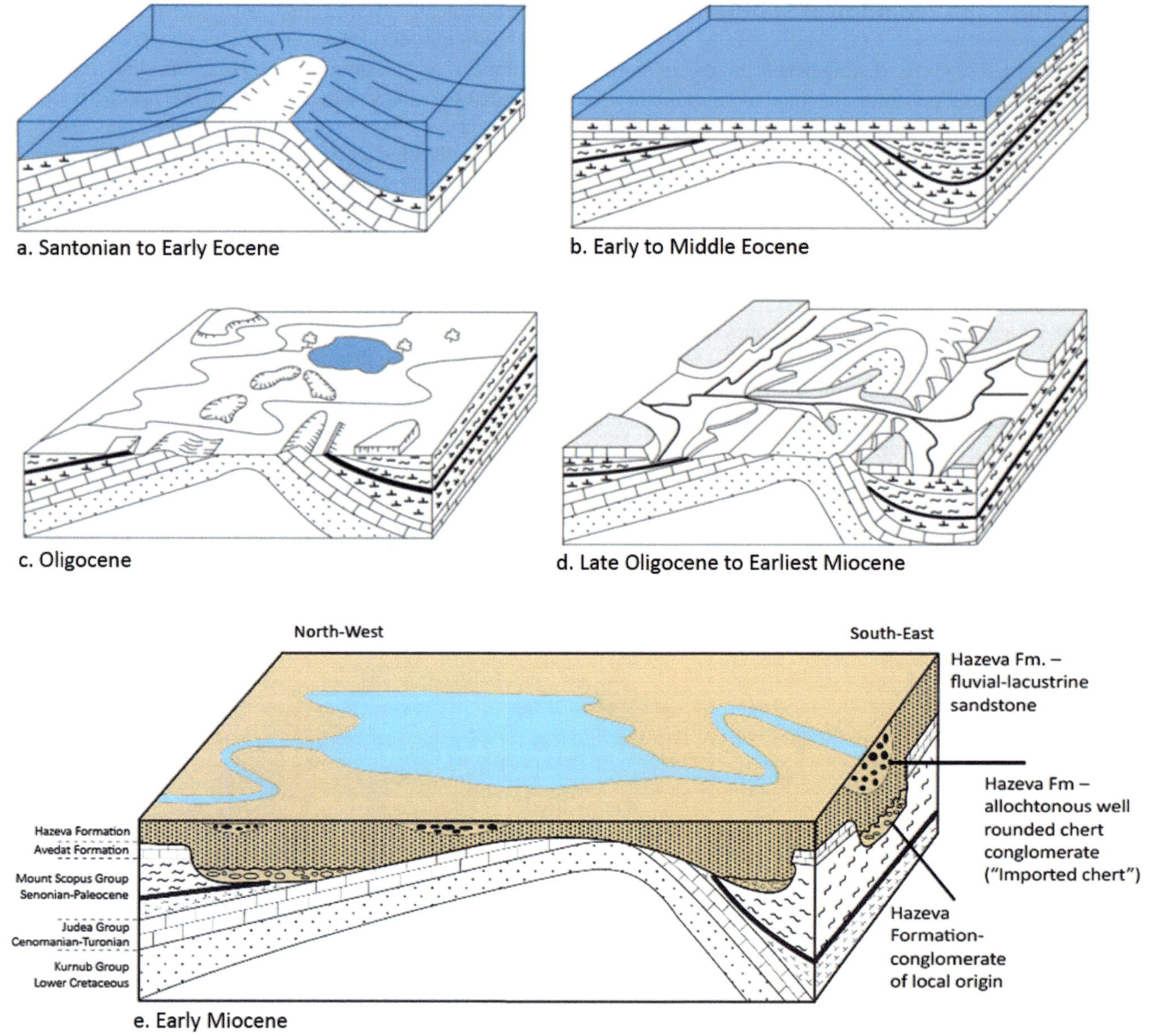

Fig. 17.5 Reconstruction of the main stages in the evolution of the erosion surfaces on a major Syrian Arc anticline: an example from the Hatira Anticline. **a** combined marine and subaerial erosion during the Senonian to Eocene transgression; **b** submergence and marine sedimentation during the Middle Eocene; **c** exposure and denudation during the Oligocene—the formation of the peneplain; **d** uplift and incision of a transverse valley during the Early Miocene tectonic phase (~23–22 my). **e** subsidence of the Negev region during the Early Miocene followed by accumulation of a thick sequence of fluvial-lacustrine sediments of Hazeva Formation, which buried the Oligocene landscape (Zilberman 1992)

Zilberman 1997), which dictated the course of the N-S oriented water divide in the Negev (Bar and Zilberman 2016b). The subsiding Dead Sea basin turned to be an erosion base level for the east-flowing streams. Erosion, which was triggered by this tectonic phase, carved most of the present relief in the Negev desert in southern Israel (Zilberman 1991; Begin and Zilberman 1997; Zilberman et al. 2011).

(4) Pliocene to Early Pleistocene—a period of major tectonic activity along the Dead Sea Transform associated with an uplift of 200–300 m of the central and southern mountain belt of Israel (Bar et al. 2013; Bar and Zilberman 2016b). Erosion of the uplifted areas was followed by deposition of a thick clastic fill in the subsiding rift and along the main drainage lines. The regional water divides were finally established along the arched mountain belt of Israel west of the rift and along the eastward-tilted, uplifted eastern margin of the rift. The present structural and morphological pattern of the Negev was finally shaped during this period

(Garfunkel and Horowitz 1966; Zilberman 1984; Garfunkel 1988; Ginat 1992; Avni 1994).

The anticlinal ridges of the Negev have been exposed to erosion processes since the Oligocene. However, it is likely that the style of erosion has varied through time, particularly from the peneplain stage to the tectonic uplift stage. In the following section, some morphological elements along the anticlinal ridges that appear to be required for the formation of the *makhteshim* will be examined in agreement with the hypotheses of Nir (1958) and Eran (1982), and evidence for the truncation of the anticlinal ridges before the formation of the *makhteshim* will be presented. In addition, the present single outlet of each *Makhtesh* that might be genetically inherited from a previous transversal stream will be discussed.

17.4 Remnants of Planation Surfaces on the Anticlinal Ridges

A planation surface is an erosion plain that cuts across various rocks and structures (Ollier 1981). In the arid climate of the Negev, where erosion is dominated by slope retreat processes, water divides are the last to be eroded and thus best preserve an ancient terrain (King 1953, 1962; Twidale 1976). Therefore, relicts of ancient planation surfaces are best demonstrated by detailed longitudinal profiles along the water divides of the anticlinal ridges, where the relicts appear as a series of peaks of similar altitudes scattered along and across the anticlinal ridges (Fig. 17.3a–d).

In the Hazera and the Halal anticlines, the upper erosion surface is well distinguished along the water divide profile (Fig. 17.3a, b) and the reconstructed amount of truncation is less than that in the Hatira and Ramon anticlines.

In the Hatira Anticline, the summits along the perimeter of the valley demonstrate relics of two extensive planation surfaces (Fig. 17.3c). The higher surface is confined to the south western part of the anticline; the lower one extends to the northwest along the entire ridge. Only the upper surface can be reconstructed to a flat planation surface. As such, it represents a mature landscape, when the top of the anticline was denuded to the level of the surrounding low-relief terrain. The Middle surface represents a very wide valley bounded in the southwest by a topographic high, which is a relict of the previous terrain. In addition, there is a low level of valleys, which represent mainly the position of the present stream valleys along the water divide. Extrapolation of the upper erosion surface across the anticline crest reveals an oval exposure of the sandy Kurnub Group at the central part of the truncated anticline (Fig. 17.5d; Roded 1982; Zilberman 1992).

The upper planation surface is best preserved around the Ramon Anticline, where it is expressed as an extensive plateau tilted to the northeast and capped by relics of the Miocene Hazeva Formation (Fig. 17.3d). The intersection between this surface and the fold structure points to a truncation of about 700 m of the fold crest (Avni 1989), exposing sandstone units of the Early Cretaceous Kurnub Group along the outer margins of the surface (Fig. 17.4) and the underlying sandy parts of the Jurassic Arad Group at its inner parts.

17.4.1 The Age of the Planation Surfaces

The history of the upper planation surfaces found on the Negev anticlines dates back to the Senonian, when the primary folds of the SAFB first emerged from the Tethys Sea, forming shallow submarine ridges or small islands (Fig. 17.5a). The summits of the folds were submitted to episodic submarine or subaerial denudation, which gradually exposed deeper parts of the growing anticlines (Picard 1951; Bentor and Vroman 1951, 1957) (Fig. 17.5a).

The amount of truncation at the anticline's crest was proportional to their structural height. However, the final truncation, which formed the present configuration of the upper erosion surfaces, occurred during the Oligocene, when the upper parts of the folds were denuded to the level of the surrounding peneplain (Fig. 17.5c) (Picard 1951; Nir 1958, 1970; Zilberman 1991). The Oligocene age of this planation surface is confirmed around *Makhtesh* Ramon, where it is capped by remnants of fluvial sediments of the Early Miocene Hazeva Formation (Fig. 17.3d) (Zilberman 1981, 1992; Calvo 2002; Avni et al. 2012; Zilberman and Calvo 2013). As will be discussed in more detail below, this formation was deposited by streams that flowed westward across the anticlinal ridges (Figs. 17.4 and 17.5d), producing the complex erosion surfaces found on the highest parts of the anticlinal ridges.

17.5 Abandoned Valleys and Their Paleogeographic Implications

Along the northwestern cliffs of the Hazera (Fig. 17.6a), Hatira (Fig. 17.7) and Halal (Fig. 17.8) erosion cirques, deep gorges are clearly evident in the water divide profiles (Fig. 17.3a–c). These gorges are situated opposite the present outlets of the drainage systems of each *makhtesh*. In the Hatira and the Hazera anticlines, the bottoms of the gorges are filled by a fluvial sequence of the Hazeva Formation, which also covers the adjacent synclinal valleys (Figs. 17.2b, 17.7b and 17.9) to the northwest.

Fig. 17.6 **a** Air photo of Makhtesh Hazera showing the present outlet (p) and the abandoned valley (**a**), which was part of the northwest-flowing Miocene transverse valley. **b** Outlet of *Makhtesh* Hazera. A view to the east

In the 42 km long valley of the *Makhtesh* Ramon, the morphology is more complex. Three deep valleys are evident along the profile of the water divide, which runs along the northwest margins of the *Makhtesh* (Fig. 17.3d). In Nahal Hawwa and Nahal Aricha, remnants of the Base Conglomerate of the Hazeva Formation occur at the level of the present stream valleys. In the upper reach of Nahal Zin valley, relicts of Miocene sandstone were found on the surrounding Avedat plateau.

The present outlet of the *Makhtesh* is located at its northeastern part, opposite the Nahal Hawwa valley, where a 60-m-thick sequence of the Base Conglomerate of the Hazeva Formation is preserved in a post-Miocene small structural depression (Zilberman 1981, 1992).

There is no information regarding the occurrence of Miocene sediments in the deep gorge of the Halal Anticline, which is more than 200 m deep. However, since this valley is not fault-controlled (Shata 1960), it is presumed that it is also a remnant of an ancient stream.

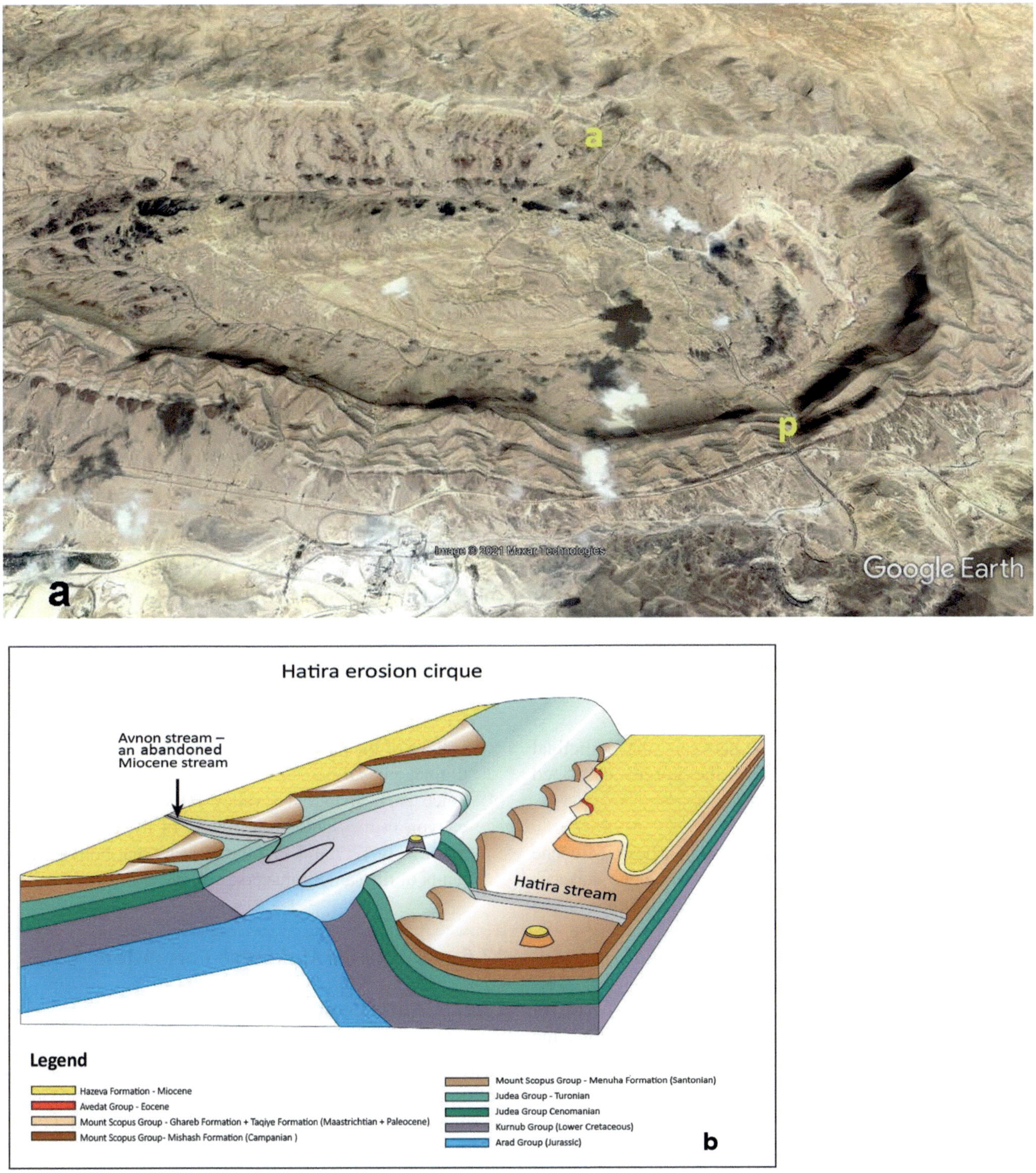

Fig. 17.7 *Makhtesh* Hatira erosion cirque. **a** air photo (from Google earth); **b** illustration of the structure of the anticline and the morphology. The abandoned Miocene stream is marked 'a', and the present outlet is marked 'p'

Fig. 17.8 *Makhtesh* in Gebel (mountain) Halal anticline. The abandoned valley is Marked; 'a' and the present outlet is marked; 'p'.

The morphological relations between the anticlinal ridges, their erosion surfaces and the gorges described above lead to the following conclusions:

(1) These abandoned gorges are of earliest Miocene age, since they are incised in the Oligocene surfaces and filled by Early Miocene fluvial sediments of the Hazeva Formation.
(2) The incision of the deep gorges in the truncated anticlines, which were formerly at the level of the surrounding terrain, reflects an Early Miocene uplift of the anticlinal ridges. Therefore, at any anticline, the amount of uplift can be estimated from the difference between the upper (Oligocene) erosion surface and the bottom of the Miocene valley along the water divide.
(3) The alluvial sediments of Hazeva Formation that fill these valleys were deposited by Miocene northwest-flowing streams that crossed the anticlinal ridges and drained toward the synclinal valleys (Figs. 17.4 and 17.5d). This is concluded from the outstanding depth of these valleys and by the alluvial fill, which is continuous with the lower local conglomerate of the Hazeva Formation in the adjacent synclines. The northwestern parts of these valleys were abandoned when the deep valleys of the *makhteshim* developed and captured their drainage basins.
(4) Since each abandoned valley faces the present outlet canyon in the southeastern flank of the anticline (e.g., Figs. 17.6, 17.7 and 17.8), it appears that both are remnants of the same transverse valley.

17.6 Model for *Makhtesh* Formation

17.6.1 Initial Conditions

There is general agreement regarding the stratigraphic control on the development of deep valleys in anticlinal ridges, as suggested by von Engeln (1942) and Picard (1951). Under arid to semi-arid conditions, the differential resistance to weathering of the hard carbonate rocks and the underlying friable sandstone enable faster erosion of the sandstone from the inner core of the folds, followed by collapse and retreat of the overlying hard carbonate cliffs, a process that declines with time (the present receding rate is 1–18 cm/ky; Finzi and Harlev 2016). In the Negev desert, the soft rocks of the Mount Scopus Group also played an imported role in this process. They were rapidly stripped

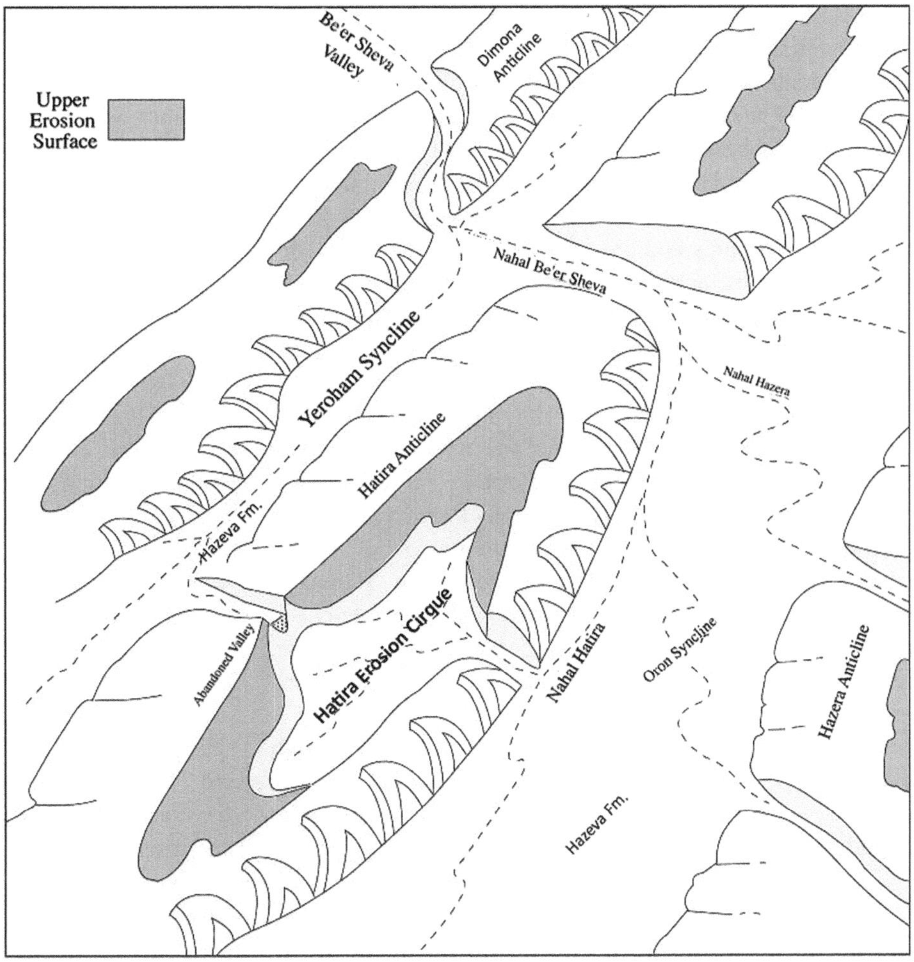

Fig. 17.9 Reconstruction of the paleogeography of the *Makhtesh* Hatira region in the Early Miocene, during the deposition of the Base Conglomerate of the Hazeva Formation. A primary *Makhtesh* is formed in the Hatira Anticline, triggered by capturing of the previously northwest-flowing transversal stream by Nahal Be'er Sheva (reconstructed after Neev 1960; Roded and Hirsch 1987)

from the folds margins, exposing structural ridges built of the underlying hard carbonate strata of the Judea Group (Fig. 17.2b).

Folding of the strata elevated the sandstone core of the folds to a high position relative to the adjacent synclinal basins in the southeast. These vertical differences determined the future depth of the *makhteshim*.

The erosion surfaces at the crests of the anticlines were also essential to the initiation of erosion of the *makhteshim*. The sandstone outcrops on the truncated anticline crests were susceptible to faster erosion as soon as the anticline was uplifted and exposed. The area and configuration of the sandstone exposures determined the dimension of the primary valley, as suggested by Nir (1958). Later on, slope retreat processes increased the size of the initial valleys to their contemporary dimensions.

Although the above initial conditions fit most of the observations at the studied *makhteshim*, they do not explain why each erosion cirque is drained through a single outlet. The model for the formation of breached anticlines in the open folds of the Jura Mountains (von Engeln 1942) predicts the incision of several deep valleys in each anticlinal ridge, which later merge into a single elongated valley. However, such abandoned valleys were not found along the anticlinal ridges of the Negev. Likewise, the explanation proposed by Picard (1951), wherein the outlet is situated in the highest part of the structure raises several problems. First, the outlets of the drainage systems of the *makhteshim* are not always located on the structurally highest part of the anticline, indicating that their location is not purely structurally controlled. Second, streams that drain other parts of the anticlines with similar or even steeper inclinations were unable to penetrate the hard carbonate envelope of the anticlines, although their drainage basins had not changed much since the exposure of the anticlinal ridges. This implies that the drainage areas of these streams are too small to produce the amount of runoff required to incise the thick, hard carbonate sequence. On the other hand, all the streams that did incise deep transverse valleys in the anticlinal ridges of the Negev are superimposed streams that have large drainage basins, which also include parts of the adjacent synclinal valleys (Picard 1951; Gvirtzman 1979).

It is therefore suggested that the development of the Miocene transverse valleys is also a prerequisite to the development of the *makhteshim.* They were especially important in the Hazera and Halal anticlines, where soft sandstone was barely exposed at the level of the Oligocene erosion surface. The incision of the Miocene transverse valley into these erosion surfaces increased the area of the sandstone exposures and exposed deeper parts of this friable unit. When the north westward-flowing streams reversed their flow direction to the present southeast trend, the streams that occupied the Miocene transverse valleys had already incised into the anticlinal ridge, and therefore had advantage on the other streams. They quickly captured the truncated area where the sandstone was exposed to become the only outlet of the developing valleys.

The Late Oligocene-Early Miocene transverse valleys are, therefore, the precursors of the present outlets of the *makhteshim*, albeit in an opposite flow direction. The abandoned valleys, which were filled by Miocene sediments and are preserved in the north western cliffs of each *Makhtesh*, represent the former downstream parts of the captured valleys.

It is therefore suggested that the following combination of initial geological and morphological conditions was prerequisite to the formation of the *makhteshim* in southern Israel:

(1) Folds, which are also prominent topographic ridges;
(2) A folded sequence of hard rocks overlying easily eroded rocks;
(3) Deep truncation of the folds' crests and exposure of the friable sandstone on a low-relief erosion surface. The outline of the contact between the hard and the friable rocks determined the primary dimension of each *Makhtesh*.
(4) Incision of transverse valleys in the erosion surfaces and the structural ridges. These valleys are the precursors of the present outlets of the *makhteshim.*

Although the above initial conditions already existed in the Early Miocene, the final opening stage of each *Makhtesh* occurred at different times triggered by the establishment of a deep erosion base to the southeast of each anticlinal ridge.

17.6.2 Stream Capture and the Evolution of the *Makhteshim*

In the Early Miocene, the Hatira Anticline was the most elevated ridge, as evident from its deep abandoned valley, which incised about 120 m in the upper erosion surface (Fig. 17.3c). This valley is the lower reach of a transverse valley, which connected the Oron synclinal valley in the southeast side of the Hatira Anticline with the Yeroham synclinal valley at its northwestern side, both covered by fluvial-lacustrine sediments of the Hazeva Formation (Fig. 17.9).

The capturing of the northwest-flowing stream and the incision of a primary deep valley in the anticlinal core must have occurred for the first time in the Early Miocene during the deposition of the Zefa Member of the Hazeva Formation (Calvo 2002). This conclusion is based on two observations: (a) a remnant of a conglomerate of the Hazeva Formation, which is preserved within *Makhtesh* Hatira 70 m above its bottom, at the same morphostratigraphic position as the conglomerate in the Oron syncline near the present outlet (Fig. 17.7), and (b) clasts, derived from Jurassic and Early Cretaceous sequences exposed within the *Makhtesh*, are found in the conglomerate in the Oron syncline. These clasts indicate exposure of the anticlinal core by a Miocene southeast-flowing stream (Fig. 17.9). The reversal of the northwest-flowing stream in the transverse valley of the Hatira Anticline is related to stream capture, by a tributary of the ancient Be'er Sheva River (Roded and Hirsch 1987), which was the major Miocene river in the northern Negev (Neev 1960; Fig. 17.9). The capture was triggered by an Early Miocene tectonic phase associated with folding of the Hatira Anticline (Calvo 2002). The configuration and the dimension of the Miocene *Makhtesh* are not entirely clear, but the relic of the conglomerate of the Hazeva formation within the *Makhtesh* indicates that its bottom was 70 m higher than at present. Toward the end of the early Miocene, this relief was buried under a thick fluvial-lacustrine sequence of the Rotem member of the Hazeva Formation, when the entire Negev subsided (Zilberman 1992; Calvo 2002; Zilberman and Calvo 2013).

Although the Hazera and the Mahmal anticlines were also crossed by transversal streams during the Oligocene and the beginning of the Early Miocene, they remained low in relation to their surrounding terrain; therefore, deep valleys did not develop in their truncated cores. The *Makhtesh* Ramon area was also low in relation to the northern Negev, as evident by 100–350-m-thick sequences of the Hazeva Formation preserved on the surrounding terrain (Bentor and Vroman 1951; Zilberman 1981; Zilberman and Calvo 2013).

The incision of *Makhtesh* Hazera and *Makhtesh* Ramon, and the re-incision of the buried Early Miocene *Makhtesh* Hatira, started in the Late Miocene or Early Pliocene, triggered by the uplift of the Negev mountains belt and the formation of the Dead Sea Rift Valley in the east as a deep erosional base level.

The uplift of the Negev was associated with a long-wave arching (upwarp) (Picard 1951; Wdowinski and Zilberman 1996, 1997), which eventually dictated the route of the regional N-S oriented water divide of Israel. The uplift started in the Early Miocene (Sneh and Buchbinder 1984; Bar et al. 2016a) and continued during the Late Miocene in the central Negev (Chaldekas 2019). As a result of this uplift and the subsidence of the Dead Sea Rift Valley, the regional northwest-flowing fluvial system, which dominated southern Israel since the Oligocene, was cut off from its drainage basin in the east. In addition, the upwarp of the SAFB reversed the regional gradient of the eastern Negev to the east. The present regional water divide was established along the uplifted SAFB, winding west of the Hatira, Hazera and Ramon anticlines, which were also tilted toward the subsiding Rift Valley. The new eastward-flowing drainage systems removed the cover of the Hazeva Formation and exposed the Early Miocene relief of the anticlinal ridges. The streams, which coincided with the ancient transverse valleys, rapidly incised into the truncated anticlinal ridges and eroded the friable sandstone from the cores of the folds.

The upwarp of the central Negev tilted the Ramon Anticline and its truncated upper part (The erosion surface) to the northeast (Zilberman 1984; Ben David and Mazor 1988; Avni 1989; Ben David 1991). Therefore, the erosion surface in the western part of *Makhtesh* Ramon is t300–400 m higher than in its north eastern part (Fig. 17.3d). As a result of this tilting, a northeast-flowing drainage system developed (Avni 1989; Ben David 1991), forming a valley, which drains toward the Dead Sea Rift (Bentor and Vroman 1957; Baer 1981), with outlet located in the northeast part of the *Makhtesh*, probably using the eastern reach of Nahal Hawwa Miocene transverse valley. The opening of *Makhtesh* Halal may be genetically related to both tectonic activity and the establishment of the post-Miocene drainage system of Sinai. This structure is located near the Sa'ad-Nafha tectonic line, which is part of the six E-W-oriented dextral strike-slip faults of the Sinai-Negev Shear Zone (Bartov 1974). This system was highly active during the Late Oligocene to Early Miocene and during the Middle to Late Miocene (Wienberger et al. 2020). The main tectonic activity along Sa'ad-Nafha fault occurred a short time after the deposition of the Early Miocene Hazeva Formation and ended before this fluvial cover was removed by erosion from the entire Negev (Zilberman 1981; Zilberman et al. 2011; Zilberman and Calvo 2013). Today *Makhtesh* Halal is drained southward, toward Nahal El-Arish, which flows to the Mediterranean Sea, via a transverse valley that crosses the eastern part of the Halal anticline.

To summarize, the deep valleys of the *makhteshim* formed after the uplift of the Negev region in the Middle to Late Miocene and the formation of the Dead Sea Rift Valley as a deep inland erosion base level, together with the establishment of the main topographic elements in the present Negev landscape.

17.7 Conclusions

Although the valleys of the *makhteshim* of the Negev seem to be merely a subset within a broader family of intra-fold valleys, their complex evolution and peculiar morphology raise questions about the simple model suggested for the evolution of similar valleys by von Engeln (1942) and Picard (1951). In the arid to semi-arid climatic regime that prevailed in southern Israel at least since the Pleistocene, small streams that drain the steep flanks of the anticlines were unable to penetrate the 400-m-thick hard carbonate envelope of the anticlines to form additional outlets for the *makhteshim*. The formation of deep outlets that enabled the rapid incision in the anticlinal cores required additional conditions. These, as described above (truncation surfaces and ancient transversal streams), were already part of the morphology before the Miocene. It was the Middle Miocene to late Pliocene tectonic phase, with the upwarping of the SAFB and the formation of the Dead Sea Rift that ultimately triggered the changes in the drainage systems that led to stream reversal and to the incision of the *makhteshim*. Unfortunately, there is not much information about the evolution of similar valleys in other parts of the world; closed valleys with only one outlet, as found in the Negev, are not yet described elsewhere.

References

Amiran DHK (1951) Geomorphology of the central Negev highlands, Israel. Isr Explor J 1:107–120. https://www.jstor.org/stable/27924432

Avni Y (1989) The geology of the western part of the Ramon Anticline. M.Sc. thesis, Hebrew Univ., Jerusalem, 160pp (in Hebrew, English abstr)

Avni Y (1994) Plio–Pleistocene tectonic deformation—A key to dividing the continental Dead Sea Group in the Negev and the Arava Valley. Isr Geol Soc Annu Mtg. Nof Genosar, p 4

Avni Y, Segev A, Ginat H (2012) Oligocene regional denudation of the northern Afar Dome: pre and syn-breakup stages of the Afro-Arabian plate. Geol Soc Am Bull 124:1871–1897. https://doi.org/10.1130/B30634.1

Baer G (1981) The geology of the Arif-Batur lineament Ma'ale Hameshar area. M.Sc. thesis, Hebrew Univ. Jerusalem, 86pp (in Hebrew, English abstr)

Bar O, Zilberman E (2016) Subsidence and conversion of the Dead Sea basin to an inland erosion base level in the early middle Miocene as inferred from geomorphological analysis of its ancient western fluvial outlet. Geomorphology 261:147–161. https://doi.org/10.1016/j.geomorph.2016.02.028

Bar O, Gvirtzman Z, Feinstein S, Zilberman E (2013) Accelerated subsidence and sedimentation in the Levant Basin during the Late Tertiary and concurrent uplift of the Arabian Platform: tectonic versus counteracting sedimentary loading effects. Tectonics 32:1–17. https://doi.org/10.1002/tect.20026

Bar O, Zilberman E, Feinstein S, Calvo R, Gvirtzman Z (2016) The uplift history of the Arabian Plateau as inferred from geomorphologic analysis of its northwestern edge. Tectonophysics 671:9–23. https://doi.org/10.1016/j.tecto.2016.01.004

Bartov Y (1974) A structural and paleogeographical study of the Central Sinai faults and domes. Ph.D. thesis, Hebrew Univ. Jerusalem, 143pp (in Hebrew, English abstr)

Ben-David R (1991) The geology of Be'erot Oded area and western *Makhtesh* Ramon, and stages in the evolution of landscape since the Late Miocene to the present. M.Sc. thesis, Ben-Gurion Univ., Be'er Sheva. 118pp (in Hebrew, English abstr)

Ben-David R, Mazor E (1988) Stages in the evolution of *Makhtesh* Ramon and its drainage system. Isr J Earth Sci 37:125–135

Begin ZB, Zilberman E (1997) The rate and stages of landscape evolution in Eretz Israel. Isr. Geol. Surv. Rep. GSI/24/97, 64 pp. (In Hebrew, Eng. Abst.)

Bentor YK, Vroman A (1951) The geological map of the Negev, 1:100,000, sheet 18: Abde (Avedat), with explanatory notes. 98pp (in Hebrew)

Bentor YK, Vroman A (1954) A structural contour map of Israel (1:250.000) with remarks on its dynamic interpretation. Geol Surv Isr Bull No. 7, 10pp

Bentor YK, Vroman A (1957) The geological map of the Negev, 1:100,000, Sheet 19: Arava Valley, with explanatory text. Geol Surv Isr 66pp

Bentor YK, Vroman A (1960) The geological map of the Negev, 1:100,000, sheet 16: Mt. Sedom, with explanatory text (revised edition). Geol Surv Isr 117pp

Calvo R (2002) Stratigraphy and petrology of the Hazeva Formation in the Arava and Negev: Implications for the development of the sedimentary basins and the morphotectonics of the Dead Sea Rift Valley. PhD thesis, Hebrew Univ (in Hebrew, English abstr) 264pp

Chaldekas O (2019) U-Pb geochronology of phreatic and vadose speleothems: recording falling groundwater levels and uplift phases along the western shoulder of Dead Sea Transform. Isr Geol Surv Rep GSI/16/2019, 64pp

Eran G (1982) Erosion surfaces in Israel. Unpublished report, Geol. Dept., Hebrew Univ., Jerusalem, 17 pp (in Hebrew)

Finzi Y, Harlev N (2016) A regional approach for modeling cliff retreat rate: the Makhteshim Country, Israel. Geomorphology 271:65–73. https://doi.org/10.1016/j.geomorph.2016.07.015

Garfunkel Z (1981) Internal structure of the Dead Sea leaky transform (rift) in relation to plate kinematics. Tectonophysics 80:81–108. https://doi.org/10.1016/0040-1951(81)90143-8

Garfunkel Z (1988) The pre-quaternary geology of Israel. In: Yom-Tov Y, Tchernov E (eds) The zoogeography of Israel. Dr W. Junk, Dordrecht, pp 7–33

Garfunkel Z, Horowitz A (1966) The upper tertiary and quaternary morphology of the Negev. Isr J Earth Sci 15:101–117

Ginat H (1992) Reversal in the flow direction of Nahal Hiyyon during the deposition of the Arava Conglomerate. Isr Geol Soc Annu Mtg Ashqelon, p 51

Gvirtzman G (1979) The geology and geomorphology of the Biq'at Be'er Sheva valley. In: Gardos Y, Shtern E (eds) The book of Be'er Sheva. Ben-Gurion Univ of the Negev, pp 333–354

Hall JK (1994) DTM map of Israel and adjacent areas 1:500,000. Geol Surv Isr

Issar A (1983) Emerging ground water, a triggering factor in the formation of the *Makhteshim* (erosion cirques) in the Negev and Sinai. Isr J Earth Sci 32:53–61

King LC (1953) Canons of landscape evolution. Geol Soc Am Bull 64:721–754. https://doi.org/10.1130/0016-7606(1953)64[721:COLE]2.0.CO;2

King LC (1962) The morphology of the earth. Edinburgh, Oliver and Boyd, 699pp

Maret RE, Coe AV Robertson (1960) Geology of Sinbad valley Anticline (Colorado-Utah). Four Corners Geol Soc 3d field Conf Guidebook, pp 43–46

Neev D (1960) A pre-Neogene erosion channel in the southern coastal plain of Isr. Geol Surv Isr Bull 25, 20pp

Nir D (1958) Les pays de bray el les *makhteshim* en Negev. Un cas de geomorphologie comparative. Rev Geomorph Dynam 9:112–118

Nir D (1970) The geomorphology of Israel. Academon Press, Hebrew Univ., Jerusalem, 404pp (in Hebrew)

Nuriel P, Weinberger R, Kylander-Clark ARC, Hacker BR, Craddoc JP (2017) The onset of the Dead Sea transform based on calcite age-strain analyses. Geology 45(7):587–590. https://doi.org/10.1130/G38903.1

Oberlander MT (1985) Origin of drainage transverse to structures in orogenes. In: Morisawa M, Hack JT (eds) Tectonic geomorphology. Allen and Unwin, Boston, pp 155–182

Ollier C (1981) Tectonics and landforms (Geomorphology texts; 6). Longman, Harlow, UK

Picard L (1951) Geomorphogeny of Israel, Part 1: the Negev. Bull Res Counc Isr 1:5–32

Quennell AM (1958) The structural and geomorphic evolution of the Dead Sea Rift (with discussion). Q J Geol Soc 114:1–24. https://doi.org/10.1144/gsjgs.114.1.0001

Roded R (1982) The geological map of Israel, Oron sheet, scale 1:50,000. Geol Surv Isr

Roded R, Hirsch F (1987) Geomorphological history of the Rotem Oron Basin in Neogene "Hazeva" time. Isr Geol Soc Annu Mtg Mizpe Ramon, p 105

Shata A (1960) The geology and geomorphology of El- Qusaima area. Bull Soc Geograph Egypt 33:95–146

Sneh A, Buchbinder B (1984) Miocene to Pleistocene surfaces and their associated sediments in the Shefela region. Israel Isr Geol Surv Current Res 1983–1984:60–64

Twidale CR (1976) On the survival of paleoforms. Am J Sci 276:77–95

Von Engeln OD (1942) Geomorphology. Macmillan, New York, p 655

Wdowinski S, Zilberman E (1996) Kinematic modelling of large-scale structural asymmetry across the Dead Sea Rift. Tectonophysics 266:187–201. https://doi.org/10.1016/S0040-1951(96)00238-7

Wdowinski S, Zilberman E (1997) Systematic analyses of the large-scale topography and structure across the Dead Sea Rift. Tectonics 16:409–424. https://doi.org/10.1029/97TC00814

Weinberger R, Nuriel P, Kylander-Clark AR, Craddock JP (2020) Temporal and spatial relations between large-scale fault systems: Evidence from the Sinai-Negev shear zone and the Dead Sea Fault. Earth Sci Rev 211:103377. https://doi.org/10.1016/j.earscirev.2020.103377

Zilberman E (1981) The geology of the central Negev-Sinai shear zone, central Negev. Part A: the Sa'ad-Nafha Lineament. Geol. Surv Isr Rep Hydrol 1/81, 59pp

Zilberman E (1984) The Tertiary and the Quaternary of the northwestern Negev. In: Begin ZB (ed) Outlines of the geology of the northwestern Negev. Isr Geol Surv Rep GSI/19/84, pp 41–76

Zilberman E (1991) Landscape evolution in the central, northern and northwestern Negev during the Neogene and the Quaternary. Isr Geol Surv Rep GSI/45/90. 164pp (In Hebrew, Eng. abstr)

Zilberman E (1992) Remnants of Miocene landscape in the central and northern Negev and their paleogeographical implications. Geol Surv Isr Bull 83, 54pp

Zilberman E (2000) The formation of the "*makhteshim*" – Unique erosion cirques in the Negev, southern Israel. Israel J Earth Sci 49(3):127–142

Zilberman E, Calvo R (2013) Remnants of Miocene fluvial sediments in the Negev Desert, Israel, and the Jordanian Plateau: evidence for an extensive subsiding basin in the northwestern margins of the Arabian plate. J African Earth Sci 82:33–53. https://doi.org/10.1016/j.jafrearsci.2013.02.006

Zilberman E, Edelman A, Avni Y, Ginat H (2011) The geology and landscape evolution of the Negev, Israel. 188 p. (In Hebrew). Published by the Non-Profit Organization Association for the Conversation of Nature in the Middle East

Makhtesh Hatzera Erosion Cirque, the Negev Desert—Landforms and Sediment Dynamics

18

Noam Greenbaum, Uri Schwartz, Motti Zohar and Judith Lekach

Abstract

Quantitative GIS-based analysis of the morpho-stratigraphic map of the Makhtesh Hatzera Erosion Cirque (MHEC) and RTL ages of the eight alluvial terraces (Plakht 2003) were used to calculate the volumes and rates of sediments transported outside the cirque through its opening, specifically during the last > 500 ka and since its initiation in the Pliocene (~5 Ma). The calculated average rates of sediment transport for the entire MHEC are 1840 m^3 $year^{-1}$. For the alluvial terraces, the rates range between 344 and 1328 m^3 $year^{-1}$, which are similar to the present-day rates as documented at a small reservoir in the MHEC. An earlier study showed that floods at the Hatzera stream such as the 2004 flood (470 m^3 s^{-1}) can transport boulders up to 2.1 m and weighing up to 15 tons at least once in 120 years. This size is about the maximum boulder size that was found on the terraces within the cirque. Field survey across the terraces shows, as expected, that (a) the desert pavement and soil parameters systematically and gradually change from the active channel and the sub-recent terrace to the oldest terrace due to the increasingly longer intervals of weathering, dust accumulation and pedogenesis; (b) the boulder size is reduced up the alluvial terraces sequence—i.e. with time, due to desert weathering and they almost completely disappear within about 100 ka. Sediment evacuation from the erosion cirque as well as the size of the opening are highly controlled by the flood regime which transports the sediments that are directly related to the climatic conditions. Periods of slow sediment aggradation coincide with the build-up of the alluvial terraces and correlate with more humid periods, whereas rapid incision phases characterize more arid conditions.

Keywords

Makhtesh Hatzera Erosion Cirque (MHEC) · Sediment evacuation · Boulders transport · Alluvial terraces · Ungauged catchment · Desert flood · Desert pavement · Reg soil

18.1 Introduction

18.1.1 Erosion Cirques ("Makhtesh") in the Negev Desert—Geological Background

"Makhtesh" ("Crater" in Hebrew) is an erosion cirque, four of which occur in the Negev Desert (Israel) and Sinai (Egypt) (Fig. 18.1). These erosion cirques are usually deep and elongated-oval depressions situated at the crest of the southwest—northeast, asymmetric Negev anticlines that form prominent topographic ridges, where the southeastern flanks of the anticlines are steeper 60°–90°, in comparison with the northwestern flanks-6°–15° (Zilberman 2000; this volume). The geology of the structures is composed of hard Upper Cretaceous carbonate rocks with occasional chert and silicified carbonate rocks (Arkin and Hamaoui 1967; Yechieli et al. 1994; Hirsch 1995) exposed at the top of the cliffs overlying relatively soft lower Cretaceous quartz sandstone (Weissbrod 1993) (Fig. 18.2c). At the bottom, along the axis of the anticline, older Jurassic dolomite and marls are exposed. The erosion cirques have one outlet through a deep gorge incised in the steeply inclined southeastern flank, which is the only opening for the fluvial systems that drain the erosion cirques (Fig. 18.2).

N. Greenbaum (✉) · U. Schwartz · M. Zohar
School of Environmental Sciences, University of Haifa, Mt. Carmel, 3498838 Haifa, Israel
e-mail: noamgr@geo.haifa.ac.il

J. Lekach
The Hebrew University of Jerusalem, 9190401 Givat Ram, Jerusalem, Israel

A. Frumkin and N. Shtober-Zisu (eds.), *Landscapes and Landforms of Israel*, World Geomorphological Landscapes, https://doi.org/10.1007/978-3-031-44764-8_18

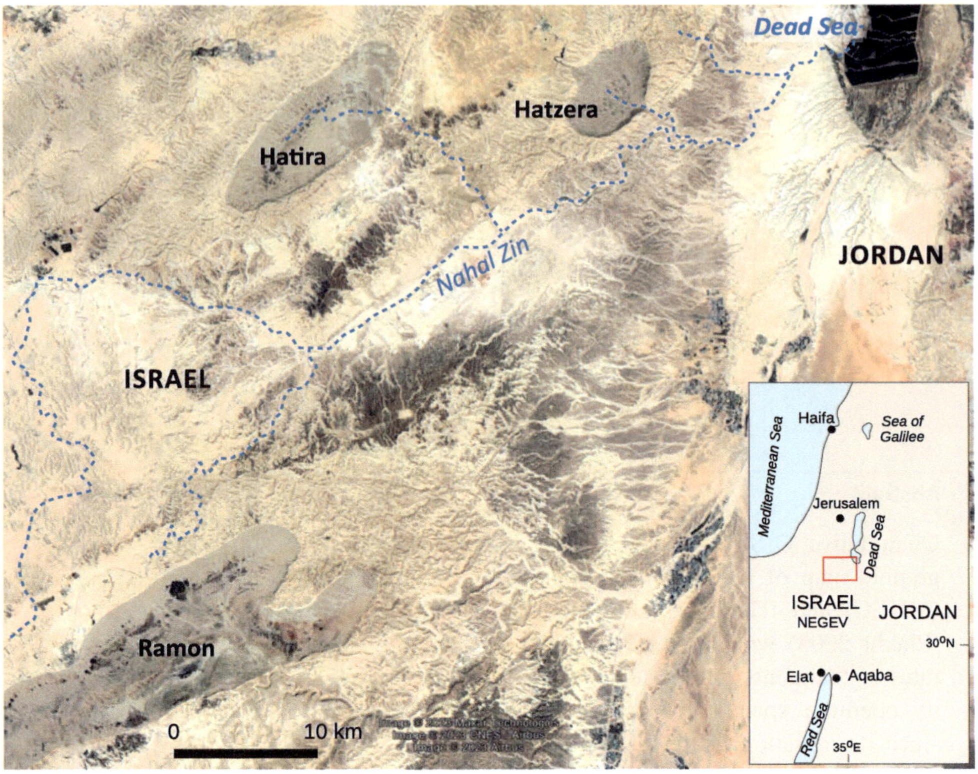

Fig. 18.1 Location map of the erosion cirques (Makhtesh) along the southwest–northeast anticlines of the Negev, and Nahal Zin—the mainstream system of the northeastern Negev which enters into the southern basin of the Dead sea covered with evaporation ponds of the Dead Sea Works, at present; the Hatira and Hatzera Makhtesh drain into Nahal Zin via Nahal Hatira and Nahal Hatzera ephemeral stream systems, respectively

Picard (1951) introduced the first model for the formation of a "Makhtesh": (1) Exposure of the fold as a topographic ridge. (2) Incision of several stream channels into the steep southeastern flanks of the anticlines, where the steepest-gradient stream developed into a deep gorge exposing the friable sandstone. (3) Rapid erosion of the sandstone, followed by collapse of the surrounding cliffs, forming a deep oval valley that developed along the fold axis. Amiran (1951) suggested a tectonic mechanism for the formation of the opening of the erosion cirque, whereas Nir (1958), Eran (1982) and Zilberman (2000) proposed a truncation phase prior to incision. In contrast, Issar (1983) proposed that emerging groundwater from the lower Cretaceous Nubian sandstone aquifer triggered the erosion of the cirque. Zilberman (2000) showed that the present single outlet of each erosion cirque is genetically inherited from a previous southeast–northwest transversal stream.

The formation stages of the "Makhtesh" as summarized by Zilberman (2000; this volume) are: (1) Asymmetric folds, which are also prominent topographic ridges. (2) A folded sequence of hard rocks overlying easily eroded rocks. (3) Deep truncation of the fold crests and exposure of the friable sandstone on a low-relief erosion surface. The outline of the contact between hard and friable rocks determined the primary dimension of the cirque. (4) Incision of transversal valleys in the erosion surfaces and the structural ridges. These valleys are the precursors of the present outlets of the "Makhtesh". (5) Reversal of the northwest-flowing systems in the transversal valleys to the southeast in response to the formation of the nearby rift valley (Picard 1951; Garfunkel and Horovitz 1966) and the arching (upwarp) of the Syrian-African Folding Belt (Picard 1951; Wdowinski and Zilberman 1996, 1997). The regional northwest-flowing fluvial system, which dominated southern Israel since the Oligocene, was cut off from its drainage basin in the east by the subsiding rift. The incision of the erosion cirques started during the Early Pliocene.

All studies, except for Issar (1983), suggest that the mechanism for the formation of these large depressions is fluvial erosion along the northwest–southeast flowing streams that eroded and evacuated the softer rocks inside the erosion cirques through the outlet toward the Dead Sea Valley. Nevertheless, none of the studies ever attempted to estimate the amounts and rates of erosion of the material that has been transported out of the erosion cirque. Fruchter et al. (2011) suggested that the size and elevation of the opening control the flows and sediment transport through the outlet outside the erosion cirque.

18.1.2 Climate

The Makhtesh Hatzera Erosion Cirque (MHEC) is located in the northeastern Negev Desert close to the southern basin of the Dead Sea. The climate at the MHEC at present is arid with average annual rainfall of about 80 mm year^{-1}. It is

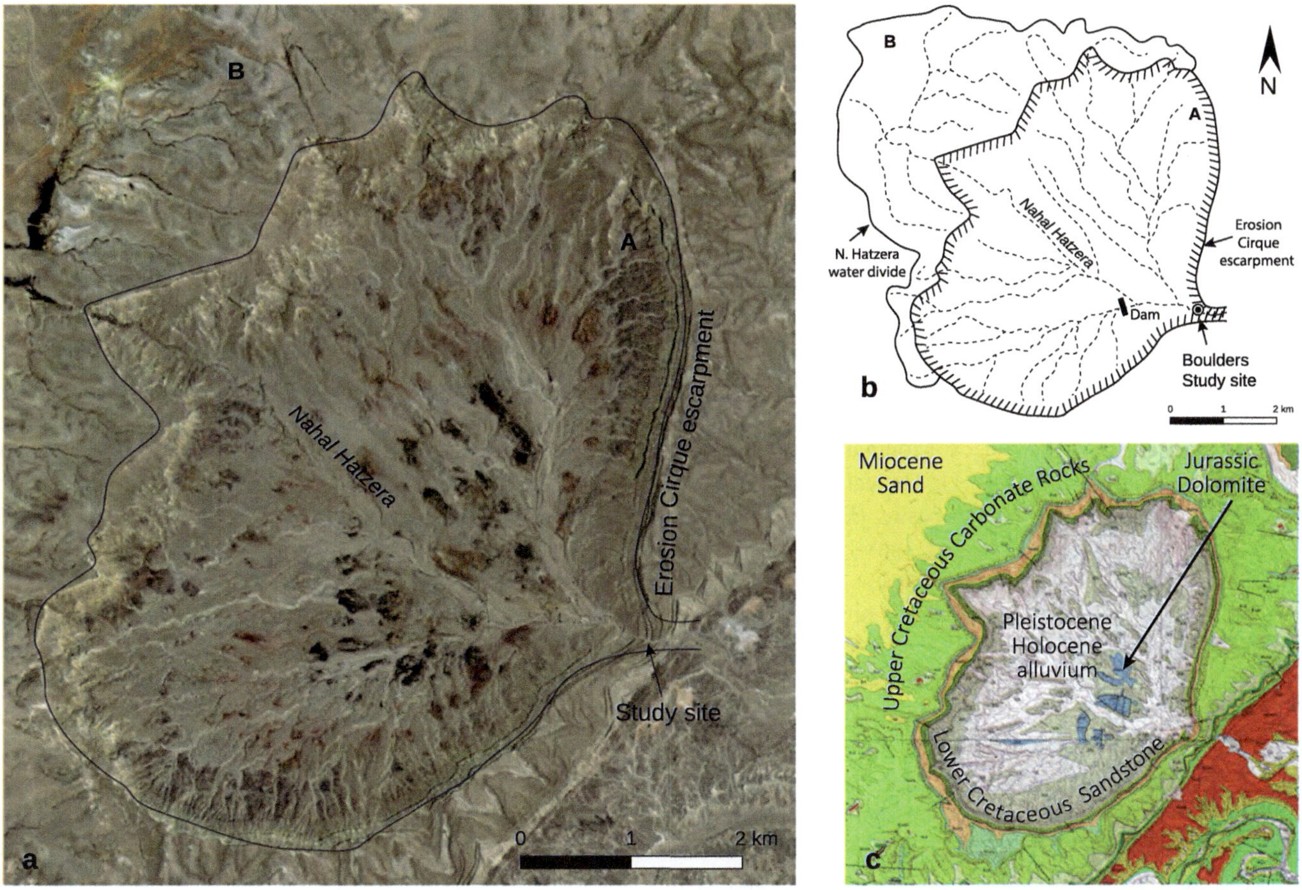

Fig. 18.2 Drainage basin of Nahal Hatzera (45 km^2) and the borders of the Makhtesh Hatzera Erosion Cirque (MHEC)—(**a**) Airphoto showing the present channels, alluvial surfaces and the rock exposures within the erosion cirque, the erosion cirque escarpment and the location of the boulders study site. (**A**) Sub-basin within the MHEC (30 km^2). (**B**) Sub-basin outside the cirque (15 km^2). (**b**) The channel network of Nahal Hatzera ephemeral stream. Note the opposite directions of the tributaries inside the MHEC (southeast) and outside the MHEC (northwest) related to the reversal of the drainage system toward the nearby Dead Sea Rift during the Plio-Pleistocene. Note the location of the dam and the boulder study site near the opening of the MHEC. (**c**) Geological map of the Makhtesh Hatzera Erosion Cirque

affected mainly by three synoptic systems: (1) Active Red Sea Trough (ARST) system (Kahana et al. 2002; Dayan and Morin 2006; Armon et al. 2018), mostly frequent during autumn and spring—September to November and March to May. The ARST is characterized by short and intensive convective rainstorms. The trough creates barometric conditions that encourage the development of strong, but limited rainstorms with localized, high rainfall intensities for short durations (minutes). This system produces the majority of the large floods in the Negev Desert (Shentsis et al. 2012). (2) During winter, the most active system is the Mediterranean Cyclones (MC), which produce cold fronts of the Mediterranean Sea, accompanied by heavy rainfall of hours to days and may extend eastward and southeast over the northeastern Negev and the Judean Desert (Dayan and Morin 2006). This system is evenly distributed between November and March. (3) Tropical plumes (TP)—elongated bands of upper- and mid-level clouds stretching from the tropics polewards and into the subtropics, typically accompanied by a subtropical jet (STJ) streak and a trough on their poleward side (Knippertz 2005). TP systems account for ~ 8% of the large floods in the Negev and usually produce relatively widespread and prolonged rainfall (Kahana et al. 2002; Armon et al. 2018).

18.2 The Landscape of Makhtesh Hatzera Erosion Cirque (MHEC)

Makhtesh Hatzera Erosion Cirque is a deep, oval-shaped depression (5 km × 7 km) with a NE–SW long-axis developed in the center of the Hatzera anticline (Figs. 18.1 and 18.2a–c). The Makhtesh is surrounded by 300–400 m high, almost vertical cliffs. The slopes are mostly composed of exposed rocks and vertical escarpments that are covered by active talus (wash and debris slopes in Fig. 18.3) at the bottom of the escarpment. Transitional colluvial–alluvial units (pediments in Fig. 18.3) connect and integrate the lower parts of slopes with alluvial terraces covering most of the area of the cirque. Except for some exposures of Jurassic brownish dolomite and reddish and yellowish Lower Cretaceous quartz sandstone, the bottom of the cirque covered by Pleistocene–Holocene alluvial sediments (Fig. 18.3).

18.2.1 Nahal Hatzera Catchment

Nahal Hatzera ephemeral stream catchment (45 km^2), which drains the MHEC, is a tributary to the major stream of Nahal Zin (catchment area of 1400 km^2), which drains the eastern flanks of the Negev Mountains to the northeast into the Dead Sea (Fig. 18.1). The catchment of Nahal Hatzera is composed of two sub-basins (Fig. 18.2b): (A) a Pliocene–Pleistocene, oval-shaped, sub-basin inside the erosion cirque (30.84 km^2), which drains the steep slopes and the bottom of the cirque to the southeast, toward the outlet of the MHEC; (B) an older, Miocene–Pliocene drainage system outside the cirque (15 km^2), with a northwest flow direction toward the Mediterranean, which was reversed and captured to the southeast by the Dead Sea Rift during the Plio-Pleistocene. This sub-basin is covered by a 180 m thick sequence of Miocene sandstone, sand and gravel of the Hazeva Formation (Calvo 2000) (Fig. 18.2a, c). This sand is occasionally transported during floods to the Nahal Hatzera system. Three major bedrock knickpoints located along the Nahal Hazera channel separate the MHEC from the direct influence of the Dead Sea base level (Plakht 2000; Fruchter et al. 2011): (a) a steep dry waterfall that connects the upper sub-basin with the bottom of the MHEC, (b) at the outlet of the erosion cirque and (c) a large dry waterfall, 5 km downstream from the erosion cirque outlet. Therefore, fluvial activity and morphology of the Nahal Hatzera within the erosion cirque are isolated from effects of changes in base level.

The drainage network of the sub-basin A within the cirque is dense (drainage density—5.3 km/km^2) (Fig. 18.1b), and the gradient ranges between 2 and 3%, which is effective in sediment transport. The sandy sub-basin B outside the cirque has a relatively sparse drainage network (drainage density—2.9 km/km^2) (Fig. 18.1b), and the average gradient is low $\leq$ 1%.

18.2.2 Quaternary Geomorphological Units Within the Makhtesh Hatzera Erosion Cirque

Detailed mapping of Quaternary deposits and RTL dating of alluvial terraces of the ancient fluvial systems in the MHEC was performed by Plakht (2003) (Fig. 18.3). This map was modified and reanalyzed using GIS procedure (Sect. 4.1). Sediments from the initial stages of the erosion cirque development were not preserved. Younger alluvial–colluvial sediments and terraces dated to about < 500 ka cover the bottom of the MHEC.

18.2.2.1 Talus/Colluvium (Pediments)

These transitional formations that are composed of sediments transported from the slopes to the fluvial system were defined at this map as pediments (9% of the MHEC area) (Fig. 18.3). These geomorphological features together with fluvial terraces constitute geomorphic pairs assumed as synchronous in time. Pediments have a parabolic profile—steep at their upper parts merging with the alluvial terraces at their lower parts having moderate gradients of about 4–2%. Their upper and steeper parts (20%) are rough and covered by coarse clasts/gravels, whereas the lower moderately sloping parts are usually finer in texture, smooth and may be covered by desert pavement and Reg soils. Three generations of these geomorphological features, which are exposed along the bottom of the erosion cirque surrounding cliffs (Fig. 18.3), were mapped. The ages of the pediments are unknown, but since they connect the slope formations—talus/colluvium with the alluvial surfaces, their ages can be related to the ages of the alluvial surfaces to which they correlate.

18.2.2.2 Alluvial Terraces

The bottom of the erosion cirque (56% of the area of the cirque) is composed of exposed bedrock (about 4% of the area of the cirque), which includes mainly Lower Cretaceous sandstone and some Jurassic dolomite. The surfaces of these bedrock relics are usually flat, in places they contain few gravels indicating that they are rock-cut surfaces created by the older fluvial systems. Most of the bottom area (43% of the area of the cirque) is covered by Pleistocene to present alluvium in the form of seven abandoned gravelly alluvial terraces located above the present channel (Fig. 18.3). The rest of the area (44%) is composed of relatively steep rocky and debris-covered slopes. The alluvial terraces are composed of sandstone at the bottom truncated and covered by a 3–12 m thick section of coarse alluvium (Figs. 18.4 and 18.5a, c). The gravel composition includes carbonate, sandstone, chert and dolomite, and the matrix is usually sandy. The sedimentological descriptions of different alluvial terraces are given in Table 18.1 and are supported by data from Plakht (2003) and Fruchter et al. (2011).

The alluvial terraces were dated by thermo-luminescence (RTL) to between > 500 ka and 36–33 ka (Plakht 2003). The oldest age—about 500 ka, is uncertain since it is at the limit of this dating method. Terrace no. 5 is covered by up to 4 m thick travertine (Fig. 18.4c) dated

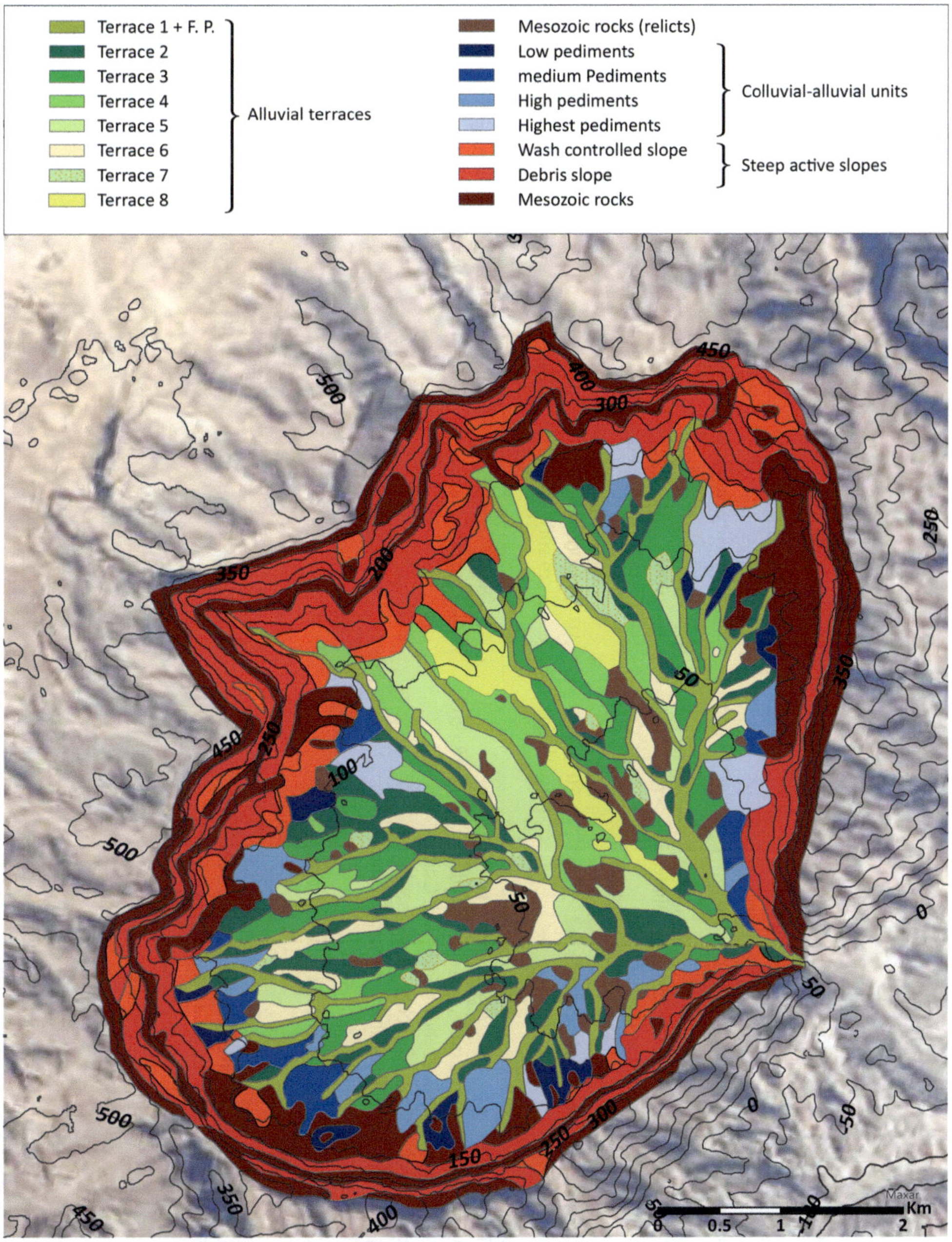

Fig. 18.3 Quaternary map of the Makhtesh Hazera Erosion Cirque—modified after Plakht (2003). The original mapping was based on field survey, identification of the sedimentary units and interpretation of color aerial photographs at a scale of 1:20,000. Hand-processed map was digitized and processed using ARCINFO. Terrace characteristics are shown in Table 18.1. The GIS modification process is explained in Sect. 18.4.1

by uranium series to 40–119 ka (Kronfeld and Livnat 1997) indicting multi-phase deposition. The older age of 119 ka fits well with the chronostratigraphy of the alluvial terraces. Three terraces (probably nos. 6, 4 and 3) were dated to 334 ± 36—278 ± 18 ka, 155 ± 22 ka and 63 ± 6—47.3 ± 9.4 ka by Fruchter et al. (2011), using OSL abandonment ages and ^{10}Be exposure ages. Their study shows that the abandonment age of each terrace is similar to the basal age of the next lower terrace suggesting that relatively rapid incision immediately followed the abandonment of each of the three terraces, whereas the depositional phases and sediment accumulation extended for 100–120 ky. Nevertheless, episodes of erosion within the sequence of the fluvial terraces indicate that the sedimentary succession in each terrace was not deposited as a single event, but rather as a long continuous process of erosion and deposition (Fig. 18.5a, c; Fruchter et al. 2011).

18.2.2.3 Alluvial Terraces Surfaces, Desert Pavement and Reg Soils

The characteristics of alluvial terraces were described through a field survey (Table 18.1), which was performed along a southwest–northeast topographical transect which cuts most of the terraces.

Similar to earlier studies (Gerson et al. 1985; Amit and Gerson 1986; Gerson and Amit 1987; Amit et al. 1993), the

Table 18.1 Sedimentological and pedological properties of the various alluvial terraces in the MHEC

Terrace No.	Alluvial cover thickness (m)	Sedimentary description (where exposed)	Reg Soil properties (Fig. 18.6)			
			Thickness (cm)	A Horizon	B horizon	C horizon
8	5–6	Alternating chert and limestone cobbles in sandy matrix and sand layers	>65 (Fig. 18.6cc)	*A0—95% Av—5 cm, silty, vesicular	B1—15 cm; g.f B2y—15 cm, gypsic,	C1y,z—30 cm, C2z—unexposed -petrosalic
7	~12	Alternating thick layers (2–2.5 m) of pebbles in sandy matrix and sand layers (0.3 m)	>75	*Ao—90% Av—3 cm silty, vesicular	B1—3 cm; g.f B2y—20 cm, gypsic	C1y,z—25 cm, C2z—24 cm petrosalic
6 (Fig. 18.4.a, d)	~8	Alternating layers of pebbles and sand	>70 (Fig. 18.6d)	*A0—80–90% Av—3 cm silty, vesicular	B1—6 cm; g.f B2y—16 cm, gypsic	C1y,z—20 cm, C2z—25 cm petrosalic
5 (Fig. 18.4a)	4–7	Fine gravel layer (1.1 m), 70% pebbles, up to 10 cm. Gravels—limestone, dolomite, sandstone, chert, Matrix—reddish fine sand. Coarse gravel layer (0.9 m), 60% cobbles and boulders up to 45 cm. Gravels—Limestone and dolomite. Matrix—reddish sand	80 (Fig. 18.6b)	*Ao—80% Av—2 cm silty, vesicular	B1—6 cm; g.f B2y—12 cm, gypsic	C1y,z—28 cm, C2z—32 cm salic
4	?	Alternating layers of pebbles in sandy matrix and sand layers				
3	~6	Alternating 4 layers of pebbles (0.4–1.5 m), 70–80% pebbles, Matrix—red sand, with 2 cobbles layers (0.5–1.4 m) 30–40% gravel. Matix-sand (Fig. 18.5a, after Fruchter et al. 2011)	70 (Fig. 18.6e)	*Ao—60% Av—2 cm silty, vesicular	B1—6 cm; g.f B2y—14 cm, gypsic	C1y,z—15 cm, C2z—32 cm salic
2 (Fig. 18.4b)	4–6	Alternating, well-bedded, fine gravel layers (0.6-2 m), 50–70% pebbles 1–10 cm, with poorly bedded coarse gravel layers (1–2 m), 60–80% cobbles and boulders, up to 60 cm. Gravels—limestone, dolomite, sandstone, chert, shales. Matrix—reddish fine sand. (Fig. 18.5c)	50	*Ao—Islands with 50%. boulders Av—0.7 cm	B—7 cm	C1y,z—5 cm, C2z—42 cm
1	Channel	Fine gravel up to 6 cm, limestone, dolomite, sandstone, chert, shales (30–40%). Occasional single cobbles and boulders, few concentrated in bars. Matrix coarse-medium sand and granules				

*A0—desert pavement stone cover

values that characterize desert pavement and soil parameters systematically and gradually change from the active channel and the sub-recent terrace 1 to the oldest terrace 8. This time-dependent soil maturation process is driven by increasingly longer intervals of weathering, dust accumulation and pedogenesis (Gerson et al. 1985; Amit and Gerson 1986; Gerson and Amit 1987; Amit et al. 1993) (Fig. 18.6). Terraces 3–8 are flat, smooth and covered by increasing relative area of desert pavement (Fig. 18.4a, c; Table 18.1), whereas terrace 2 still preserve the original bar and swale morphology of initial alluvial deposition with some "islands" of desert pavement (Fig. 18.4b). The clast density within the pavement increases from about 60–70% in terrace 3 to about 90–95% in terraces 7 and 8 (Fig. 18.4a, c; Table 18.1), whereas the average size of the clasts forming the pavement decreases with time. Greenbaum et al. (2020a) indicated inverse correlation between the density and the average size of the clasts forming the desert pavement.

The degree of pedogenesis is evident by the gradual changes in the properties of the Reg soil with time (Amit and Gerson 1986; Gerson and Amit 1987; Amit et al. 1993). For example, soil thickness increases from 50 cm on the terrace 2 to about 90 cm in terraces 7,8. The thickness of the Av horizon increases from 0.7 cm to 4 cm, respectively (Fig. 18.6; Table 18.1). B horizon is only 2–3 cm thick in terrace 2 but reaches 28–36 cm in terraces 7, 8. The texture of the Av and B horizons becomes finer with time. The surfaces of the alluvial terraces, except for terrace no.2, are usually smooth and covered by well-developed desert pavement and Reg soils (Figs. 18.4c, 18.6c), which are characterized by low permeability and therefore generate runoff (Greenbaum et al. 2020b).

18.2.2.4 Boulders Distribution and Weathering

Boulders, which are sourced mainly from the resistant capping of the Upper Cretaceous carbonate rocks, were transported along the channel of Nahal Hatzera and can be found along the channel as single boulders or concentrated in boulder bars.

A large boulder bar located at the outlet of the cirque (Figs. 18.2b, 18.5d) was documented and analyzed by Greenabum et al. (2020b). Boulders are also scattered over various terrace surfaces and their distribution and estimated

Fig. 18.4 **a** Alluvial terrace no. 5—note the smooth surface covered by well-developed desert pavement and weathered boulders **b** Alluvial terrace no. 2 in the MHEC. Note large carbonate boulders—up to about 2 m in size. **c** Alluvial terrace no. 6 covered by travertine (Kronfeld and Livnat 1997). **d** Weathered carbonate boulder over alluvial terrace no. 6

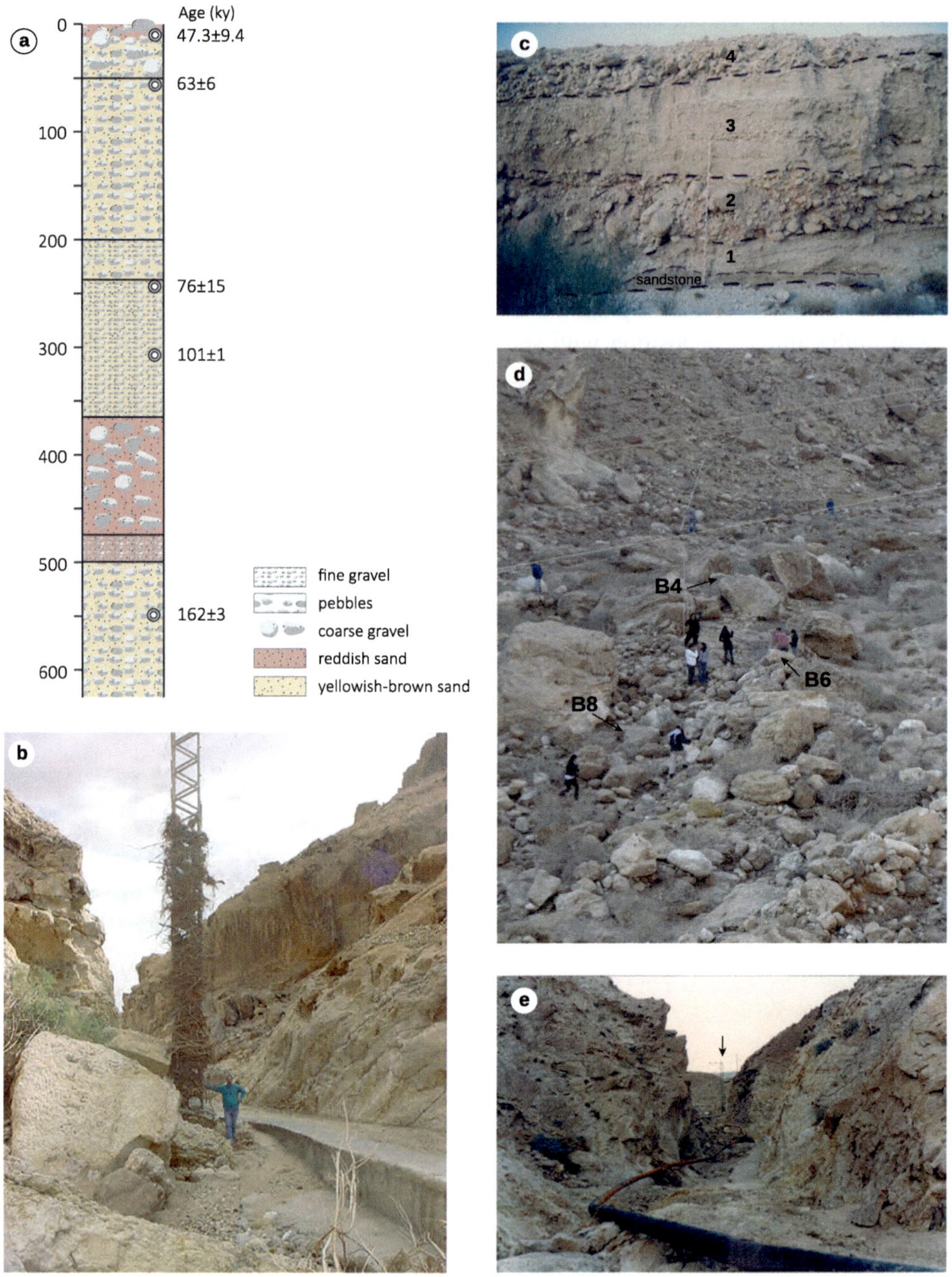

Fig. 18.5 **a** Sedimentary section of terrace no. 3 (Table 18.1; modified after Fruchter et al. 2011). Note the alternating layers of well-bedded fine gravel and pebbles and the poorly bedded coarse layers. The OSL ages span over a period of about 115–100 ky indicating long, non-continuous aggradation, which includes unconformities suggesting erosion phases among sub-periods of aggradation (Fruchter et al. 2011). **b** The electricity pole at the inlet to the gorge—the MHEC outlet with a person for scale. Note the high water marks of the 2004 flood accumulated on the pole indicating water elevation of > 8.5 m. **c** A 5 m sedimentary section of terrace no. 2 above channel bed. Note the alternating layers of well-bedded fine gravel and pebbles and the poorly bedded coarse layers with erosional contacts between the layers similar to terrace no. 3 (Fig. 18.5a). The gravel layers are deposited on a sandstone strath. **d** The boulder bar at the outlet of the of the MHEC with concrete boulders marked with arrows and numbered using letters "B." B6—intermediate axis—0.85 m, weight—0.82 t; B4—intermediate axis—1.05 m, weight—1.56 t; **e** The natural bedrock channel of the outlet gorge after the 1994 flood (230 $m^3 s^{-1}$; Greenbaum and Lekach 1997), which completely evacuated the sediments and an old British road. The electricity pole is marked in black arrow (see also Fig. 18.5b). View is looking upstream. (*Source* of panels b, d, e after Greenbaum et al. 2020b)

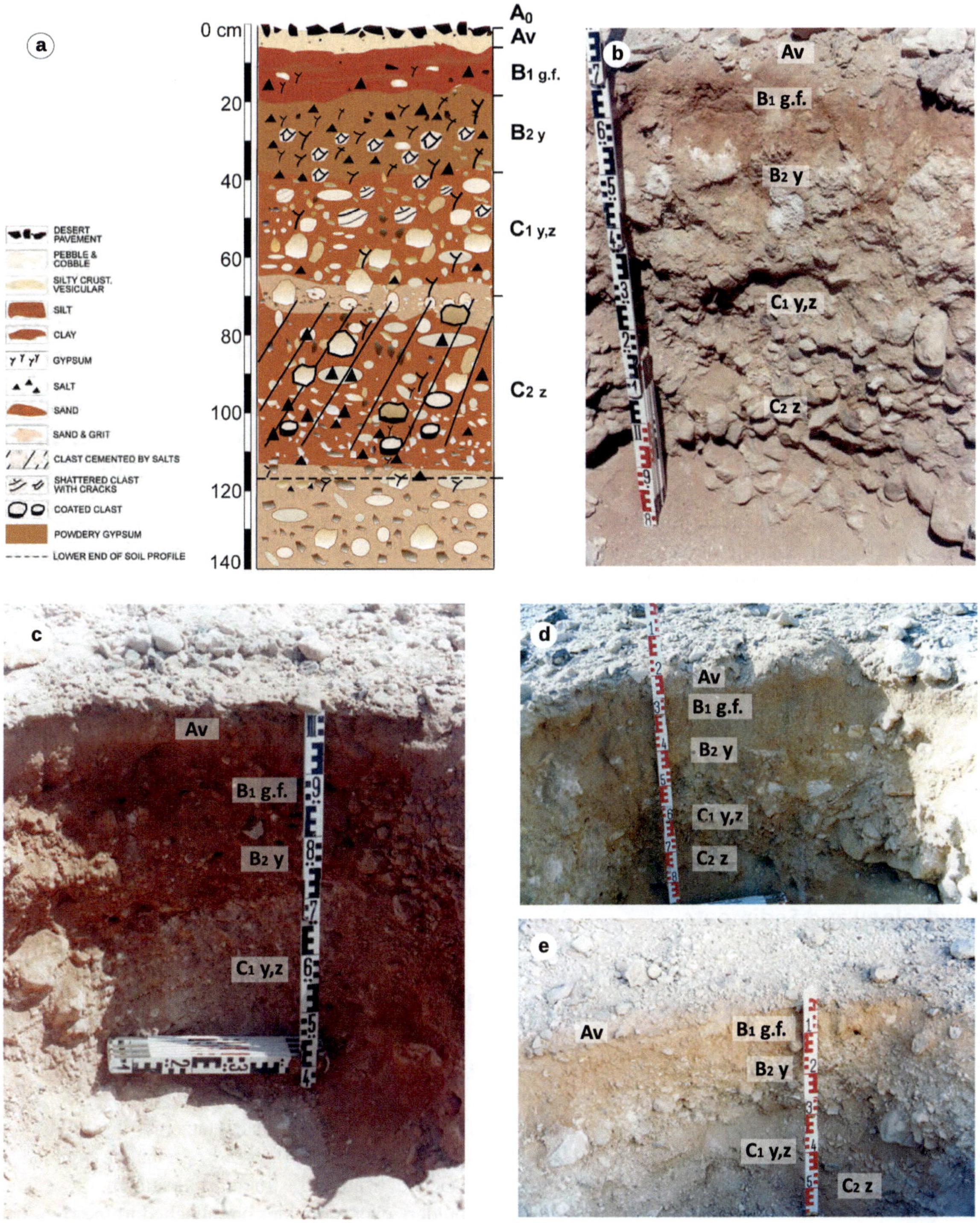

Fig. 18.6 Typical middle-late Pleistocene Reg soil profiles on the terraces of the MHEC. Soil properties are shown in Table 18.1; Note the time-dependent increase in thickness of B—horizons from 20 cm in terrace no. 3 (Fig. 18.6e) to 30 cm in terrace no. 8 (Fig. 18.6c); **a** Typical schematic late Pleistocene Reg soil profile. These soils accumulate airborne dust and salts—gypsum and halite and are covered by desert pavement; they develop with time over gravelly alluvial terraces in desert environments; g.f.—gravel-free horizon (after Greenbaum et al. 2020a). **b** Reg soil profile of terrace no. 5 dated to 155 ± 22 ka (Fruchter et al. 2011)—119 ka (Plakht 2003); profile depth 80 cm. **c** Reg soil profile of terrace no. 8 dated to > 500 ka. The depth of the profile is 65 cm only, because the maximum depth of the soil was not reached, due to the indurated C2z horizon. **d** Reg soil profile of terrace no. 6 OSL dated to 334 ± 36 ka (Fruchter et al. 2011) and IRSL to 278 ± 69—220 ± 55 ka (Plakht 2003). The depth of the profile is 70 cm only, because the maximum depth of the soil was not reached, due to the indurated C2z horizon. **e** Reg soil profile of terrace no. 3, OSL dated to 63 ± 6 ka (Fruchter et al. 2011) and IRSL to 66 ± 16—54 ± 11 ka (Plakht 2003). The depth of the profile is 70 cm

weathering condition of the different lithological types were studied. The boulders show a general trend of decrease in frequency and increase in weathering condition up the sequence of the alluvial terraces, i.e. with time. Boulder density is much higher on the lower and younger terraces and sparse on the upper and older terraces (Fig. 18.4c). Also, on the upper and older alluvial terraces, the boulders are highly weathered and hardly preserved. The most preserved boulders > 2 m in size were found on the lowest terrace no. 2 (Fig. 18.4b), most of which are carbonate boulders, but a few extremely weathered and shattered sandstone and Jurassic dolomite boulders were also found. The carbonate boulders were found to be the most frequent and resistant to weathering except for the chert gravels, which are rare in general but are relatively frequent on the higher terraces, where they are usually in size of small boulders and cobbles. Only a few remnants of carbonate boulders, extremely weathered, were found on the surface of terraces 4 and 5 (dated to about 100–140 ka) (Fig. 18.4d), indicating that preservation of the carbonate boulders in the arid climate is limited to such period, before they are almost completely weathered. Thus, at the higher and older alluvial surfaces, carbonate boulder remnants are rare. Similar results of complete weathering of boulders within a period of about 100 ka were also documented by Amit et al. (1993) and Greenbaum et al. (2020a) in the Negev Desert.

18.3 Geomorphological Processes at the Makhtesh Hatzera Erosion Cirque

18.3.1 Hydrology and Sediments

Estimating the frequency of high-magnitude floods that transport sediment, especially in arid areas where hydrological data are scarce or absent, has always caused debate and has been the subject for numerous estimations and interpretations among hydrologists, geologists and engineers (Baker 2003). The hydrological records of floods in deserts are usually short, partial or absent; therefore, determination of frequency of extreme events for basins such as Nahal Hatzera, is problematic (Zituni et al. 2021). Greenbaum et al. (2020b) used paleoflood hydrology (Kochel and Baker 1982; Baker 1987, 2003; Greenbaum et al. 2006) to reconstruct a long-term natural record of the magnitude and frequency for the largest past floods in the Nahal Hatzera at the MHEC outlet (Fig. 18.2b). Moreover, the data on sediment transport in desert ephemeral streams is even more rare. In the Negev Desert of Israel, some data on sediment transport and yield is available, mainly from field laboratories and reservoirs (Lekach and Schick 1982; Schick and Lekach 1993; Laronne et al. 1994; Greenbaum and Lekach 1997; Laronne and Wilhelm 2001; Alexandrov et al. 2003; Cohen and Laronne 2005; Schwartz and Greenbaum 2009). Boulders in general and boulder bars may serve as indirect evidence for past extreme events, but direct measurements of boulder entrainment in the Negev Desert wadis are absent. Evidence for boulder motion in nature is rare and usually cannot directly be related to a specific time, discharge or other hydraulic parameters.

18.3.2 Floods in the Ungauged Nahal Hatzera Ephemeral Stream

The similar distance, about 7 km from the escarpment to the outlet, for most tributaries within the erosion cirque due to its oval shape (Fig. 18.2b), determines similar tributary gradients resulting in similar concentration times for floods generated by rainstorms that cover the catchment. Therefore, the floods at the outlet of the MHEC are generally characterized by high peak discharges and short durations.

The paleohydrological study of Greenbaum et al. (2020b) at the outlet of the erosion cirque (Fig. 18.2b) reconstructed 23 paleofloods with discharges ranging from 200 to 760 $m^3 s^{-1}$, during the period 1400–2000 AD (about 600 years). Additional three observed floods in 1994, 2004 and 2018 with peak discharges of 230, 470 and 130 $m^3 s^{-1}$, respectively, were documented. According to the frequency analysis, the largest observed floods of 230 and 470 $m^3 s^{-1}$ are associated with recurrence intervals of 25 and 120 years, whereas the largest paleoflood (760 $m^3 s^{-1}$) attained a recurrence interval of about 800 years.

18.3.3 Sediment Transport

18.3.3.1 Boulders

The 2004 flood in Nahal Hatzera (Fig. 18.5b) transported and deposited large 0.85–2.1 m concrete boulders and slabs detached from infrastructures upstream as well as natural boulders over a boulder bar attached to the right bank of the active channel, close to the outlet of the MHEC (Greenbaum et al. 2020b) (Figs. 18.2b, 18.6d). The paleoflood analysis together with measurements of the transported boulders provided an estimate for the frequency of the motion of large boulders. The results indicate that a flood with peak discharge of 470 $m^3 s^{-1}$ (recurrence interval of 120 years), such as the 2004 flood, can transport all range of the boulders that were documented in the MHEC—at a maximum size of 2.1 m and a weight of about 15 tons (Fig. 18.6d).

18.3.3.2 Fine Sediment Yield

Greenbaum and Lekach (1997) calculated the total sediment volume and yield for a 7.3 km^2 tributary of the Nahal Hatzera, which was dammed and the sediments accumulated in a small reservoir at the back of the dam (Fig. 18.2b). These sediments are composed mainly of fine material which includes reddish quartz sand and silt mostly derived from the Lower Cretaceous sandstone exposed in the lower part of the surrounding cliffs. The coarse fraction included small carbonate pebbles and granules and relatively few sandstone gravels, but no boulders were documented. Fruchter et al. (2011) using cosmogenic isotopes showed that most of the sand in the present channel is sourced from the exposed sandstone at the lower parts of the cliffs. The sedimentological stratigraphy of the reservoir included evidence for 19 floods over a period of 30 years (1964–1994), with an average frequency of one flood per 1.6 years. The largest amount of sediment was deposited in the reservoir during the 1994 flood (estimated peak discharge of 40 $m^3\ s{-}^1$ and frequency of 10–20 years) and filled the reservoir to its maximum capacity after which it breached. For the entire 30-year period, sediment volume varied between 50 and 7800 m^3 per event and about 580 $m^3\ year^{-1}$ (880 ton $year{-}^1$) on average. Mean annual specific sediment volume was calculated to about 80 $m^3\ km{-}^2\ year^{-1}$ and mean annual sediment yield to about 120 ton $km{-}^2\ year^{-1}$.

18.4 Geomorphological Processes—Volumes and Rates

18.4.1 Sediment Transport Out of the MHEC—GIS-Based Approach

The modified morpho-stratigraphic map of Plakht (2003) (Fig. 18.3) provides a real basis for the calculations of sediment transport volumes for the alluvial surfaces inside the erosion cirque. The systematic RTL dating of Plakht (2003) that yielded ages between about 500 ka and present as well as the ages of Fruchter et al. (2011) for their studied terraces, provide the time frame for the later Quaternary phases of development of the bottom of the MHEC resulting in the formation of the present-day landscape. The major, large-scale geomorphological processes include sediment transport out of the cirque through the outlet. This process was not continuous with time, based on the observations and dating of phases of sediment aggradation of about 100 ka for each terrace (Fig. 18.6a, c) before the rapid incision to the level of a lower surface (Fruchter et al. 2011).

For the purpose of modifying the morpho-stratigraphic map of Plakht (2003) (Fig. 18.3) and the GIS analysis, the map was scanned and georeferenced to the Israel Transverse Mercator (ITM) (Mugnier 2000) coordinate reference system. Accordingly, the total area and perimeter of the erosion cirque are 30.838 km^2 and 23.087 km, respectively. Following the morphologic definitions offered by Plakht (2003) and digitalization of maps, we have calculated the volumes of the various eroded units including (a) rocks/active slopes, (b) pediments and (c) alluvial terraces using the actual area when the alluvial surface was active and the difference in elevation between each pair of terraces. The average paleo-rates of sediment transport were calculated by dividing the volumes with the age difference between each pair of terraces.

Table 18.2 present the results of the calculations. The total volume of sediment that was evacuated from the erosion cirque since the Pliocene (5 Ma) was calculated to about 9.2 km^3. For the rock section that included carbonate rocks and the underlying exposed sandstone and the active wash and debris slopes at the base of the cliffs, the calculated volume was about 7.9 km^3, whereas for the pediments, the volume was calculated to about 0.92 km^3. In contrast, the alluvial terraces covering most of the area of the bottom of the cirque account for about 0.38 km^3, 4% of the total volume, only. Calculated average rates of sediment transport for the entire 5 Ma are 1840 $m^3\ year{-}^1$, including 2028 $m^3\ year{-}^1$ for the rocks, slopes and pediments, whereas for the alluvial terraces, the range of rates is 344–1328 $m^3\ year{-}^1$. The largest sediment volumes—0.092 and 0.074 km^3 were calculated for terraces 8 and 7 due to their relatively high elevation differences—7 m and 6 m, respectively, and their long duration of activity—184 and 216 ky, respectively. The calculations for the alluvial terraces suggest that during the last 0.5 Ma, only about 4% of the total volume of the MHEC was evacuated, and most of the sediments were transported earlier, during the Pliocene and early Pleistocene. These geological calculations average different climates including wetter periods, which are documented at least during the mid-Pleistocene.

The cycles of long-term sediment aggradation and residence periods of sediment inside the MHEC followed by rapid incision as documented by Fruchter et al. (2011) also support the climatic mechanism of the changes in the erosion–deposition regime.

18.4.2 The Outlet of the Makhtesh Hatzera Erosion Cirque as a Control on Boulder Transport

Fruchter et al. (2011) concluded that the sediment transport outside the erosion cirque through the outlet is controlled only by its size and elevation (Fig. 18.6e), which is formed in accordance to the hydrological regime, i.e.

Table 18.2 Geomorphological properties and IRSL chronology for the various geomorphological formations at the MHEC, including the results of the GIS-based calculations of volumes and rates of sediment transport outside the cirque

Geomorph. Unit	Elevation above present channel bed (m)	Age (ka)	Area (km^2)	Slope (%)	Volume (km^3)	Average sediment transport rate (m^3 year^{-1})
Rocks and debris slopes			13.617	>70	7.899	2028
Pediments		>5 × 10^6—34 (??)	2.808	20–2	0.918	
Bedrock relicts	35–38		1.231			
Terrace 8	35–38	>500	0.804	4	0.092	501
Terrace 7	26–31	506 ± 126 -426 ± 106	0.356	2.3	0.074	344
Terrace 6	22–24	[a]*334 ± 36—278 ± 18* 278 ± 69 - 220 ± 55	1.153	2	0.060	925
Terrace 5	18–20	[a]*155 ± 22* [b]119	1.674	3	0.054	836
Terrace 4	13–15	140–100	1.516	?	0.037	613
Terrace 3	9–11	[a] *63 ± 6—47.3 ± 9.4* 66 ± 16 -54 ± 11	2.370	2	0.035	1328
Terrace 2	5–6	36–33	1.863	2.7	0.029	858
Terrace 1	0–1	Holocene-Recent	3.443	2		
All terraces			*13.179*		*0.381*	
Total cirque		>5 × 10^6	30.835		9.199	1840

[a]*Italic*—OSL and ^{10}Be ages, after Fruchter et al. (2011)
[b]Uranium Series ages after Kronfeld and Livnat (1997)

floods magnitude and frequency. The present study traced remnants of terrace no. 2, which contains large boulders, all the way through/within the outlet gorge and outside the erosion cirque indicating that the present narrow (<12 m) outlet already was active since about 36–33 ka. Therefore, the presence of terrace no. 2 within the MHEC upstream of the outlet occurred while the outlet was opened and its deposition was not related to any blockage of the opening. The 2004 flood proved that the geometry and gradient of this narrow outlet allow for the transport of large boulders ≥ 2 m in size through the outlet. These boulder sizes are the largest that were documented at the boulder bar near the outlet and on the alluvial terraces within the erosion cirque (Figs. 18.4b, 18.6d). We conclude that even the present narrow outlet of the MHEC (Fig. 18.6e) allows the ephemeral stream to transport the eroded material outside the MHEC including the largest boulders that were found (> 2 m in size).

18.4.3 Paleo-Climate

Fruchter et al. (2011) indicated that the size of the outlet of the MHEC serves as a downstream control for sediment aggradation. Also, they did not find any correlation to climatic periods and claimed that the aggradation phases extended to both glacial and interglacial periods as documented at the Soreq cave stalagmites (Bar-Matthews et al. 2000, 2003). Nevertheless, the mechanism that can transport the sediments downstream toward the outlet as well as control the size of the outlet are the floods that are climate-controlled.

Amit et al. (2006) showed that climate was arid throughout the entire Quaternary and therefore does not correlate to either erosional or accumulation periods (Fruchter et al. 2011). In contrast, Plakht (2000) based on the similarity in terrace morpho-stratigraphy among all the erosion cirques suggests climatic control on their formation.

Vaks et al. (2006, 2010) and Bar-Matthews et al. (2017) used the record of stalactite deposition in desert caves, inactive at present and documented the following relatively humid periods during the last 350 ka: at 350–290 ka, 220–190 ka and 142–109 ka. These moist periods fall within the range of ages of the base and the top of the terraces documented and dated by Fruchter et al. (2011)—334–278 ka, 271–155 ka and 162–63 ka, respectively. This may suggest that the protracted aggradation of the sediments of these terraces occurred during humid periods, whereas the following, relatively short incision phases occurred during dry phases. These dry phases are responsible for the rapid evacuation of the sediments outside the MHEC, and therefore, the actual evacuation time of the sediments through the outlet was much shorter. The travertine deposits covering the eastern parts of terraces 4 and 5 (Kronfeld and Livnat 1997) (Fig. 18.4c) also testify to humid conditions during which discharge of springs increased, probably related to higher groundwater levels and/or a water body that may have existed at the eastern/lower part of the erosion cirque.

18.5 Conclusions

The study of Quaternary sediments inside the cirque allowed for the following conclusions:

- The GIS-based calculations for the alluvial terraces suggest that during the last 0.5 Ma, only about 4% of the total volume of the MHEC was evacuated, whereas most of the sediments were transported away earlier, during the Pliocene and early Pleistocene. The rates of sediment transport during the last 0.5 Ma are similar to the present-day rates.
- The large boulders of the terrace no. 2 were traced all the way through/within the gorge of the outlet and outside the erosion cirque indicating that the present narrow (<12 m) outlet was already active at least since about 36–33 ka.
- Protracted aggradation of sediments of the terraces occurred during humid periods, whereas the following, relatively short incision phases occurred during dry phases, which were responsible for the evacuation of the sediments outside the MHEC.
- The paleoflood analysis and measurements of the transported boulders indicate that flood with peak discharge of 470 $m^3 s-^1$ (recurrence interval of 120 years), such as the 2004 flood, can transport all range of the boulders that were documented in the MHEC, with a maximum size of 2.1 m and a weight of about 15 ton.
- The boulder size is reduced up the alluvial terraces sequence, i.e. with time, due to the desert weathering and they almost completely disappear within about 100 ka.
- The main control on sediment evacuation from the erosion cirque is the flood regime, which transports the sediments and is highly controlled by climatic conditions as well as the size of the opening.
- The desert pavement and soil parameters systematically and gradually change from the active channel and the sub-recent terrace to the oldest terrace. This time-dependent maturation process is driven by increasingly longer intervals of weathering, dust accumulation and pedogenesis.

Acknowledgements The authors wish to thank N. Yoselevich of the Cartography laboratory, the Geography Department, University of Haifa, for drawing the figures.

References

Alexandrov Y, Laronne JB, Reid I (2003) Suspended sediment transport in flash floods of the semiarid northern Negev, Israel. In: Servat E, Najem W, Leduc C, Shakeel A (eds) Hydrology of the mediterranean and semiarid regions. IAHS Publ, vol 278, pp 346–352

Amiran DHK (1951) Geomorphology of the central Negev highlands, Israel. Israel Exploration J 1:107–120

Amit R, Enzel Y, Sharon D (2006) Permanent quaternary hyperaridity in the Negev, Israel, resulting from regional tectonic blocking Mediterranean frontal system Geology 54:509–512

Amit R, Gerson R (1986) The evolution of Holocene Reg (gravelly) soils in deserts—an example from the Dead Sea region. CATENA 13:59–79

Amit R, Gerson R, Yaalon DH (1993) Stages and rate of the gravel shattering process by salts in desert Reg soils. Geoderma 57:295–324

Arkin Y, Hamaoui M (1967) The Judea Group (Upper Cretaceous) in the central and southern Israel. Geological Society of Israel Bulletin 42:17–35

Armon M, Dente E, Smith JA, Enzel Y, Morin E (2018) Synoptic-scale control over modern rainfall and flood patterns in the Levant drylands with implications for past climates. J Hydrometeorol 19:1077–1096. https://doi.org/10.1175/JHM-D-18-0013.1

Baker VR (1987) Paleoflood hydrology and extraordinary flood events. J Hydrol 96:79–99. https://doi.org/10.1016/0022-1694(87)90145-4

Baker VR (2003) A bright future for old flows: origins, status and future of paleoflood hydrology. In: Thorndycraft VRG, Benito G, Barriendos M, Llassat MC (eds) Paleofloods, historical data and climatic variabillity: applications in flood risk assessment, pp 13–18

Bar-Matthews M, Ayalon A, Kaufman A (2000) Timing and hydrological conditions of Sapropel events in the Eastern Mediterranean, as evident from speleothems, Soreq cave, Israel. Chem Geol 169:145–156

Bar-Matthews M, Ayalon A, Gilmour M, Matthews A, Hawkesworth CJ (2003) Sea–land oxygen isotopic relationships from planktonic foraminifera and speleothems in the Eastern Mediterranean region and their implication for paleorainfall during interglacial intervals. Geochimica et Cosmochimica Acta 67:3181–3199

Bar-Matthews M, Ayalon A, Vaks A, Frumkin A (2017) Climate and environmental reconstruction based on speleothems from the Levant. In: Enzel Y, Bar-Yosef O (eds) Quaternary of the Levant Cambridge University Press, pp 151–164

Calvo R (2000) Stratigraphy and petrology of Hazeva Formation in the Arava and the Negev: implications for the development of sedimentary basins and the morphostratigraphic of the Dead Sea Rift Valley. PhD Thesis, the Hebrew University of Jerusalem Israel. (in Hebrew with English abstract).

Cohen H, Laronne JB (2005) High rates of sediment transport by flashfloods in the Southern Judean Desert, Israel. Hydrol Process 19:1687–1702

Dayan U, Morin E (2006) Flash floods-producing rainstorms over the Dead Sea: a review. In: Enzel Y, Agnon A, Stein M (eds) New frontiers in dead sea paleoenvironment research. Geol Soc Am Spec Paper 401:53–62

Eran G (1982) The geometry of the Negev monoclines. Master in Geology. The Hebrew University of Jerusalem Israel (in Hebrew with English abstract)

Fruchter N, Matmon A, Avni Y, Fink D (2011) Revealing sediment sources, mixing, and transport during erosional crater evolution in the hyper-arid Negev desert Israel. Geomorphology 134(3–4):363–377. https://doi.org/10.1016/j.geomorph.2011.07.011

Garfunkel Z, Horowitz A (1966) The upper tertiary and quaternary morphology of the Negev. Israel J Earth Sci 15:101–117

Gerson R, Amit R, Grossman S (1985) Dust availability in desert terrains, DAJA-83-C-0041. Development and Standardization Group, UK, The Institute of Earth Sciences, The Hebrew University, Jerusalem, Israel, US Army Research (111 pp)

Gerson R, Amit R (1987) Rates and modes of dust accretion and deposition in an arid region, the Negev, Israel. In: Frostick L, Reid I (eds) Desert Sediments: Ancient and Modern: Journal of the Geological Society of London Special Publication 35, pp 157–169

Greenbaum N, Lekach J (1997) The potential water resource and efficiency of detention storage of the reservoirs in the Makhtesh Katan catchment. Report to Mekorot Water Company Ltd., Hebrew University Jerusalem, 26pp (in Hebrew)

Greenbaum N, Ben-Zvi A, Haviv I, Enzel Y (2006) The hydrology and paleohydrology of the Dead Sea tributaries. In: Enzel Y, Agnon A, Stein M (eds) New frontiers in dead sea paleoenvironment research. Geolog Soc Am Special Paper 401:63–93. https://doi.org/10.1130/2006.2401(05)

Greenbaum N, Mushkin M, Porat N, Amit R (2020a) Runoff generation, rill erosion and time-scales for hyper-arid abandoned alluvial surfaces, the Negev desert Israel. Geomorphology 358:107101

Greenbaum N, Schwartz U, Carling PA, Bergman N, Mushkin A, Zituni R, Halevi R, Benito G, Porat N (2020b) Frequency of boulders transport during large floods in hyperarid areas using paleoflood analysis—an example from the Negev Desert Israel. Earth-Sci Rev 202:103086

Hirsch F (1995) Geological map of Israel 1:50,000, sheet 19-ii, HaMakhtesh HaQatan Geological Survey of Israel

Issar A (1983) Emerging ground water, a triggering factor in the formation of the makhteshim (erosion cirques) in the Negev and Sinai. Israel J Earth Sci 32:53–61

Kahana R, Ziv B, Enzel Y, Dayan U (2002) Synoptic climatology of major floods in the Negev desert, Israel. Int J Climatol 22:867–882. https://doi.org/10.1002/joc.766

Knippertz P (2005) Tropical-extratropical interactions associated with an Atlantic tropical plume and subtropical jet streak. Mon Weather Rev 133:2759–2776. https://doi.org/10.1175/MWR2999.1

Kochel RC, Baker VR (1982) Paleoflood hydrology. Science 215:353–361. https://doi.org/10.1126/science.215.4531

Kronfeld J, Livnat A (1997) Uranium series dating of rift valley travertines and lacustrine deposits: paleoclimatic and stratigraphic applications. Geological Survey of Israel, Jerusalem, 83p

Laronne JB, Wilhelm R (2001) Shifting stage-volume curves: predicting event sedimentation rate based on reservoir stratigraphy. In: Anthony D, Harvey M, Laronne JB, Mosley MP (eds) Applying geomorphology to environmental management. Water Resources Publ, Littleton, Colorado, pp 33–54

Laronne JB, Reid I, Yitshak Y, Frostick E (1994) The nonlayering of gravel streambeds under ephemeral flood regimes. J Hydrol 159:353–363

Lekach J, Schick AP (1982) Suspended sediment in desert floods in small catchments. Israel J Earth Sci 31:144–156

Mugnier CJ (2000) Grids and datums: the state of Israel. Photogramm Eng Remote Sens 66(8):915–917

Nir D (1958) Les pays de bray el les makhteshimes en Negev. un cas de geomorphologie comparative. Rev Geomorph Dynam 9:112–118

Picard L (1951) Geomorphology of Israel, Part I — the Negev. Bullet Res Counc Israel 1:5

Plakht J (2000) Quaternary units in the makhteshim, Negev Desert: similarities and peculiarities. Israel J Earth Sci 49:179–187

Plakht J (2003) Quaternary maps of Makhtesh Hatira and Makhtesh Hazera, Negev Desert. Israel J Earth Sci 52:31–38

Shentsis I, Laronne JB, Alpert P (2012) Red sea trough flood events in the Negev, Israel (1964–2007). Hydrol Sci J 57: 42–51. https://doi.org/10.1080/02626667.2011.636922

Schick AP, Lekach J (1993) An evaluation of two ten-year sediment budgets, Nahal Yael. Israel. Phys Geogr 14(3):225–238

Schwartz U, Greenbaum N (2009) Extremely high sediment yield from a small aridcatchment - Giv'at Hayil, northwestern Negev, Israel. Isr J Earth Sci 57:167–175

Vaks A, Bar-Matthews M, Ayalon A, Matthews A, Frumkin A, Dayan U, Halicz L, Almogi-Labin A, Schilman B (2006) Paleoclimate and location of the border between Mediterranean climate region and the Saharo-Arabian Desert as revealed by speleothems from the northern Negev Desert, Israel. Earth Planet Sci Lett 249:384–399

Vaks A, Bar-Matthews M, Matthews A, Ayalon A, Frumkin A (2010) Middle-Late Quaternary paleoclimate of northern margins of the Saharan-Arabian Desert: reconstruction from speleothems of Negev Desert, Israel. Quatern Sci Rev 29:2647–2662

Wdowinski S, Zilberman E (1996) Kinematic modelling of large-scale structural asymmetry across the Dead Sea Rift. Tectonophysics 266:187–201

Wdowinski S, Zilberman E (1997) Systematic analyses of the large-scale topography and structure across the Dead Sea Rift. Tectonics 16:409–424

Weissbrod T (1993) Lithology and stratigraphy of Lower Cretaceous Formations in the southern Negev. Current Research, Geological Survey of Israel 8:62–64

Yechieli Y, Elron E, Sneh A (1994) Geological map of Israel 1:50,000, sheet 19-IV, 20iii, Neot Hakikar. Geological Survey of Israel.

Zilberman E (2000) Formation of "Makhteshim"—unique erosion cirques in the Negev, southern Israel. Israel J Earth Sci 49:127–141

Zituni R, Greenbaum N, Porat N, Benito G (2021) Magnitude, frequency and hazard assessment of the largest floods in steep, mountainous bedrock channels of the Southern Judean Desert Israel. J Hydrol: Reg Stud 37:100886

Collapse into Relict Karst Voids: Examples from the Northeastern and Central Parts of the Negev

19

Amir Eyal and Amos Frumkin

Abstract

We describe collapse features, tens up to 100 m in diameter, in the north-central Negev Desert. These affect mainly the upper part of the Judea Group and the Mount Scopus Group layers. These indicate the development of isolated chamber-like karstic voids under hypogenic conditions, probably by long-range groundwater flow during the Oligocene—early Miocene. These had collapsed and affected the overlying layers.

Keywords

Hypogenic karst · Groundwater dissolution · Sinkhole filling · Isolated caves

19.1 Introduction

The Negev Desert has been under an arid climate for millions of years, with short intervals of wetter episodes (Amit et al. 2006; Vaks et al. 2010, 2013). However, relict caves in the region indicate an ancient period of karstification, probably under hypogenic conditions (Frumkin and Langford 2017). The speleologic evidence suggests past rising water across presently known areas of rising water, forming hypogene karst. This water must have been aggressive during cave formation. Analysis of the present hydrogeology suggests that the water was rising from the Kurnub Group of the lower Cretaceous, invading the Judea Group of the upper Cretaceous under confined conditions, and developing hypogene voids within Judea Group carbonate succession. The Kurnub Group groundwater was flowing northward from far-field recharge areas in the south, mixing with Judea Group groundwater which flows southward from the Judea Mountains. Here, we discuss additional evidence which may indicate deep underlying relict karstification.

It is well-known that the development of subsurface voids can, in some cases, be indicated by collapse features, forming sinkholes in overlying strata (Kranjc 2013; Frumkin et al. 2015).

The collapse of higher parts of the geological section into voids within deeper layers is known from the Negev in varying spatial densities and without known order or trend. These phenomena are mentioned and described in a number of works, mainly in mapping projects or in studies describing the development of the landscape in the Negev (Eyal 1984; Elron 1980; Hirsch and Shahar 1993; Bar 2003). These features indicate collapse which formed karstic sinkholes, usually round or elliptic, tens up to 100 m in diameter. However, the phenomenon has never been systematically described, no mapping has been made and there is no documentation on the geological maps at the scale of 1:50,000, published by the Geological Survey of Israel. In the present chapter, we discuss several indicative examples of these features in the northeastern and central Negev (Fig. 19.1).

19.2 The Landforms and Features

The most common manifestation of karstic sinkholes in the Negev is the collapse of parts of the Mount Scopus Group, mainly Menuha, Mishash, and Ghareb formations, into the upper parts of the Judea group, mainly Nezer and Gerofit formations (Figs. 19.1, 19.2, 19.3 and 19.9) as well as the collapse of Nezer and Shivta formations into the Derorim formation (Figs. 19.4, 19.5, 19.6 19.7 and 19.8). The plane that separates the sinkhole fill from the surrounding rocks might be erroneously attributed to a fault plane.

A. Eyal (✉)
Shimshon Amitzur 12/13, 8471712 Be'er Sheva, Israel
e-mail: noamir401@gmail.com

A. Frumkin
Institute of Earth Sciences, The Hebrew University of Jerusalem, 9190401 Jerusalem, Israel

A. Frumkin and N. Shtober-Zisu (eds.), *Landscapes and Landforms of Israel*,
World Geomorphological Landscapes, https://doi.org/10.1007/978-3-031-44764-8_19

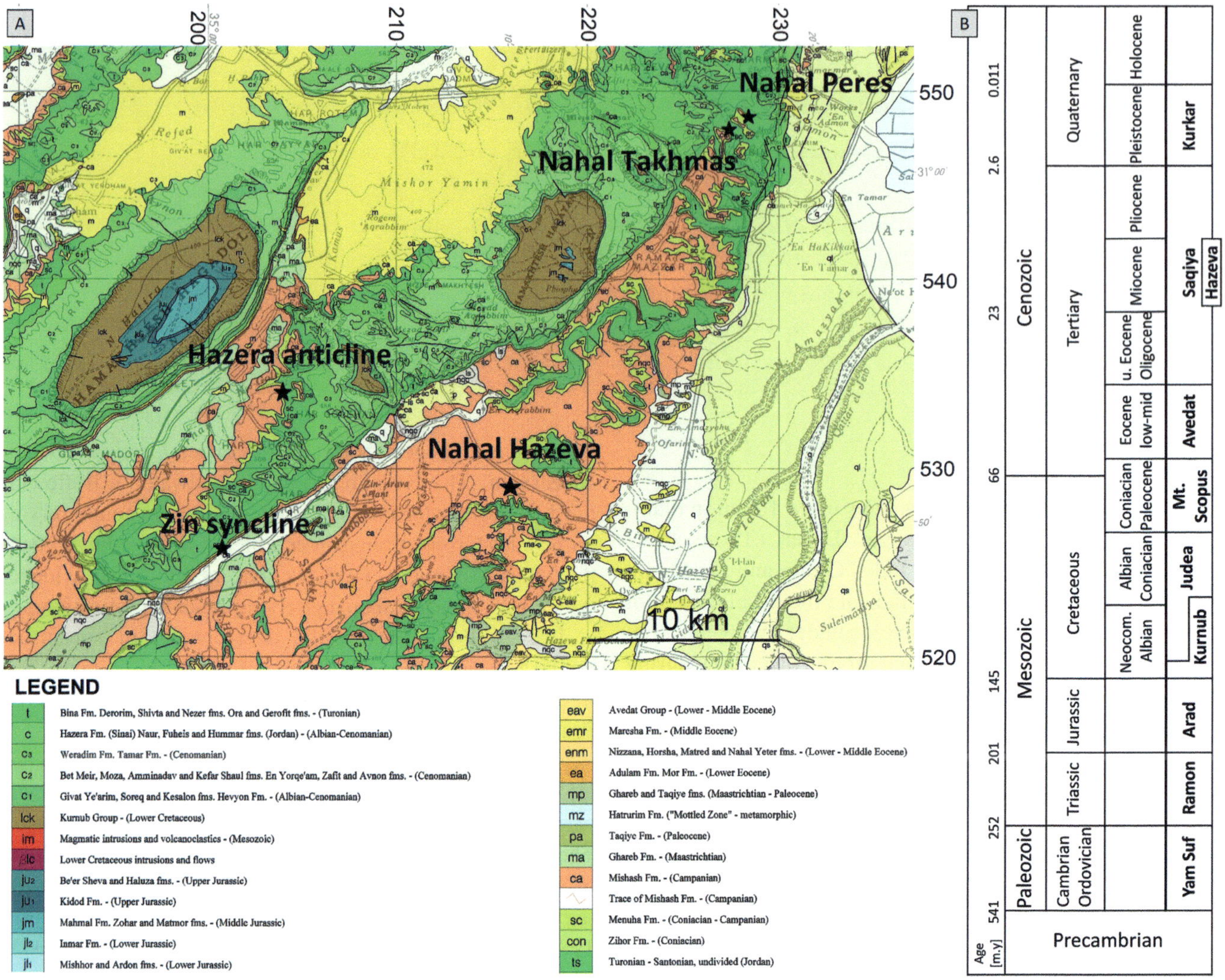

Fig. 19.1 A-Location map of the discussed paleosinkholes on the geological map at a scale of 1:200,000 (reduced), the Geological Survey of Israel (Sneh et al. 1997); B-Main lithostratigraphic groups (right column) in Israel (modified after Fleisher 2002)

A comprehensive consideration of the geometry of structures formed by the collapse clarifies that these structures are associated with paleo-karstic activity and not with faults. The sinkhole fillings differ from fault breccia, as they include a range of collapsed material, from small particles to boulders of different sizes and up to fully preserved collapsed sections of the original overlying layers.

We have not observed similar collapse features in lower formations (Tamar Formation and below) or in the Avedat Group, overlying the Mount Scopus Group, in the study area (Fig. 19.1), although small karstic cavities filled with sand from the Hazeva Formation have been identified and described (Zilberman 1977, 1991; Avni 1991).

The phenomena described above are hardly observed in outcrops of synclines, where the Judea Group is covered by the Mount Scopus and Avedat groups. However, during prospection drilling for phosphate in Rotem, Oron, and Nahal Zin phosphate mines, and during mining in these fields, sinkholes were found in abundance (Fig. 19.3) (Eyal and Bar David 2005) but, as previously stated, were never described before.

19.3 Discussion and Conclusion

Large sinkholes into which higher parts of the geological section collapsed are mainly within the Turonian Derorim, Shivta, and Nezer formations, comprising the uppermost part of the Judea Group. This group is the most karstified sedimentary succession in Israel. The compilation of the described collapse features indicates large chamber-like isolated caves, developed at the top of the Judea Group (Frumkin and Fischhendler 2005). The overlying Mount Scopus Group comprises mainly chalk, chert, phosphate,

Fig. 19.2 Paleosinkholes in different members of the sedimentary succession of the north-central Negev. A sinkhole in the Shivta Formation filled with chert fragments of the Mishash Formation is seen in the upper scarp slope, whereas a smaller sinkhole (or pipe) in the Derorim Formation, also filled with chert fragments, occurs in the foreground. Hazera anticline, ITM coordinates 20450/53213

Fig. 19.3 Sinkhole in the phosphate member of Mishash Formation and Ghareb Formation. The upper chalky member of Ghareb Formation (center) collapsed into the lower marly part of the same formation (right). Zin syncline, Zin mine, Hagor field, ITM coordinates 19987/52413

and marl, which are commonly not karstified, but this group includes collapse features associated with voids in the underlying Judea Group.

The impervious Mount Scopus Group seems to have acted as a caprock, confining the deeper Judea Group aquifer, when the regional relief was flatter than today, and the water-table was in a higher position. This configuration is in agreement with the Oligocene—early Miocene times. The karstification of the upper Judea Group layers is attributed to hypogenic water flow, associated with deep groundwater circulation (Frumkin et al. 2017). The most probable recharge area of the groundwater was the fringe of the Afar dome developed along the present Red Sea rift zone (Avni et al. 2012). In this large region, the erosion had truncated

Fig. 19.4 Sinkhole in the Derorim and Shivta Formations, Nahal Hazeva, ITM coordinates 21587/52765. The disintegrated upper formations (labeled sinkhole fill) collapsed into the underlying formations

Fig. 19.5 Closer look at the Nahal Hazeva sinkhole fill next to its wall, showing the brecciated boulders of the sinkhole fill (center), compared with the undisturbed part of the stratified Derorim Formation (right)

the dome, exposing wide areas of Paleozoic to Cretaceous pervious rocks. The uplift of the dome provided the gradient for both subaerial and subsurface flow to the northwest (Calvo 2002; Frumkin et al. 2017).

Since this type of karst is terrestrial, the earliest age that can be attributed to it in the Negev is Oligocene (Hirsch and Shahar 1993), but younger (e.g., Miocene, Pliocene, or Pleistocene) ages may also be appropriate. The collapses could have been gradual, with or without reaching the surface. The collapse of overlying layers into the karst voids occurred either contemporaneously with the dissolution, or in a later phase. Several later processes could

Fig. 19.6 Sinkhole in the Derorim and Shivta Formations, tributary of Nahal Enmar, ITM coordinates 20886/51032. The horn-like prominent structure was formed because of cementation of the sinkhole fill and faster erosion of the surrounding Derorim Formation. The sinkhole is located south of the area covered by Fig. 19.1

Fig. 19.7 Closer look at the Nahal Enmar sinkhole fill next to its wall, (center), compared with the undisturbed part of the stratified Derorim Formation (left)

induce collapse. These include falling water-table, which dewaters the void and eliminates the buoyant support of the void roof; additional loading of sediments above the void; tectonic deformation; and earthquakes. Some collapse features became better exposed following the Plio-Pleistocene uplift and erosion, including stream entrenchment.

Fig. 19.8 Sinkhole in the Derorim and Shivta Formations, Nahal Peres, ITM coordinates 22747/54639. Note that Nezer Formation behind the sinkhole is not disturbed

Fig. 19.9 Sinkhole in the Nezer Formation, filled with fragments from higher formations, mainly chert of the Mishash Formation. Nahal Takhmas, ITM coordinates. 22712/54597

References

Amit R, Enzel Y, Sharon D (2006) Permanent Quaternary hyperaridity in the Negev, Israel, resulting from regional tectonics blocking Mediterranean frontal systems. Geology 34:509–512

Avni Y (1991) The geology, paleogeography and landscape evolution in the central Negev highlands and the western Ramon structure. MSc thesis, in Hebrew

Avni Y, Segev A, Ginat H (2012) Oligocene regional denudation of the northern Afar dome: pre- and synbreakup stages of the Afro-Arabian plate. Geol Soc Am Bull 124:1871–1897

Bar O (2003) The geology, paleogeography and structural relations in the eastern Ramon area and their implications on the tectonic development of the area after the Eocene. MSc thesis, in Hebrew

Calvo R (2002) Stratigraphy and Petrology of the Hazeva Formation in the Arava and the Negev: Implications for the Development of the Sedimentary Basins and the Morphotectonics of the Dead Sea Rift Valley: Geol Surv Israel Rep GSI/22/02, 264 p. [in Hebrew with English abstract]

Elron E (1980) The geology of Nahal Zin area (Hazeva sheet, northern part) MSc thesis, in Hebrew

Eyal A (1984) The geology of the northern Arava and its western margins in the En Yahav—Hazeva region. MSc thesis, in Hebrew

Eyal A, Bar David N (2005) A visit at the phosphate mine in Nahal Zin, showing the mining activity and a variety of phenomena exposed by it. Israel Geol Soc Ann Meet—Field trip, in Hebrew

Fleisher L (2002) Stratigraphic table of Israel—outcrops and subsurface. Geophysical Institute of Israel

Frumkin A, Fischhendler I (2005) Morphometry and distribution of isolated caves as a guide for phreatic and confined paleohydrological conditions. Geomorphology 67:457–471

Frumkin A, Zaidner Y, Na'aman I, Tsatskin A, Porat N, Vulfson L (2015) Sagging and collapse sinkholes over hypogenic hydrothermal karst in a carbonate terrain. Geomorphology 229:45–57

Frumkin A, Langford B (2017) Arid hypogene karst in a multi-aquifer system: hydrogeology and speleogenesis of Ashalim Cave, Negev Desert, Israel. In: Parise M, Gabrovsek F, Kaufmann G, Ravbar N (eds) Advances in Karst Research: theory, fieldwork and applications. Geological Society, London, Special Publications 466

Frumkin A, Langford B, Lisker S, Amrani A (2017) Hypogenic Karst at the Arabian Platform Margins: Implications for Far-Field Groundwater Systems. Bull Geol Soc Am 129(11–12):1636–1659

Hirsch F, Shahar Y (1993) Post Turonian events south of Arad and different facies of Hazeva formation. Israel Geol Soc Ann Meet – Field trip, in Hebrew

Kranjc A, ZRC SAZU P (2013) Classification of closed depressions in carbonate karst. In: Frumkin A, Shroder JF (eds) 2013 Karst Geomorphology 6.10, Vol 6, Treatise on Geomorphology, Academic Press, San Diego, pp 104–111

Sneh A, Bartov Y, Weisbrod T, Rosensaft M (1997) Geological map of Israel 1:200,000, Sheet 3.

Vaks A, Bar-Matthews M, Matthews A, Ayalon A, Frumkin A (2010) Middle-late quaternary paleoclimate of northern Saharan-Arabian desert: reconstruction from speleothems of Negev Desert, Israel. Quat Sci Rev 29:1201–1211

Vaks A, Woodhead J, Bar-Matthews M, Ayalon A, Cliff R, Zilberman T, Matthews A, Frumkin A (2013) Pliocene-Pleistocene climate of the northern margin of Saharan-Arabian desert recorded in speleothems from the Negev Desert, Israel. Earth Planet Sci Lett 368:88–100

Zilberman E (1977) The geology of Sa'ad Nafha line, Avedat plateau. MSc thesis, in Hebrew

Zilberman E (1991) Landscape evolution in the central, northern and northwestern Negev during the Neogene and the Quaternary. PhD thesis, in Hebrew.

The Anthropogenic "Runoff" Landscape of the Central Negev Desert

20

Hendrik J. Bruins

Abstract

Landscapes are inherently formed by geological, climatic, and geomorphological processes. However, anthropogenic activities may also affect the shaping of the landscape. This has been the case in the hilly central Negev desert, where remarkable human-made landscape changes occurred, related to the utilization of precious runoff rainwater in an arid environment. Hundreds of cisterns were constructed in selected geological strata, accompanied by cleverly designed hillside conduits to harvest runoff for vital drinking water. Many thousands of stone terrace walls were built in wadis to capture runoff water for food production. Millions of stones were repositioned along hillslopes in stone mounds and stone strips to enhance runoff generation. These anthropogenic changes in the landscape led to beneficial soil aggradation in valleys and beautiful constructs in terms of landscape architecture.

Keywords

Negev desert · Climatic classification · Runoff harvesting · Agricultural terraces · Cisterns · Hillslope modifications

H. J. Bruins (✉)
Ben-Gurion University of the Negev, Jacob Blaustein Institutes for Desert Research, Swiss Institute for Dryland Environmental and Energy Research (SIDEER), Sede Boker Campus, 8499000 Midreshet Ben-Gurion, Israel
e-mail: hjbruins@bgu.ac.il

20.1 Introduction

> Turn our captivity, O LORD, as streams in the Negev (Psalm 126:4)

Occasional strong rainfall in the desert abruptly turns a dry riverbed (*nahal* in Hebrew; *wadi* in Arabic) into a flowing stream of water (Fig. 20.1). Such a sudden change in the physical landscape is reflected metaphorically in the above Psalm. The word "Negev" occurs in the original Hebrew (Masoretic) text and is retained as such in the above English translation by the Jewish Publication Society. The Hebrew word translated by streams is "*aphiqim*", meaning more precisely "stream-channels" in a valley. Utilization of "streams in the Negev" during ancient times—harvesting runoff rainwater and floodwater for both agriculture and drinking water (cisterns)—constituted the basis for the formation of distinctive anthropogenic features in this desert landscape.

The hilly central Negev desert is geomorphologically characterized by a dense network of hills and valleys. Late Cretaceous to Eocene rocks of marine origin, featuring limestone, dolomite, chalk, marl, and chert, are widespread in the region (Avni 1991). The altitude of the examined area ranges from about 300 m in the north-west up to 1000 m in the south. However, height differences between hill ranges and adjacent valleys are usually quite moderate, in the range of tens or scores of metres. Thus, the area is naturally endowed with a beneficial geomorphic division of hillslopes, generating runoff rainwater, and valleys, receiving runoff/floodwater, utilized by human ingenuity and engineering in antiquity during several archaeological periods (Bruins 2012).

In addition, the landscape has been dusted during the quaternary by fine loess sediments (Yaalon and Dan 1974; Singer 2007), which settled on hillslopes, but concentrated particularly in valleys, due to redeposition by runoff rainwater (Fig. 20.1). Dust storms occur also at present (Gerson and Amit 1987; Goossens and Offer 1993; Offer

A. Frumkin and N. Shtober-Zisu (eds.), *Landscapes and Landforms of Israel*, World Geomorphological Landscapes, https://doi.org/10.1007/978-3-031-44764-8_20

Fig. 20.1 "Streams in the Negev". Nahal Zin is the largest wadi in the central Negev Desert. Its catchment, having a size of 1400 km^2, begins near Mizpe Ramon at an altitude of about 900 m, descending to the Dead Sea at − 400 m (Fig. 20.4). **a** Dry stream bed of Nahal Zin (1.8 km west of the ancient Byzantine city of Avdat); a blue arrow indicates the entrance into a cistern, hewn in white Eocene chalk, designed to be filled with water during a flood. **b** Floodwater, laden with yellowish-brown loessial sediment, flowing in Nahal Zin (3.4 km north of Avdat). **c** Floodwaters falling into the Zin Canyon, slightly more downstream (*Photos* HJ Bruins, **a** 23-04-2012, **b** & **c** 18-01-2010)

and Goossens 2001; Sandler 2013; Lucke et al. 2019), usually caused by Mediterranean cyclones, which transport dust from northern Africa to Israel by strong south-westerly winds (Yaalon and Ganor 1979; Goldreich 2003).

The central Negev Highlands contain thousands of remnants of ancient agricultural field systems (Evenari et al. 1958, 1982; Mayerson 1960; Kedar 1967; Shanan and Schick 1980; Bruins 1986, 2012; Bruins et al. 2019; Ashkenazi et al. 2012; Avni et al. 2012, 2013), particularly manifested as stone terrace walls in ephemeral stream valleys (Fig. 20.2c). Modifications were also made on many hillslopes in the form of runoff conduit channels, as well as surface clearance of stones and their repositioning in stone mounds and stone strips, which look from the air (Tadmor et al. 1958) as astonishing designs in terms of landscape architecture.

The loess-derived soils in the terraced wadi fields enabled farming in the desert, because the silt-loam texture of loess has a large water holding capacity, as capillary force is related to pore size. This property is vital to keep water available in the soil for agricultural crops, because significant rainfall events that produce runoff and floods usually occur in the central Negev desert during the rainy season at large time intervals of several weeks or even more.

Drinking water was of course essential to establish settlements in this desert region and to utilize the landscape for livestock grazing, travel, and trade during ancient times. Hundreds of cisterns were constructed, as springs and wells are quite rare. Different types of cisterns (Fig. 20.2a, b) are often related to distinct geological-geomorphological landscape properties (Fig. 20.1a). Suitable geological strata were required, in which cisterns could be hewn with technologies available in different archaeological/historical periods. The geological layer itself must consist of non-permeable (aquiclude) material (marl, clay), or human-made plaster must be applied to the chalk or limestone bedrock in which the cistern was hewn to prevent leakage of water (Evenari et al. 1958, 1982; Moran and Palmach 1985; Ore et al. 2020). Most cistern types (Ore et al., 2020) required hillside conduits in sophisticated spatial landscape positions, in order to convey and direct runoff water into the cistern.

Concluding this introduction, it is important to understand that four basic landscape elements are required to enable farming in the desert based on runoff capture by

Fig. 20.2 **a** Open cistern filled with runoff water: Bor Hemet near Mizpe Ramon north of the Makhtesh Ramon erosion cirque (Figs. 20.4 and 20.5), constructed in soft Turonian marl of the Drorim/Lower Ora Formation (Ore et al. 2020). **b** Roofed rock cistern along the southern hillslope of Har Eldad, south-west of Avdat (Fig. 20.4), hewn in a chalk layer of the Middle Eocene Matred Formation; a darker coloured layer of hard limestone overlies the roof. **c** Terraced wadi field with a cereal crop in Nahal Havat Habor, a wadi of fourth stream order, south-west of Har Eldad. The stone terrace wall (partly eroded), built to capture runoff floodwater, probably during the Byzantine period, shows orderly rows of hard limestone blocks, available in the surrounding landscape. (*Photos* HJ Bruins, **a** 09-04-2010, **b** & **c** 22-04-2010)

stone terrace walls in suitable wadis (Shanan and Schick 1980; Evenari et al. 1982; Bruins 1986, 2012; Bruins et al. 2019): (1) A runoff-contributing catchment area with geomorphic surface conditions suitable for runoff generation and overland flow, i.e. rainfall intensity > soil infiltration rate. (2) Hilly catchment topography enabling sustained flow of runoff water downwards to terraced agricultural fields in adjacent valleys. (3) The runoff-receiving terraced wadi fields must have soils of suitable texture—silt loam—and depth to store sufficient runoff water in the root-zone of agricultural crops or fruit trees. (4) Rock outcrops of well-bedded hard limestone or chert in the vicinity to facilitate quarrying and transport of suitable building stones over small distances, for the durable construction of stone terrace walls. This fourth requirement, often disregarded, is emphasized here as an additional necessary landscape element.

20.2 Setting and Location

The Negev desert in southern Israel (Fig. 20.3) is part of the largest planetary desert belt on Earth, which stretches over a length of more than 8000 km from the Atlantic coast of northern Africa eastwards through the Sahara, the Arabian, and Iranian deserts to the Thar desert in India. The summer months in Israel and the Near East are usually completely dry from about May to October, as the region is under the influence of descending air masses of the northern Hadley cell circulation (Ziv et al. 2004). During autumn, winter, and spring, the westerlies prevail, and meteorological depression systems may bring rain to Israel.

The Negev is a subtropical desert. The coldest month is January, and the hottest months are July and August. Average temperatures in the more elevated central Negev range from 8 °C in winter to 26 °C in summer. Light frost may occur in winter, while in summer maximum temperatures may reach 40 °C (Goldreich 2003).

Most rains are usually concentrated in the winter months, from December to March. Early rains may occur in the autumn (October, November) and late rains in the spring (April, May). The amount of average annual rainfall ranges from 300 mm in the northern Negev to only 25 mm in the southern Negev (Goldreich 2003). Average annual precipitation in the Negev Highlands is approximately 100 mm, characterized by large yearly variations. The average annual evaporation from a class A pan in the region is about 2600 mm (Goldreich 2003).

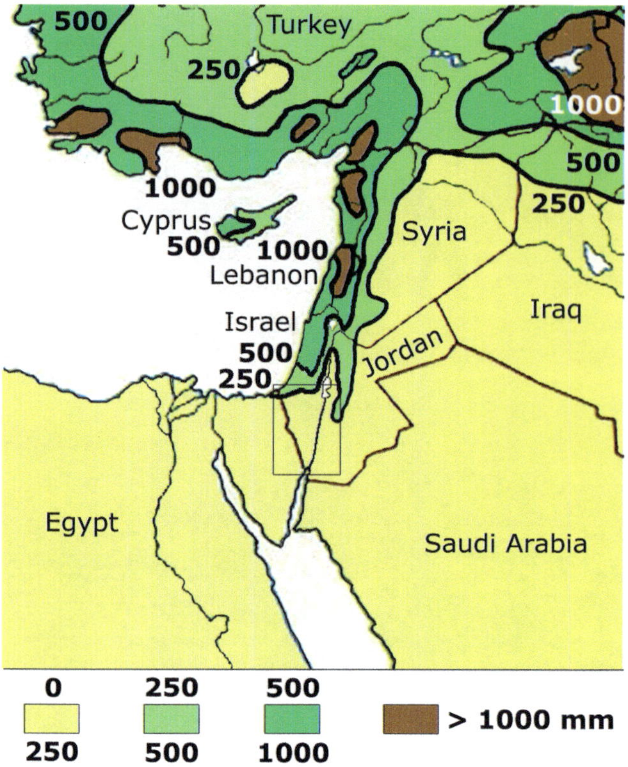

Fig. 20.3 Average annual precipitation in the Near East, showing the extensive (yellow) desert zone (0–250 mm). The location of the Negev desert in Israel is indicated by a rectangle (after Bruins 2012: 29)

The main source of winter precipitation in Israel are cyclones moving over the Mediterranean from west to east. The so-called Cyprus Low (Saaroni et al. 2010), moving over Cyprus (Fig. 20.3), produces most precipitation. When the rains from this system begin to reach Israel, the accompanying wind direction is south-west. Thus, the Negev, due to its more land-locked geographic position (Amit et al. 2006), receives dry air from the deserts of Sinai and North Africa. Only when the Cyprus Low moves further away to the east, the winds reaching Israel change direction, becoming west to north-west, so that moist air from the Mediterranean Sea may bring also rain to the northern and central Negev (Zangvil, personal communications; Zangvil and Druian 1990; Zangvil et al. 2003).

Autumn and spring rains in the Negev and other parts of Israel may relate to the so-called Red Sea Trough, a tongue of low pressure extending northwards from East Africa over the Red Sea towards the Eastern Mediterranean and the Levant. Heavy showers and thunderstorms may develop, which often lead to flash floods, particularly in the Negev region (Goldreich 2003; Tsvieli and Zangvil 2005; Hochman et al. 2021).

20.3 Climatic Classification of the Negev Based on the P/PET Aridity Index

A novel climatic map of southern Israel (Bruins 2012) is based on the numerical P/PET classification system. This method is based on the ratio between annual precipitation (P) and annual potential evapotranspiration (PET), developed by UNEP in its World Atlas of Desertification (Middleton and Thomas 1997). This numerical climatic classification system is ideally suited for time-series analysis, as used by Kafle and Bruins (2009) regarding climatic trends in Israel for the period 1970–2002.

Annual precipitation P is easily measured in rain gauges. However, potential evapotranspiration (PET) cannot be measured, but must be calculated. Based on meteorological data from the Israel Meteorological Service (IMS), available for 13 stations in southern Israel (Fig. 20.4), the annual P/PET index was calculated for each station, for each year, during the decade 1990–2000. A rainfall year in the Levant is usually taken from 1st September to 31st August, in order not to cut the rainfall season and related agricultural year in the middle, which happens with the common 1st January –31st December annual division. The methodology used to calculate PET is detailed in Bruins (2012: 30–32; see also Middleton and Thomas 1997; Kafle and Bruins 2009).

The southern and eastern borders of the ancient runoff farming region, mapped by Kedar (1967), generally coincide with the climatic boundary (P/PET = 0.05) of the hyper-arid zone (Bruins, 2012). Ancient agricultural terraced fields usually do not exist in the latter zone, considered too dry for runoff farming. Exceptions include (Fig. 20.4) a few terraced fields inside the Makhtesh Ramon erosion cirque (Map 204, Rosen 1994), and the unique Biqat Uvda valley (Avner 1998).

The studied decade 1990–2000 included both the wettest year ever recorded in Israel (1991–92) and the severe drought year 1998–99. Precipitation at the Sede Boker meteorological station during the wet year 1991–92 was 187.8 mm, while the drought year 1998–99 yielded only 34.0 mm. The position of the P/PET climatic boundaries in Israel changed dramatically during the wet year 1991–92. The humid zone (P/PET > 0.75), normally not present in Israel, moved southwards until Gaza and Hebron. However, further south, inside the Negev, climatic shifts were minor. Even in this wettest year on record, the ancient runoff farming region in the central Negev remained within the arid zone (Bruins 2012).

On the other hand, during the severe drought year 1998–99, the hyper-arid and arid zones shifted dramatically northwards (Bruins 2012), when compared with their average positions (Fig. 20.4). The entire runoff farming region

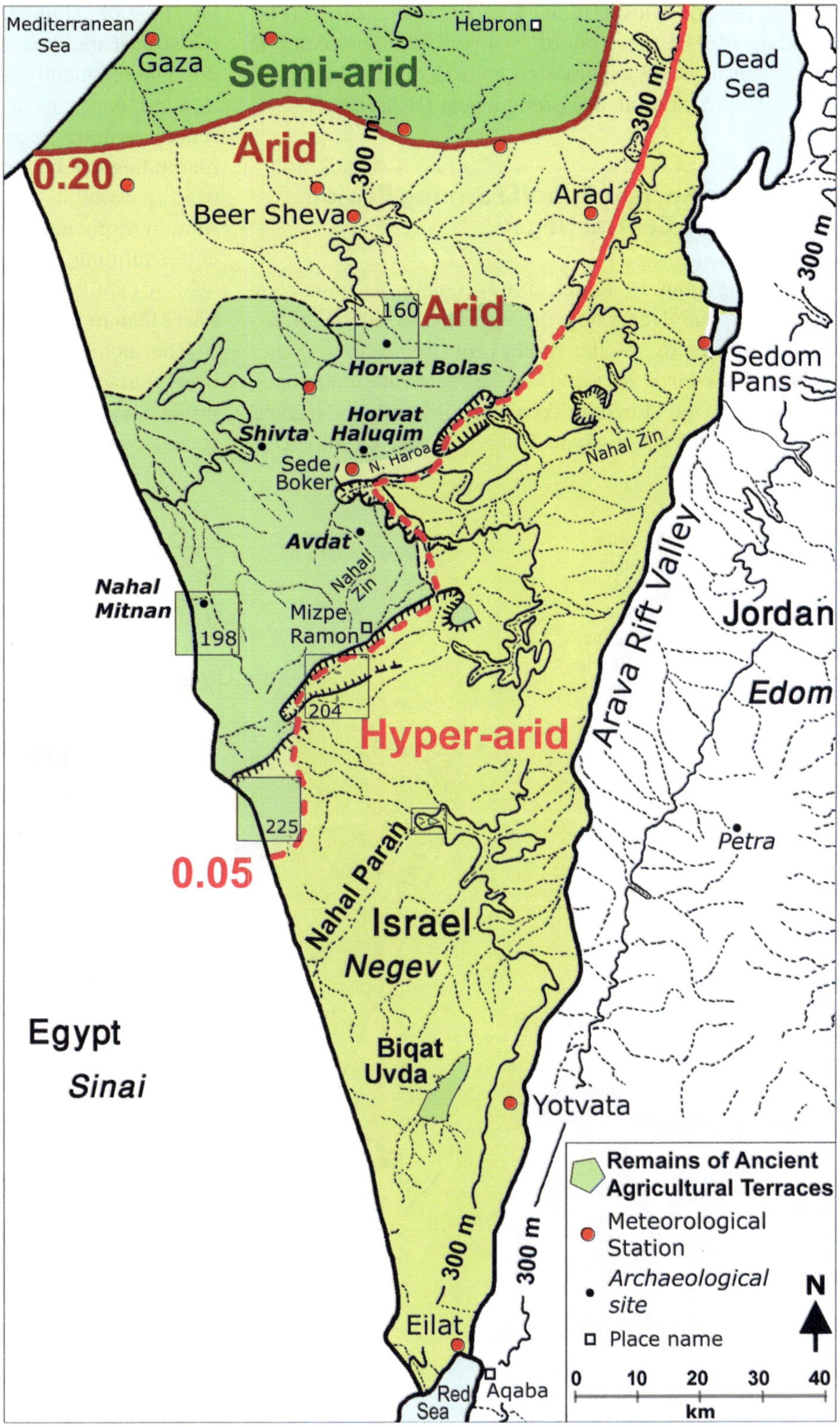

Fig. 20.4 Average position of the P/PET climatic boundaries in southern Israel for the decade 1990–2000. Dashed lines are estimates related to topography, due to a lack of meteorological stations in these areas. Rain-fed agriculture is possible in the semi-arid zone with P/PET values of 0.20 and higher. The ancient runoff farming region (olive green), characterized by stone terrace walls, is situated in the drier part of the arid zone, having P/PET values ranging between ca 0.10–0.05. The numbered rectangles are 10 km × 10 km map units surveyed in detail by the Israel Antiquities Authorities, see Fig. 20.5 for more details and references of these map numbers (modified from Bruins, 2012: 33)

had a hyper-arid climate in the year 1998–99, as the P/PET boundary of 0.05 moved north of Beer Sheva and Arad. In fact, much of Israel experienced an arid desert climate (P/PET 0.05–0.20) during that drought year (Bruins 2012).

20.4 The Ancient Runoff Farming Region in the Central Negev

The investigations by Kedar (1967) remain the most comprehensive study regarding the spatial distribution of terraced wadi fields in the valleys of the central Negev highlands. Kedar (1967) divided the region into ten zones: Haluza–Rehovot, Revivim–Yeroham, Mamshit, Nizzana, Ruth, Shivta, Ashalim–Sede Boker, Avdat, La'ana–Yeter, Horsha–Loz (Fig. 20.5). He determined for each zone the size of the area with terraced agricultural fields (Table 20.1), based mainly on aerial photographs.

Sand dunes exist in the northern part of the ancient runoff farming region (Fig. 20.5), forming here a geomorphological border. The high infiltration rate of water in sand prohibits runoff generation during rainfall events. Moreover, not enough water can be stored in sandy soils for desert farming, as the field capacity, i.e. water that can be held by capillary force against the force of gravity, is much lower than in loess-derived soils.

The archaeological survey by Haiman (1986) revealed many terraced fields in map 198, also across the border inside Sinai (Egypt). Haiman (1986: 22) noted that ancient stone terrace walls were constructed in nearly all the wadis

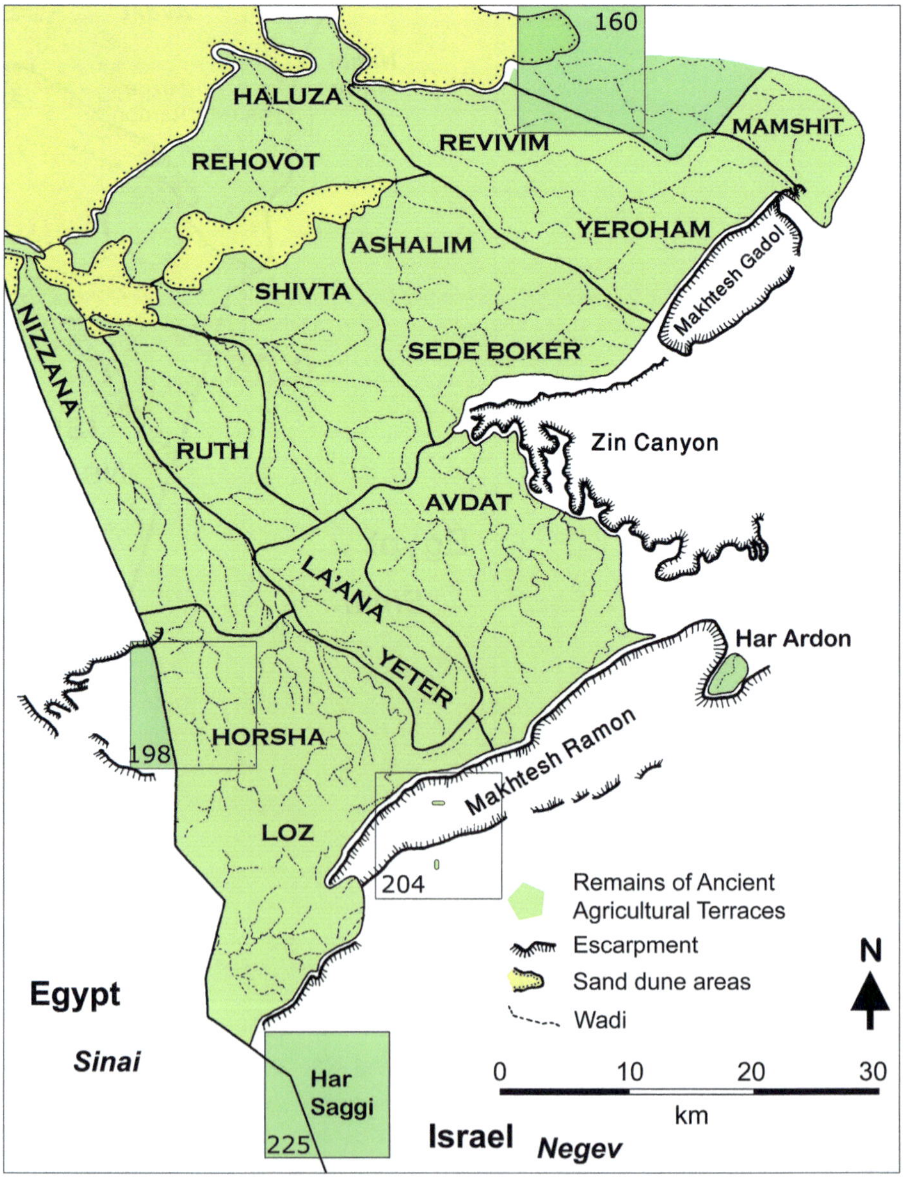

Fig. 20.5 Subdivision of the ancient agricultural region in the central Negev highlands, based on Kedar (1967). New additions at the borders are the Masaf Negev area (map 160) in the north (Eldar-Nir and Danino 2014); the Har Saggi area (map 225) in the south (Avni 1992); the Har Hamran–southwest area (map 198) surveyed by Haiman (1986); two terraced wadis in the Makhtesh Ramon area (map 204) surveyed by Rosen (1994). Some terraced fields exist in the elevated Har Ardon area, as noticed by Bruins on Google Earth satellite images

Table 20.1 Spatial data of the ancient runoff farming region in the central Negev, mainly based on Kedar (1967)

Name of zone	Total area (ha.)	Terraced fields (ha.)	Percentage of terraced fields within zone	Percentage of all terraced fields in the region
Haluza–Rehovot	15,800	239	1.5	5.72
Revivim–Yeroham	27.900	227	0.8	5.44
Mamshit	6.200	42	0.7	1.00
Nizzana	17.000	1.489	8.8	35.66
Ruth	9.200	237	2.6	5.68
Shivta	18.800	495	2.6	11.86
Ashalim–Sede Boker	25.500	303	1.2	7.26
Avdat	12.500	678	5.4	16.24
La'ana–Yeter	22.400	95	0.4	2.28
Horsha–Loz	42.300	200	0.5	4.79
Har Saggi map 225	10.000	170	1.7	4.07
Total	207.600	4.175	2.0	100.00

Added are archaeological survey data of the Har Saggi area (Avni 1992), map 225, which was not included in Kedar's investigation (Fig. 20.5)

Table 20.2 Comparison of the area of terraced agricultural fields as determined by Kedar (1967), who used aerial photographs, and Haiman (1986), who conducted an archaeological survey inside the area on foot

	Name of mapping zone	Area of mapping zone (hectare)	Area of terraced fields (hectare)	Percentage of terraced fields within the mapping zone
Kedar (1967)	Horsha-Loz	42,300	200	0.5
Haiman (1986)	Map 198	10,000	350	3.5
Avni (1992)	Map 225	10.000	170	1.7

The archaeological survey of Har Saggi (map 225, Avni 1992), situated more to the south, is also added for comparison

of this area, except for the largest stream channels. The accumulative length of all terraced fields in map 198 is more than 100 km and the total area of terraced agricultural fields is at least 3500 dunam, i.e. 350 hectares (Haiman 1986). The numerical assessment is based on determinations of both the length of individual terraced wadis and their width, measured at several places along its course (Haiman 2021, personal communication).

The Har Hamran south-west area (map 198), surveyed by Haiman (1986), overlaps with the Horsha-Loz zone (Fig. 20.5). The former mapping area is 100 km^2, of which about 27 km^2 is situated in Egyptian territory. The Horsha-Loz zone of Kedar (1967) encompasses 42,300 hectares (Table 20.1), i.e. 423 km^2, in which there are 200 hectares of ancient agricultural fields, according to Kedar's methodology. Comparing the data by Haiman (1986) with the assessments by Kedar (1967), it becomes clear (Table 20.2) that the archaeological survey methodology by Haiman yielded a much larger area of terraced agricultural fields. The difference may be explained by dissimilar methodologies. Kedar (1967) used mainly aerial photographs in his pioneering research. The archaeological surveys, conducted by Haiman (1986), as well as by Avni (1992), involved actual examination on the ground within the respective mapping area of 10 km × 10 km by jeep and on foot.

20.5 Stone Terrace Walls in Ephemeral Stream Valleys

Cereal food crops, like wheat and barley require at least 200–300 mm rainfall to obtain a reasonable yield. Fruit trees, including grapevines, olives and pomegranates need more than 400 mm of precipitation. However, the average annual rainfall in the ancient runoff farming region in the central Negev desert is only about 70–120 mm. Therefore, the building of stone terrace walls in suitable wadis was essential to capture runoff/floodwater. The addition of runoff water to the soils of the terraced fields facilitated the growth of agricultural crops during average and wet years, and sometimes even in drought years (Shanan and Schick 1980; Evenari et al. 1982; Bruins et al. 2019).

The building of many thousands of terrace walls in suitable valleys in the central Negev desert created an anthropogenic landscape. Besides capturing runoff water, the terrace walls also caused sedimentation of sediment transported by the floodwater (Fig. 20.1b, c). This had a beneficial effect for agriculture, as the aggradation of sediment in the terraced wadi fields led to thicker loess-derived soils, in which more runoff water was kept by capillary force. Sediment aggradation also necessitated the ancient farmers

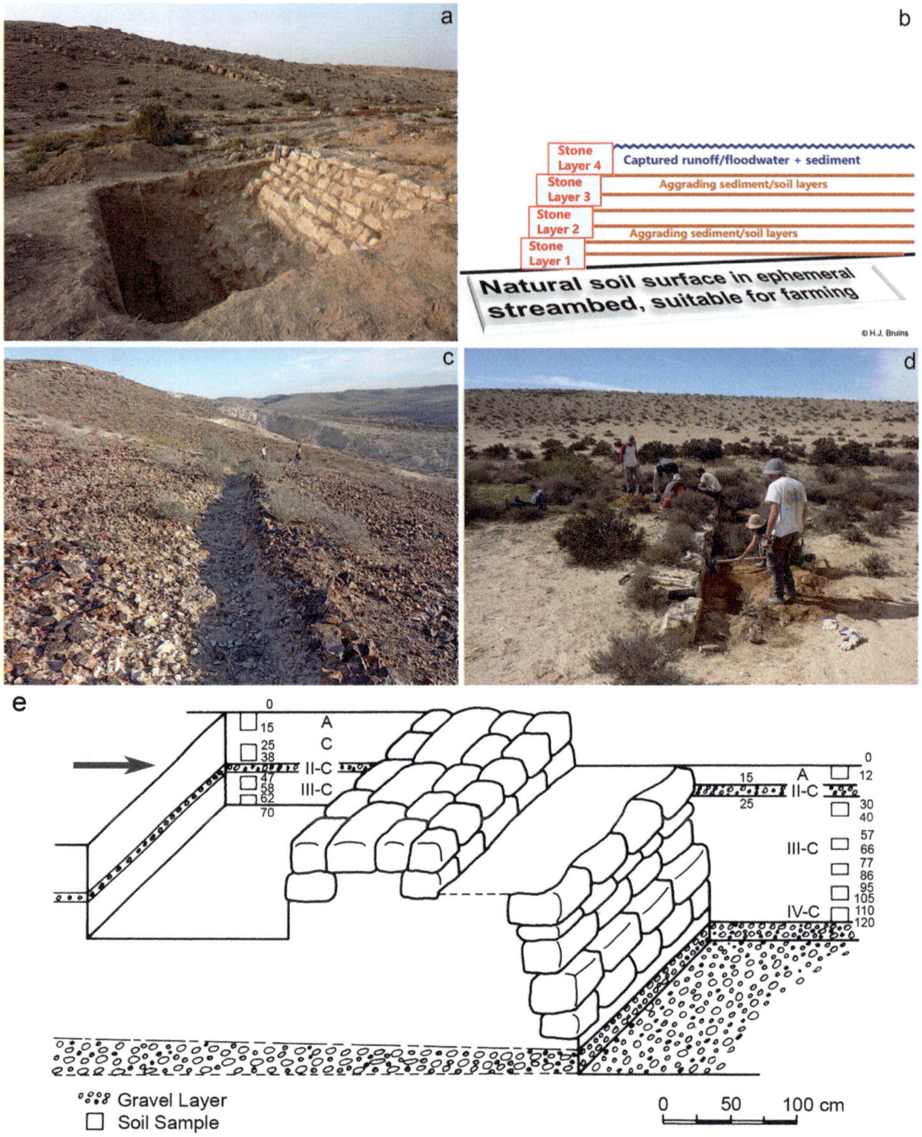

Fig. 20.6 Construction of water-retaining structures. **a** Terrace wall at the site of Horvat Bolas, built of hard limestone, probably of the adjacent Turonian Nezer Formation. Notice an older stone terrace wall at depth, displaying a dissimilar, less orderly architecture. **b** Process of gradual sediment aggradation, due to runoff floodwater capture, necessitating the building of additional stone layers to the terrace walls. **c** Hillside conduit conveying runoff water to a cistern in an area covered by chert of the Campanian Mishash Formation, north-east of Sede Boker. **d** Terrace wall in the same area built of chert blocks of the Mishash Formation. **e** Terrace wall built of hard limestone, excavated by the author in Nahal Mitnan (Bruins 1990), close to the border with Sinai (Egypt). The arrow indicates the flow direction of runoff floodwater. (*Photos* HJ Bruins, a 26-05-2010, c & d 20-01-2014)

to add, in the course of time, additional layers of stones on top of the terrace walls, so that the wall remained higher than the related soil surface of the terraced field. Otherwise, not enough runoff water would be captured. This process of gradual aggradation and subsequent raising of the terrace walls is illustrated in Fig. 20.6b (after Bruins et al. 2020).

The site of Horvat Bolas (Fig. 20.4) is situated close to the northern border of the ancient runoff farming region, in map sheet 160 (Masaf Negev, Eldar-Nir and Danino, 2014). Here a terrace wall (Fig. 20.6a) was excavated during 2012 by the Israel Antiquities Authority and examined by the author. Only the uppermost stone layer was visible before the excavations began. The wall had become almost completely buried with sediment. The neat architecture of the terrace wall shows orderly layers of hard limestone blocks, typical for the Byzantine period and its continuation in the Early Islamic period. Each layer is characterized by stones of the same height. Radiocarbon dates of sediment layers that accumulated against the terrace wall confirm this assessment. An older stone terrace wall is visible at depth (Fig. 20.6a), showing a different architecture.

Terrace walls in the ancient runoff farming region in the central Negev desert were always built of hard rocks, according to the field observations by the author. Chalk was apparently not suitable, but blocks of hard limestone, dolomite, and sometimes chert were used to construct terrace walls. The area north-east of Sede Boker, eastwards of Nahal Haro'a (Fig. 20.4), is characterized by massive chert layers of the Campanian Mishash Formation (Avni and Wieler 2013). An unnamed terraced wadi, a tributary of Nahal Haro'a, is shown in Fig. 20.6d. The terrace wall, partly excavated by the author with assistance of staff members of the Sede Boker Field School, consists of large blocks of chert. The ancient farmers could quarry the chert blocks from the adjacent hillslope. The weight of each building stone in terrace walls may be as much as 50–80 kg (Ore and Bruins 2012). Transporting such heavy stones over large distances is not easy. Therefore, hard rocks suitable for the construction of runoff-capturing terrace walls were quarried as close as possible to the wadi to be developed for farming.

Cisterns also exist in this area north-east of Sede Boker (Hatira Anticline), apparently hewn in chalk of the Santonian Menuha Formation. A hillside conduit (Fig. 20.6c), which conveyed runoff water to a distant cistern, is situated on a hillslope covered by chert gravel, which are erosive remnants of the overlying Mishash Formation.

Another example of an excavated terrace wall (Fig. 20.6e; Bruins 1990) is from Nahal Mitnan in the Har Hamran area (map 198, Haiman 1986) close to the border with Sinai, Egypt (Fig. 20.4). Nahal Mitnan is surrounded by limestone and chalk layers of the Eocene Nahal Yeter Formation and, slightly further to the north, by limestone of the Eocene Nizzana Formation (Zilberman and Avni 2004). A terrace wall built of hard limestone blocks, probably of the Nahal Yeter Formation, was excavated by the author (Bruins 1990) in 1981 during an archaeological survey in the area with Haiman who conducted excavations at an adjacent farmstead (Haiman 1982, 1995).

Nahal Mitnan is terraced over most of its length (about 3.5 km), having an average gradient of ca 2% from south-west to north-east. This fourth-order stream descends from an altitude of ca 540 to 475 m, where it joins Nahal Horsha. Among the many terraced fields (ca 150), the excavated terrace wall was selected, because it is closest to the farmhouse, situated ca 50 m south of Nahal Mitnan. The architecture of the terrace wall shows a high building quality (Fig. 20.6e), which was also noted by Haiman (1982, 1995) for the walls of the farmhouse.

The excavation by the author in 1981 at the terraced wadi field in Nahal Mitnan, was probably the first of its kind in the Negev. Such investigations in ancient agricultural fields, involving the stratigraphy of sediment/soil material, required a background in the Earth sciences, which is not a common part of the training of a typical archaeologist. Prior to the excavation, only the left part of the terrace wall was visible (Fig. 20.6e), as the right part was completely buried by sediment. The latter, older part is composed of five rows of stones, having a combined height of 110 cm. The lowest course of stones was obviously built first. Subsequent aggradation of sediment, due to runoff floodwater capture, necessitated the addition of more and more stone layers (Fig. 20.6b). The cumulative height of the terrace wall, including the left part, is 133 cm. This is also the overall depth of sediment aggradation, suggesting a significant time interval. Radiocarbon dating was not possible, due to a complete lack of organic material in the excavated terraced field. Age assessments by Haiman (1982, 1995) for the adjacent farmhouse may suggest that the first stage of terrace wall building probably began in the Byzantine period and continued during the Early Islamic period, perhaps covering a time interval from approximately the sixth- to the ninth-century CE.

The aggraded sediment consists almost entirely of fine loess-derived soil (see Bruins (1990) for detailed data). A conspicuous gravel layer, about 10 cm thick, was deposited after the building of the fourth stone course in the right part of the terrace wall. Bedload transport of gravel is rare in Nahal Mitnan. An extraordinary large rainfall event probably caused a flood of sufficient magnitude to transport gravel, notwithstanding the presence of many terrace walls in the streambed, which slow down the flow of water. A thicker gravel layer is visible in the subsoil below the bottom of the terrace wall, which was deposited in the more distant past before anthropogenic landscape changes.

At some time during the Byzantine or Early Islamic period, after the large flood that deposited the upper gravel layer, the ancient farmers at Nahal Mitnan decided to build a major addition to the investigated terrace wall. The left part of the terrace wall, being two stone courses high and four stone courses wide, was constructed about 1 m upstream of the older part (Fig. 20.6e). Later in the Early Islamic period, probably during the Abbasid Caliphate, the ancient farmers left the area (Haiman 1995). However, seasonal runoff flows continued, and alluvial sediment accumulated until reaching the uppermost part of the left terrace wall. Even without maintenance, the terrace wall remained intact until today, more than a millennium after its abandonment, which is noteworthy for a terraced wadi of the fourth stream order.

At the rural archaeological site of Horvat Haluqim (Figs. 20.4 and 20.5), near Kibbutz Sede Boker, three small first-order terraced wadis (Bruins 1986) descend along the southern hillslopes of the Haluqim Anticline (Fig. 20.7a). Horvat Haluqim was partly excavated in the 1970s by the late Rudolph Cohen (1976), who attributed most building remains to the tenth-century BCE within the Iron Age IIA. He also suggested that terraced wadi fields at the site may date to the Iron Age.

However, only geo-archaeological excavations in these terraced wadi fields can reveal their origin and history through time. The author worked in the early 1980s with the Negev Archaeological Emergency Survey and received permission from Rudolph Cohen to investigate their stratigraphy and geo-archaeological history (Bruins 1986). Charred organic material, fragments of animal bones, and remains of animal dung in sediment/soil layers of terraced fields provided not only evidence for manuring practices by the ancient farmers (Bruins and Jongmans 2012; Bruins

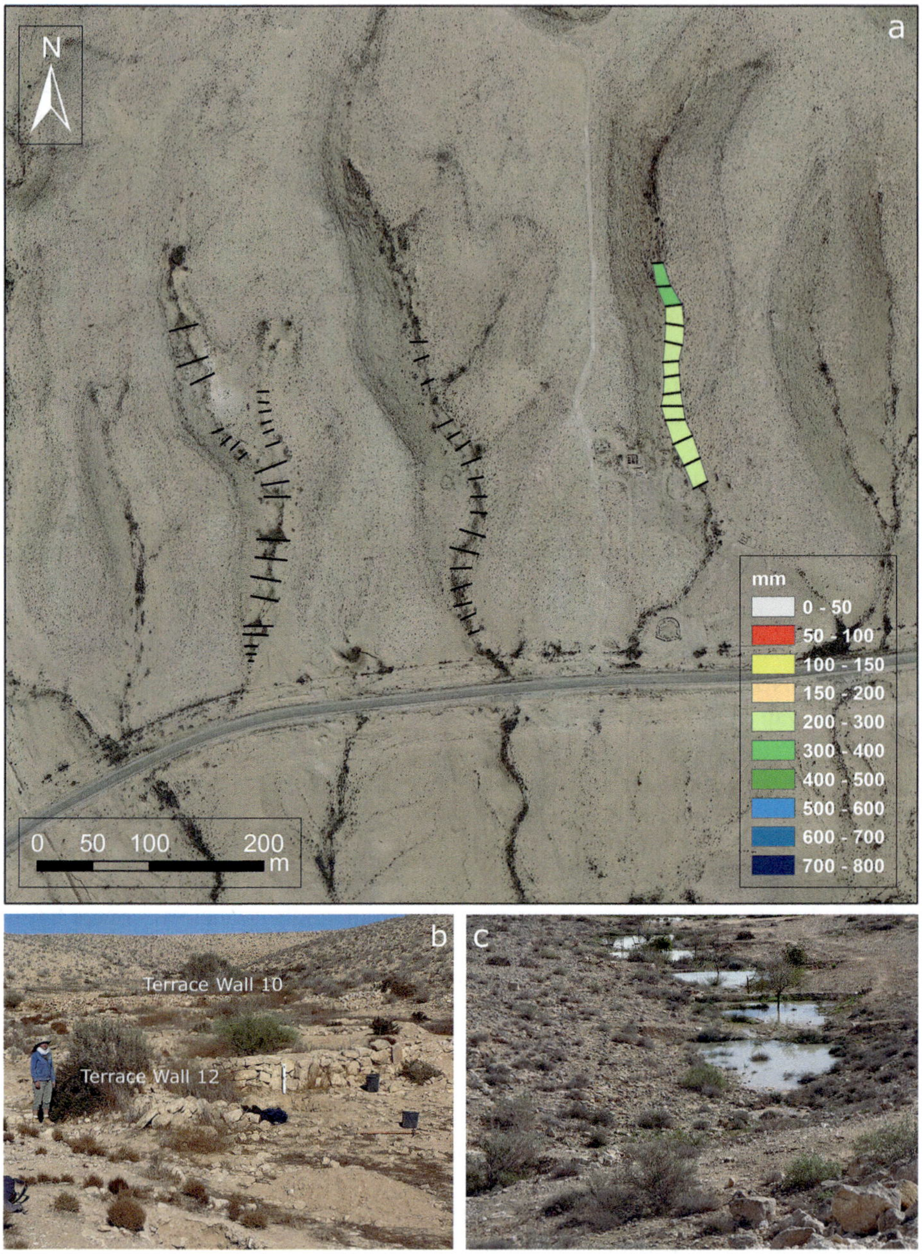

Fig. 20.7 Terraced wadis at Horvat Haluqim. **a** Distribution of stone terraces in three wadis. A GIS hydrological model calculated runoff received by each of the 13 terraced fields in the eastern wadi (Bruins et al. 2019). The average rainfall year 1989–90 (88 mm rainfall) produced 200–400 mm runoff onto the terraced fields, as indicated. For comparison, the very wet year 1991–92 (188 mm rainfall) yielded 400–800 mm runoff. The severe drought year 1999–2000 (34 mm rainfall) produced only 50–150 mm runoff. **b** "Chaotic" stone terrace walls 12 and 10 in the eastern wadi (*Photo* HJ Bruins, 02-12-2010). **c** Runoff water captured by stone terrace walls, recently constructed by the Jewish National Fund in a small wadi to facilitate tree growth in the desert, adjacent to the western wadi of Horvat Haluqim. (*Photo* HJ Bruins, 19-01-2010)

and Van der Plicht 2017a, 2017b; Bruins et al. 2020; Van Bommel et al. 2021), but also enabled radiocarbon dating (Bruins et al. 2011, 2018; Bruins and Van der Plicht 2017a, 2017b).

The ^{14}C dating results at Horvat Haluqim of soil strata in terraced fields showed that runoff farming was carried out during several archaeological periods, ca 1600–1200 BCE (Middle/Late Bronze Age), ca 1200–980 BCE (Iron Age I), and to a lesser extent in Iron Age II. One ^{14}C date of an animal bone found in the upper terrace layers (GrA-34270) has a 1 sigma calibrated age range of 139–241 CE within the Roman period, similar in age as a coin found in the nearby Roman watchtower (Cohen 1976). Finally, the Early Islamic period (Bruins 2012; Bruins et al. 2020) is represented in several terraced field, with ^{14}C dates ranging in age from the seventh to the tenth-century CE.

The Byzantine period appears so far to be absent at Horvat Haluqim. Indeed, the highest terrace walls at Horvat Haluqim, i.e. no. 12 and 10 in the Eastern Wadi (Fig. 20.7b), have a rather chaotic architecture, quite different from the orderly building style of the Byzantine period (Fig. 20.6a, e). Investigations of terraced wadis in several areas of the central Negev by Avni et al. (2006, 2012, 2013), using OSL dating, highlighted the widespread presence of runoff farming during the Byzantine and Early Islamic periods. Nevertheless, the authors also found evidence for runoff farming in the Iron Age in a tributary of Nahal Zin, ca 3 km north of Mizpe Ramon. Here the uppermost loess layer trapped behind the terrace wall gave an OSL age of 2.9 ± 0.7 ka, which corresponds to the Iron Age (Avni et al. 2012), as discussed also by Bruins and Van der Plicht (2017a).

Other examples of high-quality terrace walls of the Byzantine period are found in Nahal Havat Habor (Fig. 20.8), a terraced wadi of the fourth stream order, near Har Eldad south-west of Avdat (Ore and Bruins 2012). An adjacent Byzantine farmhouse and a rock-roofed cistern (Fig. 20.2b) gave the name to this terraced wadi, in which Bedouin nowadays grow wheat and/or barley (Fig. 20.2c). The lowermost terrace wall in Nahal Havat Habor (Fig. 20.8a, b), next to the confluence with Nahal Zin, is a substantial structure, consisting of two parallel stone walls, altogether 1 m wide. The average stone weight within this wall is 65 kg, calculated according to the dimensions of 10 randomly selected stones (Ore and Bruins 2012). The building stones were probably quarried nearby at Har Eldad from hard limestone layers of the Middle Eocene Matred Formation.

Terrace walls were not built in the streambed of Nahal Zin, being of the sixth stream order, as floods (Fig. 20.1b, c) are too powerful (Greenbaum et al. 2000). The construction of stone terrace walls in Nahal Havat Habor led to aggradation, which resulted in a topographic difference of 1.70 m at the confluence with Nahal Zin (Fig. 20.8a, b). This is a specific example of remarkable anthropogenic landscape change in the region. Normal geomorphological processes would cause Nahal Havat Habor and Nahal Zin to be at the same topographic level at the confluence, which was the case before the construction of terrace walls in Nahal Havat Habor (Fig. 20.8b).

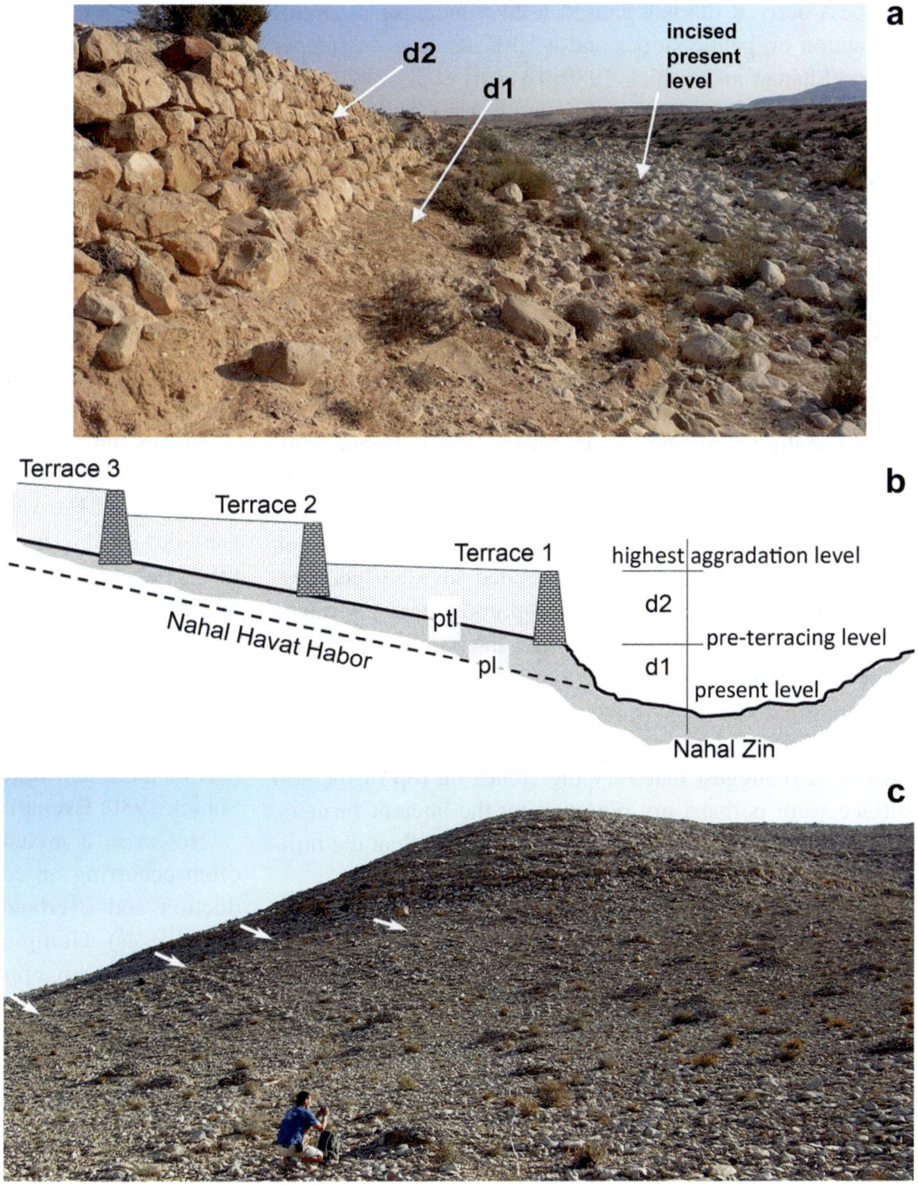

Fig. 20.8 Anthropogenic landscape changes at the confluence of a terraced fourth-order wadi (Nahal Havat Habor, left) with the large sixth-order Nahal Zin (right), next to Har Eldad, 2.3 km SSW of Avdat (Fig. 20.4). **a** and **b** Building of terrace walls caused aggradation of sediment (d2), considerably raising the former base-level (ptl) in Nahal Havat Habor. The difference between the present base-level of Nahal Zin and its former base-level in the Byzantine period is indicated by d1. Without anthropogenic landscape changes (stone terrace walls), the dashed line (pl) would probably be the present wadi surface of Nahal Havat Habor (after Ore and Bruins 2012). **c** Lower north-eastern hillslope of Har Eldad shows many individual stone mounds (having a dark shadow, one next to the person for scale) and stone strips (indicated by arrows), resulting from surface stone clearance in the past to obtain more runoff water. (*Photo* HJ Bruins, 09–07-2008)

20.6 Stone Mounds and Stone Strips on Hillslopes

The phenomenon of stone mounds on hillslopes in certain parts of the central Negev was first reported by Palmer (1871). The Bedouin name for the stone mounds—tuleilat el 'anab (= grape mounds)—led Palmer to a mistaken understanding that grapevines could have been grown on the "sunny slopes… while these little mounds would allow the vines to trail along them and would still keep the clusters off the ground" (Palmer, 1871: 367). This is not possible, because the soils on the hillslopes are very shallow and often saline, while the stone mounds do not capture runoff water. The vexing phenomenon of stone mounds resulted in various theories (see Evenari et al. 1982). The issue was investigated experimentally by Shanan along a hillslope at the Avdat farm. Clearance of stones from the surface of the loess-derived lithosol caused a 24% increase in runoff production on gentle slopes and a 50% increase on steeper slopes (Shanan and Schick 1980; Evenari et al. 1982). The stone mounds themselves were found to have no hydrological function.

Subsequent investigations (Lavee and Poesen 1991; Poesen and Lavee 1994; Lavee et al. 1997) indicated that stones already deeply embedded in the loess soil were not extracted. Such stones enhance runoff. However, stones lying on top of the soil, or only slightly embedded, increase infiltration, thus having a negative influence on runoff production. The explanation is that small and medium stones lying on top of the soil protect the soil underneath from surface sealing (crusting) by raindrop impact. Hence, runoff generated on the slope will partly infiltrate in the more porous soil below such stones. The ancient farmers somehow understood this negative effect and removed such stones. However, stones well-embedded in a sealed loess soil increased runoff, as hard non-porous stones produce usually more runoff than soil, while the runoff from such embedded stones flows directly onto the surface seal of the surrounding loess soil (Lavee and Poesen 1991; Poesen and Lavee 1994; Lavee et al. 1997). Experiments by these authors also suggest that very big stones on top of the soil surface were perhaps not removed by the ancient farmers, because their large size increased runoff more than the infiltration losses in the porous soil underneath.

Stone clearance on hillslopes required a lot of work. The phenomenon occurred mainly in areas around Shivta and Avdat, apparently related to grapevine cultivation and wine production during the Byzantine period (Erez-Edelson 2004). The growing of perennial grapevines demanded larger quantities of runoff water than seasonal cereal crops.

Bruins and Ore (2008) carried out an investigation of stone mounds and stone strips existing on the northern, lower (colluvial) slope of Har Eldad (Fig. 20.8c), situated south-west of Avdat (Fig. 20.4). The stone mounds were built systematically along a spatial grid. The stones were cleared from surface areas of similar size, ca. 180 square metres per stone mound, according to our calculations. Stone mounds were found to be larger and taller close to the steep rocky upper slope, when compared with the lower part of the colluvial slope. Evidently, more stones and comparatively larger stones accumulated by natural geomorphic processes along the upper part of the colluvial slope, because immediately above is the rocky upper slope from where rock fragments and stones eroded, moving down by gravity.

Although the stone mounds are a human-made waste product to get rid of the collected surface stones with minimal walking and carrying distances, they were built in a stable manner to prevent redispersal of the stones. Thus, the larger rock fragments were usually placed at the outer part of the stone mound or at its downslope part to improve mound stability. The following three variables seem to explain the architectural diversity of stone mound construction (Bruins and Ore 2008): (1) stone density per surface area, (2) stone size, and (3) slope angle.

The stone strips (Fig. 20.8c) do have a function to guide the flow of runoff in a certain direction, somewhat like hillslope conduits (Fig. 20.6c). However, the latter are built as ditch channels carved inside the hillslope, designed to convert slow runoff overland flow into faster channel flow (Shanan and Schick 1980). Thus, more water can be conveyed to a cistern or to a terraced field.

Bruins and Ore (2008) evaluated the geomorphic relationships of various hillslope conduits that were built on the tabular hill of Har Eldad. It became clear that most conduits were designed to harvest runoff water from gentle hillslopes (3% gradient) on the rather flat hilltop plateau (ca. 50 ha) of Har Eldad, an area covered for 98% by loess and some stones. Evidently, loess-covered surfaces were used in the past for runoff harvesting. The loess in the Negev is typically covered by a crust, which enhances runoff production during rainfall events, as fine clay dominates the upper part of the crust, reducing the infiltration rate (Shanan and Schick 1980; Evenari et al. 1982).

However, a mixed surface of loess with many stones, often occurring on colluvial slopes, diminishes runoff production and overland flow, as infiltration increases (Yair et al., 1978). Going down from the hilltop plateau of Har Eldad, the steep upper slope (130 m long) is dominated by bedrock outcrops, considered very favourable for runoff production (Yair et al. 1978; Yair and Kossovsky 2002; Yair and Raz-Yassif 2004; Wieler et al. 2016). Indeed, stone clearance was not carried out by the ancient farmers on the rocky upper slope of Har Eldad. Further down, the colluvial slope (224 m long) shows systematic patterns of stone mounds and stone strips on the northern side (Fig. 20.8c), already described, as well as long-range conduits. One such conduit received water from the loess-covered hilltop plateau, descended steeply along the rocky upper slope and continued over the colluvial lower slope towards Nahal Havat Habor (Bruins and Ore 2008).

20.7 Runoff Farming and Palaeohydrology: Some Notes

From the perspective of runoff farming in the Negev desert, there is certainly a big difference between a drought year and a wet year in terms of agricultural yields (Evenari et al. 1982), although the issue has not been studied in detail. Wet years are usually characterized by more and larger runoff floods, when compared with drought years (Shanan and Schick 1980; Greenbaum et al. 2000).

Dead Sea level variations indicate wetter and drier periods during the Holocene (Frumkin et al. 1991, 2001; Frumkin 2009; Bookman et al. 2004, 2014; Kagan et al. 2015). How wet was the central Negev during wet periods? P/PET studies by Bruins (2012) of annual climatic fluctuations in the Negev show that, even in the wettest year on record (1991–92), the central Negev remained arid and the southern Negev hyper-arid. Persisting aridity is also indicated by soil properties (Amit et al. 2006).

A detailed review of climatic changes is beyond the scope of this article. However, palaeohydrological investigations in the region, which are directly related to runoff, are compared with runoff farming periods. Nahal Mishqafaim (Frumkin and Ford 1995), draining a small catchment of 0.078 km^2 in rock salt on the Mount Sedom diapir, in a hyper-arid region (50 mm average annual rainfall) next to the Dead Sea, was investigated in terms of palaeohydrology. Rapid entrenchments of stream profiles in rock salt resulted in relict cave levels with organic matter deposited by runoff floods. Radiocarbon dating indicated wetter periods with more floods in the calibrated age ranges (Frumkin et al. 1998) of 1436–1258 cal BCE (level D), 54 cal BCE–408 cal CE (level C), and 1033–1217 cal CE (level B).

Comparing these palaeohydrological data with runoff farming periods, the oldest wet phase at Nahal Mishqafaim (level D) matches very well with the Late Bronze Age ancient agricultural soil layers in the Eastern Wadi of Horvat Haluqim. Level C relates to the Roman period, during which not much is known about runoff farming activities in the Negev, though a matching ^{14}C date is present in a terraced wadi field at Horvat Haluqim (Bruins 2012). The wet palaeohydrological phase of level B at Nahal Mishqafaim (1033–1217 cal CE, Frumkin et al. 1998) postdates runoff farming during the Early Islamic period, which ceased in the tenth- or eleventh- century CE (Bruins 2012; Avni et al. 2013). Evidently, the end of runoff farming in the Negev during the Early Islamic period was not caused by aridity, but occurred during a period with rising Dead Sea levels (Bookman et al. 2004). Factors other than climate were responsible.

Slackwater deposits along Nahal Zin, the largest wadi in the central Negev (Figs. 20.1 and 20.4) with a catchment of 1400 km^2, facilitated palaeoflood studies (Greenbaum et al. 2000). Sedimentary deposits of 28 floods were identified, ranging in magnitude from 200 to 1500 m^3/s. These flood deposits were dated to the last 2000 years, based on five radiocarbon measurements of organic debris found on top of the respective palaeoflood units. Greenbaum et al. (2000) concluded that periods with low flood frequency, ca 270–620 cal CE and ca 1120–1940 cal CE, often (not always!) correlate with low Dead Sea levels. Two periods with a high flood frequency and high magnitude floods were distinguished: (1) the period before ca 270 CE (Roman period); (2) the period of ca 620–1120 cal CE (Latest Byzantine and Early Islamic period).

Agriculture based on runoff capture in the central Negev reached the peak of its development during the Byzantine period (324–640 CE), both in areal extent and technological advances. Nevertheless, the Byzantine period correlates with *drier* palaeohydrological conditions in Nahal Zin (Greenbaum et al. 2000) and Mount Sedom (Frumkin et al. 1998). Evidently, human farming activities in the Negev were not simply determined by wetter climatic periods.

On the other hand, during the subsequent Early Islamic period a high flood frequency in Nahal Zin (ca 620–1120 cal CE, Greenbaum et al. 2000) does coincide with runoff farming in the central Negev and qanat agriculture in the Arava valley (Avner and Magness 1998; Evenari et al. 1982).

20.8 Conclusions

The ancient anthropogenic activities in the hilly central Negev desert to utilize local runoff rainwater left many amazing imprints on the arid landscape. Ephemeral valleys of first to fourth stream order were often terraced with cross-channel stone terrace walls to capture runoff/floodwater for farming. The presence of hard rock layers that could be quarried along the hillslopes was essential to provide building stones for constructing the terrace walls. The presence of silt-loam loess-derived soil in the wadi bed was another requirement to store and retain the captured runoff water in the soil profile by capillary force for usage by agricultural crops.

Terraced wadi fields were already used at Horvat Haluqim during the Late Bronze Age and Iron Age I (Bruins and Van der Plicht 2017a, 2017b) and probably even earlier (Bruins 2012). However, the peak of development in terrace wall building for agriculture in the central Negev desert occurred during the Byzantine period, typified by more advanced engineering and technology. Even along fifth- or sixth-order wadis, such as Nahal Zin, diversion channels were built to direct floodwater unto Pleistocene loess-covered valley plains next to the streambed, where terraced fields were constructed (Fig. 20.9a). The building of thousands of terrace walls in valleys had positive anthropogenic effects on the landscape, leading to sediment aggradation and soil conservation.

Conduit channels on hillslopes, requiring careful planning, were necessary to harvest and direct sufficient runoff into cisterns for drinking water. The so-called open cisterns (Fig. 20.2a) in the central Negev may date already to the Early Bronze Age in the third millennium BCE (Ore et al. 2020). The many settlements in the region during this millennium obviously needed sources of drinking water, well before the Iron Age.

Fig. 20.9 Amazing landscape architecture by the ancient inhabitants of the central Negev near the Byzantine town of Shivta (Fig. 20.4) to increase runoff production, which enabled agriculture in the desert and the collection of drinking water in cisterns. **a** Area situated 1.5 km ESE of Shivta (Google Earth Pro image). The blue arrow indicates a water diversion channel, more than 500 m long, which diverted floodwater from the main wadi bed of Nahal Drorim to ancient agricultural terraced fields. On the hillslopes above the terraced fields regularly spaced stone mounds (black dots) and stone strips are visible. **b** Aerial photograph (Tadmor et al. 1958) showing another area near Shivta. Millions of dark chert stones were repositioned into mounds and strips to enhance runoff production on the hillslopes. Thus, extraordinary anthropogenic landscape patterns were formed

Later in time, probably during the Byzantine period, hillside conduits were also built to direct more runoff water to terraced agricultural fields (Fig. 20.10a, b), apparently related to the growing of grapevines and industrial wine production. This process was accompanied by the relocation of millions of stones on hillslopes into stone mounds and stone strips to enhance runoff production in the vicinity of the Byzantine towns of Shivta (Fig. 20.9) and Avdat (Fig. 20.8c).

The restoration of ancient Byzantine runoff farms by Evenari, Shanan and Tadmor (1982) in 1958 near Shivta and since 1960 near Avdat (Fig. 20.10) has been of exceptional value for our present understanding of these ancient desert farming systems. Their hydrological and agronomic research showed that agriculture based on local runoff rainwater is generally possible in the present arid climate. They found important hydrological differences between terraced wadis related to small catchments and terraced fields receiving runoff from large catchments. Small catchments (1–7 ha) produce runoff more frequently (Fig. 20.10a, b), as the threshold rainfall is only 3–5.5 mm, but large catchments (> 300 ha) require about 7–8 mm rainfall, before producing runoff floods (Shanan and Schick 1980; Evenari et al. 1982). Therefore, there are years in which terraced agricultural fields, related to small catchments, may receive a number of runoff floods during the rainy season, while fields related to large catchments do not get any runoff at all (Evenari et al. 1982).

Palaeohydrological investigations (Frumkin et al. 1998; Greenbaum et al. 2000) indicate that the peak of development of runoff farming in the Byzantine period occurred during a comparatively dry period. A letter written in the fifth- or sixth-century CE gives evidence of both climatic stress and the growing of grapevines near Haluza, a Byzantine town situated in the northern part of the runoff farming region, surrounded by sandy areas (Fig. 20.5). Procopius of Gaza (ca 450–526 CE) wrote a letter to Jerome, who lived in Haluza, but was staying in Egypt. The letter includes the following sentence: "For there will be a day when you will see Elusa (= Haluza) again and you will weep at the sand being shifted by the wind and stripping the vines naked to their roots" (Mayerson 1983: 252).

Runoff farming is a human activity, and its feasibility is also determined by costs and benefits within a specific socio-economic and political context. Apparently, the drier climate in the Byzantine period did not reach a critical impact level on the hydrology and viability of the

Fig. 20.10 Reconstructed hillside conduit system for enhanced runoff collection near the ancient Nabatean-Byzantine town of Avdat (Fig. 20.4). **a** View from the Byzantine winepress at Avdat towards the runoff farm, restored by Evenari et al. (1982). Hillside conduits no. 1–7 drain small sub-catchments (1–7 ha), which yield runoff more frequently. This system was probably used to grow grapevines for wine production. Number 8 is not a conduit but a wadi, draining a large catchment of 3450 ha, yielding runoff less frequently. (*Photo* HJ Bruins, 29-03-2013, Bruins et al. 2019: 97). **b** Hillside conduit No. 5 conveying runoff as faster channel flow to the agricultural fields (*Photo* HJ Bruins, November 1984). **c** Terraced agricultural fields filled with runoff water at the Avdat farm, looking north to the ancient town of Avdat, situated on the flat hilltop (*Photo* HJ Bruins, March 1985)

runoff farming system. For comparison, the large Biqat Uvda Valley in the southern Negev has a hyper-arid climate (Fig. 20.4), receiving about 30 mm average annual rainfall. The valley naturally collects runoff water from the surrounding hills during rare rainstorms. Thus, Biqat Uvda has been used for occasional farming since the Chalcolithic in the fifth millennium BCE (Avner 1998). More recently, the Haiwat Bedouin cultivated wheat and barley in the Uvda Valley in the early twentieth century, but only following sufficient runoff flooding, which happened in 4 out of 10 years during the decade 1939–1948. Yields reached approximately 800 kg/ha, according to interviews with the Bedouin (Avner 1998).

A modern farmer, depending entirely on the sale of agricultural produce, would not be able to have a viable economy based only on 4 out of 10 years (six drought years without any yields). However, the Haiwat Bedouin had an economy based primarily on extensive pastoralism. Farming in the Uvda Valley, whenever possible, provided a welcome addition of cereal grains for internal consumption and some trade. Similarly, socio-economic systems in the past may have benefitted from runoff farming in the Negev desert, even within drier climatic periods.

There is generally no simple deterministic relationship between climate history and human history in the southern Levant (Bruins 1994: 310), although severe drought periods may lead to famine and population movements (Genesis 41–47). The fact that virtually no runoff-capturing terrace walls were built in wadis in the hyper-arid zone (Fig. 20.4; Bruins 2012) underlines the importance of climate as a co-determining factor.

References

Amit R, Enzel Y, Sharon D (2006) Permanent Quaternary hyperaridity in the Negev, Israel, resulting from regional tectonics blocking Mediterranean frontal systems. Geology 34(6):509–512. https://doi.org/10.1130/G22354.1

Ashkenazi E, Avni Y, Avni G (2012) A comprehensive characterization of ancient desert agricultural systems in the Negev Highlands of Israel. J Arid Environ 86:55–64. https://doi.org/10.1016/j.jaridenv.2012.02.020

Avner U (1998) Settlement, agriculture, and paleoclimate in 'Uvda valley, Southern Negev desert, 6th–3rd millennia B.C. In: Issar AS, Brown N (eds) Water. Environment and Society in Times of Climatic Change, Kluwer/Springer, Dordrecht, pp 147–202

Avner U, Magness J (1998) Early Islamic settlement in the southern Negev. Bulletin of the American Schools of Oriental Research (BASOR) 310:39–57

Avni G (1992) Map of Har Saggi—Northeast (225). Archaeological Survey of Israel, Israel Antiquities Authority, Jerusalem

Avni G, Porat N, Avni Y (2013) Byzantine-Early Islamic agricultural systems in the Negev Highlands: Stages of development as interpreted through OSL dating. J Field Archaeol 38(4):332–346. https://doi.org/10.1179/0093469013Z.00000000052

Avni Y (1991) The geology, paleogeography and landscape evolution in the central Negev Highlands and the western Ramon structure. Isr Geol Surv Rep GSI/6/91, 153p.

Avni Y, Porat N, Avni G (2012) Pre-farming environment and OSL chronology in the Negev Highlands, Israel. J Arid Environ 86:12–27. https://doi.org/10.1016/j.jaridenv.2012.01.002

Avni Y, Porat N, Plakht J, Avni G (2006) Geomorphologic changes leading to natural desertification processes versus anthropogenic land conservation in an arid environment, the Negev Highlands, Israel. Geomorphology 82:177–200. https://doi.org/10.1016/j.geomorph.2006.05.002

Avni Y, Wieler N (2013) Geological Map of Israel 1:50,000, Sede Boqer Sheet 18-IV, Geol Surv Isr, Jerusalem.

Bookman (Ken-Tor) R, Enzel Y, Agnon A, Stein M, (2004) Late Holocene lake levels of the Dead Sea. Geol Soc Am Bull 116(5/6):555–571. https://doi.org/10.1130/B25286.1

Bookman R, Filin S, Avni Y, Rosenfeld D, Marco S (2014) Possible connection between large volcanic eruptions and level rise episodes in the Dead Sea Basin. Quat Sci Rev 89:123–128. https://doi.org/10.1016/j.quascirev.2014.02.009

Bruins HJ (1986) Desert Environment and Agriculture in the Central Negev and Kadesh-Barnea during Historical Times. Published Ph.D. Dissertation, Wageningen University, Stichting Midbar Foundation, Nijkerk

Bruins HJ (1990) Ancient agricultural terraces at Nahal Mitnan. Atiqot 10:127–128 (Hebrew), 22*–28* (English)

Bruins HJ (1994) Comparative chronology of climatic and human history in the southern Levant from the late Chalcolithic to the Early Arab Period. In: Bar-Yosef O, Kra R (eds) Late Quaternary Chronology and Paleoclimates of the Eastern Mediterranean. Radiocarbon, Tucson, p 301–314.

Bruins HJ (2012) Ancient desert agriculture in the Negev and climate-zone boundary changes during average, wet and drought years. J Arid Environ 86:28–42. https://doi.org/10.1016/j.jaridenv.2012.01.015

Bruins HJ, Bithan-Guedj H, Svoray T (2019) GIS-based hydrological modelling to assess runoff yields in ancient-agricultural terraced wadi fields (central Negev desert). J Arid Environ 166:91–107. https://doi.org/10.1016/j.jaridenv.2019.02.010

Bruins HJ, Jongmans AG (2012) Micromorphology of ancient agricultural terraces in the Negev Desert, Horvat Haluqim (Israel). In: Poch RM, Casamitjana M, Francis ML (eds) Proceedings of the 14th International Working Meeting on Soil Micromorphology. International Union of Soil Sciences & University of Lleida, Lleida, p 292–295

Bruins HJ, Jongmans T, Van der Plicht J (2020) Ancient runoff farming and soil aggradation in terraced wadi fields (Negev, Israel): Obliteration of sedimentary strata by ants, scorpions and humans. Quat Int 545:87–101. https://doi.org/10.1016/j.quaint.2019.11.027

Bruins HJ, Nijboer AJ, Van der Plicht J (2011) Iron Age Mediterranean chronology: a reply. Radiocarbon 53(1):199–220. https://doi.org/10.1017/S0033822200034470

Bruins HJ, Ore G (2008) Runoff from loess or bedrock? Hillslope geoarchaeology of ancient runoff farming systems at Horvat Haluqim and Har Eldad in the central Negev desert. Isr J Earth Sci 57(3–4):231–247

Bruins HJ, Segal I, Van der Plicht J (2018) Bronze chisel at Horvat Haluqim (central Negev highlands) in a sequence of radiocarbon dated Late Bronze to Iron Age I layers. In: Ben-Yosef E (ed.) Mining for Ancient Copper: Essays in Memory of Beno Rothenberg. Monograph Series 37, Sonia and Marco Nadler Institute of Archaeology. Tel Aviv University, Tel Aviv & Eisenbrauns, Winona Lake, Indiana, p 297–308

Bruins HJ, Van der Plicht J (2017a) Iron-Age agriculture — A critical rejoinder to "Settlement oscillations in the Negev Highlands revisited: the impact of microarchaeological methods." Radiocarbon 59(1):1–16. https://doi.org/10.1017/RDC.2016.95

Bruins HJ, Van der Plicht J (2017b) Dating of Iron Age agriculture in the Negev highlands: A response to Shahack-Gross and Finkelstein. Radiocarbon 59(4):1233–1239. https://doi.org/10.1017/RDC.2017.57

Eldar-Nir I, Danino D (2014) Map of Masaf Negev (160). Arch Surv Isr, Israel Antiquities Authority, Jerusalem

Erez-Edelson R (2004) Sedentary Zone in the Negev during the Byzantine Period: Settlement Distribution and Runoff Farming. Unpublished Ph.D. dissertation. Bar-Ilan University, Ramat-Gan (Hebrew)

Evenari M, Aharoni Y, Shanan L, Tadmor NH (1958) The ancient desert agriculture of the Negev, early beginnings. Israel Explor J 8:231–268 https://www.jstor.org/stable/27924751

Evenari M, Shanan L, Tadmor N (1982) The Negev — The Challenge of a Desert, 2nd edn. Harvard University Press, Cambridge MA

Frumkin A (2009) Stable isotopes of a subfossil Tamarix tree from the Dead Sea region, Israel, and their implications for the Intermediate Bronze Age environmental crisis. Quat Res 71(3):319–328. https://doi.org/10.1016/j.yqres.2009.01.009

Frumkin A, Ford DC (1995) Rapid entrenchment of stream profiles in the salt caves of Mount Sedom, Israel. Earth Surf Proc Land 20:139–152. https://doi.org/10.1002/esp.3290200205

Frumkin A, Magaritz M, Carmi I, Zak I (1991) The Holocene climatic record of the salt caves of Mount Sedom, Israel. The Holocene 1:191–200. https://doi.org/10.1177/095968369100100301

Frumkin A, Greenbaum N, Schick AP (1998) Paleohydrology of the Northern Negev: Comparative Evaluation of Two Catchments. In: Issar AS, Brown N (eds) Water, Environment and Society in Times of Climatic Change. Kluwer/Springer, Dordrecht, pp 97–111

Frumkin A, Kadan G, Enzel Y, Eyal Y (2001) Radiocarbon chronology of the Holocene Dead Sea: attempting a regional correlation. Radiocarbon 43(3):1179–1189. https://doi.org/10.1017/S0033822200038479

Gerson R, Amit R (1987) Rates and modes of dust accretion and deposition in an arid region – the Negev, Israel. In: Frostick L, Reid I (eds) Desert Sediments: Ancient and Modern. Geological Society Special Publication 35, London, p 157–169

Goldreich Y (2003) The Climate of Israel: Observation, Research and Application. Kluwer/Plenum, New York

Goossens D, Offer ZY (1993) Eolian deposition of dust over symmetrical hills: an evaluation of wind tunnel data by means of terrain measurements. Z Geomorph 37:103–111

Greenbaum N, Schick AP, Baker VR (2000) The palaeoflood record of a hyperarid catchment, Nahal Zin, Negev desert, Israel. Earth Surf Proc Land 25:951–971https://doi.org/10.1002/1096-9837(200008)25:9<951::AID-ESP110>3.0.CO;2-8

Haiman M (1982) Nahal Mitnan (Notes and News). Israel Exploration Journal 32:265–266

Haiman M (1986) Map of Har Hamran–Southwest (198) Archaeological Survey of Israel, Department of Antiquities and Museums, Jerusalem
Haiman M (1995) An early Islamic period farm at Naḥal Mitnan in the Negev highlands. Atiqot 26:1–13
Hochman A, Rostkier-Edelstein D, Kunin P, Pinto JG (2021) Changes in the characteristics of 'wet' and 'dry' Red Sea Trough over the Eastern Mediterranean in CMIP5 climate projections. Theor Appl Climatol 143:781–794. https://doi.org/10.1007/s00704-020-03449-0
Kagan EJ, Langgut D, Boaretto E, Neumann FH, Stein M (2015) Dead Sea levels during the Bronze and Iron Ages. Radiocarbon 57(2):237–252. https://doi.org/10.2458/azu_rc.57.18560
Kafle H, Bruins HJ (2009) Climatic trends in Israel 1970–2002: warmer and increasing aridity inland. Clim Change 96:63–77. https://doi.org/10.1007/s10584-009-9578-2
Kedar Y (1967) The Ancient Agriculture in the Negev Mountains. Bialik Institute, Jerusalem (Hebrew)
Lavee H, Poesen J (1991) Overland flow generation and continuity on stone-covered soil surfaces. Hydrol Proc 5:345–360. https://doi.org/10.1002/hyp.3360050403
Lavee H, Poesen J, Yair A (1997) Evidence of high efficiency water-harvesting by ancient farmers in the Negev Desert, Israel. J Arid Environ 35:341–348. https://doi.org/10.1006/jare.1996.0170
Lucke B, Roskin J, Vanselow KA, Bruins HJ, Abu-Jaber N, Deckers K, Lindauer S, Porat N, Reimer PJ, Bäumler R, Erickson-Gini T ... Kouki P (2019) Character, rates, and environmental significance of Holocene dust accumulation in archaeological hilltop ruins in the Southern Levant. Geosciences 9:190. https://doi.org/10.3390/geosciences9040190
Mayerson P (1960) The ancient agricultural remains of the central Negeb: methodology and dating criteria. Bull Am Sch Orient Res 160:27–37
Mayerson P (1983) The city of Elusa in the literary sources of the fourth–sixth centuries. Israel Explor J 33:247–253. https://www.jstor.org/stable/27925903
Middleton N, Thomas D (eds) (1997) World Atlas of Desertification, 2nd edn. UNEP. Arnold, London
Moran O, Palmach D (1985) The Water Cisterns in the Negev Highlands. Society for the Protection of Nature, Midreshet Sede Boqer (Hebrew)
Offer ZY, Goossens D (2001) Ten years of aeolian dust dynamics in a desert region (Negev desert, Israel): analysis of airborne dust concentration, dust accumulation and the high-magnitude dust events. J of Arid Environ 47:211–249. https://doi.org/10.1006/jare.2000.0706
Ore G, Bruins HJ (2012) Design features of ancient agricultural terrace walls in the Negev desert: human-made geodiversity. Land Degrad Dev 23:409–418. https://doi.org/10.1002/ldr.2152
Ore G, Bruins HJ, Meir IA (2020) Ancient cisterns in the Negev Highlands: types and spatial correlation with Bronze and Iron Age sites. J Arch Sci Rep 30:102227
Palmer EH (1871) The Desert of the Exodus. Deighton, Cambridge
Poesen J, Lavee H (1994) Rock fragments in top soils: significance and processes. Catena 23:1–28. https://doi.org/10.1016/0341-8162(94)90050-7
Rosen SA (1994) Map of Makhtesh Ramon (204). Archaeological Survey of Israel, Israel Antiquities Authority, Jerusalem
Saaroni H, Halfon N, Ziv B, Alpert P, Kutiel H (2010) Links between the rainfall regime in Israel and location and intensity of Cyprus Lows. Int J Climatol 30:1014–1025. https://doi.org/10.1002/joc.1912
Sandler A (2013) Clay distribution over the landscape of Israel: from the hyper-arid to the Mediterranean climate regimes. Catena 110:119–132. https://doi.org/10.1016/j.catena.2013.05.016
Shanan L, Schick AP (1980) A hydrological model for the Negev desert highlands: effects of infiltration, runoff and ancient agriculture. Hydrol Sci B 25:269–282. https://doi.org/10.1080/02626668009491934
Singer A (2007) The soils of Israel. Springer, Berlin
Tadmor NH, Evenari M, Shanan L, Hillel D (1958) The ancient desert agriculture of the Negev. 1. Gravel mounds and gravel strips near Shivta (Sbeita). Ktavim 8:127–151
Tsvieli Y, Zangvil A (2005) Synoptic climatological analysis of 'wet' and 'dry' Red Sea Troughs over Israel. Int J of Climat 25:1997–2015. https://doi.org/10.1002/joc.1232
Van Bommel D, Bruins HJ, Lazarovitch N, Van der Plicht J (2021) Effect of dung, ash and runoff water on wheat and barley grain sizes and stable isotope ratios: Experimental studies in ancient desert agriculture (Negev, Israel). J Archaeol Sci: Reports 39:103172.https://doi.org/10.1016/j.jasrep.2021.103172
Wieler N, Avni Y, Rosensaft M (2016) The significance of the geological strata on desert runoff agriculture: indications for stable desert environment over the last 1600 years in southern Israel. J Arid Environ 135:147–163. https://doi.org/10.1016/j.jaridenv.2016.08.010
Yaalon DH, Dan J (1974) Accumulation and distribution of loess-derived deposits in the semi-desert and desert fringe areas of Israel. Z Geomorph Supp 20:91–105
Yaalon DH, Ganor E (1979) East Mediterranean trajectories of dust-carrying storms from the Sahara and Sinai. In: Morales C (ed) Saharan Dust. John Wiley and Sons, New York, pp 187–193
Yair A, Kossovsky A (2002) Climate and surface properties: hydrological response of small arid and semi-arid watersheds. Geomorphology 42:43–57. https://doi.org/10.1016/S0169-555X(01)00072-1
Yair A, Raz-Yassif N (2004) Hydrological processes in a small arid catchment: scale effects of rainfall and slope length. Geomorphology 61:155–169. https://doi.org/10.1016/j.geomorph.2003.12.003
Yair A, Sharon D, Lavee H (1978) An instrumented watershed for the study of partial area contribution of runoff in the arid zone. Z Geomorphol Supp 29:71–82
Zangvil A, Druian P (1990) Upper trough axis orientation and the spatial distribution of rainfall over Israel. Int J Climat 10:57–62. https://doi.org/10.1002/joc.3370100107
Zangvil A, Karas S, Sasson A (2003) Connection between Eastern Mediterranean seasonal mean 500 hPa height and sea-level pressure patterns and the spatial rainfall distribution over Israel. Int J Climat 23:1567–1576. https://doi.org/10.1002/joc.955
Zilberman E, Avni Y (2004) Geological map of Israel 1:50,000, Har Hamran Sheet 21-I, Geol Surv Israel, Jerusalem
Ziv B, Saaroni H, Alpert P (2004) The factors governing the summer regime of the Eastern Mediterranean. Int J Climat 24:1859–1871. https://doi.org/10.1002/joc.1113

21 Uvda Valley, Israel—An Interplay of Rock Control, Weathering and Debris Deposition in the Evolution of Hyper-Arid Rock Slopes

Nimrod Wieler and Hanan Ginat

Abstract

Looking at terrestrial hyper-arid rock slopes, one would expect such landforms in areas characterized by climatic and tectonic stability coupled with sparse or no vegetation to show minor geomorphic changes. The Uvda Valley, located in the southern part of the Negev Desert, Israel, is an excellent area to examine different controls on rock-slope geomorphic evolution. The presence of large boulders (> 1 m long) on a 16 ka alluvial terraces beneath the rock slopes challenges the concept of minor changes and indicates an ongoing dynamic erosion process. Hyper-arid rock-slope retreat models are mainly correlated to differential erosion processes (e.g., formation of mesas and cuestas), while models related to retreat of a single-lithology rock slope are based on limited field data. Quantifying weathering and transport processes in the Uvda Valley along sedimentary rock slopes composed of a single lithology reveals a systematic spatial pattern of erosional features. Such weathering features coupled with rock debris are suggested to play a major role in shaping rock-slope dynamics since the late Pleistocene.

Keywords

Rock slopes · Hyper-arid regions · Cavernous weathering · Rock debris

N. Wieler (✉)
Research Department, Israel Antiquities Authority, Jerusalem, Israel
e-mail: nimrodw@israntique.org.il

H. Ginat
Dead Sea and Arava Science Center, Eilat, Israel

21.1 Introduction

First-order indicators of landscape evolution are processes shaping rock slopes and their geometry (Sharp 1982; Howard and Selby 2009; Anderson and Anderson 2010). The main acting forces along a given slope include endogenic (i.e., tectonics) and exogenic (i.e., weathering and transport) processes which lead to the formation of either bedrock or detrital slopes differing in the proportion between erosion rate and regolith production (Gilbert 1909; Carson and Kirkby 1972). Modifications of arid rock-slope landforms are commonly associated with physical processes, e.g., scarp erosion, mass movement (Howard and Selby 2009), cliff retreat (Koons 1955; Boroda et al. 2014), or fluvial downcutting (Haviv et al. 2010). Yet, in the absence of such processes, no mechanism has been proposed to explain what could promote slope instability (Howard and Selby 2009; Viles 2013; Krautblatter and Moore 2014). Moreover, the existing models which are based on the diffusion equation (Culling 1960; Dietrich et al. 1995) link regolith production and regolith transport to hillslope generation. However, these concepts were not applied in hyper-arid environments, where local-scale tectonic activity is absent and the parameters controlling regolith production and transport are unclear.

Under the condition that the rock slopes in hyper-arid regions have adjusted to tectonic activity, they become mainly exposed to exogenic processes. Such processes that include chemical breakdown (i.e., salt that pries bedrock to small fragments) along with accumulation of settled dust lead to regolith production. The slopes lack vegetation, and hence the soil thickness is not plant-regulated (Owen et al. 2011). Furthermore, soil reservoir dynamics along bedrock slopes in arid environments was suggested to be locally correlated with landscape curvature (Amundson et al. 2015). Differential erosion model (i.e., different rock types erode to give different slopes) has been proposed as the main promoter of rock-slope dynamics in hyper-arid regions (Boroda

A. Frumkin and N. Shtober-Zisu (eds.), *Landscapes and Landforms of Israel*,
World Geomorphological Landscapes, https://doi.org/10.1007/978-3-031-44764-8_21

et al. 2014). However, there are gaps and uncertainties in our understanding of potential factors that may affect hyper-arid rock-slope evolution, e.g., differential erosion model is true for slopes composed of different lithologies, but cannot explain the dynamics of a single-lithology slope. Nevertheless, the existing models do not define the factors governing spatial distribution of instability areas along a slope of single lithology.

Various linkages between multi-scale weathering features and rock-slope evolution were suggested (Sharp 1982; Turkington and Phillips 2004; Howard and Selby 2009; Krautblatter and Moore 2014; Viles 2001, 2005, 2013). However, weathering features are mainly associated with small-scale features (< 1 m long/deep) rather than slope-scale features; moreover, the production of such small-scale weathering features yields local sediments (e.g., debris, reg soil, and rock crusts) which accumulate downslope and stabilize it. As a result, it decreases mass transport to lower parts of the slope. One more issue is that no quantitative analyses were conducted on a slope scale to test the possible correspondence between weathering erosional features and rock-slope dynamics.

Uvda Valley located in southern Israel, characterized by a long-term hyper-arid climate, is one of the most suitable places to analyze and reveal the complexity of cavernous weathering features and their possible impact on the rock-slope dynamics.

21.2 Setting and Location

Uvda Valley (29° 95′ N, 34° 94′ E, ~ 430 m.a.s.l; Fig. 21.1) that is located in the southern part of the Negev Desert, Israel (Fig. 21.1) is ~ 15 km long and 5 km wide, surrounded by ridges of 50–150 m in height. The area is underlain predominantly by sedimentary rock series, which give rise to diverse rock slopes, and maintains long-term hyper-arid conditions and stable tectonic setting since the late Pleistocene-Holocene (Ginat and Zilberman 1991; Amit et al. 2011). The southern parts of the Negev Desert is characterized with 20–50 mm annual precipitation and aridity index (P/PET) of 0.005 (Amit et al. 2011; Bruins 2012), in accordance with similar areas worldwide, e.g., the Namib and Atacama deserts (Goudie et al. 1997; Azua-Bustos et al. 2012). The long-term aridity enables it to maintain various erosion and weathering morphologies at a range of sizes along the rock slopes.

21.2.1 Geology

Uvda Valley is an elongated geological graben located 400–700 m above sea level, extending sub-parallel to the Arava Valley. To the east, the Uvda Valley is bordered by the Shaharut Plateau consisting mainly of hard, bedded limestone, and dolomite of the Late Turonian Gerofit and Zihor formations. To the west, the Uvda Valley is bordered by the Hiyyon Plain, which consists of chalk, chert, and marl of the Santonian age Mount Scopus Group, capped by gravels of the Pliocene Arava conglomerate.

The local landscape of the Uvda Valley results from early Pleistocene subsidence in the central Arava Valley and uplift of the central Negev plateau in the northwest that inverted the regional gradient, causing eastward tilting of the southern Negev. The prominent relief resulted from breaking up into blocks along the Milhan, Zihor, and Hiyyon faults. These tectonic lines then became the primary controls of the courses of the present drainage systems of the southern Negev (Ginat et al. 2000). Uvda Valley low-relief drainage system (few to 100 m wide) is incised 30–100 m below the table-like hilltops before joining the Hiyyon basin that drains eastward toward the central Arava Valley. Gradient of the valley beds declines from 1% in its southern edge to 0.1% in its northern edge.

The Uvda Valley is filled with Pleistocene to Holocene fluvial sediments, deposited during the last glacial period, between 70 and 16 ka (Amit and Yaalon 1996; Faershtein et al. 2016; Porat et al. 2010). These fluvial-alluvial materials become finer from south to north. In the valley's southern part, the alluvial fill mainly consists of coarse limestone and chert gravels, and a few fine-grained particles. In the valley's center, the fill is enriched with sand and silt, although some coarse gravels still exist. In the valley's northern part, the fill is predominated by sand, silt, and clay. This alluvial sequence consists of the southern source of fill supply, as is the case at present.

The rock slopes in this hyper-arid environment lack vegetation, and the soil cover thickness along them is not plant-regulated, but is the result of settled dust accumulation or chemical and physical breakdown. The bedrock exposed in the region consists of marine sediments, mainly limestone, dolomite, chalk, marl and chert of late Cretaceous to Santonian age (Menuha, Mishash, Gerofit and Zihor formations). Rock slopes consisting of various erosional features and unstable debris mantle are widespread in the Uvda Valley. Moreover, multi-scale erosion features can be found

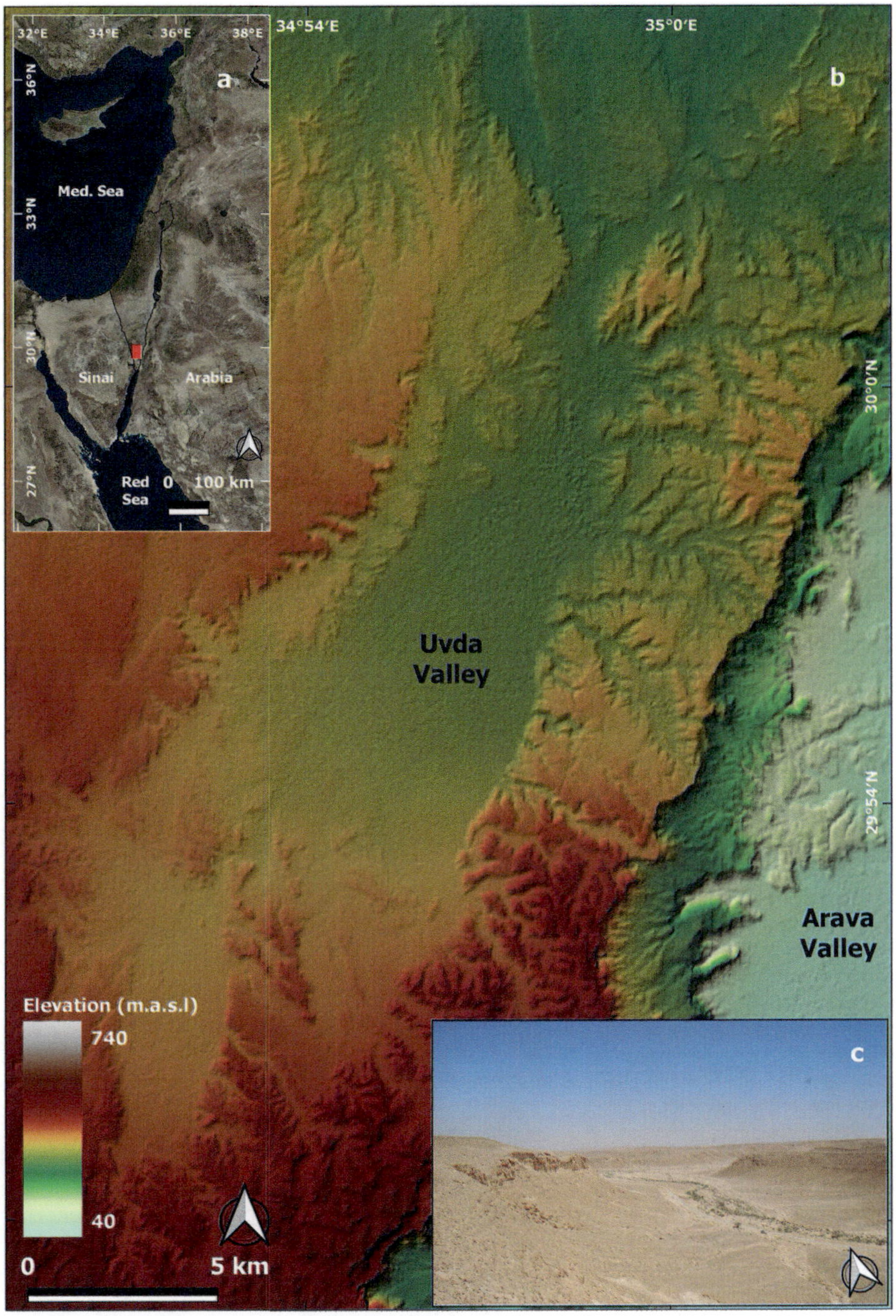

Fig. 21.1 (**a**) Geographical location of the Uvda Valley (red rectangle) in the southern Negev Desert, Israel (Google, ©2020 Landsat/Copernicus Data SIO, NOAA, U.S. Navy, NGA, GEBCO Mapa GISrael), Abbreviations: Med. Sea, Mediterranean Sea; (**b**) Shaded relief map of the Uvda Valley (SRTM 30 m/pixel; https://opentopography.org/); Uvda Valley representative slopes (**c**)

at slopes composed of single lithology, mainly dolomite (Fig. 21.2). Large boulders (> 1 m) scattered along the foot of the slopes on a 16 ka alluvial terrace (Amit and Yaalon 1996; Faershtein et al. 2016; Porat et al. 2010) manifest the young dynamic erosional processes acting along the slopes.

21.2.2 Weathering Features

The prevalence of cavernous features is observed in different areas of the Negev Desert. Their morphology does not differ between sites, despite precipitation gradient and differences in the underlying geology. In all cases, the weathering landforms are classified as either tafoni or honeycomb weathering (Goudie et al. 1997; Groom et al. 2015) (Fig. 21.3), and found to be restricted to the atmospherically exposed parts of the rock slopes.

The accepted conceptual model for the formation of cavernous rock weathering (i.e., tafoni) in hot deserts involves the presence of permeable rocks that are subjected to the movement of soluble salts and repeated episodes of drying–wetting cycles (Goudie et al. 2002; Smith 1988; Smith

Fig. 21.2 Representative 500 m long (east aspect) dolomitic rock slope (Gerofit Formation) in the Uvda Valley, containing variety of instability areas (colored rectangles) coupled with multi-scale erosional features. Increase in the instability degree along the slope can be detected as heading from the red rectangle to the blue and black rectangles. The largest cavernous features are observed at the black rectangle and the smallest ones at the red rectangle. Such differences are suggested to result from the debris coverage that limits the exposure of bedrock to atmospheric conditions; as a result, it governs the appearance of cavernous weathering features. A decrease in the gradient of debris slope was observed as heading from the red to blue and black rectangles, which correlates with the increasing appearance of cavernous weathering features

Fig. 21.3 Multi-scale cavern features detected in Uvda Valley (Gerofit Formation). Weathering is restricted to rock outcrops exposed to atmospheric conditions on all scales. The photos clearly manifest that soil acts as a 'preservative medium', in (**b**) Dashed line marks the clear-cut separation between the exposed weathered and well-preserved buried rock parts, the picture was taken at an archeological site, 6000 years old (10 cm arrow for scale). In (**a**) removal of the thin (< 10 cm) soil cover exposed an unweathered rock (30 cm hammer for scale); (**c**) and (**d**) are highly weathered boulders located on alluvial terraces below the rock slopes (30 cm hammer for scale); (**e**) and (**f**) are rock slopes, facing to the north and to the south, respectively; note that slope-scale cavernous features are also restricted to parts exposed to atmospheric conditions (cavernous features are 1.5 m high)

et al. 2005). The proposed mechanisms assume that cavernous weathering results from physicochemical processes including salt crystallization (Cooke 1979; Amit and Yaalon 1996; Scherer 2004), exfoliation (Shtober-Zisu et al. 2017), or stress erosion (Bruthans et al. 2014). Recently, Bruthans et al. (2018) conclusively demonstrated that moisture flux followed by salt crystallization at the boundary layer governs the case hardening model in temperate climates.

Rock weathering processes are typically believed to be simultaneously controlled by external factors at different scales ranging from the regional climate scale, down to local conditions of the site, and eventually the micro-scale (Smith 2009; Sperling and Cooke 1985; Viles 2001). The results of a recent study by Wieler et al. (2019) indicate that microbial colonization upon mineral surfaces mitigates evaporation and reduces water transport, thereby alleviating salt crystallization pressure in the rock pores (Scherer 2004). As a result, Wieler et al. (2019) suggested that in arid and hyper-arid environments, micro-scale conditions determine the magnitude of weathering that shape the landscape.

Elucidating the factors that govern the size and distribution of weathering features along rock slopes in the Uvda Valley, a quantitative morphometric analysis approach was used as it helps to narrow down the different variables that may be involved (e.g., slope aspect, macro-climate, and rock type). Quantifying 600 cavernous features located on a single dolomitic slope shows that their volume ranged between 0.01 m^2 and 0.1 m^3. Correlating slope parameters (e.g., inclination, aspect, and height) and cavities dimensions along the dolomitic slope reveals an increasing trend between the magnitude of weathering features and decrease of soil coverage (e.g., lithosol, rock debris). This trend is in accordance with the previous observations suggesting that cavernous features are restricted to the atmospherically exposed parts of the rock slopes. Hence, the debris resulting from weathering and erosion accumulates along the slope and thus, prevents other parts of the slope from exposure to atmospheric conditions. Therefore, understanding sediment generation processes is essential for the interpretation of spatial distribution of geomorphic features. Our findings are in agreement with studies conducted in the Atacama Desert, suggesting that geomorphic processes interact with slope and soil cover to influence spatial distribution of erosion features (Placzek et al. 2014). Other slope and geo-technical parameters (e.g., joints) tested along the rock slopes were found to play a minor role in defining the spatial distribution of caverns.

21.2.3 Role of Lithology in Shaping Rock Slope Geometry

Although weathering processes are more intense in soft than in hard lithologies (Gilbert, 1909), their influence on debris mantled slopes that consists of specific lithology in arid regions are unclear (Howard and Selby 2009). Johnstone and Hilley (2015) who revised Gilbert (1909) demonstrated that hillslopes formed in a soil-mantled landscape can be influenced by the underlying lithology through coupled processes of soil production and transport. Yet, soil-mantled bedrock slopes in hyper-arid regions are rare and therefore, the main source for cover deposits on such slopes are debris and aeolian material (i.e., accumulation of atmospheric dust) (Owen et al. 2011). In the Negev Desert, mantling slopes by soils is suggested to be lithology-dependent. Fine-bedded lithologies support slopes at moderate angles, which enable it to accumulate larger amounts of aeolian material and maintain debris material for longer periods. In comparison, coarse-bedded lithologies form steeper slopes thus accumulate less aeolian and debris material. As a result, coarse-bedded lithologies expose larger portions of rock surfaces to atmospheric conditions, thus leading to multi-scale weathering cavities. We further observed that the fine-bedded lithologies can generate only small-size weathering features (up to 50 cm in diameter), hence the impact of these formations on rock-slope geometry would be weaker than that of coarse-bedded formations. Such scenarios were detected through different lithologies in the Negev Desert.

21.3 Landscape Evolution

21.3.1 Morphodynamics of Rock Slope and Cavernous Weathering in Hyper-Arid Regions

The Negev Desert, like many other deserts throughout the world, is underlined predominantly by slopes made of sedimentary rocks (Oberlander 1994). In layered sedimentary successions, lithological alternations are common; therefore, the slope profile is nearly always associated with the change in rock strength. Hence, the resistance of beds influences the weathering and erosion pace and results in composite slope profiles typified by alternations of steep and less inclined slope units. Geomorphic features of sedimentary tablelands (e.g., scarps, cuesta, or mesa landforms) are suggested to result from the presence of a caprock of low erodibility on top of strata of higher erodibility that leads to faster erosion at the slope base, cliff development, and scarp backwasting (Boroda et al. 2014; Howard and Selby 2009; Koons 1955; Migoń and Duszyński, 2022). The main mechanisms related to escarpment retreat in layered rocks were recently reviewed by Duszyński et al. (2019) and include (1) the role of surface and underground water; (2) rock-cliff retreat by differential weathering, mass movements, lateral spreading and block gliding, landslides, block-by-block undermining, in situ degradation; and (3) aeolian processes. However, in the absence of differential weathering and seepage, these mechanisms could not allow predicting the evident retreat of a single-lithology slope.

In the Uvda Valley, dolomitic rock slopes containing instability areas are correlated with the formation of cavernous weathering features. A new suggested model of slope geometry change through time is presented below (Fig. 21.4):

a. Exposure of rock surfaces—Upper parts of soil-mantled rock slope are being exposed to atmospheric conditions. As a result, gypsum originated from the nearby soil and mobilized by dew forms chemomechanical forces in the rock pores that cause incipient weathering features accompanied by exfoliation. These small-scale cavities generate locally transported debris that accumulates in the lower parts of the slope. Thin soil (mainly built of aeolian material) accumulates downslope along with debris supply (Amit et al. 1993; Ganor and Foner 2001) and serves as an important source for rock surface burial downslope, shielding its lower parts.
b. Development of weathering processes—Chemomechanical processes lead to backwall weathering of the cavernous features, enhanced due to the small shaded

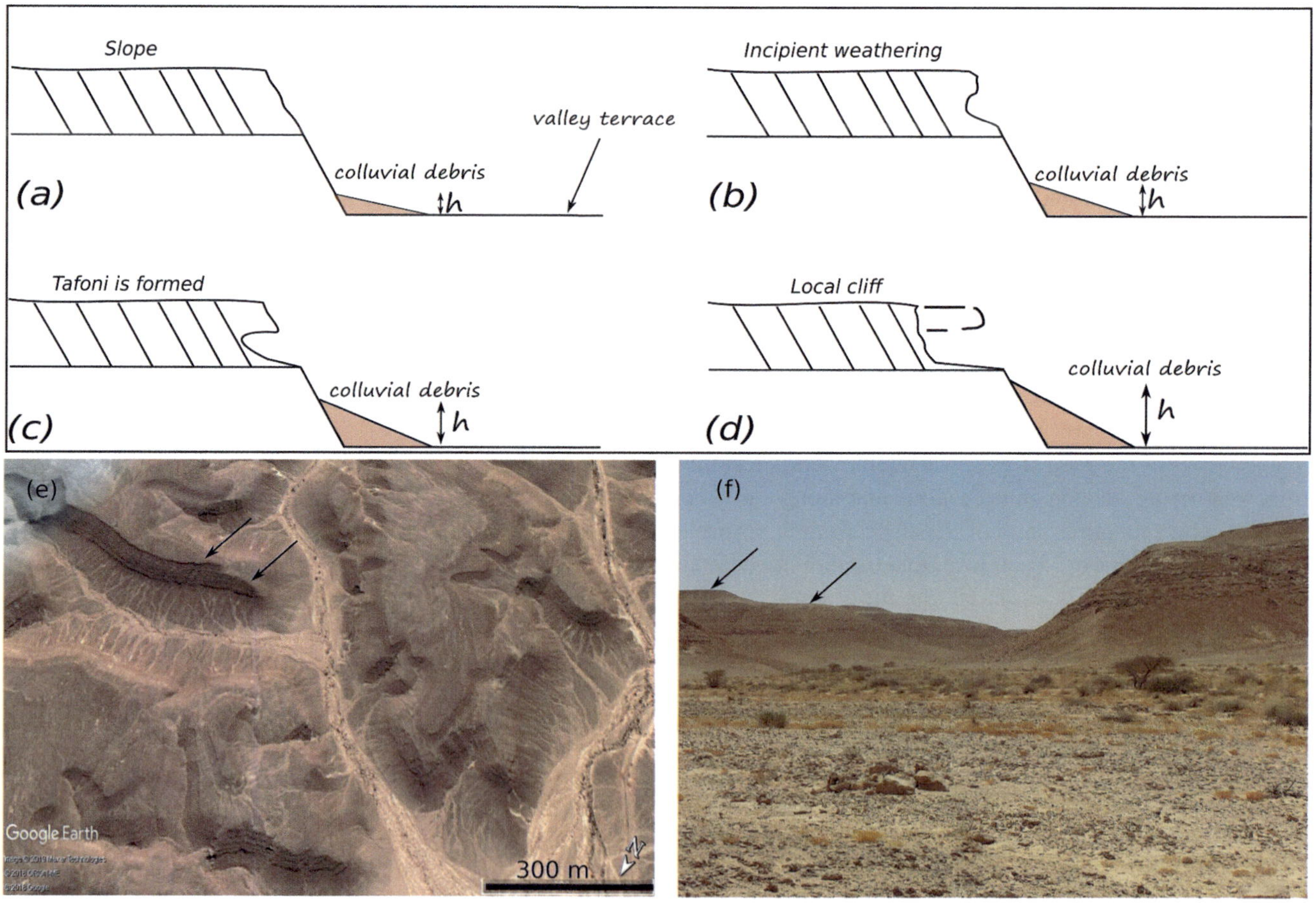

Fig. 21.4 Simplified scheme indicating the transition from slope (**a**) To local cliff (**d**) On a single-lithology slope. Both layers in the scheme are composed of similar lithology. The limiting factor is exposure of the upper layer to atmospheric conditions which leads to the origin of cavernous features; (**e**) Google Earth imagery of the surrounding slopes (Google ©2020 Landsat/Copernicus Data SIO, NOAA, U.S. Navy, NGA, GEBCO Mapa GISrael); (**f**) Side view of a mature valley in the Uvda Valley, solid arrows indicating the two-level slopes

cavities that allow for longer periods of dew condensation. As a result, rock fragments are being detached from the inside parts of cavernous features, as seen in the enlargement of exfoliation morphologies in the cavities. The detached rock fragments contribute to the debris formed at the lower parts of the slope. Supply of debris along the slope results in: (a) increased slope instability in the upper parts; (b) decreased exposure of new rock surfaces downslope, thus stabilizing the slope lower parts. This step is equally controlled by weathering- and transport-limited processes.

c. Forming new rock surfaces—Cavities keep growing as a function of strata thickness and atmospheric exposure. Due to gravitational forces, the cavity caprock collapses and turns the weathered slope into a local cliff that exposes new rock surfaces to atmospheric conditions. The collapse of boulders increases debris supply that accumulates downslope and therefore decreases the slope angle at its lower parts. This step is dominated by the transport-limited component.
d. Formation of two-level slopes—The continuous cavernous weathering in the upper slope sections maintains local cliff retreat which leads to the generation of wadis (ephemeral streams) that are wider at their top and narrow at the base. Shallow colluvial debris will start to accumulate at the base of the local cliff. This step is dominated by a weathering-limited component.

21.3.2 Rock-Slope Retreat Rates in Hyper-Arid Regions

The retreat rates along sedimentary sequences in the Negev hyper-arid regions were previously measured using cosmogenic isotopes by Boroda et al. (2011). They dated talus flatirons indicating a rate ranging between 0.006 and 0.012 m/kyr. In another study conducted at the Negev Desert, Finzi and Harlev (2016) used modeling on the basis of geometrical variables and structural parameters, and the average rates obtained ranged between 0.01 and 0.18 m/kyr. Waterfall retreat rates located beside the active Dead Sea basin in a hyper-arid climate ranged between 165–900 m per one million years (Haviv et al. 2010). Yet, exposure conditions and rock composition of cavernous weathering along sedimentary rock slopes challenge standard dating techniques such as cosmogenic isotopes or optically stimulated luminescence (Cockburn and Summerfield 2004; Viles 2016). In order to overcome the dating obstacle, large weathered boulders (Fig. 21.5) detached from the slope and placed downslope on top of young alluvial terraces, which are 16,000 years old (Porat et al. 2010) were used to constrain a maximum time frame for such processes to occur. The presence of such boulders is the result of gravitational forces operating on the upper parts of the cavernous weathering features, and marks the final step of cavern development. Although the boulder collapses might be stochastic in time, they suggest a maximum time frame for these processes, but we cannot reject the possibility that these boulders result from a recent earthquake. In order to estimate rock mass loss of these boulders, we quantitatively characterized the cavernous features. As discussed above, the median volume of the cavernous weathering features ranges between 0.01 and 0.1 m^3. Using the dimensions of the caverns and the presence of boulders on top of young alluvial terraces of known age, we could estimate the maximum retreating rate of a single-lithology sedimentary slope in hyper-arid regions to be in a range of 0.006–0.06 m/kyr. These rates are in agreement with the denudation rates (0.021–0.025 m/kyr) measured by Ryb et al. (2014) in hyper-arid carbonate terrains and with the

Fig. 21.5 **a** Typical slope (30 m high) in the Uvda Valley containing collapsed boulders in the lower parts of the slope resulting from weathering processes operating on the upper parts of the slope; the boulders are located on a ~ 16ka alluvial terrace (Porat et al. 2010); **b** Closer inspection of the weathered boulders (1.5 m high)

retreat rate (0.006–0.012 m/kyr) measured by Boroda et al. (2011) dating flatirons in the hyper-arid region. We suggest that the measured rates are attributed to chemomechanical clast contribution, resulting from cavernous weathering, as was also suggested by Ryb et al. (2014). Maximum cavernous weathering rates, based on models and cosmogenic isotope dating in granite located in humid regions, indicate that rates were faster in the initial phase (0.0055–0.0125 m/kyr) compared with the mature phase (Brandmeier et al. 2011; Matsukura and Matsuoka 1991; Norwick and Dexter 2002; Sunamura 1996).

21.4 Conclusions

Hyper-arid cavernous weathering features (i.e., tafoni) are the result of transport of dissolved salts through the rock and their eventual crystallization in surface pores, which leads to fractures and eventual flaking of rock material (Rodriguez-Navarro et al. 1999).

Field observations coupled with morphometric analyses from the hyper-arid environment of the Negev Desert permit to suggest a new model that highlights the interplay of lithological properties and rock-slope debris to form large-scale weathering erosion features, which promote rock-slope instability. We show the spatial correlation between rock-slope parameters to multi-scale cavernous weathering features. These findings demonstrate that multi-scalar cavernous weathering features can serve as reliable markers of sedimentary arid and hyper-arid rock-slope evolution. We further note that the presence of rock debris, resulting from such erosional processes and placed downslope on top of young alluvial terraces, can suggest a maximal time frame for the rock-slope weathering processes rate. Therefore, we estimate the maximum retreating rate of a single-lithology sedimentary slope in hyper-arid regions to be in a range of 0.006–0.06 m/kyr.

References

Amit R, Yaalon DH (1996) The micromorphology of gypsum and halite in Reg Soils—the Negev Desert Israel. Earth Surf Proc Land 21(12):1127–1143

Amit R, Gerson R, Yaalon DH (1993) Stages and rate of the gravel shattering process by salts in desert Reg soils. Geoderma 57(3):295–324

Amit R, Simhai O, Ayalon A, Enzel Y, Matmon A, Crouvi O, Porat N, McDonald E (2011) Transition from arid to hyper-arid environment in the southern Levant deserts as recorded by early Pleistocene cummulic Aridisols. Quat Sci Rev 30(3–4):312–323

Amundson R, Heimsath A, Owen J, Yoo K, Dietrich WE (2015) Hillslope soils and vegetation. Geomorphology 234:122–132

Anderson RS, Anderson SP (2010) Geomorphology: the mechanics and chemistry of landscapes. Cambridge University Press

Azua-Bustos A, Urrejola C, Vicuña R (2012) Life at the dry edge: microorganisms of the Atacama Desert. FEBS Lett 586(18):2939–2945

Boroda R, Amit R, Matmon A, ASTER Team, Finkel R, Porat N, Enzel Y, Eyal Y (2011) Quaternary-scale evolution of sequences of talus flatirons in the hyperarid Negev. Geomorphology 127(1–2):41–52

Boroda R, Matmon A, Amit R, Haviv I, Arnold M, Aumaître G, Bourlès DL, Keddadouche K, Eyal Y, Enzel Y (2014) Evolution and degradation of flat-top mesas in the hyper-arid Negev, Israel revealed from ^{10} Be cosmogenic nuclides: evolution and degradation of flat-top mesas in the hyper-arid Negev. Earth Surf Proc Land 39(12):1611–1621

Brandmeier M, Kuhlemann J, Krumrei I, Kappler A, Kubik PW (2011) New challenges for tafoni research. A new approach to understand processes and weathering rates. Earth Surf Proc Land 36(6):839–852

Bruins HJ (2012) Ancient desert agriculture in the Negev and climate-zone boundary changes during average, wet and drought years. J Arid Environ 86:28–42

Bruthans J, Soukup J, Vaculikova J, Filippi M, Schweigstillova J, Mayo AL, Masin D, Kletetschka G, Rihosek J (2014) Sandstone landforms shaped by negative feedback between stress and erosion. Nature Geosci 7(8):597–601

Bruthans J, Filippi M, Slavík M, Svobodová E (2018) Origin of honeycombs: testing the hydraulic and case hardening hypotheses. Geomorphology 303:68–83

Carson MA, Kirkby MJ (1972) Hillslope form and process. Cambridge University Press

Cockburn HAP, Summerfield MA (2004) Geomorphological applications of cosmogenic isotope analysis. Prog Phys Geogr: Earth Environ 28(1):1–42

Cooke RU (1979) Laboratory simulation of salt weathering processes in arid environments. Earth Surf Proc 4(4):347–359

Culling WEH (1960) Analytical theory of erosion. J Geol 68(3):336–344

Dietrich WE, Reiss R, Hsu ML, Montgomery DR (1995) A process-based model for colluvial soil depth and shallow landsliding using digital elevation data. Hydrol Proc 9(3–4):383–400

Duszyński F, Migoń P, Strzelecki MC (2019) Escarpment retreat in sedimentary tablelands and cuesta landscapes—landforms, mechanisms and patterns. Earth-Sci Rev 196:102890

Faershtein G, Porat N, Avni Y, Matmon A (2016) Aggradation–incision transition in arid environments at the end of the Pleistocene: an example from the Negev Highlands, southern Israel. Geomorphology 253:289–304

Finzi Y, Harlev N (2016) A regional approach for modeling cliff retreat rate: the Makhteshim Country, Israel. Geomorphology 271:65–73

Ganor E, Foner HA (2001) Mineral dust concentrations, deposition fluxes and deposition velocities in dust episodes over Israel. J Geophys Res: Atmosph 106(D16):18431–18437

Gilbert GK (1909) The convexity of hilltops. J Geol 17(4):344–350

Ginat H, Zilberman E (1991) Structural and morphological development of the Uvda Valley. Isr J Earth-Sci 40(1–4):209–218

Ginat H, Zilberman E, Avni Y (2000) Tectonic and paleogeographic significance of the Edom River, a Pliocene stream that crossed the Dead Sea Rift valley. Isr J Earth Sci 49:159–177

Goudie AS, Viles HA, Parker AG (1997) Monitoring of rapid salt weathering in the central Namib Desert using limestone blocks. J Arid Environ 37(4):581–598

Goudie AS, Wright E, Viles HA (2002) The roles of salt (sodium nitrate) and fog in weathering: a laboratory simulation of conditions in the northern Atacama Desert Chile. Catena 48(4):255–266

Groom KM, Allen CD, Mol L, Paradise TR, Hall K (2015) Defining tafoni: re-examining terminological ambiguity for cavernous rock decay phenomena. Prog Phys Geogr 39(6):775–793

Haviv I, Enzel Y, Whipple KX, Zilberman E, Matmon A, Stone J, Fifield KL (2010) Evolution of vertical knickpoints (waterfalls) with resistant caprock: insights from numerical modeling. J Geophys Res 115(F3)

Howard AD, Selby MJ (2009) Rock Slopes. In: Parsons AJ, Abrahams AD (eds) Geomorphology of desert environments. Springer, Netherlands, pp 189–232

Johnstone SA, Hilley GE (2015) Lithologic control on the form of soil-mantled hillslopes. Geology 43(1):83–86

Koons ED (1955) Cliff retreat in the southwestern United States. Am J Sci 253(1):44–52

Krautblatter M, Moore JR (2014) Rock slope instability and erosion: toward improved process understanding. Earth Surf Proc Land 39(9):1273–1278

Matsukura Y, Matsuoka N (1991) Rates of tafoni weathering on uplifted shore platforms in Nojima-Zaki, Boso Peninsula Japan. Earth Surf Proc Land 16(1):51–56

Migoń P, Duszyński F (2022) Landscapes and landforms in coarse clastic sedimentary tablelands—is there a unifying theme? Catena 218:106545

Norwick SA, Dexter LR (2002) Rates of development of tafoni in the Moenkopi and Kaibab formations in Meteor Crater and on the Colorado Plateau, northeastern Arizona. Earth Surf Proc Land 27(1):11–26

Oberlander TM (1994) Global deserts: a geomorphic comparison. In: Parsons AJ, Abrahams AD (eds) Geomorphology of desert environments. Springer Netherlands, pp 13–35

Owen JJ, Amundson R, Dietrich WE, Nishiizumi K, Sutter B, Chong G (2011) The sensitivity of hillslope bedrock erosion to precipitation. Earth Surf Proc Land 36(1):117–135

Placzek C, Granger DE, Matmon A, Quade J, Ryb U (2014) Geomorphic process rates in the central Atacama Desert, Chile: insights from cosmogenic nuclides and implications for the onset of hyperaridity. Am J Sci 314(10):1462–1512

Porat N, Amit R, Enzel Y, Zilberman E, Avni Y, Ginat H, Gluck D (2010) Abandonment ages of alluvial landforms in the hyper-arid Negev determined by luminescence dating. J Arid Environ 74(7):861–869

Rodriguez-Navarro C, Doehne E, Sebastian E (1999) Origins of honeycomb weathering: the role of salts and wind. GSA Bull 111(8):1250–1255

Ryb U, Matmon A, Erel Y, Haviv I, Benedetti L, Hidy AJ (2014) Styles and rates of long-term denudation in carbonate terrains under a Mediterranean to hyper-arid climatic gradient. Earth Planet Sci Lett 406:142–152

Scherer GW (2004) Stress from crystallization of salt. Cem Concr Res 34(9):1613–1624

Sharp RP (1982) Landscape evolution (a review). Proce Nat Ac Sci 79(14):4477–4486

Shtober-Zisu N, Amasha H, Frumkin A (2017) Inland notches: lithological characteristics and climatic implications of subaerial cavernous landforms in Israel. Earth Surf Proc Land 42:1820–1832

Smith BJ (1988) Weathering of superficial limestone debris in a hot desert environment. Geomorphology 1(4):355–367

Smith BJ (2009) Geomorphology of Desert Environments. In: Parsons AJ, Abrahams AD (eds) Geomorphology of desert environments. Springer, Dordrecht, pp 3–7

Smith BJ, Warke PA, McGreevy JP, Kane HL (2005) Salt-weathering simulations under hot desert conditions: agents of enlightenment or perpetuators of preconceptions? Geomorphology 67(1):211–227

Sperling CHB, Cooke RU (1985) Laboratory simulation of rock weathering by salt crystallization and hydration processes in hot, arid environments. Earth Surf Proc Land 10(6):541–555

Sunamura T (1996) A physical model for the rate of coastal tafoni development. J Geol 104(6):741–748

Turkington AV, Phillips JD (2004) Cavernous weathering, dynamical instability and self-organization. Earth Surf Proc Land 29(6):665–675

Viles HA (2001) Scale issues in weathering studies. Geomorphology 41(1):63–72

Viles HA (2005) Microclimate and weathering in the central Namib Desert Namibia. Geomorphology 67(1–2):189–209

Viles HA (2013) Linking weathering and rock slope instability: non-linear perspectives. Earth Surf Proc Land 38(1):62–70

Viles H (2016) Technology and geomorphology: Are improvements in data collection techniques transforming geomorphic science? Geomorphology 270:121–133

Wieler N, Ginat H, Gillor O, Angel R (2019) The origin and role of biological rock crusts in rocky desert weathering. Biogeosciences 16(6):1133–1145

Landscapes of Nahal Yael, Southern Negev Desert

22

Judith Lekach

Abstract

This chapter describes the landscape of a small watershed called Nahal Yael in southernmost Israel as the outcome of its climatic, geologic, hydrologic and geomorphologic conditions. The Precambrian crystalline rocks of the watershed and the hyper-arid climate of today and of the late Pleistocene create the unique conditions for weathering processes, mass movement on slopes and sediment transport along the channels by flash foods. Understanding the evolution of landscape by the slow-motion operating processes in an arid environment like the watershed of Nahal Yael requires long-term measurements of rainfall, flow, sediment transported on slopes and in channels, otherwise, one might get the impression that the landscape is stagnant and that nothing happens in such a dry climate. Over fifty years of rainfall measurements reveal its high annual and spatial variability. Due to the high intensities, only few millimeters of rain are needed to initiate runoff on slopes and in channels. The flash floods created carry mostly bedload (60–70%), and rather small amounts (30–40%) of suspended load originated from slopes and channel. The source of bedload transported by nowadays floods is remnants of Quaternary landforms–talus slopes and terraces created probably during a wetter episode (35–27 Ka) within the hyper-arid climate of the Quaternary in the Southern Negev. Such conditions create today a state of dynamic equilibrium where the amount of sediment supplied to the channel is also the amount transported out of the catchment. Evidence of this equilibrium is the fluvio-pedogenic layer (FPU) within the alluvial channel. The outcome is a watershed with unique landscape that has become a perfect field laboratory for studying and understanding the processes operating today in desert environments. As well, it enables to reveal wetter episodes within the general arid climate, with non-typical landforms in desert environments.

Keywords

Desert geomorphology · Flash flood · Sediment transport · Dynamic equilibrium · Fluvio-pedogenic units

22.1 The Nahal Yael Watershed: A Field Laboratory

22.1.1 Location

Nahal Yael is a small (0.6 km^2) watershed located ~ 5 km to the north-west of Eilat, a town along the north-west shore of the gulf of Eilat/Aqaba (with a unique landscape described by Grodek (this volume) (Figs. 22.1 and 22.2a, b). Its 160 m relief over the small area of the watershed is reflected by rough topography and steep slopes (Fig. 22.2c). The terrain is dissected by over 100 first-order channels to create a high-density drainage network (Fig. 22.2d). Nahal Yael is the last tributary of Nahal Roded (42 km^2) before it drains into the Arava Rift Valley, an area of continuous tectonics activity since 20.8–18.5 Ma (Nuriel et al. 2017).

22.1.2 History

Nahal Yael was selected in 1965 to establish a field laboratory by the late Prof. Asher Schick, to investigate long-term geomorphic and hydrologic processes in a hyper-arid region, expected to provide data that were not existent at

J. Lekach (✉)
The Fredy & Nadine Herrmann Institute of Earth Sciences, The Hebrew University of Jerusalem, The Edmond J. Safra Campus - Givat Ram, 9190401 Jerusalem, Israel
e-mail: Judith.lekach@mail.huji.ac.il

A. Frumkin and N. Shtober-Zisu (eds.), *Landscapes and Landforms of Israel*, World Geomorphological Landscapes, https://doi.org/10.1007/978-3-031-44764-8_22

Fig. 22.1 The vicinity of Nahal Yael watershed as seen in the satellite image of the southern Arava Valley—part of the Dead Sea Transform—and the regional base level. Dark rocks of the Precambrian Massif and the light Mesozoic sedimentary rocks form the watersheds of the drainage networks from Edom mountains in the east and from the southern Negev mountains in the west. The landscape of their alluvial fans shows bright strips of active channels and dark abandoned surfaces covered with desert varnish. Nahal Yael watershed (marked) drains into Nahal Roded and its alluvial fan. The satellite image is from Google Earth. *Source* Southern Arava Valley and Surrounding 29° 45′ 00″ N 35° 04′ 20″ E

that time. Currently, it holds a world record of continuous hydrological, geomorphological and ecohydrological data that represent a hyper arid region like the Southern Negev Desert. The collected data include high-resolution continuous measurements of rainfall, runoff and sediment transport that served many studies performed along the years (e.g., Schick 1968, 1970, 1977; Sharon 1970; Schick and Sharon 1974; Lekach and Schick 1982, 1983, 1995; Schick et al. 1987; Lekach et al. 1992; Clapp et al. 2000; Lekach and Enzel 2021), as well as short-time and more site-specific studies about infiltration rates and runoff generation on slopes (Yair and Klein 1973; Salmon and Schick 1980; Greenbaum 1986), desert soil formation (Amit et al. 1993, 2006), channel bed morphology (Lekach et al. 1998; Amit et al. 2007), response of Acacia trees to natural and man-induced changes in the water budget (BenDavid-Novak and Schick 1997; Armoza-Zvuloni et al. 2021).

Notwithstanding the large research body conducted in Nahal Yael, there are still unanswered questions and this unique field laboratory still suggests new frontiers in the research of hyper-arid environments.

22.1.3 Geological Features

Nahal Yael is part of the Eilat Massif, a high-grade regionally metamorphosed and complexly folded belt of meta-sedimentary, meta-igneous rocks and intrusive igneous rock, which are the northern part of the Arabian Shield of Precambrian rocks. Three well-defined lithological belts, mainly E-W trending, can be observed (Fig. 22.2e). The upper part of the watershed is dominated by amphibolite, with a small part of pelitic schist and abundant dyke-like bodies of granitic gneiss. A major belt of schist occupies the center of the watershed, and toward the outlet, there is a belt of Eilat granite. Dyke swarms are abundant in all the belts (Shimron 1974).

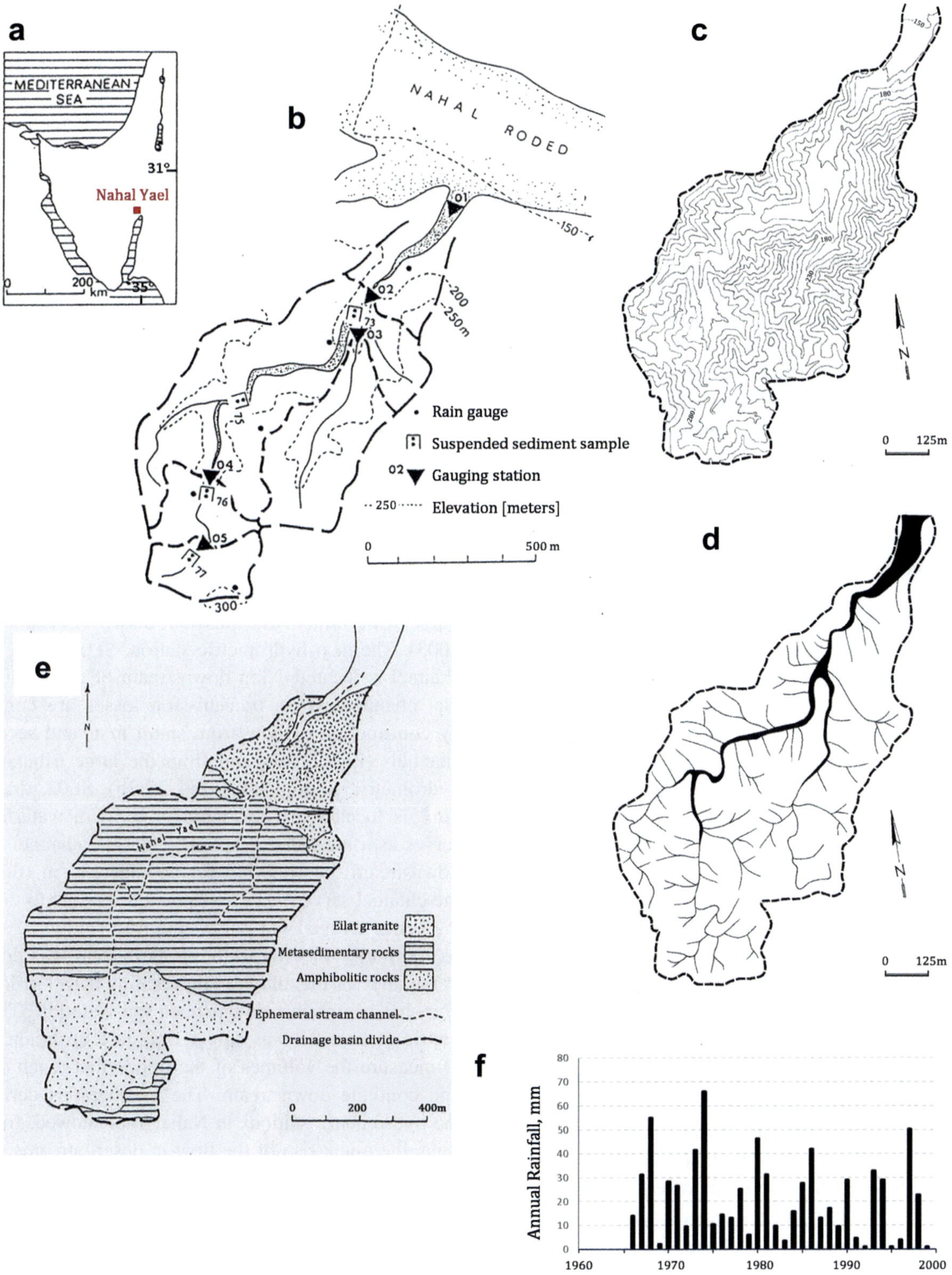

Fig. 22.2 **a** Location of Nahal Yael watershed; **b** Nahal Yael watershed selected instrumentation and location of the earth dam; **c** Topography; **d** Drainage network; **e** Geology; **f** Annual rainfall in Nahal Yael (1965–1998)

22.1.4 Climate

Nahal Yael is located in the northern part of the global desert belt, which dictates the hyper-arid climate in the area. The rain season is between October and May and the low precipitation (annual mean ~ 28 mm yr$-^1$) is characterized by high annual and spatial variability. A double and even triple amount of annual rain can fall in one year or sometimes during one event, whereas in another year, less than 1 mm of rain may be experienced (Fig. 22.2f). Three main climatic systems are responsible for the rain and its characteristics in the area: (1) the main rain contributor is the Active Red Sea Trough, that creates localized convective rain cells about 5 km in diameter, sometimes a few tens of kilometers apart, characterized by high intensities that can exceed 150 mm/hr (Sharon 1972; Sharon and Kutiel 1986; Schick 1988); (2) tropical plumes from the Intertropical Convergence Zone (ITCZ) that travel across Northern Africa and create high precipitation over the whole Negev (Rubin et al. 2007); (3) a southeastern trough of Mediterranean frontal cyclones (Kahana et al. 2002). Summer temperature can be high and reach 50 °C, while the winter temperature can be as low as 5 °C. The potential evaporation in Nahal Yael is high, ~ 2800 mm/yr (Bar-Lavee 1976). These climatic conditions dictate lack of vegetation on slopes, while perennial shrubs and Acacia trees are restricted to the active channels and the alluvial fan. After flooding events annual vegetation appears along the channels, but only after heavy rains some annual grass covers the slopes.

22.2 The Nahal Yael Watershed–Measurements and Results

22.2.1 Rain and Wind

A dense net of recorders and gauges was installed over the 0.6 km^2 of the watershed: 2 wind recorders, 13 rain recorders, 6 standard rain gauges and 40 small orifice rain gauges. High-resolution measurements of rain and wind (direction and velocity) revealed two important conclusions: (1) due to the rough topography, rain drops diverted by wind were dispersed from the rain gauges located near the watershed divide and were tunneled to the gauges located at the lower parts of the slopes near the channel. The total storm rain in the valleys can exceed the rain in higher parts of the slopes by 30% (Sharon 1970); (2) there is a high variability of precipitation even over a small area as the 0.6 km^2 of Nahal Yael (Sharon 1972).

In addition to the high spatial variability, rain storms vary in their total amounts as well as in their intensity distribution within the storm. Flood-creating rain storms vary in their amounts (5–60 mm) as well as in their intensities (Fig. 22.3). Intensities of about 50% of all rains are above 14 mm/h, 75% of those have intensities in the range of 14–35 mm/h and the intensities of the remaining 25% are above 35 mm/h (Schick 1988). A double event that occurred on 22.2.1975 is a good example of such variability. A 5-h long medium intensity rain (average less than 10 mm/h) occurred in the morning and was followed, a few hours later, by a short (5 min.) rain with intensity of 96 mm/h.

22.2.2 Hydrology

Flow data in Nahal Yael watershed have been collected by five hydrometric stations located along the channel since 1965 (Fig. 22.2b). The uppermost station is St.05 draining a sub-catchment of 0.05 km^2 which is an assemblage of three smaller sub-catchments. The rocky channel bed and the symmetrical arrangement of sub-catchment channels result in quick runoff convergence. The mostly rocky channel continues downstream to the hydrometric station St.04 with minor transmission losses (Greenbaum et al. 2003). The next hydrometric station, St.02, along the main channel is located 1 km downstream of a continuous alluvial channel, where transmission losses are compensated by contribution of flow from small first- and second-order channels (Fig. 22.2d), and from the large tributary with a hydrometric station St.03 (Fig. 22.2b). St.02, draining 0.5 km^2, is located just upstream of a 10 m waterfall which serves as a local base level. The alluvial channel continues downstream the waterfall for another 130 m (this part of the channel serves since 1978 as a reservoir) to the apex of the active alluvial fan, where the fifth station 01 existed at its toe until 1978. In 1978, an earth dam was constructed at the apex of the alluvial fan, creating a 130-m-long reservoir that begins just downstream the waterfall. The hydrometric station 01 was transferred just upstream the dam to measure the volumes of the floods that reach station 02 and continue downstream. The hydrographs derived from the hydrometric stations in Nahal Yael showed, for the first time, the quick rise of the flow in desert streams. The time span between a dry desert channel bed and a peak discharge is only few minutes (Schick 1970). A peak discharge of a given flood is not a direct function of the total amount of the rain storm. It is highly connected to the time between rain spells of a given storm and their intensity distribution, as demonstrated by a few flood events (Fig. 22.3). Floods occur only during 50% of the years on record, and about a third of these flows do not reach the outlet. These floods originated in the upper parts of the watershed, but infiltrated along the alluvial channel, are termed "preparing" or "shifting" episodes, as they only transfer sediments between

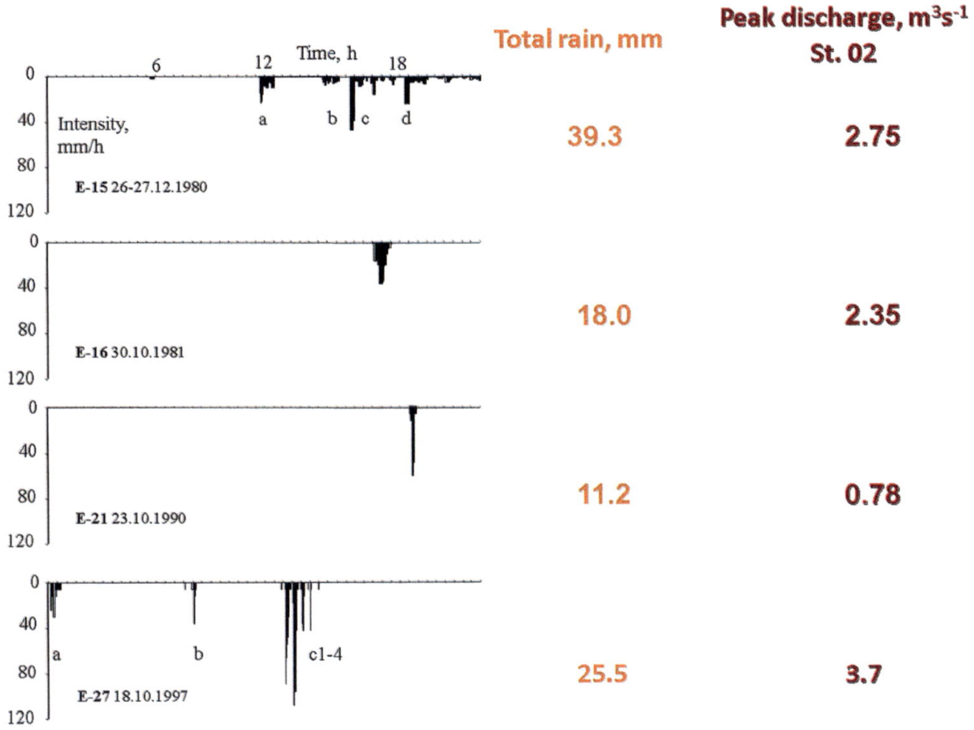

Fig. 22.3 Rain storms and their consequential floods peak discharges selected from the record of flow events (E1, E2, …) in Nahal Yael. When a flow was created by a non-continuous rain, each spell was given a letter code. Four rain spells for E15 a–d and three rain spells for E27 a–c

storage elements within the channel (Lekach et al. 1992). Floods that reach the catchment outlet occur only once in 3–3.5 years (Lekach and Enzel 2021), and until 1978 they transferred sediments from various storage elements toward the Nahal Yael active alluvial fan. All hydrological data recorded in St.02, since 1965, were published for the first time in Lekach and Enzel (2021).

22.2.3 Sediment Transport–Sampling, Tracing Programs and Total Yield Measurements

22.2.3.1 Suspended Load

Multiple stage automatic samplers constructed for the Nahal Yael project were located along the channel (Fig. 22.2b), and collected suspended load only during the rising limb of the hydrographs (Lekach and Schick 1982). This was a great disadvantage and caused an error in calculation of the sediment yields as reported by Schick in 1977. However, since the samplers were located along the channel, their data shed light on the sources of suspended load. The samplers of the rocky sub-catchments 05 and 04 provide data of suspended load contributed only by slopes, as the rocky channel's contribution to the suspended load is negligible. Most of the suspended sediments were within the silt fraction, and this coincides well with the research of Yair and Klein (1973) on slopes of the 05 sub-catchment. They found that the contribution of slopes is 95% fines and only 5% sand. The samples from samplers located further downstream, along the alluvial part of the channel, showed a double modal (silt and sand modes) grain size distribution. It reflects differential contributions of slopes and channels to the suspended load. The fine fraction represents the contribution from the slopes and the coarse fraction represents the sand brought to suspension from the alluvial channel due to high flow velocities (Fig. 27.4 in Lekach et al. 1992).

The data collected by the samplers during 35 years (1965—1980) show a close relationship between the time span separating the events which is the "preparation time", and the concentrations of fines (< 0.063mm) (Fig. 6 in Lekach and Schick 1982), pointing at the airborne dust as a preponderant source of fines in the suspended load.

22.2.3.2 Bedload

Sampling bedload is a problematic task in any environment and especially in the desert environment. Bedload in Nahal Yael was monitored by continuous measurements of travel distances (trace survey after each flow event) of tagged (painted and magnetized) bedload particles (Schick 1977; Schick et al. 1987; Lekach and Schick 1995). All the attempts to gain information about travel distance of particles was meant to get a better and reliable estimation of sediment transport using the virtual velocity approach. This approach evaluates sediment yield by longitudinal translation of the scour layer to a depth determined by scour chains over a mean distance derived from pebble tracing program. Despite its limitations (e.g., Recking et al. 2016; Mao et al. 2017), it is widely used in streams where tracing programs and scour chains are operating as in Nahal Yael (e.g., Schick 1977; Mao et al. 2017; Brenna et al. 2019). In 1985, magnetic tracers were introduced to enable detection of buried bedload particles, so more reliable bedload travel distances were determined for each flood. The results indicate that short pebble trajectories are associated with burial episodes, while the long trajectories coincide with exposure from within the active layer (Lekach and Schick 1995).

22.2.3.3 Nahal Yael Reservoir

The most distal section of the Nahal Yael alluvial channel, 130 m downstream the waterfall, is a reservoir created in 1978 by constructing an earth dam at the apex of the alluvial fan (Figs. 22.2b and 22.4). The reservoir was constructed to trap 100% of the sediment yields as part of the long-lasting program of tracing and measuring sediment transport in Nahal Yael. A detailed mapping of the reservoir was performed after every flow event, and after it dried out by evaporation (2800 mm/yr, Bar-Lavee 1976) and infiltration into its bed. It enabled the calculations of sediment yield of each individual flood and the separation into bedload and suspended load. This was possible due to the unique sedimentary sequences of event couplets that characterize an ever-emptying reservoir like the Nahal Yael reservoir (Laronne 2000): a coarse bedload layer overlain by the fining upward coarse sand to silt and clay layer deposited from suspension. Since its construction, over forty years, the reservoir has served as an almost 100% efficient trap for bedload and suspended yields of all flows that reach the reservoir. Only three floods transferred small amount of suspended load downstream the spillway.

22.2.3.4 Sediment Yields Calculations

The first sediment budget prepared for Nahal Yael was based on travel distances of a few generations of painted pebbles, thickness of the active layer derived from scour chains and data on suspended load from automatically collected samples (Schick 1977; Lekach and Schick 1982). The average annual sediment yield derived was rather high, 400-ton/km^2, with 65% suspended load and 35% bedload. These high sediment yields were not compatible with the limited geomorphic reaction of the alluvial fan which was expected to aggrade by the addition of the bulk of sediment exported to it from the watershed (Schick and Lekach 1993). Following the construction of the reservoir and its detailed mapping after every flow event, the corrected calculation of sediment yields revealed that the average annual sediment yield is 115 ton/km^2, with 70–60% bedload and 30–40% suspended load (Lekach et al. 1992; Schick and Lekach 1993; Lekach and Enzel 2021).

Fig. 22.4 Nahal Yael reservoir after a flood on 1 March 2017

22.3 Basin Morphology

22.3.1 Rock Slopes

Notwithstanding the seemingly relatively low relief of 160 m, the landscape of Nahal Yael is rugged with steep slopes exceeding 30°. Generally, bare rock characterizes the upper parts of the slopes, some of them partly covered with colluvium that thickens downslope, where substantial colluvium exists a typical talus slope is created. The variability in the colluvium cover is due to different responses to weathering of the three lithologies in Nahal Yael that create various clast sizes on the slopes (Bull and Schick 1979). The schist with the highest fracture density of 0.13–0.9 cm/cm^2, form coarse to fine gravels, which build the massive talus deposits of the central part of the basin (Fig. 22.5a). Amphibolite in the upper part of the catchment with fracture density of 0.08–0.2 cm/cm^2 weathers easily into coarse sand, but due to some gneiss and acid dykes which decompose to fine and coarse gravels, respectively, the

Fig. 22.5 Landscape of the central part of the watershed (schist strip): **a** Bare upper slopes and talus accumulated in the lower parts, attached to the alluvial terrace; **b** Accumulation of sediment from the terrace, ready to be transported by the next flood

Fig. 22.6 Landscape of sub-catchment 05. Note the automatic suspended load sampler in the channel

amphibolite slopes (Fig. 22.6) are covered by gravels from both lithologies (Yair and Klein 1973). The granitic slopes with fracture density of 0.04–0.08 cm/cm^2 in the northern basin are mostly bare rock with very thin granular colluvial cover mainly in hollows and at the base of the slopes (Fig. 22.7a). A cavernous weathering called tafone is abundant on the bare Eilat granite slopes (Fig. 22.7b) (Bull and Schick 1979).

22.3.2 Talus Slopes

Downslope of the bare rock exposed in the upper parts of schist slopes in Nahal Yael, weathered loose material accumulates to form extensive talus deposits (Fig. 22.5a). Enzel et al. (2012) in their comprehensive study of Quaternary landforms in Nahal Yael suggested that these talus deposits began to form about 45 ka, and most of their propagation downslope and the extensive thickening occurred between 35 and 20 ka. The hypothesis suggested that frequent-high intensity rainstorms of sufficient duration triggered talus formation. A study on paleofloods in Nahal Netafim (3 km west of Nahal Yael; Greenbaum et al. 2006) also suggested wetter period which lasted from 35 to 27 ka. This was probably only a wetter episode within the hyper-arid climate during the late Pleistocene in the southern Negev (Amit et al. 2006). During that wetter period, the talus slopes provided the full range of sediment sizes to the channel that triggered massive aggradation, evident today by remnants of long, ~ 2-m-high alluvial terraces along the active channel (Fig. 22.5a). What happens today? Evidence from a short period of ~ 40 years supplies an additional perspective. In 1983, the late Luna B. Leopold visited Nahal Yael and arranged a line of white quartz clasts with the B axis length within the range of 32–256 mm, across the slopes in the central part of schist lithology (Fig. 22.8). Only occasional movement of single particles has been recognized since. Abandoned research plots on the slopes in the northern and central part of the watershed (Yair and Lavee 1985) do not show substantial accumulation of clasts in their containers (Fig. 22.9). In the southern amphibolitic strip, most (95%) of the overall sediments contributed by the slopes are silt and clay, 5% is sand and there are no gravels (Yair and Klein 1973). This coincides with the longer colluvial residence time in the upper amphibolitic part of Nahal Yael as derived from cosmogenic nuclide data (Clapp et al. 2000). Data collected from the suspended load automatic samplers along the channel prove as well that the slopes are the main source of suspended load (Lekach et al. 1992). It seems that today there is no much mass movement on the talus slopes, but they still are the main contributor of coarse sediments to the active channel mainly by gully erosion (Fig. 22.10), and by scouring the toes of talus slopes that reach the active channel (Clapp et al. 2000).

Fig. 22.7 Station 02 during the recession phase of a flood: **a** Bare Eilat Granite slopes of both banks; **b** Closer look at the cavernous weathering (tafone) in the granite slopes

Fig. 22.8 The Leopold line of white quartz clasts across the schist slopes as pictured at 2019

Fig. 22.9 Abandoned research plot on a granite slope in the northern part of the watershed, not showing substantial accumulation of clasts in the containers during approximately 40 years

22.3.3 Channel Morphology

Drainage density of Nahal Yael watershed is relatively high. There are over 100 first-order channels combining into fourth-order main stream (Fig. 22.2d). The 2-km-long channel of Nahal Yael can be separated into two morphologically distinctive parts. The third part of the channel was dammed and serves as a 100% trap for sediment exported from the watershed.

22.3.3.1 Bed-Steps in a Desert Stream

The upstream 1-km-long part is a third-order, mostly a steep bedrock channel, with some shallow alluvial veneer locally more concentrated in pockets in the rock and in

Fig. 22.10 Gully developed within a talus slope. Contribution of sediments from talus slopes to the alluvial channel is by gully erosion

the form of strips along rocky depressions. Continuous alluvium, however shallow, covers the channel bed only in several parts, like a 17-m-long reach immediately upstream of the hydrometric station 05 (Fig. 22.2b) and 10-m-reach upstream of the station 04 (Fig. 22.2b). The slope of these continuous alluvial sections is 0.044 (Lekach and Schick 1983), but the overall steepness of the channel is due to bed-steps formed in both bedrock and alluvium reaches (Wohl and Grodek 1994) and due to small waterfalls, 0.5–1.0 m high, along the channel between stations 05 and 04. Morphology of bed-steps have been described mainly in perennial streams, where the clasts forming the steps are submerged by high flows (e.g., Grant et al 1990). In the hyper-arid environment of Nahal Yael, even the high flows, in the sub-catchment 05 and 04, are not so high to cover clasts of 17–25 cm length (Wohl and Grodek 1994; Table 1). Wohl and Grodek (1994) claimed that the bed-steps are not a relict feature, but rather active channel landforms in the present flow regime of Nahal Yael as no desert varnish was detected neither on bedrock or alluvial steps (Wohl and Grodek 1994). Support to their claim provide clasts of similar size to those that compose the alluvial bed-steps, that were transported by floods recorded at the station 04, as detected by the tracing program of painted pebbles operated for 22 yr (1969–1991) (Lekach 1992). The bedrock upper part of the channel ends just downstream of the station 04 with a flume-like 5 m long, 0.6 m wide and 10% steep slope bedrock reach.

22.3.3.2 The Alluvial Channel: Bars and Inner-Channels, Surficial and Subsurface Hidden Features

An abrupt change in channel morphology occurs downstream the flume-like bedrock channel and coincides with the schist belt (Fig. 22.2e). The next 1-km-long channel becomes a continuous, steep (average slope 0.05) alluvial braided stream. Channel width varies between 3 m in the upper part and 20 m further downstream. The channel narrows again (3–5 m) toward the 10-m-high waterfall just downstream of the station 02 (Fig. 22.2b). Downstream of the waterfall the alluvial channel continues for another 130 m to the apex of the long and narrow alluvial fan, active until 1978, drained into Nahal Roded (Fig. 22.11). Morphology of bars and inner-channels characterizes the braided stream of Nahal Yael (Fig. 22.12). There are two distinct populations of bars according to their surficial characteristics: (1) high and coarse-grained bars; (2) low and fine-grained bars. About 9% of the surficial material on the high bars are cobbles (> 64 mm), while on the low bars, there are no cobbles at all and pebbles (4–64 mm) compose more than 80% of the material. In the innerchannels, over 90% of the surface consists of sand and granules (0.063–4 mm) (Table 27.4 in Lekach et al. 1992).

This rather simple surficial appearance is completely different within the active layer, where scour and fill processes operate. Detection of the active layer stratigraphy was possible since 1985, when magnetic tracers were added to the

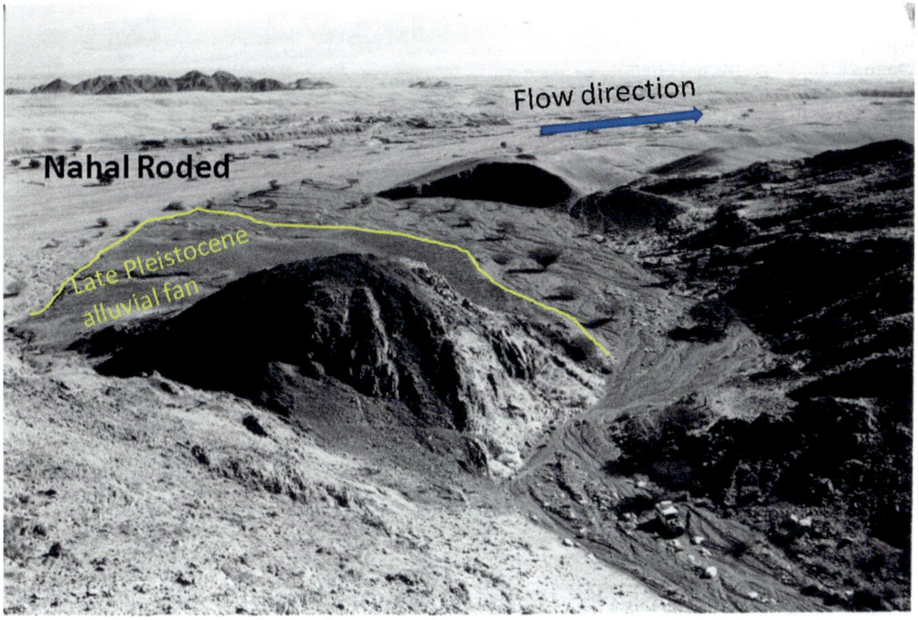

Fig. 22.11 Nahal Yael alluvial fan and its abandoned Late Pleistocene terrace as drained into Nahal Roded. Notice the unvegetated dark surface of the Late Pleistocene alluvial fan, and the acacia trees along the active alluvial fan. This photo was taken in 1975 before the construction of the earth dam at the apex (a vehicle in the lower right for scale)

Fig. 22.12 The braided alluvial channel of Nahal Yael: surficial morphology of bars and inner-channels

previous tracing program of painted pebbles. The magnetic tracers, detected after each flow event, were found deposited within the active layer beneath one or two inverse bedding units at depths of 20–30 cm (Fig. 22.13). Those units were characterized by an openwork structure, and a very well-sorted coarse surface of the inversely graded units. Such bedding is caused by the instant deposition of all sizes carried by the flow at that moment due to flow separation induced by bed disturbance (Lekach et al. 1992). They supply the proof of bedload being transported in lobes or waves, as suggested previously by Lekach and Schick (1983), who analyzed hand sampling of sediment transported during the flood of February 1975. Thus, looking into the scour layer, the surficial appearance is very illusive. Inverse bedded units can be found in newly deposited bars, on top of pre-flood fine-grained inner-channels and coarse-grained bars can be scoured to become an inner-channel with sand and granules on its surface. Thus, the stratigraphy of the active layer of Nahal Yael channel reveals the isomorphic characteristics of a braided stream.

Fig. 22.13 Two inverse bedded units deposited on top of a magnetic tracer recovered after a flow event. The scale bar is 15 cm long

22.3.3.3 The Fluvio-Pedogenic-Unit (FPU)

Closer look into the active layer raises a question about the maximum depth of the active layer during floods of the current flow regime. The next question considers the boundary between the upper mobile alluvium and the stationary alluvium beneath. Digging into the alluvial channel of Nahal Yael reveals that the upper, active, grayish and non-cohesive alluvium changes abruptly at an average depth of 50 cm into reddish, compacted distinctive layer termed a fluvio-pedogenic-unit (FPU) (Lekach et al. 1998, 2008; Amit et al. 2007). The FPU was first described beneath the active alluvial channel of Nahal Yael, and later also in other desert channels (Lekach et al. 1998; Amit et al. 2007; Enzel et al. 2012). These units are reddened due to iron oxides release, the clay content in their upper part is about 5%, which is more than twice that in the mobile alluvium above, they are continuous along and across the alluvial channel, consolidated by calcium carbonate, and their upper contact with the active layer is abrupt while the lower boundary is diffuse. The formation of FPU is connected to the cumulative influence of persistent differences in water availability to various parts of the alluvial channel before, during and after flood events during few thousands of years (Lekach et al. 1998, 2008; Amit et al. 2007). The depth of the FPU changes with the type of boundary between the channel and the adjacent slope due to variability in runoff contributed to the channel by rocky and colluvium-covered slopes. The average depth of the FPU in the channel varies between 40 and 50 cm, but it deepens to a meter and more when bare rock slopes reach the channel, due to more frequent runoff generation on the rocky slopes. On bare rocky slopes only 3 mm of rain with intensity of 20 mm/y is needed to generate flow (Yair and Klein 1973). Salmon and Schick (1980) reported runoff generated from plots on bare rocky slopes in Nahal Yael after 1–3 min of rain at average intensity of 40–25 mm/h, while on colluvial slopes more time, 3–6 min of intense rain, 50–30 mm/h, is needed to initiate runoff. When the channel is in contact with a slope covered by colluvium or a talus slope, FPU's depth is the same as in the center of the channel. Runoff generation on debris-covered slopes is created only by higher rain intensities (Salmon and Schick 1980; Greenbaum 1986; Greenbaum et al. 2003) and the source of water available for development of the FPU is the channel flow. Thus, the existence of FPU in the alluvial channel of Nahal Yael means that: (1) the current hydrologic regime is in some kind of dynamic equilibrium, all sediment supplied to the channel are transported out of the watershed, indicating that the net scour and fill in the channel is zero; (2) the maximum scour depth by floods of the current regime is the depth of the FPU.

22.3.3.4 The Alluvial Terrace

The alluvial terrace along the contemporary channel of Nahal Yael (Fig. 22.5) is a remnant of aggradation due to massive sediment transport into the channels from the talus slopes during the late Pleistocene, and following incision between 20 and 18 ka, probably immediately after the final phase of aggradation (Enzel et al. 2012). Two paleo-FPUs detected in the alluvial terrace of Nahal Yael (Fig. 10a in Enzel et al. 2012) enable a glimpse into the aggradation rates. The upper paleo-FPU was detected at the depth of ~ 30 cm from the terrace surface, representing dynamic equilibrium that existed in the catchment just before the incision to the level of the current channel. For how long this equilibrium prevailed before the incision, is still a question to be answered by further research that connects the characteristics of FPU with time of its development. The second paleo-FPU at a depth of almost 1.8 m represents some equilibrium stage before the massive aggradation between 35 and 20 ka (Enzel et al. 2012). The OSL dating and the 1.5 m distance between the two paleo-FPUs mean that aggradation was not a continuous process. Episodes of continuous aggradation were followed by episodes of dynamic equilibrium, where input of sediments into the channel was equivalent to sediment export, conditions required for development of the FPU (Lekach et al. 1998; Amit et al. 2007; Enzel et al. 2012). Dating of the terraces and their FPUs, and the characteristics of the soils developed on their surfaces support the idea that the incision in Nahal Yael occurred earlier than 13 ka (Enzel et al. 2012). The terraces serve today, by lateral accretion, as a source of sediments (Fig. 22.5b) transported along the channel and exported from the catchment (Clapp et al. 2000; Lekach and Enzel 2021).

22.3.3.5 The Alluvial Fans

Sediments exported from the watershed of Nahal Yael were partly deposited in the long alluvial fan and partly contributed to the floods in the wide braided channel of Nahal Roded (Fig. 22.11). Two alluvial fans are recognized: the non-active late Pleistocene alluvial fan and the active alluvial fan (Fig. 22.11). The surface of the late Pleistocene alluvial fan is covered with desert varnish and its reg soil development is similar to > 13 ka as the nearby alluvial surfaces of Nahal Shehoret (Amit et al. 1993, 1995, 1996; Enzel et al. 2012). A paleo-FPU is recognized about 0.5 m beneath the surface of the Pleistocene alluvial fan, which shows similar age as the upper paleo-FPU beneath the surface of the terrace in the channel of Nahal Yael. Thus, the time of incision in the catchment and creation of the alluvial terrace in Nahal Yael (20–18 ka) is compatible with the incision of the Pleistocene alluvial fan and deposition of the active alluvial fan (Enzel et al. 2012). The occurrence of FPU at depth of ~ 40 cm beneath the surface of the active alluvial fan (Fig. 10b in Enzel et al. 2012) suggests the existence of dynamic equilibrium of the current hydrological regime, where all sediments transported to the active alluvial fan are transmitted downstream toward Nahal Roded. Based on the modern sediment yields from Nahal Yael and the time needed for the late Pleistocene and Holocene alluvial fan deposition, (Enzel et al. 2012) Lekach and Enzel (2021) proposed that both alluvial fans store only a small portion of the total latest Quaternary sediment discharge out of Nahal Yael. The rest of the sediments, including probably most of the suspended sediments, were transported downstream and are part of the thick sedimentary fill of the southern Arava basin of the Dead Sea rift (Lekach and Enzel 2021).

22.4 The Holocene Hydrological Regime: A Dynamic Equilibrium

Looking at the channel of Nahal Yael, the alluvial terraces that flank the banks and the active alluvial fan raise a question regarding the current fluvial regime of Nahal Yael. An alluvial terrace is a landform formed by continuous incision into previously aggraded alluvial channel bed. Enzel et al. (2012) suggested that the incision in Nahal Yael probably occurred immediately following the final depositional age of the terraces and the Pleistocene alluvial fan at ca. 20–18 ka. Such continuous export of sediments that exceeds sediment contribution to the channels can be explained by severe flooding or exhaustion of sediments ready to be transported to the channels. Causes for this incision are unknown (Enzel et al. 2012), and they pose a challenge for next generation of researchers. By the end of the incision, the channel reached its present level, and since then it is in a dynamic equilibrium that allows the development of the FPU at the depth of maximum scour by floods of the current flow regime, and below the surface of the active alluvial fan as well. In addition, extensive talus deposits along the slopes of Nahal Yael were formed, and since the climate of the region was and still is extremely arid (Amit et al. 2006; Enzel et al. 2008) it seems that the current hydrologic regime is in some kind of dynamic equilibrium, where the amount of sediments supplied to the channel equals the amount transported downstream along the channel to the active alluvial fan to be further carried away by floods in Nahal Roded toward the Arava Valley (Lekach and Enzel 2021).

Acknowledgements The Nahal Yael Field Laboratory was established thanks to the vision and dedication of the late Prof. Asher Schick, which believed that only long-lasting research can reveal the secrets of an extremely arid environment. For more than half a century, Nahal Yael was a place of continuous research and scientific collaboration for colleagues and students from Israel and from abroad. Many researchers started their academic carrier in Nahal Yael as students. Thanks to the hard work of so many to be mentioned individually, we know today so much more about arid environments. Funding was initiated by the U.S. Army Research Grant which enabled the massive instrumentation and turning the small watershed of Nahal Yael into the first field laboratory in Israel. Financial support continues occasionally by the Israel Science Foundation, the US–Israel Binational Science Foundation, Israel Ministries of Agriculture, Defense and Infrastructure, and Central Research Fund of The Hebrew University of Jerusalem. When no financial support was available, data collection continued voluntarily by researchers and students of the Hebrew University. Today, monitoring of rain and flow continues by the Desert Floods Research Center and by The Israel Water Authority.

This final version of the chapter came to light thanks to the detailed review of Tamir Grodek who read the first version of the chapter, to Amos Frumkin, Nurit Shtober-Zisu and Prof. Piotr Migoń, whose detailed remarks help to polish this manuscript.

References

Amit R, Gerson R, Yaalon DH (1993) Stages and rate of gravel shattering processes by salts in desert Reg soils. Geoderma 57:295–324

Amit R, Harrison JBJ, Enzel Y (1995) Use of soils and colluvial deposits in analyzing tectonic events—the Southern Arava Rift, Israel. Geomorphology 12:91–107

Amit R, Harrison JBJ, Enzel Y, Porat N (1996) Soils as a tool for estimating ages of Quaternary fault scarps in a hyperarid environment—the southern Arava valley, the Dead Sea Rift, Israel. Catena 28:21–45

Amit R, Enzel Y, Sharon D (2006) Permanent quaternary aridity in the Southern Negev. Israel. Geology 34(6):509–512

Amit R, Lekach J, Ayalon A, Porat N, Grodek T (2007) New insight into pedogenic processes in extremely arid environment and its paleoclimatic implication—the Negev Desert, Israel. Quat Int 162–163:61–75

Armoza-Zvuloni R, Shlomi Y, Shem-Tov R, Stavi I, Abadi I (2021) Drought and anthropogenic effects on Acacia populations: a case study from the hyper-arid Southern Israel. Soil Syst 5:23

Bar-Lavee B (1976) Evaporation in Nahal Yael [M.Sc. thesis], Jerusalem, The Hebrew University of Jerusalem, 98 p. (in Hebrew)

BenDavid-Novak H, Schick AP (1977) The response of Acacia tree populations on small alluvial fans to changes in the hydrological regime: Southern Negev Desert, Israel. CATENA 29:341–351

Brenna A, Surian NL, Mao L (2019) Virtual velocity approach for estimating bed material transport in gravel-bed rivers: key factors and significance. Water Resour Res 55:1651–1674

Bull WB, Schick AP (1979) Impact of climatic change on an arid watershed: Nahal Yael, Southern Israel. Quat Res 11:153–171

Clapp EM, Bierman PR, Schick AP, Lekach J, Enzel Y, Caffee M (2000) Sediment yield exceeds sediment production in arid region drainage basins. Geology 28:995–998

Enzel Y, Amit R, Dayan U, Crouvi O, Kahana R, Ziv B, Sharon D (2008) The climatic and physiographic controls of the Eastern Mediterranean over the late Pleistocene climates in the southern Levant and neighbouring deserts. Global Planet Change 60:165–192

Enzel Y, Amit R, Grodek T, Ayalon A, Lekach J, Porat N, Bierman P, Blum JD, Erel Y (2012) Late quaternary weathering, erosion, and deposition in Nahal Yael, Israel: "an impact of climatic change on an arid watershed"? Geol Soc Am Bull 124:705–722

Grant GE, Swanson FJ, Wolman MG (1990) Pattern and origin of stepped-bed morphology in high-gradient streams, Western Cascades, Oregon. Geol Soc Am Bull 102:340–352

Greenbaum N (1986) Infiltration and runoff in an extremely arid region: infiltration experiments in small plots in the Southern Arava Valley and their hydrological, pedological and paleomorphological implications, [unpublished M.Sc. thesis] Jerusalem. The Hebrew University of Jerusalem, 206 p. (in Hebrew)

Greenbaum N, Salmon O, Schick AP (2003) Geomorphological implications and hydrological applications of infiltration tests in a hyperarid region. In: Greenbaum N, Lekach J, Inbar M (eds) Asher schick volume- current aspects of rainfall–runoff-sediment relations in Israel. University of Haifa, Haifa, pp 44–69 (in Hebrew)

Greenbaum N, Porat N, Rhodes E, Enzel Y (2006) Evidence of large flood episodes during oxygen isotope stage 3 in the southern Negev Desert, Israel. Quat Sci Rev 25:704–719

Kahana R, Ziv B, Enzel Y, Dayan U (2002) Synoptic climatology of major floods in the Negev Desert, Israel. Int J Climat 22(7):867–882

Laronne JB (2000) Event-based deposition in the ever-emptying Yatir Reservoir, Israel. In: Hassan MA, Slaymaker O, Berkowicz SM (eds) The hydrology-geomorphology interface: rainfall, floods, sedimentation, land use, vol. 261. IAHS Publ, Wallingford, UK, pp 285–302

Lekach J (1992) Bedload movement in a small mountain watershed in an extremely arid environment, Unpublished PhD Thesis. The Hebrew University of Jerusalem (in Hebrew with English abstract), 131p

Lekach J, Schick AP (1982) Suspended sediment in desert floods in small catchments: Isr J. Earth Sci 31:144–156

Lekach J, Schick AP (1983) Evidence for transport of bedload in waves: analysis of fluvial sediment samples in a small upland stream channel. CATENA 10:267–279

Lekach J, Schick AP, Schlesinger A (1992) Bedload yield and in-channel provenance in a flash-flood fluvial system. In: Billi P, Hey RD, Thorne CR, Tacconi P (eds) Dynamics of gravel-bed rivers. Wiley, Chichester, pp 537–554

Lekach J, Schick AP (1995) Trajectories of bed load particles within the active layer of an ephemeral stream. In: Adar EM, Leibundgut C (eds) Application of tracers in arid zone hydrology, vol 232. IAHS Publ, pp 443–452

Lekach J, Amit R, Grodek T, Schick AP (1998) Fluvio-pedogenic processes in an ephemeral stream channel, Nahal Yael, Southern Negev, Israel. Geomorphology 23:353–369

Lekach J, Amit R, Grodek T (2008) Scour envelope curve (SEC), Negev Desert, Israel. Isr J Earth Sci 57:189–197

Lekach J, Enzel Y (2021) Flood-duration integrated stream power and frequency magnitude of >50- year-long sediment discharge out of a hyperarid watershed. Earth Surf Proc Land 46(7):1348–1362

Mao L, Picco L, Lenzi MA, Surian N (2017) Bed material transport estimate in large gravel-bed rivers using the virtual velocity approach. Earth Surf Proc Land 42:595–611

Nuriel P, Weinberger R, Kylauder-Clark ARC, Hacker BR, Craddock JP (2017) The onset of the Dead Sea transform based on calcite age-strain analyses. Geology 45:587–590

Recking A, Piton G, Vazquez-Tarrio D, Parker G (2016) Quantifying the morphological print of bedload transport. Earth Surf Proc Land 41:809–822

Rubin S, Ziv B, Paldor N (2007) Tropical plumes over eastern North Africa as a source of rain in the Middle East. Month Weather Rev 135:4135–4148

Salmon O, Schick AP (1980) Infiltration tests, In: Schick AP (eds) Arid Zone geosystems, research report DA-JA-DAERO-78G-11, U.S. Army European Research Office London. Department of Physical Geography, Institute of Earth Sciences, Hebrew University, Jerusalem, pp 55–115

Schick AP (1968) The Nahal Yael Research Project in Southern Israel, problem in arid hydrology. In: Proceedings of the 21st international geographical congress, India 1968 Papers, Delhi, India, Aligarh Muslim University, Department of Geography, vol 1, pp 255–257

Schick AP (1970) Desert floods—interim results of observations in the Nahal Yael Research Watershed, Southern Israel, 1965–1970. In: IAHS—UNESCO symposium on the results of research on representative and experimental basin, Wellington, New Zealand, pp 478–493

Schick AP (1977) A tentative sediment budget for an extremely arid watershed in the Southern Negev. In: Doehring DO (ed) Geomorphology in arid regions. Allen and Unwin, London, pp 139–163

Schick AP (1988) Hydrological aspects of floods in extreme arid environments. In: Baker VR, Kochel RC, Patton PC (eds) Flood geomorphology. Wiley, NY, pp 189–203
Schick AP, Sharon D (1974) Geomorphology and climatology of arid watersheds. Technical Report, Department of Geography, Hebrew University, Jerusalem, p 161
Schick AP, Lekach J (1993) An evaluation of two ten-year sediment budgets, Nahal Yael, Israel. Phys Geog 14:225–238
Schick AP, Lekach J, Hassan MA (1987) Bedload transport in desert floods: observations in the Negev. In: Thorne CR, Bathurst JC, Hey RD (eds) Sediment transport in gravel-bed rivers. Wiley, Chichester, pp 617–636
Sharon D (1970) Areal pattern of rainfall in a small watershed as affected by wind and other meteorological conditions. Int Assoc Hydrol Sci Publ 96:3–11
Sharon D (1972) The spottiness of rainfall in a desert area. J Hydrol 17:161–175
Sharon D, Kutiel H (1986) The distribution of rainfall intensity in Israel, its regional variations and its climatological evaluation. J Climat 6:277–291
Shimron A (1974) The geology of Nahal Yael watershed, In: Schick AP, Sharon D (eds) Geomorphology and climatology of arid watersheds: Jerusalem, Israel, The Hebrew University of Jerusalem. U-S Army, Final Bi-Annual Technical Report, May 1972–September 1974 DA JA-72-C-3874, pp 103–122
Wohl EE, Grodek T (1994) Channel bed-steps along Nahal Yael, Negev desert, Israel. Geomorphology 9:117–126
Yair A, Klein M (1973) The influence of surface properties on flow and erosion processes on debris covered slopes in an arid area. CATENA 1:1–18
Yair A, Lavee H (1985) Runoff generation in arid and semi-arid zones. In: Anderson MG (ed) Hydrological Forecasting. Wiley, NY, pp 183–220

Urban Landscape and Flash-Flood Hazard on Alluvial Fans in a Hyper-Arid Zone—The Gulf of Eilat/Aqaba

23

Tamir Grodek

Abstract

Overlooking the Red Sea, the desert cities of Eilat (Israel) and Aqaba (Jordan) are located in a strategic, economic, and trade junction. The cities were built on a series of alluvial fans and a playa in-between; the only gentle slopes between the steep mountains and the sea that can accommodate large urban settings. Nevertheless, flash floods, which used to split, attenuate, and deposit sediments on the alluvial fans, are now disgorging from the mountain front into the urban areas. The attempts to mitigate the flood hazard by diversion canal/detention ponds yielded a fallacious sense of security; sedimentation reduces canal/ponds capacity and the floods breach into urbanized areas without warnings, causing destructive floods, often larger than natural flash floods. Moreover, the diversion canals only transfer the risk downwards to the Eilat playa, an area of the utmost economic importance. Today, the cities are spread over the alluvial fans and therefore, flood prevention strategies become increasingly complex and unsustainable. However, the last 70 years demonstrate that floods can cross the cities with little appreciable damage, only by minor shifts in the priorities of urban design. The incorporation of urban segments like recreation areas, football fields, main roads, parking lots, etc., into flood prevention schemes may lead to more sustainable solutions while reducing flood risk.

Keywords

Alluvial fan · Desert hydrology · Flash floods · Flood control · Flood protection · Sustainable flood management · Urban hydrology · Urban planning

23.1 Introduction—Alluvial Fans and Their Characteristics

Alluvial fans are gently sloping, cone-shaped aggradational landforms, formed by sediments eroded from the mountains and carried by mountain rivers to an open valley, where accommodation space is ample (Figs. 23.1 and 23.2; NRC 1996). At the mountain front (*fan apex*), the floodwater disgorges to an open area (*fan head*), attenuates and splits to form multiple-shallow flowpaths (Fig. 23.1b). The drop-in flow energy causes sedimentation in the flowpaths and forces flow migration to an unexpected new path. Over time, flood and sediment migration spread across the entire open valley (*main fan area*) down to the valley bottom (*fan toe*). The gentle slope, shallow flowpaths, and infrequent major floods give the inhabitants a false sense of security, but when floods do arrive, the nature of flow migration and sediment deposition interfere with the stationary urban segments (e.g., Schick 1971, 1995; NRC 1996; Larsen et al. 2001). As a result, total protection measures are implemented, including the construction of levees and diversion canals, check dams, slope stability measures, etc. However, these structures are also prone to sedimentation, making them only effective for a short period, and floods can breach and cause severe damage to the city (e.g., NRC 1996; Benito et al. 1998; Santo et al. 2015). Under such conditions, flood risk assessment cannot be reliable without considering sediment storage in mountain basins and the flood waters (Davies and McSaveney 2008, 2011). These processes and conditions are valid for both arid and humid regions, especially during extreme events.

T. Grodek (✉)
The Fredy and Nadine Herrmann Institute of Earth Sciences, The Hebrew University of Jerusalem, Jerusalem, Israel
e-mail: t.grodek@gmail.com

A. Frumkin and N. Shtober-Zisu (eds.), *Landscapes and Landforms of Israel*, World Geomorphological Landscapes, https://doi.org/10.1007/978-3-031-44764-8_23

Fig. 23.1 The cities of Eilat and Aqaba. **a** View from Eilat mountain (west) to the gulf and the cities. **b** Explanation of alluvial fan processes: Nahal Mekorot disgorged from the mountain front (*fan apex*), flooded the schoolyard sector (*fan head),* and deposited sediments. Further downstream *(the main fan area)*, the flood spreads along multiple road networks in Eilat, down to the fan margin (*fan toe)*. Note the flow paths on the active Wadi Yutum alluvial fan that inundated Aqaba. Floods from both cities frequently inundated Eilat playa. Images: a—Dafna Tal and the Israel Ministry of Tourism: b—Google Earth

In the hyper-arid Dead Sea rift valley, like in many other places in the world, alluvial fans and playas are the only gentle slopes that can accommodate large human settlements (Fig. 23.2). The monotonous dryness of the extreme desert and the shallow flowpaths on the alluvial fans leave its inhabitants mentally unprepared for floods. During the rainless years, Eilat and Aqaba expanded almost on the entire alluvial fans surface without leaving the necessary

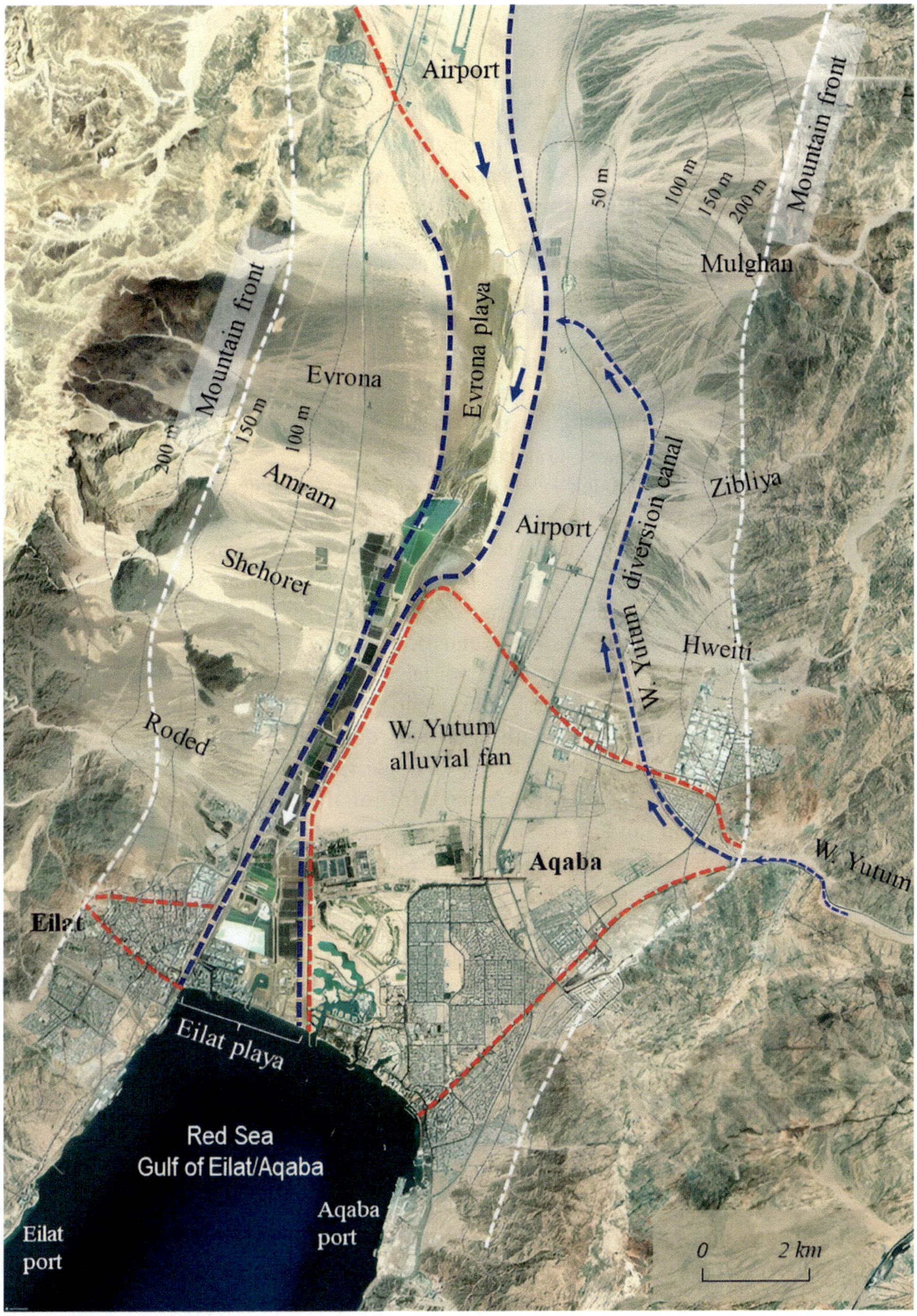

Fig. 23.2 Regional view of the southern Arava rift valley and the cities of Eilat and Aqaba on both sides of the gulf. The mountainous rivers, mountain front and various alluvial fans are visible. Alluvial fans are the only suitable place for human activities in this landscape. Urbanized alluvial fans boundaries are marked by red dashed lines. Image: ESRI

expanse for unpredictable changes in flood paths, sediment deposition, and inundation in the playa (e.g., Schick 1971; Grodek et al. 2000; Farhan and Anbar 2014). Therefore, occasional flash floods are considered as a natural 'disaster' rather than a natural course of events that we should be prepared for (Fig. 23.1b: Aqaba is frequently inundated by the active flowpaths in the Wadi Yutum alluvial fan).

23.2 Landscape and Climate

The Arava rift is a ~ 160 km long and 7–15 km wide valley, running from the Red Sea in the south to the Dead Sea in the north (Fig. 23.2). The rift is a segment of the Dead Sea Rift, the transform boundary between the African Plate (west) and the Arabian Plate (east). While the general motion is to the north-northeast, the left lateral displacement (~ 4 mm/year; total ~ 107 km) creates in some areas extension, and pull-apart basins like the Gulf of Eilat/Aqaba are formed. The intense tectonic processes at the margins of the rift have resulted in steep terrains rising at elevations 500 m asl to the west and 2000 m asl to the east. As a result of lowering the base level, alluvial fans form on both sides of the rift margins (Fig. 23.2). The continued active lateral displacement creates an offset from the feeding rivers, resulting in an anomaly between the feeding basins and the size of the alluvial fans (Bowman 2019). The mountain slopes are bare bedrock at the top and colluviated at their lower parts. The first-order steep streams are bedrock-incised, which upon transition to the third order, change into a braided pattern in coarse-grained alluvium with a moderate slope (Wohl and Grodek 1994).

Eilat in Israel and Aqaba in Jordan is part of the north-eastern margin of the north-African desert belt (Köppen classification: BWh). The arid climate is characterized by a low mean rainfall (22 mm), one rainy season (winter), high mean temperatures (38 °C; August), and low humidity (< 40%). The rainfall is highly variable; years with just a few millimeters are not infrequent (Aqaba 2.1 mm, 2016/17) as well as years with more than double the mean (56 mm, 2019/20) and even during a single rainstorm (64 mm, 1974/5; Schick 1988). About 60% of the total rainfall comes from fall and spring thunderstorms and originates in local-sporadic convective cells associated with an incursion of the Red Sea Trough yielding intensities of 30–60 mm/hr and up to 120 mm/hr for 10 min (spotty rain; Sharon 1972; Sharon and Kutiel 1986). The winter rains (December–February), which originate in southerly trajectories of Mediterranean depressions, cover large areas and typically generate longer rain spells, uniform intensities, or light intermittent showers with intensities up to 10 mm/hr. On the Wadi Yutum headwater (Ras En Naqb > 1600 m asl), the average yearly rainfall reaches 150 mm, while the measured rainfall in Eilat represents the small basin headwaters at the city margins.

Although the amount of rain in each rainstorm is low, the bare surfaces initiate slope runoff and streamflow in the small basins from ~ 6 mm of rain with intensities 30–50 mm/h (mainly granite bedrock; Yair and Lavee 1974; Schick 1995; Grodek et al. 2000). The transported sediments during flash floods comprise up to 20% of the total flood volume (depending on flood magnitudes). Of the transported sediments, ~ 66% consist of coarse sand, cobbles and boulders (> 2 mm) while the remaining ~ 34% are suspended sediments, fine sand, silt, and clay (< 2 mm). The source of the clay is partly aeolian dust, removed by the first runoff each season (Lekach and Schick 1982, 1983). In the transition to the open valley (Fig. 23.1b), the flow splits on the alluvial fan surfaces to form multiple flowpaths along the city's boundary. The process is primarily fluvial, not debris flow, based on field evidence and Melton ratio *R* (Melton 1965). The basin's *R* values range between 0.03–0.14 ($R = Hb/\sqrt{Ab}$ [km/km] where: Hb-basin relief [Δh] and Ab-basin area). In contrast debris flows, typically R values range between 0.35 and 2.0 (e.g., Wilford et al. 2004; De Scally et al. 2010; Welsh and Davies 2011; Stolle et al. 2015). In general, fluvial processes are less damaging and easier to manage when compared to debris flow (Zarn and Davies 1994). The estimated sediment transport for event in the region is 50–600 m^3/km for basins size of 0.5–13 km^2 (Schick and Lekach 1981, 1993; Lekach and Enzel 2021; Farhan 2014).

Despite the extreme aridity, the Eilat playa was inundated for weeks following widespread rain in November 1944 (Fig. 23.3; Ashbel 1975). Although not much is known about the storm, there are records of nearby exceptional rain events: at Themed (Sinai, 60 km west of Eilat; Ashbel 1938), 142 mm of rain lasted few hours in November 18, 1925, and at Nahal Tzihor (70 km north of Eilat), 120 mm of rain lasted for three hours in December 1993. Both events were likely similar types of storms that inundated the playa basins. Additionally, synthetic time series of rainfall in the region suggested 10 events over 100 mm per 1000 years (Barzilay et al. 1999). The contradictory nature of decades of rainless years and exceptional rainstorms on a century scale determines the flash-flood hazard and the style of urbanization in Eilat and Aqaba.

23.3 The Fluvial System and Urban Environment

The contemporary Eilat (pop. 52,000) and Aqaba (pop. 200,000) were first established (1949) on the lower section of the alluvial fans and the playa, and later expanded up to the mountain margins, at ~ 100 to 200 m asl (Figs. 23.1 and 23.2). Along with urban development, infrequent moderate rain and flooding have hit the region over the years, e.g., 1953, 1963, 1966, 1981, 1997, 2006, 2010, and 2016 (e.g., Schick 1971; Grodek et al. 2000; Farhan 2014; Maslamani et al. 2017). But the harsh environment and long floodless years, especially between 1981 and 1997, 'encouraged' the locals to 'forget' the flood damage shortly after its occurrence, resulting in overall neglect in the maintenance of the drainage ways (if any) which had been inactive for years (Grodek et al. 2000). Both cities divert most of the upper basin floods and the urban runoff to the playa in-between them (Fig. 23.2) which has naturally poor drainage to the sea and therefore is frequently inundated.

Fig. 23.3 Following the severe rains of 1944, flash floods inundated the Eilat playa for weeks, due to the naturally poor drainage before the establishment of the modern towns of Eilat and Aqaba. *Source* Ashbel (1975)

23.3.1 Eilat

North of Eilat (Fig. 23.1a, b) includes the fully urbanized alluvial fans, mostly developed as residential areas (5 km^2), which drain to the playa (6.1 km^2) and are occupied by mixed landuse, including industry, agriculture, residential and tourism. A belt of 20 first- and second-order streams and slopes encircle the city (total basin area of 2.3 km^2, each basin 0.05–0.8 km^2). The largest basin, Nahal Mekorot (basin 0.8 km^2) is the major threat of the 1.3 km^2 of urban alluvial fan and the playa (Fig. 23.2). The city is criss-crossed by a road network covering about 30% of the urban area (including parking lots). The down-fan streets are steep (3–7%), while the orthogonal streets, paralleling the fan contours lines, have low slopes (0–0.5%). In the older part of the city, the main road network is gulf-view oriented, leading to the boundary with the playa, an area frequently inundated by urban runoff and floods crossing the city. At the city's northern border, Nahal Roded (46 km^2) forms an active alluvial fan (1.5 km^2), invading part of the urbanized playa.

The new southern part of the city is built mostly on abandoned alluvial surfaces (5 km^2) and drained in the south-eastern direction towards the gulf by Nahal Shahamon (basin 8 km^2) and Nahal Garof (basin 2 km^2). The ports, naval, commercial, and, later, oil terminals, are situated along the narrow strip between the alluvial surfaces and the gulf.

The early period of the urban history (1949–1975) was typified by frequent floods. The main streets leading down the alluvial fan served as drainage canals, but multiple plans designed to protect the city from flooding and urban run-off have never been carried out. During the next stages of urban development (1981–1997), no significant flooding occurred, and the planning approach was to regard Eilat as a 'normal' city. This involved the construction of a sewer system, inlets and gutters based on the concept of a humid climate city. During these 16 dry years, urban waste, dust, and minor flows with sediments and debris from the mountain margins clogged the sewer system that fell into disuse shortly after its construction. In 1990, five detention ponds were first implemented (storage of 28,000 m^3) to protect the city from flash floods generated on aggregate basins covering 0.52 km^2; 30 years later, the ponds area was converted into dwellings. A new drainage master plan (1996), partly implemented, proposed to redirect floodwaters to a canal, parallel to the bypass road, which extended the flow path from 2 to 6 km through the playa to the sea.

23.3.2 Aqaba

The city of Aqaba (Jordan) covers most of the 47 km^2 of the Wadi Yutum alluvial fan including a variety of land uses such as industry, agriculture, and tourism. Among rivers contributing to the Eilat playa and the gulf, Wadi Yutum has the largest basin of 3,956 km^2, which only about 1,720 km^2, discharging directly to the city and further down across the border to Eilat playa (Fig. 23.1b). The urban area of Aqaba is highly susceptible to flood hazard. Flash floods with high sediment loads threat more often the growing city, even by low-to-moderate floods. As the city already occupies the shortest natural flowpath to the sea, a 'total protection' diversion canal extends the flow length to the sea through the Eilat playa from 12 to 27 km (Fig. 23.2).

The maximum estimated peak discharge at Wadi Yutum (February 2006), 560 m^3/s (flood volume of ~ 4.3 Mm^3; recurrence interval of 20 y), resulted from 35 mm of rain (Farhan 2014). Based on the partial available flood record, the projected peak discharges in the canyon of Wadi Yutum are in range of 1500–2100 m^3/s for recurrence interval of 25–100 year (5–1%) and flood volumes ~ 70 to 100 Mm^3 (Bany-Mustafa 2016). These projected values are far beyond the capacity of the Aqaba diversion canal, the Kinet canal (playa drainage canal), and the proposed dam/settling basin (Farhan and Anaba 2016). South of the city, smaller alluvial fans (the old city and port facilities) also face destructive floods primarily caused by coastal wadies (each basin 6–160 km^2; W. Um Jurf, W. Al-Shahbi, W. Shallalah, W. Jeishieh and W. Mubarak). These alluvial fans drain directly to the sea and therefore are not in the scope of this chapter.

23.3.3 Eilat Playa

The Eilat playa covers about 6.1 km^2 and serves as the terminal basin for accumulating flash floods from hydrologically effective basins areas of ~ 2,282 km^2 Figs. 23.1 and 23.2). These areas include the urbanized alluvial fans of Eilat (west), Aqaba (east) and the natural basins of Nahal Arava (north) and W. Yutum. Some of the northern basins drain to Yotvata and Evrona playas and together with the semi-closed basins in Wadi Yutum they contribute limited runoff to the Eilat playa. The maximum peak discharge to the playa ranges from 1,300 m^2/s (PMF estimation) to 2,100 m^2/s (hydrological model). Before the urbanization, parallel shallow flow tracks drained the playa to the gulf. The playa surface is nearly impervious, as the result of silty–clayey sediments deposited by the floods, together with salt efflorescence due to the high-water table. Low infiltration rates coupled with the minimal gradient to the sea permit frequent inundation of the playa.

Today, the playa is occupied by industry and agriculture (~3.1 km^2), mostly in the upper playa, residential use in the middle part (~0.5 km^2) and tourism along the seashore (~2.5 km^2). The development on the playa limited the Kinet canal's floodwater drainage capacity to ~100–300 m^3/s while the expected 1% flood is about three-fold higher.

23.4 Conclusions

The cities of Eilat and Aqaba are both strategically located at the intersection of three borders and the sea: Israel, Jordan, and Egypt. Between the steep mountains and the Gulf of Eilat/Aqaba, they are built on a series of alluvial fans and the playa in-between them; yet these surfaces are also prone to unforeseen extreme arid short-lived flash floods that transport abundant amounts of water and sediments to the urban areas (Fig. 23.1b). Prior to urbanization, strong flash floods split on the alluvial fans and on the playa into multiple smaller and shallower flow ways, each carrying a fraction of incoming water and sediment. These flows were evidently safer and easier to manage, but unfortunately, the urban areas extended over the naturally shortest flow routes to the sea, covering most of the alluvial fan and the playa. Therefore, in order to protect the cities from flooding, the entire flows and associated sediments have been diverted by each city to the playa through a single diversion canal. For several reasons, this strategy of total protection canal implemented in Eilat and Aqaba is confusing. First, the urban areas cover the shortest natural course of the flowpaths to the sea and the playa. Therefore, the single diversion canal in each city is at least twice the distance of the natural course. The Aqaba canal extended the flow from 12 to 27 km and the Eilat one from 3 to 6 km. The longer canal reduced the slope, resulting in enhanced sedimentation. In such condition, floods of low probability may breach into the urban area. Second, the long canal merging several stream and slope flows along the city boundary to a single canal neglects the natural benefit of flow splitting and attenuation, which would reduce risk on the alluvial surfaces. Third, the diversion canals in both cities drain urban runoff and natural basins covering about 1,765 km^2 to the playa (primarily the Aqaba canal does so), which is the most sensitive and economically important landscape in the region; that in addition to 517 km^2 of natural basins that drain also to the playa by Nahal Arava. Forth, flood protection earth dams along the mountainous, steep rivers are eventually prone to sedimentation and the loss of capacity. Consequently, a flash flood, not necessarily high, can cause a series of dam breaches and a high-impact flash flood to run over the cities and finally to the playa. Worldwide experience shows that total protection diversion canals and dams offer false sense of security and for only a short period of time.

In the city of Eilat, long-term safety could be achieved by integrating flood control schemes into the urban framework which includes roads, streets, parking lots,

Fig. 23.4 The flood of October 1997. Despite the extensive construction on the alluvial fan, the flood breached the city and caused only limited damage. The flood of Nahal Mekorot (Fig. 23.2) breached the school fence, flooding and depositing sediments on the football field and recreation area. Below the school gate (**a**), the flow split into several wide four-lane roads, each carrying a fraction of the incoming floodwaters (**b**). One flowpath has diverted through a parking lot (**c**), and over its fence, forming a 'waterfall' that further dissipated the flow energy (**d**). *Source* Grodek et al. (1997)

playgrounds, parks, etc. On the boundary between the urbanized area and the mountain front, twenty small basins and slopes disgorged floodwater onto the city with expected maximum peak discharge of ~ 10 m^3/s in each location. The disjunction of these small basins allows moderate floodwaters to enter the city and causes divergence of the flows, attenuation of their velocity, and capture of floodwaters in local storages (Fig. 23.4). The result is a fast decline of peak discharge, a longer flow duration, and little appreciable damage. Aqaba, however, present a different situation. The expected extreme floods from Wadi Yutum 1,300–2,100 m^3/s, require much larger expanses, and the current urban layout cannot accommodate floods of this magnitude. Thus, the city is faced with two options: (i) to allow sector part of the alluvial fan to remain open to unpredictable changes in flood paths and sediment deposition, or (ii) to implement a flood total protection diversion canal that is expected to burst causing destructive floods in the city. Due to rapid development on the alluvial fan, the next decade will determine which solution is preferred.

Unfortunately, in most cases, the urban design is in favor of total protection, leading to an illusion of security while ignoring the possibility of sustainable solution by integrating natural flows into the urban design scheme.

References

Ashbel D (1938) Great floods in Sinai Peninsula, Palestine, Syria and the Syrian desert, and the influence of the Red Sea on their formation. Quart J Roy Meteor Soc 64:635–639

Ashbel D (1975) Forty-five years observations on the climatology and hydrology of the Dead Sea, 1929–1974. Hebrew University Report

Bany-Mustafa M (2016) Flash flood hazard and risk maps for Wadi Yutum. In: The second international symposium on flash floods in wadi systems, ISFF 25–27 Oct 2016 Technische Universität Berlin Campus El Gouna, Egypt

Barzilay E, Enzel Y, Amit R (1999) Constructing synthetic time series of rainfall events for a hyperarid environment, southern Arava, Israel, In: Hassan MA et al (eds) The hydrology-geomorphology interface: rainfall, floods, sedimentation, land use, vol, 261. IAHS Publ. pp 29–42

Bowman D (2019) The regional approach: alluvial fans along the dead Sea-Arava rift valley. In: Principles of Alluvial fan morphology. Springer, Dordrecht. https://doi.org/10.1007/978-94-024-1558-2_19

Benito G, Grodek T, Enzel Y (1998) The geomorphic and hydrologic impacts of the catastrophic failure of flood-control-dams during the 1996-Biescas flood (Central Pyrenees, Spain). Z Geomorphologie 42:417–437

Davies TR, McSaveney MJ (2008) Principles of sustainable development on fans. J Hydrol (NZ) 47:43–65

Davies TR, McSaveney MJ (2011) Bedload sediment flux and flood risk management in New Zealand. J Hydrol (NZ) 50:181–190

de Scally FA, Owens IF, Louis J (2010) Controls on fan depositional processes in the schist ranges of the Southern Alps, New Zealand, and implications for debris-flow hazard assessment. Geomorphology 122:99–116

Farhan Y (2014) Geomorphological evaluation for Urban development using remote sensing and GIS, Southern Coast of Aqaba, Jordan. J Environ Earth Sci 4(5):104–118

Farhan Y, Anbar A (2014) Fragile landscape: impact and consequences of May 2014 flash-flood disaster in the Aqaba Area, Southern Jordan. Res J Environ Earth Sci 6(9):451–465

Farhan Y, Anaba O (2016) Flash flood risk estimation of Wadi Yutum (Southern Jordan) watershed using GIS based morphometric analysis and remote sensing techniques. Open J Mod Hydrol 6:79–100

Grodek T, Lekach J, Schick AP (2000) Urbanizing alluvial fans as flood-conveying and flood-reducing systems: lessons from the October 1997 Eilat flood. In: Hassan MA et al (eds) The hydrology-geomorphology interface: rainfall, floods, sedimentation, land use, vol 261. Proc Jerusalem Conference, May 1999. IAHS, pp 229–250

Larsen CM, Wieczorek GF, Eaton SL, Morgan BA, Torres-Sierra H (2001) Venezuelan debris flow and flash flood disaster of 1999 studied. Eos 82(47):572–573

Lekach J, Schick AP (1982) Suspended sediment in desert floods in small catchments. Israel J Earth Sci 31:144–156

Lekach J, Schick AP (1983) Evidence for transport of bedload in waves: analysis of fluvial sediment samples in a small upland stream channel. CATENA 10:267–279

Lekach J, Enzel Y (2021) Flood-duration-integrated stream power and frequency magnitude of >50-year-long sediment discharge out of a hyperarid watershed. Earth Surface Process Landforms 2021:1–15. https://doi.org/10.1002/esp.5104

Maslamani A, Zeitoun M, Almagbile A (2017) Hydro-meteorological analysis of flash flood in Southern Jordan. J Environ Sci Allied Res 2017:32–38

Melton MA (1965) The geomorphic and paleoclimatic significance of alluvial deposits in southern Arizona. J Geol 73:1–38

NRC, National Research Council (1996) Alluvial fan flooding. National Academy Press Washington DC

Santo A, Santangelo N, Di Crescenzo G, Scorpio V, De Falco M, Chirico GB (2015) Flash flood occurrence and magnitude assessment in an alluvial fan context: the October 2011 event in the Southern Apennines. Nat Hazards 78:417–442. https://doi.org/10.1007/s11069-015-1728-4

Schick AP (1971) A desert flood: physical characteristics, effects on man, geomorphic significance, human adaptation—a case study of the Southern Arava watershed. Jerusalem Stud Geogr 2:91–155

Schick AP (1988) Hydrologic aspects of floods in hyper-arid environments. In: Baker VR et al (eds) Flood geomorphology. Wiley, New York pp 189–203

Schick AP, Lekach J (1981) High bedload transport rates in relation to stream power, Wadi Mikeimin, Sinai. CATENA 8:43–47

Schick AP, Lekach J (1993) An evaluation of two ten-years sediment budgets, Nahal Yael, Israel. Phys Geogr 14:225–238

Schick AP (1995) Fluvial processes on an urbanizing alluvial fan, Eilat, Israel. In: Costa, JC et al (eds) Anthropogenic influences in fluvial geomorphology, vol 89. American Geophysical Union, pp 209–219

Sharon D (1972) The spottiness of rainfall in a desert area. J Hydrol 17:161–175

Sharon D, Kutiel H (1986) The distribution of rainfall intensity in Israel, its regional and seasonal variations and its climatological evaluation. Int J Climatol 6:277–291

Stolle A, Langer M, Blöthe JH, Korup O (2015) On predicting debris flows in arid mountain belts. Global Planet Change 126:1–13

Welsh A, Davies T (2011) Identification of alluvial fans susceptible to debris-flow hazards. Landslides 8:183–194

Wilford DJ, Sakals ME, Innes JL, Sidle RC, Bergerud WA (2004) Recognition of debris flow, debris flood and flood hazard through watershed morphometrics. Landslides 1:61–66

Wohl EE, Grodek T (1994) Channel bed-steps along Nahal Yael, Negev desert, Israel. Geomorphology 9:117–126

Yair A, Lavee H (1974) Areal contribution to runoff on scree slopes in an extreme arid environment—a simulated rainstorm experiment. Z Geomorph NF Suppl 21:106–121

Zarn B, Davies TRH (1994) The significance of processes on alluvial fans to hazard assessment. Z Geomorphol 38:487–500

Index

A. Frumkin and N. Shtober-Zisu (eds.), *Landscapes and Landforms of Israel*,
World Geomorphological Landscapes, https://doi.org/10.1007/978-3-031-44764-8